APPROXIMATE METHODS OF HIGHER ANALYSIS

L. V. KANTOROVICH
AND
V. I. KRYLOV

Translated by
Curtis D. Benster

Dover Publications, Inc.
Mineola, New York

Bibliographical Note

This Dover edition, first published in 2018, is an unabridged republication of the work originally published by Interscience Publishers, Inc., New York, and P. Noordhoff Ltd., Groningen, the Netherlands, in 1958.

Library of Congress Cataloging-in-Publication Data

Names: Kantorovich, L. V. (Leonid Vital§evich), 1912-1986, author. | Krylov, V. I. (Vladimir Ivanovich), 1902-author. | Benster, Curtis D., translator.
Title: Approximate methods of higher analysis / L.V. Kantorovich and V.I. Krylov ; translated by Curtis D. Benster.
Other titles: Metody priblizhennogo resheniëiia uraveniæi v chastnykh proizvodnykh. English
Description: Dover edition. | Mineola, New York : Dover Publications, Inc., 2018. | Originally published: New York : Interscience Publishers, 1958.
Identifiers: LCCN 2017048238| ISBN 9780486821603 | ISBN 0486821609
Subjects: LCSH: Mathematical physics. | Differential equations, Partial. | Approximation theory.
Classification: LCC QA401 .K263 2018 | DDC 530.15—dc23
LC record available at https://lccn.loc.gov/2017048238

Manufactured in the United States by LSC Communications
82160901 2018
www.doverpublications.com

TABLE OF CONTENTS

CHAPTER II. THE APPROXIMATE SOLUTION OF THE INTEGRAL EQUATIONS OF FREDHOLM

CHAPTER III. THE METHOD OF NETS

CHAPTER IV. VARIATIONAL METHODS

PREFACE TO THE THIRD EDITION

The problems of mathematical physics have found wide application in the most diverse fields of engineering. In courses of mathematical physics, general methods of solution are ordinarily expounded; these are of a purely theoretical character and do not in fact make it possible to actually find the solution of such problems; classical examples of exact solutions are also given for the simplest cases. In the practical problems of technology, however, problems are frequently encountered where the exact solution either cannot be found or has such a complicated structure that it is used only with difficulty in calculations.

Approximate methods of solving the problems of mathematical physics, particularly the method of nets and variational methods, which were developed at the beginning of this century, have therefore been met with much interest on the part of engineers, and promptly obtained wide diffusion. The basic merit of the approximate methods consists in their being universal and efficient, since they permit the determination of an approximate solution for a wide class of cases, and in application require but simple and quite realizable computations.

During the past three decades approximate methods have obtained an especially great development in the Soviet Union. Simultaneously with the elaboration of methods already known, Soviet scientists have proposed a number of new ones. In this common labor, occasioned chiefly by the requirements of technics, many representatives of mechanics and other applied sciences participated together with the mathematicians.

This book came out in first edition in 1936 under the title "Metody priblizhënnogo resheniia uravnenii̮ v chastnykh proizvodnykh" [Methods for the approximate solution of partial differential equations]. An entire range of problems was not covered by it. In it were discussed primarily boundary-value problems for linear equations, but even for them not all the known methods were expounded.

In 1941 the book was reissued under changed title: "Priblizhënnye metody vysshego analiza" [Approximate methods of higher analysis]. Some chapters had been subjected to revision.

The demand for such a book has become especially acute at the present time in view of the wide application of approximate methods in the work of the scientific and technical institutions.

More than eight years have passed since the issuance of the second edition. In these years the basic approximate methods have undergone further development and a number of new methods have appeared. Essential new works have appeared on questions of convergence and estimates of error. Considerable experience in the application of the methods to concrete problems of physics and engineering has been accumulated. General viewpoints have arisen which make it possible to treat of whole groups of problems with greater clarity and simplicity. Finally, in connection with the utilization of a calculating-machine technology for the solution of mathematical problems, the matter of estimating the efficiency and realizability of the methods has taken on quite a different aspect.

The incorporation in this book of all the material arising from these changes would require that it be radically revised and that its volume be increased substantially. To accomplish this would require several years of labor. The authors intend to undertake such a revision in the future, but do not consider it possible to postpone the reprinting for such a long time. In view of this the authors have renounced even a partial revision and have introduced only insignificant changes in the 1941 edition, which changes chiefly amount to the provision of references to new literature.

A more complete survey of the Soviet literature on approximate methods can be found in the authors' article in the book "Matematika v SSSR za tridtsat' let" [Thirty years of mathematics in the USSR].

L. V. Kantorovich
V. I. Krylov

FROM THE PREFACE TO THE SECOND EDITION

The first attempt at a systematic exposition of the principal effective approximate methods was made by the authors in the book "Metody priblizhennogo resheniia uravneniĭ v chastnykh proizvodnykh" [Methods for the approximate solution of partial differential equations]. That book, issued in 1936, is long since out of print. The present book represents a revision of it. We have found it appropriate to change its title, since, together with methods for solving partial differential equations, a significant place in it is devoted to the discussion of conformal mapping and the approximate solution of integral equations. Substantial modifications have been made in the chapter on the method of nets, the chapter on variational methods has been reworked, and the chapter on Schwarz's method has been written anew. The authors have consciously limited themselves to the consideration of methods for the approximate solution of boundary-value problems only, since the inclusion of methods for the solution of problems with initial—as well as with mixed—boundary conditions would necessitate a further increase in the volume of the book. As a rule we have confined ourselves to linear equations, and have only in certain cases given an indication that the given method could also be applied to nonlinear problems.

Chapters I, II and IV were written by L. V. Kantorovich; Chapters III, V, VI and VII, by V. I. Krylov; N. P. Stenin is the author of § 10 of Chapter V.

All chapters are independent of one another, with the exception of the sixth chapter, which relies essentially on the fifth.

L. V. Kantorovich
V. I. Krylov

TRANSLATOR'S PREFACE

The source of the present translation is the Fourth Edition.

The reader's attention is called to the appearance in this book of certain usages – unfamiliar but unobjectionable, and perhaps having a certain merit – that were adopted from the Russian original: *sh* appears instead of *sinh*, and there are like notations for the other hyperbolic functions; signs of the elementary algebraic operations are repeated when a line is split. Moreover, subdivisions of Sections are called "Nos.", in conformity with the authors' practice.

The translator feels that publication of this translation would in a sense be incomplete over his signature alone, and takes this occassion to acknowledge his indebtedness to Dr. Ivan S. Sokolnikoff and to Dr. George E. Forsythe, but for whose initiative and encouragement this translation would scarcely have been conceivable. He hastens to add that shortcomings in the translation are to be laid only at his door.

The translator also wishes to make more personal acknowledgments, for help received at many stages of his work, to his father, L. Halsey Benster, Mechanical Engineer – who understood the value of fine tools, and to his wife, Ada – who just understood.

CURTIS D. BENSTER

Chapter I

METHODS BASED ON THE REPRESENTATION OF THE SOLUTION AS AN INFINITE SERIES

§ 1. FOURIER'S METHOD

One of the fundamental and widely-met methods of solving the problems of mathematical physics is the method of Fourier. A solution found by this method is ordinarily obtained in the form of an infinite series; in many cases such series converge rapidly, and therefore Fourier's method may often be used even for the determination of the required numerical results: a segment of this series will give an approximate solution of the problem. We assume that the reader is familiar with the simplest applications of the Fourier method to the solution of problems of mathematical physics — for example, with the solution of the problem of the vibrating string by Fourier's method [1]). We shall not, therefore, give a thorough and detailed exposition of the Fourier method here, but shall consider only its application to several problems relating to equations of the elliptic type and chiefly to the Laplace equation.

1. The Dirichlet problem for the rectangle. The simplest, and the fundamental, partial differential equation of the elliptic type is the *Laplace equation*:

$$\Delta u = \frac{\partial^2 u}{\partial x^2} + \frac{\partial^2 u}{\partial y^2} = 0. \tag{1}$$

Continuous functions which have partial derivatives of the first and second order and which are solutions of this equation are called *harmonic*.

The first fundamental problem relating to the Laplace equation is the *Dirichlet problem*. It is formulated thus: *To find, a function u, harmonic in the given region D and taking assigned values on the closed contour L bounding the region D.*

If the equations of the contour L are given in parametric form: $x = x(s)$, $y = y(s)$, this condition may, for example, be written as:

$$u[x(s), y(s)] = f(s),$$

where $f(s)$ is a given function. We shall write, more briefly,

$$u = f(s) \text{ on } L.$$

---

[1]) See V. I. Smirnov, [1], vol. II (ed. 9), 1948; R. Courant and D. Hilbert, [1]; S. L. Sobolev, [1]; I. G. Petrovskiĭ, [2]; V. I. Levin and IU. I. Grosberg, [1].

We shall make an observation which will be useful to us. If the function $f(s)$ is a sum, $f(s) = f_1(s) + f_2(s)$, it is obviously sufficient to find the solutions u_1 and u_2 which on L become $f_1(s)$ and $f_2(s)$ respectively, since $u = u_1 + u_2$ gives the solution of the problem which becomes $f(s)$ on L.

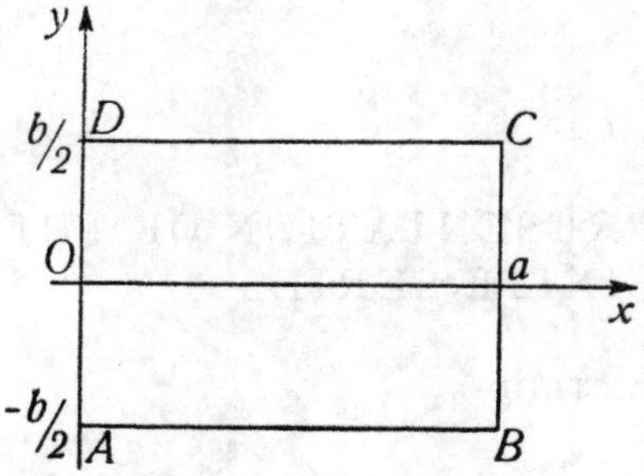

Fig. 1.

Let us now pass to the solution of the Dirichlet problem for the case when the region D is the rectangle $ABCD$ with sides a and b. We shall choose the axes of coordinates as is shown in Fig. 1, i.e., for the axis of abscissas we shall take the horizontal center line, and for the axis of ordinates, the left side.

The general Dirichlet problem will here consist in finding the harmonic function which equals the given arbitrary functions on the sides of the rectangle. Accordingly the boundary conditions may be written thus:

$$\left.\begin{array}{l} u = \begin{cases} \varphi_1(x) & \text{for } y = \dfrac{b}{2} \\ \varphi_2(x) & \text{,, } \; y = -\dfrac{b}{2} \end{cases} \\ u = \begin{cases} \psi_1(y) & \text{,, } \; x = 0 \\ \psi_2(y) & \text{,, } \; x = a \end{cases} \end{array}\right\}. \tag{2}$$

The solution of this problem can be obtained as the sum $u_1 + u_2$ of the solutions of the two simpler problems:

$$u_1 = \left\{\begin{array}{ll} \varphi_1(x) & \text{for } y = \dfrac{b}{2} \\ \varphi_2(x) & \text{,, } \; y = -\dfrac{b}{2} \\ 0 & \text{,, } \; x = 0 \text{ and } x = a \end{array}\right\} \tag{3}$$

$$u_2 = \left\{\begin{array}{ll} \psi_1(y) & \text{for } x = 0 \\ \psi_2(y) & \text{,, } \; x = a \\ 0 & \text{,, } \; y = \pm\dfrac{b}{2} \end{array}\right\}. \tag{3'}$$

These problems are simpler because their solutions are to vanish on two sides of the rectangle. These problems do not differ essentially from one another, and the solution of one of them can easily be reduced to the solution of the other; we shall consider in detail, therefore, only the problem of finding the function u_1.

Let us solve the problem by Fourier's method. We first of all seek elementary, basic solutions, in the form

$$u = X(x) \cdot Y(y), \tag{4}$$

which will satisfy the Laplace equation (1) and only the last of conditions (3).

Substituting (4) in the Laplace equation, we find

$$X''Y + Y''X = 0 \text{ or } \frac{X''}{X} = -\frac{Y''}{Y} = k, \tag{5}$$

where k is some constant, for both expressions $\frac{X''}{X}$ and $-\frac{Y''}{Y}$ must be equal to the same constant, since only in this case can a function of x alone be identically equal to a function of y alone. For the determination of the function X from (5) we obtain an equation of the second order:

$$X'' - kX = 0, \tag{6}$$

and in order that the function u, defined by equation (4), satisfy the last of conditions (3), the function $X(x)$ must satisfy the conditions:

$$X(0) = X(a) = 0. \tag{7}$$

We have to find a solution of equation (6) satisfying condition (7). If we put $k = -\lambda^2$ [1]), the general solution of equation (6) will be:

$$X(x) = C_1 \cos \lambda x + C_2 \sin \lambda x.$$

Conditions (7) give

$$C_1 = 0,\ C_1 \cos \lambda a + C_2 \sin \lambda a = 0;$$

accordingly $C_1 = 0$ invariably, and C_2 can be taken as unequal to zero only on condition that

$$\sin \lambda a = 0, \tag{8}$$

i.e., that λa be a numerical multiple of π:

$$\lambda a = n\pi \qquad (n = 1, 2, 3, \ldots);$$

non-positive n are absent here, since they do not give new solutions. Thus for $\lambda = \frac{n\pi}{a}$ we obtain the solution:

$$X(x) = C \sin \frac{n\pi x}{a} \qquad (n = 1, 2, \ldots). \tag{9}$$

---

[1]) It would be possible to consider positive values of k also, but they would not yield any solutions different from zero. Indeed, putting $k = \lambda^2$, we should have $X(x) = C_1 \operatorname{ch} \lambda x + C_2 \operatorname{sh} \lambda x$, and conditions (7) would compel the adoption here of $C_1 = C_2 = 0$, i.e., we should have obtained a trivial solution.

Substituting $k = -\lambda^2 = -\left(\frac{n\pi}{a}\right)^2$ in (5), we obtain for Y the equation

$$Y'' - \left(\frac{n\pi}{a}\right)^2 Y = 0, \tag{9'}$$

on solving which we obtain

$$Y = C_1^* e^{\frac{n\pi}{a}y} + C_2^* e^{-\frac{n\pi}{a}y} = C_1 \operatorname{ch} \frac{n\pi}{a} y + C_2 \operatorname{sh} \frac{n\pi}{a} y, \tag{10}$$

where C_1 and C_2 are arbitrary constants.

In view of (9) and (10), the basic solution (4) may be written thus:

$$u = \left(A \operatorname{ch} \frac{n\pi}{a} y + B \operatorname{sh} \frac{n\pi}{a} y\right) \sin \frac{n\pi}{a} x, \tag{11}$$

where A and B are arbitrary constants.

Now we can seek the function u_1, the solution of our problem, in the form of a sum of the fundamental solutions:

$$u_1 = \sum_{n=1}^{\infty} \left[A_n \operatorname{ch} \frac{n\pi}{a} y + B_n \operatorname{sh} \frac{n\pi}{a} y\right] \sin \frac{n\pi}{a} x. \tag{12}$$

This function satisfies the Laplace equation (1) and the last of conditions (3); it remains to select the constants A_n and B_n so as to satisfy the first two of conditions (3). On the basis of these conditions we write the following equations:

$$\left.\begin{aligned} u_1\Big|_{y=\frac{b}{2}} &= \sum_{n=1}^{\infty} \left(A_n \operatorname{ch} \frac{n\pi b}{2a} + B_n \operatorname{sh} \frac{n\pi b}{2a}\right) \sin \frac{n\pi}{a} x = \varphi_1(x), \\ u_1\Big|_{y=-\frac{b}{2}} &= \sum_{n=1}^{\infty} \left(A_n \operatorname{ch} \frac{n\pi b}{2a} - B_n \operatorname{sh} \frac{n\pi b}{2a}\right) \sin \frac{n\pi}{a} x = \varphi_2(x). \end{aligned}\right\} \tag{13}$$

On the other hand, expanding $\varphi_1(x)$ and $\varphi_2(x)$ in series involving $\sin \frac{n\pi}{a} x$ in the interval $(0, a)$, we have:

$$\varphi_1(x) = \sum_{n=1}^{\infty} \beta_n^{(1)} \sin \frac{n\pi x}{a}, \qquad \varphi_2(x) = \sum_{n=1}^{\infty} \beta_n^{(2)} \sin \frac{n\pi x}{a}, \tag{14}$$

where

$$\beta_n^{(1)} = \frac{2}{a} \int_0^a \varphi_1(x) \sin \frac{n\pi x}{a}\, dx, \qquad \beta_n^{(2)} = \frac{2}{a} \int_0^a \varphi_2(x) \sin \frac{n\pi x}{a}\, dx. \tag{15}$$

Comparing the coefficients of series (13) and (14), we obtain

$$\begin{aligned} A_n \operatorname{ch} \frac{n\pi b}{2a} + B_n \operatorname{sh} \frac{n\pi b}{2a} &= \beta_n^{(1)}, \\ A_n \operatorname{ch} \frac{n\pi b}{2a} - B_n \operatorname{sh} \frac{n\pi b}{2a} &= \beta_n^{(2)}. \end{aligned} \tag{16}$$

If A_n and B_n be determined from these equations and substituted in (12), we shall finally have obtained for u_1:

$$u_1 = \sum_{n=1}^{\infty} \left[\frac{\beta_n^{(1)} + \beta_n^{(2)}}{2 \operatorname{ch} \frac{n\pi b}{2a}} \operatorname{ch} \frac{n\pi y}{a} + \frac{\beta_n^{(1)} - \beta_n^{(2)}}{2 \operatorname{sh} \frac{n\pi b}{2a}} \operatorname{sh} \frac{n\pi y}{a} \right] \sin \frac{n\pi x}{a}. \tag{17}$$

This series will certainly be convergent at any point within the rectangle, for $\beta_n^{(1)}$ and $\beta_n^{(2)}$, as the Fourier coefficients of an integrable function, tend to zero, while the ratios

$$\frac{\operatorname{ch} \frac{n\pi y}{a}}{\operatorname{ch} \frac{n\pi b}{2a}}, \quad \frac{\operatorname{sh} \frac{n\pi y}{a}}{\operatorname{sh} \frac{n\pi b}{2a}}$$

are quantities of the order of $e^{n\frac{\pi}{a}\left(|y| - \frac{b}{2}\right)}$; since $|y| < \frac{b}{2}$, the series for u_1 converges as a geometric progression with ratio $e^{\frac{\pi}{a}\left(|y| - \frac{b}{2}\right)}$.

It should be remarked that if series (17) is to converge rapidly, especially at points close to the contour (when $|y|$ is close to $b/2$), the rate at which the coefficients $\beta_n^{(1)}$ and $\beta_n^{(2)}$ decrease is of essential significance. This circumstance offers a means of improving the convergence of series (17) (see § 5).

In view of the rapid convergence of series (17), the series for the derivatives of u_1 will converge, and a term-by-term differentiation will be admissible; therefore, since each term is a solution of the equation, it is clear that the function $u_1(x, y)$ satisfies the Laplace equation (1). Furthermore, if $\varphi_1(x)$ and $\varphi_2(x)$ are continuous, it can be shown that the limiting values for $u_1(x, y)$ when approaching the contour will be those prescribed by conditions (3). Thus the function u_1 found satisfies all the conditions laid down.

To determine the function u_2 the Laplace equation must be solved under conditions (3'). But on introducing a transformation of coordinates by means of the formulas

$$x' = y + \frac{b}{2}, \qquad y' = x - \frac{a}{2},$$

we shall have brought conditions (3') into form (3), the roles of the numbers a and b being interchanged, however, and ψ_1 and ψ_2 replacing

φ_1 and φ_2. Having, therefore, made use of formula (17) and afterwards returning to the variables x and y, we obtain the solution for the function u_2 in the following form:

$$u_2 = \sum_{n=1}^{\infty} \left[\frac{\delta_n^{(1)} + \delta_n^{(2)}}{2 \operatorname{ch} \frac{n\pi a}{2b}} \operatorname{ch} \frac{n\pi}{b}\left(x - \frac{a}{2}\right) + \frac{\delta_n^{(1)} - \delta_n^{(2)}}{2 \operatorname{sh} \frac{n\pi a}{2b}} \operatorname{sh} \frac{n\pi}{b}\left(x - \frac{a}{2}\right)\right] \sin \frac{n\pi}{b}\left(y + \frac{b}{2}\right), \qquad (18)$$

where

$$\delta_n^{(1)} = \frac{2}{b} \int_{-\frac{b}{2}}^{\frac{b}{2}} \psi_1(y) \sin \frac{n\pi}{b}\left(y + \frac{b}{2}\right) dy,$$

$$\delta_n^{(2)} = \frac{2}{b} \int_{-\frac{b}{2}}^{\frac{b}{2}} \psi_2(y) \sin \frac{n\pi}{b}\left(y + \frac{b}{2}\right) dy. \qquad (19)$$

The complete solution of the problem that has been posed — the function u — will be obtained on adding solutions (17) and (18). It should be observed that the expressions for the functions u_1 and u_2 are simplified if $\varphi_1(x)$ and $\varphi_2(x)$ [or $\psi_1(y)$ and $\psi_2(y)$ respectively] are equal. For instance, if $\varphi_1(x) = \varphi_2(x) = \varphi(x)$, we have

$$\beta_n^{(1)} = \beta_n^{(2)} = \beta_n = \frac{2}{a} \int_0^a \varphi(x) \cdot \sin \frac{n\pi x}{a}\, dx,$$

and the expression for u_1 takes the form

$$u_1 = \sum_{n=1}^{\infty} \frac{\beta_n \operatorname{ch} \frac{n\pi y}{a}}{\operatorname{ch} \frac{n\pi b}{2a}} \sin \frac{n\pi x}{a}. \qquad (20)$$

The second fundamental boundary-value problem, too — the Neumann problem — can be solved for the case of the rectangle in the same fashion.

The *Neumann problem* is generally formulated as follows: *To find, a function u, harmonic in the region D, whose normal derivative $\frac{\partial u}{\partial n}$, i.e., the derivative along the direction of the normal to the contour, has assigned values on the contour L:*

$$\frac{\partial u}{\partial n} = f_1(s) \text{ on } L. \qquad (21)$$

The Neumann problem has some peculiarities in comparison with the Dirichlet problem. First, it does not have a unique solution, for obviously the addition of a constant to some solution of the problem leads again to a solution. Second, the function $f_1(s)$ cannot be chosen arbitrarily, for in accordance with a familiar property of an harmonic function, the integral of its normal derivative along a closed contour must equal zero [1]),

$$\int_L \frac{\partial u}{\partial n} ds = \int_L f_1(s) ds = 0. \tag{22}$$

Let us now proceed to the case of the rectangle. We will preserve the notation and system of coordinates of the Dirichlet problem. Just as there, we may separate the problem into two, and moreover in each case the conditions are null on two sides.

For the function u_1 we shall have the conditions

$$\frac{\partial u_1}{\partial y} = \begin{cases} \chi_1(x) \text{ for } y = \dfrac{b}{2} \\ \chi_2(x) \text{ for } y = -\dfrac{b}{2}, \end{cases} \tag{23}$$

$$\frac{\partial u_1}{\partial x} = 0 \text{ for } x = 0,\ x = a.$$

Let us solve the problem by Fourier's method. Seeking a solution in the form $u = XY$, we have here for $X(x)$, instead of (7), the conditions

$$X'(0) = X'(a) = 0. \tag{24}$$

On the basis of these conditions we shall obtain, in the same way, $\lambda = \dfrac{n\pi}{a}$ and

$$X = \cos \frac{n\pi}{a} x \quad (n = 0, 1, 2, \ldots). \tag{25}$$

Then for u_1 we shall find, instead of (12), the series

$$u_1 = \sum_{n=1}^{\infty} \left[A_n \operatorname{ch} \frac{n\pi y}{a} + B_n \operatorname{sh} \frac{n\pi y}{a} \right] \cos \frac{n\pi}{a} x + A_0 y + B_0; \tag{26}$$

the last summand $A_0 y + B_0$ is the solution of equation (9′) corresponding to the case $n = 0$.

The constants A_n and B_n can be determined by using the first two

---

[1]) See V. I. Smirnov, [1], vol. II, 1948, p. 571. While it is true that the proof there given is for a contour lying within the region, by passing thence to the limit the same equation is obtained for the boundary contour.

of conditions (23). On the basis of these conditions, then, and series (26), we write

$$\left.\begin{aligned} \left.\frac{\partial u_1}{\partial y}\right|_{y=\frac{b}{2}} &= \sum_{n=1}^{\infty} \frac{n\pi}{a}\left[A_n \operatorname{sh}\frac{n\pi b}{2a} + B_n \operatorname{ch}\frac{n\pi b}{2a}\right]\cos\frac{n\pi}{a}x + A_0 = \chi_1(x), \\ \left.\frac{\partial u_1}{\partial y}\right|_{y=-\frac{b}{2}} &= \sum_{n=1}^{\infty} \frac{n\pi}{a}\left[-A_n \operatorname{sh}\frac{n\pi b}{2a} + B_n \operatorname{ch}\frac{n\pi b}{2a}\right]\cos\frac{n\pi}{a}x + A_0 = \chi_2(x). \end{aligned}\right\} \quad (27)$$

Let us denote by $\alpha_n^{(1)}$ and $\alpha_n^{(2)}$ the Fourier coefficients of the functions $\chi_1(x)$ and $\chi_2(x)$ in their expansion in terms of $\cos\frac{n\pi}{a}x$:

$$\left.\begin{aligned} \alpha_n^{(1)} &= \frac{2}{a}\int_0^a \chi_1(x)\cos\frac{n\pi x}{a}\,dx \\ \alpha_n^{(2)} &= \frac{2}{a}\int_0^a \chi_2(x)\cos\frac{n\pi x}{a}\,dx \end{aligned}\right\} \quad (n = 0, 1, 2, \ldots).$$

If the coefficients of $\cos\frac{n\pi x}{a}$ on both sides of equations (27) be compared, we find the system for the determination of A_n and B_n:

$$\left.\begin{aligned} A_n \operatorname{sh}\frac{n\pi b}{2a} + B_n \operatorname{ch}\frac{n\pi b}{2a} &= \frac{a}{n\pi}\alpha_n^{(1)} \\ -A_n \operatorname{sh}\frac{n\pi b}{2a} + B_n \operatorname{ch}\frac{n\pi b}{2a} &= \frac{a}{n\pi}\alpha_n^{(2)} \end{aligned}\right\} \quad (n = 1, 2, \ldots). \quad (27')$$

For $n = 0$ we shall obtain directly from (27) the equations:

$$A_0 = \frac{\alpha_0^{(1)}}{2}, \quad A_0 = \frac{\alpha_0^{(2)}}{2}.$$

The last equations can be satisfied only if $\alpha_0^{(1)} = \alpha_0^{(2)}$. Under the same condition we shall have $A_0 = \frac{\alpha_0^{(1)}}{2} = \frac{\alpha_0^{(2)}}{2} = \frac{\alpha}{2}$. On determining A_n and B_n from systems (27′) and substituting in (26) the values obtained, we finally find:

$$u_1 = \frac{\alpha}{2}y + B_0 + \sum_{n=1}^{\infty}\left[\frac{\alpha_n^{(1)} - \alpha_n^{(2)}}{\frac{2n\pi}{a}\operatorname{sh}\frac{n\pi b}{2a}}\operatorname{ch}\frac{n\pi y}{a} + \right.$$

$$\left. + \frac{\alpha_n^{(1)} + \alpha_n^{(2)}}{\frac{2n\pi}{a}\operatorname{ch}\frac{n\pi b}{2a}}\operatorname{sh}\frac{n\pi y}{a}\right]\cos\frac{n\pi x}{a}. \quad (28)$$

In the process of solution the peculiarities of the Neumann problem noted above have revealed themselves: first, an arbitrary constant B_0 has entered the solution; second, we have obtained a supplementary condition $\alpha_0^{(1)} - \alpha_0^{(2)} = 0$. This condition denotes nothing but condition (22) for the problem in hand, for indeed,

$$\alpha_0^{(1)} - \alpha_0^{(2)} = \frac{2}{a}\int_0^a \chi_1(x)dx - \frac{2}{a}\int_0^a \chi_2(x)dx =$$

$$= \frac{2}{a}\left[\int_{DC} \frac{\partial u_1}{\partial y} dx - \int_{AB} \frac{\partial u_1}{\partial y} dx\right] = -\frac{2}{a}\int_L \frac{\partial u_1}{\partial n} ds,$$

since along the remaining sides the latter integral is equal to zero. Thus the condition $\alpha_0^{(1)} - \alpha_0^{(2)} = 0$ is equivalent to (22).

The problem of the determination of the second part, u_2, may be posed and solved analogously.

We shall note in addition that the solution of the problem may be found by Fourier's method even for the case when on some sides of the rectangle the unknown function itself is given, and on others its normal derivative, and also when a linear combination of the function and its normal derivative is given.

The possibility of solving the fundamental boundary-value problems for the Laplace equation in the case of the rectangle also permits in many cases the obtaining of the solution of the same problems for the *Poisson equation*:

$$\Delta u = \frac{\partial^2 u}{\partial x^2} + \frac{\partial^2 u}{\partial y^2} = f(x, y). \tag{29}$$

Indeed, for the sake of definiteness let us speak about the Dirichlet problem: $u = \varphi(s)$ on L. Let us assume that we have succeeded in finding some particular solution or another, $u_0(x, y)$, of the Poisson equation: $\Delta u_0 = f$. We then introduce a new unknown function

$$u_1(x, y) = u(x, y) - u_0(x, y).$$

The function $u_1(x, y)$ must already satisfy the Laplace equation, for

$$\Delta u_1 = \Delta u - \Delta u_0 = f - f = 0.$$

The boundary conditions for u_1 will be, it is true, other ones, viz.: if we denote by $f_0(s)$ that known function to which $u_0(x, y)$ reduces on the contour L, then for u_1 we shall obtain the condition:

$$u_1 = f(s) - f_0(s) = f_1(s) \text{ on } L.$$

Thus the solution of the boundary-value problem for the Poisson equation always reduces to finding a particular solution of it and to the solution of the same boundary-value problem for the Laplace equation.

As regards the particular solution $u_0(x, y)$, it is often easy to find it by seeking it in form corresponding to the right member and containing undetermined coefficients. For example, if the right side $f(x, y)$ is a polynomial of the nth degree, the solution must be sought in the form of a polynomial of degree $n + 2$ with undetermined coefficients. Arbitrary values may be given to two of the coefficients of the terms of degree $k \leqslant n + 2$, the remaining ones then being uniquely determined. This arbitrariness may be utilized to attribute to the particular solution the simplest form possible.

Let us examine now one example of the solution of the Dirichlet problem, to wit, that in the case of the Poisson equation.

Example. The problem of the torsion of a rectangular prism.

By the theory of Saint-Venant, the problem of the torsion of any prismatic body whose section is the region D bounded by the contour L reduces to the following boundary-value problem: to find, the solution of the Poisson equation

$$\Delta u = -2, \tag{30}$$

that reduces to zero on the contour L:

$$u = 0 \text{ on } L. \tag{31}$$

Here the basic quantities required from the calculation are expressed in terms of the function u (called the *stress function*) as follows [1]: the components of the tangential stress

$$\tau_{zx} = G\vartheta \frac{\partial u}{\partial y}, \quad \tau_{zy} = -G\vartheta \frac{\partial u}{\partial x}, \tag{32}$$

and the torsional moment

$$M = G\vartheta \iint\limits_D u \, dx \, dy. \tag{33}$$

Here ϑ is the angle of twist per unit length, and G is the modulus of shear.

We shall now give the solution of the torsion problem for a rectangle with sides a and b. We have to find a solution of equation (30) reducing to zero on the contour. In order to reduce the problem to the solution of the Laplace equation, we seek a particular solution, u_0, of equation (30).

Let us seek it in the form $u_0 = Ax^2 + By^2$. Substituting u_0 in the

---

[1]) See P. F. Papkovich, [1]; L. S. Leĭbenzon, [1]; Filonenko-Borodich, [1]; J. W. Heckeler, [1], p. 17. The physical significance of the function u is this: if the displacement in the direction of the axis Z is $w = \vartheta\varphi(x, y)$, and if ψ is the function conjugate to φ, i.e., such that $\frac{\partial \varphi}{\partial x} = \frac{\partial \psi}{\partial y}$, $\frac{\partial \varphi}{\partial y} = -\frac{\partial \psi}{\partial x}$, then $u = \psi - \frac{x^2 + y^2}{2}$.

aforementioned equation, we obtain

$$2A + 2B = -2.$$

It is simplest to adopt $A = -1$ and $B = 0$. In addition, to the solution obtained one can add an arbitrary linear function. It is most convenient to adopt

$$u_0 = -x^2 + ax,$$

since then u_0 reduces to zero on the sides $x = 0$ and $x = a$. If one now introduces the unknown function $u_1 = u - u_0$, it will satisfy the equation $\Delta u_1 = 0$; the boundary conditions for it are

$$u_1 = -(ax - x^2) \text{ for } y = \pm \frac{b}{2}$$

$$u_1 = 0 \quad \text{,,} \quad x = 0,\ x = a.$$

For u_1 we can utilize solution (20), where in the present case $\varphi(x) = -(ax - x^2)$, and therefore

$$\beta_n = -\frac{2}{a}\int_0^a (ax - x^2)\sin\frac{n\pi x}{a}\,dx =$$

$$= -\Big[-\frac{2}{a}(ax - x^2)\frac{a}{n\pi}\cos\frac{n\pi}{a}x\Big]_0^a - \frac{2}{a}\frac{a}{n\pi}\int_0^a (a - 2x)\cos\frac{n\pi x}{a}\,dx =$$

$$= -\Big[\frac{2}{n\pi}(a - 2x)\frac{a}{n\pi}\sin\frac{n\pi x}{a}\Big]_0^a - \frac{2a}{n^2\pi^2}\int_0^a 2\sin\frac{n\pi x}{a}\,dx =$$

$$= -\frac{4a}{n^2\pi^2}\Big[-\frac{a}{n\pi}\cos\frac{n\pi x}{a}\Big]_0^a = -\frac{4a^2}{n^3\pi^3}[1 - \cos n\pi].$$

Thus for n even, $\beta_n = 0$, for n odd, $\beta_n = -\frac{8a^2}{n^3\pi^3}$. Substituting the value found for β_n in (20) we obtain

$$u_1 = -\sum_{n=1,3,5\,\ldots} \frac{8a^2 \operatorname{ch}\frac{n\pi y}{2a}}{\pi^3 n^3 \operatorname{ch}\frac{n\pi b}{a}}\sin\frac{n\pi x}{a}.$$

For u we finally obtain the solution

$$u = x(a - x) - \frac{8a^2}{\pi^3}\left[\frac{\operatorname{ch}\frac{\pi y}{a}}{1^3 \operatorname{ch}\frac{\pi b}{2a}}\sin\frac{\pi x}{a} + \frac{\operatorname{ch}\frac{3\pi y}{a}}{3^3 \operatorname{ch}\frac{3\pi b}{2a}}\sin\frac{3\pi x}{a} + \ldots\right] \tag{34}$$

By means of this solution the tangential stresses τ and moment M can be found. Thus for the moment we have

$$M = 2G\vartheta \int_0^a \int_{-\frac{b}{2}}^{\frac{b}{2}} u\,dx\,dy = 2G\vartheta \Big\{ b \Big[\frac{ax^2}{2} - \frac{x^3}{3} \Big]_0^a -$$

$$- \frac{8a^2}{\pi^3} \sum_{n=1,3,\ldots} \frac{1}{n^3 \operatorname{ch} \frac{n\pi b}{2a}} \Big[\frac{a}{n\pi} \operatorname{sh} \frac{n\pi y}{a} \Big]_{-\frac{b}{2}}^{\frac{b}{2}} \Big[-\frac{a}{n\pi} \cos \frac{n\pi x}{a} \Big]_0^a \Big\} =$$

$$= 2G\vartheta \Big\{ \frac{a^3 b}{6} - \frac{32a^4}{\pi^5} \sum_{n=1,3,\ldots} \frac{1}{n^5} \operatorname{th} \frac{n\pi b}{2a} \Big\}.$$

The series for τ_{zx} and τ_{zy} are obtained by differentiating series (34).

We subjoin some numerical data [1]). For the moment M we obtain the following values:

$$M = 0.1408 G\vartheta a^4 \quad \text{for} \quad \frac{a}{b} = 1$$

$$M = 0.1957 G\vartheta a b^3 \quad ,, \quad \frac{a}{b} = 1.5$$

$$M = 0.2287 G\vartheta a b^3 \quad ,, \quad \frac{a}{b} = 2.$$

Moreover we cite values of u at certain points for the case $a = 2$ and $b = 1$:

$$u(1, 0) = 0.91098; \; u(\tfrac{1}{2}, 0) = 0.77694$$

$$u(1, \tfrac{1}{4}) = 0.66374; \; u(\tfrac{1}{2}, \tfrac{1}{4}) = 0.58982.$$

We note that the mathematical problem just solved may be used otherwise as well in the theory of elasticity, viz.: the deflection u of a uniformly loaded membrane satisfies the equation

$$\Delta u = C,$$

where C is a constant. If the membrane is supported along the contour of the rectangle, then u must be equal to zero on the contour. It is clear, therefore, that the solution of the problem of a uniformly loaded membrane can easily be obtained from (34). In particular, the values for u exhibited above give us (apart from a constant factor) the magnitude of the deflection at various points of a membrane with sides 1 and 2.

---

[1]) See also J. W. Heckeler, [1], p. 30; P. F. Papkovich [1], p. 231.

2. The Dirichlet and Neumann problems for an annulus in the case of the Laplace equation. Let it be required to solve the Dirichlet problem for the Laplace equation

$$\Delta u = 0$$

in the case of the region between two concentric circles of radii R_1 and R_2 with center at the origin.

Introducing polar coordinates, we can rewrite the equation cited above in the form [1]):

$$r^2 \frac{\partial^2 u}{\partial r^2} + r \frac{\partial u}{\partial r} + \frac{\partial^2 u}{\partial \theta^2} = 0. \tag{35}$$

The infinite strip $(-\infty < \theta < +\infty,\ R_1 \leqslant r \leqslant R_2)$ serves as the region in polar coordinates; the boundary conditions are the periodicity, as a function of θ, of the solution sought, and the assigned values on the circumferences:

$$\left.\begin{aligned} &u(r, \theta + 2\pi) = u(r, \theta) \\ &u(R_1, \theta) = f_1(\theta),\ u(R_2, \theta) = f_2(\theta). \end{aligned}\right\} \tag{36}$$

We shall employ Fourier's method for the solution of the problem. Substituting $u = T(\theta)R(r)$, we shall obtain, after transformations,

$$\frac{T''(\theta)}{T(\theta)} = -\frac{r^2 R''(r) + rR'(r)}{R(r)} = -k^2. \tag{37}$$

Here the constant on the right side of (37) must be chosen negative, as in the contrary case it would be impossible to satisfy the first of conditions (36). The first of equations (37) gives

$$T(\theta) = A \cos k\theta + B \sin k\theta. \tag{38}$$

In order that the first of conditions (36) be satisfied, it is necessary that k be an integer: $k = 0, 1, 2, \ldots$.

The second equation takes the form:

$$r^2 R''(r) + rR'(r) - k^2 R(r) = 0.$$

For $k \neq 0$, this equation is of the Euler type, and its fundamental solutions are r^k and r^{-k}. For $k = 0$ we satisfy ourselves directly that its solutions are 1 and $\ln r$.

Taking this and (38) into consideration, we can write the solution u in the form of the series

$$u(r, \theta) = \sum_{k=1}^{\infty} [(A_k r^k + B_k r^{-k}) \cos k\theta +$$
$$+ (C_k r^k + D_k r^{-k}) \sin k\theta] + A_0 \ln r + B_0. \tag{39}$$

---

[1]) See Smirnov, [1], Vol. II, p. 342.

For the determination of the constants we expand the functions $f_1(\theta)$ and $f_2(\theta)$ in trigonometric series:

$$\left.\begin{aligned} f_1(\theta) &= \frac{\alpha_0^{(1)}}{2} + \sum_{k=1}^{\infty} (\alpha_k^{(1)} \cos k\theta + \beta_k^{(1)} \sin k\theta) \\ f_2(\theta) &= \frac{\alpha_0^{(2)}}{2} + \sum_{k=1}^{\infty} (\alpha_k^{(2)} \cos k\theta + \beta_k^{(2)} \sin k\theta). \end{aligned}\right\} \tag{40}$$

Using the last of conditions (36) and equating the coefficients of corresponding sines and cosines in (39) and (40), we obtain the following equations for the determination of the constants A, B, C and D:

$$\left.\begin{aligned} A_k R_1^k + B_k R_1^{-k} &= \alpha_k^{(1)}, \quad C_k R_1^k + D_k R_1^{-k} = \beta_k^{(1)}, \\ A_k R_2^k + B_k R_2^{-k} &= \alpha_k^{(2)}, \quad C_k R_2^k + D_k R_2^{-k} = \beta_k^{(2)}, \\ A_0 \ln R_1 + B_0 &= \frac{\alpha_0^{(1)}}{2}, \quad A_0 \ln R_2 + B_0 = \frac{\alpha_0^{(2)}}{2}. \end{aligned}\right\} \tag{41}$$

Finding the values of the constants from these equations and substituting in (39), we find, finally,

$$\begin{aligned} u = {} & \frac{\alpha_0^{(2)} - \alpha_0^{(1)}}{2(\ln R_2 - \ln R_1)} \ln r + \frac{\alpha_0^{(1)} \ln R_2 - \alpha_0^{(2)} \ln R_1}{2(\ln R_2 - \ln R_1)} + \\ & + \sum_{k=1}^{\infty} \Big[\frac{(\alpha_k^{(1)} R_2^{-k} - \alpha_k^{(2)} R_1^{-k}) r^k - (\alpha_k^{(1)} R_2^k - \alpha_k^{(2)} R_1^k) r^{-k}}{R_1^k R_2^{-k} - R_1^{-k} R_2^k} \cos k\theta + \\ & + \frac{(\beta_k^{(1)} R_2^{-k} - \beta_k^{(2)} R_1^{-k}) r^k - (\beta_k^{(1)} R_2^k - \beta_k^{(2)} R_1^k) r^{-k}}{R_1^k R_2^{-k} - R_1^{-k} R_2^k} \sin k\theta. \Big] \end{aligned} \tag{42}$$

It is readily verified directly that series (42) certainly converges for $R_1 < r < R_2$, since the quantities α_k and β_k tend to zero. Moreover, if the functions $f_1(\theta)$ and $f_2(\theta)$ satisfy the Dirichlet condition, series (40) converge, and therefore series (42) also converges for $r = R_1$ and $r = R_2$.

Let us consider the two most important limiting cases — when the annulus reduces to a circle, and when it becomes the region exterior to a circle. For the first case, adopting in (42)

$$R_1 = 0, R_2 = R, f_1(\theta) = \frac{\alpha_0}{2}, f_2(\theta) = f(\theta), \alpha_k^{(2)} = \alpha_k, \beta_k^{(2)} = \beta_k,$$

we find the solution of the Dirichlet problem for the *circle*

$$\begin{aligned} u(r, \theta) &= \frac{\alpha_0}{2} + \sum_{k=1}^{\infty} \Big(\frac{-\alpha_k r^k}{-R^k} \cos k\theta + \frac{-\beta_k r^k}{-R^k} \sin k\theta \Big) = \\ &= \frac{\alpha_0}{2} + \sum_{k=1}^{\infty} \Big(\frac{r}{R}\Big)^k (\alpha_k \cos k\theta + \beta_k \sin k\theta). \end{aligned} \tag{43}$$

In the second case, putting $R_1 = R$ and letting $R_2 \to \infty$,

$$f_1(\theta) = f(\theta),\ f_2(\theta) = \frac{\alpha_0}{2},$$

$$\alpha_k^{(1)} = \alpha_k,\ \beta_k^{(1)} = \beta_k,$$

we obtain the solution of the Dirichlet problem for the *region exterior to a circle* of radius R:

$$u(r, \theta) = \frac{\alpha_0}{2} + \sum_{k=1}^{\infty} \left(\frac{-\alpha_k r^{-k}}{-R^{-k}} \cos k\theta + \frac{-\beta_k r^{-k}}{-R^{-k}} \sin k\theta \right) =$$

$$= \frac{\alpha_0}{2} + \sum_{k=1}^{\infty} \left(\frac{R}{r}\right)^k (\alpha_k \cos k\theta + \beta_k \sin k\theta). \tag{44}$$

The solution obtained, $u(r, \theta)$, has as its limit as $r \to \infty$ the number $\frac{\alpha_0}{2}$.

The solutions (43) and (44) found here have been represented in the form of infinite series; however they can be transformed into definite integrals.

Doing this for solution (43), for instance, we obtain

$$u(r, \theta) = \frac{\alpha_0}{2} + \sum_{k=1}^{\infty} \left(\frac{r}{R}\right)^k (\alpha_k \cos k\theta + \beta_k \sin k\theta) =$$

$$= \frac{1}{2\pi} \int_0^{2\pi} f(\lambda) d\lambda + \sum_{k=1}^{\infty} \left(\frac{r}{R}\right)^k \left[\left(\frac{1}{\pi} \int_0^{2\pi} f(\lambda) \cos k\lambda \, d\lambda \right) \cos k\theta + \right.$$

$$\left. + \left(\frac{1}{\pi} \int_0^{2\pi} f(\lambda) \sin k\lambda \, d\lambda \right) \sin k\theta \right] =$$

$$= \frac{1}{2\pi} \int_0^{2\pi} \left[1 + 2 \sum_{k=1}^{\infty} \left(\frac{r}{R}\right)^k (\cos k\lambda \cos k\theta + \sin k\lambda \sin k\theta) \right] f(\lambda) d\lambda. \tag{45}$$

But the infinite series in brackets can be summed. Utilizing Euler's formulas, we find

$$1 + 2 \sum_{k=1}^{\infty} \left(\frac{r}{R}\right)^k \cos k(\lambda - \theta) = 1 + 2\mathrm{Re} \left[\sum_{k=1}^{\infty} \left(\frac{r}{R}\right)^k e^{i(\theta-\lambda)k} \right]^{1)} =$$

$$= 1 + 2\mathrm{Re} \left[\frac{\frac{r}{R} e^{i(\theta-\lambda)}}{1 - \frac{r}{R} e^{i(\theta-\lambda)}} \right] = \frac{R^2 - r^2}{R^2 - 2rR \cos(\theta - \lambda) + r^2};$$

---

1) Here and henceforth the sign Re signifies that the real part of the expression following it is to be taken.

substituting in (45) the expression obtained, we find the solution of the interior Dirichlet problem for the circle in a form known as the *Poisson integral*:

$$u(r, \theta) = \frac{1}{2\pi} \int_0^{2\pi} \frac{R^2 - r^2}{R^2 - 2rR \cos(\theta - \lambda) + r^2} f(\lambda) d\lambda. \text{ [1]} \tag{46}$$

For the exterior problem we find, analogously,

$$u(r, \theta) = \frac{1}{2\pi} \int_0^{2\pi} \frac{r^2 - R^2}{R^2 - 2rR \cos(\theta - \lambda) + r^2} f(\lambda) d\lambda. \tag{46'}$$

Such a summation and transformation of solution (42) into a definite integral for the case of an annulus cannot be effected in terms of elementary functions, but can be accomplished only by drawing upon special — elliptical — functions.

Just as the Dirichlet problem, the Neumann problem can be solved for an annulus, and for the interior and exterior of a circle. In the case of the annulus, the boundary conditions will consist in the assignment of the values of the normal derivative:

$$\left.\frac{\partial u(r, \theta)}{\partial u}\right|_{r=R_1} = f_1(\theta), \quad \left.\frac{\partial u(r, \theta)}{\partial r}\right|_{r=R_2} = f_2(\theta). \tag{47}$$

If we then seek the solution of the problem in form (39) and denote as before the Fourier coefficients of the functions $f_1(\theta)$ and $f_2(\theta)$ by $\alpha_k^{(1)}$, $\beta_k^{(1)}$, $\alpha_k^{(2)}$, $\beta_k^{(2)}$, we find

$$u(r, \theta) = \frac{\alpha_0^{(1)} R_1}{2} \ln r + B_0 +$$

$$+ \sum_{k=1}^{\infty} \frac{(\alpha_k^{(1)} R_2^{-k-1} - \alpha_k^{(2)} R_1^{-k-1}) r^k + (\alpha_k^{(1)} R_2^{k-1} - \alpha_k^{(2)} R_1^{k-1}) r^{-k}}{k(R_1^{k-1} R_2^{-k-1} - R_2^{k-1} R_1^{-k-1})} \cos k\theta +$$

$$+ \sum_{k=1}^{\infty} \frac{(\beta_k^{(1)} R_2^{-k-1} - \beta_k^{(2)} R_1^{-k-1}) r^k + (\beta_k^{(1)} R_2^{k-1} - \beta_k^{(2)} R_1^{k-1}) r^{-k}}{k(R_1^{k-1} R_2^{-k-1} - R_2^{k-1} R_1^{-k-1}} \sin k\theta, \tag{48}$$

where the solution of the problem is possible only on condition that

$$\alpha_0^{(1)} R_1 = \alpha_0^{(2)} R_2, \tag{49}$$

and moreover the solution is not unique, containing as it does the arbitrary constant B_0. Condition (49) refers to the fact that the integrals of the normal derivatives along both circumferences must be equal to each other, which is again equivalent to condition (22) of § 1 No. 1.

---

[1]) A more rigorous mathematical discussion of the solution of the Dirichlet problem for the rectangle and circle can be found, for example, in L. V. Kantorovich's book [1].

The solution found, (48), acquires a simpler form if the annulus reduces to a circle or the exterior of a circle, where as in (46) and (46'), it can be represented in the form of a definite integral [1]).

3. An example of the biharmonic problem. Problems in which a solution of the equation

$$\Delta\Delta u = \frac{\partial^4 u}{\partial x^4} + 2\frac{\partial^4 u}{\partial x^2\,\partial y^2} + \frac{\partial^4 u}{\partial y^4} = 0, \tag{50}$$

is to be found such as will satisfy on the contour of the region certain boundary conditions with respect to the function u itself and its partial derivatives of the first three orders, are generally called *biharmonic problems*. These conditions are formulated by the meaning of the problem itself. In a number of cases the Fourier method proves to be applicable to the solution of biharmonic problems. As an example of such a problem let us examine the problem of the deflection of a uniformly loaded rectangular plate supported along the four sides. The deflection of this plate, $u = u(x, y)$, is known to be given by the solution of the differential equation [2]):

$$\Delta\Delta u = \frac{p}{N}, \tag{51}$$

where p is the load per unit area and N is the rigidity of the plate. For the boundary conditions we have

$$\left.\begin{array}{lll} 1) & u = 0 & \\ & \dfrac{\partial^2 u}{\partial y^2} = 0 & \text{for } y = 0,\ y = b, \\ 2) & u = 0 & \\ & \dfrac{\partial^2 u}{\partial x^2} = 0 & \text{for } x = -\dfrac{a}{2},\ x = \dfrac{a}{2}. \end{array}\right\} \tag{52}$$

In order to reduce the problem to the solution of a homogeneous equation, we find a particular solution. For such a particular solution it is most convenient to take a polynomial of the fourth degree in y satisfying equation (51) and the first of conditions (52). This polynomial will obviously be

$$u_0 = \frac{pb^4}{24N}\left(\frac{y}{b} - 2\frac{y^3}{b^3} + \frac{y^4}{b^4}\right). \text{ [3])} \tag{53}$$

If we now denote the difference $u - u_0$ by u_1, the function $u_1(x, y)$

[1]) Numerous examples of the application of the Fourier method and a more general method developed by the author may be found in the extensive monograph of G. A. Grinberg, [1].

[2]) Heckeler, [1], p. 127; S. P. Timoshenko, [1]; B. G. Galerkin, [1].

[3]) The physical significance of this solution is that u_0 is the deflection of a freely supported beam of length b under a uniform load of intensity p/N.

will satisfy the homogeneous biharmonic equation $\Delta\Delta u_1 = 0$ and the boundary conditions

$$\left.\begin{array}{l} 1) \quad \left.\begin{array}{l} u_1 = 0 \\ \dfrac{\partial^2 u_1}{\partial y^2} = 0 \end{array}\right| \text{ for } y = 0,\ y = b, \\ 2) \quad \left.\begin{array}{l} u_1 = -\dfrac{pb^4}{24N}\left(\dfrac{y}{b} - 2\dfrac{y^3}{b^3} + \dfrac{y^4}{b^4}\right) \\ \dfrac{\partial^2 u_1}{\partial x^2} = 0 \end{array}\right| \text{ for } x = -\dfrac{a}{2},\ x = \dfrac{a}{2}. \end{array}\right\} \tag{54}$$

The function u_1 may be found by Fourier's method.

We seek fundamental solutions of homogeneous equation (50) in the form XY. Substituting in equation (50), we obtain

$$X^{IV}Y + 2X''Y'' + XY^{IV} = 0. \tag{55}$$

In order to obtain an equation involving only the function X, it is sufficient to take as Y a function satisfying the equation

$$Y'' = cY. \tag{56}$$

This function Y must in addition satisfy the first group of conditions (54). The first two of those conditions: $Y = 0$ for $y = 0$ and $y = b$, together with equation (56), already completely determine, as in the preceding cases, the sought fundamental functions Y:

$$Y_n = \sin\frac{n\pi y}{b} \quad (n = 1, 2, \ldots). \tag{57}$$

But these functions, as is readily verified, also satisfy the two other conditions of the same group, viz.:

$$\frac{\partial^2 Y}{\partial y^2} = 0 \text{ for } y = 0 \text{ and } y = b.$$

This fortunate circumstance holds only with the given special boundary conditions, not having a general character. Owing to this the given method is not applicable to any biharmonic problem but only to certain ones like that under consideration.

Continuing the investigation, let us substitute in (55) the above expressions for Y_n; for the determination of X_n we obtain the equation

$$X^{IV} - 2\frac{n^2\pi^2}{b^2}X'' + \frac{n^4\pi^4}{b^4}X = 0, \tag{58}$$

whence

$$X_n = a_n \operatorname{ch}\frac{n\pi x}{b} + b_n\frac{n\pi x}{b}\operatorname{sh}\frac{n\pi x}{b} + c_n \operatorname{sh}\frac{n\pi x}{b} + d_n\frac{n\pi x}{b}\operatorname{ch}\frac{n\pi x}{b}, \tag{59}$$

and the sought solution u_1 takes the form:

$$u_1 = \sum_{n=1}^{\infty} X_n \sin \frac{n\pi y}{b}. \tag{60}$$

For the determination of the constants a_n, b_n, c_n and d_n we must utilize the second group of boundary conditions (54). The coefficients c_n and d_n are to be taken equal to zero, since the solution must obviously be an even function of x.

For the determination of the coefficients a_n and b_n we use conditions (54). Expanding the right side of the next-to-the-last equation of (54) in a Fourier series, we shall have

$$-\frac{pb^4}{24N}\left(\frac{y}{b} - 2\frac{y^3}{b^3} + \frac{y^4}{b^4}\right) = -\frac{4pb^4}{\pi^5 N} \sum_{n=1,3,\ldots}^{\infty} \frac{1}{n^5} \sin \frac{n\pi y}{b}. \tag{61}$$

Applying now conditions (54) to expression (60), we shall obtain for the determination of a_n and b_n ($n = 1, 3, 5, \ldots$) the equations

$$a_n \operatorname{ch} \frac{n\pi a}{2b} + b_n \frac{n\pi a}{2b} \operatorname{sh} \frac{n\pi a}{2b} = -\frac{4pb^4}{\pi^5 N n^5},$$

$$a_n \frac{n^2\pi^2}{b^2} \operatorname{ch} \frac{n\pi a}{2b} + b_n \cdot \frac{n^2\pi^2}{b^2}\left[2 \operatorname{ch} \frac{n\pi a}{2b} + \frac{n\pi a}{2b} \operatorname{sh} \frac{n\pi a}{2b}\right] = 0,$$

a_n and b_n of even index proving to be equal to zero.

Solving the cited equations for a_n and b_n and substituting in (59) and (60) the values found for them, we shall finally obtain, after all transformations:

$$u = u_0 + u_1 =$$

$$= \frac{4pb^4}{\pi^5 N} \sum_{n=1,3\ldots}^{\infty} \frac{1}{n^5}\left[1 - \frac{2 \operatorname{ch}\alpha \operatorname{ch}\frac{n\pi x}{b} + \alpha \operatorname{sh}\alpha \operatorname{ch}\frac{n\pi x}{b} - \frac{n\pi x}{b} \operatorname{sh}\frac{n\pi x}{b} \operatorname{ch}\alpha}{1 + \operatorname{ch}\frac{n\pi a}{b}}\right] \sin\frac{n\pi y}{b}, \tag{62}$$

where $\alpha = \dfrac{n\pi a}{2b}$. This series converges so fast that it is ordinarily practicable to limit it to the first term,

$$u \cong \frac{4pb^4}{\pi^5 N}\left[1 - \frac{2 \operatorname{ch}\frac{\pi a}{2b} \operatorname{ch}\frac{\pi x}{b} + \frac{\pi a}{2b} \operatorname{sh}\frac{\pi a}{2b} \operatorname{ch}\frac{\pi x}{b} - \frac{\pi x}{b} \operatorname{sh}\frac{\pi x}{b} \operatorname{ch}\frac{\pi a}{2b}}{2 \operatorname{ch}^2 \frac{\pi a}{2b}}\right] \sin\frac{\pi y}{b}.$$

§ 2. INFINITE SYSTEMS OF EQUATIONS

1. Fundamental definitions. The theory of infinite systems of equations in an infinite set of unknowns began to be developed at the end of the last century; at the present time there is a vast literature devoted to it [1]). However, the theory is not in finished form as yet.

The theory of infinite systems of linear equations arose and developed in connection with its applications in problems of the integration of ordinary differential equations, in the theory of integral equations, and, most especially, in the solution of boundary-value problems of mathematical physics. In this section we shall consider the foundations of the theory of infinite systems; the application of infinite systems to the solution of some boundary-value problems is given in the section following.

The system of equations

$$\left.\begin{array}{l} a_{1,1}x_1 + a_{1,2}x_2 + \ldots = b_1 \\ a_{2,1}x_1 + a_{2,2}x_2 + \ldots = b_2 \\ \cdots\cdots\cdots\cdots \\ \cdots\cdots\cdots\cdots \end{array}\right\} \tag{1}$$

is called an *infinite system of linear equations* in an infinite set of unknowns. Here $a_{i,k}$ are the known coefficients, b_i the free members and x_i the unknowns. A set of numerical values of the quantities x_1, $x_2, \ldots$ is called a solution of system (1) if on substituting these values in the left member of equations (1) we obtain convergent series and all these equations are satisfied.

In studying and solving infinite systems the following problems arise:

1) to establish whether the system has a solution satisfying the given conditions;

2) to establish whether a solution is unique;

3) to indicate a method by means of which this solution may be found; generally speaking, to accomplish this, infinite operations are required;

4) to indicate devices for finding, by means of a finite number of operations, approximate values for the unknowns, and such as to make possible an estimate of the error of these approximations.

We shall consider here the solution of these problems for one important class of systems, those known as *regular systems*.

We note as a preliminary that by singling out x_i in the ith equation and transposing it to the left side, the given system may be given the

---

[1]) Of the literature in which this theory is expounded, we cite: V. F. Kagan, [1]; F. Riesz, [1]. The literature of the subject is referred to in an article of Hellinger and Toeplitz, [1], and in the collection "Matematika v SSSR za 30 let" [30 years of mathematics in the USSR].

following form:

$$x_i = \sum_{k=1}^{\infty} c_{i,k} x_k + b_i \quad (i = 1, 2, \ldots). \tag{2}$$

Henceforth it is this system of form (2) which we shall consider.

2. Theorems on the comparison of systems [1]). Before passing on to the study of the class of regular systems, we shall prove some general theorems concerning the comparison of two systems of type (2).

We shall call the system of equations

$$X_i = \sum_{k=1}^{\infty} C_{i,k} X_k + B_i \quad (i = 1, 2, \ldots) \tag{3}$$

majorant for the system (2) if the following inequalities hold:

$$\left.\begin{aligned} &|c_{i,k}| \leqslant C_{i,k} \ (i = 1, 2, \ldots;\ k = 1, 2, \ldots),\\ &|b_i| \leqslant B_i \ (i = 1, 2, \ldots). \end{aligned}\right\} \tag{4}$$

THEOREM I. (On the existence of a solution). *If the majorant system* (3) *has a nonnegative solution* $X_i' \geqslant 0$, *the given system* (2) *has the solution* x_i^*, *which may be found by the method of successive approximations and which is subject to the inequality*

$$|x_i^*| \leqslant X_i'. \tag{5}$$

Proof. Let us make a preliminary application of the method of successive approximations to system (3), with which aim in view we put $X_i^{(0)} = 0$ and next determine the $X_i^{(n)}$ by induction:

$$X_i^{(n+1)} = \sum_{k=1}^{\infty} C_{i,k} X_k^{(n)} + B_i \quad (i = 1, 2, \ldots). \tag{6}$$

Obviously we have $X_i^{(1)} = B_i \geqslant 0 = X_i^{(0)}$. If it has now been established already that $X_i^{(n)} \geqslant X_i^{(n-1)}$, we obtain on the basis of (6):

$$X_i^{(n+1)} = \sum_{k=1}^{\infty} C_{i,k} X_k^{(n)} + B_i \geqslant \sum_{k=1}^{\infty} C_{i,k} X_k^{(n-1)} + B_i = X_i^{(n)}.$$

Thus for all n and i we have $X_i^{(n+1)} \geqslant X_i^{(n)}$.

On the other hand $X_i^{(0)} = 0 \leqslant X_i'$. Let it have been established already that $X_i^{(n)} \leqslant X_i'$; then again on the basis of (6), using the fact that X_i' satisfies system (3), we conclude that

$$X_i^{(n+1)} = \sum_{k=1}^{\infty} C_{i,k} X_k^{(n)} + B_i \leqslant \sum_{k=1}^{\infty} C_{i,k} X_k' + B_i = X_i',$$

---

[1]) On the subject of the theorems of the No., see Pellet, [1], [2], [3] and L. V. Kantorovich, [2]. See also the literature on the theory of regular systems cited on p. 26.

where the convergence of the series on the right side of the inequality implies the convergence of that on the left side.

Thus on the basis of the induction principle for all i and n: $X_i^{(n)} \leqslant X_i'$. We have thus established that for fixed i the sequence $X_i^{(n)}$ grows with the increase of n, and remains bounded by the number X_i'; there accordingly exist limits:

$$\lim_{n\to\infty} X_i^{(n)} = X_i^* \leqslant X_i' \quad (i = 1, 2, \ldots). \tag{7}$$

We shall show that the numbers X_i^* obtained in this manner give the solution of system (3). Indeed, let us pass to the limit as $n \to \infty$ in equation (6). In the right member a term-by-term passage to the limit is admissible, for the series at the right converges uniformly as regards n, since it is majored by the series with constant terms $\sum_{k=1}^{\infty} C_{i,k} X_k'$. Thus on effecting this passage we find that

$$X_i^* = \sum_{k=1}^{\infty} C_{i,k} X_k^* + B_i, \tag{8}$$

i.e., X_i^* is indeed the solution of system (3).

Let us apply this same method of successive approximations to system (2). Accordingly we put $x_i^{(0)} = 0$ and then determine the $x_i^{(n)}$ inductively, using

$$x_i^{(n+1)} = \sum_{k=1}^{\infty} c_{i,k} x_k^{(n)} + b_i. \tag{9}$$

We shall now establish the inequality

$$|x_i^{(n+1)} - x_i^{(n)}| \leqslant X_i^{(n+1)} - X_i^{(n)}. \tag{10}$$

We have, obviously,

$$|x_i^{(1)} - x_i^{(0)}| = |b_i \leqslant B_i = X_i^{(1)} - X_i^{(0)}.$$

Let us assume that the inequality $|x_i^{(n)} - x_i^{(n-1)}| \leqslant X_i^{(n)} - X_i^{(n-1)}$ has already been established; then we have

$$|x_i^{(n+1)} - x_i^{(n)}| = |[\sum_{k=1}^{\infty} c_{i,k} x_k^{(n)} + b_i] - [\sum_{k=1}^{\infty} c_{i,k} x_k^{(n-1)} + b_i]| =$$

$$= |\sum_{k=1}^{\infty} c_{i,k}(x_k^{(n)} - x_k^{(n-1)})| \leqslant \sum_{k=1}^{\infty} C_{i,k}(X_k^{(n)} - X_k^{(n-1)}) = X_i^{(n+1)} - X_i^{(n)}.$$

Inequality (10) has thus been established. Using it, we see that the series

$$x_i^{(0)} + (x_i^{(1)} - x_i^{(0)}) + (x_i^{(2)} - x_i^{(1)}) + \ldots \tag{11}$$

converges, since it is majored by the converging series [cf. (7)]

$$X_i^{(0)} + (X_i^{(1)} - X_i^{(0)}) + (X_i^{(2)} - X_i^{(1)}) + \ldots = \lim_{n\to\infty} X_i^{(n)} = X_i^*. \tag{12}$$

Let us designate by x_i^* the sum of series (11):

$$x_i^{(0)} + (x_i^{(1)} - x_i^{(0)}) + \ldots = \lim_{n\to\infty} x_i^{(n)} = x_i^* . \tag{13}$$

From (12) and (13) it is clear that $|x_i^*| \leqslant X_i^*$. Finally, on passing to the limit in equation (9) we find that

$$x_i^* = \sum_{k=1}^{\infty} c_{i,k} x_k^* + b_i \quad (i = 1, 2, \ldots),$$

the termwise passage to the limit having been admissible since the series on the right is majored by the series $\sum_{k=1}^{\infty} C_{i,k} X_k'$ and therefore converges uniformly with respect to n. Thus x_i^* are the solutions of system (2), and $|x_i^*| \leqslant X_i^* \leqslant X_i'$, and the theorem is completely proved.

Observation 1. We will call the solutions x_i^* and X_i^* of systems (2) and (3), found by the method of successive approximations with zero initial values, the *principal solutions* of these systems.

Observation 2. We note that if the coefficients and free members of the system are nonnegative: $c_{i,k} \geqslant 0$, $b_i \geqslant 0$, so is the principal solution nonnegative: $x_i^* \geqslant 0$. Indeed, as is clear by induction from (9), all $x_i^{(n)} \geqslant 0$, and therefore $x_i^* \geqslant 0$ too.

Let us now pass to the question of the uniqueness of a solution of an infinite system. It should be stated here that the question of the uniqueness of the solution depends upon what conditions are imposed on the solution. Thus, for example, the system of equations

$$x_i = \tfrac{1}{2} x_{i+1} \quad (i = 1, 2, \ldots)$$

has no unique solution: together with the trivial solution $x_i = 0$, it has the solution $x_i = 2^i$. However, its bounded solution, i.e., the solution satisfying the condition $|x_i| < M$ $(i = 1, 2, \ldots)$, is unique; this is $x_i = 0$. In Theorem II we indicate the region in which the solution of the system is certainly unique.

THEOREM II. (On the uniqueness of the solution). *Under the conditions of, and employing the notation of the preceding theorem, the unique solution of system* (2) *satisfying the inequality* $|x_i| \leqslant PX_i^*$ ($P \geqslant 1$ *is a constant*) *is its principal solution. We arrive at this solution without fail by starting from any initial values* $\bar{x}_i^{(0)}$ *satisfying the conditions* $|\bar{x}_i^{(0)}| \leqslant PX_i^*$.

Proof. It will be more convenient to prove first the second of the statements of our theorem. It must be proved that starting with values of $\bar{x}_i^{(0)}$ satisfying the inequality $|\bar{x}_i^{(0)}| \leqslant PX_i^*$, and determining $\bar{x}_i^{(n)}$ by induction, putting

$$\bar{x}_i^{(n+1)} = \sum_{k=1}^{\infty} c_{i,k} \bar{x}_k^{(n)} + b_i , \tag{14}$$

we shall have $\lim_{n\to\infty} \bar{x}_i^{(n)} = x_i^*$. With this aim in view let us prove the inequality

$$|\bar{x}_i^{(n)} - x_i^{(n)}| \leqslant P(X_i^* - X_i^{(n)}). \tag{15}$$

For $n = 0$ the inequality is valid: $|\bar{x}_i^{(0)} - x_i^{(0)}| = |\bar{x}_i^{(0)}| \leqslant PX_i^* = P(X_i^* - X_i^{(0)})$. We shall now show the possibility of passing from n to $n + 1$. On the basis of (14), (9), (15), (6) and (8) we have

$$|\bar{x}_i^{(n+1)} - x_i^{(n+1)}| = |\sum_{k=1}^{\infty} c_{i,k}(\bar{x}_k^{(n)} - x_k^{(n)})| \leqslant P\sum_{k=1}^{\infty} C_{i,k}(X_k^* - X_k^{(n)}) =$$

$$= P[(\sum_{k=1}^{\infty} C_{i,k}X_k^* + B_i) - (\sum_{k=1}^{\infty} C_{i,k}X_k^{(n)} + B_i)] = P(X_i^* - X_i^{(n+1)}),$$

and inequality (15), with n replaced by $n + 1$, is established.

From it, on the basis of (7), we conclude that $(\bar{x}_i^{(n)} - x_i^{(n)}) \to 0$ and, since $x_i^{(n)} \to x_i^*$ [in accordance with (13)], so does $\bar{x}_i^{(n)} \to x_i^*$, and the second statement of the Theorem is established.

Now let $\bar{x}_i$ be some solution of system (2) satisfying the condition $|\bar{x}_i| \leqslant PX_i^*$. Let us put $\bar{x}_i^{(0)} = \bar{x}_i$; then all $\bar{x}_i^{(n)}$ determined successively by equation (14) will obviously be equal to $\bar{x}_i$. By what has already been proved we must have $\lim_{n\to\infty} \bar{x}_i^{(n)} = x_i^*$, but since the $\bar{x}_i^{(n)}$ do not in fact depend on n and are equal to $\bar{x}_i$, we have $\bar{x}_i = \bar{x}_i^*$, and the first part of the Theorem is also established.

For finding the solution of an infinite system of the type we have under consideration, besides the method of successive approximations, another method may be given. To provide a basis for this method we shall require a theorem concerning the passage to the limit in infinite systems, with which we shall accordingly begin.

THEOREM III (Concerning passing to the limit). *Let there be given a sequence of infinite systems of type* (2):

$$x_i = \sum_{k=1}^{\infty} \overset{s}{c}_{i,k} x_k + \overset{s}{b}_i \quad (i = 1, 2, \ldots), \tag{16}$$

$(s = 1, 2, \ldots)$, *having a common majorant system* (3) *with a nonnegative solution. Moreover let*

$$\lim_{s\to\infty} \overset{s}{c}_{i,k} = c_{i,k}, \quad \lim_{s\to\infty} \overset{s}{b}_i = b_i. \tag{17}$$

Then if x_i^* is the principal solution of system (2) *and $\overset{s}{x}_i^*$ is the principal solution of a system* (16),

$$\lim_{s\to\infty} \overset{s}{x}_i^* = x_i^*. \tag{18}$$

Proof. Let us apply to each of systems (16) the method of successive approximations, putting $\overset{s}{x}_i^{(0)} = 0$ and then by induction

$$\overset{s}{x}_i^{(n+1)} = \sum_{k=1}^{\infty} \overset{s}{c}_{i,k} \overset{s}{x}_k^{(n)} + \overset{s}{b}_i. \tag{19}$$

We shall show that

$$\lim_{s\to\infty} \overset{s}{x}_i^{(n)} = x_i^{(n)}. \tag{20}$$

For $n = 0$ this is obvious; we shall show the possibility of passing from n to $n + 1$. In equation (19) let us pass to the limit as $s \to \infty$. Thanks to the fact that the series on the right is majored by the convergent series with constant terms $\sum_k C_{i,k} X'_k$, a term-by-term passage to the limit is admissible, and on using (17), (20) and (19) we find that

$$\lim_{s\to\infty} \overset{s}{x}_i^{(n+1)} = \lim_{s\to\infty} [\sum_{k=1}^{\infty} \overset{s}{c}_{i,k} \overset{s}{x}_k^{(n)} + \overset{s}{b}_i] = \sum_{k=1}^{\infty} c_{i,k} x_k^{(n)} + b_i = x_i^{(n+1)},$$

and relation (20) is established for any n. Furthermore we have [cf. (13)]:

$$\overset{s}{x}_i^{*} = \overset{s}{x}_i^{(0)} + (\overset{s}{x}_i^{(1)} - \overset{s}{x}_i^{(0)}) + (\overset{s}{x}_i^{(2)} - \overset{s}{x}_i^{(1)}) + \ldots;$$

here the series on the right is majored by series (12) for all s, and a term-by-term passage to the limit as $s \to \infty$ is therefore admissible. Effecting it, using (20) and (13) we find

$$\lim_{s\to\infty} \overset{s}{x}_i^{*} = x_i^{(0)} + (x_i^{(1)} - x_i^{(0)}) + (x_i^{(2)} - x_i^{(1)}) + \ldots = x_i^{*},$$

and relation (18), and with it the Theorem, is proved.

THEOREM IV (Concerning the method of reduction). *The solution of system* (2), *having the majorant system* (3) *with nonnegative solution, may be found by the method of reduction, i.e., by means of a passage to the limit in the solution of the finite system obtained from the given infinite system by discarding all equations and unknowns commencing with a certain one. More specifically, if the finite system*

$$\left.\begin{aligned} x_1 &= c_{1,1}x_1 + c_{1,2}x_2 + \ldots + c_{1,N}x_N + b_1 \\ x_2 &= c_{2,1}x_1 + c_{2,2}x_2 + \ldots + c_{2,N}x_N + b_2 \\ &\ldots\ldots\ldots\ldots\ldots\ldots\ldots\ldots \\ &\ldots\ldots\ldots\ldots\ldots\ldots\ldots\ldots \\ x_N &= c_{N,1}x_1 + c_{N,2}x_2 + \ldots + c_{N,N}x_N + b_N \end{aligned}\right\} \tag{21}$$

be formed and its principal solution[1] *be denoted by* $\overset{N}{x}_i$ $(i = 1, 2, \ldots, N)$,

---

[1] As we shall see below (No. 3, Observation 2 to Theorem IIa), the finite systems of form (21) have, as a rule, a unique solution, and it would therefore be possible here to speak simply of the solution of system (21).

then

$$\lim_{N\to\infty} \overset{N}{x_i} = x_i^*, \tag{22}$$

where x_i^* denotes the principal solution of system (2).

Proof. Let us examine, along with system (21), the following infinite system of equations ($N = 1, 2, \ldots$):

$$\left.\begin{aligned} x_i &= \sum_{k=1}^{\infty} c_{i,k}x_k + b_i \quad (i = 1, 2, \ldots, N), \\ x_i &= 0 \qquad\qquad\qquad (i = N+1, N+2, \ldots). \end{aligned}\right\} \tag{23}$$

As the principal solution of this system the sequence of numbers $\overset{N}{x_1}$, $\overset{N}{x_2}, \ldots, \overset{N}{x_N}, 0, 0, \ldots$ will obviously serve, or, if $\overset{N}{x_i} = 0$ be adopted for $i > N$, it can be said more briefly that the solution of the Nth system (23) will be the sequence $\overset{N}{x_i}$ ($i = 1, 2, \ldots$). Moreover it is clear that system (2) serves as the limit for the sequence of systems (23). Indeed, as soon as $N \geqslant i$, the coefficient of system (23) standing in the (i, k)th place, $c_{i,k}^N$, is equal to $c_{i,k}$, and therefore the coefficients (and free terms) of system (23) indeed tend to the corresponding coefficients and free terms of system (2). Finally, it is quite obvious that system (3) serves as a common majorant for systems (23). Thus to the sequence of systems (23) Theorem III is applicable. Using it, we conclude that the principal solution of system (23) converges to the principal solution of system (2):

$$\lim_{N\to\infty} \overset{N}{x_i} = x_i^*, \tag{22}$$

Q.E.D.

3. Regular and fully regular systems. We shall call an infinite system of type (2) for which the sum of the moduli of the coefficients of each row is less than unity,

$$\sum_{k=1}^{\infty} |c_{i,k}| < 1 \quad (i = 1, 2, \ldots), \tag{24}$$

a *regular system*.

If this sum does not exceed a constant number less than unity:

$$\sum_{k=1}^{\infty} |c_{i,k}| \leqslant 1 - \theta < 1 \quad (i = 1, 2, \ldots), \tag{25}$$

the system will be called *fully regular* [1])

---

[1]) On the theory of regular and fully regular systems, see B. M. Koialovich, [1], R. O. Kuz'min, [1], [2].
See also Dixon [1], Koch, H. v., [1], Wintner, A., [1].

We introduce the following designation:

$$\varrho_i = 1 - \sum_{k=1}^{\infty} |c_{i,k}| \quad (i = 1, 2, \ldots). \tag{26}$$

For regular systems the numbers $\varrho_i > 0$; for fully regular systems $\varrho_i \geqslant \theta > 0$.

The theorems of the preceding No. concerning the comparison of systems permit us to at once obtain a series of results with respect to regular and fully regular systems.

Let us assume that a regular system (2) has been given, the free terms of which satisfy the condition

$$|b_i| \leqslant K\varrho_i, \tag{27}$$

where $K > 0$ is a constant. Then the system of equations

$$X_i = \sum_{k=1}^{\infty} |c_{i,k}| X_k + K\varrho_i \tag{28}$$

will be majorant for system (2); indeed, thanks to (27) it is clear that conditions (4) are observed. System (28) has an obviously positive solution $X_i = K > 0$, for on the basis of (26) it is clear that

$$K = \sum_{k=1}^{\infty} |c_{i,k}| K + K\varrho_i.$$

Thanks to this, Theorem I is applicable to system (2). On using it we obtain the following theorem concerning regular systems.

THEOREM Ia. *A regular system of equations, the free terms of which satisfy condition* (27), *has the bounded solution*

$$|x_i| \leqslant K, \tag{29}$$

which may be found by the method of successive approximations.

Corollary. In the case of a fully regular system, condition (27), since $\varrho_i \geqslant \theta$, may be replaced by the condition $|b_i| \leqslant K\theta = P$. Since the number P is arbitrary, we see that the solution of a fully regular system exists for any bounded set of free terms; and if $|b_i| \leqslant P$, then

$$|x_i| \leqslant \frac{P}{\theta} \ (= K).$$

Observation 1. By the choice of a suitable K, condition (27) can certainly be satisfied, if of the free terms of the system all except a finite number are equal to zero. Thus in this case a regular system has certainly a bounded solution.

Observation 2. We have carried out a proof of the existence of a solution for a regular system under condition (27) by using the fact that such systems represent a particular case of systems admitting a majorant with a nonnegative solution. But in reality this particular case almost coincides with the general one; for any system admitting a majorant with positive free terms $B_i > 0$ and positive solution X'_i can be reduced to a regular system. Indeed, introducing new unknowns in system (2) by putting $z_i = \frac{x_i}{X'_i}$, we may rewrite it thus:

$$z_i = \sum_{k=1}^{\infty} \left(c_{i,k} \frac{X'_k}{X'_i} \right) z_k + \frac{b_i}{X'_i}. \tag{30}$$

Then, since X'_i is the solution of system (3),

$$\sum_{k=1}^{\infty} |c_{i,k}| \frac{X'_k}{X'_i} \leqslant \frac{1}{X'_i} \sum_{k=1}^{\infty} C_{i,k} X'_k = \frac{X'_i - B_i}{X'_i} < 1$$

and

$$\frac{b_i}{X'_i} \leqslant \frac{B'_i}{X'_i} = 1 - \frac{X'_i - B_i}{X'_i} \leqslant 1 - \sum_{k=1}^{\infty} |c_{i,k}| \frac{X'_k}{X'_i} = \varrho_i,$$

i.e., system (30) is regular, and condition (27) is fulfilled.

Let us pass on to the question of the uniqueness of the solution for regular systems. Here Theorem II allows us to establish the following proposition.

THEOREM IIa. *If the principal solution of the system of equations* (28) *is bounded below by the positive number*

$$X_i^* \geqslant \alpha > 0, \tag{31}$$

the regular system (2), *under condition* (27), *has a unique bounded solution, to which the method of successive approximations leads from whatever initial values* $\bar{x}_i^{(0)}$ *(bounded in the aggregate) we start. Lastly, in this case* $X_i^* = K$ *invariably.*

Indeed, in accordance with Theorem II, the unique solution of system (2) satisfying the condition $|x_i| \leqslant PX_i^*$ is the principal solution. But for any bounded solution, for suitable P we shall have $|x_i| \leqslant P\alpha$, i.e., thanks to (31), the condition $|x_i| \leqslant PX_i^*$ will be satisfied for it, and such a solution cannot, therefore, be distinct from the principal one. The second statement is obtained in exactly the same manner from the second half of Theorem II.

Lastly, the final statement, that $X_i^* = K$, is at once obtained if the first statement of the Theorem be applied to system (28), which serves as its own majorant. We see, then, that its bounded solution $X'_i = K$ must coincide with the principal one, $K = X_i^*$.

Observation 1. For fully regular systems condition (31) is always satisfied. Indeed, in this case

$$X_i^* \geqslant X_i^{(1)} = B_i = K\varrho_i \geqslant K\theta > 0.$$

Therefore a fully regular system always has a unique bounded solution [1]), which

[1]) This fact — that a bounded solution of a fully regular system is unique — may be established directly very simply. To establish it, one must show that the bounded solution of the homogeneous system $x_i = \sum c_{i,k} x_k$ is zero. But, denoting by M the exact upper bound of $|x_i|$, we obtain $|x_i| \leqslant \sum |c_{i,k}| M \leqslant M(1 - \theta)$, and therefore $M \leqslant M(1 - \theta)$, i.e., $M = 0$.

may be found by the method of successive approximations, starting from any bounded system of initial values.

Observation 2. A regular system composed of a finite number of equations is unfailingly fully regular. Indeed, as θ the least of the numbers ϱ_i may be adopted. Therefore in accordance with the previous observation, a bounded solution of such a system is unique; but since there cannot be unbounded solutions of a finite system, the solution of such a system is unique generally. A finite system of equations having a majorant with positive free terms and a positive solution is reducible, as we saw above (Observation 3 to Theorem Ia) to a regular finite system, and therefore such a system has a unique solution. We have already referred to this fact above (Note on page 25).

Observation 3. A regular system may have even several bounded solutions. Thus by direct substitution one can convince oneself that the system

$$x_i = \left(1 - \frac{1}{(i+1)^2}\right) x_{i+1} + \frac{1}{(i+1)^2} \qquad (i = 1, 2, \ldots) \tag{32}$$

has two different solutions

$$x_i = 1 \qquad (i = 1, 2, \ldots)$$

and

$$x_i = \frac{1}{i+1} \qquad (i = 1, 2, \ldots)$$

The principal one will be the second of these solutions, which is readily verified directly and which follows from the theorem next following.

THEOREM II*b* [1]). *A regular system can have no more than one solution tending to zero, that is, such that* $\lim_{i\to\infty} x_i = 0$. *Moreover if the coefficients and free terms of a regular system are positive, the positive solution of it that tends to zero is always its principal solution.*

We shall first of all show that the homogeneous regular system cannot have a solution tending to zero and different from zero. Indeed, let x_i be such a solution of the system:

$$x_i = \sum_{k=1}^{\infty} c_{i,k} x_k, \quad \sum_{k=1}^{\infty} |c_{i,k}| < 1. \quad (i = 1, 2, \ldots)$$

Let us denote by $Q > 0$ the exact upper bound of $|x_i|$. Since $x_i \to 0$, we will have, beginning with some i $(i \geqslant i_1)$, $|x_i| < \frac{1}{2}Q$, and therefore it must be possible to find an $i_0 < i_1$ such that $|x_{i_0}| = Q$. From the equation of the system of index i_0 we have

$$Q = |x_{i_0}| \leqslant \sum_{k=1}^{\infty} |c_{i_0,k}|\, |x_k| \leqslant Q \sum_{k=1}^{\infty} |c_{i_0,k}|,$$

whence it follows at once that $Q = 0$, and therefore all $x_i = 0$.

Hence issues directly the first statement of the Theorem, since if the non-homogeneous regular system had two solutions tending to zero,

---

[1]) This theorem and its corollary belong to P. S. Bondarenko.

x'_i and x''_i, their difference $x_i = x'_i - x''_i$ would represent just such a solution of the corresponding homogeneous system, which by what has been proved is possible only if $x_i = 0$, i.e., $x'_i = x''_i$.

Moving on to proof of the second statement, we will note first of all that if the system with $c_{i,k} \geqslant 0$ and $b_i \geqslant 0$ has a positive solution $\{x_i\}$, the principal solution does not exceed it: $0 \leqslant x_i^* \leqslant x_i$. This follows at once from Theorem I, if we adopt $X'_i = x_i$ in it.

Therefore if a given solution is known to tend to zero, $x_i \to 0$, we will have $x_i^* \to 0$ too, and then by what has been proved above we must have $x_i = x_i^*$, i.e., $\{x_i\}$ is the principal solution.

Corollary. If for any substitution of the form

$$x_i = H_i z_i, \quad (H_i \neq 0), \quad \lim_{i\to\infty} H_i = \infty$$

in the regular system (2), the system in the unknowns z_i obtained as the result of this substitution,

$$z_i = \sum_{k=1}^{\infty} \frac{c_{i,k}H_k}{H_i} z_i + \frac{b_i}{H_i} \quad (i = 1, 2, \ldots) \tag{2a}$$

turns out to be regular again, i.e.,

$$\sum_{k=1}^{\infty} |c_{i,k}| \left| \frac{H_k}{H_i} \right| = 1 - \varrho'_i, \quad \varrho'_i > 0, \quad |b_i| \leqslant K' H_i \varrho'_i,$$

then the given system has a unique bounded solution.

But indeed, each bounded solution x_i of system (2) gives the solution $z_i = \frac{1}{H_i} x_i$ of system (2a), which tends to zero, and since system (2a) can have not more than one such solution, the bounded solution of system (2) is unique.

This proposition gives a valuable method of establishing the uniqueness of bounded solutions for regular systems. An example of its application is given in § 3 No. 5.

We shall not formulate Theorem III specially for the case of regular systems; in view of the importance of Theorem IV, we give a formulation of it.

THEOREM IVa. *The principal solution of a regular system*

$$x_i = \sum_{i=1}^{\infty} c_{i,k} x_k + b_i \quad (i = 1, 2, \ldots), \tag{2}$$

with free terms satisfying the condition $|b_i| \leqslant K\varrho_i$, *may be found by the method of reduction, i.e., if* $\overset{N}{x}_i$ *is the solution of the finite system*

$$x_i = \sum_{k=1}^{N} c_{i,k} x_k + b_i \quad (i = 1, 2, \ldots N), \tag{21}$$

then

$$x_i^* = \lim_{N\to\infty} \overset{N}{x_i}. \tag{22}$$

We had the right to speak simply of the solution, and not of the principal solution, of the finite system (21), since the solution of this system, as we saw above (Observation 2 to Theorem IIa) is unique.

In conclusion, in order to show the possibility of using the Theorems of this No., we conduct an investigation of one particular infinite system.

Example. Let us consider the system of equations

$$\sum_{k=1}^{\infty} \frac{x_k}{(2i+1-2k)(2i-1-2k)} + b_i = 0 \qquad (i = 1, 2, \ldots) \tag{33}$$

or, in extended form,

$$\left.\begin{aligned} x_1 &= 0\cdot x_1 + \frac{1}{1\cdot 3}x_2 + \frac{1}{3\cdot 5}x_3 + \frac{1}{5\cdot 7}x_4 + \ldots + b_1 \\ x_2 &= \frac{1}{1\cdot 3}x_1 + 0\cdot x_2 + \frac{1}{1\cdot 3}x_3 + \frac{1}{3\cdot 5}x_4 + \ldots + b_2 \\ x_3 &= \frac{1}{3\cdot 5}x_1 + \frac{1}{1\cdot 3}x_2 + 0\cdot x_3 + \frac{1}{1\cdot 3}x_4 + \ldots + b_3 \\ &\ldots\ldots\ldots\ldots\ldots\ldots\ldots\ldots \end{aligned}\right\} \tag{33'}$$

This system of equations will be regular, but not fully regular. Indeed, in the given case we have

$$\sum_{k=1}^{\infty} c_{i,k} = \sum_{k=1}^{k=i-1} \frac{1}{(2i+1-2k)(2i-1-2k)} + \sum_{k=i+1}^{\infty} \frac{1}{(2i+1-2k)(2i-1-2k)} =$$

$$= \frac{1}{1\cdot 3} + \frac{1}{3\cdot 5} + \ldots + \frac{1}{(2i-3)(2i-1)} + \left(\frac{1}{1\cdot 3} + \frac{1}{3\cdot 5} + \frac{1}{5\cdot 7} + \ldots\right) =$$

$$= \frac{1}{2}\left(1 - \frac{1}{2i-1}\right) + \frac{1}{2} = 1 - \frac{1}{4i-2}.$$

Therefore in the case given,

$$\varrho_i = \frac{1}{4i-2},$$

$\varrho_i > 0$, but it is obviously impossible to indicate such a $\theta > 0$ that $\varrho_i \geqslant \theta$. On the basis of Theorem Ia, one can draw the conclusion that the system of equations (33) certainly has a solution, and the method of successive approximations is applicable, if its free terms satisfy the conditions

$$|b_i| \leqslant \frac{K}{4i-2}. \tag{34}$$

The last condition is equivalent to this, that $|b_i| \leqslant \frac{K_1}{i}$, i.e., the free terms are not of lower order than $\frac{1}{i}$: $|b_i| = O\left(\frac{1}{i}\right)$.

Furthermore, in order to establish the uniqueness of the bounded solution for system (33), it must be established that the system of type (28) formed for the given case has a principal solution bounded from below: $X_i^* \geqslant \alpha > 0$.

The last system has just the same coefficients as does (33), and the free terms $\varrho_i = \frac{1}{4i - 2}$ (we put $K = 1$, which is inessential), so that it takes the form

$$\left.\begin{aligned} X_1 &= \frac{1}{1 \cdot 3} X_2 + \frac{1}{3 \cdot 5} X_3 + \frac{1}{5 \cdot 7} X_4 + \ldots + \frac{1}{2} \\ X_2 &= \frac{1}{1 \cdot 3} X_1 + \frac{1}{1 \cdot 3} X_3 + \frac{1}{3 \cdot 5} X_4 + \ldots + \frac{1}{6} \\ &\ldots\ldots\ldots\ldots\ldots\ldots \\ X_i &= \frac{1}{(2i-1)(2i-3)} X_1 + \frac{1}{(2i-3)(2i-5)} X_2 + \ldots + \frac{1}{1 \cdot 3} X_{i-1} + \\ &+ \frac{1}{1 \cdot 3} X_{i+1} + \frac{1}{3 \cdot 5} X_{i+2} + \ldots + \frac{1}{4i-2} \\ &\ldots\ldots\ldots\ldots\ldots\ldots \\ &\ldots\ldots\ldots\ldots\ldots\ldots \end{aligned}\right\} \tag{35}$$

We shall show that $X_i^* \geqslant \frac{1}{2}$. For this we make a direct estimate of the successive approximations.

Starting with the values $X_i^{(0)} = 0$, for the first approximation we have

$$X_i^{(1)} = \frac{1}{4i - 2}.$$

We shall show that for any nth approximation we have the valid inequality

$$X_i^{(n)} \geqslant \frac{n}{4i - 4 + 2n}. \tag{36}$$

It is true for $n = 1$. We shall show that it remains valid on passing from n to $n + 1$. Indeed, on the basis of the ith equation of system (35) and inequalities (36), we have:

$$\begin{aligned} X_i^{(n+1)} &= \frac{1}{(2i-1)(2i-3)} X_1^{(n)} + \frac{1}{(2i-3)(2i-5)} X_2^{(n)} + \ldots + \frac{1}{1 \cdot 3} X_{i-1}^{(n)} + \\ &+ \frac{1}{1 \cdot 3} X_{i+1}^{(n)} + \ldots + \frac{1}{4i-2} \geqslant \frac{1}{(2i-1)(2i-3)} \frac{n}{2n} + \\ &+ \frac{1}{(2i-3)(2i-5)} \frac{n}{2n+4} + \ldots + \frac{1}{1 \cdot 3} \frac{n}{4(i-1) - 4 + 2n} + \\ &+ \frac{1}{1 \cdot 3} \frac{n}{4(i-1) + 4 + 2n} + \ldots + \frac{1}{4i-2} \geqslant \\ &\geqslant \frac{1}{1 \cdot 3}\left(\frac{n}{4(i-1) + 2n + 4} + \frac{n}{4(i-1) + 2n - 4}\right) + \end{aligned}$$

$$+ \frac{1}{3 \cdot 5}\left(\frac{n}{4(i-1)+2n+8} + \frac{n}{4(i-1)+2n-8}\right) + \ldots$$

$$+ \frac{1}{(2i-3)(2i-1)}\left(\frac{n}{4(i-1)+2n+4(i-1)} + \right.$$

$$\left. + \frac{n}{4(i-1)+2n-4(i-1)}\right) + \frac{1}{4i-2}.$$

Moreover, in view of the concavity of the function $\frac{n}{4(i-1)+2n+x}$ we have the inequality

$$\frac{n}{4(i-1)+2n-j} + \frac{n}{4(i-1)+2n+j} > \frac{2n}{4(i-1)+2n}.$$

On utilizing it, we impart to the preceding estimate the following form:

$$X_i^{(n+1)} > \frac{2n}{4(i-1)+2n}\left(\frac{1}{1 \cdot 3} + \frac{1}{3 \cdot 5} + \ldots + \frac{1}{(2i-3)(2i-1)}\right) + \frac{1}{4i-2} =$$

$$= \frac{2n}{4i-4+2n}\left(\frac{1}{2} - \frac{1}{4i-2}\right) + \frac{1}{4i-2} = \frac{1}{4i-2} + \frac{n(4i-4)}{[(4i-4+2n)(4i-2)]} >$$

$$> \frac{1}{4i-2} + \frac{n(4i-4)}{(4i-2)(4i-2+2n)} = \frac{n+1}{4i-4+2(n-1)},$$

and inequality (36) is established. Passing to the limit in it as $n \to \infty$, we find

$$X_i^* = \lim_{n\to\infty} X_i^{(n)} \geqslant \lim_{n\to\infty} \frac{n}{4i-4+2n} = \frac{1}{2}.$$

Hence it follows, on the basis of Theorem IIa, that the bounded solution of system (33) is unique. Thus we have shown that with free terms of order not lower than $\frac{1}{i}$, system (33) has a bounded solution, and this is unique.

4. The approximate solution of regular systems. We shall now show how deficient (lower) or excessive (upper) approximations to the solution of a regular infinite system can be found by means of the solution of finite systems.

Let us consider a regular system satisfying condition (27). Let us first assume that $c_{i,k} \geqslant 0$ and $b_i \geqslant 0$. In this case the solutions $\overset{N}{x}_i$ of the finite system (21) give approximations for the unknowns from below. Indeed, the same numbers $\overset{N}{x}_i$ are solutions of infinite system (23), which is obviously majored by system (2), and therefore we have for its principal solutions $\overset{N}{x}_i \leqslant x_i^*$.

It will be more convenient here for us to denote a deficient approximation by $\tilde{x}_i$ (omitting the N), so that $\tilde{x}_i = \overset{N}{x}_i \leqslant x_i^*$. For obtaining excessive approximations we reason thus. We would have obtained exact values for the numbers $x_1, x_2, \ldots, x_N$ had we substituted in the first N equations of system (2), in place of $x_{N+1}, x_{N+2}, \ldots$, their exact values, and had then solved the resulting system of N equations in N unknowns:

$$\left.\begin{array}{l} x_1 = c_{1,1}x_1 + c_{1,2}x_2 + \ldots + c_{1,N}x_N + b_1 + \sum\limits_{k=N+1}^{\infty} c_{1,k}x_k^* \\ \ldots\ldots\ldots\ldots\ldots\ldots\ldots\ldots\ldots\ldots \\ x_N = c_{N,1}x_1 + c_{N,2}x_2 + \ldots + c_{N,N}x_N + b_N + \sum\limits_{k=N+1}^{\infty} c_{N,k}x_k^* . \end{array}\right\} \quad (37)$$

However we do not know the exact values of the unknowns x_{N+1}, $x_{N+2}, \ldots$ which figure in the free terms of system of equations (37); we do know, however, that none of them exceed K; therefore by replacing the numbers x_k^* $(k = N + 1, N + 2, \ldots)$ by K in system (37) we will obtain a finite (regular) system of equations majoring it and therefore having solutions not less than the numbers $x_1^*, x_2^*, \ldots, x_N^*$.

So if we denote by $\bar{x}_1, \ldots, \bar{x}_N$ the solutions of the system

$$\left.\begin{array}{l} x_1 = c_{1,1} + c_{1,2}x_2 + \ldots + c_{1,N}x_N + b_1 + K \sum\limits_{k=N+1}^{\infty} c_{1,k} \\ \ldots\ldots\ldots\ldots\ldots\ldots\ldots\ldots\ldots\ldots \\ x_N = c_{N,1}x_1 + c_{N,2}x_2 + \ldots + c_{N,N}x_N + b_N + K \sum\limits_{k=N+1}^{\infty} c_{N,k}, \end{array}\right\} \quad (38)$$

we will have $\bar{x}_i \geqslant x_i^*$. Thus for $i = 1, 2, \ldots, N$

$$\tilde{x}_i \leqslant x_i^* \leqslant \bar{x}_i . \quad (39)$$

If, moreover, we put $\tilde{x}_i = 0$ and $\bar{x}_i = K$ for $i > N$, inequality (39) will be valid for all values of i without exception.

Let us now pass on to the question of what happens to these approximations as $N \to \infty$. With respect to the approximations with a deficiency, $\tilde{x}_i = \overset{N}{x_i}$, it has already been established by us that as $N \to \infty$ they converge to the principal solutions of the system [see (22)]:

$$\lim_{N\to\infty} \tilde{x}_i = x_i^* . \quad (40)$$

As regards the approximations with an excess, $\bar{x}_i$, it is clear that they do not increase with an increase of N.

Indeed, if we examine a system of form (38) with $(N + 1)$ equations, we find that for the group of the first N equations of this system in the unknowns $\bar{x}_1, \bar{x}_2, \ldots, \bar{x}_N$, system (38) with N equations will serve as majorant, since the second is obtained from the first by the replacement of $\bar{x}_{N+1}$ by the number K, known to be not less. Therefore there exist the limits

$$\lim_{N\to\infty} \bar{x}_i = x_i .$$

On passing to the limit in each equation of system (38), we convince ourselves that, thanks to the fact that

$$\lim_{N\to\infty} \sum_{k=N+1}^{\infty} c_{i,k} = 0,$$

these limits x_i satisfy the system of equations (2). If for system (2), therefore, the conditions securing the uniqueness of the bounded solution (Theorem IIa) are fulfilled — in particular, if it is fully regular — we have $x_i = x_i^*$, and therefore

$$\lim_{N\to\infty} \bar{x}_i = x_i^*. \tag{41}$$

We have just given a method of finding approximations with deficiency and with excess, which approach the sought values in the case of positive systems (with unique solution). We shall now show a quite similar method for the case of systems with coefficients and free terms of arbitrary sign.

For finding the approximations one can proceed thus. Let us employ the designation

$$R_N^{(i)} = \sum_{k=N+1}^{\infty} |c_{i,k}| \tag{42}$$

and form the system of $2N$ equations in $2N$ unknowns $\tilde{x}_i$ and $\bar{x}_i$:

$$\left.\begin{aligned} \tilde{x}_i &= \sum_{k=1}^{N}{}' c_{i,k}\tilde{x}_k + \sum_{k=1}^{N}{}'' c_{i,k}\bar{x}_k + b_i - R_N^{(i)}K \quad (i = 1, 2, \ldots N) \\ \bar{x}_i &= \sum_{k=1}^{N}{}' c_{i,k}\bar{x}_k + \sum_{k=1}^{N}{}'' c_{i,k}\tilde{x}_k + b_i + R_N^{(i)}K \quad (i = 1, 2, \ldots N), \end{aligned}\right\} \tag{43}$$

where the sum Σ' on each row ranges over those k for which $c_{i,k} > 0$, and Σ'' over those for which $c_{i,k} < 0$. We observe first of all that this system has a definite solution. Indeed, subtracting the first equations from the second, and also adding the corresponding equations, we obtain two independent systems of equations in the unknowns $(\bar{x}_i - \tilde{x}_i)$ and $(\bar{x}_i + \tilde{x}_i)$:

$$\begin{aligned} \bar{x}_i - \tilde{x}_i &= \sum_{k=1}^{N}{}' c_{i,k}(\bar{x}_k - \tilde{x}_k) + \sum_{k=1}^{N}{}'' (-c_{i,k})(\bar{x}_k - \tilde{x}_k) + 2R_N^{(i)}K = \\ &= \sum_{k=1}^{N} |c_{i,k}| (\bar{x}_k - \tilde{x}_k) + 2R_N^{(i)}K \qquad (i = 1, 2, \ldots N) \end{aligned} \tag{44}$$

$$\bar{x}_i + \tilde{x}_i = \sum_{k=1}^{N} c_{i,k}(\bar{x}_k + \tilde{x}_k) + 2b_i \qquad (i = 1, 2, \ldots N). \tag{45}$$

Both systems (44) and (45) are fully regular, and therefore (see Observation 2 to Theorem IIa) $(\bar{x}_i - \tilde{x}_i)$ and $(\bar{x}_i + \tilde{x}_i)$ are uniquely defined, and hence $\bar{x}_i$ and $\tilde{x}_i$ as well. We shall show that the numbers

$\tilde{x}_1, \tilde{x}_2, \ldots, \tilde{x}_N$ so determined are deficient approximations to the sought values $x_1^*, x_2^*, \ldots, x_N^*$, and that $\bar{x}_1, \bar{x}_2, \ldots, \bar{x}_N$ are excessive approximations. For proof of this we shall solve system (43) by the method of successive approximations, in the following manner: For the initial values we shall adopt $\tilde{x}_i^{(0)} = -K$, $\bar{x}_i^{(0)} = K$; substituting these values in the right part of (43), we find $\tilde{x}_i^{(1)}$ and $\bar{x}_i^{(1)}$, and so on. The solution of system (43) by this method is equivalent to solving systems (44) and (45) independently of one another by the method of successive approximations, beginning with the values $(\bar{x}_i^{(0)} - \tilde{x}_i^{(0)}) = 2K$, $\bar{x}_i^{(0)} + \tilde{x}_i^{(0)} = 0$, and therefore as $l \to \infty$, $\bar{x}_i^{(l)} \to \bar{x}_i$, $\tilde{x}_i^{(l)} \to \tilde{x}_i$. For the initial values we obviously have $\tilde{x}_i^{(0)} \leqslant x_i^* \leqslant \bar{x}_i^{(0)}$.

Let it have been already established that $\tilde{x}_i^{(l)} \leqslant x_i^* \leqslant \bar{x}_i^{(l)}$. We shall show that $\tilde{x}_i^{(l+1)} \leqslant x_i^* \leqslant \bar{x}_i^{(l+1)}$ too. Indeed, from (43) we have, for $\tilde{x}_i^{(l+1)}$, for instance,

$$\tilde{x}_i^{(l+1)} = \sum_{k=1}^{N}{}' c_{i,k}\tilde{x}_k^{(l)} + \sum_{k=1}^{N}{}'' c_{i,k}\bar{x}_k^{(l)} + b_i - R_N^{(i)}K \leqslant$$

$$\leqslant \sum_{k=1}^{N} c_{i,k}x_k^* + b_i - K\sum_{k=N+1}^{\infty} |c_{i,k}| \leqslant \sum_{k=1}^{\infty} c_{i,k}x_k^* + b_i = x_i^*.$$

So for all l the inequality $\tilde{x}_i^{(l)} \leqslant x_i^* \leqslant \bar{x}_i^{(l)}$ is established. Thence on passing to the limit we conclude that the inequality

$$\tilde{x}_i \leqslant x_i^* \leqslant \bar{x}_i \tag{39}$$

is satisfied.

The circumstance that these approximations $\tilde{x}_i$ and $\bar{x}_i$ do in fact converge to the principal solution can be verified again on the additional assumption that the conditions of Theorem IIa are fulfilled for system (2). Indeed, as one can readily satisfy oneself by comparing the successive approximations of the solutions of systems (43) for various N, with the growth of N the numbers $\tilde{x}_i$ do not decrease, and the $\bar{x}_i$ do not increase; therefore as $N \to \infty$ they tend to definite limits. Accordingly there exist the limits

$$\lim_{N\to\infty} (\bar{x}_i - \tilde{x}_i) = z_i.$$

Passing to the limit as $N \to \infty$ in system (44), we convince ourselves that the numbers z_i satisfy the homogeneous system of equations

$$z_i = \sum_{k=1}^{\infty} |c_{i,k}|\, z_k. \qquad (i = 1, 2, \ldots)$$

In such a case $z_i = 0$. Indeed, once the conditions of Theorem IIa are satisfied for system (2), they are also satisfied for the system cited above defining z_i. On applying this theorem, we verify that any bounded solution of this system must coincide with its principal solution, obviously equal to zero. So: $z_i = 0$; $\lim\limits_{N\to\infty} (\bar{x}_i - \tilde{x}_i) = 0$, and thanks to inequality (39)

$$\lim_{N\to\infty} \bar{x}_i = \lim_{N\to\infty} \tilde{x}_i = x_i^*,$$

which was what had to be established.

As an example we shall find approximations to the solutions of system (33) with free terms $b_1 = 1$, $b_2 = 0$, $b_3 = 0, \ldots$:

$$x_1 = \frac{1}{1\cdot 3} x_2 + \frac{1}{3\cdot 5} x_3 + \ldots + 1$$

$$x_2 = \frac{1}{1\cdot 3} x_1 + \frac{1}{1\cdot 3} x_3 + \ldots$$

$$x_3 = \frac{1}{3\cdot 5} x_1 + \frac{1}{1\cdot 3} x_2 + \ldots$$

$$\cdots\cdots\cdots\cdots\cdots\cdots$$

The systems for the determination of the excessive and deficient approximations n the given case (with $N = 5$) have the form [cf. (21) and (38), $K = 2$]:

$$\tilde{x}_1 = \tfrac{1}{3}\tilde{x}_2 + \tfrac{1}{15}\tilde{x}_3 + \tfrac{1}{35}\tilde{x}_4 + \tfrac{1}{63}\tilde{x}_5 + 1, \quad \bar{x}_1 = \tfrac{1}{3}\bar{x}_2 + \tfrac{1}{15}\bar{x}_3 + \tfrac{1}{35}\bar{x}_4 + \tfrac{1}{63}\bar{x}_5 + 1 + \tfrac{1}{9},$$

$$\tilde{x}_2 = \tfrac{1}{3}\tilde{x}_1 + \tfrac{1}{3}\tilde{x}_3 + \tfrac{1}{15}\tilde{x}_4 + \tfrac{1}{35}\tilde{x}_5, \quad \bar{x}_2 = \tfrac{1}{3}\bar{x}_1 + \tfrac{1}{3}\bar{x}_3 + \tfrac{1}{15}\bar{x}_4 + \tfrac{1}{35}\bar{x}_5 + \tfrac{1}{7},$$

$$\tilde{x}_3 = \tfrac{1}{15}\tilde{x}_1 + \tfrac{1}{3}\tilde{x}_2 + \tfrac{1}{3}\tilde{x}_4 + \tfrac{1}{15}\tilde{x}_5, \quad \bar{x}_3 = \tfrac{1}{15}\bar{x}_1 + \tfrac{1}{3}\bar{x}_2 + \tfrac{1}{3}\bar{x}_4 + \tfrac{1}{15}\bar{x}_5 + \tfrac{1}{5},$$

$$\tilde{x}_4 = \tfrac{1}{35}\tilde{x}_1 + \tfrac{1}{15}\tilde{x}_2 + \tfrac{1}{3}\tilde{x}_3 + \tfrac{1}{3}\tilde{x}_5, \quad \bar{x}_4 = \tfrac{1}{35}\bar{x}_1 + \tfrac{1}{15}\bar{x}_2 + \tfrac{1}{3}\bar{x}_3 + \tfrac{1}{3}\bar{x}_5 + \tfrac{1}{3},$$

$$\tilde{x}_5 = \tfrac{1}{63}\tilde{x}_1 + \tfrac{1}{35}\tilde{x}_2 + \tfrac{1}{15}\tilde{x}_3 + \tfrac{1}{3}\tilde{x}_4, \quad \bar{x}_5 = \tfrac{1}{63}\bar{x}_1 + \tfrac{1}{35}\bar{x}_2 + \tfrac{1}{15}\bar{x}_3 + \tfrac{1}{3}\bar{x}_4 + 1.$$

On solving these systems we obtain

$$\tilde{x}_1 = 1.2112;\ \tilde{x}_2 = 0.5383;\ \tilde{x}_3 = 0.3461;\ \tilde{x}_4 = 0.2307;\ \tilde{x}_5 = 0.1346;$$

$$\bar{x}_1 = 1.7034;\ \bar{x}_2 = 1.3040;\ \bar{x}_3 = 1.3461;\ \bar{x}_4 = 1.4651;\ \bar{x}_5 = 1.6424.$$

We thus find the following bounds for the unknowns:

$$1.2112 < x_1 < 1.7034;\ 0.5383 < x_2 < 1.3040;\ 0.3461 < x_3 < 1.3461;$$

$$0.2307 < x_4 < 1.4651;\ 0.1346 < x_5 < 1.6424;\ 0 \leqslant x_i \leqslant 2\ (i = 6, 7, \ldots).$$

The bounds obtained for the unknowns with such an approximation are only very rough (they are somewhat improved below in No. 5).

We shall remark that for a practical approximate solution of an infinite system and to find both approximations to the unknowns, it proves to be more efficient (for large N) not to solve the system (43) exactly, but to solve it by the method of successive approximations, so as to obtain for $\tilde{x}_i$ a deficient approximation, and for $\bar{x}_i$ an excessive one. Moreover in the course of the computations the number N of equations taken into consideration may profitably be gradually increased simultaneously [1]).

[1]) A detailed expository solution of another numerical example is contained in the work of B. M. Koialovich, [1].

An example of the solution of a system of a type encountered in a torsion problem is also studied in a quite recent work of N. Kh. Arutiunian, [1].

5. Limitants. Diverse generalizations of regular systems. At the outset we shall indicate a concept introduced by B. M. Koialovich: that of limitants — special expressions which in certain cases permit the estimation and study of the values of the unknowns for large i.

We shall assume that the coefficients of a regular system (2) satisfying condition (27) are positive, and that the system has a unique bounded solution, i.e., that the principal solutions of system (28) are $X_i^* = K$. Let us assume in addition that we know the values $x_1^*, x_2^*, \ldots, x_p^*$ of the first p unknowns. Let us substitute these values in all equations beginning with the $(p+1)$th. We then obtain for the determination of the unknowns $x_{p+1}, x_{p+2}, \ldots$ the system of equations

$$x_i = \sum_{k=p+1}^{\infty} c_{i,k} x_k + (b_i + \sum_{k=1}^{p} c_{i,k} x_i^*) \qquad (i = p+1, p+2, \ldots). \tag{46}$$

This system has as majorant the system

$$X_i = \sum_{k=p+1}^{\infty} C_{i,k} X_k + (K\varrho_i + K\sum_{k=1}^{p} C_{i,k}) \qquad (i = p+1, p+2, \ldots) \tag{47}$$

with principal solutions $X_i^* = K$ $(i = p+1, \ldots)$. Comparing the free terms of systems (46) and (47), one can indicate known bounds in which the unknowns $x_{p+1}, x_{p+2}, \ldots$ are contained, for if the ratio of these free terms, for all $i = p+1$, $p+2, \ldots$, lies between the bounds h and H:

$$h < v_i^{(p)} = \frac{\sum_{k=1}^{p} c_{i,k} x_i^* + b_i}{\sum_{k=1}^{p} C_{i,k} X_k^* + K\varrho_i} = \frac{\sum_{k=1}^{p} c_{i,k} x_i^* + b_i}{K\sum_{k=1}^{p} C_{i,k} + K\varrho_i} < H, \tag{48}$$

the unknowns x_i of the bounded solution of system (46) lie between the limits

$$hK = hX_i^* < x_i < HX_i^* = HK \quad (i = p+1, p+2, \ldots). \tag{49}$$

Indeed, from (46) and (48) it is clear that for the unknowns $x_i - Kh$ and $KH - x_i$, systems with positive free terms are obtained; in such a case these unknowns, which are the principal and unique solutions of the corresponding system, are also non-negative (No. 2, Observation 2 to Theorem I):

$$x_i - Kh \geqslant 0, \; KH - x_i \geqslant 0, \qquad (i = p+1, p+2, \ldots)$$

and inequalities (49) are established. The expressions $v_i^{(p)}$ figuring in (48) are called *limitants* by B. M. Koialovich, who has given the following interesting application of them. During the practical solution of certain systems it had been noted that with an increase of the index of the unknown, the ratio $\frac{x_i^*}{X_i^*}$ approached a certain constant number, or putting it otherwise, x_i^* could be written in the form

$$x_i^* = (G + \varepsilon_i) X_i^*, \tag{50}$$

where G is a constant and $\varepsilon_i \to 0$. Using limitants, B. M. Koialovich has shown [1]) that this circumstance holds for regular systems satisfying a certain supplementary condition, viz., the following:

---

[1]) See B. M. Koialovich, op. cit., Chap. IV. The author there considers a system with a double series of unknowns and makes along with condition (a) given below a second assumption which is in our case superfluous, as is the last part of the author's reasoning, pp. 162–165.

(a) All coefficients lying to the left of the diagonal are of the order of the free terms of the majorant system, i.e., two numbers l and L exist such that

$$0 < l \leqslant \frac{c_{i,k}}{K\varrho_i} \leqslant L \qquad \begin{pmatrix} k = 1, 2, \ldots, i-1 \\ i = 1, 2, \ldots \end{pmatrix}. \tag{51}$$

Thus if the system is regular, with unique bounded solution, and satisfies conditions (a) and (27), then as $i \to \infty$ the unknowns have a definite finite limit

$$\lim_{i\to\infty} x_i^* . \tag{52}$$

In order to show an application of limitants, by using inequality (49), we shall narrow the bounds on the unknowns x_1^*, x_2^*, ... in the example considered at the end of No. 4. We adopt $p = 5$ in (48) and in finding H replace the numbers $x_1^*, x_2^*, x_3^*, x_4^*, x_5^*$, which are unknown to us, by the values x_1, x_2, x_3, x_4, x_5 found above. It is easily verified that the maximum ratio is attained with $i = 6$, and therefore

$$H' = \frac{\dfrac{1}{9\cdot 11}\bar{x}_1 + \dfrac{1}{7\cdot 9}\bar{x}_2 + \dfrac{1}{5\cdot 7}\bar{x}_3 + \dfrac{1}{3\cdot 5}\bar{x}_4 + \dfrac{1}{1\cdot 3}\bar{x}_5 + 0}{2\left(\dfrac{1}{9\cdot 11} + \dfrac{1}{7\cdot 9} + \dfrac{1}{5\cdot 7} + \dfrac{1}{3\cdot 5} + \dfrac{1}{1\cdot 3}\right) + 2\dfrac{1}{22}} =$$

$$= \tfrac{1}{99}\, 1.7034 + \tfrac{1}{63}\, 1.3040 + \tfrac{1}{35}\, 1.3461 + \tfrac{1}{15}\, 1.4651 + \tfrac{1}{3}\, 1.6424 = 0.7231.$$

We thus obtain a new estimate for the last unknowns

$$x_i^* \leqslant \bar{x}_i' = KH' = 2\cdot 0.7231 = 1.4462. \qquad (i = 6, 7, \ldots)$$

As is readily verified, $h = 0$ in the given case, and we do not find an estimate from below other than a trivial one. The values $\bar{x}_i'$ $(i = 6, 7, \ldots)$ obtained permit in their turn improvement of the excessive approximations for the first five unknowns, found above. The values of these excessive approximations are indeed to be found from a system obtained just as that above is, but with the replacement of x_i $(i = 6, 7, \ldots)$, not by $K = 2$ but by KH'. It is clear, therefore, that the newly found excessive approximations will equal $\bar{x}_i' = \tilde{x}_i + H'(\bar{x}_i - \tilde{x}_i)$ $(i = 1, 2, 3, 4, 5)$. By means of the values $\bar{x}_i'$ thus found we may again diminish the value of H and pass from H' to $H'' < H'$, and so on. Here two successive values of H will be connected by the equality

$$H^{(k+1)} = \frac{1}{9\cdot 11}[\tilde{x}_1 + (\bar{x}_1 - \tilde{x}_1)H^{(k)}] + \frac{1}{7\cdot 9}[\tilde{x}_2 + (\bar{x}_2 - \tilde{x}_2)H^{(k)}] +$$
$$+ \frac{1}{5\cdot 7}[\tilde{x}_3 + (\bar{x}_3 - \tilde{x}_3)H^{(k)}] + \frac{1}{3\cdot 5}[\tilde{x}_4 + (\bar{x}_4 - \tilde{x}_4)H^{(k)}] +$$
$$+ \frac{1}{1\cdot 3}[\tilde{x}_5 + (\bar{x}_5 - \tilde{x}_5)H^{(k)}].$$

The number $H = \lim_{k\to\infty} H^{(k)}$ is determined from the equation which is obtained from the preceding equation if we replace $H^{(k)}$ and $H^{(k+1)}$ in it by H. On solving the equation, we find $H = 0.4923$. Hence by the formulas $\bar{x}_i^{(0)} = \tilde{x}_i + H(\bar{x}_i - \tilde{x}_i)$ $(i = 1, 2, 3, 4, 5)$, we obtain new values for the excessive approximations. We are finally brought to the following bounds for the unknowns:

$$1.211 \leqslant x_1 \leqslant 1.331;\ 0.538 \leqslant x_2 \leqslant 0.726;\ 0.346 \leqslant x_3 \leqslant 0.592;$$

$$0.231 \leqslant x_4 \leqslant 0.534;\ 0.135 \leqslant x_5 \leqslant 0.506;\ 0 \leqslant x_i \leqslant 0.492. \ (i = 6, 7 \ldots)$$

We shall now indicate certain types of systems, the theory of which is reducible to that of regular systems or to which this theory is generalizable.

I. One sometimes has to consider the set of two infinite systems

$$\left.\begin{aligned} x_i &= \sum_{k=1}^{\infty} c_{i,k}x_k + b_i \quad (i = 1, 2, \ldots) \\ y_i &= \sum_{k=1}^{\infty} \gamma_{i,k}y_k + \beta_i \quad (i = 1, 2, \ldots) \end{aligned}\right\} \tag{53}$$

with two sets of unknowns $x_1, x_2, \ldots$ and $y_1, y_2, \ldots$ The results of Nos. 2 to 4 may obviously be extended to such systems, since, on introducing new unknowns z, putting

$$z_{2i-1} = x_i,\ z_{2i} = y_i \qquad (i = 1, 2, \ldots)$$

we will obtain one set of equations for them.

II. It is possible to consider systems of equations that are infinite on both sides, with unknowns x_i $(i = 0, \pm 1, \pm 2, \ldots)$, namely systems of the form

$$x_i = \sum_{k=-\infty}^{+\infty} c_{i,k}x_k + b_i. \qquad (i = 0, \pm 1, \pm 2, \ldots) \tag{54}$$

The theory of regular systems may be extended to such systems too; in particular, if

$$\sum_{k=-\infty}^{+\infty} |c_{i,k}| \leqslant 1 - \theta,\ \theta > 0, \qquad (i = 0, \pm 1, \pm 2, \ldots)$$

then for system (54) all that has been said above concerning fully regular systems remains true.

III. In many cases a system that is not regular is reducible to the form of a regular one with the aid of certain transformations, for example, by means of the substitution $z_i = \alpha_i x_i$ or by adding and subtracting the equations of the system.

IV. We shall in addition make mention of *quasi-regular systems*. Such a system is one in which the condition of regularity is satisfied only in some rows beginning with a certain one, i.e.,

$$\sum_{k=1}^{\infty} |c_{i,k}| < 1 \qquad (i = N + 1; N + 2, \ldots) \tag{55}$$

and in addition

$$\left.\begin{aligned} &\sum_{k=1}^{\infty} |c_{i,k}| < +\infty \qquad (i = 1, 2, \ldots N \\ &|b_i| \leqslant K\varrho_i = K(1 - \sum_{k=N+1}^{\infty} |c_{i,k}|) \qquad (i = N + 1 \ldots) \end{aligned}\right\} \tag{56}$$

The question of the existence of a solution of such a system is reducible to the question of the existence of a solution of a finite system. Indeed, let us consider the regular system of equations

$$x_i = \sum_{k=N+1}^{\infty} c_{i,k}x_k + (b_i + \sum_{k=1}^{N} c_{i,k}x_i). \qquad (i = N + 1, N + 2, \ldots) \tag{57}$$

A similar system with free terms b_i has a bounded solution $|\xi_i| \leqslant K$; with free terms $c_{i,k}$ $(k = 1, 2, \ldots, N)$ it also has the bounded solutions $a_i^{(k)}$, $(|a_i^{(k)}| \leqslant 1$, since $|c_{i,k}| \leqslant 1 - \sum_{k=N+1}^{\infty} |c_{i,k}|)$, on the strength of (55). Therefore the solution of system (57) is finally expressible as

$$x_i = a_i^{(1)}x_1 + a_i^{(2)}x_2 + \ldots + a_i^{(N)}x_N + \xi_i. \quad (i = N+1, N+2, \ldots) \tag{58}$$

Substituting these values in the equation

$$x_i = \sum_{k=1}^{\infty} c_{i,k}x_k + b_i \qquad (i = 1, 2, \ldots N)$$

[the substitution makes sense on the strength of the first of conditions (56)], we shall obtain a system of N equations in N unknowns for the determination of $x_1, x_2, \ldots, x_N$. If this finite system has a solution, then we shall find the solution of the original system by means of equations (58). If the solution of this finite system is unique and satisfies the condition for the uniqueness of the solution of the regular system (57), then so is the solution of the original system unique.

V. Lastly, the theory carried through above for linear systems is in considerable measure extensible to *nonlinear systems*. In particular, in this case the theorem on comparison of systems is extensible *in toto* [1]). Let us consider an infinite system of nonlinear equations,

$$x_i = f_i(x_1, x_2, \ldots) = b^{(i)} + \sum_{k_1=1}^{\infty} c_{k_1}^{(i)}x_{k_1} + \sum_{k_1=1}^{\infty}\sum_{k_2=1}^{\infty} c_{k_1,k_2}^{(i)}x_{k_1}x_{k_2} +$$

$$+ \sum_{k_1=1}^{\infty}\sum_{k_2=1}^{\infty}\cdots\sum_{k_j=1}^{\infty} c_{k_1,k_2,\ldots k_j}^{(i)}x_{k_1}x_{k_2}\ldots x_{k_j} + \ldots \quad (i = 1, 2, \ldots) \tag{59}$$

and the system majorant for it, i.e., the system of the same form,

$$X_i = F_i(X_1, X_2, \ldots) = B^{(i)} + \sum_{k_1=1}^{\infty} C_{k_1}^{(i)}X_{k_1} + \sum_{k_1=1}^{\infty}\sum_{k_2=1}^{\infty} C_{k_1,k_2}^{(i)}X_{k_1}X_{k_2} +$$

$$+ \ldots + \sum_{k_1=1}^{\infty}\sum_{k_2=1}^{\infty}\cdots\sum_{k_j=1}^{\infty} C_{k_1,k_2,\ldots k_j}^{(i)}X_{k_1}X_{k_2}\ldots X_{k_j} + \ldots, \quad (i = 1, 2, \ldots) \tag{60}$$

where the functions F_i are majorants for the f_i, that is, it is always true that

$$|b^{(i)}| \leqslant B^{(i)}, \quad |c_{k_1,k_2,\ldots k_j}^{(i)}| \leqslant C_{k_1,k_2,\ldots k_j}^{(i)}. \tag{61}$$

Then it can be shown that if the majorant system has a non-negative solution $X_i' \geqslant 0$, the given system has a solution x_i^* (the principal one) satisfying the condition $|x_i^*| \leqslant X_i'$. Furthermore if X_i^* is the principal solution of system (60), the solution of system (59) satisfying the condition $|x_i| \leqslant X_i^*$ is unique. Finally, the principal solution of system (59) can be obtained not only by the method of successive approximations but also by means of the method of reduction, viz.: if we designate by $\overset{N}{x_i}$ the principal solution of the finite system of equations

$$x_i = \sum_{j=1}^{N}\sum_{k_1,\ldots k_j=1}^{N} c_{k_1,\ldots k_j}^{(i)}x_{k_1}\ldots x_{k_j} + b^{(i)}, \qquad (i = 1, 2, \ldots N), \tag{62}$$

---

[1]) See the works of Pellet, [1] (for the case of finite systems) and L. V. Kantorovich, [2].

the limit of these solutions gives the principal solution of system (59):

$$\lim_{N\to\infty} \overset{N}{x_i} = x_i^*.$$

With this we conclude the theory of regular systems. We have allotted them so much space in view of their importance for applications. In the No. following we shall give a survey of other investigations relating to infinite systems of equations; there, notwithstanding the theoretical value of these investigations, we will limit ourselves to an exposition of the basic results without proofs, for which we refer the interested reader to the special textbooks of V. F. Kagan and F. Riesz cited above.

6. Brief survey of other investigations relating to infinite systems. I. Infinite determinants. Infinite determinants were first applied in 1877 by Hill in the integration of a differential equation obtained in studying the motion of the moon. Soon afterwards, in the same connection, Poincaré developed the beginnings of a theory of them. After this their systematic investigation was carried out in a series of works of Koch.

Let us consider the infinite matrix:

$$\left|\begin{matrix} 1 + c_{11}, & c_{12}, & c_{13}, & \dots \\ c_{12}, & 1 + c_{22}, & c_{23}, & \dots \\ c_{31}, & c_{32}, & 1 + c_{33}, & \dots \\ \dots & \dots & \dots & \dots \end{matrix}\right| \tag{63}$$

and assume that the double series $\sum_{i,k=1}^{\infty} |c_{i,k}|$ converges; it can then be shown that the determinant of n^2 of the elements of this matrix, Δ_n, tends to a limit as $n \to \infty$; this limit Δ is called the *infinite determinant* of the matrix (63). Under the assumption made, it is called the *normal* determinant.

If in matrix (63) we now replace the kth column by a series of bounded numbers $b_1, b_2, \dots$, the limit of the finite determinants $\Delta_n^{(k)}$ as $n \to \infty$ will again exist; it is denoted by $\Delta^{(k)}$.

Upon the introduction of such determinants for the system of equations whose coefficients constitute matrix (63):

$$x_i + \sum_{k=1}^{\infty} c_{i,k} x_k = b_i, \tag{64}$$

a theory of solution can be developed analogous to the theory of solution of finite systems by means of determinants, in particular, theorems can be proved that are analogous to the theorems of Cramer and Rouchet.

Explicitly, the following theorems can be proved:

1) If the determinant of the system $\Delta \neq 0$, the system has a unique, bounded solution, which is given by

$$x_i = \frac{\Delta^{(i)}}{\Delta}. \tag{65}$$

2) If the determinant $\Delta = 0$, the homogeneous system obtained from (64) has non-zero solutions, and moreover has r independent solutions, where r is the rank of the matrix (63) (the number is always finite); thus in particular r unknowns of the homogeneous system may be fixed so that the system becomes determinate.

3) If $\Delta = 0$, the non-homogeneous system (64) does not, generally speaking, have a solution. In order that it have a solution, the free terms b_i of the system must satisfy certain definite relations (r in number). If these relations are satisfied, a solution exists, but is not unique.

Koch showed that these results remain true for markedly wider classes of determinants than the normal ones. The most interesting generalization consists in the fact that the convergence requirement of the series $\sum_{i,k=1}^{\infty} |c_{i,k}|$ may be replaced by a convergence requirement of the series

$$\sum_{i=1}^{\infty} |c_{i,i}|, \quad \sum_{i,k=1}^{\infty} c_{i,k}^2 \text{ and } \sum_{i=1}^{\infty} b_i^2,$$

and that the solution will then satisfy the condition that $\sum_{i=1}^{\infty} x_i^2$ converges. Instead of the conditions stipulated above, it is also possible to require, for instance, that the series $\sum_{i=1}^{\infty} |c_{i,i}|$ converge and that $|c_{i,k}| \leqslant \alpha_i$, where the series $\sum_{i=1}^{\infty} \alpha_i$ converges.

II. Systems with a completely continuous form [1]). In connection with the construction of the theory of integral equations, a special theory of infinite systems of equations was developed by Hilbert.

Let us consider, together with system (64), the bilinear form

$$K(x, y) = \sum_{i,k=1}^{\infty} c_{i,k} x_i y_k. \tag{66}$$

This form is completely continuous if the condition

$$\left| \sum_{i,k=1}^{n} c_{i,k} x_i y_k - \sum_{i,k=1}^{m} c_{i,k} x_i y_k \right| < \varepsilon \text{ for } n, m \geqslant N$$

is satisfied uniformly with respect to x_i and y_k lying in the region

$$\sum_{i=1}^{\infty} x_i^2 \leqslant 1, \quad \sum_{k=1}^{\infty} y_k^2 \leqslant 1.$$

A sufficient condition for this is, for example, the convergence of the series $\sum_{i,k=1}^{\infty} c_{i,k}^2$, but it is even possible to specify other, weaker sufficient conditions, which are, however, of a quite complicated character.

For systems (64) with completely continuous form (66), the following proposition, generalizing the like result for finite systems, was proved by Hilbert: given convergence of the series $\sum_{i=1}^{\infty} b_i^2$, either the given system has a unique solution satisfying the condition that $\sum_{i=1}^{\infty} x_i^2$ converges, or the homogeneous system corresponding to system (64) has a non-zero solution. So also are extended the remaining theorems regarding finite systems [2]).

III. Schmidt has considered [3]) systems of form (64) under the sole condition that $\sum_{k=1}^{\infty} c_{i,k}^2$ converges. For such systems he has given a necessary and sufficient

---

[1]) F. Riesz, [1], Chap. IV, V; D. Hilbert, [1].

[2]) The question of the applicability of the method of reduction and of the approximate solution of systems with completely continuous form and systems satisfying the Koch condition is examined in a work of L. V. Kantorovich, [3], Chap. II, § 3.

[3]) E. Schmidt, [1]. Some of the methods indicated by Schmidt, and also by T. Kötterizsch, may prove to be useful also for the numerical solution of infinite systems. See the dissertation of E. Goldschmidt, [1].

condition that the solution exist. Thus, for example, the condition that the homogeneous system (64) not have a solution different from zero with converging series $\sum_{i=1}^{\infty} x_i^2$ is the vanishing of the following expression:

$$\lim_{\alpha\to\infty}\left\{\begin{vmatrix} s_{11} & \cdots & s_{1\alpha} & c_{1k} \\ \cdot & \cdot & \cdot & \cdot \\ \cdot & \cdot & \cdot & \cdot \\ s_{\alpha 1} & \cdots & s_{\alpha\alpha} & c_{\alpha k} \\ c_{1k} & \cdots & c_{\alpha k} & 1 \end{vmatrix} : \begin{vmatrix} s_{11} & \cdots & s_{1\alpha} \\ \cdot & \cdot & \cdot \\ \cdot & \cdot & \cdot \\ s_{\alpha 1} & \cdots & s_{\alpha\alpha} \end{vmatrix}\right\}, \text{ where } s_{\alpha\beta} = \sum_{k=1}^{\infty} c_{\alpha k} c_{\beta k}.$$

$$(k = 1, 2, \ldots).$$

A condition of analogous form can be given for the non-homogeneous system, too, and even the unknowns themselves can be given expressions of similar form; however these expressions are scarcely algorithmic and it is therefore doubtful that they could be useful for a practical solution.

§ 3. THE SOLUTION OF BOUNDARY-VALUE PROBLEMS BY MEANS OF NON-ORTHOGONAL SERIES

1. General principles. In § 1 we considered the solution of boundary problems by Fourier's method using series of fundamental functions, i.e., series involving functions that form on the contour an orthogonal system of functions (in the case of the Dirichlet problem; in general, though, the contour operators on them form an orthogonal system). The discovery of such a system of functions generally represents, however, a matter of much difficulty; we shall therefore consider here methods of finding the solution in the form of a series composed of arbitrary solutions not forming an orthogonal system on the contour. Here infinite systems of equations will find application.

So let it be required to find the solution of a homogeneous equation of the elliptic type:

$$M(u) = 0 \tag{1}$$

under the boundary condition

$$P(u) = f(s) \text{ on the contour } L, \tag{2}$$

where P is a homogeneous linear operation on u.

Let some infinite series of solutions of equation (1) be known to us: $u_1(x, y)$, $u_2(x, y)$, ... Operation $P(u)$ on these solutions gives on the contour a certain system of functions of the arc coordinate s:

$$P[u_n(x, y)] = \varphi_n(s) \text{ on } L \quad (n = 1, 2, \ldots). \tag{3}$$

If we could expand the given function $f(s)$ in a series involving the functions $\varphi_n(s)$,

$$f(s) = \sum_{n=1}^{\infty} c_n \varphi_n(s), \tag{4}$$

we would be able to write the sought solution in the form of the series [1])

$$u(x, y) = \sum_{n=1}^{\infty} c_n u_n(x, y). \tag{5}$$

Thus we have before us the problem of expanding an arbitrary function $f(s)$ in terms of a certain system of non-orthogonal functions. In the following No. we shall examine methods of solving this problem, and also examples of the solution of various boundary problems by means of such expressions.

2. Solution of the problem of the expansion of an arbitrary function in terms of preassigned functions by means of orthogonalization. Let there be given an infinite system of functions

$$\varphi_1(x), \varphi_2(x), \ldots, \tag{6}$$

defined and continuous in the interval (a, b).

At the outset we can exclude from the given system of functions those that represent linear combinations of the preceding ones, since by their nature they do not extend the system. Let us now carry through the orthogonalization of system (6), i.e., let us construct the system of functions

$$\psi_1(x), \psi_2(x), \ldots, \tag{7}$$

orthogonalized and normalized; each function of system (7) is to represent some linear combination of the functions of system (6); more exactly, $\psi_n(x)$ is to have the form

$$\psi_n(x) = a_1^{(n)}\varphi_1(x) + a_2^{(n)}\varphi_2(x) + \ldots + a_n^{(n)}\varphi_n(x). \tag{8}$$

Let us carry out this orthogonalization step by step. The first function $\psi_1(x)$ must have the form $c\varphi_1(x)$. Determining the constant c from the condition $\int_a^b \psi_1^2 dx = 1$, we find

$$\psi_1(x) = \frac{\varphi_1(x)}{\sqrt{\int_a^b \varphi_1^2(x)dx}}.$$

Let the first n functions $\psi_1(x), \psi_2(x), \ldots, \psi_n(x)$ have been determined.

---

[1]) We shall observe that for this purpose there is even no need to expand $f(s)$ in a series involving the $\varphi_n(s)$ themselves; it is sufficient to expand it in a series involving linear combinations of them, i.e., to write $f(s)$ in the form

$$f(s) = \sum_{j=1}^{\infty} [\sum_{n=1}^{n_j} c_n^j \varphi_n(s)],$$

since then $u(x, y)$ can be formed analogously:

$$u(x, y) = \sum_{j=1}^{\infty} [\sum_{n=1}^{n_j} c_n^j u_n(x, y)].$$

The function $\psi_{n+1}(x)$ must be a linear combination of these and the function $\varphi_{n+1}(x)$, i.e., must have the form:

$$\psi_{n+1}(x) = c_1\psi_1(x) + c_2\psi_2(x) + \ldots + c_n\psi_n(x) + c\varphi_{n+1}(x). \tag{9}$$

We determine the constants c_i from the condition of orthogonality of $\psi_{n+1}(x)$ to $\psi_1(x), \ldots, \psi_n(x)$: multiplying (9) by $\psi_i(x)$ and integrating, we obtain, using the orthogonality and normality of $\psi_1(x), \ldots, \psi_n(x)$:

$$c_i + c\int_a^b \varphi_{n+1}(x)\psi_1(x)dx = 0.$$

Determining c_i from this, substituting in (9) and then finding the constant c from the condition $\int_a^b \psi_{n+1}^2 dx = 1$, we obtain

$$\psi_{n+1}(x) = \frac{\varphi_{n+1}(x) - \sum_{i=1}^{n}\left(\int_a^b \varphi_{n+1}\psi_i dx\right)\psi_i(x)}{\left\{\int_a^b \left[\varphi_{n+1}(x) - \sum_{i=1}^{n}\left(\int_a^b \varphi_{n+1}\psi_i dx\right)\psi_i(x)\right]^2 dx\right\}^{\frac{1}{2}}}. \tag{10}$$

We observe that the denominator here is not zero, thanks to the linear independence of the functions $\varphi_1(x), \varphi_2(x), \ldots, \varphi_{n+1}(x)$. By means of this equation the functions $\psi_1(x), \psi_2(x), \ldots$ may be found in succession.

We note that by using determinants, one could write directly the expression for the nth function $\psi_n(x)$:

$$\psi_n(x) = \frac{\begin{vmatrix} \int_a^b \varphi_1^2 dx & \int_a^b \varphi_1\varphi_2 dx \ldots & \int_a^b \varphi_1\varphi_{n-1}dx & \varphi_1(x) \\ \int_a^b \varphi_2\varphi_1 dx & \int_a^b \varphi_2^2 dx \ldots & \int_a^b \varphi_2\varphi_{n-1}dx & \varphi_2(x) \\ \cdots & \cdots & \cdots & \cdots \\ \int_a^b \varphi_n\varphi_1 dx & \int_a^b \varphi_n\varphi_2 dx \ldots & \int_a^b \varphi_n\varphi_{n-1}dx & \varphi_n(x) \end{vmatrix}}{\sqrt{\Delta_{n-1}\cdot\Delta_n}} \tag{10'}$$

where Δ_n is known as the *Gramm determinant* of the functions $\varphi_1, \ldots, \varphi_n$,

$$\Delta_n = \begin{vmatrix} \int_a^b \varphi_1^2 dx & \int_a^b \varphi_1\varphi_2 dx \ldots & \int_a^b \varphi_1\varphi_n dx \\ \int_a^b \varphi_2\varphi_1 dx & \int_a^b \varphi_2^2 dx \ldots & \int_a^b \varphi_2\varphi_n dx \\ \cdots & \cdots & \cdots \\ \int_a^b \varphi_n\varphi_1 dx & \int_a^b \varphi_n\varphi_2 dx \ldots & \int_a^b \varphi_n^2 dx \end{vmatrix}.$$

This determinant is certainly different from zero if the functions $\varphi_1, \ldots, \varphi_n$ are linearly independent.

Of the fact that the functions $\psi_1, \ldots, \psi_n$ defined by equation (10') are orthogonal and normalized, one may satisfy oneself by a direct check.

Just as we carried through the orthogonalization for functions given in the interval, the orthogonalization can be carried through for functions given on the contour L; it is necessary only to take $\int_L$ in place of $\int_a^b$ everywhere.

After having constructed the orthogonal system, one can write, for any function $f(s)$, its series expansion in terms of these functions:

$$f(x) = \sum_{n=1}^{\infty} A_n \psi_n(x); \quad A_n = \int_a^b f(x)\psi_n(x)dx. \tag{11}$$

We remark that it is of course generally impossible to guarantee the convergence of the series in the right side of equation (11), and it would therefore be more correct to say that this series corresponds to the function $f(x)$ rather than that it equals it. Knowledge of the series involving the functions $\psi_n(x)$ and corresponding to the function $f(x)$ permits us, since ψ_n is a linear combination (8) of the functions φ_i, to write a series involving linear combinations of the functions φ_i that corresponds to $f(x)$:

$$f(x) = \sum_{n=1}^{\infty} A_n [a_1^{(n)}\varphi_1(x) + a_2^{(n)}\varphi_2(x) + \ldots + a_n^{(n)}\varphi_n(x)]. \tag{12}$$

As regards the question of the convergence of series (11) and (12), the so-called completeness of the system of functions $\{\varphi_n(x)\}$ is of essential significance. A system of functions $\{\varphi_n(x)\}$ is said to be *complete* if no function $\Phi(x)$ exists with $\int_a^b |\Phi(x)|^2 dx > 0$ and which is orthogonal to all the functions $\varphi_n(x)$ simultaneously, i.e., such that

$$\int_a^b \varphi_n(x)\Phi(x)dx = 0 \qquad (n = 1, 2, \ldots). \tag{13}$$

We note the following nearly obvious property of complete systems: two continuous functions all of whose Fourier coefficients are equal are identical. Indeed, the difference of two such functions is orthogonal to all $\varphi_i(x)$, and is therefore equal to zero.

Obviously the system of functions $\varphi_n(x)$ will be complete simultaneously with the system of functions $\psi_n(x)$. Many systems of functions are known to be complete, e.g., the trigonometric functions $\cos nx$, $\sin nx$ $(n = 0, 1, 2, \ldots)$ form a complete system in the interval $(0, 2\pi)$, as do the positive integral powers of x: $1, x, x^2, \ldots$ in any interval [1]. It can be shown that the general condition for completeness of an orthogonal, normalized system of functions is the satisfaction, for any square-integrable function $f(x)$, of the so-called *closure equation* of V. A. Steklov:

$$\sum_{n=1}^{\infty} A_n^2 = \int_a^b f^2(x)dx, \qquad A_n = \int_a^b f(x)\psi_n(x)dx. \tag{14}$$

If the system of functions $\psi_n(x)$ is complete, it is nonetheless impossible to affirm that the series on the right side of (11) converges for any function $f(x)$; one can, however, affirm that a convergence of this series to $f(x)$ obtains in a more general sense.

Let us consider the following integral:

$$I_n = \int_a^b [f(x) - \sum_{i=1}^{n} A_i\psi_i(x)]^2dx.$$

On integrating, using the orthogonality and normality of the $\psi_i(x)$ and the definition of the A_i, we easily satisfy ourselves that

$$I_n = \int_a^b [f(x) - \sum_{i=1}^{n} A_i\psi_i(x)]^2dx =$$
$$= \int_a^b f^2(x)dx - 2\sum_{i=1}^{n} A_i \int_a^b f(x)\psi_i(x)dx + \int_a^b [\sum_{i=1}^{n} A_i\psi_i(x)]^2dx =$$
$$= \int_a^b f^2(x)dx - \sum_{i=1}^{n} A_i^2. \tag{15}$$

Hence on the basis of the closure equation it is clear that the quantity I_n tends to zero as $n \to \infty$. For large n, therefore, the sum of n terms of the series (11), $\sum_{i=1}^{n} A_i\psi_i(x)$, although it may at some points be markedly different from $f(x)$, will differ very little from $f(x)$ in the mean — integrally. The sequence of functions $F_n(x)$ is generally said to *converge in the mean* to the function $F(x)$ if

$$\lim_{n\to\infty} \int_a^b [F(x) - F_n(x)]^2dx = 0. \tag{16}$$

In view of (15), therefore, we can say that the partial sums of the series on the right side of equation (11) converge in the mean to the function

---

[1]) See, for example, I. I. Privalov, [1], pp. 41 and 47, and also V. L. Goncharov, [1], p. 174. It was established by G. M. Müntz that any system of powers, 1, x^{λ_1}, $x^{\lambda_2}, \ldots$ forms a complete system in the interval (0, 1) only if the series $\sum \frac{1}{\lambda_n}$ diverges. See I. P. Natanson, [1]. On the subject of the generalization of this theorem see H. Steinhaus, [1], p. 142. The recent investigations of N. K. Bari, [1], are also devoted to questions of the completeness of systems of functions.

$f(x)$; the same thing relates analogously to series (12), too. We shall promptly avail ourselves of this fact. Let the function $f(s)$ represent the boundary values of the posed problem, and $\varphi_n(s)$ the contour values of the basic solutions $u_n(x, y)$. Moreover let the system of functions $\varphi_n(s)$ be complete. On having orthogonalized this system, we obtain for the function $f(s)$ expansions (11) and (12).

Form the function

$$u(x,y) = \sum_{n=1}^{\infty} A_n[a_1^{(n)}u_1(x, y) + a_2^{(n)}u_2(x, y) + \ldots + a_n^{(n)}u_n(x, y)]. \tag{17}$$

This function $u(x, y)$ satisfies formally all the conditions of the problem. We shall show that at least in the case of the Dirichlet problem for the Laplace equation, series (17) converges and represents the actual solution of the problem. We shall prove, to wit, the following theorem.

THEOREM [1]). *Let $u_n(x, y)$ be harmonic functions in the region D, bounded by the contour Γ consisting of a finite number of analytic arcs. Let the contour values of the functions $u_n(x, y)$ form a complete system of functions $\varphi_n(s)$. Then if $f(s)$ is a continuous function, series* (17) *converges and gives the function $u(x, y)$, harmonic in the region D, whose values, in an approach to the contour, approach $f(s)$.*

We shall first of all prove this Theorem for the case when the region D is the unit circle. Let us introduce polar coordinates. The function $u_n(x, y)$ reduces to $\varphi_n(\theta)$ on the contour; it is therefore expressible by the Poisson integral [§ 1, No. 2, (46)]:

$$u_n(r, \theta) = \frac{1}{2\pi}\int_0^{2\pi} \frac{(1 - r^2)\varphi_n(\lambda)}{1 - 2r\cos(\theta - \lambda) + r^2}\, d\lambda, \tag{18}$$

thanks to which the general term of series (17) — the function v_n — is likewise expressible as a Poisson integral

$$v_n(r, \theta) = A_n[a_1^{(n)}u_1(r, \theta) + a_2^{(n)}u_2(r, \theta) + \ldots + a_n^{(n)}u_n(r, \theta)] =$$

$$= \frac{A_n}{2\pi}\int_0^{2\pi} \frac{(1 - r^2)\psi_n(\lambda)}{1 - 2r\cos(\theta - \lambda) + r^2}\, d\lambda. \tag{19}$$

Let us define — by means of the Poisson integral and the series — the harmonic function

$$u(r, \theta) = \frac{1}{2\pi}\int_0^{2\pi} \frac{1 - r^2}{1 - 2r\cos(\theta - \lambda) + r^2} f(\lambda) d\lambda =$$

$$= \frac{a_0}{2} + \sum_{n=1}^{\infty} (a_n \cos n\theta + b_n \sin n\theta) r^n, \tag{20}$$

---

[1]) M. Picone, [1].

where a_n and b_n are the Fourier coefficients of the function $f(\theta)$. This function represents, as we know (§ 1, No. 2), the solution of the Dirichlet problem for the circle with boundary values defined by the function $f(\theta)$. Therefore

$$\lim_{r\to 1} u(r, \theta) = f(\theta).$$

It is not difficult to verify that this same limit will equal $f(\theta)$ if, in an approach of a point to the contour from within, r and θ be varied simultaneously. We shall now prove that within the unit circle the series $\sum_{n=1}^{\infty} v_n(r, \theta)$ converges and gives as its sum $u(r, \theta)$. For this let us consider the difference

$$|u(r, \theta) - \sum_{i=1}^{n} v_i(r, \theta)| =$$

$$= \left| \frac{1}{2\pi} \int_0^{2\pi} \frac{1 - r^2}{1 - 2r\cos(\theta - \lambda) + r^2} [f(\lambda) - \sum_{i=1}^{n} A_i \psi_i(\lambda)] d\lambda \right|. \quad (21)$$

For the integral standing on the right side of (21), on making use of Buniakovsky's [Schwarz's] inequality:

$$|\int_a^b f_1 \cdot f_2 dx| \leqslant [\int_a^b f_1^2 dx \cdot \int_a^b f_2^2 dx]^{\frac{1}{2}}, \quad (22)$$

we find that

$$|u(r, \theta) - \sum_{i=1}^{n} v_i(r, \theta)| \leqslant$$

$$\leqslant \frac{1}{2\pi} \Big[\int_0^{2\pi} \Big(\frac{1 - r^2}{1 - 2r\cos(\theta - \lambda) + r^2} \Big)^2 d\lambda \Big]^{\frac{1}{2}} \Big\{ \int_0^{2\pi} \Big[f(\lambda) - \sum_{i=1}^{n} A_i \psi_i(\lambda) \Big]^2 d\lambda \Big\}^{\frac{1}{2}} \cdot \quad (23)$$

In the first integral on the right side the integrand obviously does not exceed $\left[\frac{1 - r^2}{(1 - r)^2}\right]^2 = \left(\frac{1 + r}{1 - r}\right)^2 < \frac{4}{(1 - r)^2}$, and the first factor itself does not exceed the quantity $\frac{\sqrt{2\pi} \cdot 2}{1 - r}$, i.e., is a bounded quantity. The second integral, inasmuch as the series $\sum_{n=1}^{\infty} A_n \psi_n(\lambda)$ converges in the mean to $f(\lambda)$ [cf. (15)], has the limit zero. Thus the convergence of the series $\sum_{i=1}^{\infty} v_i(r, \theta)$ and the fact that its sum is equal to $u(r, \theta)$, have been established, and hence equality (17) is proved.

In case the given region is not a circle, we could have established equation (17) by analogous reasoning, with this sole distinction, that instead of Poisson's formula we should have had to utilize the general formula giving the solution of the Dirichlet problem in the case of any region

$$u(x, y) = \int_L G(x, y, s)f(s)ds, \tag{24}$$

which can be constructed by means of the Green function.

Example. We shall again solve the Dirichlet problem for a rectangle as our example: for definiteness let us take the square $(-1, 1; -1, 1)$. This time we shall choose the fundamental solutions otherwise than we did in No. 1, § 1, to wit: we shall adopt as the fundamental solutions for the circle $r^n \cos n\varphi$, $r^n \sin n\varphi$. On the contour of the square these functions will now not be orthogonal, and we must therefore carry through their orthogonalization.

If we go over to Cartesian coordinates $x = r \cos \varphi$ and $y = r \sin \varphi$ and expand $\cos n\varphi$ and $\sin n\varphi$ by the familiar Moivre formulas, we shall find the so-called *harmonic polynomials*:

$$r^n \cos n\varphi = r^n \left[\cos^n \varphi - \frac{n(n-1)}{1\cdot 2}\cos^{n-2}\varphi \sin^2\varphi + \right.$$

$$\left. + \frac{n(n-1)(n-2)(n-3)}{1\cdot 2\cdot 3\cdot 4}\cos^{n-4}\varphi \sin^4\varphi - \ldots\right] =$$

$$= x^n - \frac{n(n-1)}{1\cdot 2}x^{n-2}y^2 + \frac{n(n-1)(n-2)(n-3)}{1\cdot 2\cdot 3\cdot 4}x^{n-4}y^4 - \ldots$$

$$r^n \sin n\varphi = r^n\left[n\cos^{n-1}\varphi \sin\varphi - \frac{n(n-1)(n-2)}{1\cdot 2\cdot 3}\cos^{n-3}\varphi \sin^3\varphi + \ldots\right] =$$

$$= nx^{n-1}y - \frac{n(n-1)(n-2)}{1\cdot 2\cdot 3}x^{n-3}y^3 + \ldots$$

Hence the first harmonic polynomials will be: 1; x; y; $x^2 - y^2$; $2xy$; $x^3 - 3xy^2$; $3x^2y - y^3$; $x^4 - 6x^2y^2 + y^4$; $4x^3y - 4xy^3$; ... We orthogonalize these polynomials; here, to keep the calculations brief, we shall limit ourselves to the even polynomials: 1; $x^2 - y^2$; $x^4 - 6x^2y^2 + y^4$; ... We observe that the preservation of only this partial sequence of polynomials will make possible the solution of the Dirichlet problem with symmetric boundary conditions only.

Let us denote the contour of the square by L. The first function, 1, need only be normalized. We have, obviously,

$$\int_L 1^2\, ds = 8;$$

therefore the first function $\psi_1(x) = 1/\sqrt{8}$. For the second function we first of all compute

$$\int_L 1(x^2 - y^2)ds = \int_{(1,-1)}^{(1,1)} + \int_{(1,1)}^{(-1,1)} + \int_{(-1,1)}^{(-1,-1)} + \int_{(-1,-1)}^{(1,-1)} =$$

$$= \int_{-1}^{1}(1 - y^2)dy + \int_{1}^{-1}(x^2 - 1)(-dx) + \int_{1}^{-1}(1 - y^2)(-dy) + \int_{-1}^{1}(x^2 - 1)dx = 0.$$

Accordingly the function $x^2 - y^2$ is orthogonal to 1, and we need therefore only normalize it:

$$\int_L (x^2 - y^2)^2 ds = 4 \int_{-1}^{1} (1 - x^2)^2 dx = \tfrac{64}{15},$$

whence finally $\psi_2(x) = \sqrt{15/64}\,(x^2 - y^2)$. Next for finding $\psi_3(x)$ by (10) we seek

$$\int_L (x^4 - 6x^2y^2 + y^4) \frac{1}{\sqrt{8}}\, ds =$$

$$= \frac{1}{\sqrt{8}} \left[4 \int_{-1}^{1} (x^4 + 1) dx - 24 \int_{-1}^{1} x^2 dx \right] = -\, 6.4 \cdot \frac{1}{\sqrt{8}},$$

$$\int_L (x^4 - 6x^2y^2 + y^4) \sqrt{\tfrac{15}{64}} (x^2 - y^2) ds =$$

$$= \sqrt{\tfrac{15}{64}} \int_L (x^6 - 7x^4y^2 + 7x^2y^4 - y^6) ds = 0.$$

Then

$$\psi_3 = C[\varphi_3 - (\int_L \varphi_3\psi_1 ds)\psi_1 - (\int_L \varphi_3\psi_2 ds)\psi_2] =$$

$$= C\left(x^4 - 6x^2y^2 + y^4 + 6.4 \cdot \frac{1}{\sqrt{8}} \cdot \frac{1}{\sqrt{8}}\right).$$

Lastly on normalizing the function ψ_3 by the condition $\int_L \psi_3^2 ds = 1$, we obtain finally

$$\psi_3 = \frac{x^4 - 6x^2y^2 + y^4 + 0.8}{[\int_L (x^4 - 6x^2y^2 + y^4 + 0.8)^2 ds]^{\frac{1}{2}}} = 0.2244(x^4 - 6x^2y^2 + y^4 + 0.8).$$

By means of the polynomials $\psi_1, \psi_2, \psi_3, \ldots$, the first of which we have found, one can solve the Dirichlet problem for the square with arbitrary (symmetric) conditions on the contour [1]). As an example we shall solve the Dirichlet problem with the boundary conditions

$$u = 1 - x^2 \text{ for } y = \pm 1,$$
$$u = 1 - y^2 \text{ for } x = \pm 1,$$

or, briefly, $u = f(s)$ on L. According to the method indicated above, we must expand $f(s)$ on the contour L into a series involving $\psi_1, \psi_2, \psi_3, \ldots$. For this we seek

$$a_1 = \int_L f(s)\psi_1(s) ds = 4 \int_{-1}^{1} \frac{1}{\sqrt{8}} (1 - x^2) dx = \frac{16}{3\sqrt{8}} = \tfrac{4}{3}\sqrt{2},$$

---

[1]) This follows from the proposition that harmonic polynomials form a complete system (on the contour) for any simply connected region bounded by a rectifiable contour. The last proposition itself may be easily deduced from the theorem that an analytic function in a simply connected region may be uniformly approximated by polynomials of a complex variable. See V. L. Goncharov, [1], Chap. II.

$$a_2 = \int_L f(s)\psi_2(s)ds = 4\int_{-1}^{1} \sqrt{\tfrac{15}{64}}(x^2 - 1)(1 - x^2)dx +$$
$$+ 2\int_{-1}^{1} \sqrt{\tfrac{15}{64}}(1 - y^2)(1 - y^2)dy = 0.$$

$$a_3 = \int_L f(s)\psi_3(s)ds = 0.2244 \cdot 4\int_{-1}^{1} (x^4 - 6x^2 + 1 + 0.8)(1 - x^2)dx = 0.8206.$$

In this case
$$f(s) \approx a_1\psi_1(s) + a_2\psi_2(s) + a_3\psi_3(s),$$

and we find the harmonic function itself approximately in the form

$$u(x, y) \approx \tfrac{4}{3}\sqrt{2} \cdot \frac{1}{\sqrt{8}} + 0.8206 \cdot 0.2244(x^4 - 6x^2y^2 + y^4 + 0.8) =$$
$$= \tfrac{2}{3} + 0.1841(x^4 - 6x^2y^2 + y^4 + 0.8).$$

Hence, for example, $u(0, 0) \approx 0.814$; exactly, however, by the method of No. 1, § 1, we would have found $u(0, 0) = 0.816$.

Observation 1. We shall note that the method of solving the Dirichlet problem utilized in the example for solving that problem for a square can be applied for the Laplace equation in the case of other regions too. In fact, it is always possible to take these same harmonic polynomials $r^n \cos n\varphi$ and $r^n \sin n\varphi$ and orthogonalize them on the contour of the given region. This method is most convenient in the case of regions bounded by polygons. True, this method requires very substantial computations; but these computations are nevertheless practically realizable, particularly if one can succeed in introducing simplifications into them by utilizing considerations of symmetry, etc.

Observation 2. It should be remarked that along with the usual orthogonalized and normalized systems that we considered above,

$$\int_a^b \varphi_i\varphi_k dx = \begin{cases} 1, & \text{if } i = k \\ 0, & \text{,, } i \neq k, \end{cases}$$

one also finds useful systems orthogonalized and normalized in respect of a weight $p(x)$, where $p(x) \geqslant 0$ is some integrable function, namely, such that

$$\int_a^b \varphi_i\varphi_k p dx = \begin{cases} 1 & \text{if } i = k \\ 0, & \text{,, } i \neq k. \end{cases}$$

In the case of the usual orthogonalization, obviously the weight $p(x) = 1$. When orthogonalizing by weight there may also be formed for any function the series expansion in the functions φ_i:

$$f(x) = \sum_{i=1}^{\infty} a_i\varphi_k(x), \quad a_i = \int_a^b \varphi_i f p dx.$$

An arbitrary system of functions may be orthogonalized and normalized by weight just as in the usual case; one need only add the factor $p(x)$ to the integrand in all of the integrals figuring in (10).

Other generalizations of orthogonal systems sometimes find application, for example bi-orthonormal systems: these are two systems of functions $\{\varphi_i(x)\}$, $\{\psi_i(x)\}$ satisfying the conditions

$$\int_a^b \varphi_i\psi_k dx = \begin{cases} 1, & \text{if } i = k \\ 0, & \text{,, } i \neq k. \end{cases}$$

Here too, for any function $f(x)$, there may be formed the expansion

$$f(x) = \sum_{i=1}^{\infty} a_i\varphi_i(x), \quad a_i = \int_a^b f\psi_i dx.$$

Observation 3. We have considered in this No. boundary conditions of the form

$$P(u) = f(s) \text{ on the contour } L,$$

where $P(u)$ is some linear operator on the contour. The same way of solving is applicable, however, in a more general case, when on different parts of the contour different operators are given:

$$P_1(u) = f_1(s) \text{ on } L_1; \quad P_2(u) = f_2(s) \text{ on } L_2,$$

where L_1 and L_2 are two parts of the contour L. For example, we have such a problem if on one part of the contour the function itself is given, and on another its normal derivative. In case of such a mixed problem one must form the functions $\varphi_n(s)$ in terms of the fundamental solutions $u_n(x, y)$, where the definitions are as follows:

$$\varphi_n(s) = \begin{cases} P_1[u_n(x, y)] \text{ on } L_1 \\ P_2[u_n(x, y)] \text{ on } L_2 \end{cases}$$

and one must orthonormalize this system of functions, after which one must form the expansion, in terms of this system, of the function $f(s)$, which is equal to $f_1(s)$ on L_1 and $f_2(s)$ on L_2.

3. Solution of the problem of the expansion of an arbitrary function in terms of given functions by means of infinite systems of equations. Let us consider the same problem of expanding an arbitrary function $f(x)$ in the interval (a, b) in terms of the functions $\varphi_n(x)$ in another way. Let us write the sought expansion with undetermined coefficients A_k in the form

$$f(x) = \sum_{k=1}^{\infty} A_k\varphi_k(x). \tag{25}$$

Let us now take any complete system of functions $\{\psi_i(x)\}$, multiply both sides of equation (25) by $\psi_i(x)$ and integrate it, whereupon we shall have obtained a system of equations

$$\sum_{k=1}^{\infty} c_{i,k}A_k = b_i, \qquad (i = 1, 2, \ldots), \tag{26}$$

where

$$c_{i,k} = \int_a^b \varphi_k(x)\psi_i(x)dx, \quad b_i = \int_a^b f(x)\psi_i(x)dx. \tag{27}$$

System (26) serves for the determination of the constants A_i. If we can succeed in determining values for A_i which satisfy system (26) and are such that the series on the right side of (25) converges to some continuous function $f^*(x)$ (even though the series converge only in the mean), and if, moreover, a term-by-term integration of series (25) proves to be admissible, then the sum of this series $f^*(x) = f(x)$. Indeed, the integrals for the functions $f(x)$ and $f^*(x)$ are identical: $\int \psi_i f dx = \int \psi_i f^* dx$; in such a case for the orthogonal system obtained by orthogonalization from the functions ψ_i, f and f^* have all their Fourier coefficients the same, and by the observation made above [p. 47], they are identical.

It will be useful to indicate also another method of obtaining system (26), which is doubtless even more convenient. We have in mind that one can arrive at system (26) in another way.

Let us expand in a series in terms of any complete orthogonal system of functions $\{\psi_i(x)\}$ — for example, in an ordinary Fourier series — the function $f(x)$ and all $\varphi_k(x)$; let

$$f(x) = \sum_{i=1}^{\infty} b_i \psi_i(x), \quad \varphi_k(x) = \sum_{i=1}^{\infty} c_{i,k} \psi_i(x), \tag{28}$$

where $c_{i,k}$ and b_i are defined by formulas (27).

Let us now substitute series (28) in equation (25) and equate the coefficients of $\psi_i(x)$ on both sides of this equation, whereupon we shall obviously obtain system (26). We note that the second method of obtaining system (26) presupposes, as distinct from the first, a system of functions $\psi_i(x)$ that is orthogonal and normalized.

It is impossible to give more or less general conditions under which system (26) has a solution. However, in a number of practically important cases, system (26) proves to be regular, as we shall see later in the examples. We mention one such case.

If the functions $\psi_n(x)$ are trigonometric and the functions $\varphi_n(x)$ differ only very little from them, the system proves to be regular. More than that: let $\psi_n(x)$ be any bounded functions orthogonal and normalized in the interval (a, b):

$$|\psi_n(x)| \leqslant \frac{M}{b-a},$$

and let

$$|\varphi_n(x) - \psi_n(x)| \leqslant \frac{\varepsilon_n}{M},$$

where

$$\sum_{n=1}^{\infty} \varepsilon_n = q < 1. \tag{29}$$

System (26) will then be fully regular. Indeed

$$\left.\begin{aligned} |c_{i,k}| &= \left|\int_a^b \varphi_k(x)\psi_i(x)dx\right| \leqslant \left|\int_a^b \psi_k(x)\psi_i(x)dx\right| + \\ &+ \left|\int_a^b [\varphi_k(x)-\psi_k(x)]\psi_i(x)dx\right| \leqslant \frac{\varepsilon_k}{M}\cdot\frac{M}{b-a}(b-a) = \varepsilon_k \text{ for } i \neq k, \\ |c_{i,i} - 1| &= \left|\int_a^b \varphi_i(x)\psi_i(x)dx - 1\right| \leqslant \\ &\leqslant \int_a^b |\varphi_i(x) - \psi_i(x)|\,|\psi_i(x)|\,dx \leqslant \varepsilon_i. \end{aligned}\right\} \tag{30}$$

On the strength of this, if we introduce

$$c^*_{i,k} = -c_{i,k},\ i \neq k;\ c^*_{i,i} = 1 - c_{i,i}, \tag{31}$$

system (26) may be rewritten as

$$A_i = \sum_{k=1}^{\infty} c^*_{i,k} A_k + b_i, \quad (i = 1, 2, \ldots), \tag{32}$$

and it is obviously fully regular, since in view of inequalities (30)

$$\sum_{k=1}^{\infty} |c^*_{i,k}| = |c_{i,i} - 1| + \sum_{\substack{k=1\\k\neq i}}^{\infty} |c_{i,k}| < \varepsilon_i + \sum_{\substack{k=1\\k\neq i}}^{\infty} \varepsilon_k = q < 1.$$

We now exhibit some examples of problems of mathematical physics that are soluble by means of infinite systems.

4. Example 1. A mixed boundary-value problem for the Laplace equation. Let us consider the following boundary-value problem. To be found: a function harmonic in the unit circle (Fig. 2), for which on half the contour of the circumference the values of the function itself are given, and on the other half, the values of its normal derivative:

$$u|_{r=1} = f(\theta) \text{ for } -\frac{\pi}{2} \leqslant \theta \leqslant \frac{\pi}{2}$$

$$\left.\frac{\partial u}{\partial r}\right|_{r=1} = \varphi(\theta) \text{ for } \frac{\pi}{2} < \theta < \frac{3\pi}{2}.$$

To simplify the analysis we shall assume the conditions to be symmetric with respect to the axis Ox, i.e.,

$$f(-\theta) = f(\theta),$$
$$\varphi(\pi + \theta) = \varphi(\pi - \theta).$$

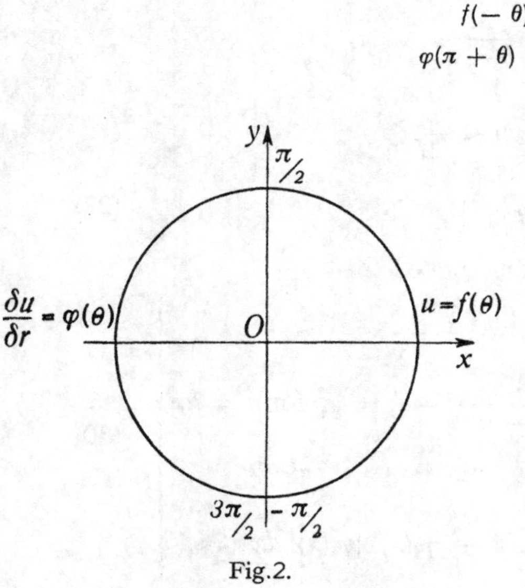

Fig.2.

We seek the solution, in view of the symmetry with respect to the axis Ox, in the form

$$u = \sum_{n=0}^{\infty} A_n r^n \cos n\theta.$$

On the basis of the first condition we must have

$$\sum_{n=0}^{\infty} A_n \cos n\theta = f(\theta), \quad -\frac{\pi}{2} \leqslant \theta \leqslant \frac{\pi}{2}.$$

The functions 1, $\cos 2\theta$, $\cos 4\theta$, ... form a complete system in the interval $(-\pi/2, \pi/2)$ (complete in respect of even functions); therefore on multiplying the equation above by each of these functions and integrating from $-\pi/2$ to $\pi/2$, we find the required system for the determination of the unknown coefficients A_n. For setting up this system we first of all compute

$$\int_{-\pi/2}^{\pi/2} \cos k\theta \cdot \cos 2i\theta d\theta = \tfrac{1}{2} \int_{-\pi/2}^{\pi/2} [\cos (k + 2i)\theta + \cos (k - 2i)\theta]d\theta =$$

$$= \begin{cases} \pi & \text{for } k = 2i = 0 \\ \dfrac{\pi}{2} & ,, \quad k = 2i \qquad (i > 0) \\ 0 & ,, \quad k = 2j \qquad (j \neq i) \\ \dfrac{(-1)^{i+j}2(2j+1)}{(2j+1)^2 - 4i^2} & ,, \quad k = 2j + 1. \end{cases}$$

Let us now introduce the designation

$$\alpha_{2i} = \int_{-\pi/2}^{\pi/2} f(\theta) \cdot \cos 2i\theta d\theta; \qquad (i = 0, 1, 2, \ldots)$$

the system of equations obtained may then be written in the form

$$\left.\begin{array}{l} \pi A_0 + \dfrac{2\cdot 1}{1\cdot 1} A_1 - \dfrac{2\cdot 3}{3\cdot 3} A_3 + \dfrac{2\cdot 5}{5\cdot 5} A_5 - \dfrac{2\cdot 7}{7\cdot 7} A_7 + \ldots = \alpha_0 \\ \dfrac{\pi}{2} A_2 - \dfrac{2\cdot 1}{-1\cdot 3} A_1 + \dfrac{2\cdot 3}{1\cdot 5} A_3 - \dfrac{2\cdot 5}{3\cdot 7} A_5 + \dfrac{2\cdot 7}{5\cdot 9} A_7 - \ldots = \alpha_2 \\ \dfrac{\pi}{2} A_4 + \dfrac{2\cdot 1}{-3\cdot 5} A_1 - \dfrac{2\cdot 3}{-1\cdot 7} A_3 + \dfrac{2\cdot 5}{1\cdot 9} A_5 - \dfrac{2\cdot 7}{3\cdot 11} A_7 + \ldots = \alpha_4 \\ \ldots\ldots\ldots\ldots\ldots\ldots\ldots\ldots\ldots\ldots\ldots\ldots \\ \dfrac{\pi}{2} A_{2i} + \dfrac{(-1)^i 2\cdot 1}{(1-2i)(1+2i)} A_1 + \dfrac{(-1)^{i+1} 2\cdot 3}{(3-2i)(3+2i)} A_3 + \ldots + \\ \qquad + \dfrac{(-1)^{i+j}2(2j+1)}{(2j+1-2i)(2j+1+2i)} A_{2j+1} + \ldots = \alpha_{2i} \\ \ldots\ldots\ldots\ldots\ldots\ldots\ldots\ldots\ldots\ldots\ldots\ldots \end{array}\right\} \quad \text{(a)}$$

Next the condition on the second part of the boundary

$$\left.\frac{\partial u}{\partial r}\right|_{r=1} = \sum_{n=1}^{\infty} nA_n \cos n\theta = \varphi(\theta) \text{ for } \frac{\pi}{2} < \theta < \frac{3\pi}{2}.$$

Here let us multiply both sides of the equation by $\cos 2i(\theta - \pi) = \cos 2i\theta$; the corresponding integral will be distinguished by sign only from the preceding one:

$$\int_{\pi/2}^{3\pi/2} \cos k\theta \cdot \cos 2i\theta d\theta = \int_{3\pi/2}^{5\pi/2} \cos (k\theta - k\pi) \cos (2i\theta - 2i\pi) d\theta =$$

$$= (-1)^k \int_{-\pi/2}^{\pi/2} \cos k\theta \cos 2i\theta d\theta.$$

If we now introduce in addition the designation

$$\beta_{2i} = \int_{\pi/2}^{3\pi/2} \varphi(\theta) \cos 2i\theta d\theta,$$

the system of equations obtained will be

$$\left.\begin{aligned}
&-\frac{2\cdot 1^2}{1\cdot 1}A_1 + \frac{2\cdot 3^2}{3\cdot 3}A_3 - \frac{2\cdot 5^2}{5\cdot 5}A_5 + \frac{2\cdot 7^2}{7\cdot 7}A_7 - \ldots = \beta_0 \\
&2\cdot\frac{\pi}{2}A_2 + \frac{2\cdot 1^2}{-1\cdot 3}A_1 - \frac{2\cdot 3^2}{1\cdot 5}A_3 + \frac{2\cdot 5^2}{3\cdot 7}A_5 - \frac{2\cdot 7^2}{5\cdot 9}A_7 + \ldots = \beta_2 \\
&\cdots\cdots\cdots\cdots\cdots\cdots\cdots\cdots \\
&2i\cdot\frac{\pi}{2}A_{2i} + \frac{(-1)^{i+1}2\cdot 1^2}{(1-2i)(1+2i)}A_1 + \frac{(-1)^{i+2}2\cdot 3^2}{(3-2i)(3+2i)}A_3 + \ldots + \\
&\qquad + \frac{(-1)^{i+j+1}2(2j+1)^2}{(2j+1-2i)(2j+1+2i)}A_{2j+1} + \ldots = \beta_{2i} \\
&\cdots\cdots\cdots\cdots\cdots\cdots\cdots\cdots
\end{aligned}\right\} \quad \text{(b)}$$

We shall transform this system somewhat. Let us first eliminate the unknowns $A_2, A_4, \ldots$, to accomplish which each (ith) equation of system (b) will be divided by $2i$ and the corresponding equation of system (a) subtracted from it; the first equation of system (b) will be left unchanged. As the result we arrive at the system

$$\left.\begin{aligned}
&-\frac{2\cdot 1}{1}A_1 + \frac{2\cdot 3}{3}A_3 - \frac{2\cdot 5}{5}A_5 + \frac{2\cdot 7}{7}A_7 - \ldots = \beta_0 \\
&\frac{2\cdot 1}{-1\cdot 2}A_1 - \frac{2\cdot 3}{1\cdot 2}A_3 + \frac{2\cdot 5}{3\cdot 2}A_5 - \frac{2\cdot 7}{5\cdot 2}A_7 + \ldots = \frac{\beta_2}{2} - \alpha_2 \\
&-\frac{2\cdot 1}{-3\cdot 4}A_1 + \frac{2\cdot 3}{-1\cdot 4}A_3 - \frac{2\cdot 5}{1\cdot 4}A_5 + \frac{2\cdot 7}{3\cdot 4}A_7 - \ldots = \frac{\beta_4}{4} - \alpha_4 \\
&\cdots\cdots\cdots\cdots\cdots\cdots\cdots\cdots \\
&\frac{(-1)^{i+1}2\cdot 1}{(1-2i)2i}A_1 + \frac{(-1)^{i+2}2\cdot 3}{(3-2i)2i}A_3 + \ldots + \\
&\qquad + \frac{(-1)^{i+j+1}2\cdot(2j+1)}{(2j+1-2i)2i}A_{2j+1} + \ldots = \frac{\beta_{2i}}{2i} - \alpha_{2i} \\
&\cdots\cdots\cdots\cdots\cdots\cdots\cdots\cdots
\end{aligned}\right\} \quad \text{(c)}$$

Now let us multiply the ith equation ($i \neq 0$) by i, divide the first by 2, and introduce new unknowns:

$$B_1 = -A_1\cdot 1,\ B_3 = A_3\cdot 3,\ B_5 = -A_5\cdot 5,\ \ldots,$$

$$B_{2j-1} = (-1)^j A_{2j-1}(2j-1), \ldots$$

for these we shall have obtained the system

$$\left.\begin{array}{l}
B_1 + \dfrac{B_3}{3} + \dfrac{B_5}{5} + \dfrac{B_7}{7} + \ldots = \dfrac{\beta_0}{2} \\
-\dfrac{B_1}{-1} - \dfrac{B_3}{1} - \dfrac{B_5}{3} - \dfrac{B_7}{5} - \ldots = \dfrac{\beta_2}{2} - \alpha_2 \\
\dfrac{B_1}{-3} + \dfrac{B_3}{-1} + \dfrac{B_5}{1} + \dfrac{B_7}{3} + \ldots = \dfrac{\beta_4}{2} - 2\alpha_4 \\
-\dfrac{B_1}{-5} - \dfrac{B_3}{-3} - \dfrac{B_5}{-1} - \dfrac{B_7}{1} - \ldots = \dfrac{\beta_6}{2} - 3\alpha_6 \\
\ldots\ldots\ldots\ldots\ldots\ldots\ldots\ldots\ldots\ldots
\end{array}\right\} \quad \text{(d)}$$

In this system let us take half the sum of the first and second equations, of the second and third, and so on; together with this let us change the signs in some of the equations obtained; we shall then obtain, finally,

$$\left.\begin{array}{l}
B_1 - \dfrac{B_3}{1\cdot 3} - \dfrac{B_5}{3\cdot 5} - \dfrac{B_7}{5\cdot 7} - \ldots = \tfrac{1}{4}(\beta_0 + \beta_2 - 2\alpha_2) \\
-\dfrac{B_1}{1\cdot 3} + B_3 - \dfrac{B_5}{1\cdot 3} - \dfrac{B_7}{3\cdot 5} - \ldots = -\tfrac{1}{4}(\beta_2 + \beta_4 - 2\alpha_2 - 4\alpha_4) \\
-\dfrac{B_1}{3\cdot 5} - \dfrac{B_3}{1\cdot 3} + B_5 - \dfrac{B_7}{1\cdot 3} - \ldots = \tfrac{1}{4}(\beta_4 + \beta_6 - 4\alpha_4 - 6\alpha_6) \\
-\dfrac{B_1}{5\cdot 7} - \dfrac{B_3}{3\cdot 5} - \dfrac{B_5}{1\cdot 3} + B_7 - \ldots = -\tfrac{1}{4}(\beta_6 + \beta_8 - 6\alpha_6 - 8\alpha_8) \\
\ldots\ldots\ldots\ldots\ldots\ldots\ldots\ldots\ldots\ldots
\end{array}\right\} \quad \text{(e)}$$

Such a transition from system (d) to system (e) is only half legitimate: while any solution of system (d) will obviously satisfy system (e) too, we cannot, however, vouch for the converse.

The latter system, (e), we have examined as the example of No. 3, § 2 (33) and established there that this system has a solution that is certainly bounded if the free terms are of orders not lower than $1/n$. Since we have on the right the quantities β_{2i} and $2i\alpha_{2i}$, i.e., the Fourier coefficients of the functions $\varphi(\theta)$ and $\dfrac{df}{d\theta}$, it is sufficient to require that these functions be of bounded variation. After the variables $B_1, B_3, \ldots$ have been determined from the last system, the values of $A_1, A_3, \ldots$ can be found in accordance with them, and by these, $A_0, A_2, A_4, \ldots$ can be found from system (a). However to prove that the values obtained for A_i actually give the solution of the initial problem, the boundedness of the numbers B_{2i-1} is insufficient. Indeed, we obtain, for example, the solution $B_{2i-1} = 1$ with free terms $\dfrac{1}{4i-2}$ in the ith equation; it is obvious, however, that this solution does not satisfy system (d), since on substituting it, the series on the right side turn out to be divergent. It is therefore necessary to place a stronger limitation on the free terms of system (e).

Let us assume that the free terms of the system are of orders not lower than $\dfrac{1}{n\sqrt{n}}$, i.e., $|\beta_{2n-2}|$ and $|2n\alpha_{2n-2}|$ do not exceed $\dfrac{K}{n\sqrt{n}}$; in such a case it can be shown that the solutions will be of order not lower than $\frac{1}{\sqrt{n}}$, i.e., $|B_{2n-1}| < \frac{K_1}{\sqrt{n}}$ [1]). We shall prove that these solutions of system (e) satisfy system (d) too. To begin with, it is clear that their substitution in the equations of system (d) gives on the left sides convergent series with general terms of order not lower than $1/n^{3/2}$. Let us consider the nth equation of system (d) and estimate its left side, using the fact that $|B_{2n-1}| < \frac{K_1}{\sqrt{n}}$. Here in making the estimate we replace the sum by an integral — as is done in deducing Cauchy's integral criterion of convergence of a series We have, taking absolute values,

$$\left|\frac{B_1}{-(2n-3)} + \frac{B_3}{-(2n-5)} + \ldots + \frac{B_{2n-3}}{1} + \frac{B_{2n-1}}{1} + \frac{B_{2n+1}}{3} + \ldots\right| <$$

$$\leqslant K_1\left|\frac{1}{(2n-3)\sqrt{1}} + \frac{1}{(2n-5)\sqrt{2}} + \ldots + \frac{1}{\sqrt{n-1}} + \frac{1}{\sqrt{n}} + \frac{1}{3\sqrt{n+1}} + \ldots\right|$$

$$\leqslant K_1\left[\int_0^{n-2}\frac{dx}{(2n-3-2x)\sqrt{x}} + \int_n^{\infty}\frac{dx}{\sqrt{x}(2x+1-2n)} + \frac{1}{\sqrt{n-1}} + \frac{1}{\sqrt{n}}\right]$$

$$= K_1\left[\frac{1}{\sqrt{2(2n-3)}}\ln\frac{\sqrt{2n-3}+\sqrt{2n-4}}{\sqrt{2n-3}-\sqrt{2n-4}} + \right.$$

$$\left. + \frac{1}{\sqrt{2(2n-1)}}\ln\frac{\sqrt{2n}+\sqrt{2n-1}}{\sqrt{2n}-\sqrt{2n-1}} + \frac{1}{\sqrt{n-1}} + \frac{1}{\sqrt{n}}\right] < \frac{K^* \ln n}{\sqrt{n}}.$$

The right member of the nth equation of system (d) is also arbitrarily small, under the assumptions made. This shows that for any $\varepsilon > 0$, one can find a sufficiently distant equation of system (d) that is true accurate to ε. It could be briefly said that an infinitely distant equation of system (d) is satisfied. Let us take the nth equation of system (d), which is satisfied with an accuracy to ε. The equation obtained by the addition of the nth and $(n-1)$th equations of system (d) is the doubled $(n-1)$th equation of system (e), and is therefore satisfied exactly. Hence we conclude that the $(n-1)$th equation is satisfied with an accuracy to ε, and continuing in the same manner, we shall establish the same thing for all equations right up to the first. In view of the arbitrariness of ε, one can conclude that all equations of system (d) are satisfied exactly. After this we find

$$A_{2n-1} = \frac{(-1)^n B_{2n-1}}{2n-1},$$

which will be of order not lower than $1/n^{3/2}$ and will give the solution of system (c). Next we can find $A_0, A_2, A_4, \ldots$ from system (a); it is easy to establish by means of an estimate analogous to that carried through above that the numbers A_{2n}

---

[1]) To prove this it is sufficient to verify that the substitution of $B_{2n-1} = \frac{1}{\sqrt{n}}$ n system (e) yields free terms of order $\frac{1}{n\sqrt{n}}$, and then to compare solutions (§ 2, No. 2 Theorem 1 (5)).

will be of order not lower than $n^{-3/2} \ln n$ and satisfy together with A_{2n-1} the equations of systems (a) and (b). Furthermore, if for the functions $u(r, \theta)$ and $\frac{\partial u(r, \theta)}{\partial r}$ we consider series in $\cos 2n\theta$ in the intervals $-\frac{\pi}{2} \leqslant \theta \leqslant \frac{\pi}{2}$ and $\frac{\pi}{2} \leqslant \theta \leqslant \frac{3\pi}{2}$, these series will converge absolutely and uniformly [1]), and a term-by-term passage to the limit as $r \to 1$ will be admissible, which will give, on the strength of the equations of systems (a) and (b):

$$\lim_{r\to 1} u(r, \theta) = f(\theta) \quad -\frac{\pi}{2} \leqslant \theta \leqslant \frac{\pi}{2}, \qquad \lim_{r\to 1} \frac{\partial u(r, \theta)}{\partial r} = \varphi(\theta) \quad \frac{\pi}{2} \leqslant \theta \leqslant \frac{3\pi}{2}.$$

In this example we have had in view how the solution of such a problem can be carried out by means of infinite systems, with a complete investigation of the systems obtained. We note that the problem we have considered can be solved in another way without employing infinite systems [2]).

The solution of the following more general problem can also be brought to an infinite system, but of a more complex form:

To find: u, under the condition

$$a_1 u + b_1 \frac{\partial u}{\partial r}\bigg|_{r=1} = f_1(\theta), \quad \left(-\frac{\pi}{2} \leqslant \theta \leqslant \frac{\pi}{2}\right),$$
$$a_2 u + b_2 \frac{\partial u}{\partial r}\bigg|_{r=1} = f_2(\theta), \quad \left(\frac{\pi}{2} \leqslant \theta \leqslant \frac{3\pi}{2}\right).$$

5. Example 2. The clamped plate. It is known that the deflection of a thin rectangular plate with sides $2a$ and $2b$ which is clamped along the sides and subjected to uniform loading, is a function $u(x, y)$ satisfying the equation

$$\Delta\Delta u = \frac{\partial^4 u}{\partial x^4} + 2\frac{\partial^4 u}{\partial x^2 \partial y^2} + \frac{\partial^4 u}{\partial y^4} = c; \quad c = \frac{P}{N}, \tag{33}$$

(where P is the unit load, and N is a constant connected with the elastic properties of the plate), and satisfying also the boundary conditions

$$u = 0 \text{ for } x = \pm a, \; y = \pm b,$$
$$\frac{\partial u}{\partial x} = 0 \text{ for } x = \pm a, \qquad \frac{\partial u}{\partial y} = 0 \text{ for } y = \pm b. \tag{34}$$

A particular solution of equation (33) will obviously be

$$u_0 = \frac{c}{8}(a^2 - x^2)(b^2 - y^2). \tag{35}$$

Then for the function $u_1 = u - u_0$ we have the equation $\Delta\Delta u_1 = 0$, and the

---

[1]) In respect of the series for $\frac{\partial u}{\partial r}$, this is, granted, valid only with certain additional limitations.

[2]) This problem is the limiting case of Hilbert's problem, which is examined in V. I. Smirnov's textbook, [1], Vol. III, p. 382 (1933), or below, Chapter VI, § 4, No. 2.

boundary conditions

$$\left.\begin{aligned} &u_1 = 0 \text{ for } x = \pm a,\ y = \pm b, \\ &\frac{\partial u_1}{\partial x} = \pm \frac{ca}{4}(b^2 - y^2) \text{ for } x = \pm a, \\ &\frac{\partial u_1}{\partial y} = \pm \frac{cb}{4}(a^2 - x^2) \text{ ,, } y = \pm b. \end{aligned}\right\} \tag{36}$$

In view of the symmetry of the boundary conditions, u_1 represents an even function of the variables x and y. It will therefore be sufficient to seek the solution in the form of a series of even functions, and to satisfy the boundary conditions with $x = +a$ and $y = +b$, since for $x = -a$ and $y = -b$ they will be automatically satisfied. Let us now expand the functions in the boundary conditions in trigonometric series, whereupon the boundary conditions may be written thus:

$$\left.\begin{aligned} &u_1 = 0 \text{ for } x = a,\ u_1 = 0 \text{ for } y = b, \\ &\left.\frac{\partial u_1}{\partial x}\right|_{x=a} = \frac{a^3 c \cdot 8\varphi^2}{\pi^3} \sum_{n=1}^{\infty} \frac{\sin n\dfrac{\pi}{2}}{n^3} \cos\frac{n\pi}{2b} y, \\ &\left.\frac{\partial u_1}{\partial y}\right|_{y=b} = \frac{a^3 c \cdot 8\varphi}{\pi^3} \sum_{n=1}^{\infty} \frac{\sin n\dfrac{\pi}{2}}{n^3} \cos\frac{n\pi}{2a} x, \end{aligned}\right\} \tag{37}$$

where $\varphi = b/a$.

We shall try to find the solution in the form

$$u_1 = -\frac{ca^3}{2\pi^2}\left\{\sum_{n=1,3,5,\ldots} A_n \frac{H_1(y, n\pi)}{n^2 \operatorname{ch}^2 \dfrac{n\pi\varphi}{2}} \cos\frac{n\pi}{2a} x + \right.$$

$$\left. + \varphi^3 \sum_{n=1,3,5,\ldots} B_n \frac{H(x, n\pi)}{n^2 \operatorname{ch}^2 \dfrac{n\pi}{2\varphi}} \cos\frac{n\pi}{2b} y\right\}, \tag{38}$$

where the functions H and H_1 are the even solutions of differential equation (58) which would occur in a study of the deflection of a freely supported plate (§ 1, No. 3); they satisfy the conditions $H(a, n\pi) = 0$ and $H_1(b, n\pi) = 0$ respectively:

$$\left.\begin{aligned} H(x, n\pi) &= (a + x) \operatorname{sh}\frac{n\pi}{2b}(a - x) + (a - x) \operatorname{sh}\frac{n\pi}{2b}(a + x), \\ H_1(y, n\pi) &= (b + y) \operatorname{sh}\frac{n\pi}{2a}(b - y) + (b - y) \operatorname{sh}\frac{n\pi}{2a}(b + y). \end{aligned}\right\} \tag{39}$$

It is obvious that solution (38) formally satisfies the equation $\Delta\Delta u_1 = 0$ and the conditions $u_1 = 0$ for $x = a$ and $y = b$. We must now so choose the coefficients A_n and B_n as to satisfy the last two conditions of (37).

It is readily verified that

$$\left.\begin{aligned} H'(x, n\pi)\,|_{x=a} &= -\left(\frac{n\pi}{\varphi} + \operatorname{sh}\frac{n\pi}{\varphi}\right), \\ H_1'(y, n\pi)\,|_{y=b} &= -(n\pi\varphi + \operatorname{sh} n\pi\varphi). \end{aligned}\right\} \tag{40}$$

Hence we have

$$\left.\begin{aligned}
\left.\frac{\partial u_1}{\partial x}\right|_{x=a} &= \frac{ca^3}{2\pi^2}\left\{\sum_{n=1,3,\ldots} A_n \frac{H_1(y, n\pi)}{n^2\,\mathrm{ch}^2\dfrac{n\pi\varphi}{2}}\frac{n\pi}{2a}\sin\frac{n\pi}{2} + \right. \\
&\left. + \varphi^3 \sum_{n=1,3,\ldots} B_n \frac{\dfrac{n\pi}{\varphi} + \mathrm{sh}\dfrac{n\pi}{\varphi}}{n^2\,\mathrm{ch}^2\dfrac{n\pi}{2\varphi}}\cos\frac{n\pi}{2b}y\right\}, \\
\left.\frac{\partial u_1}{\partial y}\right|_{y=b} &= \frac{ca^3}{2\pi^2}\left\{\sum_{n=1,3,\ldots} A_n \frac{n\pi\varphi + \mathrm{sh}\,n\pi\varphi}{n^2\,\mathrm{ch}^2\dfrac{n\pi\varphi}{2}}\cdot\cos\frac{n\pi}{2a}x + \right. \\
&\left. + \varphi^3 \sum_{n=1,3,\ldots} B_n \frac{H(x, n\pi)}{n^2\,\mathrm{ch}^2\dfrac{n\pi}{2\varphi}}\frac{n\pi}{2b}\sin\frac{n\pi}{2}\right\}.
\end{aligned}\right\} \tag{41}$$

Now in order to compare expressions (41) and (37), we must expand the functions $H(x, n\pi)$ and $H_1(y, n\pi)$ in terms of $\cos\dfrac{n\pi}{2a}x$ and $\cos\dfrac{n\pi}{2b}y$. We have

$$\left.\begin{aligned}
\frac{n\pi}{2a}H_1(y, n\pi) &= \sum_{i=1,3,\ldots} C_i \cos\frac{i\pi}{2b}y, \\
\frac{n\pi}{2b}H(x, n\pi) &= \sum_{i=1,3,\ldots} D_i \cos\frac{i\pi}{2a}x,
\end{aligned}\right\} \tag{42}$$

where, in conformity with the formulas for the coefficients of series of orthogonal functions,

$$\left.\begin{aligned}
C_i &= \frac{n\pi}{2a}\,\frac{\int_0^b H_1(y, n\pi)\cos\dfrac{i\pi}{2b}y\,dy}{\int_0^b \cos^2\dfrac{i\pi}{2b}y\,dy} = \frac{16\varphi^2}{n\pi\left(\dfrac{i^2}{n^2}+\varphi^2\right)^2}\cdot\frac{i}{n}\,\mathrm{ch}^2\frac{n\pi\varphi}{2}\sin\frac{i\pi}{2}, \\
D_i &= \frac{n\pi}{2b}\,\frac{\int_0^a H(x, n\pi)\cos\dfrac{i\pi}{2a}x\,dx}{\int_0^a \cos^2\dfrac{i\pi}{2a}x\,dx} = \frac{16\varphi^2}{n\pi\left(\dfrac{i^2}{n^2}\varphi^2+1\right)^2}\cdot\frac{i}{n}\,\mathrm{ch}^2\frac{n\pi}{2\varphi}\sin\frac{i\pi}{2}.
\end{aligned}\right\} \tag{43}$$

Substituting for H and H_1 series (42) in expressions (41) and equating them to expressions (37), we arrive, on dividing by ca^3, at the following equations:

$$-\frac{8\cdot\varphi^2}{\pi^3}\sum_{n=1,3,\ldots}\frac{\sin n\frac{\pi}{2}}{n^3}\cos\frac{n\pi}{2b}y+$$

$$+\sum_{n=1,3,\ldots}\sum_{i=1,3,\ldots}\left\{\frac{8}{n^3\pi^3}\cdot\frac{i}{n}\frac{A_n}{\left(\frac{i^2}{n^2}+\varphi^2\right)^2}\varphi^2\sin\frac{n\pi}{2}\cdot\sin\frac{i\pi}{2}\cdot\cos\frac{i\pi}{2b}y\right\}+$$

$$+\sum_{n=1,3,\ldots}\frac{\varphi^3B_n}{2n^2\pi^2}\cdot\frac{\frac{n\pi}{\varphi}+\operatorname{ch}\frac{n\pi}{\varphi}}{\operatorname{ch}^2\frac{n\pi}{2\varphi}}\cos\frac{n\pi}{2b}y=0,$$

$$-\frac{8\varphi}{\pi^3}\sum_{n=1,3,\ldots}\frac{\sin n\frac{\pi}{2}}{n^3}\cos\frac{n\pi}{2a}x+$$

$$+\sum_{n=1,3,\ldots}\sum_{i=1,3,\ldots}\left\{\frac{8}{n^3\pi^3}\frac{i}{n}\frac{B_n}{\left(\frac{i^2}{n^2}\varphi^2+1\right)^2}\varphi^5\sin\frac{n\pi}{2}\sin\frac{i\pi}{2}\cos\frac{i\pi}{2a}x\right\}+$$

$$+\sum_{n=1,3,\ldots}\frac{A_n}{2n^2\pi^2}\frac{n\pi\varphi+\operatorname{sh}n\pi\varphi}{\operatorname{ch}^2\frac{n\pi\varphi}{2}}\cdot\cos\frac{n\pi}{2a}x=0. \tag{44}$$

Converting the double series into simple ones in terms of $\cos\frac{n\pi}{2b}y$ and $\cos\frac{n\pi}{2a}x$ (replacing i by n and n by ϱ) and equating to zero all coefficients of these functions, we arrive at the system of equations

$$\left.\begin{aligned}&\beta_nB_n\sin\frac{n\pi}{2}+\sum_{\varrho=1,3,\ldots}A_\varrho\left(\frac{n^2}{n^2+\varrho^2\varphi^2}\right)^2\sin\frac{\varrho\pi}{2}=1,\\&\alpha_nA_n\sin\frac{n\pi}{2}+\sum_{\varrho=1,3,\ldots}B_\varrho\left(\frac{n^2}{\varphi^2n^2+\varrho^2}\right)^2\sin\frac{\varrho\pi}{2}=\frac{1}{\varphi^4},\end{aligned}\quad(n=1,3,5,7,\ldots)\right\} \tag{45}$$

where to simplify the writing we have used the designations

$$\alpha_n=\varphi^{-5}\frac{n\pi}{16}\frac{n\pi\varphi+\operatorname{sh}n\pi\varphi}{\operatorname{ch}^2\frac{n\pi\varphi}{2}},\quad\beta_n=\varphi\frac{n\pi}{16}\frac{\frac{n\pi}{\varphi}+\operatorname{sh}\frac{n\pi}{\varphi}}{\operatorname{ch}^2\frac{n\pi}{2\varphi}}. \tag{46}$$

The coefficients A_n and B_n must be determined from system (45); the solution (38) can then be formed in terms of them.

Let us linger in more detail upon the case of a square plate. In this case

$$b=a,\quad\varphi=\frac{b}{a}=1,\quad B_n=A_n,\ \beta_n=\alpha_n.$$

Of the two systems of equations (45), one may be preserved. To write this system completely, we compute the α_n; we have:

$$\alpha_1=\frac{\pi^2}{16\operatorname{ch}^2\frac{\pi}{2}}+\frac{2\pi}{16}\operatorname{th}\frac{\pi}{2}=0.45814,$$

$$\alpha_3 = \frac{9\pi^2}{16\,\mathrm{ch}^2\dfrac{3\pi}{2}} + \frac{6\pi}{16}\,\mathrm{th}\,\tfrac{3}{2}\pi = 1.17989,$$

$$\alpha_5 = 1.96355,\ \alpha_7 = 2.74889,\ \alpha_9 = 3.53417,\ \alpha_{11} = 4.31969.$$

For large n ($n \geqslant 10$) we may adopt

$$\alpha_n = \frac{n\pi}{16}\,\frac{n\pi + \mathrm{sh}\,n\pi}{\mathrm{ch}^2\dfrac{n\pi}{2}} = \frac{n\pi}{16}\,\frac{n\pi + \mathrm{sh}\,n\pi}{\frac{1}{2} + \frac{1}{2}\,\mathrm{ch}\,n\pi} \approx \frac{n\pi}{8}.$$

If in system (45) the term of the sum containing A_n be combined with the term standing outside the sum, the system will acquire the form

$$\left.\begin{array}{l}
0.70814A_1 - (\frac{1}{10})^2A_3 + (\frac{1}{26})^2A_5 - (\frac{1}{50})^2A_7 + (\frac{1}{82})^2A_9 - (\frac{1}{122})^2A_{11} + \ldots = 1\\
(\frac{9}{10})^2A_1 - 1.42989A_3 + (\frac{9}{34})^2A_5 - (\frac{9}{58})^2A_7 + (\frac{9}{90})^2A_9 - (\frac{9}{130})^2A_{11} + \ldots = 1\\
(\frac{25}{26})^2A_1 - (\frac{25}{34})^2A_3 + 2.21355A_5 - (\frac{25}{74})^2A_7 + (\frac{25}{106})^2A_9 - (\frac{25}{146})^2A_{11} + \ldots = 1\\
(\frac{49}{50})^2A_1 - (\frac{49}{58})^2A_3 + (\frac{49}{74})^2A_5 - 2.99889A_7 + (\frac{49}{130})^2A_9 - (\frac{49}{170})^2A_{11} + \ldots = 1\\
(\frac{81}{82})^2A_1 - (\frac{81}{90})^2A_3 + (\frac{81}{106})^2A_5 - (\frac{81}{130})^2A_7 + 3.78417A_9 - (\frac{81}{202})^2A_{11} + \ldots = 1\\
(\frac{121}{122})^2A_1 - (\frac{121}{130})^2A_3 + (\frac{121}{146})^2A_5 - (\frac{121}{170})^2A_7 + (\frac{121}{202})^2A_9 - 4.51969A_{11} + \ldots = 1\\
\ldots\ldots\ldots\ldots\ldots\ldots\ldots\ldots\ldots\ldots\ldots\ldots
\end{array}\right\} \quad (47)$$

This system, as we shall show below, is regular and has a unique bounded solution for the given free terms. This solution may be found not only by the method of successive approximations indicated in § 2 but also by a somewhat modified method of successive approximations, namely the following one. Separate the terms beyond the principal diagonal and transpose them to the right side; now first discarding these terms and solving the remaining recurrent system, we obtain a first approximation to the unknowns $A_1, A_3, \ldots$ Using these first approximations to calculate the omitted terms and solving anew the recurrent system obtained, we obtain second approximations for the unknowns, and so forth. It can be established without difficulty that these successive approximations will tend to the same principal solution as do those obtained by the usual method. Utilizing this method Hencky [1]) found the following values for the unknowns:

$$\begin{array}{llll}
A_1 = & 1.4138 & A_{15} = & 0.0193\\
A_3 = & 0.0945 & A_{17} = & -0.0141\\
A_5 = & -0.1093 & A_{19} = & 0.0103
\end{array}$$

---

[1]) The method of solving this problem of the clamped plate by a reduction to an infinite system of equations was first developed by B. M. Koialovich, [2], 1902. On the subject of the solution presented and numerical data see B. G. Galerkin, [2], and his book [1], p. 90. In addition see S. P. Timoshenko, [1], and H. Hencky, [1].

Examples of the application of infinite systems in other problems of the theory of elasticity are contained in articles of: D. M. Volkov and A. A. Nazarov, [1]. D. I. Luchinin, [1], N. I. Muskhelishvili, [1], M. I. Gorbunov-Posadov, [1], S. S. Gurevich, [1], C. B. Biezeno and J. J. Koch, [1].

$$A_7 = 0.0770 \qquad A_{21} = -0.0077$$
$$A_9 = -0.0510 \qquad A_{23} = 0.0051$$
$$A_{11} = 0.0371 \qquad A_{25} = -0.0041.$$
$$A_{13} = -0.0264$$

Now, when the constants $A_1, A_3, \ldots$ have been determined, the deflection of the plate at any place, $u = u_0 + u_1$ is also determined, being found from equations (35) and (38); by means of it the moments and the other quantities required of the calculation can also be found. Thus, for instance, the deflection at the center:

$$u(0, 0) = \frac{ca^4}{8} - \frac{ca^3}{2\pi^2} \sum_{n=1,3,\ldots} \frac{4a \operatorname{sh} \frac{n\pi}{2}}{n^2 \operatorname{ch}^2 \frac{n\pi}{2}} A_n;$$

the largest moment, in the middle of a side:

$$M_x \bigg|_{\substack{x=a \\ y=0}} = \frac{a^2 p}{4} \left(1 - \frac{4}{\pi} \sum_{n=1}^{\infty} \frac{A_n}{n}\right).$$

The solution for a rectangular plate can be carried through analogously, except that in that case one has to take into consideration both systems of equations (45).

On the basis of the solution he obtained, Hencky gives the following numerical results, exhibiting for comparison the results for a fixed-end beam ($\varphi = \infty$):

$\varphi = \dfrac{b}{a}$	In the center of the plate		At the middle of a side	
	M_y	M_x	M_y	M_x
1.0	$0.092\, pa^2$	$0.092\, pa^2$	$0.205\, pa^2$	$0.205\, pa^2$
1.5	$0.148\, pa^2$	$0.080\, pa^2$	$0.306\, pa^2$	$0.226\, pa^2$
∞	$0.167\, pa^2$	—	$0.333\, pa^2$	—

Let us now conduct an investigation of the system of equations (47). We shall prove that this system is regular and has a bounded solution for the given free terms. For this let us estimate the sum of the moduli of the coefficients (excluding the diagonal one) in each equation of the system. We have, for $n = 1, 3, 5, \ldots$

$$\left(\frac{n^2}{n^2+1}\right)^2 + \left(\frac{n^2}{n^2+3^2}\right)^2 + \ldots + \left[\frac{n^2}{n^2+(n-2)^2}\right]^2 + \left[\frac{n^2}{n^2+(n+2)^2}\right]^2 + \ldots <$$

$$< \int_{-\frac{1}{2}}^{\frac{n}{2}-1} \left[\frac{n^2}{n^2+(2x+1)^2}\right]^2 dx + \frac{8}{3n} + \int_{\frac{n}{2}}^{\infty} \left[\frac{n^2}{n^2+(2x+1)^2}\right]^2 dx <$$

$$< \frac{n}{2} \int_0^{\infty} \frac{du}{(u^2+1)^2} = \frac{n}{2} \left[\frac{u}{2(1+u^2)} + \tfrac{1}{2} \operatorname{arctg} u\right]_0^{\infty} = \frac{\pi n}{8}.$$

The first sum has been replaced by an integral and a remainder term, using the tangential rule. The second infinite sum has been replaced directly by the integral, since the curve is already concave upward.

As regards the diagonal coefficient of the unknown A_n, it is equal in absolute value to

$$\alpha_n + \left(\frac{n^2}{n^2+n^2}\right)^2 = \alpha_n + \tfrac{1}{4} > \frac{n\pi}{8} + \tfrac{1}{4};$$

[of the fact that $\alpha_n > \frac{n\pi}{8}$ one can readily satisfy oneself by using (46)]. Dividing the equations of system (47) by this diagonal coefficient $\alpha_n + \frac{1}{4}$ and transposing to the right all terms except the diagonal one, we bring system (47) into normal form. The sum of the moduli of the coefficients of the unknowns on the right side will then be

$$\sum_{\varrho=1}^{\infty} |c_{n,\varrho}| = \frac{1}{\alpha_n + \frac{1}{4}}\left[\left(\frac{n^2}{n^2+1}\right)^2 + \left(\frac{n^2}{n^2+3^2}\right)^2 + \ldots + \left(\frac{n^2}{n^2+(n-2)^2}\right)^2 + \right.$$

$$\left. + \left(\frac{n^2}{n^2+(n+2)^2}\right)^2 + \ldots\right] < \frac{\frac{n\pi}{8}}{\frac{n\pi}{8}+\frac{1}{4}} < 1 - \frac{1}{n\pi} < 1.$$

This inequality shows that the system obtained is regular and that it has certainly a bounded solution for free terms of order not lower than $\frac{1}{n\pi}$, i.e., $< \frac{K}{n\pi}$. But after division by $\alpha_n + \frac{1}{4}$, the free terms 1 are replaced by

$$\frac{1}{\alpha_n + \frac{1}{4}} < \frac{1}{\frac{n\pi}{8}} = 8 \cdot \frac{1}{n\pi},$$

i.e., satisfy this condition for $K = 8$. Thus system of equations (47) does indeed have a bounded solution and the method of successive approximations is a convergent process.

For proof of the uniqueness [1]) of the bounded solution of system (47), we shall apply the method indicated in the Corollary to Theorem IIb (p. 30), viz.: the introduction, in place of the unknowns A_n, of new unknowns C_n, putting

$$A_n = nC_n \qquad (n = 1, 3, 5, \ldots).$$

With respect to the unknowns C_n system (47) [see also (45)] acquires the form

$$C_n = \frac{n^2}{\alpha_n + \frac{1}{4}} \sum_{k=1,3,\ldots}{}' \frac{k(-1)^{\frac{k+n+2}{2}}}{(n^2+k^2)^2} C_k + \frac{1}{n(\alpha_n + \frac{1}{4})} \qquad (n = 1, 3, 5, \ldots),$$

where the prime on the sum sign shows that k should not be given the value $k = n$.

It is necessary that the regularity of the system obtained be verified. For this

---

[1]) The direct proof of the uniqueness of the solution expounded below was suggested by P. S. Bondarenko. R. O. Kuz'min, in his work [1], had previously given a proof of the uniqueness of the solution which relied on the uniqueness theorem for the biharmonic problem.

let us estimate the sum of the moduli of its coefficients. We have:

$$\sum_{k=1,3,5,\dots}' \frac{k}{(n^2+k^2)^2} = \tfrac{1}{2}\frac{1}{(n^2+1)^2} + \left[\tfrac{1}{2}\frac{1}{(n^2+1)^2} + \right.$$

$$\left. + \sum_{k=3,5,\dots,n-2} \frac{k}{(n^2+k^2)^2} + \tfrac{1}{2}\frac{1}{4n^3}\right] - \tfrac{1}{2}\cdot\frac{1}{4n^3} +$$

$$+ \sum_{k=n+2,\dots} \frac{k}{(n^2+k^2)^2} \leqslant \tfrac{1}{2}\frac{1}{(n^2+1)^2} + \tfrac{1}{2}\int_1^n \frac{xdx}{(n^2+x^2)^2} +$$

$$+ \tfrac{1}{2}\int_n^\infty \frac{xdx}{(n^2+x^2)^2} \leqslant \tfrac{1}{2}\frac{1}{(n^2+1)^2} + \tfrac{1}{2}\int_0^\infty \frac{x}{(n^2+x^2)^2} = \tfrac{1}{2}\cdot\frac{1}{(n^2+1)^2} + \tfrac{1}{4}\cdot\frac{1}{n^2}.$$

Here, utilizing the fact that the function $\dfrac{x^2}{(n^2+x^2)^2}$ is concave downwards in the interval $(0, n)$, we have replaced by $\int_1^n$ the sum in square brackets corresponding to the trapezoidal formula (with interval $h = 2$), and have replaced the infinite sum by $\int_n^\infty$, utilizing the fact that this function is a decreasing one in the interval $(n, +\infty)$. Therefore on introducing the factor standing in front of the sum sign, we find:

$$\frac{n^3}{\alpha_n + \frac{1}{4}} \sum_{k=1,3,5,\dots} \frac{k}{(n^2+k^2)^2} \leqslant \frac{\frac{1}{4}n + \frac{1}{2}\dfrac{n^3}{(n^2+1)^2}}{\alpha_n + \frac{1}{4}} \leqslant$$

$$\leqslant \frac{\frac{1}{4}n + \frac{1}{8}}{\dfrac{n\pi}{8} + \frac{1}{4}} = \frac{\frac{1}{4}(n+\frac{1}{2})}{\dfrac{\pi}{8}\left(n + \dfrac{2}{\pi}\right)} \leqslant \frac{2}{\pi}.$$

The inequality we have obtained shows that the transformed system has even proved to be fully regular, and therefore it — and consequently the initial system too — has a unique bounded solution.

§ 4. THE APPLICATION OF DOUBLE SERIES TO THE SOLUTION OF BOUNDARY-VALUE PROBLEMS

1. Statement of the problem. The foundations of the method. Let us pose the problem: to find, the solution of an equation of elliptic type,

$$M(u) = f(x, y), \tag{1}$$

with null boundary conditions:

$$P(u) = 0 \text{ on the contour } \Gamma \tag{2}$$

and various free terms $f(x, y)$. We note that the requirement of zero boundary conditions does not reduce the generality. Indeed, if the boundary conditions are not zero: $P(u) = \varphi(s)$ on Γ, on constructing an arbitrary function $u_0(x, y)$ satisfying those boundary conditions, we would obviously have for the difference $u_1 = u - u_0$ the null boundary conditions $P(u_1) = P(u) - P(u_0) = 0$ on Γ, and the equation

$$M(u_1) = M(u) - M(u_0) = f(x, y) - M[u_0(x, y)],$$

i.e., one of the same type as (1) but with a different free term.

For solving the problem posed, double series often prove useful. Let us assume that we have found the functions $u_{i,k}(x, y)$ satisfying the null boundary conditions

$$P[u_{i,k}(x, y)] = 0 \text{ on } \Gamma, \tag{3}$$

and let

$$M[u_{i,k}(x, y)] = v_{i,k}(x, y). \tag{4}$$

We shall assume that we can expand the function $f(x, y)$ in a series in terms of the functions $v_{i,k}$:

$$f(x, y) = \sum_{i,k=1}^{\infty} a_{i,k} v_{i,k}(x, y), \tag{5}$$

the sought solution of the boundary problem being then formally written as

$$u(x, y) = \sum_{i,k=1}^{\infty} a_{i,k} u_{i,k}(x, y). \tag{6}$$

We shall now indicate a method of finding the functions $u_{i,k}(x, y)$. In many cases one can successfully adopt for them the fundamental functions of the equation

$$M(u) = \lambda u, \tag{7}$$

under the boundary condition $P(u) = 0$. These functions, generally speaking, are orthogonal to each other and form a complete system of functions in the given region. Let us assume that functions $u_{i,k}(x, y)$ and numbers $\lambda_{i,k}$ have been found such that

$$M(u_{i,k}) = \lambda_{i,k} u_{i,k}. \tag{8}$$

If the function $f(x, y)$ be then expanded in a series

$$f(x, y) = \sum_{i,k=1}^{\infty} A_{i,k} u_{i,k}(x, y), \tag{9}$$

the sought solution, $u(x, y)$, can be written in the form:

$$u(x, y) = \sum_{i,k=1}^{\infty} \frac{A_{i,k}}{\lambda_{i,k}} u_{i,k}(x, y). \tag{10}$$

It frequently proves to be possible to find the fundamental functions $u_{i,k}$ by Fourier's method.

2. Poisson's equation for a rectangle. Let us pose the following problem: to find, the solution of the Poisson equation

$$\Delta u = f(x, y), \tag{11}$$

vanishing on the boundaries of some rectangle

$$\left.\begin{aligned} u = 0 \text{ for } x = 0, \quad y = 0, \\ x = a, \quad y = b. \end{aligned}\right\} \tag{12}$$

In § 1 we presented the solution of the Dirichlet problem by Fourier's method for the Laplace equation $\Delta u = 0$ with arbitrary boundary conditions. This solution made it possible in certain cases to find the solution even of the problem of the Poisson equation now before us. In this No. we give, by means of double series, the solution of the problem in the general case with an arbitrary function $f(x, y)$.

Proceeding as has been stated above, we seek by Fourier's method the fundamental functions of the equation

$$\Delta u = \lambda u.$$

Substituting in it $u = X(x) \cdot Y(y)$, we have

$$X''Y + XY'' = \lambda XY$$

or

$$\frac{X''}{X} + \frac{Y''}{Y} = \lambda,$$

for which it is necessary that

$$\frac{X''}{X} = p, \quad \frac{Y''}{Y} = q,$$

where p and q are constants. Next determining the functions $X(x)$ and $Y(y)$ from these equations and from the boundary conditions

$$X(a) = X(0) = 0, \quad Y(b) = Y(0) = 0,$$

we find that a non-trivial solution exists only on condition that

$$p = -\frac{m^2\pi^2}{a^2}, \quad q = -\frac{n^2\pi^2}{b^2} \quad \begin{pmatrix} m = 1, 2, 3, \ldots \\ n = 1, 2, 3, \ldots \end{pmatrix},$$

and that this solution will be

$$X_m(x) = \sin\frac{m\pi x}{a}, \quad Y_n(y) = \sin\frac{n\pi y}{b}.$$

The functions

$$u_{m,n}(x, y) = X_m(x) \cdot Y_n(y) = \sin\frac{m\pi x}{a} \cdot \sin\frac{n\pi y}{b} \tag{13}$$

form a complete system of orthogonal functions in the rectangle (0, a; 0, b). Furthermore,

$$\Delta X_m Y_n = -\pi^2\left(\frac{m^2}{a^2} + \frac{n^2}{b^2}\right) X_m Y_n. \tag{14}$$

The function $f(x, y)$ can be expanded, generally speaking, in a double Fourier series [1]), i.e., in a series involving the functions $u_{m,n}$:

$$f(x, y) = \sum_{m,n=1}^{\infty} a_{m,n} \sin\frac{m\pi x}{a} \sin\frac{n\pi y}{b}, \tag{15}$$

where

$$a_{m,n} = \frac{4}{ab}\int_0^a\int_0^b f(\xi, \eta) \sin\frac{m\pi\xi}{a} \sin\frac{n\pi\eta}{b}\, d\xi d\eta. \tag{16}$$

In view of (10), the solution u can then be stated thus:

$$u(x, y) = -\sum_{m,n=1}^{\infty} \frac{a_{m,n}}{\pi^2\left(\dfrac{m^2}{a^2} + \dfrac{n^2}{b^2}\right)} \sin\frac{m\pi x}{a} \sin\frac{n\pi y}{b}. \tag{17}$$

The solution obtained, (17), may be given a somewhat different form.

On substituting in (17) for the Fourier coefficients $a_{m,n}$ their expressions (16), we obtain

$$u(x, y) = -\sum_{m,n=1}^{\infty} \frac{4\sin\dfrac{m\pi x}{a}\sin\dfrac{n\pi y}{b}}{\pi^2\left(\dfrac{m^2}{a^2} + \dfrac{n^2}{b^2}\right)ab}\int_0^a\int_0^b f(\xi, \eta) \sin\frac{m\pi\xi}{a} \sin\frac{n\pi\eta}{b}\, d\xi d\eta =$$

$$= \int_0^a\int_0^b \left[-\frac{4}{\pi^2 ab}\sum_{m,n=1}^{\infty} \frac{\sin\dfrac{m\pi x}{a}\sin\dfrac{n\pi y}{b}\sin\dfrac{m\pi\xi}{a}\sin\dfrac{n\pi\eta}{b}}{\dfrac{m^2}{a^2} + \dfrac{n^2}{b^2}}\right] f(\xi, \eta) d\xi d\eta. \tag{18}$$

If the function of four variables in the brackets be considered separately,

$$G(x, y; \xi, \eta) = -\frac{4}{\pi^2 ab}\sum_{m,n=1}^{\infty} \frac{\sin\dfrac{m\pi x}{a}\sin\dfrac{n\pi y}{b}\sin\dfrac{m\pi\xi}{a}\sin\dfrac{n\pi\eta}{b}}{\dfrac{m^2}{a^2} + \dfrac{n^2}{b^2}}, \tag{19}$$

by means of this function — which does not depend on the free term — the solution of the problem with an arbitrary free term can be very

---

[1]) See, e.g., I. Privalov, [1], § 27.

simply expressed:

$$u(x, y) = \int_0^a \int_0^b G(x, y; \xi, \eta) f(\xi, \eta) d\xi d\eta. \tag{20}$$

Concerning the function $G(x, y; \xi, \eta)$ we note that the series for it always converges and gives a finite value, with the exception of the case $\xi = x$, $\eta = y$, when it turns out to be infinite of logarithmic order. Thus the double integral giving $u(x, y)$ is generally improper; the integrand is certainly, however, absolutely integrable. The function $G(x, y; \xi, \eta)$ is, apart from a constant factor, the Green function for a rectangle [1]).

We note that by means of double trigonometric series the Poisson equation can be solved for the rectangle not only in the case of the Dirichlet problem, but also under other boundary conditions, for instance in the case of the Neumann problem. Moreover, by means of double series boundary problems for the rectangle can be solved not only in the case of the Poisson equation, but also for all equations of the second order of elliptic type with constant coefficients and not containing mixed nor first derivatives, i.e., for equations of the form:

$$a^2 \frac{\partial^2 u}{\partial x^2} + b^2 \frac{\partial^2 u}{\partial y^2} + cu = f(x, y). \tag{21}$$

3. Application to equations of the fourth order. The general principles of the application of double series to the solution of equations of the fourth order remain the same, and we shall therefore limit ourselves to the consideration of two particular problems only.

1. *A rectangular plate freely mounted along the edges.* The differential quation for its deflection is:

$$N \cdot \Delta\Delta u = N\left(\frac{\partial^4 u}{\partial x^4} + 2\frac{\partial^4 u}{\partial x^2 \partial y^2} + \frac{\partial^4 u}{\partial y^4}\right) = p(x, y), \tag{22}$$

where $p(x, y)$ is the unit load and the boundary conditions are

$$\left.\begin{aligned} u = 0, \frac{\partial^2 u}{\partial x^2} = 0 \text{ for } x = 0, \quad x = a, \\ u = 0, \frac{\partial^2 u}{\partial y^2} = 0 \text{ for } y = 0, \quad y = b. \end{aligned}\right\} \tag{23}$$

The particular case of this problem with p constant has already been examined in No. 3 of § 1. The basic functions satisfying the boundary conditions will here be, as may readily be verified,

$$\sin \frac{m\pi x}{a} \sin \frac{n\pi y}{b} \quad (m, n = 1, 2, 3, \ldots).$$

---

[1]) See Goursat, [1], Vol. III, Part 1, p. 184.

We therefore seek the solution in double series form:

$$\sum_{m=1}^{\infty} \sum_{n=1}^{\infty} A_{m,n} \sin \frac{m\pi x}{a} \sin \frac{n\pi y}{b}.$$

On applying the operation $N\Delta\Delta$ to one term we see that

$$N\Delta\Delta A_{m,n} \sin \frac{m\pi x}{a} \sin \frac{n\pi y}{b} =$$

$$= N \left(\frac{m^2}{a^2} + \frac{n^2}{b^2} \right)^2 \pi^4 A_{m,n} \sin \frac{m\pi x}{a} \sin \frac{n\pi y}{b}, \tag{24}$$

i.e., on substitution in the right side this term acquires the factor

$$N\pi^4 \left(\frac{m^2}{a^2} + \frac{n^2}{b^2} \right)^2.$$

This shows that the Fourier coefficients of $u(x, y)$ must differ from the Fourier coefficients of $p(x, y)$ by a reciprocal factor.

Thus if

$$p(x, y) = \sum_{m=1}^{\infty} \sum_{n=1}^{\infty} a_{m,n} \sin \frac{m\pi x}{a} \sin \frac{n\pi y}{b}, \tag{25}$$

then

$$u(x, y) = \sum_{m=1}^{\infty} \sum_{n=1}^{\infty} \frac{a_{m,n} \sin \frac{m\pi x}{a} \sin \frac{n\pi y}{b}}{N\pi^4 \left(\frac{m^2}{a^2} + \frac{n^2}{b^2} \right)^2}, \tag{26}$$

where

$$a_{m,n} = \frac{4}{ab} \int_0^a \int_0^b p(\xi, \eta) \sin \frac{m\pi \xi}{a} \sin \frac{n\pi \eta}{b} \, d\xi d\eta. \tag{27}$$

The series for $u(x, y)$ will certainly be absolutely convergent, and the series for $\frac{\partial^2 u}{\partial x^2}, \frac{\partial^2 u}{\partial y^2}$, used for the determination of the moments, will also converge.

Let us consider as an example the deflection under the action of a concentrated force of magnitude P applied at the point $x = \xi$, $y = \eta$. Such a force may be regarded as the limit of a uniform load of intensity $\frac{P}{\delta^2}$ applied to the rectangle $(\xi, \xi + \delta;\ \eta, \eta + \delta)$, as $\delta \to 0$. Let us find the limit of the coefficient $a_{m,n}$ as $\delta \to 0$. For the case in hand we have, using the theorem of the mean:

$$a_{m,n} = \frac{4}{ab} \int_{\xi}^{\xi+\delta} \int_{\eta}^{\eta+\delta} \frac{P}{\delta^2} \sin \frac{m\pi x}{a} \sin \frac{n\pi y}{b} \, dx \, dy =$$

$$= \frac{4}{ab} \frac{P}{\delta^2} \sin \frac{m\pi}{a} (\xi + \theta\delta) \sin \frac{n\pi}{b} (\eta + \theta_1\delta) \cdot \delta^2$$

$$\to \frac{4P}{ab} \sin \frac{m\pi\xi}{a} \sin \frac{n\pi\eta}{b}. \tag{28}$$

Thus the deflection under the action of a concentrated force P at the point (ξ, η) is given by the series

$$u(x, y) = \sum_{m=1}^{\infty} \sum_{n=1}^{\infty} \frac{4P}{N\pi^4 ab} \frac{\sin \frac{m\pi x}{a} \sin \frac{n\pi y}{b} \sin \frac{m\pi\xi}{a} \sin \frac{n\pi\eta}{b}}{\left(\frac{m^2}{a^2} + \frac{n^2}{b^2}\right)^2}. \tag{29}$$

We note that for $u(x, y)$ we here too have a convergent series, and that passing to the limit has been permissible; however had we replaced $a_{m,n}$ by its limiting value in the series for $p(x, y)$, we would have obtained a divergent series — but we had no need of passing to the limit in that series.

Let us substitute expression (27) for the coefficient $a_{m,n}$ in the solution $u(x, y)$; we can then represent it in the form:

$$u(x, y) = \int_0^a \int_0^b \left[\frac{4}{N\pi^4 ab} \sum_{m=1}^{\infty} \sum_{n=1}^{\infty} \frac{\sin \frac{m\pi x}{a} \sin \frac{m\pi\xi}{a} \sin \frac{n\pi y}{b} \sin \frac{n\pi\eta}{b}}{\left(\frac{m^2}{a^2} + \frac{n^2}{b^2}\right)^2} \right] \times$$

$$\times \, p(\xi, \eta) d\xi \, d\eta. \tag{30}$$

Or, if the function of four variables in brackets be denoted by $G(x, y, \xi, \eta)$, we have

$$u(x, y) = \int_0^a \int_0^b G(x, y, \xi, \eta) p(\xi, \eta) d\xi \, d\eta. \tag{31}$$

The function $G(x, y, \xi, \eta)$ is called the *Green function* for the problem given; the solution for any form of the free term — i.e., for any loading — can be obtained, in the form cited, by means of this function. This function represents, as a comparison with (29) will convince one, the deflection for the case of a unit force, $P = 1$, applied at the point (ξ, η), and it is therefore natural that the deflection under a load of intensity $p(x, y)$ be given by formula (31). Indeed, the deflection at the point (x, y) under the action of a force $p(\xi, \eta) d\xi \, d\eta$ applied at the point (ξ, η) will equal $G(x, y, \xi, \eta) p(\xi, \eta) d\xi d\eta$, in accordance with (29), and therefore the deflection under the action of the whole load of intensity $p(\xi, \eta)$ is given by integral (31).

The biharmonic equation can be solved under diverse other conditions as well by means of double series. Examples of such solutions in application to beamless floors may be found in a book by Leve [1]). It should be observed, however, that to obtain definitive results by means of double series requires very considerable computational labor.

2. *A rectangular plate on an elastic base.* Its equation differs from the usual equation for a plate in that there is added to the load, with a minus sign, the reaction of the base — proportional to the deflection — and the equation will thus be

$$N\Delta\Delta u = p(x, y) - kNu$$

or

$$N\left(\frac{\partial^4 u}{\partial x^4} + 2\frac{\partial^4 u}{\partial x^2 \partial y^2} + \frac{\partial^4 u}{\partial y^4} + ku\right) = p(x, y).$$

The boundary conditions and the loading are adopted in conformity with the case of an infinite beamless floor (slab floor).

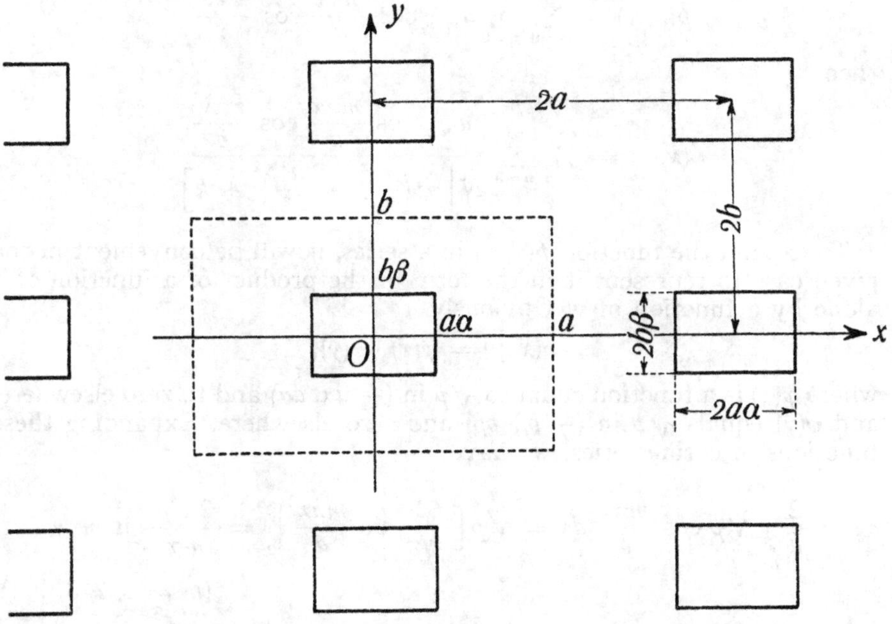

Fig. 3

The floor is an infinite slab on which columns rest at definite intervals (Fig. 3). At the base of each column, therefore, on a given rectangle, a uniform load of intensity p acts on the slab. The columns must be spaced symmetrically in both directions, i.e., u is, with respect to the

[1]) D. Leve, [1].

center lines of this spacing, an even function, and thus the boundary conditions for u will be:

$$\frac{\partial u}{\partial x} = 0, \ \frac{\partial^3 u}{\partial x^3} = 0 \text{ for } x = \pm a; \ \frac{\partial u}{\partial y} = 0, \ \frac{\partial^3 u}{\partial y^3} = 0 \text{ for } y = \pm b.$$

In solving the problem it is obviously sufficient to limit ourselves to the rectangle $(-a, a; -b, b)$.

The right side — the intensity of load, $p(x, y)$ — is a function equal to p in the rectangle $(-a\alpha, a\alpha; -b\beta, b\beta)$ and zero elsewhere.

For the fundamental solutions, in view of the evenness of the sought function u, we should adopt

$$\cos\frac{m\pi x}{a}\cos\frac{n\pi y}{b} \qquad (m, n = 0, 1, 2, \ldots),$$

since the boundary conditions are obviously satisfied for these functions.

Just as in the case preceding, we find that if $a_{m,n}$ are the coefficients of $p(x, y)$ in the expansion in terms of these functions, i.e., if

$$p(x, y) = \sum_{m=0}^{\infty}\sum_{n=0}^{\infty} a_{m,n}\cos\frac{m\pi x}{a}\cos\frac{n\pi y}{b},$$

then

$$u(x, y) = \sum_{m=0}^{\infty}\sum_{n=0}^{\infty}\frac{a_{m,n}\cos\dfrac{m\pi x}{a}\cos\dfrac{n\pi y}{b}}{N\left[\pi^4\left(\dfrac{m^2}{a^2}+\dfrac{n^2}{b^2}\right)^2 + k\right]}.$$

To expand the function $p(x, y)$ in a series, it will be convenient in the given case to represent it in the form of the product of a function of x alone by a function of y. Obviously

$$p(x, y) = \varphi(x)\cdot\psi(y),$$

where $\varphi(x)$ is a function equal to $\sqrt{p}$ in $(-a\alpha, a\alpha)$ and to zero elsewhere, and $\psi(y)$ equals $\sqrt{p}$ in $(-b\beta, b\beta)$ and zero elsewhere. Expanding these functions in cosine series, we have:

$$c_m = \frac{2}{a}\int_0^{a\alpha}\sqrt{p}\cos\frac{m\pi x}{a}\,dx = \sqrt{p}\left[\frac{2}{m\pi}\sin\frac{m\pi x}{a}\right]_0^{a\alpha} = \frac{2\sqrt{p}}{m\pi}\sin m\pi\alpha$$

$$(m = 1, 2, 3, \ldots),$$

$$c_0 = \frac{2}{a}\int_0^{a\alpha}\sqrt{p}\,dx = 2\sqrt{p}\alpha,$$

and consequently

$$\varphi(x) = \sqrt{p}\alpha + 2\sqrt{p}\sum_{m=1}^{\infty}\frac{\sin m\pi\alpha}{m\pi}\cos\frac{m\pi x}{a};$$

analogously

$$\psi(y) = \sqrt{p}\beta + 2\sqrt{p}\sum_{n=1}^{\infty} \frac{\sin n\pi\beta}{n\pi} \cos \frac{n\pi y}{b}.$$

Then

$$p(x, y) = \varphi(x)\psi(y) = p\alpha\beta + 2p\beta \sum_{m=1}^{\infty} \frac{\sin m\pi\alpha}{m\pi} \cos \frac{m\pi x}{a} +$$

$$+ 2p\alpha \sum_{n=1}^{\infty} \frac{\sin n\pi\beta}{n\pi} \cos \frac{n\pi y}{b} +$$

$$+ \frac{4p}{\pi^2} \sum_{m=1}^{\infty} \sum_{n=1}^{\infty} \frac{\sin m\pi\alpha \sin n\pi\beta \cos \frac{m\pi x}{a} \cos \frac{n\pi y}{b}}{m \cdot n}.$$

Finally:

$$u(x, y) = \frac{p\alpha\beta}{kN} + \frac{2p\beta}{N} \sum_{m=1}^{\infty} \frac{\sin m\pi\alpha \cos \frac{m\pi x}{a}}{m\pi \left[\pi^4 \left(\frac{m}{a}\right)^4 + k\right]} +$$

$$+ \frac{2p\alpha}{N} \sum_{n=1}^{\infty} \frac{\sin n\pi\beta \cos \frac{n\pi y}{b}}{n\pi \left[\pi^4 \left(\frac{n}{b}\right)^4 + k\right]} +$$

$$+ \frac{4p}{\pi^2 N} \sum_{m=1}^{\infty} \sum_{n=1}^{\infty} \frac{\sin m\pi\alpha \sin n\pi\beta}{mn \left\{\left(\frac{m^2}{a^2} + \frac{n^2}{b^2}\right)^2 \pi^4 + k\right\}} \cos \frac{m\pi x}{a} \cos \frac{n\pi y}{b}.$$

We note that in general one can successfully solve boundary-value problems for the rectangle by means of double trigonometric series in the case of equations of the form:

$$A \frac{\partial^4 u}{\partial x^4} + B \frac{\partial^4 u}{\partial x^2 \partial y^2} + C \frac{\partial^4 u}{\partial y^4} + D \frac{\partial^2 u}{\partial x^2} + E \frac{\partial^2 u}{\partial y^2} + Fu = f(x, y),$$

where A, B, C, D, E, F are constants and $B^2 - 4AC < 0$ [1]).

§ 5. THE IMPROVEMENT OF THE CONVERGENCE OF SERIES

1. The general principles on which methods for improving convergence are founded. In the approximate solution of many prob-

---

[1]) For the solution of boundary-value problems, and especially in the case of infinite regions, the apparatus of Fourier integrals can be employed. A number of examples of its successful utilization in biharmonic problems can be found in a monograph of S. S. Golushkevich, [1]. An efficient method for the numerical determination of the values of oscillating integrals of the type of the Fourier integral has been proposed by M. V. Nikolaeva, [1].

lems of analysis it frequently happens that the given process — most often a series — proves to converge very slowly, which makes it almost — and sometimes completely — unsuitable for actually obtaining the solution. Such a slow convergence is usually occasioned by the presence of a singularity in the given problem. The general method for overcoming the slow convergence consists in removing the singularity.

Let us consider first its application to a simpler problem — that of computing the sum of a slowly convergent series of constant terms.

Let it be required, for example, that we determine the sum of the series

$$\sum_{n=1}^{\infty} \frac{1}{n^2+1} = \tfrac{1}{2} + \tfrac{1}{5} + \tfrac{1}{10} + \tfrac{1}{17} + \dots$$

For the direct determination of the sum of this series with an accuracy of 0.001, one must take 1000 terms of it. Let us proceed otherwise: let us extract from the general term the lower powers of $\frac{1}{n}$, which are producing the slow convergence. We have:

$$\sum_{n=1}^{\infty} \frac{1}{n^2+1} = \sum_{n=1}^{\infty} \left(\frac{1}{n^2} - \frac{1}{n^4} + \frac{1}{n^4(n^2+1)} \right) =$$

$$= \sum_{n=1}^{\infty} \frac{1}{n^2} - \sum_{n=1}^{\infty} \frac{1}{n^4} + \sum_{n=1}^{\infty} \frac{1}{n^4+n^6}.$$

The first two series, which are slowly convergent, are exactly summable, having the sums $\frac{\pi^2}{6}$ and $\frac{\pi^4}{90}$; the last series converges so rapidly that in computing its sum we limit ourselves to four terms:

$$\sum_{n=1}^{\infty} \frac{1}{n^4+n^6} = \tfrac{1}{2} + \tfrac{1}{80} + \tfrac{1}{810} + \tfrac{1}{4352} \approx 0.514.$$

Finally

$$\sum_{n=1}^{\infty} \frac{1}{n^2+1} = \frac{\pi^2}{6} - \frac{\pi^4}{90} + 0.514 = 1.076.$$

To improve the convergence in the given case and in others analogous, when the general term is a rational function, rather than the series $\sum_{n=1}^{\infty} \frac{1}{n^2}$ and $\sum_{n=1}^{\infty} \frac{1}{n^4}$ one can employ the following still more elementary series:

$$\sum_{n=1}^{\infty} \frac{1}{n(n+1)} = 1, \quad \sum_{n=1}^{\infty} \frac{1}{n(n+1)(n+2)} = \tfrac{1}{4},$$

$$\sum_{n=1}^{\infty} \frac{1}{n(n+1)(n+2)(n+3)} = \tfrac{1}{18}.$$

Thus on successively removing from the general term of the given series the terms of the series cited above with appropriately chosen coefficients, we have:

$$\frac{1}{n^2+1} = \frac{1}{n(n+1)} + \frac{n-1}{(n^2+1)(n^2+n)} = \frac{1}{n(n+1)} +$$
$$+ \frac{1}{n(n+1)(n+2)} + \frac{n-3}{(n^2+n)(n+2)(n^2+1)}.$$

The first two series give as a sum $1 + \frac{1}{4} = \frac{5}{4}$; we find the sum of the last approximately:

$$\sum_{n=1}^{\infty} \frac{n-3}{(n^2+n)(n+2)(n^2+1)} \approx -\tfrac{2}{12} - \tfrac{1}{120} + \tfrac{1}{2040} + \tfrac{2}{5460} \approx -0.174,$$

and, finally, the sum of the complete series will be equal to

$$\sum_{n=1}^{\infty} \frac{1}{n^2+1} = \tfrac{5}{4} - 0.174 = 1.076 \text{ [1]}.$$

2. The method of Acad. A. N. Krylov for improving the convergence of trigonometric series [2]. Any function satisfying the so-called Dirichlet conditions is known to have a convergent Fourier series:

$$f(x) = \frac{a_0}{2} + \sum_{n=1}^{\infty} (a_n \cos nx + b_n \sin nx),$$

$$a_n = \frac{1}{\pi} \int_{-\pi}^{\pi} f(x) \cos nx \, dx, \quad b_n = \frac{1}{\pi} \int_{-\pi}^{\pi} f(x) \sin nx \, dx.$$

Nevertheless the coefficients a_n and b_n do not, generally speaking, decrease very rapidly, the Fourier series being slowly convergent and therefore of little utility for the representation of the function. The Fourier series will be rapidly convergent if the function $f(x)$ itself and a certain number of its first derivatives are continuous and periodic. Indeed, let — for example — $f(x)$, $f'(x)$, $f''(x)$ and $f'''(x)$ be continuous and periodic, and $f^{IV}(x)$ integrable. Denote by a_n and b_n the Fourier coefficients of $f(x)$; by $a_n^{(1)}$, $b_n^{(1)}$ those of $f'(x)$; by $a_n^{(2)}$, $b_n^{(2)}$ those of $f''(x)$; by $a_n^{(3)}$, $b_n^{(3)}$ those of $f'''(x)$; and by $a_n^{(4)}$, $b_n^{(4)}$ those of $f^{IV}(x)$.

On integrating by parts the expressions for a_n and b_n and utilizing the periodicity of $f(x)$, we obtain:

$$a_n = \frac{1}{\pi} \int_{-\pi}^{\pi} f(x) \cos nx \, dx = \left[\frac{1}{\pi} f(x) \frac{\sin nx}{n}\right]_{-\pi}^{\pi} -$$

---

[1] For more detail concerning the improvement of the convergence of series of constant terms see A. A. Markov, [1], and IA. S. Bezikovich, [1].

[2] A. N. Krylov, [1], Chap. VI.

$$-\frac{1}{n\pi}\int_{-\pi}^{\pi} f'(x)\sin nx\,dx = -\frac{b_n^{(1)}}{n},$$

$$b_n = \frac{1}{\pi}\int_{-\pi}^{\pi} f(x)\sin nx\,dx = \left[-\frac{1}{\pi} f(x)\frac{\cos nx}{n}\right]_{-\pi}^{\pi} -$$

$$-\frac{1}{n\pi}\int_{-\pi}^{\pi} f'(x)\cos nx\,dx = \frac{a_n^{(1)}}{n}.$$

Now employing the same formulas successively for $f'(x)$, $f''(x)$, $f'''(x)$, we find:

$$a_n = -\frac{b_n^{(1)}}{n} = -\frac{a_n^{(2)}}{n^2} = \frac{b_n^{(3)}}{n^3} = \frac{a_n^{(4)}}{n^4},$$

$$b_n = \frac{a_n^{(1)}}{n} = -\frac{b_n^{(2)}}{n^2} = -\frac{a_n^{(3)}}{n^3} = \frac{b_n^{(4)}}{n^4}.$$

The numbers $a_n^{(4)}$ and $b_n^{(4)}$, as the Fourier coefficients of an integrable function, tend to zero as $n \to \infty$, and therefore the Fourier series for $f(x)$ will in this case converge more rapidly than the series $\Sigma \frac{1}{n^4}$ [1]).

Thus in order to have a quickly converging Fourier series for the function, its continuity and that of its first derivatives must be attained. This may frequently be accomplished by removing from the given function of elementary form a function having the same discontinuities, and such that its first derivatives have the same discontinuities as do the first derivatives of the given function.

The construction of this type of function is effected in a very elementary fashion when the discontinuities of the given function and its first derivatives are of a simple character. Let us start from a familiar and easily verifiable expansion:

$$\sigma(x) = \sum_{n=1}^{\infty} \frac{\sin nx}{n} = \begin{cases} \dfrac{-\pi - x}{2} & \text{for } -2\pi < x < 0, \\ \dfrac{\pi - x}{2} & \text{,, } \quad 0 < x < 2\pi, \\ 0 & \text{,, } \quad x = 0,\ 2\pi,\ -2\pi, \end{cases}$$

which gives a periodic function that undergoes a saltus, with a jump

[1]) If $f^{IV}(x)$ is a function with bounded variation (in particular, if it satisfies the Dirichlet condition), its Fourier coefficients are of order $\frac{1}{n}$, and the series for $f(x)$ converges as $\Sigma \frac{1}{n^5}$

of magnitude π at the point $x = 0$:

$$\sigma(+0) - \sigma(-0) = \pi,$$

and is elsewhere linear. The function $\sigma(x - x_0)$ is a function of this same form, but having a discontinuity at some $x = x_0$ $(0 < x_0 < 2\pi)$ rather than at $x = 0$. The last function is defined by the equations:

$$\sigma(x - x_0) = \begin{cases} \dfrac{-\pi + (x_0 - x)}{2} & 0 \leqslant x < x_0, \\ \dfrac{\pi + (x_0 - x)}{2} & x_0 < x \leqslant 2\pi, \\ 0 & x = x_0 \end{cases}$$

and the Fourier series for it will be:

$$\sigma(x - x_0) = \sum_{n=1}^{\infty} \frac{\sin n(x - x_0)}{n} = \sum_{n=1}^{\infty} \frac{\cos nx_0 \sin nx - \sin nx_0 \cos nx}{n}.$$

In order to obtain a function of elementary form having a discontinuity in the first derivative, we integrate term by term the series for the function $\sigma(x)$, finding

$$\int_0^x \sigma(x)dx = -\frac{(\pi - x)^2}{4} + \frac{\pi^2}{4} =$$

$$= \sum_{n=1}^{\infty} \left[\frac{-\cos nx}{n^2} \right]_0^x = \sum_{n=1}^{\infty} \frac{1}{n^2} - \sum_{n=1}^{\infty} \frac{\cos nx}{n^2} \qquad 0 \leqslant x \leqslant 2\pi.$$

Since $\sum \frac{1}{n^2}$ proves to be the free term of the Fourier series of the function standing on the left, we have

$$\sum_{n=1}^{\infty} \frac{1}{n^2} = \frac{1}{\pi} \int_0^x \left[\frac{\pi^2}{4} - \frac{(\pi - x)^2}{4} \right] dx = \frac{\pi^2}{6}.$$

Let us therefore finally adopt the periodic function

$$\sigma_1(x) = \int_0^x \sigma(x)dx - \frac{\pi^2}{6} = -\sum_{n=1}^{\infty} \frac{\cos nx}{n^2} =$$

$$= \begin{cases} \dfrac{\pi^2}{12} - \dfrac{(\pi + x)^2}{4} & \text{for } -2\pi \leqslant x \leqslant 0, \\ \dfrac{\pi^2}{12} - \dfrac{(\pi - x)^2}{4} & \text{,, } \; 0 \leqslant x \leqslant 2\pi, \end{cases}$$

for which it is obvious that

$$\sigma_1'(+0) - \sigma_1'(-0) = \pi.$$

Just as for $\sigma(x)$, a function $\sigma_1(x - x_0)$ can be introduced, the derivative of which has a jump of magnitude π at the point $x = x_0$. Integrating once again, we introduce the periodic function

$$\sigma_2(x) = -\sum_{n=1}^{\infty} \frac{\sin nx}{n^3} = \frac{3\pi x^2 - 2\pi^2 x - x^3}{12} \qquad (0 \leqslant x \leqslant 2\pi),$$

having a jump in the second derivative at the point $x = 0$. In just the same manner the subsequent functions $\sigma_s(x)$ can be constructed.

Now let $f(x)$ be a function having everywhere, with the exception of a finite number of points, bounded derivatives of the kth order, and let it and these derivatives be everywhere continuous except at a finite number of points, where they may have a discontinuity of the first kind. The derivative of the $(k + 1)$th order satisfies the Dirichlet condition. Let the points of discontinuity of the function $f(x)$ be $x_1^{(0)}, x_2^{(0)}, \ldots, x_{m_0}^{(0)}$, and let the magnitude of the jumps be

$$h_s = f(x_s^{(0)} + 0) - f(x_s^{(0)} - 0) \qquad (s = 1, 2, \ldots, m_0).$$

Let the jth derivative have discontinuities at the points $x_1^{(j)}, \ldots, x_{mj}^{(j)}$ with the magnitude of the jumps

$$h_s^{(j)} = f^{(j)}(x_s^{(j)} + 0) - f^{(j)}(x_s^{(j)} - 0) \quad \begin{pmatrix} s = 1, 2, \ldots, m_j \\ j = 1, 2, \ldots, k \end{pmatrix}.$$

We can now represent the function $f(x)$ in the form:

$$f(x) = \sum_{s=1}^{m_0} \frac{1}{\pi} h_s \sigma(x - x_s^{(0)}) + \sum_{s=1}^{m_1} \frac{1}{\pi} h_s^{(1)} \sigma_1(x - x_s^{(1)}) + \ldots + \\ + \sum_{s=1}^{m_k} \frac{1}{\pi} h_s^{(k)} \sigma_k(x - x_s^{(k)}) + \varphi(x).$$

Here the function $\varphi(x)$ will now be continuous together with its first k derivatives, and its $(k + 1)$th derivative will satisfy the Dirichlet condition. Let us verify, for instance, that $\varphi(x)$ is continuous. Only the point $x_s^{(0)}$ may occasion doubt, but at that point all the terms are continuous with the exception of $f(x)$ and $\dfrac{1}{\pi} h_s \sigma(x - x_s^{(0)})$. The difference of the two latter functions is nevertheless continuous, since on subtracting from the limit from the right the limit from the left, we have:

$$\left[f(x_s^{(0)} + 0) - \frac{1}{\pi} h_s \sigma(x_s^{(0)} + 0 - x_s^{(0)})\right] -$$

$$-\left[f(x_s^{(0)}-0)-\frac{1}{\pi}h_s\sigma(x_s^{(0)}-0-x_s^{(0)})\right]=$$

$$=f(x_s^{(0)}+0)-f(x_s^{(0)}-0)-\frac{1}{\pi}h_s[\sigma(+0)-\sigma(-0)]=0.$$

Thus the difference in question is continuous and $\varphi(x)$ is therefore continuous at the point $x_s^{(0)}$.

One can satisfy oneself analogously of the continuity of the first k derivatives of the function $\varphi(x)$, and lastly, that its $(k+1)$th derivative, as does $f^{(k+1)}(x)$, satisfies the Dirichlet condition. Therefore the Fourier series for the function $\varphi(x)$:

$$\varphi(x)=\frac{\alpha_0}{2}+\sum_{n=1}^{\infty}(\alpha_n\cos nx+\beta_n\sin nx)$$

will converge very rapidly, in fact α_n and β_n will be of the order $\frac{1}{n^{k+2}}$. The Fourier series for the function $f(x)$, if we take into account the series for $\sigma(x)$, $\sigma_1(x)$, etc., will be

$$f(x)=\sum_{s=1}^{m_0}\frac{1}{\pi}h_s\sum_{n=1}^{\infty}\frac{\sin n(x-x_s^{(0)})}{n}-\sum_{s=1}^{m_1}\frac{1}{\pi}h_s^{(1)}\sum_{n=1}^{\infty}\frac{\cos n(x-x_s^{(1)})}{n^2}-$$

$$-\sum_{s=1}^{m_2}\frac{1}{\pi}h_s^{(2)}\sum_{n=1}^{\infty}\frac{\sin n(x-x_s^{(2)})}{n^3}+\ldots+\sum_{n=1}^{\infty}(\alpha_n\cos nx+\beta_n\sin nx).$$

We have in this manner removed from the Fourier series for the function $f(x)$ the slowly convergent parts, which are summable in an elementary way, and in the remainder have obtained a rapidly converging series.

One must more often, however, solve the problem in reverse order, to wit: without knowing the function itself, we obtain by one method or another its Fourier series, usually as the result of the solution of some problem of mathematical physics. In many cases it turns out that this series converges slowly, and that it is very poorly suited to the computation of the values of the function itself, and is entirely unsuitable for the determination of its derivatives, since divergent series are obtained for them. In this case one can often succeed in carrying out an improvement of the convergence on the series itself, without reverting to the generating function. This is most conveniently done by removing the slowly convergent parts from the series and summing them on the basis of the known series for $\sigma(x-x_0)$, $\sigma_1(x-x_0)$, etc. We shall analyse two examples.

1. Let us examine the series

$$\frac{\sin x}{1}+\frac{\sin 3x}{3}+\frac{\sin 5x}{5}+\ldots.$$

It can be transformed thus:

$$\sum_{k=1}^{\infty} \frac{\sin (2k+1)x}{2k+1} = \sum_{n=1}^{\infty} \frac{1-\cos n\pi}{2} \frac{\sin nx}{n} =$$

$$= \tfrac{1}{2} \sum_{n=1}^{\infty} \frac{\sin nx}{n} - \tfrac{1}{2} \sum_{n=1}^{\infty} \frac{\sin n(x-\pi)}{n} = \tfrac{1}{2}\sigma(x) - \tfrac{1}{2}\sigma(x-\pi).$$

We see that in the case given this method has made it possible to sum the series right to the end, viz.: if the expression for $\sigma(x)$ be utilized, we find that the sum of this series is equal to

$$S(x) = \tfrac{1}{2}[\sigma(x) - \sigma(x-\pi)] =$$

$$= \begin{cases} \tfrac{1}{2}\left[\dfrac{\pi - x}{2} - \left(-\dfrac{x}{2}\right)\right] = \dfrac{\pi}{4} & \text{for } 0 < x < \pi, \\ \tfrac{1}{2}\left[\dfrac{\pi - x}{2} - \dfrac{2\pi - x}{2}\right] = -\dfrac{\pi}{4} & ,, \quad \pi < x < 2\pi, \\ 0 & ,, \quad x = 0,\ \pi,\ 2\pi. \end{cases}$$

2. Let there be given the Fourier series

$$f(x) = -\frac{2}{\pi} \sum_{n=1}^{\infty} \frac{n \cos n \dfrac{\pi}{2}}{n^2 - 1} \sin nx \qquad (0 \leqslant x \leqslant \pi)$$

Removing the lower powers of $\dfrac{1}{n}$ from the coefficient, we have

$$\frac{n}{n^2-1} = \frac{1}{n} + \frac{1}{n^3} + \frac{1}{n^5 - n^3}.$$

After the substitution of this expression, the Fourier series is separable into three parts:

$$f(x) = -\frac{2}{\pi} \sum_{n=1}^{\infty} \frac{\cos n \dfrac{\pi}{2} \sin nx}{n} - \frac{2}{\pi} \sum_{n=1}^{\infty} \frac{\cos n \dfrac{\pi}{2}}{n^3} \sin nx -$$

$$- \frac{2}{\pi} \sum_{n=1}^{\infty} \frac{\cos \dfrac{n\pi}{2}}{n^5 - n^3} \sin nx.$$

Using the series for $\sigma(x)$ and $\sigma_2(x)$ as in the preceding example, we can sum the first two series in finite form:

$$S_1(x) = -\frac{2}{\pi}\sum_{n=1}^{\infty}\frac{\cos n\frac{\pi}{2}\sin nx}{n} =$$

$$= -\frac{1}{\pi}\sum_{n=1}^{\infty}\frac{\sin n\left(x+\frac{\pi}{2}\right)+\sin n\left(x-\frac{\pi}{2}\right)}{n} =$$

$$= -\frac{1}{\pi}\left[\sigma\left(x+\frac{\pi}{2}\right)+\sigma\left(x-\frac{\pi}{2}\right)\right],$$

$$S_2(x) = -\frac{2}{\pi}\sum_{n=1}^{\infty}\frac{\cos n\frac{\pi}{2}\sin nx}{n^3} =$$

$$= -\frac{1}{\pi}\sum_{n=1}^{\infty}\frac{\sin n\left(x+\frac{\pi}{2}\right)+\sin n\left(x-\frac{\pi}{2}\right)}{n^3} =$$

$$= \frac{1}{\pi}\left[\sigma_2\left(x+\frac{\pi}{2}\right)+\sigma_2\left(x-\frac{\pi}{2}\right)\right].$$

Using the definition of the functions $\sigma(x)$ and $\sigma_2(x)$ and their periodicity, we easily find that

$$S_1(x) = \begin{cases} \dfrac{x}{\pi} & \left(0 \leqslant x < \dfrac{\pi}{2}\right), \\ \dfrac{x-\pi}{\pi} & \left(\dfrac{\pi}{2} < x \leqslant \pi\right), \\ 0 & \left(x = \dfrac{\pi}{2}\right). \end{cases}$$

$$S_2(x) = \begin{cases} -\dfrac{x^3}{6\pi} + \dfrac{\pi}{24}x & \left(0 \leqslant x \leqslant \dfrac{\pi}{2}\right), \\ -\dfrac{(x-\pi)^3}{6\pi} + \dfrac{\pi}{24}(x-\pi) & \left(\dfrac{\pi}{2} \leqslant x \leqslant \pi\right). \end{cases}$$

Thus we finally obtain for the function $f(x)$:

$$f(x) = S_1(x) + S_2(x) - \frac{2}{\pi}\sum_{n=1}^{\infty}\frac{\cos n\frac{\pi}{2}}{n^3(n^2-1)}\sin nx$$

$$(0 \leqslant x \leqslant \pi).$$

This representation of the function is of course many times more convenient than the original one, since it allows one to compute directly the values of the function $f(x)$ and its first derivatives.

We shall observe that an improvement of the convergence of Fourier series can be achieved not only with Fourier series of the form we have considered, i.e., Fourier series whose generating function, with its first derivatives, is bounded, but with the Fourier series of other functions as well. Then, however, for the summation of the slowly convergent parts the series adduced above for $\sigma(x)$, $\sigma_1(x)$, etc., may prove insufficient. Here in many cases one can employ the series

$$\cos x + \tfrac{1}{2}\cos 2x + \tfrac{1}{3}\cos 3x + \ldots = \\ = -\operatorname{Re}\,[\ln(1 - e^{ix})] = -\ln\left(2\sin\frac{x}{2}\right)$$

and series obtained analogously, for instance:

$$\frac{\cos 2x}{1\cdot 2} + \frac{\cos 3x}{2\cdot 3} + \frac{\cos 4x}{3\cdot 4} + \ldots = \\ = (1 - \cos x)\ln 2\sin\frac{x}{2} + \left(\frac{\pi}{2} - \frac{x}{2}\right)\sin x + \cos x.$$

By means of these series and the series indicated above, the convergence can be improved for any series of the form:

$$\sum_{n=1}^{\infty}\Big[\Big(A\left(\frac{1}{n}\right)\sin nx_0 + B\left(\frac{1}{n}\right)\cos nx_0\Big)\sin nx + \\ + \Big(C\left(\frac{1}{n}\right)\sin nx_0 + D\left(\frac{1}{n}\right)\cos nx_0\Big)\cos nx\Big],$$

where $A(1/n)$, $B(1/n)$, $C(1/n)$, $D(1/n)$ are analytic functions of $1/n$ for small values of the argument.

3. Fourier series with strengthened convergence (A. S. Maliev). As was indicated in the preceding No, a periodic function having $(k-1)$ continuous derivatives and the kth satisfying the Dirichlet condition, has Fourier coefficients of the order $1/n^{k+1}$. A non-periodic function, even one having derivatives of all orders within the interval $(0, 2\pi)$, but such that either the function itself or one of its derivatives has different values at $x = 0$ and $x = 2\pi$, has a slowly convergent expansion. A. S. Maliev [1]) has proposed a method of obtaining rapidly convergent trigonometric series which are powerful aids in representing functions of this form.

Let the function $f(x)$ be given in the interval $(0, \pi)$ and have there continuous derivatives up to the $(k-1)$th order, the kth derivative satisfying the Dirichlet condition. This function can be expanded in a series involving both $\sin 2nx$ and $\cos 2nx$, or it can be expanded in a series involving $\sin nx$ or $\cos nx$ only, which corresponds to an odd or an even continuation into the interval $(-\pi, 0)$. Generally speaking, however, these series will converge slowly. If we are not interested in the special character of the trigonometric series by means of which we want to represent $f(x)$, we can continue it into the interval $(-\pi, 0)$ in any manner. We shall attain an order of $1/n^{k+1}$ for its Fourier coefficients if we define it in the interval $(-\pi, \pi)$ in such a way that with a periodic continuation of it to $(-\infty, \infty)$ a function would be obtained whose derivatives up to the kth order exist and whose kth satisfies the Dirichlet condition. To realize such a continuation, it is sufficient to take $f(x)$ equal to $\varphi(x)$ in the interval $(-\pi, 0)$, $\varphi(x)$ being a polynomial of the $(2k-1)$th degree which at the points 0 and $-\pi$ has, together with its derivatives of up to the $(k-1)$th order, values equal to the corresponding values

[1]) A. S. Maliev, [1], [2].

of $f(x)$ and its derivatives at $x = 0$ and $x = \pi$, namely:

$$\begin{aligned} &\varphi(0) = f(0) && \varphi(-\pi) = f(\pi) \\ &\varphi'(0) = f'(0) && \varphi'(-\pi) = f'(\pi) \\ &\dots\dots && \dots\dots \\ &\varphi^{(k-1)}(0) = f^{(k-1)}(0) && \varphi^{(k-1)}(-\pi) = f^{(k-1)}(\pi). \end{aligned}$$

The polynomial $\varphi(x)$ required here is most conveniently sought in the form:

$$\varphi(x) = (x+\pi)^k\left(A_0 + A_1x + \dots + \frac{A_{k-1}}{(k-1)!}x^{k-1}\right) + \\ + x^k\left[B_0 + B_1(x+\pi) + \dots + \frac{B_{k-1}}{(k-1)!}(x+\pi)^{k-1}\right],$$

the coefficients $A_0, A_1, \dots, A_{k-1}$; $B_0, B_1, \dots, B_{k-1}$ then being determined, obviously, from the following two systems of equations:

$$\begin{aligned} \varphi(0) &= \pi^k A_0 = f(0), \\ \varphi'(0) &= k\pi^{k-1}A_0 + \pi^k A_1 = f'(0), \\ \varphi''(0) &= C_2^0 k(k-1)\pi^{k-2}A_0 + C_2^1 k\pi^{k-1}A_1 + C_2^2\pi^k A_2 = f''(0), \\ &\dots\dots\dots\dots\dots\dots\dots\dots \\ \varphi^{(k-1)}(0) &= C_{k-1}^0 k(k-1)\dots 2\pi A_0 + C_{k-1}^1 k(k-1)\dots 3\pi^2 A_1 + \dots + \\ &\quad + C_{k-1}^{k-2}\, k\pi^{k-1}A_{k-2} + C_{k-1}^{k-1}\,\pi^k A_{k-1} = f^{(k-1)}(0), \end{aligned}$$

$$\begin{aligned} \varphi(-\pi) &= (-\pi)^k B_0 = f(\pi), \\ \varphi'(-\pi) &= k(-\pi)^{k-1}B_0 + (-\pi)^k B_1 = f'(\pi), \\ &\dots\dots\dots\dots\dots\dots\dots\dots \\ \varphi^{(k-1)}(-\pi) &= C_{k-1}^0 k(k-1)\dots 2(-\pi)B_0 + \\ &\quad + C_{k-1}^1 k(k-1)\dots 3(-\pi)^2 B_1 + \dots + C_{k-1}^{k-1}(-\pi)^k B_{k-1} = f^{(k-1)}(\pi). \end{aligned}$$

If the polynomial $\varphi(x)$ be defined in the manner indicated, the function

$$\psi(x) = \begin{cases} f(x) & (0 \leqslant x \leqslant \pi) \\ \varphi(x) & (-\pi \leqslant x \leqslant 0), \end{cases}$$

which is continued elsewhere periodically, has $(k-1)$ continuous derivatives and the kth satisfying the Dirichlet condition. Its Fourier series converges as $1/n^{k+1}$, and since in the interval $(0, \pi)$ it gives the function $f(x)$, we shall indeed have obtained for the latter the requisite representation.

Example. Let us construct a rapidly converging Fourier series for the function

$$f(x) = x - \frac{\pi}{2} \quad (0 \leqslant x \leqslant \pi).$$

Desiring to obtain convergence of the order $1/n^4$, we must adopt $k = 3$.

For the given case the system of equations takes the form

$$\begin{aligned} &\pi^3 A_0 = -\frac{\pi}{2}, && -\pi^3 B_0 = \frac{\pi}{2}, \\ &3\pi^2 A_0 + \pi^3 A_1 = 1, && 3\pi^2 B_0 - \pi^3 B_1 = 1, \\ &6\pi A_0 + 6\pi^2 A_1 + \pi^3 A_2 = 0, && -6\pi B_0 + 6\pi^2 B_1 - \pi^3 B_2 = 0. \end{aligned}$$

Hence we find

$$A_0 = -\frac{1}{2\pi^2}, \quad A_1 = \frac{5}{2\pi^3}, \quad A_2 = -\frac{12}{\pi^4},$$

$$B_0 = -\frac{1}{2\pi^2}, \quad B_1 = -\frac{5}{2\pi^3}, \quad B_2 = -\frac{12}{\pi^4},$$

and thus

$$\varphi(x) = (x+\pi)^3\left(-\frac{1}{2\pi^2} + \frac{5}{2\pi^3}x - \frac{6}{\pi^4}x^2\right) + \\ + x^3\left[-\frac{1}{2\pi^2} - \frac{5}{2\pi^3}(x+\pi) - \frac{6}{\pi^4}(x+\pi)^2\right].$$

Now expanding in a Fourier series the function

$$\psi(x) = \begin{cases} x - \dfrac{\pi}{2} & (0 \leqslant x \leqslant \pi), \\ \varphi(x) & (-\pi \leqslant x \leqslant 0), \end{cases}$$

we shall obtain for $f(x)$ in the interval $(0, \pi)$ the required expansion:

$$f(x) = \frac{240}{\pi^3} \sum_{n=1,3,5\,\ldots} \left(\frac{1}{n^4} - \frac{12}{\pi^2 n^6}\right)\cos nx + \frac{1440}{\pi^4} \sum_{n=2,4,6\,\ldots} \frac{1}{n^5}\sin nx.$$

This method of improving the convergence can find application to the solution of the problem of the expansion of a function in a series involving given functions that do not form an orthogonal system — an important problem, as we have seen in § 3, for some methods of solution of boundary problems. Let it be required to expand a function $g(x)$ in the interval $(0, \pi)$ in terms of assigned functions $f_n(x)$. In § 3 the method of infinite systems of equations was considered for finding the coefficients of such an expansion. If an improvement of convergence is sought, the system can be formed thus: Continue the functions $g(x)$ and $f_n(x)$ to $(-\pi, \pi)$ and form for them the rapidly convergent series:

$$g(x) = \sum_{n=0}^{\infty} (a_n \cos nx + b_n \sin nx),$$

$$f_m(x) = \sum_{n=0}^{\infty} (\alpha_n^{(m)} \cos nx + \beta_n^{(m)} \sin nx).$$

Then seeking the expansion of $g(x)$ in the form

$$g(x) = A_1 f_1(x) + A_2 f_2(x) + \ldots,$$

we obtain, after substituting the series and equating the coefficients, the system of equations

$$\sum_{m=1}^{\infty} \alpha_n^{(m)} A_m = a_n, \qquad \sum_{m=1}^{\infty} \beta_n^{(m)} A_m = b_n.$$

This system of equations may prove to be more convenient for solution than that formed in the ordinary way, thanks to the fact that its coefficients diminish very much more rapidly.

4. General methods for improving the convergence of approximate solutions of boundary-value problems [1]). The infinite series that we found in solving various boundary-value problems were obtained as the result of a certain transformation of the Fourier series of the boundary function: the coefficients of the series giving the solution were determined in terms of the Fourier coefficients of the boundary function. Thus for rapid convergence of the series giving the solution, the order of the Fourier coefficients of the boundary function — i.e., in the final analysis (as we saw in No. 2), the regularity of the boundary function, or more exactly speaking, its continuity and that of its first derivatives — is of essential significance. In the same way, in solving by means of double series non-homogeneous equations — for example $\Delta u = f(x, y)$, $\Delta\Delta u = f(x, y)$ — the coefficients of the double Fourier series of the function $f(x, y)$ figure in the series giving the solution, and therefore in order to secure rapid convergence the regularity of the function $f(x, y)$ is necessary.

Thus for rapid convergence of the series that give the solution, the existence and continuity of the first derivatives: 1) of the boundary function; 2) of the function on the right side of the equation — i.e., the regularity of the problem itself, has an essential significance. As we shall see below, the satisfaction of these two conditions is necessary for rapid convergence even with other methods for the approximate solution of boundary-value problems, and not only in using the method of infinite series.

The considerations expounded above lead to this thought: rather than improving the convergence of the series obtained as the result of the solution, let us proceed otherwise — by conducting a preliminary, so to speak, "prophylactic" improvement of the original boundary-value problem, ensuring the satisfaction of the two conditions indicated above, whereupon the series obtained as the result of the solution will converge rapidly of themselves.

A general device for bringing the problem to regularity consists in the following. Let us discuss finding the solution of an equation of the second order of elliptic type:

$$M[u(x, y)] = f(x, y),$$

which is to become a given function $\varphi(s)$ on the contour L that bounds the region D.

We shall seek a function $u_0(x, y)$ such that

$$M[u_0(x, y)] = f_0(x, y), \qquad u_0(x, y) = \varphi_0(s) \text{ on } L,$$

where $f_0(x, y)$ and $\varphi_0(s)$ are any functions having only those singularities that $f(x, y)$ and $\varphi(s)$ have, i.e., such that the differences:

$$f(x, y) - f_0(x, y) = f_1(x, y), \qquad \varphi(s) - \varphi_0(s) = \varphi_1(s)$$

---

[1]) See L. V. Kantorovich, [4], [7].

are continuous together with a certain number of their first derivatives. If such a function $u_0(x, y)$ be constructed, then on introducing a new unknown function

$$u_1(x, y) = u(x, y) - u_0(x, y),$$

we have for its determination the problem, now regular:

$$M[u_1(x, y)] = f_1(x, y), \qquad u_1(x, y) = \varphi_1(s) \text{ on the contour } L.$$

After having found $u_1(x, y)$, we shall obtain the sought u by adding the known function $u_0(x, y)$. We remark further that the elimination of the singularities in $f(x, y)$ and in $\varphi(s)$ can be done successively, viz.: one can first eliminate the singularities in $f(x, y)$; the boundary data may then be altered, after which the singularities in the boundary function are eliminated; but then, to leave the right side unchanged, $u_0(x, y)$, which is the solution of the homogeneous equation $M(u) = 0$, must be sought. We shall now give devices for constructing the required function $u_0(x, y)$ in order to eliminate singularities of one sort or another.

I. *The removal of singularities in a function given on the contour.* Let the function $\varphi(s)$, given on the contour, have singular points. Let us assume it to be continuous, together with several of its first derivatives, everywhere except at isolated points, namely points of a discontinuity of the first kind of the function $\varphi(s)$ itself, and angular points of the curve $\sigma = \varphi(s)$, i.e., points of a discontinuity of the first kind of the derivative of $\varphi(s)$, and so on.

Let it be necessary for us to remove some singularity of the function $\varphi(s)$. Let us assume for definiteness that at the point M_1 corresponding to $s = s_1$, the function has a discontinuity of the first kind, with saltus

$$\varphi(s_1 + 0) - \varphi(s_1 - 0) = \sigma.$$

Let us take a region D_0 of simple form, wholly containing the given region D and bounded by some contour L_0 having near the point M_1 a part in common with the contour L. Having in the simplest way assigned on the contour L_0 the boundary values forming a saltus of magnitude σ at the point M_1, let us find in the region D_0 a solution of the equation $M(u) = 0$ satisfying these conditions. Then taking as the contour values of u_1 the difference between $\varphi(s)$ and the values of $u_0(x, y)$ on L, we shall have eliminated the discontinuity at the point M. We note that the requirement that a common part of the contours L and L_0 be present close to M_1 may, with certain provisos, be replaced by the weaker requirement that at the point M_1 the contour L_0 be tangent to L.

As the region D_0 it is most convenient to use: a circle, the exterior of a circle, or a half-plane, placed appropriately with respect to the given contour.

This device is most conveniently used in the case when the operator $M(u)$ is the Laplace operator Δu. In this case one can construct solutions regular at all points except one, where it has a given singularity.

Let the point M_1 on the contour L be a point of discontinuity of $\varphi(s)$ of the first kind. Let us assume that at the point M_1 there is a tangent, and that near M_1 the contour L lies along one side of the tangent. Without limiting the generality, we can consider the point M_1 to be (0, 0), that the tangent is the axis of ordinates, and that the contour L (near M_1) lies to the right of it. Then the function

$$U_0(x, y) = \frac{\sigma}{\pi} \mathrm{I}(\ln z) = \frac{\sigma}{\pi} \mathrm{I}[\ln(x + iy)] = \frac{\sigma}{\pi} \operatorname{arctg} \frac{y}{x}\ ^{1)}$$

is a regular harmonic function everywhere except at the origin of coordinates; at the origin of coordinates in moving along the contour L it undergoes a discontinuity of the first kind with saltus σ. Indeed:

$$\text{for } y < 0 \quad \lim_{\substack{y\to 0\\ x\to 0}} \frac{\sigma}{\pi} \operatorname{arctg} \frac{y}{x} = \frac{\sigma}{\pi} \operatorname{arctg} (-\infty) = -\frac{\sigma}{2},$$

$$\text{for } y > 0 \quad \lim_{\substack{y\to 0\\ x\to 0}} \frac{\sigma}{\pi} \operatorname{arctg} \frac{y}{x} = \frac{\sigma}{\pi} \operatorname{arctg} (+\infty) = \frac{\sigma}{2},$$

since $x > 0$ and x is an infinitesimal of a higher order than y in an approach along the curve L. Thus $U_0(x, y)$ has a saltus of magnitude σ on the curve L and, subtracting this function, we eliminate the corresponding discontinuity of $\varphi(s)$. In just the same way the discontinuity in the derivative can be removed, but there one must utilize the function:

$$U_0^*(x,y) = -\frac{\sigma^*}{\pi} \mathrm{Re}\{(x + iy)\,[\ln (x + iy) - 1]\} =$$

$$= -\frac{\sigma^*}{\pi} \left[x(\ln \sqrt{x^2 + y^2} - 1 - y \operatorname{arctg} \frac{y}{x}\right].$$

This function $U_0^*(x, y)$ has at the origin of coordinates a discontinuity in the derivative along the arc of the contour L, with a saltus of magnitude σ^*. True, to establish this it is necessary to assume that the tangent to the contour L changes continuously near M_1 and that $\lim xy^{-1} \ln y = 0$. The second of these conditions is certainly satisfied if at the point M_1 the curvature of the curve L is finite. Subtracting the function $U_0^*(x, y)$, we can remove the saltus of magnitude σ^* in the derivative. Thus, discontinuities of the first kind in the contour function itself and in its derivative along the arc can always be removed (on the assumptions made above) in the case of a boundary problem for the Poisson equation $\Delta u = f(x, y)$ by means of the functions $U_0(x, y)$ and $U_0^*(x, y)$.

We note that on the axis of ordinates itself, the contour values of the functions $U_0(x, y)$ and $U_0^*(x, y)$ in an approach from the side of

---

[1]) I(ln z) denotes the imaginary part of ln z.

the right half-plane will be respectively

$$U_0(0, y) = \begin{cases} \frac{\sigma}{2} & y > 0 \\ 0 & y = 0 \\ -\frac{\sigma}{2} & y < 0 \end{cases} \qquad U_0^*(0, y) = \begin{cases} \frac{\sigma^*}{2} y & y \geqslant 0 \\ \\ -\frac{\sigma^*}{2} y & y \leqslant 0. \end{cases}$$

This shows that even in the case when the contour has a straight part near the point M_1, one can use the functions $U_0(x, y)$ and $U_0^*(x, y)$ for eliminating a singularity.

It is not hard to construct functions that make possible the removal of discontinuities in the higher-order derivatives as well.

Let us now consider the application of the method we have expounded to two particular cases, namely to the circle and the annulus.

1. *The Dirichlet problem for the circle.* It is required to find a function harmonic in the circle $x^2 + y^2 \leqslant R^2$ and reducing to a given function $\varphi(\theta)$ on the contour. The solution of the problem is given, as we know, in the form of the Poisson integral or the series

$$u(r, \theta) = \frac{1}{2\pi} \int_0^{2\pi} \frac{R^2 - r^2}{R^2 - 2rR \cos(\theta - \lambda) + r^2} \varphi(\lambda) d\lambda =$$

$$= \frac{a_0}{2} + \sum_{n=1}^{\infty} (a_n \cos n\theta + b_n \sin n\theta) \left(\frac{r}{R}\right)^n, \qquad (*)$$

where a_n and b_n are the Fourier coefficients of the function $\varphi(\theta)$. Let us assume that the function $\varphi(\theta)$ has a singular point; for definiteness let this be a discontinuity with saltus σ for $\theta = 0$. The Fourier coefficients of the function $\varphi(\theta)$, a_n and b_n, will then be of order $\frac{1}{n}$, and the series which gives the solution, cited above, is not suitable for computations; the computation of the integral is also inconvenient here. Following the method indicated above, we must subtract from $u(x, y)$ the function $U_0(x, y)$, which must be adjusted to the point $(R, 0)$. Thus x must be replaced by $(R - x)$ in it and we must use the function

$$\frac{\sigma}{\pi} \operatorname{arctg} \frac{y}{R - x}.$$

Let us denote by $\varphi_0(\theta)$ its contour values; then, introducing $x = R \cos \theta$, $y = R \sin \theta$, we find:

$$\varphi_0(\theta) = \frac{\sigma}{\pi} \operatorname{arctg} \frac{R \sin \theta}{R - R \cos \theta} = \frac{\sigma}{\pi} \operatorname{arctg} \left(\operatorname{ctg} \frac{\theta}{2}\right) = \frac{\sigma}{\pi} \frac{\pi - \theta}{2}$$

$$\text{for } 0 < \theta < 2\pi.$$

The Fourier series for the function $\varphi_0(\theta)$ will be:

$$\varphi_0(\theta) = \frac{\sigma}{\pi}\sum_{n=1}^{\infty}\frac{\sin n\theta}{n}.$$

Now it is necessary to determine the harmonic function $u(r, \theta)$ from the contour function,

$$\varphi_1(\theta) = \varphi(\theta) - \varphi_0(\theta) = \frac{a_0}{2} + \\ + \sum_{n=1}^{\infty}(a_n \cos n\theta + b_n \sin n\theta) - \frac{\sigma}{\pi}\sum_{n=1}^{\infty}\frac{\sin n\theta}{n}$$

and to add to the function $u_1(r, \theta)$ obtained, the function

$$U_0(x, y) = \frac{\sigma}{\pi}\operatorname{arctg}\frac{y}{R - x} = \frac{\sigma}{\pi}\operatorname{arctg}\frac{r\sin\theta}{R - r\cos\theta}.$$

We see that in the case in hand the employment of the device for improving the boundary conditions has essentially reduced to this, that we have withdrawn from the series (*) the slowly convergent part

$$\frac{\sigma}{\pi}\sum_{n=1}^{\infty}\frac{\sin n\theta}{n}\left(\frac{r}{R}\right)^n$$

and have summed it in finite form, i.e., have carried out an improvement of the convergence of the trigonometric series.

Thus in this simplest case the method of improving the boundary conditions is equivalent to improving the convergence of the trigonometric series.

Analogously if for $\theta = 0$ the function $\varphi(\theta)$ has a saltus in the first derivative of magnitude σ^*, one must subtract from the function $u(x, y)$ the function $U_0^*(x, y)$, in which x is replaced by $(R - x)$, i.e.,

$$-\frac{\sigma^*}{\pi}\left[(R - x)(\ln\sqrt{(R - x)^2 + y^2} - 1 - y\operatorname{arctg}\frac{y}{R - x}\right] = \\ = -\frac{\sigma^*}{\pi}\operatorname{Re}[(R - x + iy)(\ln(R - x + iy) - 1)].$$

In such a case it will be necessary to subtract from the contour function the function $\varphi_0^*(\theta)$, obtained from that cited above by the substitution $x = R\cos\theta$, $y = R\sin\theta$:

$$\varphi_0^*(\theta) = -\frac{\sigma^*}{\pi}\left[R(1 - \cos\theta)(\ln R\sqrt{2 - 2\cos\theta} - 1) - \right. \\ \left. - \frac{\pi - \theta}{2}R\sin\theta\right] = -\frac{\sigma^*}{\pi}\operatorname{Re}[(R - Re^{-i\theta})(\ln(R - Re^{-i\theta}) - 1)].$$

2. *The Dirichlet problem for the annulus.* A function harmonic in an annulus $R_1 < r < R_2$ and reducing on the circumferences $r = R_1$ and $r = R_2$ to given functions $\varphi_1(\theta)$ and $\varphi_2(\theta)$ may be given, as we know from § 1, No. 2, in the form of the series:

$$u = \frac{\alpha_0^{(1)} - \alpha_0^{(2)}}{2(\ln R_1 - \ln R_2)} \ln r + \frac{\alpha_0^{(1)} \ln R_2 - \alpha_0^{(2)} \ln R_1}{2(\ln R_2 - \ln R_1)} +$$

$$+ \sum_{k=1}^{\infty} \Big[\frac{(\alpha_k^{(1)} R_2^{-k} - \alpha_k^{(2)} R_1^{-k}) r^k + (\alpha_k^{(1)} R_2^k - \alpha_k^{(2)} R_1^k) r^{-k}}{R_1^k R_2^{-k} - R_1^{-k} R_2^k} \cos k\theta +$$

$$+ \frac{(\beta_k^{(1)} R_2^{-k} - \beta_k^{(2)} R_1^{-k}) r^k + (\beta_k^{(1)} R_2^k - \beta_k^{(2)} R_1^k) r^{-k}}{R_1^k R_2^{-k} - R_1^{-k} R_2^k} \sin k\theta \Big]. \quad (**)$$

Here $\alpha_k^{(1)}$ and $\beta_k^{(1)}$ are the Fourier coefficients of $\varphi_1(\theta)$ and $\alpha_k^{(2)}$ and $\beta_k^{(2)}$ are those of $\varphi_2(\theta)$. The convergence near the circumferences $r = R_1$ and $r = R_2$ will be the same as for the trigonometric series of the functions $\varphi_1(\theta)$ and $\varphi_2(\theta)$, i.e., very slow, if the functions φ_1 and φ_2 have singularities.

In the given case, we shall not conduct an improvement of the order of convergence by eliminating separate singular points, but shall eliminate all the singularities with a single device in the following manner. Let us form two functions: $u_1^*(r, \theta)$ — the solution of the Dirichlet problem for the exterior of a circle of radius $r = R_1$ with contour values $\varphi_1(\theta)$, and $u_2^*(r, \theta)$ — the solution of the Dirichlet problem for the interior of a circle of radius $r = R_2$ for the boundary function $\varphi_2(\theta)$:

$$u_1^*(r, \theta) = \frac{1}{2\pi} \int_0^{2\pi} \frac{r^2 - R_1^2}{R_1^2 - 2rR_1 \cos(\theta - \lambda) + r^2} \varphi_1(\lambda) d\lambda =$$

$$= \frac{\alpha_0^{(1)}}{2} + \sum_{k=1}^{\infty} (\alpha_k^{(1)} \cos k\theta + \beta_k^{(1)} \sin k\theta) \left(\frac{R_1}{r}\right)^k,$$

$$u_2^*(r, \theta) = \frac{1}{2\pi} \int_0^{2\pi} \frac{R_2^2 - r^2}{R_2^2 - 2rR_2 \cos(\theta - \lambda) + r^2} \varphi_2(\lambda) d\lambda =$$

$$= \frac{\alpha_0^{(2)}}{2} + \sum_{k=1}^{\infty} (\alpha_k^{(2)} \cos k\theta + \beta_k^{(2)} \sin k\theta) \left(\frac{r}{R_2}\right)^k.$$

Let us subtract them from the function $u(r, \theta)$; then for the difference $v(r, \theta) = u(r, \theta) - u_1^*(r, \theta) - u_2^*(r, \theta)$ the contour conditions will be:

$$v(R_1, \theta) = -\frac{\alpha_0^{(2)}}{2} - \sum_{k=1}^{\infty} (\alpha_k^{(2)} \cos k\theta + \beta_k^{(2)} \sin k\theta) \left(\frac{R_1}{R_2}\right)^k,$$

$$v(R_2, \theta) = -\frac{\alpha_0^{(1)}}{2} - \sum_{k=1}^{\infty} (\alpha_k^{(1)} \cos k\theta + \beta_k^{(1)} \sin k\theta) \left(\frac{R_1}{R_2}\right)^k.$$

For the determination of the function $v(r, \theta)$ we have obtained a problem of the same form as previously for $u(r, \theta)$, but here the Fourier coefficients of the contour functions have at least the order of the progression $\left(\frac{R_1}{R_2}\right)^k$, and therefore the series (**) for the function $v(r, \theta)$ will certainly converge as a progression with the ratio $\frac{R_1}{R_2}$. Explicitly, it will be the series:

$$v(r, \theta) = -\frac{\alpha_0^{(2)} - \alpha_0^{(1)}}{2(\ln R_1 - \ln R_2)} \ln r - \frac{\alpha_0^{(2)} \ln R_2 - \alpha_0^{(1)} \ln R_1}{2(\ln R_2 - \ln R_1)} -$$

$$- \sum_{k=1}^{\infty} \frac{(\alpha_k^{(2)} R_2^{-k} - \alpha_k^{(1)} R_1^{-k}) r^k - (\alpha_k^{(2)} R_2^k - \alpha_k^{(1)} R_1^k) r^{-k}}{R_1^k R_2^{-k} - R_1^{-k} R_2^k} \left(\frac{R_1}{R_2}\right)^k \cos k\theta -$$

$$- \sum_{k=1}^{\infty} \frac{(\beta_k^{(2)} R_2^{-k} - \beta_k^{(1)} R_1^{-k}) r^k - (\beta_k^{(2)} R_2^k - \beta_k^{(1)} R_1^k) r^{-k}}{R_1^k R_2^{-k} - R_1^{-k} R_2^k} \left(\frac{R_1}{R_2}\right)^k \sin k\theta.$$

Of the fact that this series converges as a geometrical progression with ratio $\frac{R_1}{R_2}$, one can convince oneself directly.

After the function $v(r, \theta)$ is found, we obtain $u(r, \theta)$ by adding $u_1^*(r, \theta)$ and $u_2^*(r, \theta)$ to it; as regards the functions $u_1^*(r, \theta)$ and $u_2^*(r, \theta)$, in determining them by means of the series or the Poisson integral, one can employ the devices for improving the convergence indicated above for the circle — in the simplest cases they will be found in finite form.

The device here expounded is applicable not only for an annulus but also for other multiply connected regions, and permits the transition from non-analytic to analytic boundary conditions [1]).

II. *The removal of the singularities in the right member*. The elimination of the singularities of the function $f(x, y)$ can be accomplished by means of devices constructed on the same principles as those above. Without going into detail, we shall indicate two such devices.

1. Let the function $f(x, y)$ or its partial derivatives have discontinuities. Let us remove a function $f_0(x, y)$ of simple form, which is such that the difference $f(x, y) - f_0(x, y)$ is continuous together with its first partial derivatives. Let us continue the function $f_0(x, y)$ into some region D_0, of simple form and containing D; let the function obtained be $f_0^*(x, y)$. For this function let us solve the equation $M(u) = f_0^*$ in the region D_0 under any boundary conditions; this solution will give the required function $u_0(x, y)$.

It is desirable that the region D_0 be so chosen that we can solve the equation $M(u) = f$ for it with any free term. For example, for the Poisson equation one can take a rectangle, upon which one obtains a problem with a regular right side for finding the difference $u_1 = u - u_0$. As the function $f_0(x, y)$ one can take, in particular, the function $f(x, y)$ itself, thereby obtaining a homogeneous equation for u_1.

---

[1]) Other devices for improving the convergence of series encountered in the solution of the problems of mathematical physics, with examples, can be found in Chapter XII of the monograph of G. A. Grinberg, [1].

2. One can also proceed differently. To eliminate the given singularity of f it is sufficient to construct any function $u_0(x, y)$ for which the result of the operation $M(u_0)$ has the same singularity as does f. On subtracting u_0 from u, we shall have eliminated this singularity.

For example, if $M(u) = \Delta u$, and the function $f(x, y)$ undergoes along the line $x = x_1$ a saltus of magnitude $\sigma(y)$, one can adopt as such a function:

$$u_0(x, y) = \begin{cases} \frac{1}{2}(x - x_1)^2\sigma(y) & \text{for } x \geqslant x_1, \\ 0 & \text{,, } x \leqslant x_1. \end{cases}$$

Both devices alter the boundary values, and therefore after employing them there may appear, in the boundary conditions, singularities of the first kind. These singularities can be eliminated by the methods considered above (p. 92). We note that the free term $f(x, y)$ will not thereby be altered.

CHAPTER II

THE APPROXIMATE SOLUTION OF THE INTEGRAL EQUATIONS OF FREDHOLM

§ 1. THE REPLACEMENT OF AN INTEGRAL EQUATION BY A SYSTEM OF LINEAR EQUATIONS

1. Fundamental definitions. The equation:

$$\varphi(x) - \lambda \int_a^b K(x, y)\varphi(y)dy = f(x),$$

where $\varphi(x)$ is an unknown function, $K(x, y)$ is the kernel of the equation, $f(x)$ the free term and λ a numerical parameter, is called a Fredholm integral equation of the second kind. If the free term is absent, we obtain the homogeneous equation

$$\varphi(x) = \lambda \int_a^b K(x, y)\varphi(y)dy.$$

An equation of the form

$$\int_a^b K(x, y)\varphi(y)dy = f(x).$$

is called a Fredholm equation of the first kind. A majority of the boundary-value problems of mathematical physics [1]) can be brought into the form of integral equations of the second kind.

Two basic problems arise in the solution of an equation of the second kind.

1. The direct determination of the solution of a non-homogeneous equation for a definite value of the parameter λ and for a definite or arbitrary right side.

2. The problem of the solution of the homogeneous equation. The homogeneous equation has, as a rule, only the trivial null solution $\varphi(x) = 0$, having a non-trivial solution only for exceptional values of the parameter λ. These exceptional values of λ are called the *characteristic* (or *proper*) *values* of the given integral equation, and the corresponding

---

[1]) On the theory of integral equations, see I. I. Privalov, [2]; E. Goursat, [1], vol. III, part 2; V. I. Smirnov, [1], vol. IV, 1941; I. G. Petrovskiĭ, [1]; S. G. Mikhlin, [1].

solutions of the homogeneous equation are called the *characteristic* (or *proper*) *functions*. The second problem then is that of the determination of the proper values and proper functions of the integral equation.

In practice we encounter the first problem on applying integral equations to equations of elliptic type, and the second problem chiefly on applying them to equations of hyperbolic or parabolic type. It should be remarked that the second problem is the more difficult and general one — for if all proper numbers and the corresponding proper functions are known, the solution of the first problem may be easily obtained, as follows from the general theory of integral equations.

2. The replacement of an integral equation by a finite system of linear algebraic equations. Let there be given an integral equation of the second kind:

$$\varphi(x) - \lambda \int_a^b K(x, y)\varphi(y)dy = f(x). \tag{1}$$

Using any formula for approximate integration, we can approximately replace the integral figuring in this equation by some simple form of expression not involving the integral sign. Indeed, any linear formula for approximate integration has the form:

$$\int_a^b \psi(x)dx = \sum_{k=1}^n A_k\psi(x_k) + \varrho, \tag{2}$$

where A_k and x_k are numbers which are constant for the given interval and for a given formula, and ϱ is the error. Moreover, ordinarily [1]) $A_k \geqslant 0$ and $\sum_{k=1}^n A_k = b - a$.

For example, for the *rectangular formula*:

$$x_1 = a,\ x_2 = a + \frac{b-a}{n},\ \ldots,\ x_n = a - (n-1)\frac{b-a}{n},\quad A_k = \frac{b-a}{n};$$

for the *trapezoidal formula*:

$$x_1 = a,\ x_2 = a + \frac{b-a}{n-1},\ \ldots,\ x_n = b,$$

$$A_1 = A_n = \frac{b-a}{2n-2},\ A_2 = A_3 = \ldots = A_{n-1} = \frac{b-a}{n-1};$$

[1]) For the formulas which we are going to use henceforth, we shall assume these conditions to have been met.

On the subject of formulas for approximate integration, see A. N. Krylov, [2], IA. S. Bezikovich, [2] and also [1].

for the *tangential formula*:

$$x_1 = a + \frac{b-a}{2n},\quad x_2 = a + 3\frac{b-a}{2n}, \ldots,$$

$$x_n = a + (2n-1)\frac{b-a}{2n},$$

$$A_k = \frac{b-a}{n};$$

for *Chebyshev's formula*:

$$x_k = \frac{b-a}{2} + \frac{b-a}{2} x_k^{(n)},$$

$$A_k = \frac{b-a}{n},$$

where $x_k^{(n)}$ are the Chebyshev points, the roots of the Chebyshev polynomials.
Lastly, for *Gauss's formula:*

$$x_k = a + (b-a)x_k^{(n)},$$

$$A_k = (b-a)A_k^{(n)},$$

where $x_k^{(n)}$ are the Gaussian points (the roots of the Legendre polynomial), and $A_k^{(n)}$ are the Gaussian coefficients for the interval (0, 1).

After applying formula (2) to the integral on the left side of equation (1) we arrive at the equation:

$$\varphi(x) - \lambda \sum_{k=1}^{n} A_k K(x, x_k)\varphi(x_k) = f(x) + \lambda\varrho(x). \tag{3}$$

In particular, successively putting

$$x = x_1, x_2, \ldots, x_n,$$

in equation (3), we arrive at the following system of equations, which the numbers $\varphi(x_i)$ — the values of the sought function at the points x_i — satisfy:

$$\varphi(x_i) - \lambda \sum_{k=1}^{n} A_k K(x_i, x_k)\varphi(x_k) = f(x_i) + \lambda\varrho_i, \qquad (i = 1, 2, \ldots, n), \tag{4}$$

where $\varrho_i = \varrho(x_i)$. Neglecting the small quantity ϱ_i in the right side of equations (4), the exact value of which is unknown to us, we obtain the following system of n equations in the n unknowns $\tilde{\varphi}(x_1), \tilde{\varphi}(x_2), \ldots, \tilde{\varphi}(x_n)$:

$$\tilde{\varphi}(x_i) - \lambda \sum_{k=1}^{n} A_k K(x_i, x_k)\tilde{\varphi}(x_k) = f(x_i) \qquad (i = 1, 2, \ldots, n) \tag{5}$$

or *in extenso*,

$$\left.\begin{aligned}
&\tilde{\varphi}(x_1)[1 - \lambda A_1 K(x_1, x_1)] - \lambda\tilde{\varphi}(x_2)A_2K(x_1, x_2) - \ldots - \\
&\qquad\qquad - \lambda\tilde{\varphi}(x_n)A_nK(x_1, x_n) = f(x_1) \\
&- \lambda\tilde{\varphi}(x_1)A_1K(x_2, x_1) + \tilde{\varphi}(x_2)[1 - \lambda A_2K(x_2, x_2)] - \ldots - \\
&\qquad\qquad - \lambda\tilde{\varphi}(x_n)A_nK(x_2, x_n) = f(x_2) \\
&\ldots\ldots\ldots\ldots\ldots\ldots\ldots\ldots\ldots \\
&- \lambda\tilde{\varphi}(x_1)A_1K(x_n, x_1) - \lambda\tilde{\varphi}(x_2)A_2K(x_n, x_2) - \ldots + \\
&\qquad\qquad + \tilde{\varphi}(x_n)[1 - A_n\lambda K(x_n, x_n)] = f(x_n).
\end{aligned}\right\} \tag{5'}$$

On solving this system of equations, we find approximations $\tilde{\varphi}(x_1)$, $\tilde{\varphi}(x_2), \ldots, \tilde{\varphi}(x_n)$ to the values of the sought function [1] $\varphi(x_1), \varphi(x_2), \ldots, \varphi(x_n)$. In terms of these values the approximate value of the function itself can be found by means of one device or another for interpolation. In the special case above, it is most convenient to obtain this value, $\tilde{\varphi}(x)$, by starting from equation (3), discarding $\varrho(x)$ in it and replacing $\varphi(x_i)$ by $\tilde{\varphi}(x_i)$. Then $\tilde{\varphi}(x)$ takes the form

$$\tilde{\varphi}(x) = f(x) + \lambda \sum_{k=1}^{n} A_k K(x, x_k)\tilde{\varphi}(x_k). \tag{6}$$

It is obvious that the accuracy of the result obtained by the replacement of integral equation (1) by system (5) of linear equations will be the higher the less error we make in replacing the integral by the sum. The precise evaluation of the error introduced by employing this method constitutes the contents of the following No.; here we shall only make a few remarks of a general character.

Let us first of all dwell on the choice of the formula for mechanical quadrature. In view of the fact that enlarging the number of the ordinates of the unknown function to be employed increases the difficulty of solving system of equations (5), it is generally desirable to use the most exact of the formulas, namely Chebyshev's and Gauss's; of these formulas that of Gauss is somewhat the more exact, but in Chebyshev's formula all A_k are equal, and therefore by utilizing it system (5) is made simpler and more convenient. We further note that in the special case where the functions $f(x)$ and $K(x, y)$ are periodic and of period $b - a$ [the solution $\varphi(x)$ is then periodic too, of course], the same accuracy as the Gauss formula gives is yielded by the rectangular formula, and it is of course more convenient to use the latter. The system (5) thereupon acquires the following extremely simple form, if we take $a = 0$,

[1] The replacement of an integral equation by a system of linear equations was already employed in the fundamental works of Fredholm and Hilbert for theoretical inquiries. There the rectangular formula is always used. In his works [1], [2], the Finnish mathematician Nyström systematically made this substitution for the approximate solution; he also proposed the application to this question of the other more exact formulas of approximate integration.

$b = 2\pi$ for definiteness:

$$\tilde{\varphi}\left(i\frac{2\pi}{n}\right) - \frac{2\pi\lambda}{n}\sum_{k=0}^{n-1} K\left(i\frac{2\pi}{n}, k\frac{2\pi}{n}\right)\tilde{\varphi}\left(k\frac{2\pi}{n}\right) = f\left(i\frac{2\pi}{n}\right)$$

$$(i = 0, 1, \ldots, n-1).$$

In case the interval of integration is of the form $(0, b)$, if the kernel is an even or an odd function of its variables, an application of special formulas for the integration of even and odd functions is expedient [1]). If the unknown function or the kernel is known to vanish at the ends of the interval, it is expedient to apply A. A. Markov's quadrature formula [2]).

The application of the higher quadrature formulas can be justified only in case the integrand is regular (has a certain number of derivatives). Accordingly in the case given, the regularity of $K(x, y)$ and $\varphi(y)$ is necessary, but the regularity of the latter will be guaranteed if, besides the kernel, the free term also, $f(x)$, is regular [3]).

It is useful to point out that in some cases where there is no regularity in the problem as it is posed, this may nevertheless be contrived. For example, if the kernel is regular but the free term has a singularity, by introducing a new unknown function

$$\psi(x) = \varphi(x) - f(x), \tag{7}$$

we shall obtain for it the equation

$$\psi(x) - \lambda\int_a^b K(x, y)\psi(y)dy = \lambda\int_a^b K(x, y)f(y)dy, \tag{8}$$

i.e., an equation of the same type as for $\varphi(x)$, but in which the free term is now regular.

Let us yet consider the case where the kernel has a singularity for $y = x$, as this often happens; that is, the kernel itself, or its derivative $K_y'(x, y)$, undergoes a break of continuity at $y = x$. In this case it is advantageous, before replacing the integral equation by the system of linear equations, to transform it in the following manner:

$$\varphi(x)[1 - \lambda\int_a^b K(x, y)dy] - \lambda\int_a^b K(x, y)[\varphi(y) - \varphi(x)]dy = f(x). \tag{9}$$

Indeed, if we now examine the second integral on the left side, we see that thanks to the presence of the factor $[\varphi(y) - \varphi(x)]$, which vanishes for $y = x$, its integrand turns out to be more regular

---

[1]) See L. V. Kantorovich, [6].

[2]) IA. S. Bezikovich, [1], p. 290.

[3]) Indeed, if, for example, there exist both continuous $f^{(n)}(x)$ and $K_{x^n}^{(n)}(x, y)$, then on differentiating integral equation (1) n times with respect to x, we verify that the nth derivative of $\varphi(x)$ exists:

$$\varphi^{(n)}(x) = \lambda\int_a^b K_{x^n}^{(n)}(x, y)\varphi(y)dy + f^{(n)}(x).$$

than $K(x, y)\varphi(y)$, and the replacement of this integral by a finite sum is more justified than for the original one. As regards $\int_a^b K(x, y)dy$, inasmuch as it does not contain the unknown function, it can be computed without difficulty and will give some definite function $\chi(x)$. Constructing for equation (9) a system of equations just as we did system (5) for equation (1), we arrive at the system:

$$\tilde{\varphi}(x_i)[1 - \lambda\chi(x_i)] - \lambda \sum_{k=1}^{n} A_k K(x_i, x_k)[\tilde{\varphi}(x_k) - \tilde{\varphi}(x_i)] = f(x_i) \quad (10)$$

$$(i = 1, 2, \ldots, n),$$

the solution of which gives a better approximation in this case than the solution [1]) of system (5).

If the kernel $K(x, y)$ becomes infinite for $y = x$, as happens for singular integral equations, the ith term on the left side of (10) loses sense. In this case it is expedient to replace it by the expression which is obtained if one or other interpolation formula be applied to the function $K(x_i, y)[\varphi(y) - \varphi(x_i)]$. In particular, if linear interpolation be utilized, the expression $K(x_i, x_i)[\tilde{\varphi}(x_i) - \tilde{\varphi}(x_i)]$ must be replaced by

$$\frac{x_{i+1} - x_i}{x_{i+1} - x_{i-1}} K(x_i, x_{i-1})[\tilde{\varphi}(x_{i-1}) - \tilde{\varphi}(x_i)] +$$

$$+ \frac{x_i - x_{i-1}}{x_{i+1} - x_{i-1}} K(x_i, x_{i+1}) \cdot [\tilde{\varphi}(x_{i+1}) - \tilde{\varphi}(x_i)].$$

Lastly it should be remarked that the method of replacement of an integral equation by a system of algebraic equations is suitable also for the approximate solution of the problem of the determination of the proper numbers and proper functions, to wit: for the solution of this problem, instead of system of equations (5) one must consider the corresponding homogeneous system

$$\tilde{\varphi}(x_i) - \lambda \sum_{k=1}^{n} A_k K(x_i, x_k)\tilde{\varphi}(x_k) = 0 \qquad (i = 1, 2, \ldots, n). \quad (11)$$

The determinant of this system will be

$$\Delta(\lambda) = \begin{vmatrix} 1 - \lambda A_1 K(x_1, x_1), & -\lambda A_2 K(x_1, x_2), & \ldots, & -\lambda A_n K(x_1, x_n) \\ -\lambda A_1 K(x_2, x_1), & 1 - \lambda A_2 K(x_2, x_2), & \ldots, & -\lambda A_n K(x_2, x_n) \\ \cdot & \cdot & \cdot & \cdot \\ \cdot & \cdot & \cdot & \cdot \\ -\lambda A_1 K(x_n, x_1), & -\lambda A_2 K(x_n, x_2), & \ldots, & 1 - \lambda A_n K(x_n, x_n) \end{vmatrix}. \quad (12)$$

---

[1]) See L. V. Kantorovich, [7]. There still another method of annulling a singularity in a kernel for $y = x$ is shown.

Another possibility of the application of the given method in case of the presence of singularities in the kernel, based on the utilization of the quadrature formulas of V. A. Steklov, has been elaborated by N. A. Artmeladze, [1]. On this question see also Nyström, [3], and Hildebrand, [1].

Equating this determinant to zero and solving the corresponding equation

$$\Delta(\lambda) = 0. \tag{13}$$

we find those values of λ for which the system of equations (11) has a solution $\{\tilde{\varphi}(x_i)\}$ not identically equal to zero [1]). These values of λ will then be approximate values for the proper numbers. Furthermore, taking values of λ equal to the roots of equation (13), and forming for these values of λ the respective independent solutions of system (11), we can obtain approximate expressions for the proper functions of the equation [2]).

It is necessary to note that even in case the kernel is symmetric, $K(x, y) = K(y, x)$, the matrix of system of equations (11) may prove to be asymmetric, which makes the solution of equation (13) somewhat more difficult. In such a case it is expedient to render the matrix symmetric, for which it is sufficient to multiply the ith equation of system (11) by $\sqrt{A_i}$ and introduce, instead of $\tilde{\varphi}(x_i)$, new unknowns $z_i = \sqrt{A_i}\tilde{\varphi}(x_i)$.

3. The evaluation of the error incurred by replacing the integral equation by a system of linear equations. We shall consider in detail the evaluation of the error arising when the first problem — the determination of the function $\varphi(x)$ — is solved.

If by $\Delta_{i,k}$ be designated the algebraic complement of the element of the ith row and kth column in the determinant $\Delta(\lambda)$, (12), then the solution of the system of equations (5) can be written in the form:

$$\tilde{\varphi}(x_i) = \frac{\sum_{k=1}^{n} \Delta_{i,k} f(x_k)}{\Delta(\lambda)} \qquad (i = 1, 2, \ldots, n). \tag{14}$$

Let us estimate how much these approximate values differ from the exact values of the sought function $\varphi(x)$ at these same points:

$$\varphi(x_1), \varphi(x_2), \ldots, \varphi(x_n).$$

---

1) A method of expanding the determinant of the "secular" equation (12), considerably surpassing those known, was proposed by A. N. Krylov in 1931, [3]. Another interesting method with this same objective has been given by A. M. Danilevskii, [1].

Indirect methods of finding the proper numbers and proper vectors of a matrix without expanding the secular determinant are also efficient in applications.

Finally, the method of steepest descent is applicable to this problem. See L. V. Kantorovich, [8] or [3].

A detailed exposition of the different numerical methods for the determination of the proper numbers of matrices is given in the book of V. N. Faddeeva, [1]. See also I. M. Gel'fand [2].

2) Another way to find the proper numbers and functions, convenient in practice, is given in a work of O. Kellogg, [1]. See also S. G. Mikhlin, [1]. A device permitting one to indicate the bounds between which a proper number is contained, is given in a work of L. Collatz, [1].

Let us assume that the functions $f(x)$ and $K(x, y)$ have a certain number p of continuous derivatives; $\varphi(x)$ then has just as many derivatives [1]).

Denote by $H^{(s)}$, $N^{(s)}$, $M_x^{(s)}$, $M_y^{(s)}$ the upper bounds of

$$|\varphi^{(s)}(x)|, |f^{(s)}(x)|, \left|\frac{d^s}{dx^s} K(x, y)\right|, \left|\frac{d^s}{dy^s} K(x, y)\right|$$

and by $H^{(0)}$, $N^{(0)}$, $M^{(0)}$ the upper bounds of the moduli of the functions $\varphi(x)$, $f(x)$ and $K(x, y)$ themselves. In such a case the derivatives of the integrand $K(x, y)\varphi(y)$ can also be evaluated easily, for, as is obvious on the strength of Leibnitz's rule:

$$\left.\begin{aligned}\left|\frac{d^s}{dy^s}[K(x, y)\varphi(y)]\right| \leqslant H^{(0)}M_y^{(s)} + \\ + C_s^1 H^{(1)} M_y^{(s-1)} + \ldots + H^{(s)}M^{(0)} = T^{(s)}.\end{aligned}\right\} \tag{15}$$

If we now denote by σ the maximum error that we make in replacing $\int_a^b K(x, y)\varphi(y)dy$ by the sum, i.e., the maximum modulus of the quantity

$$\int_a^b K(x, y)\varphi(y)dy - \sum_{k=1}^{n} A_n K(x, x_k)\varphi(x_k) = \varrho(x), \tag{16}$$

so that

$$|\varrho(x)| \leqslant \sigma, \quad |\varrho_i| \leqslant \sigma, \tag{17}$$

this quantity σ may be easily evaluated by a certain expression containing $T^{(s)}$.

This evaluation can be made by means of the familiar expressions for the remainder terms which we have for all the formulas of mechanical quadrature. In particular we have, for instance:

for the trapezoidal formula: $\sigma \leqslant \dfrac{1(b-a)^3}{12(n-1)^2} T^{(2)}$;

for Simpson's formula: $\sigma \leqslant \frac{1}{90}(b-a)^5 \dfrac{T^{(4)}}{(n-1)^4}$;

for Gauss's formula:

$$\sigma \leqslant \frac{(b-a)^{2n+1}}{2n+1}\left\{\frac{1\cdot 2\cdot 3\ldots n}{(n+1)\ldots 2n}\right\}^2 \cdot \frac{T^{(2n)}}{1\cdot 2\cdot 3\ldots 2n}.$$

In all these formulas this estimate has the form $\sigma \leqslant k_n T^{(s)}$, where k_n is a definite factor for a given formula, and does not depend on the form of the integrand.

We shall now derive the required estimate. Let $\varphi(x)$ be the solution

---

[1]) See footnote [3]) to page 101.

of integral equation (1). Its values at the points $x_1, x_2, \ldots, x_n$ satisfy the system of equations:

$$\varphi(x_i) - \lambda \sum_{k=1}^{n} A_k K(x_i, x_k)\varphi(x_k) = f(x_i) + \lambda \varrho_i.$$

For the solution of this system let us utilize formulas (14), but taking $f(x_k) + \lambda\varrho_k$ instead of $f(x_k)$; we then find:

$$\varphi(x_i) = \frac{\sum_{k=1}^{n} \Delta_{i,k}[f(x_k) + \lambda\varrho_k]}{\Delta(\lambda)} \qquad (i = 1, 2, \ldots, n). \tag{18}$$

Comparing with (14), we obtain:

$$|\varphi(x_i) - \tilde{\varphi}(x_i)| = \left| \frac{\sum_{k=1}^{n} \Delta_{(i,k)}\lambda\varrho_k}{\Delta(\lambda)} \right| \leqslant \sigma\,|\lambda|\, \frac{\sum_{k=1}^{n} |\Delta_{i,k}|}{\Delta(\lambda)} \leqslant |\lambda|\,\sigma B, \tag{19}$$

if we denote by B the upper bound of the ratio of $\sum_{k=1}^{n} |\Delta_{i,k}|$ to $|\Delta(\lambda)|$, i.e., a number such that

$$\frac{\sum_{k=1}^{n} |\Delta_{i,k}|}{|\Delta(\lambda)|} \leqslant B. \tag{20}$$

Estimate (19) is quite handy, since the quantities $\Delta_{i,k}$ and $\Delta(\lambda)$ required for this estimate are determined automatically when system of equations (5) is actually solved. It still remains to show the limit of error which we admit at the remaining points by using the interpolation formula (6) for $\tilde{\varphi}(x)$. Indeed, the approximate value of $\varphi(x)$ is

$$\tilde{\varphi}(x) = f(x) + \lambda \sum_{k=1}^{n} A_k K(x, x_k)\tilde{\varphi}(x_k). \tag{6}$$

On the other hand, the exact value of $\varphi(x)$, by (3), is

$$\varphi(x) = f(x) + \lambda \sum_{k=1}^{n} A_k K(x, x_k)\varphi(x_k) + \lambda\varrho(x),$$

where $|\rho(x)| \leqslant \sigma$. Subtracting the one equation from the other and using (19), we obtain:

$$|\varphi(x) - \tilde{\varphi}(x)| \leqslant |\lambda| \sum_{k=1}^{n} A_k\, |K(x, x_k)| \cdot |\varphi(x_k) - \tilde{\varphi}(x_k)| + \\ + |\lambda| \cdot \sigma \leqslant |\lambda|\,\sigma[1 + |\lambda|\, BM^{(0)}(b - a)], \tag{21}$$

since [1] $\sum_{k=1}^{n} A_k = b - a$. The sole shortcoming of the estimate obtained,

[1] This equation is true for all of the quadrature formulas enumerated on pp. 98–99 and generally is for those that give an exact value for a function equal to a constant.

(21), consists in the fact that σ is evaluated in terms of $T^{(s)}$, and in the composition of $T^{(s)}$ there figure $H^{(0)}, H^{(1)}, \ldots, H^{(s)}$, the bounds of the unknown function $\varphi(x)$ and its derivatives. This inconvenience can be removed successfully, since these bounds can be expressed in terms of the known quantities $M_x^{(s)}$, $N^{(s)}$ and B. First differentiating equation (1) s times, we find:

$$\varphi^{(s)}(x) = \lambda \int_a^b \left[\frac{d^s}{dx^s} K(x, y) \right] \varphi(y) dy + f^{(s)}(x),$$

whence

$$H^{(s)} \leqslant H^{(0)} \cdot |\lambda| \, (b - a) M_x^{(s)} + N^{(s)} \qquad (s = 0, 1, 2, \ldots). \quad (22)$$

Now the expression for $T^{(s)}$, (15), can be evaluated thus:

$$T^{(s)} = H^{(0)} M_y^{(s)} + C_s^1 H^{(1)} M_y^{(s-1)} + \ldots + H^{(s)} M^{(0)} \leqslant N^{(0)} M_y^{(s)} + \\ + C_s^1 N^{(1)} M_y^{(s-1)} + \ldots + N^{(s)} M^{(0)} + |\lambda| \, (b - a)(M^{(0)} M_y^{(s)} + \\ + C_s^1 M_x^{(1)} M_y^{(s-1)} + \ldots + M_x^{(s)} M^{(0)}) H^{(0)} = P_s + Q_s H^{(0)},$$

where P_s and Q_s are known quantities. Next σ is evaluated, as has been indicated above, in terms of $T^{(s)}$, and consequently the estimate of it has the form:

$$\sigma \leqslant k_n T^{(s)} \leqslant k_n (P_s + Q_s H^{(0)}), \quad (23)$$

where k_n is the coefficient of the derivative figuring in the estimate of the remainder term of the quadrature formula applied.

Using inequality (23), estimate (21) can be rewritten thus:

$$|\varphi(x) - \tilde{\varphi}(x)| \leqslant [M^{(0)} \, |\lambda| \, B(b - a) + 1] k_n (P_s + Q_s H^{(0)}) \, |\lambda|. \quad (24)$$

Let us denote by S the upper bound of the modulus of the approximate solution [1]) $\tilde{\varphi}(x)$. We then obtain:

$$|\varphi(x)| \leqslant S + [M^{(0)} \, |\lambda| \, B(b - a) + 1] k_n (P_s + Q_s H^{(0)}) \, |\lambda|.$$

---

[1]) The quantity S can be found directly from (6), but it can also easily be evaluated in terms of the constants introduced earlier. Indeed, by (14) and (20):

$$|\tilde{\varphi}(x_i)| \leqslant B \cdot N^{(0)}.$$

Moreover, in view of (6):

$$|\tilde{\varphi}(x)| \leqslant |f(x)| + |\lambda| \sum_{k=1}^{n} A_k \, |K(x, x_k) \cdot |\tilde{\varphi}(x_k)| \leqslant N^{(0)} + M^{(0)} \, |\lambda| \, (b - a) N^{(0)} B.$$

Hence it is evident that the expression last obtained can be adopted for S.

Hence also

$$H^{(0)} \leqslant S + [M^{(0)} |\lambda| B(b-a) + 1] k_n (P_s + Q_s H^{(0)}) |\lambda|.$$

Next, transforming this inequality, we find an estimate for [1] $H^{(0)}$:

$$H^{(0)} \leqslant \frac{S + [M^{(0)} |\lambda| B(b-a) + 1] |\lambda| k_n P_s}{1 - [M^{(0)} |\lambda| B(b-a) + 1] |\lambda| k_n Q_s}, \tag{25}$$

which is true if the denominator of the right side is positive. For this, since k_n is very small, it is required only that the remaining quantities B, $M^{(0)}$, Q_s, $|\lambda|$ prove not to be too large.

In case this denominator is positive, the solution $\varphi(x)$ turns out to be bounded, and this at once permits us to affirm that the given value of λ is not a proper number. Indeed, in the contrary case, to $\varphi(x)$ could be added an arbitrary solution of the homogeneous equation, which could be taken arbitrarily large, and inequality (25) would then not be observed.

Substituting the estimate of $H^{(0)}$ from inequality (25) in (24), we can obtain a definitive estimate of $|\varphi(x) - \tilde{\varphi}(x)|$ — of the accuracy of the approximation — in which only known quantities figure.

The exhibited estimates also permit the evaluation of the error made in finding the proper numbers and functions by the method expounded. Without going into details, we shall limit ourselves to an indication of the principle only.

Let us take a certain region of values of λ, for example, the circle $|\lambda| < \Lambda_0$. Evaluate $\sum_{k=1}^{n} |\Delta_{i,k}|$ in this region; let this be the number R. Then in inequality (25) we can adopt for B the quantity $\frac{R}{\Delta(\lambda)}$. If now the value of λ be such that

$$1 - \left[M^{(0)} |\lambda| \cdot \frac{R}{\Delta(\lambda)} (b-a) + 1\right] |\lambda| k_n Q_s > 0, \tag{26}$$

λ is not a proper number. The proper values can therefore be situated only in a region of values of λ not satisfying inequality (26). This observation indeed makes possible an estimate of the accuracy of the approximate values of the proper numbers found by the solution of the equation $\Delta(\lambda) = 0$. The estimate of the error in the determination of the proper functions can be made on the basis of like considerations, on which we shall not dwell in detail.

---

[1]) A less thorough effort at an estimation of the error that is due to the replacement of an integral equation by a system of linear equations was made by I. Akbergenov, [1], 1935.

Another estimate of the error, based on an investigation of the Fredholm determinant, has been given by Ostrowski, [1]. This estimate relates to the application of the rectangular formula and is of a more immediate theoretical interest. From a general point of view the question has been considered in Chapter II, § 4 of L. V. Kantorovich's work [3].

4. Example. Let us apply the method of approximate solution of integral equations which has been considered above to the solution of the equation

$$\varphi(x) - \tfrac{1}{2}\int_0^1 e^{xy}\varphi(y)dy = 1 - \frac{1}{2x}(e^x - 1),$$

using the Gauss quadrature formula.

Replacing this integral equation by a system with $n = 2$, and noting that here $\lambda = \frac{1}{2}$, $A_1 = A_2 = \frac{1}{2}$, we obtain system (5) for the case in hand in the form

$$(1 - \tfrac{1}{4}K_{1,1})\tilde{\varphi}(x_1) - \tfrac{1}{4}K_{1,2}\tilde{\varphi}(x_2) = f_1,$$
$$-\tfrac{1}{4}K_{2,1}\tilde{\varphi}(x_1) + (1 - \tfrac{1}{4}K_{2,2})\tilde{\varphi}(x_2) = f_2.$$

In accordance with what has been stated above, for x_1 and x_2 are adopted the Gaussian abscissas for the interval (0, 1): $x_1 = 0.2113$; $x_2 = 0.7887$. Computing the values of $K_{i,k} = K(x_i, x_k)$, $f_i = f(x_i)$ and substituting them in the system, we bring it into the form:

$$0.7386\tilde{\varphi}(x_1) - 0.2954\tilde{\varphi}(x_2) = 0.4434,$$
$$-0.2954\tilde{\varphi}(x_1) + 0.5343\tilde{\varphi}(x_2) = 0.2384.$$

The values of the required determinants will be:

$$\Delta = \begin{vmatrix} 0.7386 & -0.2954 \\ -0.2954 & 0.5343 \end{vmatrix} = 0.3074,$$

$$\Delta_1 = \begin{vmatrix} 0.4434 & -0.2954 \\ 0.2384 & 0.5343 \end{vmatrix} = 0.3073, \quad \Delta_2 = \begin{vmatrix} 0.7386 & 0.4434 \\ -0.2954 & 0.2384 \end{vmatrix} = 0.3071,$$

whence

$$\tilde{\varphi}(x_1) = \frac{\Delta_1}{\Delta} = 0.9997, \quad \tilde{\varphi}(x_2) = \frac{\Delta_2}{\Delta} = 0.9990.$$

The approximate solution at the other points is given, in accordance with formula (6), by the equation:

$$\tilde{\varphi}(x) = \tfrac{1}{4}(e^{0.2113x}0.9997 + e^{0.7887x}0.9990) + 1 - \frac{1}{2x}(e^x - 1).$$

The exact solution of the given equation, as is easily verified by substitution, is $\varphi(x) = 1$. Comparing the values of these solutions at the points x_1, x_2, we see that the difference is respectively 0.0003 and 0.0010. One can, for example, find $\tilde{\varphi}(0) = 0.9997$, and again the difference is 0.0003. As we see, the approximate solution nearly coincides with the exact.

We shall now give a precise estimate of the error of the approximate solution found, utilizing the method indicated in No. 3.

Let us find some necessary quantities. B designates the estimate

$$\frac{\sum_{k=1}^{n} |\Delta_{i,k}|}{\Delta}.$$

In the case in hand $\Delta_{i,k}$ are the elements of the determinant Δ, and the maximum ratio will be

$$\frac{0.7386 + 0.2954}{0.3074} < 3.4,$$

and accordingly we may adopt $B = 3.4$.

Expanding the free term and its derivatives in series, we find:

$$f(x) = 1 - \frac{1}{2x}(e^x - 1) = \frac{1}{2} - \frac{x}{4} - \frac{x^2}{12} - \frac{x^3}{48} - \frac{x^4}{240} - \frac{x^5}{1440} - \frac{x^6}{10080} - \ldots,$$

$$f'(x) = -\left(\tfrac{1}{4} + \tfrac{1}{6}x + \tfrac{1}{16}x^2 + \tfrac{1}{60}x^3 + \tfrac{1}{288}x^4 + \tfrac{1}{1680}x^5 + \ldots\right),$$

$$f''(x) = -\left(\tfrac{1}{6} + \tfrac{1}{8}x + \tfrac{1}{20}x^2 + \tfrac{1}{72}x^3 + \tfrac{1}{336}x^4 + \ldots\right),$$

$$f'''(x) = -\left(\tfrac{1}{8} + \tfrac{1}{10}x + \tfrac{1}{24}x^2 + \tfrac{1}{84}x^3 + \ldots\right),$$

$$f^{(IV)}(x) = -\left(\tfrac{1}{10} + \tfrac{1}{12}x + \tfrac{1}{28}x^2 + \ldots\right).$$

Hence it is evident that for the bounds of the function f and its derivatives in (0, 1) we may adopt

$$N^{(0)} = 0.5,\ N^{(1)} = 0.5,\ N^{(2)} = 0.36,\ N^{(3)} = 0.29,\ N^{(4)} = 0.25.$$

Next the kernel and its derivatives will be

$$K = e^{xy},\ K'_x = ye^{xy},\ K''_x = y^2e^{xy},\ K'''_x = y^3e^{xy},\ K_x^{(IV)} = y^4e^{xy};$$

the derivatives with respect to y have analogous expressions. For their bounds in (0, 1) we may therefore adopt:

$$M^{(0)} = M_x^{(1)} = \ldots = M_x^{(4)} = M_y^{(1)} = \ldots = M_y^{(4)} = e.$$

Hence

$$P_4 = N^{(0)}M_y^{(4)} + C_4^1N^{(1)}M_y^{(3)} + C_4^2N^{(2)}M_y^{(2)} + C_4^3N^{(3)}M_y^{(1)} + N^{(4)}M^{(0)} < 8e < 22,$$

$$Q_4 = |\lambda|\ (b - a)\ (M^{(0)}M_y^{(4)} + C_4^1M_x^{(1)}M_y^{(3)} + C_4^2M_x^{(2)}M_y^{(2)} + C_4^3M_x^{(3)}M_y^{(1)} + M_x^{(4)}M^{(0)}) = 8e^2 < 60.$$

Then for the derivative $\frac{d^4}{dy^4}K(x, y)\varphi(y)$ we obtain the estimate

$$T^{(4)} = P_4 + Q_4H^{(0)},$$

where $H^{(0)}$ is the upper bound of $\varphi(y)$. Next we find for the estimate of the error due to the replacement of the integral by the sum, using the expression of the estimate for the Gauss formula (p. 104), for $n = 2$:

$$\sigma = k_2T^{(4)} = \frac{1}{5}\left\{\frac{1\cdot 2}{3\cdot 4}\right\}^2 \frac{T^{(4)}}{1\cdot 2\cdot 3\cdot 4} = \frac{1}{4320}T^{(4)},$$

i.e., $k_2 = \frac{1}{4320} = 0.00024$.

Lastly S, the upper bound of $\tilde{\varphi}(x)$, may be adopted as equal to 1.4, Indeed, the maximum value of the first term of $\tilde{\varphi}(x)$ is $\frac{1}{4}(e^{0.212} + e^{0.789}) \approx 0.9$, and the maximum value of the difference following is $\frac{1}{2}$.

Substituting the values found, we find for $H^{(0)}$ the following estimate:

$$H^{(0)} \leqslant \frac{S + [M^{(0)}B(b-a)\,|\lambda| + 1]\,|\lambda|\,k_2P_4}{1 - [M^{(0)}B(b-a)\,|\lambda| + 1]\,|\lambda|\,k_2Q_4} =$$

$$= \frac{1.4 + (2.72 \cdot 3.4 \cdot 0.5 + 1) \cdot 0.5 \cdot 0.00024 \cdot 22}{1 - (2.72 \cdot 3.4 \cdot 0.5 + 1) \cdot 0.5 \cdot 0.00024 \cdot 60} = 1.48.$$

Finally, for the difference which is of interest to us, $|\varphi = \tilde{\varphi}|$, we obtain the estimate

$$|\varphi(x) - \tilde{\varphi}(x)\,| < |\lambda|\,[1 + BM^{(0)}(b-a)\,|\lambda|]\,k_2(P_4 + Q_4H^{(0)}) =$$

$$= 0.5 \cdot 5.62 \cdot 0.00024(22 + 60 \cdot 1.48) = 0.07.$$

As we saw, in reality the error is of the order of 0.001, which is considerably [1]) lower than that which we just now obtained by the estimate, 0.070. However, the making of the estimate has established, for example, with complete definiteness, that $\lambda = 0.5$ is not a proper value of the given integral equation and that it accordingly has a unique solution [2]).

§ 2. THE METHOD OF SUCCESSIVE APPROXIMATIONS AND ANALYTIC CONTINUATION

1. The method of successive approximations. Let us consider the integral equation of the second kind:

$$\varphi(x) - \lambda\int_a^b K(x, y)\varphi(y)dy = f(x). \tag{1}$$

The method of successive approximations for it consists in the following. We seek the solution in the form of a series arrayed in powers of λ:

$$\varphi(x) = \varphi_0(x) + \lambda\varphi_1(x) + \lambda^2\varphi_2(x) + \dots \tag{2}$$

Substituting this series in equation (1), we find:

$$\varphi_0(x) + \lambda\varphi_1(x) + \lambda^2\varphi_2(x) + \dots =$$

$$= f(x) + \lambda\int_a^b K(x, y)\,[\varphi_0(y) + \lambda\varphi_1(y) + \lambda^2\varphi_2(y) + \dots]dy.$$

---

[1]) The exceptional coarseness of the estimate in the given case is accidental and is occasioned by the fact that the exact solution is a constant. Had we taken this into account, we would have found that in the given case $T^{(4)} = H^{(0)} \cdot M_y^{(4)} = e$. Therefore the estimate of the error turns out to equal 0.5(5.62) (0.00024) (2.72) = = 0.0018, i.e., is sufficiently close to the actual error.

[2]) Here we have considered the approximate solution of integral equations of the Fredholm type; methods for the solution of equations with variable limits (of the Volterra type) can be found in the works: Sh. E. Mikeladze, [1]; V. I. Krylov, [1]; Whittaker and Robinson, [1]; G. M. Müntz, [1].

Equating the coefficients of like powers of λ, we obtain:

$$\left.\begin{aligned} \varphi_0(x) &= f(x), \\ \varphi_1(x) &= \int_a^b K(x, y)\varphi_0(y)dy, \\ \varphi_2(x) &= \int_a^b K(x, y)\varphi_1(y)dy. \\ &\ldots\ldots\ldots\ldots \end{aligned}\right\} \tag{3}$$

From these equations there may successively be determined all the functions $\varphi_1(x)$, $\varphi_2(x)$, ...; explicitly, if we introduce the so-called iterated kernels:

$$\left.\begin{aligned} K_1(x, y) &= K(x, y), \\ K_2(x, y) &= \int_a^b K(x, t)K_1(t, y)dt, \\ K_3(x, y) &= \int_a^b K(x, t)K_2(t, y)dt, \\ &\ldots\ldots\ldots\ldots \end{aligned}\right\} \tag{4}$$

we can then write the following expressions for the functions $\varphi_n(x)$:

$$\varphi_0(x) = f(x), \quad \varphi_n(x) = \int_a^b K_n(x, y)f(y)dy \quad (n = 1, 2, \ldots). \tag{5}$$

Series (2) can then be written in the form:

$$\begin{aligned} \varphi(x) &= f(x) + \lambda \int_a^b K_1(x, y)f(y)dy + \lambda^2 \int_a^b K_2(x, y)f(y)dy + \ldots = \\ &= f(x) + \lambda \int_a^b [K_1(x, y) + \lambda K_2(x, y) + \ldots]f(y)dy = \\ &= f(x) + \lambda \int_a^b \Gamma(x, y, \lambda)f(y)dy, \end{aligned} \tag{6}$$

if by Γ we denote the series in brackets:

$$\Gamma(x, y, \lambda) = K_1(x, y) + \lambda K_2(x, y) + \ldots \tag{7}$$

The function Γ is called the resolvent of equation (1). It can be proved, assuming a bounded kernel $K(x, y)$: $|K(x, y)| \leqslant M$, that series (7), (6) and (2) will be uniformly convergent and $\varphi(y)$ will be the solution of equation (1) if λ is sufficiently small — more exactly, if λ satisfies the condition:

$$|\lambda| < \frac{1}{M(b-a)}\ ^{1)}. \tag{8}$$

If λ satisfies inequality (8), series (2) can successfully be used for an approximate solution. Often the required quadratures (3) can be performed exactly; then the series (2) obtained will converge at least as a progression with ratio $|\lambda|\ M(b-a)$. The error which we admit by limiting ourselves to n terms in series (2) can be evaluated without any special labor, for, assuming $|f(x)| \leqslant N$, we easily obtain in succession:

$$|\varphi_n(x)| < NM^n(b-a)^n, \tag{9}$$

and the remainder of series (2) after n terms is therefore

$$|\lambda^n\varphi_n(x) + \lambda^{n+1}\varphi_{n+1}(x) + \dots | < |\lambda|^n NM^n(b-a)^n +$$
$$+ |\lambda|^{n+1}NM^{n+1}(b-a)^{n+1} + \dots = \frac{NM^n(b-a)^n\,|\lambda|^n\ ^{2)}}{1 - M(b-a)\,|\lambda|}. \tag{10}$$

As a rule, however, quadratures (3) cannot be performed, and in such cases a formula for approximate integration must be employed. We shall indicate a most convenient order of arrangement for the computations in such a case.

We choose a definite quadrature formula:

$$\int_a^b \psi(x)dx = \sum_{k=1}^{m} A_k\psi(x_k).$$

Introduce the abbreviated notations:

$$K_{i,k} = K(x_i, x_k), \quad \varphi_n^{(i)} = \varphi_n(x_i), \quad \varphi^{(i)} = \varphi(x_i), \quad f^{(i)} = f(x_i), \tag{11}$$

along with which the approximate values for $\varphi^{(i)}$ and $\varphi_n^{(i)}$ will be denoted

---

1) More precise estimates for the radius of convergence of series (2), (6) and (7) can be given. Estimate (8) remains true, for instance, if M is not an upper bound of $K(x, y)$ but is any number such that one of the following inequalities is satisfied:

$$\int_a^b |K(x, y)|\, dy < M(b-a), \tag{a}$$

$$\{\int_a^b\int_a^b K^2(x, y)dxdy\}^{\frac{1}{2}} \leqslant M(b-a), \tag{b}$$

$$|K_p(x, y)| < M^p(b-a)^{p-1}. \tag{c}$$

2) For the solution of the integral equation and also for finding its proper numbers and functions, another method of successive approximations may also be employed — the method of steepest descent. On this see Chap. III, § 6 of L. V. Kantorovich's work [3].

by $\tilde{\varphi}^{(i)}$ and $\tilde{\varphi}_n^{(i)}$. Equations (3) then give:

$$\varphi_0^{(i)} = f^{(i)},$$

$$\varphi_1^{(i)} = \int_a^b K(x_i, y)\varphi_0(y)dy \cong \sum_{k=1}^m A_k K_{i,k}\varphi_0^{(k)}, \quad \tilde{\varphi}_1^{(i)} = \sum_{k=1}^m A_k K_{i,k}\varphi_0^{(k)},$$

and generally

$$\tilde{\varphi}_{n+1}^{(i)} = \sum_{k=1}^m A_k K_{i,k}\tilde{\varphi}_n^{(k)}. \tag{12}$$

The computations by these formulas are most conveniently arranged in a table:

	1	2		m	Con-trol	$\varphi_0 = f$	$\tilde{\varphi}_1$	$\tilde{\varphi}_2$		$\tilde{\varphi}$
x_1	$\lambda A_1 K_{1,1}$	$\lambda A_1 K_{2,1}$	...	$\lambda A_1 K_{m,1}$	K_1^*	$\varphi_0^{(1)}$	$\lambda\tilde{\varphi}_1^{(1)}$	$\lambda^2\tilde{\varphi}_2^{(1)}$	...	$\tilde{\varphi}^{(1)}$
x_2	$\lambda A_2 K_{1,2}$	$\lambda A_2 K_{2,2}$	...	$\lambda A_2 K_{m,2}$	K_2^*	$\varphi_0^{(2)}$	$\lambda\tilde{\varphi}_1^{(2)}$	$\lambda^2\tilde{\varphi}_2^{(2)}$	...	$\tilde{\varphi}^{(2)}$
..	....	....	...	....	...	...	...	...	...	..
..	....	....	...	....	...	...	...	...	...	..
x_m	$\lambda A_m K_{1,m}$	$\lambda A_m K_{2,m}$	...	$\lambda A_m K_{m,m}$	K_m^*	$\varphi_0^{(m)}$	$\lambda\tilde{\varphi}_1^{(m)}$	$\lambda^2\tilde{\varphi}_2^{(m)}$	...	$\tilde{\varphi}^{(m)}$
							$\lambda\varphi_1^*$	$\lambda^2\varphi_2^*$	...	

This table is formed as follows: 1. Computing the values of the kernel at the points (x_i, x_k) and multiplying them by the factor λA_k, we form a square table. After it comes a control column, about which we shall speak below. 2. The values of the function f at the points x_i give us the first column following the square table. 3. To compute the elements of the second column we proceed thus: we multiply the elements of the first column after the control column by the elements of the first column of the table, and add these products, finding:

$$\lambda\tilde{\varphi}_1^{(1)} = \sum_{k=1}^m \lambda A_k K_{1,k}\varphi_0^{(k)}. \tag{13}$$

Multiplying in the same fashion the elements of this same column — the first after the control — by the elements of the second column of the table, we find

$$\lambda\tilde{\varphi}_1^{(2)} = \sum_{k=1}^m \lambda A_k K_{2,k}\varphi_0^{(k)}. \tag{13'}$$

Continuing thus, we compute all the second column after the control column. 4. The third column after the control (column $\tilde{\varphi}_2$) is computed just like the second, but instead of the first column following the table, the second must be used, for example:

$$\lambda^2\tilde{\varphi}_2^{(1)} = \sum_{k=1}^{m} \lambda A_k K_{1,k}(\lambda\tilde{\varphi}_1^{(k)}), \qquad \lambda^2\tilde{\varphi}_2^{(2)} = \sum_{k=1}^{m} \lambda A_k K_{2,k}(\lambda\tilde{\varphi}_1^{(k)}). \tag{14}$$

5. The computations must be carried on until the moment when the elements of the last column become so small that they may be neglected. After this, by adding the elements of the columns found, row by row, we obtain the last column — the approximate values of the sought function φ at the points x_i:

$$\tilde{\varphi}^{(i)} = \tilde{\varphi}_0^{(i)} + \lambda\tilde{\varphi}_1^{(i)} + \lambda^2\tilde{\varphi}_2^{(i)} + \dots \tag{15}$$

Of the fact that the series converges under condition (8), one can satisfy oneself effortlessly. Let $|\varphi_0^{(i)}| = |f_1| \leqslant N$. Then

$$|\tilde{\varphi}_1^{(i)}| = |\sum_{k=1}^{n} \lambda A_k K_{i,k}\varphi_0^{(k)}| \leqslant NM\,|\lambda| \sum_{k=1}^{n} A_k = NM\,|\lambda|\,(b-a).$$

Continuing thus, we find $|\tilde{\varphi}_n^{(i)}| \leqslant N[M\,|\lambda|\,(b-a)]^n$, i.e., series (15) does indeed converge under condition (8) [1]).

The control of the computations is most conveniently carried out on the basis of the following considerations. In view of the preceding formulas, we have:

$$\sum_{i=1}^{m} \lambda\tilde{\varphi}_1^{(i)} = \sum_{i=1}^{m}\sum_{k=1}^{m} \lambda A_k K_{i,k}\tilde{\varphi}_0^{(k)} = \sum_{k=1}^{m} \tilde{\varphi}_0^{(k)} \,[\sum_{i=1}^{m} (A_k\lambda K_{i,k})],$$

$$\sum_{i=1}^{m} \lambda^2\tilde{\varphi}_2^{(i)} = \sum_{k=1}^{m} \lambda\tilde{\varphi}_1^{(k)} \,[\sum_{i=1}^{m} (A_k\lambda K_{i,k})].$$

. . . .

If we therefore form a supplementary column of the quantities

$$K_k^* = \sum_{i=1}^{m} (A_k\lambda K_{i,k}),$$

obtained by the summation of the rows of the square table (the control column), and form the sums of the vertical columns following the square table,

$$\lambda\varphi_1^* = \sum_{i=1}^{m} \lambda\varphi_1^{(i)}, \quad \lambda^2\varphi_2^* = \sum_{i=1}^{m} \lambda^2\varphi_2^{(i)}, \quad \dots,$$

then on multiplying the control column by the columns following the

---

1) The convergence of the process of successive approximations to the solution, with the employment of special quadrature formulas and given a simultaneous increase in the number of points used within the limits of the region defined by inequality (8) or by more exact inequalities (see the footnote on page 112) was established in a work of N. Krylov and IA. Tamarkin, [1].

An application of the method of successive approximations especially to the numerical solution of an integral equation for the dispersion of light is given in quite recent works of E. S. Kuznetsov, [1], and B. V. Ovchinskiĭ, [1].

table, we must obtain the quantities $\lambda\varphi_1^*$, $\lambda^2\varphi_2^*$, ..., for example:

$$\sum_{k=1}^{m} K_k^*\varphi_0^{(k)} = \lambda\varphi_1^*, \quad \sum_{k=1}^{m} K_k^*\lambda\varphi_1^{(k)} = \lambda^2\varphi_2^*, \; \ldots . \tag{16}$$

This method consequently makes possible the check of each of the columns after computation.

The check of the correctness of the solution obtained can be effected by still another method. Using the expression for $\tilde{\varphi}^{(i)}$ and $\tilde{\varphi}_s^{(i)}$, we find:

$$\tilde{\varphi}^{(i)} - f^{(i)} = \lambda\tilde{\varphi}_1^{(i)} + \lambda^2\tilde{\varphi}_2^{(i)} + \ldots =$$

$$= \sum_{k=1}^{m} \lambda A_k K_{i,k}\tilde{\varphi}_0^{(k)} + \lambda\sum_{k=1}^{m} \lambda A_k K_{i,k}\tilde{\varphi}_1^{(k)} + \ldots =$$

$$= \lambda\sum_{k=1}^{m} A_k K_{i,k}(\tilde{\varphi}_0^{(k)} + \lambda\tilde{\varphi}_1^{(k)} + \ldots) = \lambda\sum_{k=1}^{m} A_k K_{i,k}\tilde{\varphi}^{(k)}.$$

Accordingly the values found for $\tilde{\varphi}^{(i)}$ must be solutions of a system of ordinary equations of the first degree:

$$\tilde{\varphi}^{(i)} - \sum_{k=1}^{m} (\lambda A_k K_{i,k})\tilde{\varphi}^{(k)} = f^{(i)}. \tag{17}$$

In order to verify that the $\tilde{\varphi}^{(i)}$ actually satisfy this system, one must compute, by multiplying the column $\tilde{\varphi}^{(k)}$ by the square table (as we did with each column), the column of quantities $\sum_{k=1}^{m} \lambda A_k K_{i,k}\tilde{\varphi}^{(k)}$, which must turn out to equal $\tilde{\varphi}^{(i)} - f^{(i)}$.

System (17) coincides with the system (5) that we considered in § 1. Thus the computational method that we have used here is none other than an application of a special method of solving system (17) — a method of iteration, which is usually the more convenient in practice the more direct its method.

Since the method of successive approximations gives as its result the solution of system (17), a strict evaluation of the error obtained as the result of a solution by this method can be made by means of the formulas of No. 3, § 1.

Practically, the evaluation of the error of the solution obtained can be made otherwise, too. First of all, by means of one or another interpolation formula — for instance by formula (6) No. 2, § 1 — one can, by using the numbers $\tilde{\varphi}^{(k)}$ found, obtain an approximate expression for $\varphi(x)$ in the entire interval: $\tilde{\varphi}(x) = f(x) + \lambda\sum_{k=1}^{m} A_k K(x, x_k)\tilde{\varphi}^{(k)}$. To check how well $\tilde{\varphi}(x)$ satisfies the integral equation, it may be substituted in the integral equation, the integral, however, being replaced by a sum in accordance with another quadrature formula:

$$\tilde{\varphi}(x) - \lambda\sum_{k=1}^{m^*} A_k^* K(x, x_k^*)\tilde{\varphi}(x_k^*) \cong f(x),$$

the test being of the errors for several values of x.

We have just considered the question of the approximate solution of the equation with a definite right member $f(x)$. In case the same equation must be solved for different right members, however, it is advantageous to construct the resolvent. For this purpose the iterated kernels $K_s(x, y)$ must be formed. If we introduce the designation $K_{i,k}^{(s)} = K_s(x_i, x_k)$ and denote by $\tilde{K}_{i,k}^{(s)}$ their approximate values, we shall have the following formulas for the successive approximate computation of these quantities:

$$\tilde{K}_{i,k}^{(1)} = K_{i,k}, \quad \lambda^{s+1}\tilde{K}_{i,k}^{(s+1)} = \sum_{j=1}^{m} \lambda A_j K_{i,j}(\lambda^s \tilde{K}_{j,k}^{(s)}).$$

This formula shows that if the values of $\lambda^s \tilde{K}_{i,k}^{(s)}$ are copied in matrix form, each matrix is obtained from the preceding by multiplication by the matrix $||\lambda A_k K_{j,k}||$.

This makes it possible to construct a computational scheme analogous to that considered above, and to give just such rules for checks. Adding all the matrices constructed, we shall obtain the matrix of values for the resolvent:

$$\tilde{\Gamma}_{j,k} = K_{i,k} + \lambda \tilde{K}_{i,k}^{(2)} + \lambda^2 \tilde{K}_{i,k}^{(3)} + \ldots.$$

By means of the matrix $||\tilde{\Gamma}_{i,k}||$ the solution for any free member $f(x)$ can be constructed by the formula:

$$\tilde{\varphi}_i = f^{(i)} + \lambda \sum_{k=1}^{m} A_k \tilde{\Gamma}_{i,k} f^{(k)}.$$

We shall not linger here on further details [1]).

2. The application of analytic continuation for the approximate solution of integral equations. Series (7) for the resolvent Γ represents its expansion in powers of the parameter λ about the point $\lambda = 0$, and will therefore converge up to the first singular point of this function. Accordingly the series will be convergent as long as $|\lambda| < |\lambda_1|$, where λ_1 is the first proper number of the kernel. Because of this, series (7) cannot be used if $|\lambda| > |\lambda_1|$, since it diverges, and is used with difficulty if $|\lambda|$ is close to $|\lambda_1|$, since then the series converges slowly. In these cases the employment of analytic continuation often proves convenient, particularly in those cases where the distribution of the proper numbers of the integral equation is to a certain degree known to us. We shall indicate some devices for analytic continuation which are

---

[1]) A work of D. IU. Panov, [1], is also devoted to the approximate solution of integral equations, for small values of the parameter. In this work the application of the method of S. A. Chaplygin to non-linear equations as well is considered.

The approximate solution of non-linear integral equations is considered also in works of V. P. Vetchinkin, [1]; D. M. Zagadskiĭ, [1]; and L. V. Kantorovich, [9] and [3], Chap. IV, § 3.

most suitable for the case given. In so doing we shall throughout consider for convenience that $\lambda_1 = -1$; it is obvious that this can always be contrived by replacing λ by $\bar{\lambda} = -\frac{\lambda}{\lambda_1}$.

1. *Direct continuation.* The solution for $|\lambda| < 1$ is given by the convergent series

$$\varphi = \varphi_0 + \lambda\varphi_1 + \lambda^2\varphi_2 + \ldots$$

By means of this series we can compute the values of $\varphi(x, \lambda)$ and of its derivatives with respect to λ: $\frac{d^k}{d\lambda^k}\varphi(x, \lambda)$, for $\lambda = \frac{1}{2}$, for example. This will permit us to form the expansion of the solution $\varphi(x, \lambda)$ in terms of powers of $(\lambda - \frac{1}{2})$. This expansion can also be obtained directly, replacing λ by $(\lambda' + \frac{1}{2})$ in the given series (having in view $\lambda' = \lambda - \frac{1}{2}$), and re-expanding the series in powers of λ'. The distance from the point $\lambda = \frac{1}{2}$ to the proper value $\lambda_1 = -1$ is equal to $1\frac{1}{2}$, and therefore if the following proper values do not hinder this, the power series in λ' which is obtained will have the radius of convergence $1\frac{1}{2}$ and will make it possible, for example, to compute the solution for $1 \leqslant \lambda < 2$, which would be impossible by means of the original series. Independently of the distribution of the proper values other than λ_1, the reconstructed series will certainly converge for $\lambda = 1$, i.e., $\lambda' = \frac{1}{2}$, provided this value is not proper itself.

The reconstructed series has the following form:

$$\varphi = \varphi_0 + (\lambda' + \tfrac{1}{2})\varphi_1 + (\lambda' + \tfrac{1}{2})^2\varphi_2 + \ldots =$$
$$= (\varphi_0 + \tfrac{1}{2}\varphi_1 + \tfrac{1}{4}\varphi_2 + \tfrac{1}{8}\varphi_3 + \ldots) + (\varphi_1 + \varphi_2 + \tfrac{3}{4}\varphi_3 + \ldots)\lambda' +$$
$$+ (\varphi_2 + \tfrac{3}{2}\varphi_3 + \ldots)\lambda'^2 + \ldots$$

This same series can also be used for the approximate computation of the values of φ at separate points — the numbers $\tilde{\varphi}^{(i)}$. Here one must use the successive approximations $\tilde{\varphi}_n^{(i)}$ (see the previous No.).

2. *Elimination of a pole by premultiplication.* Let us assume that the first proper number of the kernel is $\lambda_1 = -1$, and that this is a simple pole for the resolvent. For instance, the proper number will always be a simple pole if the kernel is symmetric, $K(x, y) = K(y, x)$. The remaining proper values we assume to be greater than 1 in modulus. In this case the functions $(\lambda + 1)\Gamma(x, y, \lambda)$ and $(\lambda + 1)\varphi(x, \lambda)$ no longer have $\lambda = -1$ as a singular point, and therefore the series in powers of λ for these functions will be convergent for $|\lambda| < |\lambda_2|$; in particular, they will certainly converge if $|\lambda| = 1$. The series, then, for these functions are found directly, being

$$(\lambda + 1)\Gamma(x, y, \lambda) = [K(x, y) + \lambda K_2(x, y) + \ldots](\lambda + 1) =$$
$$= K(x, y) + \lambda[K(x, y) + K_2(x, y)] + \lambda^2[K_2(x, y) + K_3(x, y)] + \ldots,$$
$$(\lambda + 1)\varphi(x, \lambda) = \varphi_0(x) + \lambda[\varphi_0(x) + \varphi_1(x)] + \ldots$$

In particular, for $\lambda = 1$:

$$\Gamma(x, y, 1) = \tfrac{1}{2}[K + (K + K_2) + (K_2 + K_3) + \ldots],$$

for which a regrouping is impermissible, since the series might then be converted into a divergent one.

These same series can be used for the computation of the values of the resolvent at separate points if the iterated kernels have been found numerically.

3. *Transformation of the series by change of variable.* The general idea consists in the following. Let us assume that we can introduce in place of λ a new variable η, putting $\eta = \omega(\lambda)$, $\lambda = \omega^*(\eta)$, where ω and ω^* are functions regular in any case in the circle $|\eta| < 1$, so that: 1) the origin $\lambda = 0$ corresponds to the origin $\eta = 0$; 2) the points $\eta_1, \eta_2, \ldots$, to which $\lambda_1, \lambda_2, \ldots$ move, are $\geqslant 1$ in modulus: $|\eta_1| \geqslant 1$, $|\eta_2| \geqslant 1, \ldots$; 3) for the value of λ of interest to us (for which the equation must be solved) we have the corresponding $|\eta| = |\omega(\lambda)| < 1$.

If we know such a function $\eta = \omega(\lambda)$, then on replacing λ by $\omega^*(\eta)$ in the series for the resolvent or the solution and expanding the series obtained in powers of η, we will obtain a series having, on the strength of condition 2, a radius of convergence $\geqslant 1$, and therefore certainly convergent for the value of η of interest to us, for which $|\eta| < 1$.

We shall give two cases of such a transformation.

1. Let it be known that all the proper values are situated in the half-plane Re $[\lambda] \leqslant -1$. In this case let us transform the variable by putting:

$$\eta = \frac{\lambda}{\lambda + 2}, \quad \lambda = \frac{2\eta}{1 - \eta} = 2[\eta + \eta^2 + \eta^3 + \ldots],$$

i.e., by taking for $\omega(\lambda)$ the function mapping the half-plane Re $[\lambda] \geqslant -1$ on the unit circle $|\eta| \leqslant 1$.

On introducing this substitution, we find for Γ, for instance, the following series:

$$\begin{aligned}
\lambda\Gamma(x, y, \lambda) &= \lambda K(x, y) + \lambda^2 K_2(x, y) + \ldots = \\
&= \frac{2\eta}{1-\eta} K + \left(\frac{2\eta}{1-\eta}\right)^2 K_2 + \left(\frac{2\eta}{1-\eta}\right)^3 K_3 + \ldots = \\
&= 2(\eta + \eta^2 + \eta^3 + \ldots)K + 4(\eta^2 + 2\eta^3 + 3\eta^4 + \ldots)K_2 + \\
&\quad + 8(\eta^3 + 3\eta^4 + 6\eta^5 + \ldots)K_3 + \ldots = \\
&= 2K\eta + (2K + 4K_2)\eta^2 + (2K + 8K_2 + 8K_3)\eta^3 + \ldots
\end{aligned}$$

This series converges for $|\eta| < 1$, which corresponds to all λ of the half-plane Re $[\lambda] > -1$. In particular, for the determination of $\Gamma(x, y, 1)$ one must take $\lambda = 1$, and therefore $\eta = \dfrac{\lambda}{\lambda + 2} = \tfrac{1}{3}$; accordingly we obtain a rapidly converging series.

2. Let us assume that all the proper numbers are real and $\leqslant -1$. In this case as the function $\omega(\lambda)$ we adopt the function mapping the plane with a cut from $\lambda = -1$ to $-\infty$ on the unit circle, viz.:

$$\eta = \omega(\lambda) = \frac{\sqrt{\lambda+1}-1}{\sqrt{\lambda+1}+1}.$$

In such a case the power series in η obtained after substitution will converge still more rapidly than the preceding; for example $\lambda = 1$ will correspond to $\eta = \dfrac{\sqrt{2}-1}{\sqrt{2}+1} = 3 - 2\sqrt{2} \approx \frac{1}{6}$. For the construction of this series, as the solution of the preceding equation with respect to λ shows, it turns out to be necessary to substitute for λ in the power series in λ, the expression:

$$\lambda = \left(\frac{\eta+1}{1-\eta}\right)^2 - 1 = (\eta+1)^2(1 + 2\eta + 3\eta^2 + \ldots) - 1 =$$
$$= 4(\eta + 2\eta^2 + 3\eta^3 + \ldots),$$

and to expand in powers of η the series obtained as the result.

While employing the above methods one may facilitate the computation of the coefficients of the series by the utilization of the table exhibited below (p. 139, Table 2).

§ 3. THE APPLICATION OF INTEGRAL EQUATIONS TO THE SOLUTION OF THE DIRICHLET PROBLEM

1. An integral equation of potential theory. At the outset we shall set forth the derivation of this integral equation.

Let us note as a preliminary that if we denote by r the distance between the points $P(x, y)$ and $M(\xi, \eta)$, then $\ln r$, regarded as a function of the variables x and y, is a harmonic function. Of this one can convince oneself, verifying by direct substitution that the function

$$u = \ln r = \ln \sqrt{(x-\xi)^2 + (y-\eta)^2} = \tfrac{1}{2} \ln [(x-\xi)^2 + (y-\eta)^2] \quad (1)$$

satisfies the Laplace equation

$$\Delta u = \frac{\partial^2 u}{\partial x^2} + \frac{\partial^2 u}{\partial y^2} = 0. \quad (2)$$

The derivatives of the function u with respect to the parameters ξ and η will also be harmonic; for instance, for $\dfrac{\partial u}{\partial \xi}$ we have:

$$\frac{\partial^2}{\partial x^2}\left(\frac{\partial u}{\partial \xi}\right) + \frac{\partial^2}{\partial y^2}\left(\frac{\partial u}{\partial \xi}\right) = \frac{\partial}{\partial \xi}\left(\frac{\partial^2 u}{\partial x^2} + \frac{\partial^2 u}{\partial y^2}\right) = 0.$$

Hence it is clear that the derivative of this function along any constant direction L with respect to the variables ξ and η, being a linear combination of $\frac{\partial u}{\partial \xi}$ and $\frac{\partial u}{\partial \eta}$, is also a harmonic function.

If we denote by φ the angle formed by the vector $\overrightarrow{PM}$ and the direction L, then this derivative, as is evident geometrically, will be

$$\frac{\partial u}{\partial L} = \frac{1}{r}\frac{\partial r}{\partial L} = \frac{\cos \varphi}{r}.$$

Thus $\frac{\cos \varphi}{r}$ is an harmonic function.

After this preliminary observation let us proceed to the consideration of the Dirichlet problem. Let a finite region D be given, bounded by a simple closed contour L. Let us assume that the equation of the contour is given in parametric form; for the parameter let the arc s be adopted:

$$x = x(s), \quad y = y(s) \quad (0 \leqslant s \leqslant s_0);$$

and let it moreover be assumed that the functions $x(s)$ and $y(s)$ have continuous derivatives $x'(s)$, $y'(s)$ not vanishing simultaneously. It is required to find a function u, harmonic in the region D, and which at the contour becomes an assigned continuous function of the arc, $f(s)$:

$$u = f(s) \text{ on } L.$$

Let us take any point $M(\xi, \eta)$ on the contour L and denote by φ the angle formed by the normal constructed at the point M and the vector $\overrightarrow{PM}$ produced from the point $P(x, y)$. Then, by what has been stated above, the function $\frac{\cos \varphi}{r}$ will be an harmonic function of the variables x and y in the region D. Now let $\mu(s)$ be an arbitrary continuous function; in such a case the integral

$$V(x, y) = \int_0^{s_0} \frac{\cos \varphi}{r}\mu(s)ds \quad (3)$$

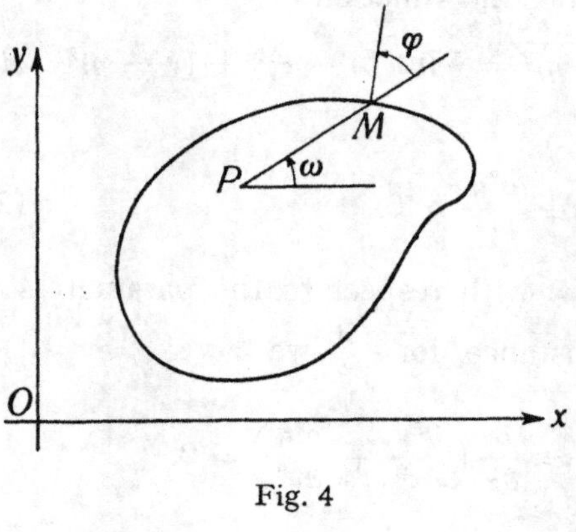

Fig. 4

is an harmonic function of the variables x and y in the region D. For indeed, in calculating ΔV, we can perform the differentiation under the integral sign; but then, since $\mu(s)$ does not depend on x and y and $\frac{\cos \varphi}{r}$ is an harmonic function, we will obtain $\Delta V = 0$.

The function V is called the potential of a double layer, and $\mu(s)$ is called its density. We shall attribute another form to the expression for V in order to make its structure clearer geometrically. Denote by ω the angle that the vector $\overrightarrow{PM}$ (Fig. 4) forms with the positive direction of the axis OX. If the position of the point M be varied, angle ω is a function of s, and $d\omega$ is the angle subtended by the element of arc ds at the point P. But the projection of ds on the perpendicular to the vector $\overrightarrow{PM}$ is $ds \cos\varphi$, and therefore the angle subtended by ds at the point P is $\frac{\cos\varphi\, ds}{r}$; so:

$$\frac{\cos\varphi}{r}\, ds = d\omega.$$

The expression for V may therefore be given the form:

$$V = \int_0^{s_0} \frac{\cos\varphi}{r}\, \mu(s) ds = \int_L \mu(s) d\omega = \int_0^{s_0} \frac{d\omega}{ds}\, \mu(s) ds. \tag{4}$$

ω is thereby a function of s and of x and y as well: $\omega = \omega(s, x, y)$; s_0 denotes the total length of the contour L.

Let us take, in particular, the density $\mu(s) = 1$; then $V = \int_L d\omega$ obviously gives the angle subtended by the entire curve L at the point P. This angle is clearly equal to 2π if the point P is within D, is equal to π if the point P lies on the contour L, and, lastly, is equal to zero if P is outside the contour L:

$$\int_0^{s_0} \frac{d\omega}{ds}\, ds = \begin{cases} 2\pi & \text{if } P \text{ is within } D, \\ \pi & \text{,, } P \text{ is on } L, \\ 0 & \text{,, } P \text{ is outside } L. \end{cases} \tag{5}$$

We shall try to find the solution of the Dirichlet problem in the form of the potential of a double layer, that is, we shall try to find a density $\mu(s)$ such that the function

$$u(x, y) = \int_0^{s_0} \mu(s)\, \frac{d\omega}{ds}\, ds \tag{6}$$

will have, on approaching the contour, the value $f(s)$.

Let us determine the limit of $u(x, y)$ when the point $P(x, y)$ tends to a point P_0 on the contour corresponding to an arc equal to σ, i.e., to a point $P_0[x(\sigma), y(\sigma)]$. We have:

$$\lim_{P\to P_0} u(x, y) = \lim_{P\to P_0} \int_0^{s_0} \mu(s)d\omega =$$

$$= \lim_{P\to P_0} \left[\int_0^{s_0} [\mu(s) - \mu(\sigma)] \frac{d\omega(s, x, y)}{ds} ds + \mu(\sigma) \int_0^{s_0} \frac{d\omega(s, x, y)}{ds} ds \right] =$$

$$= \int_0^{s_0} [\mu(s) - \mu(\sigma)] \frac{d\omega(s, \sigma)}{ds} ds + 2\pi\mu(\sigma) =$$

$$= \int_0^{s_0} \mu(s) \frac{d\omega(s, \sigma)}{ds} ds + \pi\mu(\sigma) \text{ [1]}, \tag{7}$$

since $\int_0^{s_0} \mu(\sigma)d\omega = \pi\mu(\sigma)$. Thus we must have:

$$\pi\mu(\sigma) + \int_0^{s_0} \mu(s) \frac{d\omega}{ds} ds = f(\sigma). \tag{8}$$

The equation obtained is indeed the integral equation for the determination of the function μ. If we yet divide it by π, we bring it into the form:

$$\mu(\sigma) - \int_0^{s_0} K(s, \sigma)\mu(s)ds = \frac{1}{\pi} f(\sigma), \tag{9}$$

where the kernel $K(s, \sigma)$, if we utilize the geometrically obvious expression for the angle $\omega(s, \sigma)$, may be written in the form

$$K(s, \sigma) = -\frac{1}{\pi} \frac{d\omega}{ds} = -\frac{1}{\pi} \frac{d}{ds} \operatorname{arctg} \frac{y(s) - y(\sigma)}{x(s) - x(\sigma)} . \tag{10}$$

We observe that an integral equation could have been constructed in exactly the same manner had we taken for the parameter, in the equation of the curve L, not the arc s but some other parameter t. The equation would then have had the form:

$$\mu(\tau) + \int_0^{t_0} \frac{1}{\pi} \frac{d}{dt} \operatorname{arctg} \frac{y(t) - y(\tau)}{x(t) - x(\tau)} \mu(t)dt = \frac{1}{\pi} f(t). \tag{11}$$

If we introduce the parameter in front of the integral in equation (9),

$$\mu(\sigma) - \lambda \int_0^{s_0} K(s, \sigma)\mu(s)ds = \frac{1}{\pi} f(\sigma), \tag{12}$$

---

[1]) Here $\omega(s, \sigma)$ is a brief notation for $\omega(s, x(\sigma), y(\sigma))$.

then it can be shown that the value $\lambda = 1$ which is of interest to us is not a proper number for the given integral equation. Therefore on the basis of the general theory of integral equations, equation (9) has a unique solution for any free term $f(\sigma)$.

For the approximate solution of the equation an essential role is played by the distribution of the proper values of this equation. First of all it can be shown that they are all real. Moreover $\lambda = -1$ is a proper value for this equation, since, obviously, a constant $\mu(s) = C$ satisfies the homogeneous equation corresponding to equation (9), for, as we saw in (5) above:

$$\pi C - \int_0^{s_0} C \frac{d\omega(s, \sigma)}{ds} ds = 0.$$

Lastly it can be established that in the interval $(-1, 1)$ there are no proper values [1]).

The fact that $\lambda = 1$ is not a proper value of the kernel, and that between the points -1 and 1 there are no proper values, permits us to affirm that all the devices for analytic continuation indicated in the preceding section are applicable.

In particular, the device for continuation by means of a premultiplication leads to the following series for the sought solution:

$$\mu(\sigma) = \tfrac{1}{2}\mu_0(\sigma) + \tfrac{1}{2}[\mu_0(\sigma) + \mu_1(\sigma)] + \tfrac{1}{2}[\mu_1(\sigma) + \mu_2(\sigma)] + \ldots, \quad (13)$$

where

$$\mu_0(\sigma) = \frac{1}{\pi} f(\sigma),$$

$$\mu_{n+1}(\sigma) = \int_0^{s_0} K(s, \sigma)\mu_n(s)ds = -\frac{1}{\pi}\int_0^{s_0} \frac{d\omega(s, \sigma)}{ds} \mu_n(s)ds. \quad (14)$$

We shall say a few words more about the third device. According to this device, we must introduce into the series giving the expansion of the solution in powers of λ, a new parameter η, setting $\lambda = \varphi(\eta)$ and re-expanding the series in powers of η.

There were indicated there as two possible substitutions

$$\lambda = \frac{2\eta}{1-\eta} \text{ and } \lambda = \left(\frac{\eta+1}{1-\eta}\right)^2 - 1.$$

Both of these substitutions give a satisfactory result for the given problem. Their application presents no peculiarities whatever in comparison with the general case, and in addition we shall encounter them below in an example; therefore we do not dwell in greater detail on them here. In the following No. we shall consider in detail another substitution.

---

[1]) See Goursat, [1], Vol. III, Part 2, p. 172.

2. Neumann's method. Here the substitution $\lambda = \dfrac{\eta}{1-\eta}$ is of interest to us. This substitution carries $\lambda = 1$ to $\eta = \frac{1}{2}$; $\lambda = -1$ to $\eta = \infty$. Finally to the interval $-\frac{1}{2} \leqslant \eta \leqslant \frac{1}{2}$ will correspond $-\frac{1}{3} \leqslant \lambda \leqslant 1$.

As is clear from what has been said above, in the latter interval there are no proper values; therefore the power series in η will have a radius of convergence $> \frac{1}{2}$, and will certainly converge for the value of interest to us, $\eta = \frac{1}{2}$. To obtain this power series in η we shall not use the power series in λ, but shall introduce the parameter η directly into the given integral equation. We then have:

$$\mu(\sigma) - \frac{\eta}{1-\eta} \int_0^{s_0} K(s, \sigma)\mu(s)ds = \frac{1}{\pi} f(\sigma),$$

$$(1-\eta)\mu(\sigma) - \eta \int_0^{s_0} K(s, \sigma)\mu(s)ds = \frac{1-\eta}{\pi} f(\sigma)$$

or, using the fact that $\int_0^{s_0} K(s, \sigma)ds = -1$ [see (5) and (10)], we can rewrite it once again still differently:

$$\mu(\sigma) - \eta \int_0^{s_0} K(s, \sigma)[\mu(s) - \mu(\sigma)]ds = \frac{1-\eta}{\pi} f(\sigma)\,^{1)}, \tag{15}$$

where $\eta = \frac{1}{2}$ must be taken. If we seek the solution of this equation by the method of successive approximations, we shall obtain it in the following form:

$$\mu(\sigma) = \mu_0(\sigma) + \mu_1(\sigma) + \mu_2(\sigma) + \ldots, \tag{16}$$

where

$$\left.\begin{aligned} \mu_0(\sigma) &= \frac{1}{2\pi} f(\sigma), \\ \mu_{n+1}(\sigma) &= \frac{1}{2} \int_0^{s_0} K(s, \sigma)[\mu_n(s) - \mu_n(\sigma)]ds = \\ &= \frac{1}{2\pi} \int_0^{s_0} \frac{d\omega(s, \sigma)}{ds} [\mu_n(\sigma) - \mu_n(s)]ds. \end{aligned}\right\} \tag{17}$$

[1]) One can arrive at equation (15) even more simply, viz.: by equating expression (7) to $f(s)$ before the last transformation, we would have obtained equation (15) for $\eta = \frac{1}{2}$.

Series (16) for the solution of the Dirichlet problem was proposed by Neumann [1]) as long ago as 1877. For the case of a convex contour its convergence was also established by Neumann himself, by completely elementary means [2]).

We shall present briefly this proof. As we said above, we shall consider the contour to be convex, and this moreover in the narrow sense, i.e., without straight segments. We shall make some preliminary observations.

1. All the functions $\mu_n(\sigma)$ are continuous.

For $\mu_0(\sigma) = \dfrac{1}{2\pi} f(\sigma)$ this is valid on assumption. Next let $\mu_n(\sigma)$ be continuous; in order to satisfy ourselves of the continuity of the function $\mu_{n+1}(\sigma)$ for $\sigma = \sigma_0$, let us consider its expression:

$$\mu_{n+1}(\sigma) = \frac{1}{2\pi}\int_0^{s_0} \frac{d\omega(s,\sigma)}{ds}[\mu_n(\sigma) - \mu_n(s)]ds =$$

$$= \frac{1}{2\pi}\int_0^{s_0} [\mu_n(\sigma) - \mu_n(\sigma_0)]\frac{d\omega(s,\sigma)}{ds} ds +$$

$$+ \frac{1}{2\pi}\int_0^{s_0} [\mu_n(\sigma_0) - \mu_n(s)]\frac{d\omega(s,\sigma)}{ds} ds =$$

$$= \tfrac{1}{2}[\mu_n(\sigma) - \mu_n(\sigma_0)] + \frac{1}{2\pi}\int_0^{s_0} [\mu_n(\sigma_0) - \mu_n(s)]\frac{d\omega(s,\sigma)}{ds} ds.$$

Here the first term is continuous on assumption, and the second is also continuous for $\sigma = \sigma_0$, thanks to the fact that the first factor $[\mu_n(\sigma_0) - \mu_n(s)]$ of the integrand vanishes for $s = \sigma_0$.

2. Let the function $f(s)$ be positive and bounded: $0 \leqslant f(s) \leqslant K$; then

$$|J| = \left|\int_L f(s)\left[\frac{d\omega(s,\sigma_1)}{ds} - \frac{d\omega(s,\sigma_2)}{ds}\right]ds\right| \leqslant h\pi K, \tag{18}$$

where $h < 1$ is a constant independent of the choice of the points σ_1, σ_2, and the function $f(s)$. Indeed, let us separate the contour C into two

---

[1]) See Goursat, [1], Vol. III, Part I, p. 170.

[2]) It is interesting to note that the series that is obtained if the solution of equation (15) is sought in the form of a power series in η, and the terms of which are linear combinations of two neighboring terms of series (16), will be convergent in the case of any region with a smooth contour, as follows from the reasoning set forth above.

parts: L', where the inequality

$$\frac{d\omega(s, \sigma_1)}{ds} \geqslant \frac{d\omega(s, \sigma_2)}{ds}$$

is fulfilled, and L'', where

$$\frac{d\omega(s, \sigma_1)}{ds} < \frac{d\omega(s, \sigma_2)}{ds}.$$

Then it is obvious that $|J|$ will turn out to be maximum if we take the function $f(s)$ equal to K on L' and to zero on L'', or conversely. For definiteness let us dwell on the first case; we have

$$|J| = K \int_{L'} \left[\frac{d\omega(s, \sigma_1)}{ds} - \frac{d\omega(s, \sigma_2)}{ds} \right] ds.$$

The integral on the right denotes geometrically the difference of the angles subtended by the part L' of the contour L at the points corresponding to $s = \sigma_1$ and $s = \sigma_2$. It is obvious that each term of this difference lies between 0 and π, simultaneous equality signs being excluded. Thus for each given σ_1 and σ_2:

$$\int_{L'} \left[\frac{d\omega(s, \sigma_1)}{ds} - \frac{d\omega(s, \sigma_2)}{ds} \right] ds < \pi.$$

Moreover, the quantity standing on the left is a continuous function of σ_1 and σ_2, and its upper bound is therefore $< \pi$. Let us denote this upper bound by $h\pi$ $(h < 1)$; then by the foregoing,

$$|J| \leqslant Kh\pi,$$

and inequality (18) is established.

We now present proof of the convergence of series (16), for which we shall estimate $\mu_i(\sigma)$. Let $m \leqslant \mu_0(\sigma) \leqslant M \left[\mu_0(\sigma) = \frac{1}{2\pi} f(\sigma) \right]$, then

$$\mu_1(\sigma_2) - \mu_1(\sigma_1) = \frac{1}{2\pi} \int_0^{s_0} \left\{ [\mu_0(\sigma_2) - \mu_0(s)] \frac{d\omega(s, \sigma_2)}{ds} - \right.$$

$$\left. - [\mu_0(\sigma_1) - \mu_0(s)] \cdot \frac{d\omega(s, \sigma_1)}{ds} \right\} ds = \frac{1}{2\pi} \left[\mu_0(\sigma_2) \int_0^{s_0} \frac{d\omega(s, \sigma_2)}{ds} ds - \right.$$

$$\left. - \mu_0(\sigma_1) \int_0^{s_0} \frac{d\omega(s, \sigma_1)}{ds} ds \right] + \frac{1}{2\pi} \int_0^{s_0} [\mu_0(s) - m] \times$$

$$\times \left[\frac{d\omega(s, \sigma_1)}{ds} - \frac{d\omega(s, \sigma_2)}{ds} \right] ds + \frac{m}{2\pi} \int_0^{s_0} \left[\frac{d\omega(s, \sigma_1)}{ds} - \frac{d\omega(s, \sigma_2)}{ds} \right] ds.$$

Here the first term is equal to $\frac{1}{2}[\mu_0(\sigma_2) - \mu_0(\sigma_1)]$ and does not exceed in modulus $\frac{1}{2}(M - m)$. In the second term the function $[\mu_0(s) - m]$ can play the role of $f(s)$ in proposition 2, and therefore the second integral does not exceed $\frac{1}{2}h(M - m)$; lastly, the third integral is equal to zero. Thus

$$|\mu_1(\sigma_2) - \mu_1(\sigma_1)| \leqslant \tfrac{1}{2}(M - m) + \tfrac{1}{2}h(M - m) = \varrho(M - m),$$

where

$$\varrho = \frac{1 + h}{2} < 1.$$

In exactly the same fashion we can obtain in succession:

$$|\mu_i(\sigma_2) - \mu_i(\sigma_1)| \leqslant (M - m)\varrho^i. \tag{19}$$

Hence we conclude:

$$|\mu_i(\sigma)| = \left| \frac{1}{2\pi} \int_0^{s_0} \frac{d\omega(s, \sigma)}{ds} [\mu_{i-1}(\sigma) - \mu_{i-1}(s)] ds \right| <$$

$$< (M - m)\varrho^{i-1} \left| \frac{1}{2\pi} \int_0^{s_0} \frac{d\omega(s, \sigma)}{ds} \right| = \frac{M - m}{2} \varrho^{i-1}. \tag{20}$$

On the basis of the inequality obtained, both the uniform convergence of series (16) and the fact that the solution obtained, $\mu(\sigma)$, is continuous, are now clear. Of the fact that $\mu(\sigma)$ satisfies the stipulated equation, we can satisfy ourselves by direct substitution, in any case.

We shall yet note the following inequality:

$$|\mu(\sigma) - \mu_0(\sigma)| = |\mu_1(\sigma) + \mu_2(\sigma) + \ldots| \leqslant$$

$$\leqslant \frac{M - m}{2}(1 + \varrho + \varrho^2 + \ldots) = \frac{M - m}{2(1 - \varrho)}. \tag{21}$$

We shall now show that if we assume the curve L to have everywhere a finite radius of curvature, it will be possible to make a quite precise evaluation of the quantities h and ϱ in terms of known geometrical elements [1]).

For this purpose we shall first of all estimate from below $\frac{d\omega(s, \sigma)}{ds}$, which for a convex region represents a positive quantity. Let us denote by $\theta = \theta(s)$ the angle between the tangent to the curve L and the posi-

---

[1]) See N. M. Krylov and N. N. Bogoliubov, [1], where the reasoning set forth below is contained in part.

tive direction of the axis OX. First, it is clear that

$$\left[\frac{d\omega(s,\sigma)}{ds}\right]_{s=\sigma} = \frac{1}{2}\frac{d\theta}{ds} = \frac{1}{2R_\sigma},$$

where R_σ is the radius of curvature of the curve L at the point corresponding to $s = \sigma$. Indeed, geometrically $\left[\frac{d\omega}{ds}\right]_{s=\sigma}$ is the rate of turn of the secant with respect to the arc; thus it is obviously equal to half the curvature.

Moreover if the minimum of $\frac{d\omega}{ds}$ is not attained for $s = \sigma$, at such a point we must have $\frac{d^2\omega}{ds^2} = 0$. Let us calculate this second derivative. Denote by $r_{s,\sigma}$ the length of the vector from the point σ to s, and by $\omega_{s,\sigma} = \omega(s, \sigma)$, as before, the angle formed by it and the axis OX; we have:

$$\omega_{s,\sigma} = \operatorname{arctg}\frac{y(s) - y(\sigma)}{x(s) - x(\sigma)},$$

$$\frac{d\omega_{s,\sigma}}{ds} = \frac{1}{1 + \left[\frac{y(s) - y(\sigma)}{x(s) - x(\sigma)}\right]^2} \cdot \frac{[x(s) - x(\sigma)]\frac{dy}{ds} - [y(s) - y(\sigma)]\frac{dx}{ds}}{[x(s) - x(\sigma)]^2} =$$

$$= \frac{[x(s) - x(\sigma)]\sin\theta - [y(s) - y(\sigma)]\cos\theta}{r^2_{s,\sigma}} =$$

$$= \frac{1}{r_{s,\sigma}}(\cos\omega_{s,\sigma}\sin\theta - \sin\omega_{s,\sigma}\cos\theta) = \frac{\sin(\theta - \omega_{s,\sigma})}{r_{s,\sigma}},$$

$$\frac{d^2\omega}{ds^2} =$$

$$= \frac{r\cos(\theta-\omega)\left[\frac{d\theta}{ds} - \frac{d\omega}{ds}\right] - \sin(\theta-\omega)\frac{1}{r}\{[x(s)-x(\sigma)]\cos\theta + [y(s)-y(\sigma)]\sin\theta\}}{r^2} =$$

$$= \frac{r\cos(\theta-\omega)\frac{d\theta}{ds} - \cos(\theta-\omega)\sin(\theta-\omega) - \sin(\theta-\omega)\cos(\theta-\omega)}{r^2} =$$

$$= \frac{\cos(\theta-\omega)}{r^2}\left[r\frac{1}{R_s} + 2\sin(\omega - \theta)\right].$$

Hence it is clear that $\frac{d^2\omega_{s,\sigma}}{ds^2}$ vanishes only if $\cos(\omega_{s,\sigma} - \theta) = 0$ or $\sin(\omega_{s,\sigma} - \theta) = -\frac{r_{s,\sigma}}{2R_s}$; but then $\frac{d\omega_{s,\sigma}}{ds}$ reduces to $\frac{1}{r_{s,\sigma}}$ or $\frac{1}{2R_s}$

respectively. The last value, as we saw, is taken by $\frac{d\omega_{s,\sigma}}{ds}$ at $s = \sigma$ in particular. Thus it is clear that if we denote by d_0 the maximum distance between two points of the curve, and by R_0 the maximum radius of curvature, the minimum $\frac{d\omega}{ds}$ will be $\geqslant$ the lesser of the two numbers $\frac{1}{d_0}$ and $\frac{1}{2R_0}$. It is geometrically obvious, however, that $\frac{1}{2R_0} \leqslant \frac{1}{d_0}$, and therefore, setting $\frac{1}{2R_0} = \alpha$, we can affirm that for all values of s and σ under consideration:

$$\frac{d\omega_{s,\sigma}}{ds} \geqslant \alpha. \tag{22}$$

After having obtained the estimate for $\frac{d\omega}{ds}$, it is easy to give the number $h < 1$ about which we spoke in proposition 2.

Indeed, let L' be some part of the contour L. Denote by L'' that part of L which remains after removing L'. Moreover let s' be the total length of L' and $s'' = s_0 - s'$ the length of L''. Then we have:

$$J = \int_{L'} \left[\frac{d\omega(s, \sigma_1)}{ds} - \frac{d\omega(s, \sigma_2)}{ds}\right] ds =$$

$$= \int_{L} \frac{d\omega(s, \sigma_1)}{ds} ds - \int_{L''} \frac{d\omega(s, \sigma_1)}{ds} - \int_{L'} \frac{d\omega(s, \sigma_2)}{ds} ds \leqslant \pi - s''\alpha - s'\alpha =$$

$$= \pi\left(1 - \frac{\alpha s_0}{\pi}\right) = \pi\left(1 - \frac{s_0}{2\pi R_0}\right).$$

In quite the same manner the analogous inequality for $(-J)$ can be established, and therefore

$$|J| \leqslant \pi\left(1 - \frac{s_0}{2\pi R_0}\right) = \pi h. \tag{23}$$

Consequently the number $1 - \frac{s_0}{2\pi R_0}$ can be taken as h; the subtrahend here has a simple geometric signification: it is the ratio of the length of the curve L to the maximum length of the circumference of the circle of curvature.

For the Dirichlet problem there may be employed the method considered in § 1, by which an integral equation is replaced by a system of ordinary linear equations. In so doing, one can successfully give an estimate for the error incurred owing to the application of the method, by using the estimate arrived at here for the upper bound of the sought function (21).

3. The method of N. M. Krylov and N. N. Bogoliubov. We shall consider in greater detail here one of the possible methods of replacing an integral equation by a system of algebraic equations, namely the method proposed by N. M. Krylov and N. N. Bogoliubov [1]). This method consists in replacing the integral figuring in the integral equation

$$\mu(\sigma) + \frac{1}{\pi} \int_0^{s_0} \mu(s) \frac{d\omega(s, \sigma)}{ds} ds = f(\sigma) \tag{24}$$

approximately by the following sum:

$$\frac{1}{\pi} \sum_{j=1}^{n} \mu(s_j) \int_{s_j - \frac{\Delta s}{2}}^{s_j + \frac{\Delta s}{2}} \frac{d\omega(s, \sigma)}{ds} ds =$$

$$= \frac{1}{\pi} \sum_{j=1}^{n} \mu(s_j) \left[\omega\left(s_j + \frac{\Delta s}{2}, \sigma\right) - \omega\left(s_j - \frac{\Delta s}{2}, \sigma\right)\right] =$$

$$= \frac{1}{\pi} \sum_{j=1}^{n} \mu(s_j) \Delta\omega(s_j, \sigma).$$

Here Δs is the nth part of the interval $(0, s_0)$: $\Delta s = \frac{s_0}{n}$; s_i are any points separated from each other by the distance Δs, for example:

$$s_1 = \frac{\Delta s}{2}, \quad s_2 = \frac{3}{2}\Delta s; \; \ldots, \quad s_n = \frac{(2n-1)}{2}\Delta s = s_0 - \frac{\Delta s}{2}.$$

This replacement reduces to separating the whole integral into integrals over the intervals $\left(s_j - \frac{\Delta s}{2}, s_j + \frac{\Delta s}{2}\right)$, taking out of each of these integrals the mean value of the function $\mu(s)$, and taking the value of μ at precisely the central point of the interval. Thus we replace the integral in integral equation (24) by a sum, and successively set $\sigma = s_1, s_2, \ldots, s_n$ in the equation obtained; we arrive then at a system of equations for determining the approximate values of μ at the points s_j:

$$\tilde{\mu}(s_i) + \frac{1}{\pi} \sum_{j=1}^{n} \tilde{\mu}(s_j) \Delta_j \omega(s_j, s_i) = f(s_i) \qquad (i = 1, 2, \ldots, n), \tag{25}$$

where we have for brevity put

$$\Delta_j \omega(s_j, s_i) = \int_{s_j - \frac{\Delta s}{2}}^{s_j + \frac{\Delta s}{2}} \frac{d\omega(s, s_i)}{ds} ds.$$

---

[1]) [1]. The estimate of error is there made somewhat differently.

System of equations (25) is convenient in that its coefficients $\Delta_j\omega(s_j, s_i)$ can easily be determined, even from a diagram.

On solving this system, we find approximate values for the sought function at the points s_i: $\tilde{\mu}(s_i)$. Let us denote the respective exact values by $\mu(s_i)$. If we denote by z_i the difference between them,

$$z_i = \mu(s_i) - \tilde{\mu}(s_i),$$

then the numbers z_i, as is easily seen, satisfy the system of equations

$$z_i + \frac{1}{\pi}\sum_{j=1}^{n} \Delta_j\omega(s_j,s_i)z_j = R(s_i), \tag{26}$$

where

$$R(\sigma) = \frac{1}{\pi}\left[\sum_{j=1}^{n} \mu(s_j)\Delta_j\omega(s_j, \sigma) - \int_0^{s_0} \mu(s)\frac{d\omega(s, \sigma)}{ds}\,ds\right]. \tag{27}$$

Let us denote by R^* the maximum of the quantity $R(\sigma)$; we shall concern ourselves below with the estimation of this quantity, and shall estimate the numbers z_i.

Let us first of all transform system (26). By using the equation

$$1 = \frac{1}{\pi}\sum_{j=1}^{n} \Delta_j\omega(s_j, s_i)\ \left(= \frac{1}{\pi}\int_0^{s_0} \frac{d\omega(s, \sigma)}{ds}\,ds\right),$$

we shall rewrite it in the form:

$$z_i = \frac{1}{2\pi}\sum_{j=1}^{n} \Delta_j\omega(s_j, s_i)(z_i - z_j) + \tfrac{1}{2}R(s_i). \tag{28}$$

This system of equations is analogous to the Neumann equation and can be solved by the same method of successive approximations, viz.: we find the solution z_i in the form of a series:

$$z_i = z_i^{(0)} + z_i^{(1)} + \ldots, \tag{29}$$

where

$$z_i^{(0)} = \frac{R(s_i)}{2}, \qquad z_i^{(l+1)} = \frac{1}{2\pi}\sum_{j=1}^{n} \Delta_j\omega(s_j, s_i)(z_i^{(l)} - z_j^{(l)}). \tag{30}$$

The convergence of series (29) can be established by the same method as was used for the Neumann series. Indeed, first of all the inequality analogous to (23) is valid:

$$\left|\sum_{j=1}^{n}{}' \Delta_j\omega(s_j, s_i) - \sum_{j=1}^{n}{}' \Delta_j\omega(s_j, s_k)\right| \leqslant \pi h = \pi\left(1 - \frac{s_0}{2\pi R_0}\right), \tag{31}$$

if Σ' denotes the sum, taken on some j — chosen in any manner — from 1 to n. This inequality is valid, since if we take for L' part of L consisting of the intervals $\left(s_j - \frac{\Delta s}{2}, s_j + \frac{\Delta s}{2}\right)$ corresponding to those j on which the summation is conducted, then inequality (31) is none other than (23). Moreover, analogously to proposition 2, it can be shown that if f_j are any numbers and $0 \leqslant f_j \leqslant K$, then

$$|\sum_{j=1}^{n} [\Delta_j\omega(s_j, s_i) - \Delta_j\omega(s_j, s_k)]f_j| \leqslant \pi hK. \tag{32}$$

To establish this it is sufficient to separate the sum into two, Σ' and Σ'', assigning to the first those terms for which the difference in brackets is positive, and to the second those for which it is negative. Furthermore in evaluating each sum f_i can be replaced by K and inequality (31) used.

Estimates for establishing the convergence of the series can moreover be made just as for the Neumann method, (19) and (20), if we now operate, however, with sums as we did there with integrals. These estimates will give

$$|z_i^{(0)}| \leqslant \frac{R^*}{2}, \ |z_i^{(l)}| \leqslant \frac{R^*}{2}\rho^{l-1}, \ \varrho = \frac{1+h}{2},$$

$$|z_i - z_i^{(0)}| \leqslant \frac{R^*}{2}(1 + \varrho + \varrho^2 + \dots) = \frac{R^*}{2(1-\varrho)},$$

$$|z_i| \leqslant \frac{R^*}{2} + \frac{R^*}{2(1-\varrho)} = \frac{3-h}{2(1-h)}R^* = \frac{2 + \dfrac{s_0}{2\pi R_0}}{2\cdot\dfrac{s_0}{2\pi R_0}}R^* = \tau R^*$$

where τ is a definite constant.

Thus at the points s_i we have

$$|\mu(s_i) - \tilde{\mu}(s_i)| \leqslant \tau R^*. \tag{33}$$

We note in addition that since the $\tilde{\mu}(s_i)$ satisfy just such a system as do the z_i, only with other free terms $f(s_i)$ instead of $R(s_i)$, it is clear that, if N is the upper bound of $|f(s)|$, then

$$|\tilde{\mu}(s_i)| \leqslant \tau N. \tag{34}$$

Furthermore for the exact solution $\mu(s)$, as follows from the reasoning set forth in the exposition of Neumann's method, an analogous estimate will be satisfied:

$$|\mu(s)| \leqslant \tau N. \tag{35}$$

As the result of the solution of system (25), we shall obtain approximate values of the function μ at the points s_i; at the remaining points it is natural to determine it approximately on the basis of (24), as

$$\tilde{\mu}(s) = f(s) - \frac{1}{\pi}\sum_{j=1}^{n} \tilde{\mu}(s_j)\Delta_j\omega(s_j, s). \tag{36}$$

Let us estimate the difference between this approximate solution and the exact, $\mu(s)$. For this purpose we shall write the exact equation

$$\mu(s) = f(s) - \frac{1}{\pi}\sum_{j=1}^{n} \mu(s_j)\Delta_j\omega(s_j, s) + R(s), \tag{37}$$

where $R(s)$ is the remainder term, which in modulus is $\leqslant R^*$. Comparing this equation with the preceding, (36), and using (33), we find:

$$|\mu(s) - \tilde{\mu}(s)| \leqslant R^* + \tau R^*. \tag{38}$$

Let us now occupy ourselves with the estimation of R^*.

For any function $F(s)$ we have:

$$\int_{a-h/2}^{a+h/2} F(s)ds - F(a)\cdot h = \int_{a-h/2}^{a+h/2} \Big(F(a) + (s-a)F'(a) + \\ + \frac{F''(\zeta)}{2}(s-a)^2\Big)\,ds - F(a)h = \tfrac{1}{24}F''(s^*)h^3,$$

where s^* is some mean value: $a - \dfrac{h}{2} \leqslant s^* \leqslant a + \dfrac{h}{2}$. Applying this formula to the function

$$F(s) = [\mu(s) - \mu(s_j)]\,\frac{d\omega(s, \sigma)}{ds},$$

we obtain:

$$\left|\int_{s_j-\Delta s/2}^{s_j+\Delta s/2} [\mu(s) - \mu(s_j)]\,\frac{d\omega(s, \sigma)}{ds}\,ds\right| \leqslant \\ \leqslant \frac{(\Delta s)^3}{24}\Big\{\max\left|\mu''(s)\cdot\frac{d\omega}{ds}\right| + 2\max\left|\mu'(s)\,\frac{d^2\omega}{ds^2}\right| + \\ + 2\max\left|\frac{d^3\omega}{ds^3}\,\mu(s)\right|\Big\} \leqslant \frac{(\Delta s)^3}{24}\,Q.$$

Here the constant Q can easily be determined on the basis of the data of the problem. Indeed, using (35) and then differentiating the original equation, (24), we find

$$|\mu(\sigma)| \leqslant \tau N,$$

$$|\mu'(\sigma)| < \max|f'| + \frac{1}{\pi}\,[\max|\mu(s)|]\int_0^{s_0}\left|\frac{d^2\omega}{ds\,d\sigma}\right|ds = M_1,$$

$$|\mu''(\sigma)| \leqslant \max |f''| + \frac{1}{\pi} [\max |\mu(s)|] \int_0^{s_0} \left| \frac{d^3\omega}{ds\, d\sigma^2} \right| ds = M_2,$$

where M_1, M_2 are, obviously, easily determined constants; moreover, the values of

$$\frac{d^2\omega}{ds^2}, \frac{d^3\omega}{ds^3}$$

can also easily be estimated, and thus the entire expression for Q can be found. On the basis of this,

$$|R(s)| = \left| \frac{1}{\pi} \sum_{j=1}^{n} \left[\int_{s_j - \frac{\Delta s}{2}}^{s_j + \frac{\Delta s}{2}} \mu(s) \frac{d\omega}{ds} ds - \mu(s_j) \int_{s_j - \frac{\Delta s}{2}}^{s_j + \frac{\Delta s}{2}} \frac{d\omega}{ds} ds \right] \right| \leqslant$$

$$\leqslant \frac{1}{\pi} \frac{(\Delta s)^2}{24} Q \cdot n\Delta s = \frac{1}{\pi} \frac{(\Delta s)^2}{24} Q s_0.$$

The last quantity can indeed be adopted for the estimate of R^*, and thus we have:

$$|\mu(s) - \tilde{\mu}(s)| \leqslant R^*(1 + \tau) \leqslant \frac{1 + \tau}{\pi} \frac{(\Delta s)^2}{24} Q s_0.$$

After having found the density $\mu(s)$ by means of the potential of a double layer, we can write the required solution of the Dirichlet problem:

$$\tilde{u}(x, y) = \frac{1}{\pi} \int_0^{s_0} \tilde{\mu}(s) \frac{d}{ds} \omega(s, x, y) ds,$$

where

$$\omega(s, x, y) = \operatorname{arctg} \frac{y(s) - y}{x(s) - x}.$$

Since the exact and the approximate values of u are harmonic functions, their maximum difference is attained when the point (x, y) is on the contour. But the value of the approximate solution on the contour is

$$\tilde{\mu}(\sigma) + \frac{1}{\pi} \int_0^{s_0} \tilde{\mu}(s) \frac{d\omega(s, \sigma)}{ds} ds,$$

and the value of the exact solution is

$$\mu(\sigma) + \frac{1}{\pi} \int_0^{s_0} \mu(s) \frac{d\omega(s, \sigma)}{ds} ds = f(s);$$

hence the difference on the contour will equal

$$|u(x, y) - \tilde{u}(x, y)| \leqslant |\mu(\sigma) - \tilde{\mu}(\sigma)| +$$

$$+ \frac{1}{\pi} \int_0^{s_0} |\tilde{\mu}(s) - \mu(s)| \frac{d\omega(s, \sigma)}{ds} ds \leqslant \frac{1 + \tau}{\pi} \frac{(\Delta s)^2}{12} Q s_0.$$

We shall show, however, that the approximate solution, obtained in the form of an integral, can be approximately replaced by a sum, to wit:

$$\tilde{u}(x, y) = \frac{1}{\pi} \int_0^{s_0} \tilde{\mu}(s) \frac{d}{ds} \omega(s, x, y) =$$

$$= \frac{1}{\pi} \sum_{j=1}^{n} \int_{s_j - \frac{\Delta s}{2}}^{s_j + \frac{\Delta s}{2}} \tilde{\mu}(s) \frac{d}{ds} \omega(s, x, y) ds \cong \frac{1}{\pi} \sum_{j=1}^{n} \tilde{\mu}(s_j) \Delta_j \omega(s_j, x, y),$$

where $\mu(s_j)$ are the approximate values found.

4. Example [1]**.** As an example of the application of the methods expounded in this and in the preceding section, let us consider the Dirichlet problem for the ellipse:

$$x = a \cos t, \ y = b \sin t \ (a \geqslant b), \ 0 \leqslant t \leqslant 2\pi,$$

i.e., the problem of determining a function $u(x, y)$, harmonic within the ellipse, which on the contour becomes an assigned function $g(t)$: $u(a \cos t, b \sin t) = g(t)$. This problem, if the solution is sought in the form of the potential of a double layer

$$u(x, y) = \int_0^{2\pi} \frac{d\omega(t, x, y)}{dt} \mu(t) dt,$$

leads to equation (11):

$$\mu(\tau) + \int_0^{2\pi} \frac{1}{\pi} \frac{d}{dt} \operatorname{arctg} \frac{y(t) - y(\tau)}{x(t) - x(\tau)} \mu(t) dt = \frac{1}{\pi} g(\tau).$$

Let us for brevity set

$$K(t, \tau) = \frac{d}{dt} \operatorname{arctg} \frac{y(t) - y(\tau)}{x(t) - x(\tau)}.$$

---

[1]) This example is considered in the work of Nyström, [1]. See also Bairstow-Berry, [1]; S. G. Mikhlin, [1], p. 35.

Developing this expression, we have:

$$K(t, \tau) = \frac{d}{dt} \operatorname{arctg} \frac{b \sin t - b \sin \tau}{a \cos t - a \cos \tau} = \frac{d}{dt} \operatorname{arctg} \frac{b}{a} \frac{2 \cos \frac{t+\tau}{2} \cdot \sin \frac{\tau - t}{2}}{-2 \sin \frac{t+\tau}{2} \sin \frac{\tau - t}{2}} =$$

$$= -\frac{d}{dt} \operatorname{arctg} \left(\frac{b}{a} \operatorname{ctg} \frac{t+\tau}{2} \right) = \frac{\frac{1}{2}ab}{a^2 \sin^2 \frac{t+\tau}{2} + b^2 \cos^2 \frac{t+\tau}{2}}.$$

Or, assuming $b \neq a$, i.e., excluding the case of the circle, we can give $K(t, \tau)$ the following form, which is convenient for numerical calculations:

$$K(t, \tau) = \frac{\frac{ab}{a^2 - b^2}}{\frac{a^2 + b^2}{a^2 - b^2} - \cos (t + \tau)}.$$

We shall for simplicity assume that the conditions are symmetric with respect to the coordinate axes, i.e., $g(-t) = g(t)$, $g\left(\frac{\pi}{2} + t\right) = g\left(\frac{\pi}{2} - t\right)$; the sought function $\mu(t)$ will then satisfy the same condition. In the integral equation we shall replace the integral by a sum in the following manner. For definiteness we choose 16 points in the interval $(0, 2\pi)$: 4 points from the interval $\left(0, \frac{\pi}{2}\right)$ (let these be t_1, t_2, t_3, t_4); the rest of the points, $t_5, t_6, \ldots, t_{16}$ we take symmetric with respect to the axes:

$$\begin{aligned} t_8 &= \pi - t_1, & t_9 &= \pi + t_1, & t_{16} &= 2\pi - t_1, \\ t_7 &= \pi - t_2, & t_{10} &= \pi + t_2, & t_{15} &= 2\pi - t_2, \\ t_6 &= \pi - t_3, & t_{11} &= \pi + t_3, & t_{14} &= 2\pi - t_3, \\ t_5 &= \pi - t_4, & t_{12} &= \pi + t_4, & t_{13} &= 2\pi - t_4. \end{aligned}$$

By virtue of the symmetry of the problem:

$$\begin{aligned} g(t_1) &= g(t_8) = g(t_9) = g(t_{16}), & g(t_2) &= g(t_7) = g(t_{10}) = g(t_{15}), \\ g(t_3) &= g(t_6) = g(t_{11}) = g(t_{14}), & g(t_4) &= g(t_5) = g(t_{12}) = g(t_{13}), \\ \mu(t_1) &= \mu(t_8) = \mu(t_9) = \mu(t_{16}), & \mu(t_2) &= \mu(t_7) = \mu(t_{10}) = \mu(t_{15}), \\ \mu(t_3) &= \mu(t_6) = \mu(t_{11}) = \mu(t_{14}), & \mu(t_4) &= \mu(t_5) = \mu(t_{12}) = \mu(t_{13}). \end{aligned}$$

We shall consider two quadrature formulas: the Gauss formula and the rectangular formula. We apply the Gauss formula for the interval $(0, \pi)$ with eight ordinates. Then the abscissas t_5, t_6, t_7 and t_8 are themselves symmetric with respect to the points t_1, t_2, t_3, t_4. For the points t_1, t_2, t_3, t_4, having transformed the respective Gaussian abscissas to the interval $(0, \pi)$, we must adopt:

$$t_1 = 3°34'26''.09, \quad t_2 = 18°18'0''.06, \quad t_3 = 42°42'7''.50, \quad t_4 = 73°29'27''.18.$$

For the interval $(\pi, 2\pi)$ we take this same formula of Gauss, but instead of the points $t_1, \ldots, t_8$ we use the points $t_9, \ldots, t_{16}$. The Gauss formula in application to the integral $\int_0^{2\pi} \frac{1}{\pi} K(t, \tau)\mu(t)dt$ gives:

$$\frac{1}{\pi}\int_0^{2\pi} K(t, \tau)\mu(t)dt \approx \sum_{k=1}^{16} a_k K(t_k, \tau)\mu(t_k),$$

where the coefficients a_k, thanks to the presence of the factor $\frac{1}{\pi}$, must be taken the same as for the interval (0, 1) for 8 ordinates, viz.:

$$a_1 = a_8 = a_9 = a_{16} = 0.0506143, \quad a_2 = a_7 = a_{10} = a_{15} = 0.1111905,$$

$$a_3 = a_6 = a_{11} = a_{14} = 0.1568533, \quad a_4 = a_5 = a_{12} = a_{13} = 0.1813419.$$

Substituting the sum for the integral and setting $\tau = t_1, t_2, \ldots$, we obtain the system of equations

$$\tilde{\mu}(t_i) + \sum_{k=1}^{16} a_k K(t_k, t_i)\tilde{\mu}(t_k) = \frac{1}{\pi} g(t_i) \quad (i = 1, 2, \ldots, 16).$$

We have obtained a system of 16 equations for the determination of the 16 unknowns $\tilde{\mu}(t_i)$ — the approximate values of the function μ at the points t_i. However on the strength of the symmetry mentioned above, we can keep the 4 unknowns $\tilde{\mu}(t_1), \tilde{\mu}(t_2), \tilde{\mu}(t_3), \tilde{\mu}(t_4)$ and keep only the 4 first equations, since the remaining equations repeat them.

Grouping the terms in each unknown, we can write the system in the form:

$$\tilde{\mu}(t_i) + \sum_{k=1}^{4} B_{i,k}\tilde{\mu}(t_k) = \frac{1}{\pi} g(t_i) \qquad (i = 1, 2, 3, 4),$$

where by using the expressions for $t_5, t_6, \ldots, t_{16}$ in terms of t_1, t_2, t_3, t_4, we can give the coefficients $B_{i,k}$ the form

$$B_{i,k} = \frac{ab}{a^2 - b^2} a_k \left\{ \frac{1}{\frac{a^2+b^2}{a^2-b^2} - \cos(t_k + t_i)} + \frac{1}{\frac{a^2+b^2}{a^2-b^2} - \cos(\pi - t_k + t_i)} + \right.$$

$$\left. + \frac{1}{\frac{a^2+b^2}{a^2-b^2} - \cos(\pi + t_k + t_i)} + \frac{1}{\frac{a^2+b^2}{a^2-b^2} - \cos(2\pi - t_k + t_i)} \right\} =$$

$$= \frac{ab}{a^2 - b^2} a_k \frac{1}{\frac{a^2+b^2}{a^2-b^2} \mp \cos(t_k \pm t_i)},$$

the last factor here being an abbreviated notation for the sum of the four terms in the braces.

After having determined $\tilde{\mu}(t_1), \tilde{\mu}(t_2), \tilde{\mu}(t_3), \tilde{\mu}(t_4)$ as the result of the solution of the system, we can obtain the approximate value $\tilde{\mu}(t)$ at the rest of the points by replacing the integral in the given integral equation by the sum. We obtain:

$$\tilde{\mu}(\tau) = \frac{1}{\pi} g(\tau) - \frac{ab}{a^2 - b^2} \sum_{i=1}^{4} \frac{a_i \tilde{\mu}(t_i)}{\frac{a^2+b^2}{a^2-b^2} \mp \cos(\tau \pm t_i)}.$$

For numerical calculation we shall dwell on the case $a = 1$, $b = \frac{1}{2}$. Computing the coefficients $B_{i,k}$ for the given case and using the values of t_i and a_i indicated above, we write the system of equations for the determination of the $\tilde{\mu}(t_i)$ in the form:

$$\tilde{\mu}(t_1) \cdot 1.125990 + \tilde{\mu}(t_2) \cdot 0.263037 + \tilde{\mu}(t_3) \cdot 0.311729 + \tilde{\mu}(t_4) \cdot 0.299244 = \frac{1}{\pi} g(t_1),$$

$$\tilde{\mu}(t_1) \cdot 0.119735 + \tilde{\mu}(t_2) \cdot 1.254816 + \tilde{\mu}(t_3) \cdot 0.315974 + \tilde{\mu}(t_4) \cdot 0.309480 = \frac{1}{\pi} g(t_2),$$

$$\tilde{\mu}(t_1) \cdot 0.100591 + \tilde{\mu}(t_2) \cdot 0.223987 + \tilde{\mu}(t_3) \cdot 1.321839 + \tilde{\mu}(t_4) \cdot 0.353576 = \frac{1}{\pi} g(t_3),$$

$$\tilde{\mu}(t_1) \cdot 0.083522 + \tilde{\mu}(t_2) \cdot 0.189758 + \tilde{\mu}(t_3) \cdot 0.305828 + \tilde{\mu}(t_4) \cdot 1.420914 = \frac{1}{\pi} g(t_4).$$

In order to carry the computations through to the end, we shall settle upon definite boundary conditions as well. We shall seek a function $u(x, y)$ satisfying

	$t_1 = 11°15'$	$t_2 = 33°45'$	$t_3 = 56°15'$	$t_4 = 78°45'$	$\mu_0 = g$	μ_1	μ_2
$t_1 = 11°15'$	0.300608	0.266309	0.227517	0.205566	0.6184474	0.4223938	0.4006101
$t_2 = 33°45'$	0.266309	0.261815	0.244358	0.227517	0.4892463	0.4080378	0.3990146
$t_3 = 56°15'$	0.227517	0.244358	0.261815	0.266309	0.3065284	0.3877361	0.3967589
$t_4 = 78°45'$	0.205566	0.227517	0.266309	0.300608	0.1773273	0.3733809	0.3951642

the condition [1])

$$u = \tfrac{1}{2}(x^2 + y^2) \text{ on the ellipse,}$$

or putting the matter differently, in the given case

$$g(t) = \tfrac{1}{2}\,(\cos^2 t + \tfrac{1}{4} \sin^2 t) = \tfrac{1}{16}(5 + 3 \cos 2t).$$

In such a case one must substitute in the right side

$$\frac{1}{\pi} g(t_1) = 0.158691, \quad \frac{1}{\pi} g(t_2) = 0.147386, \quad \frac{1}{\pi} g(t_3) = 0.104254,$$

$$\frac{1}{\pi} g(t_4) = 0.049428.$$

Solving directly the system of four equations cited above, we find:

$$\tilde{\mu}(t_1) = 0.103038, \quad \tilde{\mu}(t_2) = 0.092859, \quad \tilde{\mu}(t_3) = 0.054038, \quad \tilde{\mu}(t_4) = 0.004698.$$

---

[1]) To this condition leads the problem of the torsion of a rod of elliptical section, i.e., of the determination of the solution of the equation $\Delta u = -2$ with zero conditions on the contour. Indeed, introducing $u_1 = u + \frac{1}{2}(x^2 + y^2)$, we see that the function u_1 satisfies the equation $\Delta u_1 = 0$. We recall that for the torsion problem before us the exact solution is easily given, namely:

$$u = -\frac{1}{5}\left(\frac{x^2}{1} + \frac{y^2}{\frac{1}{4}} - 1\right).$$

By substituting these values in the expression exhibited above for $\tilde{\mu}(t)$, $\tilde{\mu}$ can be determined at the remaining points as well.

In this problem the exact solution of the integral equation can be obtained; it is

$$\mu(t) = \frac{1}{160\pi}(25 + 27\cos 2t).$$

Computing, for comparison, the values of μ at the points t_1, t_2, t_3, t_4, we find:

$$\mu(t_1) = 0.103034,\quad \mu(t_2) = 0.092859,\quad \mu(t_3) = 0.054040,\quad \mu(t_4) = 0.004696.$$

Comparing the values of μ and $\tilde{\mu}$ at other points, we obtain a discrepancy of the same order, for example

$$\tilde{\mu}(0) = 0.103451,\quad \mu(0) = 0.103451,$$

$$\tilde{\mu}\left(\frac{\pi}{2}\right) = -0.003985,\ \mu\left(\frac{\pi}{2}\right) = -0.003979.$$

Table 1

μ_3	μ_4	μ_5	μ_6	μ_7	μ_8	μ_9	μ_{10}
).3891895	0.3979204	0.3978903	0.3978868	0.3978860	0.3978860	0.3978858	0.3978856
).3980119	0.3979003	0.3978877	0.3978861	0.3978858	0.3978856	0.3978854	0.3978852
).3977612	0.3978724	0.3978846	0.3978858	0.3978857	0.3978855	0.3978853	0.3978851
).3975844	0.3978531	0.3978828	0.3978859	0.3978861	0.3978859	0.3978857	0.3978855

Let us employ for the solution of this same example the series given by the method of successive approximations and analytic continuation. Together with this we shall do the requisite quadratures approximately. This method will then be — as we indicated above (p. 113) — none other than the application of the method of iterations for the solution of that system of linear equations to which the solution of the integral equation reduces approximately. We shall here, however, utilize the rectangular rather than the Gauss formula, which is allowable since the functions $K(t, \tau)$ and $g(\tau)$ are periodic. Then the coefficients $B_{i,k}$ can be

Table 2

	$\lambda = \frac{2\eta}{1-\eta}$, $\eta = \frac{1}{3}$		$\lambda = \left(\frac{\eta+1}{1-\eta}\right)^2 - 1$, $\eta = 3 - 2\sqrt{2}$		$\lambda = \frac{\eta}{1-\eta}$, $\eta = \frac{1}{2}$	
	up to η^{10}	up to η^5	up to η^{10}	up to η^5	up to η^{10}	up to η^5
μ_0	1.000000	1.000000	1.000000	1.000000	1.000000	1.000000
μ_1	0.999983	0.995885	0.999998	0.999233	0.999023	0.968750
μ_2	0.999664	0.954732	0.999979	0.980458	0.989258	0.812500
μ_3	0.996596	0.790123	0.999570	0.855813	0.945312	0.500000
μ_4	0.980338	0.460905	0.995129	0.526324	0.828125	0.187500
μ_5	0.923436	0.131687	0.968748	0.152244	0.623047	0.031250
μ_6	0.786872		0.875178		0.376953	
μ_7	0.559264		0.668391		0.171875	
μ_8	0.299141		0.381319		0.054687	
μ_9	0.104049		0.138075		0.010742	
μ_{10}	0.017341		0.023178		0.000976	

found by the same formulas as above, t_i and a_i having, however, other values, viz.:

$$t_1 = \tfrac{1}{16}\pi = 11°15', \quad t_2 = \tfrac{3}{16}\pi = 33°45', \quad t_3 = \tfrac{5}{16}\pi = 56°15',$$

$$t_4 = \tfrac{7}{16}\pi = 78°45', \quad a_1 = a_2 = a_3 = a_4 = \tfrac{1}{8} = 0.125.$$

The successive computation of the functions $\tilde{\mu}_1, \tilde{\mu}_2, \ldots$ will be performed by the scheme indicated on p. 113. We first of all compute the values $\lambda K_{i,k} A_i = B_{i,k}$, and next determine $\mu_1^{(k)}, \mu_2^{(k)}, \ldots$ in succession (we use the notations employed in § 2, No. 1, except that the role of φ is played by μ). The results of the computations are set forth in Table 1 (pp. 138–139).

As we see, the $\mu_n^{(k)}$ do not diminish with increase of n; therefore the series formed directly from the successive approximations cannot be utilized for the computations. Let us employ the various devices for analytic continuation. In all cases, after the respective substitution, we reconstruct the series $\mu = \mu_0 + \lambda\mu_1 + \lambda^2\mu_2 + + \ldots$ in powers of η and afterwards, preserving a certain number of the terms

Table 3

	$\mu^1 = \mu^8 = \mu^9 = \mu^{16}$	$\mu^2 = \mu^7 = \mu^{10} = \mu^{15}$	$\mu^3 = \mu^6 = \mu^{11} = \mu^{14}$	$\mu^4 = \mu^5 = \mu^{12} = \mu^{13}$
$\mu(t) = \frac{1}{160\pi}(25 + 27\cos 2t)$ true solution	0.0993620	0.0702917	0.0291802	0.0001099
Substitution				
$\eta = \frac{\sqrt{\lambda+1}-1}{\sqrt{\lambda+1}+1}$, $\eta = 3-2\sqrt{2}$ up to η^{10}	0.0993621	0.0702919	0.0291803	0.0001099
up to η^5	0.09931	0.07024	0.02913	0.00006
Premultiplication method				
$\frac{1}{2}[\mu_0 + (\mu_0 + \mu_1) + + (\mu_1 + \mu_2) + \ldots + (\mu_4 + \mu_5)]$,	0.0993623	0.0702919	0.0291800	0.0001095
$\frac{1}{2}[\mu_0 + (\mu_0 + \mu_1) + \ldots + (\mu_9 + \mu_{10})]$	0.0993620	0.0702917	0.0291802	0.0001098
Substitution $\lambda = \frac{2\eta}{1-\eta}$, $\eta = \frac{1}{3}$,				
up to η^{10}	0.0993628	0.0702926	0.0291810	0.0001107
up to η^5	0.09916	0.07009	0.02897	0.00010
$\lambda = \frac{\eta}{1-\eta}$, $\eta = \frac{1}{2}$,				
up to η^{10}	0.0993637	0.0702925	0.0291796	0.0001073
up to η^5	0.09946	0.07033	0.02914	0.000014

of the series, we substitute for η the numerical value corresponding to $\lambda = 1$.

We shall keep terms up to η^5 or η^{10}. The expression obtained as the result of the substitution of the value of η is the sum of the μ_i, provided with various numerical coefficients. In Table 2 are set forth the values of the coefficients which are obtained upon making the several substitutions.

With the method of premultiplication, all μ_i must be taken with coefficient 1, with the exception of the last summand, which must be taken with the factor $\frac{1}{2}$. Employing these devices in the given problem and using the computed values of μ_n^j, we obtain the results which are given in Table 3.

As we see, all the devices for continuation yield quite satisfactory results; the best result is given by the simplest device for continuation, namely continuation by premultiplication. We note that the values of the coefficients exhibited in Table 2 are not confined to the given particular example and can be used in other cases too.

§ 4. THE SOLUTION OF INTEGRAL EQUATIONS BY REPLACING AN ARBITRARY KERNEL BY A DEGENERATE ONE

1. The integral equation with degenerate kernel. A kernel is said to be degenerate if it can be represented in the form of a finite sum of products:

$$K(x, y) = \sum_{s=1}^{n} \alpha_s(x)\beta_s(y). \tag{1}$$

For such a kernel it is not hard to give the complete solution of the Fredholm equation

$$\varphi(x) - \lambda \int_a^b K(x, y)\varphi(y)dy = f(x). \tag{2}$$

At the outset we can assume the functions $\alpha_s(x)$ to be linearly independent, for in the contrary case the number of terms in the expression for the kernel could have been reduced.

The solution is naturally sought in the form:

$$\varphi(x) = f(x) + \sum_{i=1}^{n} A_i\alpha_i(x), \tag{3}$$

where the A_i are constants, unknown for the time being.

Let us introduce the designations

$$\int_a^b f(x)\beta_i(x)dx = f_i, \qquad \int_a^b \alpha_i(x)\beta_j(x)dx = \beta_{j,i}. \tag{4}$$

Substituting the expression (3) for $\varphi(x)$ in the equation, we must have

$$\sum_{i=1}^{n} A_i\alpha_i(x) - \lambda \int_a^b \left[\sum_{s=1}^{n} \alpha_s(x)\beta_s(y)\right] f(y)dy -$$
$$- \lambda \int_a^b \sum_{s=1}^{n} \sum_{j=1}^{n} A_j\alpha_s(x)\beta_s(y)\alpha_j(y)dy = 0. \tag{5}$$

Hence, equating the coefficients of $\alpha_i(x)$, we find:

$$A_i - \lambda \sum_{j=1}^{n} A_j B_{i,j} = \lambda f_i \quad (i = 1, 2, \ldots, n). \tag{6}$$

If we denote by $\Delta(\lambda)$ the determinant of system (6):

$$\Delta(\lambda) = \begin{vmatrix} 1 - \lambda\beta_{1,1}, & -\lambda\beta_{1,2}, & \ldots, & -\lambda\beta_{1,n} \\ -\lambda\beta_{2,1}, & 1 - \lambda\beta_{2,2}, & \ldots, & -\lambda\beta_{2,n} \\ \cdot & \cdot & \cdot & \cdot \\ -\lambda\beta_{n,1}, & -\lambda\beta_{n,2}, & \ldots & 1 - \lambda\beta_{n,n} \end{vmatrix}, \tag{7}$$

and by $\Delta_{i,k}$ the algebraic complement of the element of the kth row and ith column, the solution of the system is expressible in the form:

$$A_i = \lambda \frac{\sum_{k=1}^{n} \Delta_{i,k} f_k}{\Delta(\lambda)}, \tag{8}$$

$$\varphi(x) = f(x) + \lambda \sum_{k=1}^{n} \frac{\sum_{k=1}^{n} \Delta_{i,k} f_k}{\Delta(\lambda)} \alpha_i(x) =$$

$$= f(x) + \lambda \frac{\sum_{k=1}^{n} \left(\sum_{i=1}^{n} \Delta_{i,k}\alpha_i(x)\right) \int_a^b \beta_k(y) f(y) dy}{\Delta(\lambda)} =$$

$$= f(x) + \lambda \int_a^b \frac{\sum_{i=1}^{n} \sum_{k=1}^{n} \Delta_{i,k}\alpha_i(x)\beta_k(y)}{\Delta(\lambda)} f(y) dy =$$

$$= f(x) + \lambda \int_a^b \frac{\Delta(x, y, \lambda)}{\Delta(\lambda)} f(y) dy, \tag{9}$$

where

$$\Delta(x, y, \lambda) = \sum_{i=1}^{n} \sum_{k=1}^{n} \Delta_{i,k}\alpha_i(x)\beta_k(y).$$

Accordingly the resolvent for an equation of this type is obtainable in explicit form:

$$\Gamma(x, y, \lambda) = \frac{\Delta(x, y, \lambda)}{\Delta(\lambda)}. \tag{10}$$

The proper numbers of the kernel are found by solving the equation

$$\Delta(\lambda) = 0.$$

2. Replacement of an arbitrary kernel by a degenerate one. The possibility of conveniently solving equations with a degenerate kernel naturally leads to the thought that in solving an equation with an arbitrary kernel we might replace this kernel approximately by a degenerate one, and solve the resulting equation rather than the original one.

A suitable degenerate kernel close to the given one can be found in many ways. In particular, as such a kernel one can adopt a segment of the Taylor series (see No. 3, Example), a segment of the Fourier series, which as we shall see below (No. 5) is equivalent to the application of the method of moments or a special interpolation device (No. 6). Right now we shall concern ourselves with estimating the error that is obtained owing to the replacement of the given kernel by another kernel close to it, in particular, by a degenerate one.

THEOREM. *Let there be given two kernels, $k(x, y)$ and $K(x, y)$, and let it be known that*

$$\int_a^b |K(x, y) - k(x, y)|\, dy < h, \tag{11}$$

and that the resolvent $\gamma(x, y, \lambda)$ of the equation with kernel $k(x, y)$ satisfies the inequality

$$\int_a^b |\gamma(x, y, \lambda)|\, dy < B, \tag{12}$$

and also that

$$|f(x) - f_1(x)| < \eta.$$

Then if the following condition is satisfied:

$$1 - |\lambda|\, h(1 + |\lambda|\, B) > 0, \tag{13}$$

the equation

$$\varphi(x) - \lambda \int_a^b K(x, y)\varphi(y)dy = f(x) \tag{14}$$

has a unique solution $\varphi(x)$, and the difference between this solution and the solution $\tilde{\varphi}(x)$ of the equation

$$\tilde{\varphi}(x) - \lambda \int_a^b k(x, y)\tilde{\varphi}(y)dy = f_1(x) \tag{15}$$

does not exceed

$$|\varphi(x) - \tilde{\varphi}(x)| < \frac{N\,|\lambda|\, h(1 + |\lambda|\, B)^2}{1 - |\lambda|\, h(1 + |\lambda|\, B)} + \eta(1 + |\lambda|\, B), \tag{16}$$

where N is the upper bound of $|f(x)|$.

PROOF. Let M be the upper bound of $|\varphi(x)|$ and $\tilde{M}$ the upper bound of $|\tilde{\varphi}(x)|$. From (14) we have:

$$\varphi(x) - \lambda \int_a^b k(x, y)\varphi(y)dy =$$

$$= f(x) + \lambda \int_a^b [K(x, y) - k(x, y)]\varphi(y)dy = f^*(x). \quad (17)$$

The right side of this equation does not exceed

$$|f^*(x)| = |f(x) + \lambda \int_a^b [K(x, y) - k(x, y)]\varphi(y)dy| \leqslant$$

$$\leqslant N + |\lambda|\, M \int_a^b |K(x, y) - k(x, y)|\, dy < N + |\lambda|\, Mh:$$

hence its solution $\varphi(x)$ does not exceed the following quantity:

$$|\varphi(x)| = |f^*(x) + \lambda \int_a^b \gamma(x, y, \lambda) f^*(y)dy| <$$

$$< N + |\lambda|\, Mh + |\lambda|\, B(N + |\lambda|\, Mh).$$

Consequently

$$M < N + |\lambda|\, hM + |\lambda|\, B(N + |\lambda|\, Mh),$$

whence

$$M < \frac{N(1 + |\lambda|\, B)}{1 - |\lambda|\, h(1 + |\lambda|\, B)}, \quad (18)$$

the last step being legitimate if the condition is satisfied that

$$1 - |\lambda|\, h(1 + |\lambda|\, B) > 0. \quad (13)$$

Thus, this condition being observed, all solutions of equation (14) are bounded, and therefore equation (14) has a unique solution; the last shows, by the way, that λ is not a proper value of this equation.

Now subtracting equation (15) from (17), we obtain the following equation for the difference $\psi(x) = \varphi(x) - \tilde{\varphi}(x)$:

$$\psi(x) - \lambda \int_a^b k(x, y)\psi(y)dy = f^*(x) - f_1(x).$$

Hence, since

$$|f^*(x) - f_1(x)| \leqslant |f^*(x) - f(x)| + |f(x) - f_1(x)| \leqslant |\lambda|\, Mh + \eta,$$

we find

$$|\psi(x)| = |f^*(x) - f_1(x) + \lambda \int_a^b \gamma(x, y, \lambda)\, [f^*(y) - f_1(y)]dy| \leqslant$$

$$\leqslant |\lambda|\, Mh + \eta + |\lambda|\, B(|\lambda|\, Mh + \eta) = (|\lambda|\, Mh + \eta)(1 + |\lambda|\, B). \quad (19)$$

Now substituting for M its estimate (18), we have

$$|\psi(x)| = |\varphi(x) - \tilde{\varphi}(x)| < \frac{N\,|\lambda|\,h(1 + |\lambda|\,B)^2}{1 - |\lambda|\,h(1 + |\lambda|\,B)} + \eta(1 + |\lambda|\,B),$$

and the required estimate (16) is obtained. We shall remark that it is possible to refine the estimate of the difference $|\varphi(x) - \tilde{\varphi}(x)| = |\psi(x)|$ in the following way. Denoting its maximum by δ, we have, obviously,

$$M \leqslant \tilde{M} + \delta,$$

and, by (19),

$$\delta \leqslant [|\lambda|\,(\tilde{M} + \delta)h + \delta](1 + |\lambda|\,B). \tag{20}$$

Solving this inequality for δ, we find

$$\delta \leqslant \frac{(|\lambda|\,\tilde{M}h + \eta)(1 + |\lambda|\,B)}{1 - |\lambda|\,h(1 + |\lambda|\,B)}. \tag{21}$$

Another method of making the estimate is given below, in No. 4 [1]).

Observation. From the proved theorem it follows that if λ is not a proper number of the kernel $K(x, y)$, and the sequence of kernels $K_n(x, y) \to K(x, y)$ and $f_n(x) \to f(x)$ (uniformly), then

$$\tilde{\varphi}_n(x) \to \varphi(x).$$

Indeed, it is sufficient to utilize estimate (16), formed for the pair of kernels $K_n(x, y)$ and $K(x, y)$ rather than for $K(x, y)$ and $k(x, y)$ respectively.

The proved theorem also gives, in particular, a method of estimating the error arising from the replacement of a kernel by an algebraic kernel. We shall illustrate this in the following simple example.

3. Example. To be solved, the equation

$$\varphi(x) - \int_0^{\frac{1}{2}} \sin xy\,\varphi(y)dy = f(x).$$

As regards $f(x)$ we shall for the time being make no particular assumptions.

Expanding $\sin xy$ in a series, we have:

$$\sin xy = xy - \frac{x^3y^3}{6} + \frac{x^5y^5}{120} - \ldots$$

---

[1]) The method of estimating the error expounded here was first published in the first edition of this monograph. An estimation of the error made on other assumptions than are set forth here is to be found in the dissertation of I. Akbergenov, completed in 1935 at Leningrad NIIMM, [3]. In the same place is given an estimate of the error in the determination of the proper values. The basic results of the dissertation can be found in Akbergenov's work [2]. See also L. V. Kantorovich, [2].

Let us replace sin xy by the first two terms of the series, and consider the corresponding equation with an algebraic kernel:

$$\tilde{\varphi}(x) - \int_0^{\frac{1}{2}} \left(xy - \frac{x^3y^3}{6}\right) \tilde{\varphi}(y)dy = f(x).$$

We seek the solution in the form

$$\tilde{\varphi}(x) = Ax + Bx^3 + f(x).$$

Substituting this expression in the equation and denoting by f_1 and f_2 the integrals

$$f_1 = \int_0^{\frac{1}{2}} yf(y)dy, \quad f_2 = -\tfrac{1}{6}\int_0^{\frac{1}{2}} y^3f(y)dy,$$

we find

$$Ax + Bx^3 - f_1x - f_2x^3 - \tfrac{1}{24}Ax - \tfrac{1}{160}Bx + \tfrac{1}{960}Ax^3 + \tfrac{1}{5376}Bx^3 = 0.$$

Hence by equating to zero the coefficients of x and x^3, we obtain:

$$\tfrac{23}{24}A - \tfrac{1}{160}B = f_1, \quad \tfrac{1}{960}A + (1 + \tfrac{1}{5376})B = f_2.$$

On solving this system, we find

$$A = 1.043277\,[\tfrac{5377}{5376}f_1 + \tfrac{1}{160}f_2], \quad B = 1.043277\,[-\tfrac{1}{960}f_1 + \tfrac{23}{24}f_2].$$

Therefore we have the solution

$$\tilde{\varphi}(x) = f(x) + Ax + Bx^3 =$$

$$= f(x) + \int_0^{\frac{1}{2}} 1.043277\,[1.0001860xy - 0.0010416x^3y - 0.0010416xy^3 -$$

$$- 0.1597222x^3y^3]f(y)dy,$$

and the resolvent:

$$\gamma(x, y, 1) = 1.043277\,[1.0001860xy - 0.0010416x^3y -$$

$$- 0.0010416xy^3 - 0.1597222x^3y^3].$$

Hence it is clear that

$$\int_0^{\frac{1}{2}} |\gamma(x, y, 1)|\, dy < \tfrac{1}{12}.$$

Accordingly in the estimate exhibited above, we can adopt $B = \dfrac{1}{12}$.

Next

$$\int_0^{\frac{1}{2}} |K(x, y) - k(x, y)|\, dy \leqslant \int_0^{\frac{1}{2}} \frac{x^5y^5}{120}\, y = \frac{1}{64 \cdot 720}\, x^5$$

and, since $x \leqslant \frac{1}{2}$, we can adopt $h = \dfrac{1}{64 \cdot 32 \cdot 720} = \dfrac{1}{1474560}$ or, more roughly, $h = \dfrac{3}{4 \cdot 10^6}$.

Then

$$|\tilde{\varphi}(x) - \varphi(x)| < N \frac{|\lambda|\, h(1 + |\lambda|\, B)^2}{1 - |\lambda|\, h(1 + |\lambda|\, B)} = N \frac{\dfrac{3}{4 \cdot 10^6}(1 + \frac{1}{12})^2}{1 - \dfrac{3}{4 \cdot 10^6}(1 + \frac{1}{12})} < \frac{N}{10^6},$$

where N is the upper bound of $|f(x)|$.

In particular, we shall adopt

$$f(x) = 1 + \frac{1}{x}\left(\cos\frac{x}{2} - 1\right) = 1 - \frac{x}{8} + \frac{x^3}{384} - \frac{x^5}{46080} + \ldots$$

The corresponding exact solution is $\varphi(x) = 1$. For the upper bound of $f(x)$ in $(0, \frac{1}{2})$ we can adopt $N = 1$.

The approximate solution, therefore, must differ from the exact by not more than 10^{-6}.

In fact the approximate solution will be

$$\varphi(x) = 1 + 0.0000009x - 0.0000002x^3.$$

4. Another estimate of the error. We have considered in No. 2 a method of estimating the error in the solution of an integral equation that is due to the replacement of the given kernel by another one close to it. Another method of estimating the error had previously been proposed by F. Tricomi [1]). This method gives, though, a considerably less exact estimate of the error than the method of No. 2. It again permits of the establishment of the convergence of the approximate solution to the exact if the difference between the kernels tends to zero.

Thus let us consider the two equations

$$\left.\begin{aligned} \varphi(x) - \lambda \int_0^1 K(x, y)\varphi(y)dy &= f(x), \\ \tilde{\varphi}(x) - \lambda \int_0^1 \tilde{K}(x, y)\tilde{\varphi}(y)dy &= f(x). \end{aligned}\right\} \quad (22)$$

We shall estimate the difference $|\varphi(x) - \tilde{\varphi}(x)|$, on the assumption that in the square $(0, 1;\ 0, 1)$:

$$|K(x, y) - \tilde{K}(x, y)| < \varepsilon. \quad (23)$$

We shall prove as a preliminary a lemma concerning determinants.

LEMMA. *Let there be given two determinants*

$$\Delta = \begin{vmatrix} a_{1,1}, & a_{1,2}, & \ldots, & a_{1,n} \\ a_{2,1}, & a_{2,2}, & \ldots, & a_{2,n} \\ \cdot & \cdot & \cdot & \cdot \\ a_{n,1}, & a_{n,2}, & \ldots, & a_{n,n} \end{vmatrix}, \quad \Delta' = \begin{vmatrix} a'_{1,1}, & a'_{1,2}, & \ldots, & a'_{1,n} \\ a'_{2,1}, & a'_{2,2}, & \ldots, & a'_{2,n} \\ \cdot & \cdot & \cdot & \cdot \\ a'_{n,1}, & a'_{n,2}, & \ldots, & a'_{n,n} \end{vmatrix},$$

the elements of the second determinant being close to those of the first: $a'_{i,k} = a_{i,k} + \delta_{i,k}$. *Then if*

$$|a_{i,k}| < N \text{ and } |\delta_{i,k}| < \varepsilon,$$

[1]) F. Tricomi, [1].

we have

$$|\Delta - \Delta'| < n^{\frac{n}{2}}[(N + \varepsilon)^n - N^n]. \tag{24}$$

PROOF. Let us resolve the determinant Δ' into a sum of determinants:

$$\Delta' = \begin{vmatrix} a_{11} + \delta_{11}, & a_{12} + \delta_{12}, & \ldots, & a_{1n} + \delta_{1n} \\ a_{21} + \delta_{21}, & a_{22} + \delta_{22}, & \ldots, & a_{2n} + \delta_{2n} \\ \cdot & \cdot & \cdot & \cdot \\ a_{n1} + \delta_{n1}, & a_{n2} + \delta_{n2}, & \ldots, & a_{nn} + \delta_{nn} \end{vmatrix} =$$

$$= \begin{vmatrix} a_{11}, & a_{12}, & \ldots, & a_{1n} \\ a_{21}, & a_{22}, & \ldots, & a_{2n} \\ \cdot & \cdot & \cdot & \cdot \\ a_{n1}, & a_{n2}, & \ldots, & a_{nn} \end{vmatrix} + \sum \begin{vmatrix} \delta_{11}, & a_{12}, & \ldots, & a_{1n} \\ \delta_{21}, & a_{22}, & \ldots, & a_{2n} \\ \cdot & \cdot & \cdot & \cdot \\ \delta_{n1}, & a_{n2}, & \ldots, & a_{nn} \end{vmatrix} +$$

$$+ \sum\sum \begin{vmatrix} \delta_{11}, & \delta_{12}, & a_{13}, & \ldots, & a_{1n} \\ \delta_{21}, & \delta_{22}, & a_{23}, & \ldots, & a_{2n} \\ \cdot & \cdot & \cdot & \cdot & \cdot \\ \delta_{n1}, & \delta_{n2}, & a_{n3}, & \ldots, & a_{nn} \end{vmatrix} + \ldots + \begin{vmatrix} \delta_{11}, & \delta_{12}, & \ldots, & \delta_{1n} \\ \delta_{21}, & \delta_{22}, & \ldots, & \delta_{2n} \\ \cdot & \cdot & \cdot & \cdot \\ \delta_{n1}, & \delta_{n2}, & \ldots, & \delta_{nn} \end{vmatrix},$$

where in each sum the typical summand is cited, the columns with δ being located in other places in the remaining summands. In the first sum the number of summands is C_n^1; each does not exceed, by Hadamard's estimate [1]), $n^{\frac{n}{2}}\varepsilon N^{n-1}$; in the second, the number of summands is C_n^2; each does not exceed $n^{\frac{n}{2}}\varepsilon^2 N^{n-2}$, etc. Therefore

$$|\Delta' - \Delta| < C_n^1 n^{\frac{n}{2}} \varepsilon N^{n-1} + C_n^2 n^{\frac{n}{2}} \varepsilon^2 N^{n-2} + \ldots + C_n^n n^{\frac{n}{2}} \varepsilon^n = n^{\frac{n}{2}} [(N + \varepsilon)^n - N^n]$$

and the lemma is proved.

We shall now give the estimate of the difference $|\varphi(x) - \tilde{\varphi}(x)|$. By the formulas of Fredholm, the solution of equations (22) may be obtained in the form:

$$\left.\begin{aligned} \varphi(x) &= f(x) - \lambda \int_0^1 \frac{\Delta(x, y, \lambda)}{\Delta(\lambda)} f(y)dy, \\ \tilde{\varphi}(x) &= f(x) - \lambda \int_0^1 \frac{\tilde{\Delta}(x, y, \lambda)}{\tilde{\Delta}(\lambda)} f(y)dy, \end{aligned}\right\} \tag{25}$$

where $\Delta(\lambda)$ and $\tilde{\Delta}(\lambda)$ are the so-called Fredholm determinants for equations (22), and $\Delta(x, y, \lambda)$ and $\tilde{\Delta}(x, y, \lambda)$ are their minors. We shall first give an estimate of the

1) Privalov, [2], 1933, p. 46.

difference $|\Delta(\lambda) - \tilde{\Delta}(\lambda)|$. This difference can be written in the form

$$\Delta(\lambda) - \tilde{\Delta}(\lambda) = \sum_{n=1}^{\infty} (-1)^n \frac{\lambda^n}{n!} \int_0^1 \int_0^1 \ldots \int_0^1 \left[K\begin{pmatrix} z_1, z_2, \ldots, z_n \\ z_1, z_2, \ldots, z_n \end{pmatrix} - \right.$$

$$\left. - \tilde{K}\begin{pmatrix} z_1, z_2, \ldots, z_n \\ z_1, z_2, \ldots, z_n \end{pmatrix} \right] dz_1 dz_2 \ldots dz_n, \tag{26}$$

where in the brackets under the integral sign are the nth-order determinants of the values of the kernels $K(x, y)$ and $\tilde{K}(x, y)$:

$$K\begin{pmatrix} z_1, z_2, \ldots, z_n \\ z_1, z_2, \ldots, z_n \end{pmatrix} = \begin{vmatrix} K(z_1, z_1), K(z_1, z_2), \ldots, K(z_1, z_n) \\ K(z_2, z_1), K(z_2, z_2), \ldots, K(z_2, z_n) \\ \cdot \quad \cdot \quad \cdot \quad \cdot \quad \cdot \quad \cdot \quad \cdot \quad \cdot \\ K(z_n, z_1), K(z_n, z_2), \ldots, K(z_n, z_n) \end{vmatrix},$$

$$\tilde{K}\begin{pmatrix} z_1, z_2, \ldots, z_n \\ z_1, z_2, \ldots, z_n \end{pmatrix} = \begin{vmatrix} \tilde{K}(z_1, z_1), \tilde{K}(z_1, z_2), \ldots, \tilde{K}(z_1, z_n) \\ \tilde{K}(z_2, z_1), \tilde{K}(z_2, z_2), \ldots, \tilde{K}(z_2, z_n) \\ \cdot \quad \cdot \quad \cdot \quad \cdot \quad \cdot \quad \cdot \quad \cdot \quad \cdot \\ \tilde{K}(z_n, z_1), \tilde{K}(z_n, z_2), \ldots, \tilde{K}(z_n, z_n) \end{vmatrix}.$$

If N is the upper bound of $K(x, y)$, then on utilizing the lemma regarding the difference of the two determinants (24), we find at once that

$$\left| K\begin{pmatrix} z_1, z_2, \ldots, z_n \\ z_1, z_2, \ldots, z_n \end{pmatrix} - \tilde{K}\begin{pmatrix} z_1, z_2, \ldots, z_n \\ z_1, z_2, \ldots, z_n \end{pmatrix} \right| < n^{\frac{n}{2}} [(N + \varepsilon)^n - N^n].$$

Hence

$$|\Delta(\lambda) - \tilde{\Delta}(\lambda)| < \sum_{n=1}^{\infty} \frac{|\lambda|^n}{n!} n^{\frac{n}{2}} [(N + \varepsilon)^n - N^n] =$$

$$= \sum_{n=1}^{\infty} \frac{n^{\frac{n}{2}}}{n!} [|\lambda|(N + \varepsilon)]^n - \sum_{n=1}^{\infty} \frac{n^{\frac{n}{2}}}{n!} (|\lambda| N)^n =$$

$$= \Omega[|\lambda| (N + \varepsilon)] - \Omega(|\lambda| N), \tag{27}$$

if we put

$$\Omega(u) = \sum_{n=1}^{\infty} \frac{n^{\frac{n}{2}}}{n!} u^n. \tag{28}$$

In exactly the same manner the difference $|\Delta(x, y, \lambda) - \tilde{\Delta}(x, y, \lambda)|$ can be estimated, proceeding from the expansion

$$\Delta(x, y, \lambda) = - \sum_{n=0}^{\infty} (-1)^n \frac{\lambda^n}{n!} \int_0^1 \int_0^1 \ldots \int_0^1 K\begin{pmatrix} x z_1 \ldots z_n \\ y z_1 \ldots z_n \end{pmatrix} dz_1 \ldots dz_n. \tag{29}$$

We then obtain:

$$|\Delta(x, y, \lambda) - \tilde{\Delta}(x, y, \lambda)| < \sum_{n=0}^{\infty} \frac{|\lambda|^n}{n!} (n+1)^{\frac{n+2}{2}} [(N+\varepsilon)^{n+1} - N^{n+1}] =$$

$$= (N+\varepsilon)\Omega'[|\lambda|(N+\varepsilon)] - N\Omega'(|\lambda| N).$$

We can now estimate the difference $|\varphi(x) - \tilde{\varphi}(x)|$; we have

$$|\varphi(x) - \tilde{\varphi}(x)| \leqslant \lambda \int_0^1 \left| \frac{\Delta(x, y, \lambda)}{\Delta(\lambda)} - \frac{\tilde{\Delta}(x, y, \lambda)}{\tilde{\Delta}(\lambda)} \right| |f(y)| \, dy. \tag{30}$$

Let us designate as F the bound of $|f(x)|$ and as $\delta\Delta$ and $\delta\Delta_{x,y}$ the estimates obtained above for $|\Delta(\lambda) - \tilde{\Delta}(\lambda)|$ and $|\Delta(x, y, \lambda) - \tilde{\Delta}(x, y, \lambda)|$; then after simplifications, having taken it into account that $|\Delta(\lambda)| < \Omega(|\lambda| N)$ and $|\Delta(x, y, \lambda)| < N\Omega'(N|\lambda|)$, we find

$$|\varphi(x) - \tilde{\varphi}(x)| < |\lambda| F \frac{N\Omega'[|\lambda| N]\delta\Delta + \Omega(|\lambda| N)\delta\Delta_{x,y}}{|\Delta(\lambda)| \, |\tilde{\Delta}(\lambda)|}.$$

Let us yet transform $\delta\Delta$ and $\delta\Delta_{x,y}$, using the theorem of the mean, and taking it into account that Ω, Ω', Ω'' are increasing functions. We then obtain, definitively,

$$|\varphi(x) - \tilde{\varphi}(x)| < \varepsilon |\lambda| F \frac{L\Omega'(L)\Omega'(\tilde{L}) + \Omega(L)\Omega'(\tilde{L}) + \tilde{L}\Omega(L)\Omega''(L)}{|\Delta(\lambda)| \, |\tilde{\Delta}(\lambda)|}, \tag{31}$$

where we have for brevity put $N|\lambda| = L$, $(N + \varepsilon)|\lambda| = \tilde{L}$.

The estimates obtained show that if the kernels $\tilde{K}(x, y)$ converge uniformly to the kernel $K(x, y)$, i.e., $\varepsilon \to 0$, then by (27) so will $\tilde{\Delta}(\lambda) \to \Delta(\lambda)$. Furthermore, if $\Delta(\lambda) \neq 0$, then for ε sufficiently small, $\tilde{\Delta}(\lambda) \neq 0$, and as is clear from estimate (33), $\tilde{\varphi}(x) \to \varphi(x)$. It should, however, be observed that the presence in Tricomi's estimate of the quantity $\Delta(\lambda)$, which is usually difficult to estimate without knowing the solution of the equation, makes it not very convenient for the actual estimation of the error. However if an equation with kernel $\tilde{K}$ has been solved and $\tilde{\Delta}(\lambda)$ determined, a lower bound for $\Delta(\lambda)$ can be given by using the estimate given above, (27), after which the estimate obtained now proves to be applicable. We will yet mention what is readily established, that for series (28) there will be majorant the series of functions

$$\Omega_0(u) = (1 + u)e^{\frac{eu^2}{2}},$$

and therefore in all the estimates above it will be possible to take Ω_0, Ω_0', Ω_0'' instead of Ω, Ω', Ω''.

5. The method of moments. This method consists in seeking the approximate solution of a given integral equation

$$L(\varphi) = \varphi(x) - \lambda \int_a^b K(x, y)\varphi(y)dy - f(x) = 0$$

in a determinate form — as a linear combination of certain functions chosen previously, with undetermined coefficients:

$$\tilde{\varphi}(x) = f(x) + \sum_{i=1}^{n} c_i \varphi_i(x). \tag{32}$$

We can obviously always consider that the functions $\varphi_i(x)$ are orthogonal, normalized, and that they are the first n functions of some complete system $\varphi_i(x)$ $(i = 1, 2, \ldots, n, \ldots)$. The coefficients c_i figuring in $\tilde{\varphi}(x)$ are to be determined from the following consideration.

In order that the function $\tilde{\varphi}(x)$ be the exact solution of the equation, $L(\tilde{\varphi})$ must be identically zero, and this is equivalent (if $L(\tilde{\varphi})$ is continuous) to its being orthogonal to all of the functions $\varphi_i(x)$ $(i = 1, 2, \ldots, n, \ldots)$. However as it has in its array only n constants c_i, we have the possibility of satisfying only the first n of these conditions:

$$\int_a^b [\tilde{\varphi}(x) - \lambda \int_a^b K(x, y)\tilde{\varphi}(y)dy - f(x)]\varphi_i(x)dx =$$

$$= \int_a^b L(\tilde{\varphi})\varphi_i(x)dx = 0. \tag{33}$$

We can therefore anticipate that the function $\tilde{\varphi}(x)$ constructed on the values of c_i found from system (33) will only be an approximate solution of the equation.

This method bears the appellation: the method of moments, since equations (33) require that the first n generalized moments of the function $L(\tilde{\varphi})$ be equal to zero, and in case $\varphi_k(x) = x^{k-1}$ $(k = 1, 2, \ldots)$ this reduces to the requirement that the first n classical Chebyshev moments be equal to zero. It has been developed chiefly by Acad. N. M. Krylov and his school.

We also note that this method of solving integral equations represents the direct analog of Galerkin's method for the approximate solution of differential equations (see Chap. IV, § 2). We shall now establish the connection between this method and the method whereby the kernel is replaced by a degenerate one.

Let us construct the degenerate kernel $K_n(x, y)$ — the nth segment of the expansion of $K(x, y)$ in a Fourier series in terms of the functions $\varphi_i(x)$:

$$K_n(x, y) = \sum_{i=1}^{n} u_i(y)\varphi_i(x), \qquad u_i(y) = \int_a^b K(x, y)\varphi_i(x)dx,$$

and form the corresponding integral equation

$$L_n(\varphi) = \varphi(x) - \lambda \int_a^b K_n(x, y)\varphi(y)dy - f(x) = 0.$$

We shall first of all remark that if we apply the method of moments to this integral equation, we shall obtain the same solution as for the original equation.

Indeed, the sole difference is that in equations of form (33) we have

$K(x, y)$ instead of $K_n(x, y)$ in the second term; however one is easily convinced that even these terms will be equal in both cases; it is sufficient merely to perform an integration with respect to x at the outset, and to consider that

$$\int_a^b K_n(x, y)\varphi_i(x)dx = \int_a^b K(x, y)\varphi_i(x)dx = u_i(y).$$

Therefore the application of the method of moments to the original equation and to the equation with degenerate kernel lead to one and the same approximate solution. But for the equation with degenerate kernel $K_n(x, y)$, as is clear from its structure, the exact solution will be of form (32); indeed

$$\varphi(x) = f(x) + \lambda \int_a^b K_n(x, y)\varphi(y)dy = f(x) + \sum_{n=1}^{n} (\lambda \int_a^b u_i(y)\varphi(y)dy)\varphi_i(x).$$

Therefore the application of the method of moments leads to the exact solution of the equation with degenerate kernel; but since this same solution is obtained on applying the method of moments to the original equation, we are led to the following conclusion:

The application of the method of moments is equivalent to the replacement of the kernel $K(x, y)$ by a degenerate kernel $K_n(x, y)$.

Thanks to this, when employing the method of moments, one can use the method of estimating the error given in No. 2, and also, by relying on the results of No. 2 or No. 4, one can under certain conditions — for example, if $K_n(x, y) \to K(x, y)$ uniformly — affirm the convergence of the approximate solutions to the exact, provided λ is not a proper value of the original equation. The question of the convergence of the method of moments, and the rate of its convergence, has been investigated in works of N. M. Krylov and M. F. Kravchuk [1]).

We further note that exclusion of $f(x)$ from expression (32) for $\tilde{\varphi}(x)$ would have necessitated replacement, in the equation with degenerate kernel, of the free term $f(x)$ by $f_n(x)$, the nth partial sum of its Fourier series in terms of the functions $\varphi_i(x)$. This modification does not require, for its validity, that the application of the method of moments to both equations lead to one and the same result, but without it we could not affirm that the exact solution of the equation with degenerate kernel has the required form $\sum_{i=1}^{n} c_i\varphi_i(x)$.

Thus in this case we may use the estimate of No. 2, but we will have $\eta \neq 0$.

In our reasoning we have assumed that the functions have been normalized and orthogonalized; however in the actual application of the

---

[1]) N. M. Krylov, [1], [2]; Kravchuk, M. F., [1]. On this question see also Oberg, E., [1], and L. V. Kantorovich, [3].

method there is no need to effect this, since this is not reflected in the form of the solution.

Lastly, it is necessary to point out that Ritz's method, in application to integral equations, is a particular case of the method of moments.

The possibility of applying Ritz's method to integral equations is connected with the fact that the problem of solving an integral equation with a symmetric kernel is in a certain sense equivalent to the problem of the minimum of the functional

$$J(\varphi) = \int_a^b \varphi^2(x)dx - \lambda \int_a^b \int_a^b K(x, y)\varphi(x)\varphi(y)dxdy - 2\int_a^b \varphi(x)f(x)dx.$$

The application of Ritz's method to this problem consists in seeking a function of form (32) giving the minimum J. But if we substitute here for φ the expression $\tilde{\varphi}$ and equate to zero the derivatives with respect to c_i, we arrive at the system of equations

$$\int_a^b \tilde{\varphi}(x)\varphi_i(x)dx - \lambda \int_a^b \int_a^b K(x, y)\tilde{\varphi}(x)\varphi_i(y)dydx -$$

$$- \int_a^b f(x)\varphi_i(x)dx = 0 \qquad (i = 1, 2, \ldots, N),$$

which obviously coincides with system (33).

We shall consider one example of the application of the method of moments.

Example. The problem of the vibration of a string, as is known, can be reduced to an integral equation with kernel

$$K(x, y) = \begin{cases} x(y - 1), & \text{if } 0 \leqslant x \leqslant y \leqslant 1 \\ y(x - 1), & \text{if } 1 \geqslant x \geqslant y \geqslant 0. \end{cases}$$

Let us employ the method of moments for the determination of the first proper number and the first proper function of this equation. The exact solution of this problem is known to be

$$\lambda_1 = \pi^2 = 9.8696, \quad \varphi_1(x) = \sqrt{2} \sin \pi x.$$

We seek the first approximation in the form of a constant:

$$\varphi(x) = A.$$

Substituting A for $\varphi(x)$ in the integral equation

$$\varphi(x) - \lambda \int_0^1 K(x, y)\varphi(y)dy = 0,$$

after multiplying by 1 and integrating with respect to x in accordance with the method of moments, we arrive at the equation

$$\int_0^1 \{A - \lambda \int_0^1 K(x, y)Ady\}dx = 0.$$

We work out beforehand

$$\int_0^1 K(x, y)dy = \int_0^x y(1 - x)dy + \int_x^1 x(1 - y)dy =$$

$$= \frac{x^2(1 - x)}{2} + \frac{x(1 - x)^2}{2} = \frac{x(1 - x)}{2}.$$

On substituting this value, we give the equation the form

$$A\int_0^1 \left[1 - \lambda \frac{x(1 - x)}{2} \right] dx = A\left(1 - \frac{\lambda}{12}\right) = 0.$$

A non-zero solution is possible only if $\lambda = 12$, which thus represents the first, very rough approximation: $\lambda_1^{(1)} = 12$.

Passing on to the determination of the next approximation, we note that it is evident from considerations of symmetry that the first proper function $\varphi_1(x)$ must be symmetric with respect to the point $x = \frac{1}{2}$. Such a first-degree polynomial, different from a constant, does not exist, and the general form of second-degree polynomial subject to this condition is

$$\varphi(x) = A + Bx(1 - x).$$

This is the form in which we shall seek $\varphi(x)$. Substituting this expression in the integral equation and premultiplying by the basic functions 1 and $x(1 - x)$, then integrating with respect to x, we arrive at the system of equations:

$$\int_0^1 \{A + Bx(1 - x) - \lambda \int_0^1 K(x, y) [A + By(1 - y)]dy\}dx = 0,$$

$$\int_0^1 \{A + Bx(1 - x) - \lambda \int_0^1 K(x, y)[A + By(1 - y)]dy\}x(1 - x)dx = 0,$$

or

$$A(1 - \tfrac{1}{12}\lambda) + B \cdot \tfrac{1}{6}(1 - \tfrac{1}{10}\lambda) = 0,$$
$$A \cdot \tfrac{1}{6}(1 - \tfrac{1}{10}\lambda) + B \cdot \tfrac{1}{30}(1 - \tfrac{17}{168}\lambda) = 0.$$

Equating the determinant of this system to zero, we obtain the equation

$$\lambda^2 - 180\lambda + 1680 = 0,$$
$$\lambda = 9.8751 \text{ and } 170.1249;$$

the lesser of these then represents the second approximation to the first proper value, $\lambda_1^{(2)} = 9.8751$. Substituting this value in the system connecting A and B, we find that $A = -0.01176B$, or, determining B from the condition $\int_0^1 \varphi^2(x)dx = 1$, we obtain finally the approximation to the proper function, of the form

$$\varphi_1^{(2)}(x) = -0.0684 + 5.817x(1 - x).$$

On comparing this with the exact value, we see that the first proper number has been determined with an error of 0.06%, and the proper function with an error of the order of 4% (from its maximum).

6. Bateman's method. It frequently may prove useful not only to replace the given kernel by a degenerate one (algebraic), but also to represent it approximately in the form of the sum of a kernel for which the resolvent is known and of a degenerate kernel. For kernels of the last form, the resolvent may also be given in finite form. A method of obtaining this resolvent has been suggested by Bateman [1]). We shall expound it in the form of the following lemma:

LEMMA. *For the equation*

$$\varphi(x) - \lambda \int_a^b k(x, y)\varphi(y)dy = f(x) \tag{34}$$

with kernel $k(x, y)$, *let there be known the resolvent* $\gamma(x, y, \lambda)$, *i.e., the function by means of which the solution of equation* (34) *is given in the form*

$$\varphi(x) = f(x) + \lambda \int_a^b \gamma(x, y, \lambda)f(y)dy. \tag{35}$$

Then for the integral equation with kernel

$$K(x, y) = \frac{\begin{vmatrix} k(x, y) & f_1(x) & \ldots & f_n(x) \\ g_1(y) & a_{11} & \ldots & a_{1,n} \\ \cdot & \cdot & \cdot & \cdot \\ g_n(y) & a_{n1} & \ldots & a_{nn} \end{vmatrix}}{\begin{vmatrix} a_{11} & a_{12} & \ldots & a_{1n} \\ a_{21} & a_{22} & \ldots & a_{2n} \\ \cdot & \cdot & \cdot & \cdot \\ a_{n1} & a_{n2} & \ldots & a_{nn} \end{vmatrix}}, \tag{36}$$

where $f_r(x)$ *and* $g_m(y)$ *are arbitrary functions and* $a_{i,k}$ *are arbitrary numbers, the resolvent is given in the form:*

$$\Gamma(x, y, \lambda) = \frac{\begin{vmatrix} \gamma(x, y, \lambda) & \varphi_1(x) & \ldots & \varphi_n(x) \\ \chi_1(y) & a_{11} + \lambda\tau_{11} & \ldots & a_{1n} + \lambda\tau_{1n} \\ \cdot & \cdot & \cdot & \cdot \\ \cdot & \cdot & \cdot & \cdot \\ \chi_n(y) & a_{n1} + \lambda\tau_{n1} & \ldots & a_{nn} + \lambda\tau_{nn} \end{vmatrix}}{\begin{vmatrix} a_{11} + \lambda\tau_{11} & \ldots & a_{1n} + \lambda\tau_{1n} \\ \cdot & \cdot & \cdot \\ \cdot & \cdot & \cdot \\ a_{n1} + \lambda\tau_{n1} & \ldots & a_{nn} + \lambda\tau_{nn} \end{vmatrix}}, \tag{37}$$

where

$$\varphi_r(x) = f_r(x) + \lambda \int_a^b \gamma(x, y, \lambda)f_r(y)dy, \tag{38}$$

---

[1]) Bateman, H., [1], [2]; cf. Goursat, E. [2].

$$\chi_m(x) = g_m(x) + \lambda \int_a^b \gamma(y, x, \lambda) g_m(y) dy, \tag{39}$$

$$\tau_{m,r} = \int_a^b \varphi_r(x) g_m(x) dx, \tag{40}$$

i.e., $\varphi_r(x)$ *and* $\chi_m(x)$ *are the solutions of the integral equations*

$$\varphi(x) - \lambda \int_a^b k(x, y) \varphi(y) dy = f_r(x), \tag{41}$$

$$\chi(x) - \lambda \int_a^b k(y, x) \chi(y) dy = g_m(x). \tag{42}$$

PROOF. In order that $\Gamma(x, y, \lambda)$ be the resolvent of the equation

$$\varphi(x) - \int_a^b K(x, y) \varphi(y) dy = f(x), \tag{43}$$

it is sufficient to establish that $\Gamma(x, y, \lambda)$ satisfies the equation, which is characteristic for the resolvent,

$$K(x, y) = \Gamma(x, y, \lambda) - \lambda \int_a^b K(x, t) \Gamma(t, y, \lambda) dt. \tag{44}$$

If at the outset we designate as Δ the determinant in the denominator of (36) and by $\Delta_{m,r}$ the algebraic complement of the element $a_{m,r}$, we will obtain, on expanding the determinant (36) by the elements of the first row and next by the elements of the first column,

$$K(x, y) = k(x, y) - \sum_{m=1}^{n} \sum_{r=1}^{n} \frac{\Delta_{m,r}}{\Delta} f_r(x) g_m(y). \tag{45}$$

In order to verify that the function $\Gamma(x, y, \lambda)$ defined by equation (37) satisfies equation (44), we shall compute the integral standing on the right side.

In view of (45) we first find

$$\lambda \int_a^b k(x, t) \Gamma(t, y, \lambda) dt.$$

Denoting by θ the determinant standing in the denominator of (37), we have

$$\lambda \int_a^b k(x, t) \Gamma(t, y, \lambda) dt =$$

$$= \frac{\lambda}{\theta} \begin{vmatrix} \int_a^b \gamma(t, y, \lambda) k(x, t) dt, & \int_a^b k(x, t) \varphi_1(t) dt, & \ldots, & \int_a^b k(x, t) \varphi_n(t) dt \\ \chi_1(y), & a_{11} + \lambda \tau_{11} & \ldots, & a_{1n} + \lambda \tau_{1n} \\ \cdots & \cdots & \cdots & \cdots \\ \chi_n(y), & a_{n1} + \lambda \tau_{n1}, & \ldots, & a_{nn} + \lambda \tau_{nn} \end{vmatrix}.$$

Next, using equations (41), and also the fact that the resolvent $\gamma(x, y, \lambda)$ satisfies the characteristic equation:

$$\gamma(x, y, \lambda) - \lambda \int_a^b k(x, t)\gamma(t, y, \lambda)dt = k(x, y),$$

we obtain

$$\lambda \int_a^b k(x, t)\Gamma(t, y, \lambda)dt =$$

$$= \frac{1}{\theta} \begin{vmatrix} \gamma(x, y, \lambda) - k(x, y), & \varphi_1(x) - f_1(x), & \ldots, & \varphi_n(x) - f_n(x) \\ \chi_1(y), & a_{11} + \lambda\tau_{11}, & \ldots, & a_{1n} + \lambda\tau_{1n} \\ \ldots & \ldots & \ldots & \ldots \\ \chi_n(y), & a_{n1} + \lambda\tau_{n1}, & \ldots, & a_{nn} + \lambda\tau_{nn} \end{vmatrix}.$$

In quite the same manner, using (40) and (42), we find:

$$\lambda \int_a^b g_m(t)\Gamma(t, y, \lambda)dt = \frac{1}{\theta} \begin{vmatrix} \chi_m(y) - g_m(y), & \lambda\tau_{m1}, & \ldots, & \lambda\tau_{mn} \\ \chi_1(y) & a_{11} + \lambda\tau_{11}, & \ldots, & a_{1n} + \lambda\tau_{1n} \\ \ldots & \ldots & \ldots & \ldots \\ \chi_n(y), & a_{n1} + \lambda\tau_{n1}, & \ldots, & a_{nn} + \lambda\tau_{nn} \end{vmatrix} =$$

$$= -\frac{1}{\theta} \begin{vmatrix} g_m(y), & a_{m1}, & \ldots, & a_{mn} \\ \chi_1(y), & a_{11} + \lambda\tau_{11}, & \ldots, & a_{1n} + \lambda\tau_{1n} \\ \ldots & \ldots & \ldots & \ldots \\ \chi_n(y), & a_{n1} + \lambda\tau_{n1}, & \ldots, & a_{nn} + \lambda\tau_{nn} \end{vmatrix},$$

if the mth row be subtracted from the first.

Furthermore, premultiplying the last equations by $-\dfrac{\Delta_{m,r}}{\Delta}$ and adding them, using for the algebraic complements the usual relations:

$$\frac{1}{\Delta} \sum_{m=1}^{n} a_{mp}\Delta_{mr} = \begin{cases} 0 & \text{for } p \neq r \\ 1 & \text{,, } p = r, \end{cases}$$

we find that

$$-\lambda \int_a^b \sum_{m=1}^{n} \frac{\Delta_{mr}}{\Delta} g_m(t)\Gamma(t, y, \lambda) =$$

$$= \frac{1}{\theta} \begin{vmatrix} \sum_{m=1}^{n} \frac{\Delta_{mr}}{\Delta} g_m(y), & 0, & \ldots, & 1, & \ldots, & 0 \\ \chi_1(y), & a_{11} + \lambda\tau_{11}, & & \ldots, & & a_{1n} + \lambda\tau_{1n} \\ \ldots & \ldots & & \ldots & & \ldots \\ \chi_n(y), & a_{n1} + \lambda\tau_{n1}, & & \ldots, & & a_{nn} + \lambda\tau_{nn} \end{vmatrix},$$

where 1 stands in the 1st row in the $(r + 1)$th place. Now multiplying this equation by $f_r(x)$, summing on r and afterwards adding the expression obtained earlier for $\lambda\int_a^b k(x, t)\Gamma(t, y, \lambda)dt$, using (45), we obtain finally:

$$\lambda\int_a^b K(x, t)\Gamma(t, y, \lambda)dt =$$

$$= \frac{1}{\theta}\begin{vmatrix} \gamma(x, y, \lambda) - K(x, y) & \varphi_1(x), & \ldots, & \varphi_n(x) \\ \chi_1(y), & a_{11} + \lambda\tau_{11}, & \ldots, & a_{1n} + \lambda\tau_{1n} \\ \cdot & \cdot & \cdot & \cdot \\ \chi_n(y), & a_{n1} + \lambda\tau_{n1}, & \ldots, & a_{nn} + \lambda\tau_{nn} \end{vmatrix} =$$

$$= \frac{1}{\theta}\begin{vmatrix} \gamma(x, y, \lambda), & \varphi_1(x), & \ldots, & \varphi_n(x) \\ \chi_1(y), & a_{11} + \lambda\tau_{11}, & \ldots, & a_{1n} + \lambda\tau_{1n} \\ \cdot & \cdot & \cdot & \cdot \\ \chi_n(y), & a_{n1} + \lambda\tau_{n1}, & \ldots, & a_{nn} + \lambda\tau_{nn} \end{vmatrix} -$$

$$- \frac{1}{\theta}\begin{vmatrix} K(x, y) & \varphi_1(x), & \ldots, & \varphi_n(x) \\ 0, & a_{11} + \lambda\tau_{11}, & \ldots, & a_{1n} + \lambda\tau_{1n} \\ \cdot & \cdot & \cdot & \cdot \\ 0, & a_{n1} + \lambda\tau_{n1}, & \ldots, & a_{nn} + \lambda\tau_{nn} \end{vmatrix} =$$

$$= \Gamma(x, y, \lambda) - K(x, y),$$

if the determinant obtained be separated into two with respect to the first column.

We have thus verified that the $\Gamma(x, y, \lambda)$ given by equation (37) actually satisfies equation (44), and is therefore the resolvent for equation (43).

Let us consider several particular cases of the general result that has been established:

1. Let us set

$$a_{mr} = \begin{cases} 0 & m \neq r \\ 1 & m = r, \end{cases}$$

i.e., putting it otherwise, in view of (36) let us assume that the kernel $K(x, y)$ has the form

$$K(x, y) = k(x, y) - \sum_{m=1}^{n} f_m(x)g_m(y).$$

The resolvent for this case will take the form:

$$\Gamma(x, y, \lambda) = \frac{\begin{vmatrix} \gamma(x, y, \lambda), & \varphi_1(x), & \varphi_2(x), & \ldots, & \varphi_n(x) \\ \chi_1(y), & 1 + \lambda\tau_{11}, & \lambda\tau_{12}, & \ldots, & \lambda\tau_{1n} \\ \cdot & \cdot & \cdot & \cdot & \cdot \\ \chi_n(y), & \lambda\tau_{n1}, & \lambda\tau_{n2}, & \ldots, & 1 + \lambda\tau_{nn} \end{vmatrix}}{\begin{vmatrix} 1 + \lambda\tau_{11}, & \lambda\tau_{12}, & \ldots, & \lambda\tau_{1n} \\ \lambda\tau_{21}, & 1 + \lambda\tau_{22}, & \ldots, & \lambda\tau_{2n} \\ \cdot & \cdot & \cdot & \cdot \\ \lambda\tau_{n1}, & \lambda\tau_{n2}, & \ldots, & 1 + \lambda\tau_{nn} \end{vmatrix}},$$

where $\varphi_i(x)$, $\chi_i(y)$ and $\tau_{m,r}$ have the same values as before.

Let us now make the additional assumption that $k(x, y) = 0$, i.e., that the kernel $K(x, y)$ is purely algebraic:

$$K(x, y) = -\sum_{i=1}^{n} f_m(x) g_m(y).$$

It is obvious that in the given case $\gamma(x, y, \lambda) = 0$; furthermore, in view of (38) and (39), $\varphi_r(x) = f_r(x)$ and $\chi_m(x) = g_m(x)$; finally

$$\tau_{mr} = \int_a^b f_r(x) g_m(x) dx.$$

The resolvent therefore acquires the form

$$\Gamma(x, y, \lambda) = \frac{\begin{vmatrix} 0, & f_1(x), & \dots, & f_m(x) \\ g_1(y), & 1 + \lambda\tau_{11}, & \dots, & \lambda\tau_{1n} \\ \cdot & \cdot & \cdot & \cdot \\ g_n(y), & \lambda\tau_{n1}, & \dots, & 1 + \lambda\tau_{nn} \end{vmatrix}}{\begin{vmatrix} 1 + \lambda\tau_{11}, & \dots, & \lambda\tau_{1n} \\ \cdot & \cdot & \cdot \\ \lambda\tau_{n1}, & \dots, & 1 + \lambda\tau_{nn} \end{vmatrix}}.$$

We considered this particular case in No. 1 and there obtained the expression for Γ in a somewhat different form, which can easily be brought into the one given here.

2. Let us consider an integral equation with a certain kernel $\bar{k}(x, y)$. In the interval (a, b) let us arbitrarily chose points $x_1, x_2, \dots, x_n$ and $y_1, y_2, \dots, y_n$. Now in equation (36) put

$$k(x, y) = 0, \quad f_r(x) = \bar{k}(x, y_r),$$
$$g_m(y) = -\bar{k}(x_m, y), \quad a_{mr} = \bar{k}(x_m, y_r).$$

Then, obviously, $\gamma(x, y, \lambda) = 0$ and the kernel $K(x, y)$ takes the form:

$$K(x, y) = -\frac{\begin{vmatrix} 0 & \bar{k}(x, y_1), & \dots, & \bar{k}(x, y_n) \\ \bar{k}(x_1, y), & \bar{k}(x_1, y_1), & \dots, & \bar{k}(x_1, y_n) \\ \cdot & \cdot & \cdot & \cdot \\ \bar{k}(x_n, y), & \bar{k}(x_n, y_1), & \dots, & \bar{k}(x_n, y_n) \end{vmatrix}}{\begin{vmatrix} \bar{k}(x_1, y_1), & \dots, & \bar{k}(x_1, y_n) \\ \cdot & \cdot & \cdot \\ \cdot & \cdot & \cdot \\ \bar{k}(x_n, y_1), & \dots, & \bar{k}(x_n, y_n) \end{vmatrix}} =$$

$$= \bar{k}(x, y) - \frac{\begin{vmatrix} \bar{k}(x, y), & \bar{k}(x, y_1), & \dots, & \bar{k}(x, y_n) \\ \bar{k}(x_1, y), & \bar{k}(x_1, y_1), & \dots, & \bar{k}(x_1, y_n) \\ \cdot & \cdot & \cdot & \cdot \\ \bar{k}(x_n, y), & \bar{k}(x_n, y_1), & \dots, & \bar{k}(x_n, y_n) \end{vmatrix}}{\begin{vmatrix} \bar{k}(x_1, y_1), & \dots, & \bar{k}(x_1, y_n) \\ \cdot & \cdot & \cdot \\ \cdot & \cdot & \cdot \\ \bar{k}(x_n, y_1), & \dots, & \bar{k}(x_n, y_n) \end{vmatrix}}.$$

This kernel is degenerate and possesses the property that it coincides with the kernel $\bar{k}(x, y)$ on the lines $x = x_m$, $y = y_r$; indeed, if we set $x = x_m$ or $y = y_r$, the determinant in the numerator of the second term will have two identical rows or columns, and will accordingly vanish; therefore

$$K(x_m, y) = \bar{k}(x_m, y), \quad K(x, y_r) = \bar{k}(x, y_r).$$

This coincidence on $2n$ lines permits us to consider generally that $K(x, y)$ is close to $\bar{k}(x, y)$, and that the solution of the equation with kernel $K(x, y)$ is close to the solution of the equation with kernel $\bar{k}(x, y)$. It should be observed that if $\bar{k}(x, y)$ is degenerate, i.e., has the form:

$$\bar{k}(x, y) = \sum_{m=1}^{n} \alpha_m(x)\beta_m(y),$$

the determinant in the numerator is identically equal to zero, in view of the familiar theorem on the resolution of a determinant into the sum of determinants, and therefore in this case

$$K(x, y) \equiv \bar{k}(x, y).$$

If $\bar{k}(x, y)$ consists of less than n summands, then the determinant in the denominator is also equal to zero, but in case this is not so, it is generally speaking different from zero, being equal to zero only for a chance choice of the points $x_1, \ldots, x_n; y_1, \ldots, y_n$.

For the kernel $K(x, y)$ the resolvent can be set up on the basis of the lemma, namely, in the case given,

$$\gamma(x, y, \lambda) = 0, \quad \varphi_r(x) = f_r(x) = \bar{k}(x, y_r)$$

$$\chi_m(y) = g_m(y) = -\bar{k}(x_m, y),$$

$$\tau_{mr} = -\int_a^b \bar{k}(x, y_r)\bar{k}(x_m, x)dx = -k^*(x_m, y_r),$$

where $k^*(x, y)$ is the second integrated kernel for $\bar{k}(x, y)$:

$$k^*(x, y) = \int_a^b \bar{k}(x, t)\bar{k}(t, y)dt,$$

and accordingly

$$\Gamma(x, y, \lambda) =$$

$$= -\frac{\begin{vmatrix} 0 & \bar{k}(x, y_1) & \ldots, & \bar{k}(x, y_n) \\ \bar{k}(x_1, y), & \bar{k}(x_1, y_1) - \lambda k^*(x_1, y_1), & \ldots, & \bar{k}(x_1, y_n) - \lambda k^*(x_1, y_n) \\ \cdot & \cdot & \cdot & \cdot \\ \bar{k}(x_n, y), & \bar{k}(x_n, y_1) - \lambda k^*(x_n, y_1), & \ldots, & \bar{k}(x_n, y_n) - \lambda k^*(x_n, y_n) \end{vmatrix}}{\begin{vmatrix} \bar{k}(x_1, y_1) - \lambda k^*(x_1, y_1), & \ldots, & \bar{k}(x_1, y_n) - \lambda k^*(x_1, y_n) \\ \cdot & \cdot & \cdot \\ \cdot & \cdot & \cdot \\ \bar{k}(x_n, y_1) - \lambda k^*(x_n, y_1), & \ldots, & \bar{k}(x_n, y_n) - \lambda k^*(x_n, y_n) \end{vmatrix}}.$$

By means of the resolvent $\Gamma(x, y, \lambda)$ we can, then, obtain an approximate solution of the equation with kernel $\bar{k}(x, y)$. In particular, we find the approximate values for the proper values λ of this kernel by equating to zero the determinant in the denominator of Γ. The estimate of the error of this method can be made on the basis of the theorems of No. 2 and No. 4. We subjoin an illustrative example, which is borrowed from the article by Bateman.

Example. Let us consider as an example the kernel of the integral equation to which the equation of the string reduces, and which in the square (0, 1; 0, 1) is given by the equations:

$$\bar{k}(x, y) = \begin{cases} x(y-1) & x \leqslant y \\ y(x-1) & x \geqslant y. \end{cases}$$

To form Γ, $k^*(x, y)$ is still needed, but

$$k^*(x, y) = \int_0^1 \bar{k}(x, t)\bar{k}(t, y)dt = \begin{cases} \frac{1}{6}x(1-y)(2y-x^2-y^2), & x \leqslant y, \\ \frac{1}{6}y(1-x)(2x-x^2-y^2), & x \geqslant y. \end{cases}$$

The points x_m and y_r will be chosen equidistant, and we shall adopt $n = 5$; then

$$x_1 = y_1 = \tfrac{1}{6},\ x_2 = y_2 = \tfrac{2}{6},\ x_3 = y_3 = \tfrac{3}{6},\ x_4 = y_4 = \tfrac{4}{6},\ x_5 = y_5 = \tfrac{5}{6}.$$

We shall concern ourselves only with the determination of the proper numbers, forming the determinant which appears in the denominator of Γ. Then, putting $\lambda = 216x$ and making the cancellation, we find

$$\begin{vmatrix} 5(1-10x) & 4(1-19x) & 3(1-26x) & 2(1-31x) & 1-34x \\ 4(1-19x) & 8(1-16x) & 6(1-23x) & 4(1-28x) & 2(1-31x) \\ 3(1-26x) & 6(1-23x) & 9(1-18x) & 6(1-23x) & 3(1-26x) \\ 2(1-31x) & 4(1-28x) & 6(1-23x) & 8(1-16x) & 4(1-19x) \\ 1-34x & 2(1-31x) & 3(1-26x) & 4(1-19x) & 5(1-10x) \end{vmatrix} = 0.$$

Upon development we arrive at the equation

$$130x^5 - 441x^4 + 488x^3 - 206x^2 + 30x - 1 = 0$$

or

$$(x-1)(2x-1)(5x-1)(13x^2-22x+1) = 0,$$

whence after solving for λ we find the following values:

$$\lambda_1 = 10.02,\ \lambda_2 = 43.2,\ \lambda_3 = 108,\ \lambda_4 = 216,\ \lambda_5 = 355.2.$$

The exact quantities for the proper values are here, as we know, π^2, $(2\pi)^2$, $(3\pi)^2$, ... i.e.,

$$\lambda_1 = 9.8696,\ \lambda_2 = 39.4784,\ \lambda_3 = 88.8264,\ \ldots,$$

consequently the error in the determination of the first proper value is 2%, in the second, 9%, in the third, 20%.

Somewhat more exact results would have been obtained had we chosen other points $x_1, x_2, \ldots,$ for instance:

$$x_1 = y_1 = \tfrac{1}{10},\ x_2 = y_2 = \tfrac{3}{10},\ x_3 = y_3 = \tfrac{5}{10},\ x_4 = y_4 = \tfrac{7}{10},\ x_5 = y_5 = \tfrac{9}{10},$$

or had we taken for x_i and y_i the five Gaussian points for the interval (0, 1). However a very high degree of accuracy cannot be attained with such a number of ordinates, since the kernel itself, $k(x, y)$, has a singularity: its derivative is discontinuous at $y = x$, in consequence of which a good approximation of it by algebraic kernels is impossible.

Chapter III

THE METHOD OF NETS

§ 1. EXPRESSIONS FOR THE DERIVATIVES IN TERMS OF DIFFERENCE RATIOS. RELATIONS BETWEEN THE VALUES OF A FUNCTION AT THE NODES OF A NET, AND THE HARMONIC AND BIHARMONIC OPERATORS

1. Expressions for the derivatives in terms of difference ratios. In applying the method of nets, the differential equation under consideration is replaced approximately by some equation in finite differences, which is, as a rule, obtained from the differential equation by replacing in it the derivatives and the differential operations of a different kind by approximate expressions for them in terms of difference ratios or the values of the function at separate points of the net.

We shall therefore commence the exposition of the method of nets by recalling the expressions for derivatives in terms of difference ratios. Owing to reasons that will be elucidated later, the formulas relating to equidistant values of the argument will be insufficient for us. We therefore cite all formulas for the general case, in which the values of the argument are not equidistant.

The expression for derivatives in terms of the values of the function at the nodes of the net is usually obtained as follows. An interpolating polynomial is constructed which at the points of the net takes the same values that the given function does, and then the derivatives of the given function are considered to be approximately equal to the corresponding derivatives of the interpolating polynomial.

All the formulas for parabolic interpolation that we have in view here can be obtained from the fundamental Newton formula by transforming it. The majority of them are, from a theoretical point of view, equivalent, and one is obtainable from the other by a regrouping of its terms. This regrouping is to be done in conformity with the aim the computations pursue; it is ordinarily done in such a way as to obtain the best accuracy using the least possible number of terms of the interpolation formula.

The theory of interpolation formulas is widely known. In our treatment we shall not pursue the aim of a detailed exposition, setting ourselves only the task of recalling it in brief outlines, meanwhile referring the reader who would have a more detailed acquaintance with it to special manuals on approximate computations [1]).

Let there be given arbitrary values $a_0, a_1, a_2, \ldots, a_n$ of the independent variable x and the values of the function corresponding to them: $y_0 = f(a_0), y_1 = f(a_1), \ldots, y_n = f(a_n)$.

One considers what are known as difference ratios, formed according to the following law:

$$f(a_1, a_0) = \frac{f(a_1) - f(a_0)}{a_1 - a_0}, \quad f(a_2, a_1) = \frac{f(a_2) - f(a_1)}{a_2 - a_1},$$

$$f(a_3, a_2) = \frac{f(a_3) - f(a_2)}{a_3 - a_2}, \ \ldots \tag{1}$$

These are difference ratios of the first order. In terms of them difference ratios of the second order are formed:

$$f(a_2, a_1, a_0) = \frac{f(a_2, a_1) - f(a_1, a_0)}{a_2 - a_0},$$

$$f(a_3, a_2, a_1) = \frac{f(a_3, a_2) - f(a_2, a_1)}{a_3 - a_1}, \ \ldots \tag{2}$$

Next the difference ratios of the third order:

$$f(a_3, a_2, a_1, a_0) = \frac{f(a_3, a_2, a_1) - f(a_2, a_1, a_0)}{a_3 - a_0}, \tag{3}$$

and so forth.

It is not hard to see that for equidistant values of the argument,

$a_0, \ a_1 = a_0 + h, \ a_2 = a_0 + 2h, \ \ldots,$

$$\left.\begin{aligned} &f(a_1, a_0) = \frac{\Delta y_0}{h}, \quad f(a_2, a_1) = \frac{\Delta y_1}{h}, \quad f(a_3, a_2) = \frac{\Delta y_2}{h}, \ \ldots \\ &f(a_2, a_1, a_0) = \frac{\Delta^2 y_0}{2! h^2}, \quad f(a_3, a_2, a_1) = \frac{\Delta^2 y_1}{2! h^2}, \ \ldots \\ &f(a_3, a_2, a_1, a_0) = \frac{\Delta^3 y_0}{3! h^3}, \ \ldots \\ &\ldots\ldots\ldots\ldots\ldots\ldots\ldots\ldots \end{aligned}\right\} \tag{4}$$

The difference ratios may be presented in a more symmetric form,

---

[1]) See, for example, Whittaker, E., and Robinson, G., [1]; A. N. Krylov, [2]; J. Scarborough, [1].

to wit:

$$
\left.\begin{aligned}
f(a_1, a_0) &= \frac{f(a_0)}{a_0 - a_1} + \frac{f(a_1)}{a_1 - a_0},\\
f(a_2, a_1, a_0) &= \frac{1}{a_2 - a_0}\left\{\frac{f(a_2)}{a_2 - a_1} + \frac{f(a_1)}{a_1 - a_2}\right\} + \\
&+ \frac{1}{a_0 - a_2}\left\{\frac{f(a_1)}{a_1 - a_0} + \frac{f(a_0)}{a_0 - a_1}\right\} = \frac{f(a_0)}{(a_0 - a_1)(a_0 - a_2)} + \\
&\qquad + \frac{f(a_1)}{(a_1 - a_0)(a_1 - a_2)} + \frac{f(a_2)}{(a_2 - a_0)(a_2 - a_1)},\\
&\dots\dots\dots\dots\dots\dots\dots\\
f(a_n, a_{n-1}, \dots, a_1, a_0) &= \\
= \sum_{k=1}^{n} & \frac{f(a_k)}{(a_k - a_0)\dots(a_k - a_{k-1})(a_k - a_{k+1})\dots(a_k - a_n)}.
\end{aligned}\right\} \quad (5)
$$

It is perfectly obvious from these equations that the right sides are symmetric with respect to $a_0, a_1, \dots, a_n$, and therefore that $f(a_n, a_{n-1}, \dots, a_0)$ is not altered by any permutation of the values a_k of the argument.

Let us seek the polynomial of degree n that will take, for values of the argument $a_0, a_1, \dots, a_n$, the respective values $y_0 = f(a_0)$, $y_1 = f(a_1), \dots, y_n = f(a_n)$.

We shall exhibit the solution of the problem in a form convenient for us, which was given by Newton in 1687 [1]). We shall rely on the obvious statement that the difference ratios of order n of a polynomial of degree n are constant [2]). Let us assume for the sake of simplicity that only four values of the argument, a_0, a_1, a_2, a_3, are given, and that we seek the polynomial of third degree acquiring for these values the assigned values $f(a_k)$.

On the strength of the constancy of the difference ratios of the third order, we have

$$f(x, a_0, a_1, a_2) = f(a_0, a_1, a_2, a_3).$$

But by definition of the difference ratios of the third order

$$f(x, a_0, a_1, a_2) = \frac{f(x, a_0, a_1) - f(a_0, a_1, a_2)}{x - a_2},$$

---

[1]) Newton, [1]: The mathematical principles of natural philosophy, Book III, Lemma V.

[2]) In order to satisfy oneself of this, it is sufficient to show that the difference ratios of the first order of any power $f(x) = x^k$ will be polynomials of a degree lower by one.

But this can be verified quite readily, for

$$\frac{f(x) - f(a)}{x - a} = \frac{x^k - a^k}{x - a} = x^{k-1} + x^{k-2}a + \dots + a^{k-1}.$$

therefore

$$f(x, a_0, a_1) = f(a_0, a_1, a_2) + (x - a_2)f(x, a_0, a_1, a_2) = \\ = f(a_0, a_1, a_2) + (x - a_2)f(a_0, a_1, a_2, a_3).$$

Furthermore, in accordance with the definition of the second difference ratio:

$$f(x, a_0, a_1) = \frac{f(x, a_0) - f(a_0, a_1)}{x - a_1},$$

$$f(x, a_0) = f(a_0, a_1) + (x - a_1)f(x, a_0, a_1) = \\ = f(a_0, a_1) + (x - a_1)f(a_0, a_1, a_2) + (x - a_1)(x - a_2)f(a_0, a_1, a_2, a_3).$$

Finally, since

$$f(x) = f(a_0) + (x - a_0)f(x, a_0),$$

we obtain:

$$f(x) = f(a_0) + (x - a_0)f(a_0, a_1) + (x - a_0)(x - a_1)f(a_0, a_1, a_2) + \\ + (x - a_0)(x - a_1)(x - a_2)f(a_0, a_1, a_2, a_3).$$

In the derivation of this interpolation formula we have carried through the reasoning for the case of four points. But the like argument is obviously applicable to the case of any other number of points. If we employ it in connection with the interpolating polynomial of degree n in a problem with $n + 1$ interpolation nodes, we obtain the following general formula of Newton's:

$$f(x) = f(a_0) + (x - a_0)f(a_0, a_1) + (x - a_0)(x - a_1)f(a_0, a_1, a_2) + \dots \\ + \dots + (x - a_0)(x - a_1) \dots (x - a_{n-1})f(a_0, a_1, \dots, a_n). \quad (6)$$

This same interpolation polynomial was obtained by Lagrange in another form, which is less convenient for computations but more convenient for theoretical investigations. We shall cite the Lagrangian form of the polynomial, since it will be useful to us in determining the remainder terms in the approximate formulas that we shall obtain later for the derivatives of different orders of the given function. In order to distinguish in symbols the given function $f(x)$ and the polynomial (6), its interpolant based upon the points $a_0, a_1, a_2, \dots$, let us agree to designate the interpolating polynomial everywhere below by $P_n(x)$.

If in (6) we collect the terms containing like values $f(a_0), \dots, f(a_n)$ of the function $f(x)$ being interpolated, the polynomial (6) can be brought to the form

$$P_n(x) = \sum_{k=0}^{n} \frac{(x - a_0) \dots (x - a_{k-1})(x - a_{k+1}) \dots (x - a_n)}{(a_k - a_0) \dots (a_k - a_{k-1})(a_k - a_{k+1}) \dots (a_k - a_n)} \cdot f(a_k). \quad (7)$$

This is indeed the interpolation polynomial in Lagrange's form. To simplify writing it we shall introduce the following notation:

$$\omega(x) = (x - a_0)(x - a_1) \dots (x - a_n).$$

$\omega(x)$ is obviously a polynomial of degree $n + 1$, having the coefficient of the leading term equal to unity and vanishing at the nodes of the interpolation.

If we call attention to the fact that

$$(x - a_0) \dots (x - a_{k-1})(x - a_{k+1}) \dots (x - a_n) = \frac{\omega(x)}{x - a_k}$$

and

$$(a_k - a_0) \dots (a_k - a_{k-1})(a_k - a_{k+1}) \dots (a_k - a_n) = \omega'(a_k),$$

it becomes clear that the Lagrange polynomial can be represented in the form

$$P_n(x) = \omega(x) \sum_{k=0}^{n} \frac{f(a_k)}{\omega'(a_k)(x - a_k)}. \tag{7$_1$}$$

If the function $f(x)$ is a polynomial of degree not higher than n, the interpolating polynomial $P_n(x)$ coincides identically with the function $f(x)$ being interpolated. In the contrary case $f(x)$ and $P_n(x)$ have the same values at the interpolation nodes, but elsewhere they diverge, generally speaking. The difference $R_{n+1}(x) = f(x) - P_n(x)$ is called the remainder term of the interpolation formula:

$$f(x) = P_n(x) + R_{n+1}(x). \tag{7$_2$}$$

We shall straightway obtain an expression for the remainder term $R_{n+1}(x)$ in a form convenient for our purposes. Let us observe at the outset that if $f(x) = 1$, the polynomial $P_n(x)$ will be identically equal to unity and therefore $\omega(x) \sum_{k=0}^{n} \frac{1}{\omega'(a_k)(x - a_k)} = 1$. Multiplying both sides of this equation by $f(x)$, we shall have put this function in the form $f(x) = \omega(x) \sum_{k=0}^{n} \frac{f(x)}{\omega'(a_k)(x - a_k)}$. If in (7$_2$) we now introduce for $f(x)$ this expression, and instead of $P_n(x)$ its value from equality (7$_1$), we obtain $R_{n+1}(x)$ in a form convenient for investigations:

$$R_{n+1}(x) = \omega(x) \sum_{k=0}^{n} \frac{f(a_k) - f(x)}{\omega'(a_k)(a_k - x)}, \tag{8}$$

$$f(x) = P_n(x) + \omega(x) \sum_{k=0}^{n} \frac{f(a_k) - f(x)}{\omega'(a_k)(a_k - x)}. \tag{9}$$

Usually in studying the remainder term it is brought into a form different from (8). Let us assume that the function $f(x)$ has derivatives of up to the $(n + 1)$th order, inclusive, and let us consider the auxiliary function of the argument z:

$$F(z) = f(z) - P_n(z) - [f(x) - P_n(x)]\frac{\omega(z)}{\omega(x)}.$$

It reduces to zero for the $n + 2$ values of the argument z: $a_0, a_1, \ldots, a_n$ and x. Therefore a point ξ can be found, lying among $a_0, a_1, \ldots, a_n, x$, at which the derivative of the $(n + 1)$th order of the function under consideration vanishes.

But since $P_n(z)$ is a polynomial of degree n and consequently its derivative of the $(n + 1)$th order is equal to zero, and $\omega(z)$ is a polynomial of degree $n + 1$ with the leading term having unit coefficient, and with its $(n + 1)$th derivative equalling $(n + 1)!$, we have

$$F^{(n+1)}(z) = f^{(n+1)}(z) - [f(x) - P_n(x)]\frac{(n + 1)!}{\omega(x)}.$$

At the point ξ we must have $F^{(n+1)}(\xi) = 0$, and therefore

$$P_{n+1}(x) = f(x) - P_n(x) = \omega(x)\frac{f^{(n+1)}(\xi)}{(n + 1)!}. \tag{8_1}$$

Here ξ is a number lying among $a_0, a_1, \ldots, a_n, x$.

We shall note one result which follows from this. From a comparison of equations (8) and (8_1) it is evident that if $f(x)$ has derivatives of up to the $(n + 1)$th order, there must exist a ξ mean between $a_0, \ldots, a_n, x$, such that the following equation holds:

$$\sum_{k=0}^{n}\frac{f(a_k) - f(x)}{\omega'(a_k)(a_k - x)} = \frac{f^{(n+1)}(\xi)}{(n + 1)!}. \tag{8_2}$$

Let us now consider particular cases of the general interpolation formula (9) for equidistant values of the argument.

Newton's formula for forward interpolation. Let us assume the function $f(x)$ to be given for the following equidistant values of the argument:

$$a_0 = a, \quad a_1 = a + h, \quad a_2 = a + 2h, \ldots, a_n = a + nh.$$

For the construction of an interpolation formula it is sufficient to calculate the difference ratios figuring in (6) and corresponding to the given interpolation nodes. The difference ratios here are at once expressible in terms of the differences of the function and the magnitude h of the step of the net, by equations (4), so that in our case the interpolation formula takes the form:

$$f(x) = f(a) + \frac{(x - a)}{1!h}\Delta f(a) + \frac{(x - a)(x - a - h)}{2!h^2}\Delta^2 f(a) + \ldots$$
$$\ldots + \frac{(x - a)(x - a - h)\ldots(x - a - (n - 1)h)}{n!h^n}\Delta^n f(a) + R_{n+1}(x). \tag{10}$$

As the name itself of this formula shows, it is convenient in computing values of the function and its derivatives for values of the argument close to the initial value a.

Newton's formula for backward interpolation. We shall mention a formula analogous to (10), but used in computing $f(x)$ and its derivatives for x close to the maximum of the given values of the argument.

Let $f(x)$ be known for the following values of x:

$$a_0 = a, \quad a_1 = a - h, \quad a_2 = a - 2h, \quad \ldots, \quad a_n = a - nh.$$

For the construction of a formula it is again sufficient for us to calculate, for our arrangement of interpolation nodes, the difference ratios in equation (6). Using the symmetry of the difference ratios and equations (4), we obtain

$$f(a_0, a_1) = f(a - h, a) = \frac{1}{h} \Delta f(a - h)$$

$$f(a_0, a_1, a_2) = f(a - 2h, a - h, a) = \frac{1}{2!h^2} \Delta^2 f(a - 2h)$$

$$\cdots\cdots\cdots\cdots\cdots\cdots\cdots\cdots\cdots$$

$$f(a_0, a_1, \ldots, a_n) = f(a - nh, \ldots, a - h, a) = \frac{1}{n!h^n} \Delta^n f(a - nh).$$

Hence it is clear that the interpolation formula in this case will have the form:

$$f(x) = f(a) + \frac{x - a}{1!h} \Delta f(a - h) + \frac{(x - a)(x - a + h)}{2!h^2} \Delta^2 f(a - 2h) +$$

$$+ \ldots + \frac{(x-a)(x-a+h) \ldots (x-a+(n-1)h)}{n!h^n} \Delta^n f(a - nh) + R_{n+1}(x). \quad (11)$$

The Newton-Gauss interpolation formula for values of the argument close to the middle of the interval. As we have already said above, the majority of the formulas for parabolic interpolation with which the discussion of this No. is concerned are in essence equivalent and differ only in the arrangement of the terms. In some problems — for example, in computing integrals — it is immaterial in what order the interpolation nodes are taken, since the order of the nodes obviously has no influence on the magnitude of the integral, for an identical role is played in each by the value of the function at each of the points of the interval of integration. But if the matter is one of the computation of the values of the function itself, or of its derivatives, at separate points of the interval, then for computational convenience it is natural to arrange the interpolation nodes so that the values of the function at the nodes closest to the value of the argument of concern may figure in the first terms of the interpolation formula, and the values of the function at distant nodes in its last terms. The formulas (10) and (11) which we have cited satisfy this requirement for the case when x is close to one of the ends of the interval of interpolation.

If x lies close to the middle of the interval, the interpolation formula is transformed to advantage by having in the first terms only the values of the function at points close to the middle of the interval. This is attainable by the following arrangement of the nodes.

Let the function $f(x)$ be given for the $n + 1$ values of x:

$$\ldots a - 2h, \quad a - h, \quad a + h, \quad a + 2h, \ldots$$

We shall not decide in advance whether $n + 1$ is an odd or an even number. If $n + 1$ is an odd number, a will be the middle node of the net; if however $n + 1$ is an even number, let the middle of the interval be located between a and $a + h$.

In formula (6) set

$$a_0 = a, \quad a_1 = a + h, \quad a_2 = a - h, \quad a_3 = a + 2h, \quad a_4 = a - 2h, \ldots$$

and calculate all the difference ratios which figure in it. If we again make use of the symmetry of the difference ratios with respect to the values of the argument, and of formulas (4), we obtain

$$\left.\begin{aligned}
&f(a_0, a_1) = f(a, a + h) = \frac{1}{1!h} \Delta f(a) \\
&f(a_0, a_1, a_2) = f(a - h, a, a + h) = \frac{1}{2!h^2} \Delta^2 f(a - h) \\
&f(a_0, a_1, a_2, a_3) = f(a - h, a, a + h, a + 2h) = \frac{1}{3!h^3} \Delta^3 f(a - h) \\
&\cdots\cdots\cdots\cdots\cdots\cdots\cdots\cdots\cdots\cdots \\
&f(x) = f(a) + \frac{x - a}{1!h} \Delta f(a) + \frac{(x - a)(x - a - h)}{2!h^2} \Delta^2 f(a - h) + \\
&+ \frac{(x - a + h)(x - a)(x - a - h)}{3!h^3} \Delta^3 f(a - h) + \\
&+ \frac{(x - a + h)(x - a)(x - a - h)(x - a - 2h)}{4!h^4} \Delta^4 f(a - 2h) + \ldots \\
&+ \ldots + R_{n+1}(x).
\end{aligned}\right\} \qquad (12)$$

Which is indeed the Newton-Gauss formula.

The Stirling formula. Let us assume that we have to find the values of the function or of its derivatives for several x very close to one of the interpolation nodes, for example a. It is then natural to take an odd number of nodes in formula (12), namely as many points to the left of a as to the right of it. The scheme of nodes will be

$$a - ph, \ \ldots a - h, \ a, \ a + h, \ \ldots a + ph.$$

In order to give (12) a more symmetric form, corresponding to the symmetry of the disposition of the nodes, we effect in it the rearrangement of terms which is indicated below:

$$f(x) = f(a) + \frac{x-a}{1!h}[\Delta f(a) - \tfrac{1}{2}\Delta^2 f(a-h)] + \frac{(x-a)^2}{2!h^2}\Delta^2 f(a-h) +$$
$$+ \frac{(x-a)[(x-a)^2-h^2]}{3!h^3}[\Delta^3 f(a-h) - \tfrac{1}{2}\Delta^4 f(a-2h)] +$$
$$+ \frac{(x-a)^2[(x-a)^2-h^2]}{4!h^4}\Delta^4 f(a-2h) + \ldots + R_{2p+1}(x).$$

Next we replace the differences of even order which stand in brackets by their expressions in terms of the odd differences, using the identities

$$\Delta^2 f(a-h) = \Delta f(a) - \Delta f(a-h)$$
$$\Delta^4 f(a-2h) = \Delta^3 f(a-h) - \Delta^3 f(a-2h)$$
$$\cdots\cdots\cdots\cdots\cdots\cdots$$

As the result we obtain Stirling's formula

$$f(x) = f(a) + \frac{x-a}{1!h}\,\frac{\Delta f(a) + \Delta f(a-h)}{2} + \frac{(x-a)^2}{2!h^2}\Delta^2 f(a-h) +$$
$$+ \frac{(x-a)[(x-a)^2-h^2]}{3!h^3}\,\frac{\Delta^3 f(a-h) + \Delta^3 f(a-2h)}{2} +$$
$$+ \frac{(x-a)^2[(x-a)^2-h^2]}{4!h^4}\Delta^4 f(a-2h) + \ldots + R_{2p+1}(x). \quad (13)$$

We shall utilize it below for determining an approximate expression for the derivatives at the nodes of the net in terms of the values of the function.

The Bessel formula. Let us now assume that we have to find the value of the function or of its derivatives either exactly at the middle of two nodes, or at points close to the middle. Here it is natural to take an even number of interpolation nodes and arrange them symmetrically about the point near which the required values are located. For example, let it be required to carry out computations for values of x near the middle of the interval between the nodes a and $a + h$. Let us take $n + 1 = 2p$ nodes, disposed symmetrically with respect to the point $a + \frac{h}{2}$:

$$a - (p-1)h, \ \ldots, \ a - h, \ a, \ a + h, \ \ldots, \ a + ph,$$

and in Newton's formula (6) let us set

$$a_0 = a, \ a_1 = a + h, \ a_2 = a - h, \ a_3 = a + 2h, \ \ldots$$

It is reducible, as we explained in deriving the Newton-Gauss formula, to form (12).

For transforming it into symmetric form we extract from it the quantities $\frac{1}{2}f(a)$, $\frac{1}{2}\Delta^2 f(a-h)$, $\frac{1}{2}\Delta^4 f(a-2h)$, ... and substitute for them their expressions from the equations

$$\begin{aligned} f(a) &= f(a+h) - \Delta f(a) \\ \Delta^2 f(a-h) &= \Delta^2 f(a) - \Delta^3 f(a-h) \\ \Delta^4 f(a-2h) &= \Delta^4 f(a-h) - \Delta^5 f(a-2h) \\ &\cdots\cdots\cdots\cdots \end{aligned}$$

Formula (12) will then take the form:

$$\left.\begin{aligned} f(x) = {} & \frac{f(a)+f(a+h)}{2} + \frac{x-a-\frac{h}{2}}{1!h}\Delta f(a) + \\ & + \frac{(x-a)(x-a-h)}{2!h^2}\,\frac{\Delta^2 f(a-h)+\Delta^2 f(a)}{2} + \\ & + \frac{(x-a)(x-a-h)\left(x-a-\frac{h}{2}\right)}{3!h^3}\Delta^3 f(a-h) + \\ & + \frac{(x-a+h)(x-a)(x-a-h)(x-a-2h)}{4!h^4} \times \\ & \times \frac{\Delta^4 f(a-2h)+\Delta^4 f(a-h)}{2} + \ldots + R_{2p}(x). \end{aligned}\right\} \tag{14}$$

And this is Bessel's formula. For us it is of interest in the following connection. Let a boundary-value problem for a differential equation have been solved by the method of nets, for a net with step h, and let the values found fail to give the necessary accuracy. We must then reduce the step of the net and solve the same problem with a finer net. Ordinarily one reduces the step of the net by half and solves the problem for a net of step $\frac{h}{2}$. In order to begin the solution of the problem for the new net by the method of iteration (see No. 2, § 3), one must at all of its nodes assign initial approximate values for the sought solution. At those nodes of the new net which are simultaneously also nodes of the old net, one may adopt as the initial approximations those that were obtained on solving the problem with a net of step h. It still remains to indicate the initial values at the nodes of the new net which were not nodes of the old net. Generally speaking, these values may be assigned arbitrarily, but the number of computational cycles that must be carried through to arrive at the result with the necessary accuracy depends on how much the assigned values differ from the true ones. The greater the error in the initial values, the longer the computations. To reduce this error to the utmost, one can utilize the previously found values at the nodes of the old net to assign the initial values, by performing an interpolation with them for the intermediate nodes of the new net. And for such an interpolation the Bessel formula, (14), may prove useful.

In making the net twice as fine, one has to interpolate right at the middle of the interval between two nodes. We shall therefore give the particular case of the Bessel formula for $x = a + \frac{h}{2}$. The coefficients of the odd differences contain the factor $x - a - \frac{h}{2}$, which vanishes for our choice of x. All such terms will fall out, and we shall obtain as the result a simple formula which is called *Bessel's formula for interpolation at the middle of an interval*:

$$f\left(a+\frac{h}{2}\right)=\frac{f(a)+f(a+h)}{2}-\frac{1}{8}\,\frac{\Delta^2 f(a-h)+\Delta^2 f(a)}{2}+$$
$$+\frac{3}{128}\,\frac{\Delta^4 f(a-2h)+\Delta^4 f(a-h)}{2}-\frac{5}{1024}\,\frac{\Delta^6 f(a-3h)+\Delta^6 f(a-2h)}{2}+$$
$$+\ldots+R_{2p}\left(a+\frac{h}{2}\right). \tag{14$_1$}$$

Let us determine more exactly the magnitude of the remainder term $R_{2p}\left(a+\frac{h}{2}\right)$. The general form of it is indicated by equation (8_1). In our case $n + 1 = 2p$, and the nodes of the interpolation are $a-(p-1)h$, $\ldots, a - h, a, a + h, \ldots, a + ph$. The polynomial $\omega(x)$ will have the form

$$\omega(x) = (x-a)(x-a-h)(x-a+h)\ldots(x-a+(p-1)h)\,(x-a-ph),$$

$$\omega\left(a+\frac{h}{2}\right)=(-1)^p\,\frac{h^{2p}(1\cdot 3\ldots(2p-1))^2}{2^{2p}},$$

and consequently:

$$R_{2p}\left(a+\frac{h}{2}\right)=(-1)^p\,\frac{h^{2p}(1\cdot 3\ldots(2p-1))^2}{2^{2p}}\cdot\frac{f^{(2p)}(\xi)}{(2_p)!}, \tag{14$_2$}$$

where ξ is some number intermediate between $a-(p-1)h$ and $a+ph$.

Let us now turn to the problem of the computation of the derivatives. Let there be given the $n + 1$ values of the independent variable $x = a_0$, $a_1, \ldots, a_n$ and the values of the function $f(a_0), f(a_1), \ldots, f(a_n)$ corresponding to them. We pose the problem of finding an expression for the derivatives of the function $f(x)$ in terms of the given values of the function itself. It is usually solved in the following manner. By the known values of the function at the points a_k is formed the interpolation polynomial $P_n(x)$:

$$f(x) = P_n(x) + R_{n+1}(x).$$

Then the remainder term $R_{n+1}(x)$ of the interpolation formula is neglected, by $P_n(x)$ replacing the function being interpolated, and the derivative of the necessary order is calculated from this. The following approximate expressions for the derivatives in terms of the values of the function at the nodes of the interpolation are obtained:

$$f'(x) \approx P'_n(x),\ f''(x) \approx P''_n(x), \ldots$$

The exact values of the derivatives, however, will be equal to

$$f'(x) = P'_n(x) + R'_{n+1}(x),\ f''(x) = P''_n(x) + R''_{n+1}(x).$$

In these equations the quantities $R'_{n+1}(x)$, $R''_{n+1}(x)$, ... signify the errors in the approximate expressions obtained for the derivatives.

The choice of the interpolation formula depends on how the system of nodes of the net is given, and for what values of x the derivatives must be calculated.

In the general case, if the values of a_k are not equidistant, Newton's general formula (6) may be used. It gives the expressions for the derivatives in terms of the difference ratios of the given function. Having designated $x - a_k = \alpha_k$ to ease the notation, we write the Newton formula in the form:

$$f(x) = f(a_0) + \alpha_0 f(a_0, a_1) + \alpha_0\alpha_1 f(a_0, a_1, a_2) + \ldots + \\ + \alpha_0\alpha_1 \ldots \alpha_{n-1} f(a_0, a_1, \ldots, a_n) + R_{n+1}(x).$$

Differentiating this equation with respect to x term by term, we obtain:

$$\left.\begin{aligned} f'(x) &= f(a_0, a_1) + (\alpha_0 + \alpha_1) f(a_0, a_1, a_2) + \\ &\quad + (\alpha_0\alpha_1 + \alpha_0\alpha_2 + \alpha_1\alpha_2) f(a_0, a_1, a_2, a_3) + \\ &\quad + \ldots + R'_{n+1}(x), \\ \frac{1}{2!} f''(x) &= f(a_0, a_1, a_2) + (\alpha_0 + \alpha_1 + \alpha_2) f(a_0, a_1, a_2, a_3) + \\ &\quad + (\alpha_0\alpha_1 + \alpha_0\alpha_2 + \alpha_0\alpha_3 + \alpha_1\alpha_2 + \\ &\quad + \alpha_1\alpha_3 + \alpha_2\alpha_3) f(a_0, a_1, a_2, a_3, a_4) + \\ &\quad + \ldots + R''_{n+1}(x), \\ \frac{1}{3!} f'''(x) &= f(a_0, a_1, a_2, a_3) + (\alpha_0+\alpha_1+\alpha_2+\alpha_3) f(a_0, a_1, a_2, a_3, a_4) + \\ &\quad + \ldots + R'''_{n+1}(x), \\ \frac{1}{4!} f^{\mathrm{IV}}(x) &= f(a_0, a_1, a_2, a_3, a_4) + (\alpha_0 + \alpha_1 + \alpha_2 + \alpha_3 + \alpha_4) \times \\ &\quad \times f(a_0, a_1, a_2, a_3, a_4, a_5) + \ldots + R^{\mathrm{IV}}_{n+1}(x), \end{aligned}\right\} \quad (15)$$

etc.

The coefficients of the difference ratios in these formulas are symmetric functions of $\alpha_0, \alpha_1, \ldots$. In the first of formulas (15) as the coefficient of $f(a_0, a_1, \ldots, a_r)$ stands a symmetric function of order $r - 1$ in $\alpha_0, \alpha_1, \ldots, \alpha_{r-1}$. In all the succeeding formulas, after each differentiation its order diminishes by one.

Equations (15), if we discard the remainder terms in them, give approximate expressions for the derivatives for any values of the argument x.

We have a particular interest in obtaining expressions for the derivatives at the nodes of the net, a_k.

Let x coincide, for example, with a_0. Then $\alpha_k = a_0 - a_k$ and $\alpha_0 = 0$, and it follows from (15) that

$$\left.\begin{aligned}
f'(a_0) &= f(a_0, a_1) + \alpha_1 f(a_0, a_1, a_2) + \alpha_1\alpha_2 f(a_0, a_1, a_2, a_3) + \\
&\quad + \ldots + R'_{n+1}(a_0),\\
\frac{1}{2!}f''(a_0) &= f(a_0, a_1, a_2) + (\alpha_1 + \alpha_2) f(a_0, a_1, a_2, a_3) + \\
&\quad + (\alpha_1\alpha_2 + \alpha_1\alpha_3 + \alpha_2\alpha_3) f(a_0, a_1, a_2, a_3, a_4) + \\
&\quad + \ldots + R''_{n+1}(a_0),\\
\frac{1}{3!}f'''(a_0) &= f(a_0, a_1, a_2, a_3) + (\alpha_1+\alpha_2+\alpha_3) f(a_0, a_1, a_2, a_3, a_4) + \\
&\quad + \ldots + R'''_{n+1}(a_0),\\
\frac{1}{4!}f^{\mathrm{IV}}(a_0) &= f(a_0, a_1, a_2, a_3, a_4) + (\alpha_1 + \alpha_2 + \alpha_3 + \alpha_4) \times \\
&\quad \times f(a_0, a_1, a_2, a_3, a_4, a_5) + \ldots + R^{\mathrm{IV}}_{n+1}(a_0),
\end{aligned}\right\} \quad (16)$$

etc.

We shall in addition find an expression for the remainder terms of the general equations (15) which will permit us to make an estimate of the errors of the approximate expressions for the derivatives. The magnitude of the remainder term $R_{n+1}(x)$ of the interpolation formula has been reduced by us to form (8). It follows from it that

$$\left.\begin{aligned}
R'_{n+1}(x) &= \omega'(x)\sum_{k=0}^{n}\frac{f(a_k)-f(x)}{\omega'(a_k)(a_k-x)} + \omega(x)\sum_{k=0}^{n}\frac{f(a_k)-f(x)-\dfrac{a_k-x}{1!}f'(x)}{\omega'(a_k)(a_k-x)^2},\\
R''_{n+1}(x) &= \omega''(x)\sum_{k=0}^{n}\frac{f(a_k)-f(x)}{\omega'(a_k)(a_k-x)} + 2\omega'(x)\sum_{k=0}^{n}\frac{f(a_k)-f(x)-\dfrac{a_k-x}{1!}f'(x)}{\omega'(a_k)(a_k-x)^2} + \\
&\quad + \omega(x)\sum_{k=0}^{n} 2\,\frac{f(a_k)-f(x)-\dfrac{(a_k-x)}{1!}f'(x)-\dfrac{(a_k-x)^2}{2!}f''(x)}{\omega'(a_k)(a_k-x)^3},\\
&\ldots\ldots\ldots\ldots\ldots\ldots\ldots\ldots\ldots\ldots\\
R^{(r)}_{n+1}(x) &= \sum_{s=0}^{r} C_r^s\omega^{(r-s)}(x)\sum_{k=0}^{n} s!\,\times\\
&\quad \times \frac{f(a_k)-f(x)-\dfrac{a_k-x}{1!}f'(x)-\ldots-\dfrac{(a_k-x)^s}{s!}f^{(s)}(x)}{\omega'(a_k)(a_k-x)^{s+1}}.
\end{aligned}\right\} \quad (17)$$

In case x coincides with one of the nodes of the interpolation, these expressions are simplified. In particular, the terms in them containing $\omega(x)$ will be absent, since for $x = a_0, a_1, \ldots,$ the polynomial $\omega(x)$ vanishes.

When the values of the function are given for equidistant values of the argument, one can utilize one of the interpolation formulas of Newton, Stirling, or Bessel. They will give expressions for the derivatives in terms of the differences of the functions.

The choice between them is made on the same bases as in the problem of interpolation. Thus, for instance, if we are given the values of the function at the points $a, a + h, \ldots, a + nh$, and we must find the values of the derivatives near the point a, we should utilize the Newton formula (10) for forward interpolation. Below we need expressions for the derivatives only at the nodes of the net, and we shall therefore consider the case $x = a$.

Calculating the derivative from (10), and setting $x = a$ in the result, we find the following expressions for the derivatives at the point a in terms of the values of the function at points lying to the right of a:

$$\left.\begin{aligned}
f'(a) &= \frac{1}{h}\Big[\Delta f(a) - \tfrac{1}{2}\Delta^2 f(a) + \tfrac{1}{3}\Delta^3 f(a) - \ldots + \\
&\qquad + (-1)^{n-1}\frac{1}{n}\Delta^n f(a)\Big] + R'_{n+1}(a),\\
f''(a) &= \frac{1}{h^2}\,[\Delta^2 f(a) - \Delta^3 f(a) + \tfrac{11}{12}\Delta^4 f(a) - \tfrac{5}{6}\Delta^5 f(a) + \\
&\qquad + \tfrac{137}{180}\Delta^6 f(a) - \ldots] + R''_{n+1}(a),\\
f'''(a) &= \frac{1}{h^3}\,[\Delta^3 f(a) - \tfrac{3}{2}\Delta^4 f(a) + \tfrac{7}{4}\Delta^5 f(a) - \tfrac{15}{8}\Delta^6 f(a) + \ldots] + \\
&\qquad + R'''_{n+1}(a),
\end{aligned}\right\}\quad (18)$$

etc.

If, on the other hand, we are given the values of the function at the points $a, a - h, a - 2h, \ldots$ and need to find the derivatives at $x = a$, we utilize the Newton formula (11) for backward interpolation. Differentiating (11) with respect to x and putting $x = a$ in the result, we find

$$\left.\begin{aligned}
f'(a) &= \frac{1}{h}\Big[\Delta f(a-h) + \tfrac{1}{2}\Delta^2 f(a-2h) + \tfrac{1}{3}\Delta^3 f(a-3h) + \ldots + \\
&\qquad + \frac{1}{n}\Delta^n f(a-nh)\Big] + R'_{n+1}(a),\\
f''(a) &= \frac{1}{h^2}\,[\Delta^2 f(a-2h) + \Delta^3 f(a-3h) + \tfrac{11}{12}\Delta^4 f(a-4h) + \\
&\qquad + \tfrac{5}{6}\Delta^5 f(a-5h) + \tfrac{137}{180}\Delta^6 f(a-6h) + \ldots] + R''_{n+1}(a),\\
f'''(a) &= \frac{1}{h^3}\,[\Delta^3 f(a-3h) + \tfrac{3}{2}\Delta^4 f(a-4h) + \tfrac{7}{4}\Delta^5 f(a-5h) + \\
&\qquad + \tfrac{15}{8}\Delta^6 f(a-6h) + \ldots] + R'''_{n+1}(a),\\
f^{\mathrm{IV}}(a) &= \frac{1}{h^4}\,[\Delta^4 f(a-4h) + 2\Delta^5 f(a-5h) + \tfrac{17}{6}\Delta^6 f(a-6h) + \ldots] + \\
&\qquad + R^{\mathrm{IV}}_{n+1}(a).
\end{aligned}\right\}\quad (19)$$

We shall dwell in more detail on the case — most important for us — when we must find expressions for the derivatives at some node of the net in terms of the values of the function at the nodes situated symmetrically on both sides of the point under consideration. Here we shall make use of Stirling's formula. Let the values of the function at the points $a - ph, \ldots, a - h, a, a + h, \ldots, a + ph$ be considered, and let it be required to find the derivatives at the point a. Let us take the derivatives of both sides of equation (13) and put $x = a$. We then find:

$$\left.\begin{aligned}
f'(a) &= \frac{1}{h}\Big[\frac{\Delta f(a) + \Delta f(a-h)}{2} - \frac{1}{3!}\,\frac{\Delta^3 f(a-h) + \Delta^3 f(a-2h)}{2} + \\
&\qquad + \frac{4}{5!}\,\frac{\Delta^5 f(a-2h) + \Delta^5 f(a-3h)}{2} - \ldots + \\
&+ \frac{(-1)^{p-1}[(p-1)!]^2}{(2p-1)!}\,\frac{\Delta^{2p-1} f(a-(p-1)h) + \Delta^{2p-1} f(a-ph)}{2}\Big] + R'_{2p+1}(a), \\
f''(a) &= \frac{1}{h^2}\Big[\Delta^2 f(a-h) - \frac{2}{4!}\Delta^4 f(a-2h) + \frac{8}{6!}\Delta^6 f(a-3h) - \ldots\Big] + \\
&\qquad + R''_{2p+1}(a), \\
f'''(a) &= \frac{1}{h^3}\Big[\frac{\Delta^3 f(a-h) + \Delta^3 f(a-2h)}{2} - \frac{30}{5!}\,\frac{\Delta^5 f(a-2h) + \Delta^5 f(a-3h)}{2} + \\
&\qquad + \ldots\Big] + R'''_{2p+1}(a), \\
f^{\mathrm{IV}}(a) &= \frac{1}{h^4}\Big[\Delta^4 f(a-2h) - \frac{120}{6!}\Delta^6 f(a-3h) + \ldots\Big] + R^{\mathrm{IV}}_{2p+1}(a),
\end{aligned}\right\} \quad (20)$$

etc.

We shall study the remainder terms of these formulas and shall establish estimates of them. Let us take the first of them, $R'_{2p+1}(a)$. Its expression for any interpolation nodes is given by the first equation of (17).

In our case $a_0 = a,\ a_1 = a + h,\ a_2 = a - h, \ldots,$

$$\omega(x) = (x - a)[(x - a)^2 - h^2] \ldots [(x - a)^2 - p^2h^2],$$

$$\omega(a) = 0,$$

$$\omega'(a) = (-1)^p (p!)^2 h^{2p}.$$

Therefore the remainder term will be equal to

$$R'_{2p+1}(a) = (-1)^p (p!)^2 h^{2p} \left[\sum_{k=0}^{2p} \frac{f(a_k) - f(x)}{\omega'(a_k)(a_k - x)}\right]_{x=a}. \quad (21)$$

If we also take into consideration equation (8_2), giving the value of the function in brackets for any values of x, it becomes evident that the remainder term in the expression for the first derivative can be presented in the form

$$R'_{2p+1}(a) = \frac{(-1)^p (p!)^2 h^{2p}}{(2p+1)!} f^{(2p+1)}(\xi) \quad (a - ph < \xi < a + ph). \tag{22}$$

We note particular cases of the values of the first derivative:

for $p = 1$:

$$f'(a) = \frac{f(a+h) - f(a-h)}{2h} - \frac{h^2}{3!} f'''(\xi), \tag{20$_1$}$$

for $p = 2$:

$$f'(a) = \frac{-f(a+2h) + 8f(a+h) - 8f(a-h) + f(a-2h)}{12h} + \frac{h^4}{30} f^{\mathrm{V}}(\xi). \tag{20$_2$}$$

Let us in addition estimate the remainder term in the expression for the second derivative. If we expand the polynomial $\omega(x)$ above in powers of $x - a$, it will become evident that only odd powers of $x - a$ figure in the expansion. Hence it follows that all the derivatives of even order of $\omega(x)$ at the point $x = a$ are equal to zero. In particular $\omega''(a) = 0$, and it is clear from the second equation of (17) that

$$R''_{2p+1}(a) = 2\omega'(a) \left[\sum_{k=0}^{2p} \frac{f(a_k) - f(x) - \frac{a_k - x}{1!} f'(x)}{\omega'(a_k)(a_k - x)^2} \right]_{x=a}.$$

We shall make an estimate of the expression standing in brackets. To do this we shall establish a property of the coefficients of the Lagrange interpolation formula. Let us consider a polynomial in the argument z:

$$\Pi_s(z) = (z - x)^s \qquad (s = 1, 2, \ldots, n).$$

We shall call the interpolation nodes $a_0, a_1, \ldots, a_n$. Let us form the interpolating polynomial for $\Pi_s(z)$. Since $\Pi_s(z)$ is a polynomial of degree not higher than n, its interpolating polynomial will be equal to the polynomial $\Pi_s(z)$ itself, and the following equation will hold, in accordance with (7_1):

$$\Pi_s(z) = (z - x)^s = \omega(z) \sum_{k=0}^{n} \frac{(a_k - x)^s}{\omega'(a_k)(z - a_k)}.$$

If we put $z = x$ here, in view of the fact that $\Pi_s(x) = 0$, it will follow that

$$\sum_{k=0}^{n} \frac{(a_k - x)^s}{\omega'(a_k)(x - a_k)} = 0 \qquad (s = 1, 2, \ldots, n). \tag{23}$$

Let us revert to the remainder term $R''_{2p+1}(a)$. Let us develop the numerator of the expression in brackets by Taylor's formula:

$$f(a_k) - f(x) - \frac{a_k - x}{1!} f'(x) = \frac{(a_k - x)^2}{2!} f''(x) + \ldots +$$

$$+ \frac{(a_k - x)^{2p+1}}{(2p+1)!} f^{(2p+1)}(x) + \frac{(a_k - x)^{2p+2}}{(2p+2)!} f^{(2p+2)}(\xi_k),$$

ξ_k being an intermediate value between x and a_k.

$$\sum_{k=0}^{2p} \frac{f(a_k) - f(x) - \frac{a_k - x}{1!} f'(x)}{\omega'(a_k)(a_k - x)^2} =$$

$$= \frac{f''(x)}{2!} \sum_{k=0}^{2p} \frac{(a_k - x)^2}{\omega'(a_k)(a_k - x)^2} + \ldots +$$

$$+ \frac{f^{(2p+1)}(x)}{(2p+1)!} \sum_{k=0}^{2p} \frac{(a_k - x)^{2p+1}}{\omega'(a_k)(x - a_k)^2} +$$

$$+ \frac{1}{(2p+2)!} \sum_{k=0}^{2p} \frac{(a_k - x)^{2p+2}}{\omega'(a_k)(x - a_k)^2} f^{(2p+2)}(\xi_k).$$

All sums on the right side of the equation, with the exception of the last, vanish, on the strength of (23). Therefore

$$R''_{2p+1}(a) = 2\omega'(a) \frac{1}{(2p+2)!} \sum_{k=0}^{2p} \frac{(a_k - a)^{2p}}{\omega'(a_k)} f^{(2p+2)}(\xi_k).$$

The quantity $\omega'(a)$ has been calculated by us above:

$$\omega'(a + kh) = (-1)^{p-k} h^{2p} (p+k)!(p-k)!$$

In $R''_{2p+1}(a)$ let us combine terms corresponding to points which are symmetric with respect to a:

$$R''_{2p+1}(a) = 2(-1)^p h^{2p} (p!)^2 \frac{1}{(2p+2)!} \times$$

$$\times \sum_{k=1}^{p} \frac{k^{2p} h^{2p} [f^{(2p+2)}(\xi_{p+k}) + f^{(2p+2)}(\xi_{p-k})]}{(-1)^{p-k} h^{2p} (p+k)!(p-k)!}.$$

Let us in addition introduce the following designation:

$$\max_{a - ph \leqslant x \leqslant a + ph} |f^{(s)}(x)| = M_s.$$

Then

$$R''_{2p+1}(a) = 4 \frac{(p!)^2 h^{2p}}{(2p+2)!} \sum_{k=1}^{p} \frac{k^{2p}}{(p+k)!(p-k)!} \vartheta M_{2p+2}, \quad |\vartheta| \leqslant 1.$$

We have for particular cases of the second derivative: for $p = 1$:

$$f''(a) = \frac{f(a+h) - 2f(a) + f(a-h)}{h^2} + \frac{h^2}{12} \vartheta M_4, \qquad (20_3)$$

for $p = 2$:

$$f''(a) = \frac{-f(a+2h) + 16f(a+h) - 30f(a) + 16f(a-h) - f(a-2h)}{12h^2} + \\ + \frac{h^4}{54}\vartheta M_6. \qquad (20_4)$$

Estimates of the remainder terms in the expressions for the derivatives of order higher than the second can be obtained in analogous fashion.

We subjoin a summary of approximate expressions for the first six derivatives [1]).

$$f_0 = f(a),$$
$$\delta_\nu = f(a+\nu h) - f(a-\nu h),$$
$$\sigma_\nu = f(a+\nu h) + f(a-\nu h),$$
$$|\vartheta| \leqslant 1;$$

$$\left.\begin{aligned}
f'(a) &= \frac{\delta_1}{2h} + \frac{h^2}{6}\vartheta M_3 = \frac{-\delta_2 + 8\delta_1}{12h} + \frac{h^4}{30}\vartheta M_5 = \\
&= \frac{\delta_3 - 9\delta_2 + 45\delta_1}{60h} + \frac{h^6}{140}\vartheta M_7, \\
f''(a) &= \frac{\sigma_1 - 2f_0}{h^2} + \frac{h^2}{12}\vartheta M_4 = \frac{-\sigma_2 + 16\sigma_1 - 30f_0}{12h^2} + \frac{h^4}{54}\vartheta M_6 = \\
&= \frac{2\sigma_3 - 27\sigma_2 + 270\sigma_1 - 490f_0}{180h^2} + \frac{47}{8480}h^6\vartheta M_8, \\
f'''(a) &= \frac{\delta_2 - 2\delta_1}{2h^3} + \frac{17h^2}{60}\vartheta M_5 = \frac{-\delta_3 + 8\delta_2 - 13\delta_1}{8h^3} + \frac{403\,h^4}{2520}\vartheta M_7, \\
f^{(4)}(a) &= \frac{\sigma_2 - 4\sigma_1 + 6f_0}{h^4} + \frac{17h^2}{90}\vartheta M_6 = \\
&= \frac{-\sigma_3 + 12\sigma_2 - 39\sigma_1 + 56f_0}{6h^4} + \frac{403\,h^4}{5040}\vartheta M_8, \\
f^{(5)}(a) &= \frac{\delta_3 - 4\delta_2 + 5\delta_1}{2h^5} + \frac{169\,h^2}{315}\vartheta M_7, \\
f^{(6)}(a) &= \frac{\sigma_3 - 6\sigma_2 + 15\sigma_1 - 20f_0}{h^6} + \frac{169\,h^2}{420}\vartheta M_8.
\end{aligned}\right\} \qquad (24)$$

2. Relations between the values of a function at the nodes of a net, and the Laplace and biharmonic operators. In the numer-

---

[1]) We borrow the estimates of the remainder terms from a brochure of L. Collatz, [2].

ical solution of the Laplace and Poisson equations we shall require certain relations between the Laplace operator and the values of the function at the nodes of a net. For an arbitrary rectangular net in the xy plane an approximate expression of the Laplace operation $\Delta u = \frac{\partial^2 u}{\partial x^2} + \frac{\partial^2 u}{\partial y^2}$ can be easily obtained if for the second derivatives one sets their expressions in terms of the second central difference ratios. Let the net in the plane be formed by the lines

$$x = x_0, x_1, x_2, \ldots$$

$$y = y_0, y_1, y_2, \ldots$$

and let it be required to find an approximate expression for the Laplace operation at the node (x_i, y_k). To compute the second partial derivative with respect to x we shall assign the argument y the value y_k and vary the argument x, compelling it to run over the values of the nodes of the net. Let us arrange these values in order of their remoteness from x_i. For instance let this take place for the following arrangement of points

$$x_i, x_{i+1}, x_{i-1}, \ldots$$

We compute the second derivative of the function u with respect to x by means of formula (15), setting $a_0 = x_i$, $a_1 = x_{i+1}$, $a_2 = x_{i-1}$, ... in it. The number of points drawn into the computation is determined by the accuracy with which we wish to compute this derivative. Next, neglecting the remainder term, we replace the second derivative $\frac{\partial^2 u}{\partial x^2}$ by the expression found for it in terms of the difference ratios. We compute analogously the approximate second partial derivative with respect to y, and lastly, on adding both of the results, we find the approximate expression for the Laplace operation.

For example, let us limit ourselves, in computing the second derivative with respect to x, to the points x_i, x_{i+1}, x_{i-1} only. For the second difference ratio, if we put $u(x_i, y_k) = u_{i,k}$, $\Delta x_i = x_{i+1} - x_i$, $\Delta y_k = y_{k+1} - y_k$, we shall have

$$\frac{\dfrac{u_{i+1,k} - u_{i,k}}{\Delta x_i} - \dfrac{u_{i,k} - u_{i-1,k}}{\Delta x_{i-1}}}{\Delta x_i + \Delta x_{i-1}}.$$

For the second difference ratio with respect to y we obtain similarly

$$\frac{\dfrac{u_{i,k+1} - u_{i,k}}{\Delta y_k} - \dfrac{u_{i,k} - u_{i,k-1}}{\Delta y_{k-1}}}{\Delta y_k + \Delta y_{k-1}}.$$

On the strength of the second of formulas (15), if we preserve only the first terms on the right side of it, we shall have the following approximate expression for the Laplacian:

$$\Delta u = 2\left[\frac{\dfrac{u_{i+1,k}-u_{i,k}}{\Delta x_i} - \dfrac{u_{i,k}-u_{i-1,k}}{\Delta x_{i-1}}}{\Delta x_i + \Delta x_{i-1}} + \frac{\dfrac{u_{i,k+1}-u_{i,k}}{\Delta y_k} - \dfrac{u_{i,k}-u_{i,k-1}}{\Delta y_{k-1}}}{\Delta y_k + \Delta y_{k-1}}\right].$$

The error of this equation can be evaluated if we estimate the magnitude of the remainder terms in formulas (15).

One can also obtain an approximate expression for the Laplace operation Δu at the point (x_i, y_k) by considering not only the nodes of the net which lie on the lines $x = x_i$, $y = y_k$, but also other nodes of the net. These expressions have the simplest form in the commonest case, of equidistant values of the arguments, i.e., when the net is square.

We shall limit ourselves to the consideration of this particular case only. Let us draw two systems of parallel lines on the plane:

$$x = x_0 + ih = x_i,$$

$$y = y_0 + kh = y_k.$$

Let us consider the node (i, k) of the net, and take the four nodes closest to it: $(i+1, k)$, $(i, k+1)$, $(i-1, k)$, $(i, k-1)$. We shall try to find an approximate expression for Δu at the node (i, k). For this we form the differences of the values of u at the point (i, k) and at the four closest points. Using Taylor's formula, we obtain the following expressions for them:

$$\left.\begin{aligned}
u_{i+1,k} - u_{i,k} &= hu_x + \frac{h^2}{2!}u_{x^2} + \frac{h^3}{3!}u_{x^3} + \frac{h^4}{4!}u_{x^4} + \dots,\\
u_{i-1,k} - u_{i,k} &= -hu_x + \frac{h^2}{2!}u_{x^2} - \frac{h^3}{3!}u_{x^3} + \frac{h^4}{4!}u_{x^4} - \dots,\\
u_{i,k+1} - u_{i,k} &= hu_y + \frac{h^2}{2!}u_{y^2} + \frac{h^3}{3!}u_{y^3} + \frac{h^4}{4!}u_{y^4} + \dots,\\
u_{i,k-1} - u_{i,k} &= -hu_y + \frac{h^2}{2!}u_{y^2} - \frac{h^3}{3!}u_{y^3} + \frac{h^4}{4!}u_{y^4} - \dots
\end{aligned}\right\} \quad (25)$$

We shall seek Δu in the form of a linear combination of these differences. In view of the obviously equal importance of these differences, all of them enter with like coefficients, and we shall have to construct the following combination of values of the function:

$$\diamond\, u_{i,k} = u_{i+1,k} + u_{i,k+1} + u_{i-1,k} + u_{i,k-1} - 4u_{i,k}. \quad (26)$$

We find the expression for it in terms of the derivatives readily enough by adding all of the equations (25) term by term:

$$\diamondsuit\, u_{i,k} = 2\left[\frac{h^2}{2!}(u_{x^2}+u_{y^2}) + \frac{h^4}{4!}(u_{x^4}+u_{y^4}) + \frac{h^6}{6!}(u_{x^6}+u_{y^6}) + \ldots\right]. \tag{27}$$

Hence we obtain:

$$\left.\begin{aligned} \frac{1}{h^2}\diamondsuit\, u_{i,k} &= \Delta u + R_{i,k}, \\ R_{i,k} &= \frac{2h^2}{4!}(u_{x^4}+u_{y^4}) + \frac{2h^4}{6!}(u_{x^6}+u_{y^6}) + \ldots \end{aligned}\right\} \tag{28}$$

The quantity $R_{i,k}$ here has the meaning of a remainder term. We note that if we had represented the differences of the functions in equations (25) by the Taylor formula with remainder term, taking the values of the derivatives of up to the third order at the point (x_i, y_k), and derivatives of the fourth order at the mean points, we would have obtained for $R_{i,k}$ an expression of the following type:

$$R_{i,k} = \frac{4h^2}{4!}\vartheta M_4, \tag{29}$$

where M_4 is the greatest value of the moduli of the derivatives of the fourth order u_{x^4} and u_{y^4} in the region where the function u is under consideration.

If the function u is harmonic or satisfies the Poisson equation $\Delta u = f(x, y)$, then on neglecting the remainder term $R_{i,k}$ in (28), we will obtain for u an equation in finite differences, either $\frac{1}{h^2}\diamondsuit\, u_{i,k} = 0$, or $\frac{1}{h^2}\diamondsuit\, u_{i,k} = f_{i,k}$, by which the exact equation (28) is replaced approximately. $R_{i,k}$ will be the error of the finite-difference equation comparable with (28). If we solve the exact equation (28) and the approximate equation in finite differences for the same boundary conditions, the two solutions will not coincide. The discrepancy between them will be the greater, generally speaking, the greater the error $R_{i,k}$ of the approximate equation. It is therefore of advantage, to obtain greater accuracy in the result, to try to reduce the magnitude of the remainder term. This can be accomplished as follows. Along with the values of the function at the nodes of the net (i, k), $(i+1, k)$, $(i, k+1)$, $(i-1, k)$, $(i, k-1)$, which figure in the combination $\diamondsuit\, u_{i,k}$, we take the values of u at the nodes $(i+1, k+1)$, $(i+1, k-1)$, $(i-1, k-1)$, $(i-1, k+1)$. Form the differences between these values and $u_{i,k}$, and expand them by Taylor's formula:

$$\left.\begin{aligned}
u_{i+1,k+1} - u_{i,k} &= h\left(\frac{\partial}{\partial x} + \frac{\partial}{\partial y}\right)u + \frac{h^2}{2!}\left(\frac{\partial}{\partial x} + \frac{\partial}{\partial y}\right)^2 u + \\
&\quad + \frac{h^3}{3!}\left(\frac{\partial}{\partial x} + \frac{\partial}{\partial y}\right)^3 u + \ldots, \\
u_{i-1,k+1} - u_{i,k} &= h\left(-\frac{\partial}{\partial x} + \frac{\partial}{\partial y}\right)u + \frac{h^2}{2!}\left(-\frac{\partial}{\partial x} + \frac{\partial}{\partial y}\right)^2 u + \\
&\quad + \frac{h^3}{3!}\left(-\frac{\partial}{\partial x} + \frac{\partial}{\partial y}\right)^3 u + \ldots, \\
u_{i-1,k-1} - u_{i,k} &= h\left(-\frac{\partial}{\partial x} - \frac{\partial}{\partial y}\right)u + \frac{h^2}{2!}\left(-\frac{\partial}{\partial x} - \frac{\partial}{\partial y}\right)^2 u + \\
&\quad + \frac{h^3}{3!}\left(-\frac{\partial}{\partial x} - \frac{\partial}{\partial y}\right)^3 u + \ldots, \\
u_{i+1,k-1} - u_{i,k} &= h\left(\frac{\partial}{\partial x} - \frac{\partial}{\partial y}\right)u + \frac{h^2}{2!}\left(\frac{\partial}{\partial x} - \frac{\partial}{\partial y}\right)^2 u + \\
&\quad + \frac{h^3}{3!}\left(\frac{\partial}{\partial x} - \frac{\partial}{\partial y}\right)^3 u + \ldots.
\end{aligned}\right\} \tag{30}$$

On the strength of the equal importance of which we spoke in forming the combination $\diamondsuit\, u_{i,k}$, in using the above differences for the formation of an approximate expression for Δu, we form the sum of all these differences. Denote it by $\Box\, u_{i,k}$ and note that on adding equations (30) all odd powers of the parentheses are cancelled, since each of them occurs twice with opposite signs. Moreover, in summing the even powers of the parentheses there are preserved as the result of the addition only those terms which contain even orders of the operators $\frac{\partial}{\partial x}$ and $\frac{\partial}{\partial y}$. Simple calculations show that

$$\begin{aligned}
\Box\, u_{i,k} &= u_{i+1,k+1} + u_{i-1,k+1} + u_{i-1,k-1} + u_{i+1,k-1} - 4u_{i,k} = \\
&= 4\left\{\frac{h^2}{2!}(u_{x^2} + u_{y^2}) + \frac{h^4}{4!}(u_{x^4} + 6u_{x^2y^2} + u_{y^4}) + \right. \\
&\quad \left. + \frac{h^6}{6!} u_{x^6} + 15u_{x^4y^2} + 15u_{x^2y^4} + u_{y^6}) + \ldots\right\}.
\end{aligned} \tag{31}$$

Let us form, lastly, the combination $a \diamondsuit\, u_{i,k} + b\, \Box\, u_{i,k}$. We recall that the remainder term in equation (28), when we used only $\diamondsuit\, u_{i,k}$, contained a term with derivatives of the fourth order, and for small h these terms were the principal ones. Choosing a and b in the combinations under study such that the terms with fourth derivatives vanish is impossible, because in $\Box\, u_{i,k}$ there appears a mixed derivative of the fourth order whereas in $\diamondsuit\, u_{i,k}$ mixed derivatives are completely absent.

But for the particular problem that we are considering now, the exclusion of the terms with fourth derivatives from the unknown remainder term and a transference of them to the category of known quantities is possible in another way. If u satisfies the Laplace or Poisson equation, then the Laplace operation is known for it, and therefore any operation derivative from it will also be known. In particular, the biharmonic operation $\Delta\Delta u = \frac{\partial^4 u}{\partial x^4} + 2\frac{\partial^4 u}{\partial x^2 \partial y^2} + \frac{\partial^4 u}{\partial y^4}$ will also be known. In accordance with this, we shall try to choose a and b such that the coefficient of Δu in the expression $a \Diamond u_{i,k} + b \square u_{i,k}$ is equal to unity and such that the terms with the fourth derivatives form a biharmonic operation $\Delta\Delta u$. For this, obviously, there must be observed the equations

$$2\frac{h^2}{2!}a + 4\frac{h^2}{2!}b = 1, \quad 2\left(2\frac{h^4}{4!}a + 4\frac{h^4}{4!}b\right) = 4\frac{h^4}{4!}6b.$$

From them we find

$$a = \frac{2}{3h^2}, \quad b = \frac{1}{6h^2}.$$

We also note that with such a choice of a and b the terms with derivatives of the sixth order can be expressed in terms of the Laplace operation Δu, for simple calculations show that if in $a \Diamond u_{i,k} + b \square u_{i,k}$ we collect all such terms, we will obtain

$$\frac{2h^4}{6!}(u_{x^6} + 5u_{x^4y^2} + 5u_{x^2y^4} + u_{y^6}) = \frac{2h^4}{6!}\left(\Delta^3 u + 2\frac{\partial^4}{\partial x^2 \partial y^2}\Delta u\right).$$

The sum of the terms with derivatives of the eighth order cannot, however, be expressed solely in terms of Δu, as is clear from the equation given above. Because of this we consign them to the remainder term:

$$\left.\begin{aligned} \frac{1}{6h^2}(4 \Diamond u_{i,k} + \square u_{i,k}) &= \Delta u + \frac{2h^2}{4!}\Delta^2 u + \\ &+ \frac{2h^4}{6!}\left(\Delta^3 u + 2\frac{\partial^4}{\partial x^2 \partial y^2}\Delta u\right) + R_{i,k}, \\ R_{i,k} = \frac{2}{3}\frac{h^6}{8!}\bigg[3\Delta^4 u + 16\frac{\partial^4}{\partial x^2 \partial y^2}&\Delta^2 u + 20\frac{\partial^8 u}{\partial x^4 \partial y^4}\bigg] + \ldots \end{aligned}\right\} \tag{32}$$

The first three terms of the right side are expressed solely in terms of the Laplace operation. For functions satisfying the equations $\Delta u = 0$ or $\Delta u = f(x, y)$, they may be regarded as known. As regards the remainder term $R_{i,k}$, we observe that if in equations (30) we had expanded by Taylor's formula with remainder term, taking derivatives of up to the seventh order at the point (i, k), and derivatives of the eighth order at some mean points, including them in the remainder term of the formula, we would have obtained for $R_{i,k}$ an expression of the following type:

$$R_{i,k} = \frac{520h^6}{3 \cdot 8!} \vartheta M_8 \qquad (|\vartheta| \leqslant 1). \tag{33}$$

M_8 is the greatest value of the modulus of the eighth derivatives in the region under consideration.

The estimate of the remainder $R_{i,k}$ provided by (33) is evidently exaggerated, and it can hardly be used efficiently in practical computations. The second equation of (32) gives the expansion of $R_{i,k}$ in a power series of the step h of the net. For small h the first term of the expansion will play the principal role, and $R_{i,k}$ may be regarded as approximately equal to this first term.

From this, then, there may be obtained an approximate evaluation of $R_{i,k}$ that is likely to be sufficient for many practical purposes. For example, if u is harmonic and consequently $\Delta u = 0$ and $\Delta^2 u = 0$ for it, we obtain the following expression for $R_{i,k}$ from (32):

$$R_{i,k} = \frac{40}{3} \frac{h^6}{8!} \frac{\partial^8 u}{\partial x^4 \partial y^4} + \text{terms of higher order.}$$

The approximate evaluation obtained from this, $|R_{i,k}| \leqslant \frac{40 \cdot h^6}{3 \cdot 8!} M_8$ will, in problems of a particular type, probably not depart greatly from the true value. It has the same order of smallness as the estimate obtained from (33), but its coefficient $\frac{40}{3 \cdot 8!}$ is one-thirteenth as large as the coefficient of the preceding estimate. It stands to reason that the new estimate has, in view of its approximate character, only an orientative value, and the possibility of applying it is still subject to investigation.

We shall require one more transformation of equation (32). We shall first discuss its purpose. Let the function u satisfy the Poisson equation $\Delta u = f(x, y)$.

If on the right side of (32) we substitute in place of Δu its value $f(x, y)$ and neglect the remainder term $R_{i,k}$, we obtain for u the following equation in finite differences:

$$\frac{1}{6h^2} [4 \diamondsuit u_{i,k} + \square u_{i,k}] = f_{i,k} + \frac{2h^2}{4!} \Delta f_{i,k} + \frac{2h^4}{6!} \left(\Delta^2 f_{i,k} + 2 \frac{\partial^4 f_{i,k}}{\partial x^2 \partial y^2} \right).$$

When the function $f(x, y)$ has been given analytically, the use of this equation presents no difficulties. But if the values of $f(x, y)$ are known only at the nodes of the net, the differential operations on it which figure in the right side of the equation cannot be calculated exactly, and are computed approximately in terms of the function itself.

Let the computations be performed accurate to quantities of the third order of smallness with respect to h, inclusive. We shall discard the last term on the right side of the equation cited above, as it contains the factor h^4, and then replace $\Delta f_{i,k}$ by its value from (28). The remainder term in it is of the order of h^2. In addition we must multiply the operation $\Delta f_{i,k}$ by $\frac{2h^2}{4}$. We therefore discard $R_{i,k}$ in (28) and set, approximately,

$$\Delta f_{i,k} = \frac{1}{h^2} \diamond f_{i,k} = \frac{f_{i+1,k} + f_{i,k+1} + f_{i-1,k} + f_{i,k-1} - 4f_{i,k}}{h^2}.$$

After simple transformations, our equation, if we substitute for $\Delta f_{i,k}$ its approximate value, takes the form

$$\frac{1}{h^2}[4 \diamond u_{i,k} + \square u_{i,k}] = \frac{u_{i+1,k+1} + u_{i-1,k-1} + u_{i+1,k-1} + u_{i-1,k+1}}{h^2} + $$
$$+ \frac{4(u_{i+1,k} + u_{i-1,k} + u_{i,k+1} + u_{i,k-1}) - 20u_{i,k}}{h^2} =$$
$$= \frac{8f_{i,k} + f_{i+1,k} + f_{i-1,k} + f_{i,k+1} + f_{i,k-1}}{2}. \tag{34}$$

The error of this new approximate equation by comparison with the exact equation (32) will be of the order of h^4.

We shall not go into the details of this type of transformation of equation (32) but shall refer the reader who wishes to familiarize himself with it to an article by Mikeladze, [2] [1]).

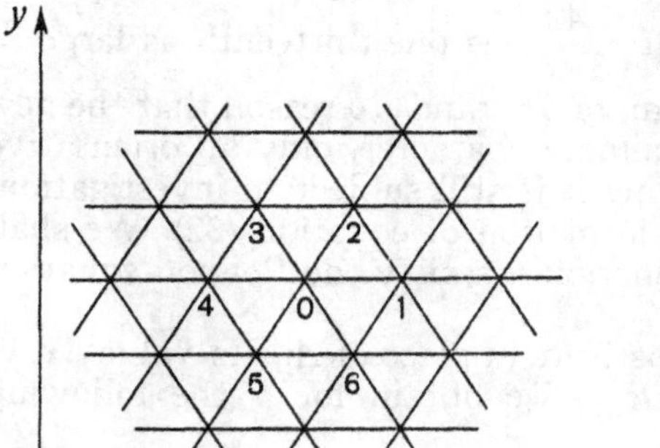

Fig. 5.

In deriving relations (28) and (32) we have used nodes of the net which are the vertices of the squares adjacent to the point (i, k). If one draws into the computations values of the function at more and more distant nodes of the net, one can obtain relations analogous to (28) and (32) involving remainder terms of higher and higher orders of smallness. As we are not striving for an exposition of the problem of the approximate representation of the Laplace and Poisson equations in all its details, having set ourselves the objective only of throwing light on the basic ideas of the method of nets, we are limiting ourselves to the simplest relations of the type we have indicated in equations (28) and (32).

---

[1]) See also other works of Sh. E. Mikeladze: [3], [4], [5], [6], [7], [8], [9].

Together with the square net we shall consider other possible regular nets in the plane — triangular and hexagonal [1]).

The first of them is depicted in Fig. 5. Let us take an inner point, identified by the number 0, and the six nearest nodal points surrounding it, 1, 2, 3, 4, 5, 6. If the coordinates of the zero point are x, y, and the length of the side of the triangle h, the coordinates of the adjoining points will be respectively:

$$(x+h, y), \left(x+\frac{1}{2}h, y+\frac{\sqrt{3}}{2}h\right), \left(x-\frac{1}{2}h, y+\frac{\sqrt{3}}{2}h\right),$$

$$(x-h, y), \left(x-\frac{1}{2}h, y-\frac{\sqrt{3}}{2}h\right), \left(x+\frac{1}{2}h, y-\frac{\sqrt{3}}{2}h\right).$$

For the differences between the values of the function u at the points $1, 2, \ldots, 6$ and at the node 0, we obtain by Taylor's formula:

$$u_1 - u_0 = hu_x + \frac{h^2}{2!}u_{x^2} + \frac{h^3}{3!}u_{x^3} + \frac{h^4}{4!}u_{x^4} + \ldots,$$

$$u_2 - u_0 = h\left(\frac{1}{2}\frac{\partial}{\partial x} + \frac{\sqrt{3}}{2}\frac{\partial}{\partial y}\right)u + \frac{h^2}{2!}\left(\frac{1}{2}\frac{\partial}{\partial x} + \frac{\sqrt{3}}{2}\frac{\partial}{\partial y}\right)^2 u +$$
$$+ \frac{h^3}{3!}\left(\frac{1}{2}\frac{\partial}{\partial x} + \frac{\sqrt{3}}{2}\frac{\partial}{\partial y}\right)^3 u + \frac{h^4}{4!}\left(\frac{1}{2}\frac{\partial}{\partial x} + \frac{\sqrt{3}}{2}\frac{\partial}{\partial y}\right)^4 u + \ldots,$$

$$u_3 - u_0 = h\left(-\frac{1}{2}\frac{\partial}{\partial x} + \frac{\sqrt{3}}{2}\frac{\partial}{\partial y}\right)u + \frac{h^2}{2!}\left(-\frac{1}{2}\frac{\partial}{\partial x} + \frac{\sqrt{3}}{2}\frac{\partial}{\partial y}\right)^2 u +$$
$$+ \frac{h^3}{3!}\left(-\frac{1}{2}\frac{\partial}{\partial x} + \frac{\sqrt{3}}{2}\frac{\partial}{\partial y}\right)^3 u + \frac{h^4}{4!}\left(-\frac{1}{2}\frac{\partial}{\partial x} + \frac{\sqrt{3}}{2}\frac{\partial}{\partial y}\right)^4 u + \ldots,$$

$$u_4 - u_0 = -hu_x + \frac{h^2}{2!}u_{x^2} - \frac{h^3}{3!}u_{x^3} + \frac{h^4}{4!}u_{x^4} - \ldots,$$

$$u_5 - u_0 = h\left(-\frac{1}{2}\frac{\partial}{\partial x} - \frac{\sqrt{3}}{2}\frac{\partial}{\partial y}\right)u + \frac{h^2}{2!}\left(-\frac{1}{2}\frac{\partial}{\partial x} - \frac{\sqrt{3}}{2}\frac{\partial}{\partial y}\right)^2 u +$$
$$+ \frac{h^3}{3!}\left(-\frac{1}{2}\frac{\partial}{\partial x} - \frac{\sqrt{3}}{2}\frac{\partial}{\partial y}\right)^3 u + \frac{h^4}{4!}\left(-\frac{1}{2}\frac{\partial}{\partial x} - \frac{\sqrt{3}}{2}\frac{\partial}{\partial y}\right)^4 u + \ldots,$$

$$u_6 - u_0 = h\left(\frac{1}{2}\frac{\partial}{\partial x} - \frac{\sqrt{3}}{2}\frac{\partial}{\partial y}\right)u + \frac{h^2}{2!}\left(\frac{1}{2}\frac{\partial}{\partial x} - \frac{\sqrt{3}}{2}\frac{\partial}{\partial y}\right)^2 u +$$
$$+ \frac{h^3}{3!}\left(\frac{1}{2}\frac{\partial}{\partial x} - \frac{\sqrt{3}}{2}\frac{\partial}{\partial y}\right)^3 u + \frac{h^4}{4!}\left(\frac{1}{2}\frac{\partial}{\partial x} - \frac{\sqrt{3}}{2}\frac{\partial}{\partial y}\right)^4 u + \ldots.$$

---

[1]) The dependence of the Laplace operator Δu upon the values of the function at the center and vertices of a regular polygon has been considered in a work of S. A. Gershgorin, [1].

Let us add all these equations. In the summation, all the derivatives obtained as the result of differentiating an odd number of times with respect to one of the variables x and y must obviously cancel. After simple calculations we find:

$$\sum_{i=1}^{6} u_i - 6u_0 = \frac{3h^2}{2!}\Delta u + \tfrac{9}{4}\frac{h^4}{4!}\Delta\Delta u +$$
$$+\frac{h^6}{6!}\left(\tfrac{33}{16}u_{x^6} + 15\cdot\tfrac{3}{16}u_{x^4y^2} + 15\cdot\tfrac{9}{16}u_{x^2y^4} + \tfrac{27}{16}u_{y^6}\right) + \dots,$$

whence follows:

$$\left.\begin{aligned} &\frac{2}{3h^2}\left(\sum_{i=1}^{6} u_i - 6u_0\right) = \Delta u + \tfrac{1}{16}h^2\Delta^2 u + R_0,\\ &R_0 = \frac{h^4}{17280}\left(33u_{x^6} + 45u_{x^4y^2} + 135u_{x^2y^4} + 27u_{y^6}\right) + \dots \end{aligned}\right\} \quad (35)$$

If the differences $u_i - u_0$ of the values of the function u be expanded by Taylor's formula with remainder term, calculating at the point x, y the values of derivatives of up to the fifth order inclusive, and derivatives of the sixth order at some mean point of the side of the triangle which joins 0 and i, we obtain for R_0, as can easily be seen, an estimate of the form

$$|R_0| \leqslant \tfrac{4}{3}\cdot\frac{h^4}{6!}\left[1 + 2\left(\frac{1+\sqrt{3}}{2}\right)^6\right]M_6 < \frac{7h^4}{270}M_5. \quad (36)$$

The application of (35) to the construction of an equation in finite differences replacing approximately the Laplace and Poisson equations will be given by us below in No. 2, § 2.

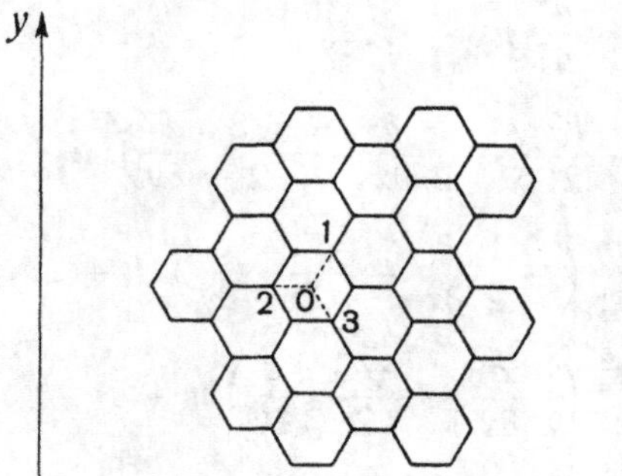

Fig. 6.

Let us now turn to a hexagonal net. Let us take the nodal point identified in Fig. 6 by the number 0, and three adjacent points 1, 2, 3. If h is the length of the side of the hexagonal, their coordinates are

$$\left(x = \tfrac{1}{2}h,\ y + \frac{\sqrt{3}}{2}h\right),$$
$$(x - h,\ y),$$
$$\left(x + \tfrac{1}{2}h,\ y - \frac{\sqrt{3}}{2}h\right).$$

For the differences of the values of the function u at the nodes we have:

$$u_1 - u_0 = h\left(\frac{1}{2}\frac{\partial}{\partial x} + \frac{\sqrt{3}}{2}\frac{\partial}{\partial y}\right)u + \frac{h^2}{2!}\left(\frac{1}{2}\frac{\partial}{\partial x} + \frac{\sqrt{3}}{2}\frac{\partial}{\partial y}\right)^2 u + \\ + \frac{h^3}{3!}\left(\frac{1}{2}\frac{\partial}{\partial x} + \frac{\sqrt{3}}{2}\frac{\partial}{\partial y}\right)^3 u + \ldots,$$

$$u_2 - u_0 = -hu_x + \frac{h^2}{2!}u_{x^2} - \frac{h^3}{3!}u_{x^3} + \ldots,$$

$$u_3 - u_0 = h\left(\frac{1}{2}\frac{\partial}{\partial x} - \frac{\sqrt{3}}{2}\frac{\partial}{\partial y}\right)u + \frac{h^2}{2!}\left(\frac{1}{2}\frac{\partial}{\partial x} - \frac{\sqrt{3}}{2}\frac{\partial}{\partial y}\right)^2 u + \\ + \frac{h^3}{3!}\left(\frac{1}{2}\frac{\partial}{\partial x} - \frac{\sqrt{3}}{2}\frac{\partial}{\partial y}\right)^3 u + \ldots.$$

After the addition of these equations we obtain:

$$\sum_{i=1}^{3} u_i - 3u_0 = \tfrac{3}{2}\frac{h^2}{2!}\Delta u - \frac{h^3}{3!}\left(\tfrac{3}{4}u_{x^3} - \tfrac{9}{4}u_{xy^2}\right) + \ldots.$$

Whence

$$\frac{4}{3h^2}\left(\sum_{i=1}^{3} u_i - 3u_0\right) = \Delta u + R_0, \quad R_0 = -\frac{h}{18}\left(3u_{x^3} - 9u_{xy^2}\right) + \ldots. \tag{37}$$

Had we represented by Taylor's formula with remainder term the above differences between the values of the function, limiting ourselves to three terms of the right side of the equation, after replacement of all third-order partial derivatives by their maximum value M_3 in the region under consideration we should have obtained the following representation for the remainder R_0 of formula (37):

$$|R_0| = \frac{2h}{9}\left[1 + 2\left(1 + \frac{\sqrt{3}}{2}\right)^3\right]\cdot\vartheta M_3 < 1.36\, hM_3. \tag{38}$$

We shall conclude this No. by obtaining an approximate expression for the biharmonic operation. Just as for the Laplace operation, approximate expressions for it can be constructed in terms of the values of the function at the nodes of a net, the accuracy depending on the number of nodes drawn into the construction of this expression. Of all the possible expressions of such a type we shall consider the simplest and least exact of them. The reader desiring to familiarize himself with more exact expressions is referred to the collection of Collatz, [2].

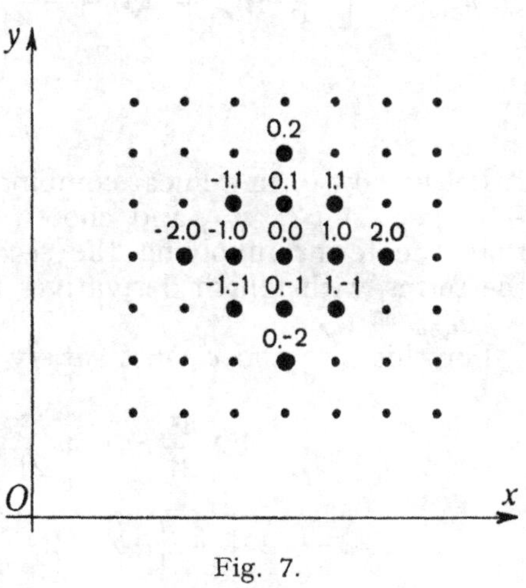

Fig. 7.

Let there be constructed on the plane a square net with side h. Let us take any node of the net. In Fig. 7 we have identified it by the numbers (0, 0). Along with it let us consider the 12 nodes closest to it, identified in the figure by bold-face points. Form the difference $u_{i,k} - u_{0,0}$ between the values of the function u at these 12 nodes and its value at (0, 0). Owing to considerations of symmetry, we shall take not the differences themselves, but their following sums:

$$\Diamond\, u_{0,0} = u_{1,0} + u_{0,1} + u_{-1,0} + u_{0,-1} - 4u_{0,0}$$

$$\Box\, u_{0,0} = u_{1,1} + u_{-1,1} + u_{-1,-1} + u_{1,-1} - 4u_{0,0}$$

$$\Diamond'\, u_{0,0} = u_{2,0} + u_{0,2} + u_{-2,0} + u_{0,-2} - 4u_{0,0}.$$

Let us expand each of the differences $u_{i,k} - u_{0,0}$ in powers of h, using Taylor's formula, and introduce their expressions into our combinations. Above, in equations (27) and (31), we showed what form $\Diamond\, u_{0,0}$ and $\Box\, u_{0,0}$ take after the substitution.

$$\Diamond\, u_{0,0} = 2\left[\frac{h^2}{2!}(u_{x^2} + u_{y^2}) + \frac{h^4}{4!}(u_{x^4} + u_{y^4}) + \frac{h^6}{6!}(u_{x^6} + u_{y^6}) + \ldots\right]$$

$$\Box\, u_{0,0} = 4\left[\frac{h^2}{2!}(u_{x^2} + u_{y^2}) + \frac{h^4}{4!}(u_{x^4} + 6u_{x^2y^2} + u_{y^4}) + \right.$$
$$\left. + \frac{h^6}{6!}(u_{x^6} + 15u_{x^4y^2} + 15u_{x^2y^4} + u_{y^6}) + \ldots\right].$$

As regards the expansion of $\Diamond'\, u_{0,0}$, it is obtainable from the expansion of $\Diamond\, u_{0,0}$ if we replace h in the latter by $2h$:

$$\Diamond'\, u_{0,0} = 2\left[\frac{4h^2}{2!}(u_{x^2} + u_{y^2}) + \frac{16h^4}{4!}(u_{x^4} + u_{y^4}) + \right.$$
$$\left. + \frac{64h^6}{6!}(u_{x^6} + u_{y^6}) + \ldots\right].$$

Let us now form a linear combination of these expressions: $a \Diamond u_{0,0} + b \Box u_{0,0} + c \Diamond' u_{0,0}$ and choose the coefficients of it in such a way that the terms involving the second derivatives u_{x^2}, u_{y^2} vanish, and the terms with fourth derivatives form the biharmonic operation $u_{x^4} + 2u_{x^2y^2} + u_{y^4}$.

For this a, b and c must satisfy the following system:

$$2\frac{h^2}{2!}a + 4\frac{h^2}{2!}b + 2\frac{4h^2}{2!}c = 0,$$

$$2\frac{h^4}{4!}a + 4\frac{h^4}{4!}b + 2\frac{16h^4}{4!}c = 1, \quad 4\frac{h^4}{4!}6b = 2,$$

on solving which we find for their values

$$a = -\frac{8}{h^4}, \quad b = \frac{2}{h^4}, \quad c = \frac{1}{h^4}.$$

Lastly, substituting in the combination under consideration the values of a, b and c we have obtained, we arrive at the following equation:

$$\frac{20u_{0,0} - 8(u_{1,0} + u_{0,1} + u_{-1,0} + u_{0,-1}) + 2(u_{1,1} + u_{-1,1} + u_{-1,-1} + u_{1,-1})}{h^4} +$$

$$+ \frac{(u_{2,0} + u_{0,2} + u_{-2,0} + u_{0,-2})}{h^4} = \Delta^2 u + R_{0,0},$$

$$R_{0,0} = \frac{h^2}{6}(u_{x^6} + u_{x^4y^2} + u_{x^2y^4} + u_{y^6}) + \ldots . \tag{39}$$

For small h the principal role in $R_{0,0}$ will be played by the first term of its expansion, which we have cited explicitly. Discarding all the rest of the terms and denoting by M_6 the greatest value of the sixth derivatives in the region under consideration, we obtain the following estimate of the remainder:

$$|R_{0,0}| \leqslant \tfrac{2}{3}h^2 M_6.$$

For obtaining not an approximate but the correct value, we would have had to use, in expanding the differences $u_{i,k} - u_{0,0}$, not a series, but Taylor's formula with remainder term, computing the sixth derivatives not at the point (0, 0) but at some mean points of the rays joining (0, 0) with the nodes (i, k). Then all the derivatives in the remainder terms would be replaced by their maximum absolute value M_6. There would then be obtained for $R_{0,0}$, as simple calculations show:

$$|R_{0,0}| \leqslant \tfrac{10}{9}h^2 M_6. \tag{40}$$

§ 2. DIFFERENTIAL EQUATIONS AND THE FINITE-DIFFERENCE EQUATIONS THAT CORRESPOND TO THEM

1. Ordinary differential equations. As in theoretical researches the art of the investigator consists rather the more in posing new problems, so in applied mathematics less art is required to carry out computations than to make a preliminary study of the problem and to bring it into most convenient form for the actual computations. The possibilities of simplifying the problem, however, are so numerous and so multifarious that it would be impossible with anything like completeness to expound them in the small number of pages that we have at our disposal. Moreover, a detailed enumeration of the methods of simplification would indeed be in considerable measure useless, since many of them are based on very simple facts and can easily be noticed by the majority of attentive and sufficiently experienced computers.

In studying the method of nets, therefore, we shall expound only the basic ideas of the replacement of differential equations by equations in finite differences, and shall elucidate them in problems the simplest.

Let us consider an equation of the second order

$$L(u) = y'' + p(x)y' + q(x)y = f(x), \tag{1}$$

whose coefficients $p(x)$, $q(x)$ and free term $f(x)$ we shall regard as continuous in the interval $a \leqslant x \leqslant b$. Let it be required to find a solution of the equation in this interval which satisfies at its ends one of the following boundary conditions:

$$y(a) = \alpha, \quad y(b) = \beta, \tag{2_1}$$

or

$$y'(a) - h_1 y(a) = \alpha, \quad y'(b) + h_2 y(b) = \beta. \tag{2_2}$$

We shall consider $p(x)$, $q(x)$, h_1, h_2 to be such that the above problem has one and only one solution.

Let us divide the interval (a, b) into N equal parts by the points

$$x_0 = a, \quad x_1 = a + h, \quad x_2 = a + 2h, \ldots, x_N = b; \quad h = \frac{b-a}{N}.$$

The net of equidistant points which we have taken is the simplest, but generally speaking not the best, it being possible to adduce examples of differential equations where a net of points not equally distant is more suitable to the nature of the case.

Let us now replace, in differential equation (1), the derivatives by their expressions in terms of difference ratios. The choice of these expressions depends on the degree of accuracy that we wish to obtain. We shall now analyse two cases of such a replacement. Let us take the ith node of the net $x_i = a + ih$. In the approximate computation of the second derivative y'' we must know the value of y at three points at least. Let us take the nodes closest to x_i: $x_{i-1} = a + (i-1)h$ and $x_{i+1} = a + (i+1)h$ and for the computation of y'' utilize formula (20_3) of the preceding section. In application to our case it gives

$$\frac{\Delta^2 y_{i-1}}{h^2} = \frac{y_{i+1} - 2y_i + y_{i-1}}{h^2} = y'' + \frac{h^2}{12}\vartheta M_4.$$

For the approximate expression of the first derivative we use the same three nodes of the net. For y' we may adopt any of the three following expressions: $\frac{y_{i+1} - y_i}{h}$, $\frac{y_i - y_{i-1}}{h}$, or (20_1) § 1. The error of the first two of them will be of the order of $\frac{h}{2}\vartheta M_2$. The error of the last, however, is $\frac{h^2}{2}\vartheta M_3$. It is more exact than either of the preceding and we shall therefore give it preference:

$$\frac{y_{i+1} - y_{i-1}}{2h} = y' + \frac{h^2}{6}\vartheta M_3.$$

Let us now form the expression

$$l_i^{(1)}(y) = \frac{y_{i+1} - 2y_i + y_{i-1}}{h^2} + p_1 \frac{y_{i+1} - y_{i-1}}{2h} + q_i y_i =$$
$$= L(y) + R_i^{(1)*}(y), \tag{3}$$

where

$$R_i^{(1)*}(y) = \frac{h^2}{12}(M_4 + 2PM_3)\vartheta,$$

$$p_i = p(a + ih), \quad q_i = q(a + ih), \quad P = \operatorname*{Max}_{a \leqq x \leqq b} |p(x)|, \quad |\vartheta| \leqslant 1.$$

If the function y satisfies the differential equation $L(y) = f(a)$, it follows from the foregoing equation that it will likewise satisfy the following finite-difference equation:

$$l_i^{(1)}(y) = f_i + R_i^{(1)*}(y).$$

Discarding here the remainder term $R_i^{(1)*}(y)$, we obtain an equation in finite differences which approximately replaces the given differential equation (1):

$$l_i^{(1)}(y) = \frac{y_{i+1} - 2y_i + y_{i-1}}{h^2} + p_i \frac{y_{i+1} - y_{i-1}}{2h} + q_i y_i = f_i. \tag{4}$$

The error $R_i^{(1)*}(y)$ of this approximate equation has the estimate

$$|R_i^{(1)*}(y)| \leqslant \frac{h^2}{12}(M_4 + 2PM_3). \tag{5}$$

Equation (4) must be satisfied at all interior points of our net.

Let us now consider the boundary points of our net, a and b. If the conditions at the ends of the interval have form (2_1), then the values y_0, y_N of the function y at the first and last points of the net are given, and there is no need to set up any equations at the boundary points. When, however, the conditions at the ends of the interval have form (2_2), the values of y at the points a and b are unknown and to determine them we must set up equations having the same error as (4) and approximately replacing the boundary conditions (2_2). Let us take the first of them; in it $y'(a)$ appears. For an approximate expression of it in terms of the values of the function at the nodes we take three points a, $a + h$, $a + 2h$ — as for the interior nodes of the net. The distinction will consist in the fact that for the interior nodes we took two additional points, one of which lay to the right and the other to the left of the node adopted; for the point a, however, both additional points $a + h$ and $a + 2h$ will be to the right of the node in question. In accordance with this, in the approximate computation of the derivatives we must utilize, not Stirling's formula, but Newton's formula for forward interpolation.

By the first of formulas (18) § 1, with $n = 2$, we have:

$$y'(a) = \frac{1}{h}(\Delta y_0 - \tfrac{1}{2}\Delta^2 y_0) + R_0^{(1)*}(a) = \frac{-y_2 + 4y_1 - 3y_0}{2h} + R_0^{(1)*}(a).$$

The remainder term $R_0^{(1)*}(a)$ on the right side of the equation can be evaluated if we proceed from the explicit expression for it given by the first of equations (17) § 1. In the previous notation it has the form $R_3'(x)$. We must put $n = 2$, $a_0 = a$, $a_1 = a + h$, $a_2 = a + 2h$. In view of the fact that $\omega(a) = 0$, it will be:

$$R_0^{(1)*}(a) = \omega'(a)\left\{\sum_{k=0}^{2} \frac{f(a_k) - f(x)}{\omega'(a_k)(a_k - x)}\right\}_{x=a}.$$

If besides this we utilize equation (8_2) § 1 for the expression standing in braces, and notice the fact that $\omega'(a) = 2h^2$, we shall obtain for the remainder term

$$R_0^{(1)*}(a) = 2h^2 \frac{f'''(\xi)}{3!} = \frac{h^2}{3}\vartheta M_3,$$

ϑ having the same meaning as before.

In the boundary condition $y'(a) - h_1 y(a) = \alpha$, let us replace the derivative by its approximate value

$$\frac{-y_2 + 4y_1 - 3y_0}{2h} - h_1 y_0 = y'(a) - h_1 y(a) - R_0^{(1)*}(a) = \alpha - R_0^{(1)*}(a).$$

Lastly, if we discard the remainder here, $R_0^{(1)*}(a)$, we will obtain the equation in finite differences we require, which approximately replaces the first of conditions (2_2):

$$\frac{-y_2 + 4y_1 - 3y_0}{2h} - h_1 y_0 = \alpha. \tag{6_1}$$

The error of this equation has the estimate

$$|R_0^{(1)*}(a)| \leqslant \frac{h^2}{3} M_3. \tag{7_1}$$

In just the same way, if in the boundary condition (2_2) for the right end we replace the derivative $y'(b)$ by its approximate expression, which is obtained from the Newton formula for backward interpolation and is given by the first equation (19) § 1, we see that it takes the form

$$\frac{3y_N - 4y_{N-1} + y_{N-2}}{2h} + h_2 y_N = y'(b) + h_2 y(b) - R_N^{(1)*}(b) = \beta - R_N^{(1)*}(b),$$

$$R_N^{(1)*}(b) = \frac{h^2}{3} f'''(\xi) = \frac{h^2}{3}\vartheta M_3.$$

And when we discard the remainder term $R_N^{(1)*}(b)$, we will obtain the approximate equation that we need,

$$\frac{3y_N - 4y_{N-1} + y_{N-2}}{2h} + h_2 y_N = \beta, \tag{6$_2$}$$

the error of which is estimated by the following inequality:

$$|R_N^{(1)*}(b)| \leqslant \frac{h^2}{3} M_3. \tag{7$_2$}$$

Thus for the approximate determination of the function y at the nodes of the net we will have to solve a system consisting of equations (4), which must be satisfied at all interior nodes of the net ($i = 1, 2, \ldots, N - 1$) and the boundary conditions of the form (2_1). When the boundary conditions have the form (2_2), however, equations (6_1) and (6_2) for the extremal points of the net must yet be adjoined to them.

Now let us construct more exact finite-difference equations than (4) and (6_1), (6_2). Let us take the ith node of the net and in computing $y''(x)$ at this point let us use the values of y not at three points, as we have just done, but at five: $a + (i - 2)h$, $a + (i - 1)h$, $a + ih$, $a + (i + 1)h$, $a + (i + 2)h$.

By formula (20_4) § 1,

$$\frac{-(y_{i+2} + y_{i-2}) + 16(y_{i+1} + y_{i-1}) - 30y_i}{12h^2} = y'' + \frac{h^4}{54}\vartheta M_6.$$

For computing the first derivative in terms of the values of the function at these same interpolation nodes we shall utilize equality (20_2) § 1:

$$\frac{-y_{i+2} + 8y_{i+1} - 8y_{i-1} + y_{i-2}}{12h} = y' + \frac{h^4}{30} M_5\vartheta.$$

Form the expression

$$l_i^{(2)}(y) = \frac{-(y_{i+2} + y_{i-2}) + 16(y_{i+1} + y_{i-1}) - 30y_i}{12h^2} +$$
$$+ p_i \frac{-y_{i+2} + 8y_{i+1} - 8y_{i-1} + y_{i-2}}{12h} + q_i y_i = L(y) + R_i^{(2)*}(y),$$

where

$$R_i^{(2)*}(y) = \frac{h^4}{6}\vartheta[\tfrac{1}{9}M_6 + \tfrac{1}{5}PM_5].$$

When the function y satisfies the given equation, $L(y) = f(x)$, it is evident from this that it will satisfy the following equation in finite differences:

$$l_i^{(2)}(y) = f_i + R_i^{(2)*}(y).$$

If we discard the remainder term $R_i^{(2)*}(y)$ here, we will obtain an equation in finite differences which approximately replaces differential equation (1):

$$l_i^{(2)}(y) = \frac{-(y_{i+2} + y_{i-2}) + 16(y_{i+1} + y_{i-1}) - 30y_i}{12h^2} +$$
$$+ p_i \frac{-y_{i+2} + 8y_{i+1} - 8y_{i-1} + y_{i-2}}{12h} + q_i y_i = f_i, \quad (8)$$

the error of which is estimated thus:

$$|R_i^{(2)*}(y)| \leqslant \frac{h^4}{6}\left(\tfrac{1}{9}M_6 + \tfrac{1}{5}PM_5\right). \quad (9)$$

The remainder term here has a fourth order of smallness by comparison with h. In the case preceding, however, when we used only three points in computing the derivatives approximately, the error of equation (4) was of the second order of smallness [see (5)].

Equation (8) must be satisfied at all interior nodes of the interval with the exception of $a + h$ and $b - h$. At these points, as also at the boundary points a and b, it must be replaced by another having an error of approximately the same order.

Let us consider the point $a + h$. Let us add to it four more of the nearest nodes, a, $a + 2h$, $a + 3h$, $a + 4h$, and find expressions for the first and second derivatives at $x = a + h$ in terms of the values at these five points. We shall utilize the Newton formula (10) § 1 for forward interpolation. If we set it up with $n = 4$, computing the first derivative from it and putting $x = a + h$ there, we find

$$y'(a + h) = \frac{y_4 - 6y_3 + 18y_2 - 10y_1 - 3y_0}{12h} + R_5'(a + h).$$

The remainder term in this equation can be estimated by proceeding from its general expression (17) § 1 by the same path that we took in evaluating the remainder terms in expressions (20) § 1 for the derivatives obtained from Stirling's formula. If in the first of equations (17) § 1 we put $x = a + h$ and take into consideration the fact that for all values of x, by (8_2) § 1,

$$\sum_{k=0}^{4} \frac{f(a_k) - f(x)}{\omega'(a_k)(a_k - x)} = \frac{f^{(5)}(\xi)}{5!} \qquad (a_k = a + kh),$$

and

$$\omega(a + h) = 0, \quad \omega'(a + h) = -6h^4,$$

we obtain for $R_5'(a + h)$ the following value:

$$R_5'(a + h) = -\frac{6h^4}{5!} f^{(5)}(\xi) = \frac{h^4}{20}\vartheta M_5.$$

The order of the remainder in the expression for the first derivative is just the same as for the points $a + 2h, \ldots$. The worsening of the estimate is expressed only in our having obtained here the numerical coefficient $\frac{1}{20}$ rather than the $\frac{1}{30}$ of the previous estimate.

The second derivative is calculated with scarcely more complications. If one bases the computation of it on these same interpolation nodes $a, a + h, \ldots, a + 4h$, the following approximate expression for it is obtained:

$$y''(a + h) = \frac{-y_4 + 4y_3 + 6y_2 - 20y_1 + 11y_0}{12h^2} + R_5''(a + h),$$

the remainder term having the estimate

$$|R_5''(a + h)| \leqslant \frac{h^3}{360}(30M_5 + 19hM_6).$$

It will, generally speaking, be of the third order of smallness with respect to h, which is worse than the estimates of any of the remainders in the approximate expressions for the derivatives which we have computed up till now, inasmuch as all the remainders have been of the fourth order of smallness. Here the asymmetric arrangement of the interpolation points with respect to the node $a + h$ where the derivative was computed has told. In order to make identical the order of all the remainder terms without exception, we prefer to involve one more node $a + 5h$ in the computation of the derivative $y''(a + h)$, trying by this means to raise the order of smallness of the remainder by one.

Let us set up the Newton formula (10) § 1 for the points $a, a + h, \ldots, a + 5h$, calculate the second derivative of both sides of it, and put $x = a + h$ in the result. Then we obtain:

$$y''(a + h) = \frac{y_5 - 6y_4 + 14y_3 - 4y_2 - 15y_1 + 10y_0}{12h^2} + R_6''(a + h),$$

where

$$R_6''(a + h) = \frac{481h^4}{900}\vartheta M_6.$$

Let us consider $L(y) = y'' + p(x)y' + q(x)y$ at the point $a + h$. From the approximate expressions obtained by us for $y''(a + h)$ and $y'(a + h)$ it is clear that the following equality will hold:

$$l_1^{(2)}(y) = \frac{y_5 - 6y_4 + 14y_3 - 4y_2 - 15y_1 + 10y_0}{12h^2} +$$

$$+ p_1\frac{y_4 - 6y_3 + 18y_2 - 10y_1 - 3y_0}{12h} + q_1y_1 = L(y) + R_1^{(2)*}(y), \qquad (10)$$

where

$$R_1^{(2)*}(y) = -R_6''(a + h) - p_1R_5'(a + h) = \frac{h^4}{900}\vartheta(481M_6 + 45PM_5).$$

If we consider that y satisfies the equation $L(y) = f(x)$ and discard the remainder term in the last equation, we will obtain the equation below, which replaces approximately the given differential equation at the point $a + h$:

$$l_1^{(2)}(y) = \frac{y_5 - 6y_4 + 14y_3 - 4y_2 - 15y_1 + 10y_0}{12h^2} +$$

$$+ p_1 \frac{y_4 - 6y_3 + 18y_2 - 10y_1 - 3y_0}{12h} + q_1 y_1 = f_1. \quad (11)$$

The error of this equation is given by the second equation of (10).

In exactly the same fashion one effects the construction of the equation analogous to (11) for the point $b - h$. If we apply, for example, Newton's formula (11) § 1 for backward interpolation and in computing $y'(b - h)$ utilize the nodes $b, b - h, \ldots, b - 4h$, and in computing $y''(b - h)$, the nodes $b, b - h, \ldots, b - 5h$, we shall obtain the equation

$$l_{N-1}^{(2)}(y) = \frac{y_{N-5} - 6y_{N-4} + 14y_{N-3} - 4y_{N-2} - 15y_{N-1} - 10y_N}{12h^2} +$$

$$+ p_{N-1} \frac{-y_{N-4} + 6y_{N-3} - 18y_{N-2} + 10y_{N-1} + 3y_N}{12h} +$$

$$+ q_{N-1} y_{N-1} = f_{N-1}, \quad (12)$$

which approximately replaces differential equation (1) at the point $b - h$. Its error $R_{N-1}^{(2)*}(y)$ has the same expression as at the point $a + h$:

$$R_{N-1}^{(2)*}(y) = \frac{h^4}{900} \vartheta(481 M_6 + 45 P M_5). \quad (13)$$

Let us pass to the ends of the interval. If the boundary conditions have the form (2_1), their application cannot occasion doubts nor questions. We shall concern ourselves only with boundary conditions of the form (2_2).

Let us take the first of them, relating to the left end a of the interval. In order to obtain, on replacing it by an approximate equation, an error of the fourth order of smallness with respect to h, which we have allowed everywhere in equations (8), (11) and (12), we shall take for the determination of the approximate expression for $y'(a)$ five nodes: $a, a + h, \ldots, a + 4h$ of the net, and utilize the first of equations (18) § 1 with $n = 4$:

$$y'(a) = \frac{1}{h} [\Delta y_0 - \tfrac{1}{2}\Delta^2 y_0 + \tfrac{1}{3}\Delta^3 y_0 - \tfrac{1}{4}\Delta^4 y_0] + R_5'(a) =$$

$$= \frac{-3y_4 + 16y_3 - 36y_2 + 48y_1 - 25y_0}{12h} + R_5'(a).$$

The remainder $R_5'(a)$ is easily estimated on the basis of its general expression (17) § 1. Here $\omega(x) = (x - a)(x - a - h) \ldots (x - a - 4h)$,

$\omega(a) = 0$, $\omega'(a) = 4!h^4$, and since

$$\sum_{k=0}^{4} \frac{f(a_k) - f(x)}{\omega'(a_k)(a_k - x)} = \frac{1}{5!} f^{(5)}(\xi),$$

we have

$$R_5'(a) = \frac{h^4}{5} f^{(5)}(\xi).$$

If in $y'(a) - h_1 y(a) = \alpha$ we now replace $y'(a)$ by the value found for it and discard the remainder $R_5'(a)$, we obtain the boundary condition in this approximate form:

$$\frac{-3y_4 + 16y_3 - 36y_2 + 48y_1 - 25y_0}{12h} - h_1 y_0 = \alpha. \quad (14)$$

Its error has the estimate

$$|R_5'(a)| \leqslant \frac{h^4}{5} M_5. \quad (15)$$

Likewise is obtained the approximate boundary condition at the point b, and with just the same estimate of its error:

$$\frac{3y_{N-4} - 16y_{N-3} + 36y_{N-2} - 48y_{N-1} + 25y_N}{12h} + h_2 y_N = \beta \quad (16)$$

$$|R_5'(b)| \leqslant \frac{h^4}{5} M_5.$$

Thus our problem for the differential equation under boundary conditions (2_1) is reduced to the solution of a system of $N-1$ equations (8), (11) and (12). For the case of boundary conditions (2_2) however, equations (14) and (16) must still be added to them.

Essentially the method of nets is one of the methods of reducing the problem of differential equations to the solution of an algebraic system. At its foundation lie two facts. The equation itself is not considered at all points of the interval of integration, but only at a finite number of them. These latter are, as a rule, taken equidistant — which, it goes without saying, is not obligatory and is done only for simplicity of computation.

The transition from the entire interval to a finite net of points does not imply errors in the equation, but it does destroy the possibility of an exact calculation of the derivatives, since they cannot be found without error on the basis of knowledge of only a finite number of the values of the function, and it compels us to replace the derivatives in the equation by some expressions or other for them in terms of the values of the function itself. The accuracy of the method of nets is therefore determined in the first place by the accuracy of the interpolation of the derivatives, since it is precisely this which is the chief source of the error occasioned by replacing the differential equation by the algebraic system.

Above, in explaining the idea of the method of nets, we have used only the simplest approximate representations of the derivatives in terms of the values of the function. A sufficiently complete catalog of this type of formula can be found, for example, in Sh. E. Mikeladze's article [10].

In principle, one could reduce without limit the error of the finite-difference equation by which the given differential equation is replaced, by taking instead of the derivatives their expressions in terms of the values of the function at the nodes of the net for an ever greater number of nodes. The complexity of these equations, however, grows rapidly, and in the majority of cases the results of the computations are probably more advantageously refined not by complicating the equations but by diminishing the step of the net.

2. Partial differential equations of elliptic type. Quite as in the preceding No., in the exposition we shall not study the question in all its generality and completeness, but shall set ourselves the task of describing the main ideas of the method for equations of a particular type. This is sufficient for an understanding of how it is to be applied, not only in practical questions, but in theoretical investigations as well. Moreover, we shall expound only the fundamental ideas and shall not undertake an investigation of the simplifications that may occur in solving equations of different special types [1]).

We shall consider the following partial differential equation of the second order, of elliptic type, which will be sufficient for our purposes:

$$L(u) = A\frac{\partial^2 u}{\partial x^2} + C\frac{\partial^2 u}{\partial y^2} + D\frac{\partial u}{\partial x} + E\frac{\partial u}{\partial y} + qu = f(x, y), \qquad (17)$$

$$(A > 0,\ C > 0).$$

Let it be required to find the solution of this equation in some region G, which solution is to satisfy on the boundary Γ of the region this or that requirement, the nature of which is at present immaterial to us.

Draw on the plane a system of straight lines parallel to the coordinate axes:

$$\left.\begin{array}{l} x = x_0,\quad x = x_1, \ldots, x = x_n,\quad x_0 < x_1 < \ldots < x_n \\ y = y_0,\quad y = y_1, \ldots, y = y_m,\quad y_0 < y_1 < \ldots < y_m \end{array}\right\}. \qquad (18)$$

They form a rectangular net, the form of which is determined by the quantities x_i, y_k. Let us construct a closed broken line C, consisting

1) We refer readers who wish to acquaint themselves with some of the simplifications of the problems for the Laplace and Poisson equations to the collection of Sh. E. Mikeladze, [4], and to G. H. Shortley and Weller, [1].

of sides and diagonals of the rectangular net, and which fits Γ as well as possible. We shall remark that the vertices of the broken line C do not necessarily lie within G or on Γ: they may even lie outside of G. In Fig. 8 is given an example of the construction of C for the case where the curve Γ is the circumference of a circle.

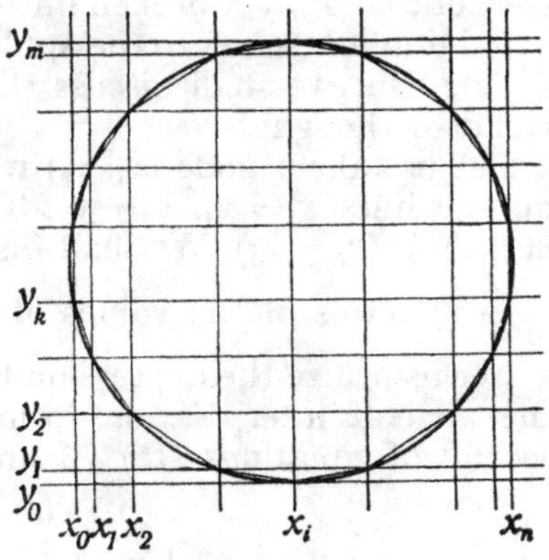

Fig. 8.

The simplest net in the plane is a square one, where the nodes of the divisions along both axes are equidistant, with the same step on both: $x_{i+1} - x_i = h$, $y_{k+1} - y_k = h$.

The equation in finite differences with which we shall later approximately replace differential equation (17) will, generally speaking, have the simplest form for a square net. In this sense the square net is preferable to a rectangular one with unequal divisions along the axes. However application of it has one shortcoming, which compels us to keep the rectangular net along with the square. By replacing the curve Γ bounding the region G, by the broken line C, we shall have to transfer from Γ to C, in some way or another, the boundary condition for the solution being determined. For definiteness we shall consider to be given the simplest of the boundary-value problems — the Dirichlet problem, where on Γ the values of the sought function u itself are known. In the transfer of the boundary values of u from Γ to C, if not all of the vertices of the broken line C lie on Γ, a certain error is unavoidable. Its magnitude will be the greater the farther C departs from Γ. Here it often happens that the order of smallness of the error obtained owing to this shift is lower than the order of smallness of the error obtained owing to the replacement of the differential equation (17) by an approximate equation in finite differences.

It is therefore important to adopt a net such that the broken line C constructed upon it will be best fitted to Γ. If the contour Γ is such that a square net leads to a broken line C whose vertices either lie exactly on Γ or are very close to Γ, then it goes without saying that it is simplest to use the square net. If this is not the case, however, to reduce the error owing to the transfer of the boundary values, it will perhaps be more advantageous to employ a rectangular net not square, upon which a broken line C closer to Γ than was the case with the square net can be constructed. This it is that compels us to preserve in the exposition the arbitrary rectangular net (18).

Later, at the end of this No., we shall indicate a method of setting up the equations in finite differences which requires no replacement of the contour Γ by a broken line C. Were we to proceed by that course, we could limit ourselves to the application of square nets only, it goes without saying. But we shall discuss that somewhat later, reverting now to our trend of thought.

Let us take a node (x_i, y_k) within C and the four nearest nodes lying on the lines $x = x_i$, $y = y_k$. Their coordinates are (x_{i+1}, y_k), (x_i, y_{k+1}), (x_{i-1}, y_k), (x_i, y_{k-1}). We shall find an expression for the second derivative $\frac{\partial^2 u}{\partial x^2}$ in terms of the values of u at the adopted nodal points.

Let us utilize the expression for the derivatives which is obtained from the general interpolation formula of Newton, namely by taking the second of equations (15) § 1 and there setting $a_0 = x_i$, $a_1 = x_{i-1}$, $a_2 = x_{i+1}$.

$$\frac{\partial^2 u}{\partial x^2} = 2\,\frac{\dfrac{\Delta_x u_{i,k}}{\Delta x_i} - \dfrac{\Delta_x u_{i-1,k}}{\Delta x_{i-1}}}{\Delta x_i + \Delta x_{i-1}} + 2R_3''(x_i) =$$

$$= 2\left[\frac{u_{i+1,k}}{\Delta x_i(\Delta x_i + \Delta x_{i-1})} - \frac{u_{i,k}}{\Delta x_i \Delta x_{i-1}} + \frac{u_{i-1,k}}{\Delta x_{i-1}(\Delta x_i + \Delta x_{i-1})}\right] + 2R_3''(x_i).$$

We shall determine the remainder term. Applying Taylor's formula to the difference $\Delta_x u_{i,k} = u_{i+1,k} - u_{i,k}$, we see that the difference ratio $\frac{\Delta_x u_{i,k}}{\Delta x_i}$ has, if we stop with terms having derivatives of up to the fourth order, the following expression:

$$\frac{\Delta_x u_{i,k}}{\Delta x_i} = \frac{\partial u}{\partial x} + \frac{\Delta x_i}{2!}\frac{\partial^2 u}{\partial x^2} + \frac{\Delta x_i^2}{3!}\frac{\partial^3 u}{\partial x^3} + \frac{\Delta x_i^3}{4!}\left(\frac{\partial^4 u}{\partial x^4}\right)_{i,i+1;k}, \tag{19$_1$}$$

where by $\left(\frac{\partial^4 u}{\partial x^4}\right)_{i,i+1;k}$ is denoted the value of the fourth derivative at some intermediate point of the interval (x_i, x_{i+1}). Analogously, for the difference ratio $\frac{\Delta_x u_{i-1,k}}{\Delta x_{i-1}}$:

$$\frac{\Delta_x u_{i-1,k}}{\Delta x_{i-1}} = \frac{\partial u}{\partial x} - \frac{\Delta x_{i-1}}{2!}\frac{\partial^2 u}{\partial x^2} + \frac{\Delta x_{i-1}^2}{3!}\frac{\partial^3 u}{\partial x^3} - \frac{\Delta x_{i-1}^3}{4!}\left(\frac{\partial^4 u}{\partial x^4}\right)_{i-1,i;k}. \tag{19$_2$}$$

If both these expressions for the difference ratios are substituted in (19), we shall obtain for the remainder term:

$$-2R_3''(x_i) = \frac{\Delta x_i - \Delta x_{i-1}}{3}\frac{\partial^3 u}{\partial x^3} + \frac{1}{12(\Delta x_i + \Delta x_{i-1})}\left[\Delta x_i^3\left(\frac{\partial^4 u}{\partial x^4}\right)_{i,i+1;k} + \right.$$

$$\left. + \Delta x_{i-1}^3\left(\frac{\partial^4 u}{\partial x^4}\right)_{i-1,i;k}\right].$$

When the divisions along the axis OX are equidistant, $\Delta x_i = h$, the expression for the remainder term is simplified, since then the term in the third derivative drops out:

$$-2R_3''(x_i) = \frac{h^2}{24}\left[\left(\frac{\partial^4 u}{\partial x^4}\right)_{i,i+1;k} + \left(\frac{\partial^4 u}{\partial x^4}\right)_{i-1,i;k}\right],$$

whence is obtained the estimate of the remainder $2R_3''(x_i)$. If the maximum of the quantities $|\Delta x_i - \Delta x_{i-1}|$, $|\Delta y_k - \Delta y_{k-1}|$ be designated as ε, and the maximum of the quantities

$$\frac{\Delta x_i^3}{\Delta x_i + \Delta x_{i-1}}, \quad \frac{\Delta x_{i-1}^3}{\Delta x_i + \Delta x_{i-1}}, \quad \frac{\Delta y_k^3}{\Delta y_k + \Delta y_{k-1}}, \quad \frac{\Delta y_{k-1}^3}{\Delta y_k + \Delta y_{k-1}}$$

as η, then

$$|2R_3''(x_i)| \leqslant \frac{\varepsilon}{3} M_3 + \frac{\eta}{6} M_4,$$

where M_3 and M_4 are the greatest values of the third and fourth derivatives of u with respect to x and y in the region where we are considering the function u.

For the case of equidistant divisions, the estimate takes the form:

$$|2R_3''(x_i)| \leqslant \frac{h^2}{12} M_4.$$

Analogously, for the second partial derivative with respect to y we find:

$$\frac{\partial^2 u}{\partial y^2} = 2\frac{\dfrac{\Delta_y u_{i,k}}{\Delta y_k} - \dfrac{\Delta_y u_{i,k-1}}{\Delta y_{k-1}}}{\Delta y_k + \Delta y_{k-1}} + 2R_3''(y_k) =$$

$$= 2\left[\frac{u_{i,k+1}}{\Delta y_k(\Delta y_k + \Delta y_{k-1})} - \frac{u_{i,k}}{\Delta y_k \Delta y_{k-1}} + \frac{u_{i,k-1}}{\Delta y_{k-1}(\Delta y_k + \Delta y_{k-1})}\right] + 2R_3''(y_k),$$

$$-2R_3''(y_k) = \frac{\Delta y_k - \Delta y_{k-1}}{3}\frac{\partial^3 u}{\partial y^3} + \frac{1}{12(\Delta y_k + \Delta y_{k-1})}\left[\Delta y_k^3\left(\frac{\partial^4 u}{\partial y^4}\right)_{i;k,k+1} + \right.$$

$$\left. + \Delta y_{k-1}^3\left(\frac{\partial^4 u}{\partial y^4}\right)_{i;k-1,k}\right], \tag{20}$$

$$|2R_3''(y_k)| \leqslant \frac{\varepsilon}{3} M_3 + \frac{\eta}{6} M_4.$$

In case of equidistant divisions along the axis OY, however, $\Delta y_k = l$,

$$-2R_3''(y_k) = \frac{l^2}{24}\left[\left(\frac{\partial^4 u}{\partial y^4}\right)_{i;k,k+1} + \left(\frac{\partial^4 u}{\partial y^4}\right)_{i;k-1,k}\right], \quad |2R_3''(y_k)| \leqslant \frac{l^2}{12} M_4.$$

Let us pass on to the formulation of approximate expressions for the first derivatives. We shall take the first of equations (15) § 1 and in it set $a_0 = x_i$, $a_1 = x_{i-1}$, $a_2 = x_{i+1}$:

$$\frac{\partial u}{\partial x} = u(a_0, a_1) + (\alpha_0 + \alpha_1)u(a_0, a_1, a_2) + R_3'(a_0) =$$

$$= \frac{\Delta x_{i-1}\dfrac{\Delta_x u_{i,k}}{\Delta x_i} + \Delta x_i \dfrac{\Delta_x u_{i-1,k}}{\Delta x_{i-1}}}{\Delta x_i + \Delta x_{i-1}} + R_3'(x_i) = \frac{\Delta x_{i-1}}{\Delta x_i(\Delta x_i + \Delta x_{i-1})} u_{i+1,k} +$$

$$+ \frac{\Delta x_i - \Delta x_{i-1}}{\Delta x_i \Delta x_{i-1}} u_{i,k} - \frac{\Delta x_i}{\Delta x_{i-1}(\Delta x_i + \Delta x_{i-1})} u_{i-1,k} + R_3'(x_i). \qquad (21)$$

In order to obtain an expression for the remainder term $R_3'(x_i)$, let us represent the difference ratios $\dfrac{\Delta_x u_{i,k}}{\Delta x_i}$ and $\dfrac{\Delta_x u_{i-1,k}}{\Delta x_{i-1}}$ by the Taylor formula, as for (19_1) and (19_2), with the difference that we shall retain terms of up to the third order. Then in the preceding equation we shall substitute for the difference ratios their expansions. Simple calculations show that

$$R_3'(x_i) = \frac{\Delta x_i \Delta x_{i-1}}{6(\Delta x_i + \Delta x_{i-1})}\left[\Delta x_i \left(\frac{\partial^3 u}{\partial x^3}\right)_{i,i+1;k} + \Delta x_{i-1}\left(\frac{\partial^3 u}{\partial x^3}\right)_{i-1,i;k}\right],$$

whence follows this estimate for $R_3'(x_i)$:

$$|R_3'(x_i)| \leqslant \frac{\eta}{3} M_3.$$

In the case of equal intervals along the axis OX,

$$|R_3'(x_i)| \leqslant \frac{h^2}{6} M_3.$$

For the derivative in the direction of the other coordinate axis we have, on the same basis, the following value:

$$\frac{\partial u}{\partial y} = \frac{\Delta y_{k-1}\dfrac{\Delta_y u_{i,k}}{\Delta y_k} + \Delta y_k \dfrac{\Delta_y u_{i,k-1}}{\Delta y_{k-1}}}{\Delta y_k + \Delta y_{k-1}} + R_3'(y_k) = \frac{\Delta y_{k-1}}{\Delta y_k(\Delta y_k + \Delta y_{k-1})} u_{i,k+1} +$$

$$+ \frac{\Delta y_k - \Delta y_{k-1}}{\Delta y_k \Delta y_{k-1}} u_{i,k} - \frac{\Delta y_k}{\Delta y_{k-1}(\Delta y_k + \Delta y_{k-1})} u_{i,k-1} + R_3'(y_k), \qquad (22)$$

$$R_3'(y_k) = \frac{\Delta y_k \Delta y_{k-1}}{6(\Delta y_k + \Delta y_{k-1})}\left[\Delta y_k \left(\frac{\partial^3 u}{\partial y^3}\right)_{i;k,k+1} + \Delta y_{k-1}\left(\frac{\partial^3 u}{\partial y^3}\right)_{i;k-1,k}\right],$$

$$|R_3'(y_k)| \leqslant \frac{\eta}{3} M_3.$$

Now let us form the following combination of the values of the function

u at the nodes of the net which have been adopted by us:

$$l_{i,k}(u) = A_{i,k} 2 \frac{\frac{\Delta_x u_{i,k}}{\Delta x_i} - \frac{\Delta_x u_{i-1,k}}{\Delta x_{i-1}}}{\Delta x_i + \Delta x_{i-1}} + C_{i,k} 2 \frac{\frac{\Delta_y u_{i,k}}{\Delta y_k} - \frac{\Delta_y u_{i,k-1}}{\Delta y_{k-1}}}{\Delta y_k + \Delta y_{k-1}} +$$

$$+ D_{i,k} \frac{\Delta x_{i-1} \frac{\Delta_x u_{i,k}}{\Delta x_i} + \Delta x_i \frac{\Delta_x u_{i-1,k}}{\Delta x_{i-1}}}{\Delta x_i + \Delta x_{i-1}} +$$

$$+ E_{i,k} \frac{\Delta y_{k-1} \frac{\Delta_y u_{i,k}}{\Delta y_k} + \Delta y_k \frac{\Delta_y u_{i,k-1}}{\Delta y_{k-1}}}{\Delta y_k + \Delta y_{k-1}} + q_{i,k} u_{i,k}. \quad (23)$$

By $A_{i,k}$, $C_{i,k}$, ... we understand here the values of the coefficients A, C, ... of equation (17) at the node (x_i, y_k) of the net. If we introduce the designations

$$a_{ik} = 2\left(\frac{A_{ik}}{\Delta x_{i-1} \Delta x_1} + \frac{C_{ik}}{\Delta y_{k-1} \Delta y_k}\right) - \frac{\Delta x_i - \Delta x_{i-1}}{\Delta x_{i-1} \Delta x_i} D_{ik} -$$

$$- \frac{\Delta y_k - \Delta y_{k-1}}{\Delta y_k \Delta y_{k-1}} E_{ik} - q_{ik},$$

$$\left.\begin{aligned} b_{ik}^{(1)} &= 2 \frac{A_{ik}}{\Delta x_i (\Delta x_i + \Delta x_{i-1})} + \frac{\Delta x_{i-1}}{\Delta x_i (\Delta x_i + \Delta x_{i-1})} D_{ik}, \\ b_{ik}^{(2)} &= 2 \frac{C_{ik}}{\Delta y_k (\Delta y_k + \Delta y_{k-1})} + \frac{\Delta y_{k-1}}{\Delta y_k (\Delta y_k + \Delta y_{k-1})} E_{ik}, \\ b_{ik}^{(3)} &= 2 \frac{A_{ik}}{\Delta x_{i-1} (\Delta x_i + \Delta x_{i-1})} - \frac{\Delta x_i}{\Delta x_{i-1} (\Delta x_i + \Delta x_{i-1})} D_{ik}, \\ b_{ik}^{(4)} &= 2 \frac{C_{ik}}{\Delta y_{k-1} (\Delta y_k + \Delta y_{k-1})} - \frac{\Delta y_k}{\Delta y_{k-1} (\Delta y_k + \Delta y_{k-1})} E_{ik}, \end{aligned}\right\} \quad (24)$$

$l_{i,k}(u)$ can be represented in the form:

$$l_{i,k} = b_{i,k}^{(1)} u_{i+1,k} + b_{i,k}^{(2)} u_{i,k+1} + b_{i,k}^{(3)} u_{i-1,k} + b_{i,k}^{(4)} u_{i,k-1} - a_{i,k} u_{i,k}. \quad (23_1)$$

The multipliers of $A_{i,k}$, $C_{i,k}$, $D_{i,k}$ are the finite-difference expressions which figured in equations (19), (20), (21), (22). If in $l_{i,k}(u)$ we substitute the values found there for them, we obtain

$$l_{i,k}(u) = L(u) + R_{i,k}(u).$$

Moreover

$$R_{i,k}(u) = - A_{i,k} 2R_3''(x_i) - C_{i,k} 2R_3''(y_k) - D_{i,k} R_3'(x_i) - E_{i,k} R_3'(y_k). \quad (25)$$

The relation we have obtained between the differential operation $L(u)$ and the difference operation $l_{i,k}(u)$ serves, indeed, for the construction of a difference equation which approximately replaces the differential equation under discussion. Let us assume that u satisfies the equation $L(u) = f$. Then it will also satisfy the equation

$$l_{i,k}(u) = f_{i,k} + R_{i,k}(u). \tag{26}$$

For small Δx_i, Δy_k, the remainder term $R_{i,k}(u)$ will, generally speaking, be a small quantity. Discarding it, we obtain an equation in finite differences,

$$l_{i,k}(u) = f_{i,k}, \tag{26$_1$}$$

approximately replacing differential equation (17).

By such a replacement we make an error in the equation itself equal to $R_{i,k}(u)$. One can anticipate that a small error in the equation will occasion a small error in its solution, too. It is therefore natural to think that the solutions of equations (26) and (26_1) satisfying the same boundary condition will differ little from each other. Just how true these anticipations are will be seen below, where we shall occupy ourselves with the investigation of the convergence of approximations constructed in accordance with the method of nets.

The estimate of the error $R_{i,k}(u)$ of the equation is obtained from the known estimates of the errors $R_3''(x_i)$, $R_3''(y_k)$ which figure in (25), and has the form:

$$|R_{i,k}(u)| \leqslant \Big[\Big(\frac{\varepsilon}{3} M_3 + \frac{\eta}{6} M_4\Big)(A_{i,k} + C_{i,k}) + \\ + \frac{\eta}{3} M_3(|D_{i,k}| + |E_{i,k}|)\Big]_{\max}. \tag{26$_2$}$$

Equation (26_1) is considerably simplified when the net is square: $\Delta x_i = \Delta y_k = h$. Its coefficients will have the following expressions:

$$a_{ik} = -\frac{2}{h^2}(A_{ik} + C_{ik}) - q_{ik},$$

$$b_{ik}^{(1)} = \frac{A_{ik}}{h^2} + \frac{D_{ik}}{2h}, \quad b_{ik}^{(2)} = \frac{C_{ik}}{h^2} + \frac{E_{ik}}{2h},$$

$$b_{ik}^{(3)} = \frac{A_{ik}}{h^2} - \frac{D_{ik}}{2h}, \quad b_{ik}^{(4)} = \frac{C_{ik}}{h^2} - \frac{E_{ik}}{2h},$$

and equation (26_1) takes the form:

$$\Big(A_{ik} + \frac{h}{2} D_{ik}\Big) u_{i+1,k} + \Big(C_{ik} + \frac{h}{2} E_{ik}\Big) u_{i,k+1} + \\ + \Big(A_{ik} - \frac{h}{2} D_{ik}\Big) u_{i-1,k} + \Big(C_{ik} - \frac{h}{2} E_{ik}\Big) u_{i,k-1} - \\ - 2\Big(A_{ik} + C_{ik} - \frac{h^2}{2} q_{ik}\Big) u_{ik} = h^2 f_{ik}. \tag{26$_3$}$$

The much greater simplicity of this expression in comparison with the general form (26_1) compels one in practical calculations to give preference to square nets over rectangular, even if they have to be taken with a smaller step to approximate the contour Γ of the region G by the broken line C as closely as in the case of the rectangular net.

Example. Let us consider the Poisson equation

$$\frac{\partial^2 u}{\partial x^2} + \frac{\partial^2 u}{\partial y^2} = f(x, y).$$

Here $A = C = 1$, $D = E = q = 0$, $a_{ik} = \dfrac{2}{\Delta x_{i-1}\Delta x_i} + \dfrac{2}{\Delta y_{k-1}\Delta y_k}$,

$$b_{ik}^{(1)} = \frac{2}{\Delta x_i(\Delta x_{i-1} + \Delta x_i)}, \qquad b_{ik}^{(2)} = \frac{2}{\Delta y_k(\Delta y_{k-1} + \Delta y_k)},$$

$$b_{ik}^{(3)} = \frac{2}{\Delta x_{i-1}(\Delta x_{i-1} + \Delta x_i)}, \qquad b_{ik}^{(4)} = \frac{2}{\Delta y_{k-1}(\Delta y_{k-1} + \Delta y_k)}.$$

Equation (26) will have the form

$$\frac{2}{\Delta x_i + \Delta x_{i-1}}\left(\frac{u_{i+1,k}}{\Delta x_i} + \frac{u_{i-1,k}}{\Delta x_{i-1}}\right) + \frac{2}{\Delta y_k + \Delta y_{k-1}}\left(\frac{u_{i,k+1}}{\Delta y_k} + \frac{u_{i,k-1}}{\Delta y_{k-1}}\right) -$$

$$- 2\left(\frac{1}{\Delta x_{i-1}\Delta x_i} + \frac{1}{\Delta y_{k-1}\Delta y_k}\right) u_{ik} = f_{ik}. \tag{27}$$

In the case of a square net we obtain:

$$u_{i+1,k} + u_{i,k+1} + u_{i-1,k} + u_{i,k-1} - 4u_{ik} = h^2 f_{ik}. \tag{28}$$

The estimate of the error of this equation is obtained from (26_2) if we introduce there for A, C, D, E their values, and for ε, η the values they take for a square net: $\varepsilon = 0$, $\eta = \dfrac{h^2}{2}$,

$$|R_{i,k}(u)| \leqslant \frac{h^2}{6} M_4. \tag{29}$$

Let us remind ourselves that we have contemplated using a rectangular net solely owing to the consideration that by adopting such a net it could be contrived that the broken line C by which we replace the contour Γ, fit the latter better than it would with a square net.

One can completely avoid considering rectangular nets and limit oneself to square nets only, by resorting to the following device. We shall explain the idea of it on the example of a very simple problem. Let us take the Laplace equation $\Delta u = 0$ and consider the Dirichlet problem for it in the region G. Let us construct in the plane xy a square net

$$x_i = x_0 + ih, \quad y_k = y_0 + kh.$$

Let us pick out nodes of the net which either lie within G or on its boundary Γ. The values of u at the nodes that lie on Γ are known. The values of it at nodes of the net which belong to the interior of G are subject to determination, however. For the determination of them we shall form, as above, an equation in finite differences which approximately replaces

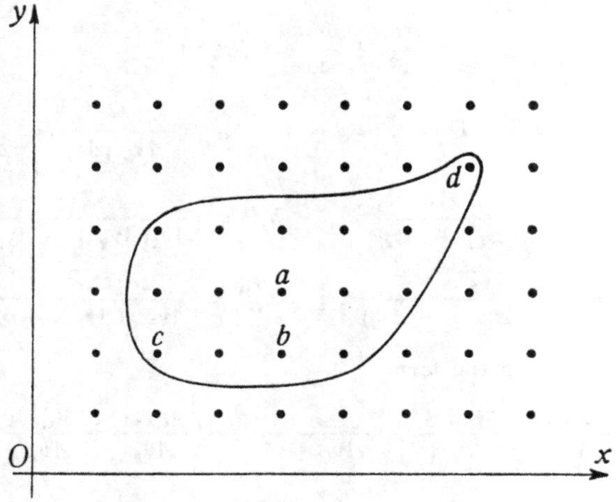

Fig. 9.

the Laplace equation, and for the formation of the approximate value of the derivative $\frac{\partial^2 u}{\partial x^2}$ $\left(\text{or } \frac{\partial^2 u}{\partial y^2}\right)$ at the node (i, k) of the net, we shall draw upon the values of u both at the node (i, k) itself and at two points lying on the line $x = x_i$ (or $y = y_k$), which points are either nodes of our net or points of the boundary Γ and adjacent to the node (i, k). This will lead us either to an equation of type (27) or to one of type (28), depending upon what internal point we select. In both cases $f_{i,k} = 0$. Let us consider the setting up of the equations in greater detail.

We shall divide into two groups the nodes of the net which lie within Γ. To the first group we assign those nodes for which all four of the nodes of the net closest to them lie either inside G or on the boundary Γ. Such will be the node a in Fig. 9, for instance. At all such points we shall replace the Laplace equation by the following equation involving the values of u at the nodes:

$$u_{i+1,k} + u_{i,k+1} + u_{i-1,k} + u_{i,k-1} - 4u_{i,k} = 0. \tag{30}$$

To the second group we shall assign all of the nodes for which at least one of the four nodes of the net closest to it is located outside Γ. Such are the points b, c, d in the Figure.

We adopt one of the points of the second group and denote it by the number 0 (Fig. 10). The four neighbouring points of the net we will denote by the numbers 1, 2, 3, 4. Some of them will lie outside G. Let us consider the case when all four neighboring points lie outside G. Let the contour Γ intersect the straight lines joining 0 with 1, 2, 3, 4 at the points $\alpha, \beta, \gamma, \delta$, the distances from which to 0 are respectively $t_1 h$, $t_2 h$, $t_3 h$, $t_4 h$. The numbers t_i are positive and do not exceed unity.

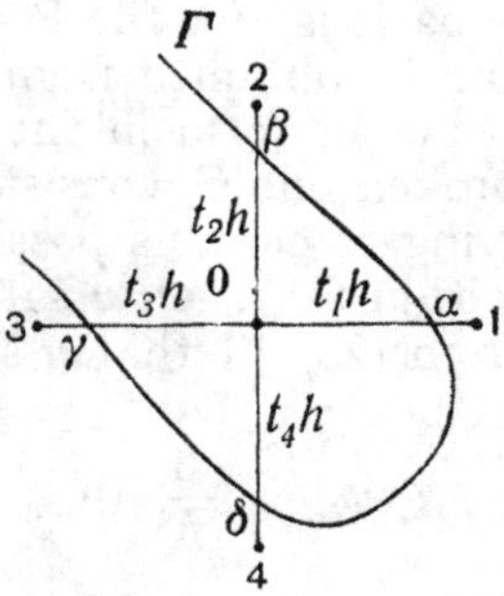

Fig. 10.

The case of the distribution of the points 1, 2, 3, 4 under consideration is general in the following sense. If it turns out that one of the points, for example 1, lies on Γ, or within G, then this new case is obtainable from that under consideration by letting $t_1 = 1$. The same thing can be said of the remaining points.

An equation in finite differences for any disposition of the points 1, 2, 3, 4 can be obtained from the equation (31), which we construct below, with a suitable choice of the values t_1, t_2, t_3, t_4.

The reasoning with regard to setting up an equation in finite differences for the point 0 is quite the same as that which we carried through in deriving equations (26) and (27); for our case, we have $A = C = 1$, $D = E = q = f = 0$,

$$\Delta x_i = t_1 h, \quad \Delta x_{i-1} = t_3 h,$$
$$\Delta y_k = t_2 h, \quad \Delta y_{k-1} = t_4 h,$$
$$u_{i,k} = u_0, \quad u_{i+1,k} = u_1, \quad u_{i,k+1} = u_2,$$
$$u_{i-1,k} = u_3, \quad u_{i,k-1} = u_4.$$

We therefore obtain for the point 0 an equation of the following form, which emerges from (27), if we substitute there for $u_{i,k}, \ldots$ and $\Delta x_i, \ldots$ their values cited above, and afterwards multiply everything by $\frac{1}{2}h^2$:

$$\frac{1}{t_1+t_3}\left(\frac{u_1}{t_1}+\frac{u_3}{t_3}\right)+\frac{1}{t_2+t_4}\left(\frac{u_2}{t_2}+\frac{u_4}{t_4}\right)-\left(\frac{1}{t_1 t_3}+\frac{1}{t_2 t_4}\right)u_0 = 0. \quad (31)$$

On setting up equations of types (30) and (31) for every node within G, we obtain a system of as many equations as there are unknown values of u at these nodes.

Equation (31) is of course more complex than (30), but the use of it has the advantage of freeing us of shifting the boundary values of u from Γ to the nodes of the net which belong to the auxiliary line C replacing the contour Γ, which line we should have to construct if we wished to limit ourselves to equations of type (30).

We shall now pass on to a more detailed consideration of the "finite equations" for the Poisson and Laplace equations. The simplest of them can be obtained from relations (28) § 1.

Let us form in the plane xy a square net of step h and construct a broken line C, consisting of the sides and diagonals of the squares, and lying as close as possible to the contour Γ. Let us take any node (i, k) within C. When u satisfies the Poisson equation $\Delta u = f(x, y)$, it follows from (28) § 1 that u satisfies the following equation:

$$\frac{1}{h^2} \Diamond u_{i,k} = \frac{1}{h^2} (u_{i+1,k} + u_{i,k+1} + u_{i-1,k} + u_{i,k-1} - 4u_{i,k}) = f_{i,k} + R_{i,k}(u).$$

If we discard the remainder term here, we obtain the finite-difference equation already known to us, (28) § 1, which approximately replaces the Poisson equation $\Delta u = f(x, y)$. Its error has the estimate (29) § 1.

A more exact equation can be obtained on the basis of relations (32) § 1. If it be set up for some node (i, k) of the net, there will appear in it, in addition to the values $u_{i,k}$ of the function at the point (i, k) itself, the values at all the vertices of the squares with a corner at the node (i, k). We can therefore consider to be an interior node any node of the net region, if together with it there also belong to the net region or to its boundary, the eight nodes which are the vertices of the squares that border upon the node in question. In conformity with this we must alter the structure of the curve C that bounds the net region and by which we replace G. On the net we mark a contour C passing only along sides of squares and fitting Γ as well as possible. Let us take any point (i, k) of the net, lying within C. When u satisfies the Poisson equation $\Delta u = f$, it will also, as is clear from (32) § 1, satisfy the equation:

$$\frac{1}{6h^2} (4 \Diamond u_{i,k} + \Box u_{i,k}) = f_{i,k} + \frac{2h^2}{4!} \Delta f_{i,k} + \\ + \frac{2h^4}{6!} \left[\Delta f_{i,k} + 2\left(\frac{\partial^4 f}{\partial x^2 \partial y^2}\right)_{i,k}\right] + R_{i,k}(u).$$

On discarding the remainder term here, $R_{i,k}(u)$, we obtain an approximate equation in finite differences:

$$\frac{1}{6h^2} (4 \Diamond u_{i,k} + \Box u_{i,k}) = \frac{u_{i+1,k+1} + u_{i-1,k+1} + u_{i-1,k-1} + u_{i+1,k-1}}{6h^2} + \\ + \frac{4(u_{i+1,k} + u_{i,k+1} + u_{i-1,k} + u_{i,k-1}) - 20u_{i,k}}{6h^2} = \\ = f_{i,k} + \frac{2h^2}{4!} \Delta f_{i,k} + \frac{2h^4}{6!} \left[\Delta^2 f_{i,k} + 2\left(\frac{\partial^4 f}{\partial x^2 \partial y^2}\right)_{i,k}\right], \quad (32)$$

the error of which has the estimate [§ 1 (33)]

$$|R_{i,k}(u)| \leqslant \frac{520}{3 \cdot 8!} h^6 M_8.$$

This approximate equation looks especially simple in the case of the Laplace equation, when $f = 0$; then, after multiplying (32) by $6h^2$, we have:

$$4 \Diamond u_{i,k} + \Box u_{i,k} = u_{i+1,k+1} + u_{i-1,k+1} + u_{i-1,k-1} + u_{i+1,k-1} + \\ + 4(u_{i+1,k} + u_{i,k+1} + u_{i-1,k} + u_{i,k-1}) - 20u_{i,k} = 0. \quad (33)$$

Equation (32) is somewhat more complicated in point of structure than (28) is, but possesses a considerably greater accuracy, since its remainder term will be a small quantity, of the sixth order with respect to h, whereas the remainder term of (28) is of but the second order of smallness.

In case the free member, $f(x, y)$, of the Poisson equation is given tabularly, or the computation of its derivatives presents difficulties owing to one reason or another, equation (32) becomes inconvenient and instead of it one can use equation (34) § 1, which is less exact than (32) since its error is of the fourth order of smallness with respect to h, and not, like (32), of the sixth; but it has by comparison with the latter the advantage of not containing derivatives of $f(x, y)$ and of requiring knowledge of f only at the nodes of the net. It stands to reason that one can construct an equation requiring, like (34) § 1, knowledge only of the function itself at the nodes of the net, the error of which will be of the same order as in (32). We shall not dwell on this, since we are not pursuing the aim of constructing all the types of equations that can be encountered in practical applications.

Let us now turn to other regular nets in the plane. Assume that a triangular net of step h has been constructed in the plane. In § 1 we established that for any function u having continuous derivatives of up to the sixth order, relation (35) § 1 holds between the values of u at the nodes of the net and the Laplace operation upon the function.

In particular when u satisfies the Poisson equation $\Delta u = f$, it will satisfy the following equation

$$\frac{2}{3h^2}\left(\sum_{i=1}^{6} u_i - 6u_0\right) = f_0 + \tfrac{1}{16}h^2 \Delta f_0 + R_0(u). \quad (34)$$

The remainder $R_0(u)$, the error of the equation obtained from (34) by discarding $R_0(u)$ from it, has the following estimate [see (36) § 1]:

$$|R_0(u)| \leqslant \frac{7h^4}{270} M_6.$$

For harmonic functions, in view of the fact that $f(x, y) = 0$, equation

(34) simplifies, acquiring the form

$$\sum_{i=1}^{6} u_i - 6u_0 = 0. \tag{35}$$

Lastly, when the net is formed of regular hexagons, equation (37) § 1 is fulfilled for any function having continuous derivatives of up to the third order. For functions that satisfy the Poisson equation $\Delta u = f$, however, it gives

$$\frac{4}{3h^2}\left(\sum_{i=1}^{3} u_i - 3u_0\right) = f_0 + R_0(u),$$

in which the remainder term, we recall, is estimated as follows:

$$|R_0(u)| \leqslant 1.36\, hM_3.$$

Discarding the remainder, we obtain a "finite equation" of the form:

$$\frac{4}{3h^2}\left(\sum_{i=1}^{3} u_i - 3u_0\right) = f_0. \tag{36}$$

The error of this equation is of the first order of smallness with respect to h. In view of the low accuracy of the equation, it is hardly of any practical value.

3. Boundary conditions for finite-difference equations. The simplest case will be that where on Γ the boundary values of the function $u(x, y)$ are assigned:

$$u|_\Gamma = \varphi(x, y).$$

These values must by one means or another be transferred to the contour C and adopted as the end values of $u_{i,k}$. It will be most natural to transfer the end values from Γ to the boundary nodes of the net via the shortest distance. Often the shift is made along straight lines parallel to the coordinate axes.

The boundary values for $u_{i,k}$ which are obtained as the result of such a shift will generally not coincide with the actual values that the function $u(x, y)$ takes on at the bounding nodes.

A computer who is experienced in solving this sort of problem can often indicate approximately, judging by the boundary conditions, the corrections that have to be assigned to the quantities being shifted in order to obtain as the result quantities which are closer to the actual values of the sought function at the boundary nodes.

A rough method of determining that part of the error that is due to the error in the end conditions is given by us below in § 4.

The correction of the errors r_{ik} can be effected, for example, in the following manner. After determining u_{ik} at all the nodes of the net, the u_{ik} are extrapolated to those points of the contour Γ from which the values were shifted to C, and the magnitude of the error is judged by the magnitude of the deviation from φ of the values obtained.

If the results obtained are unsatisfactory as regards accuracy, the equation for the error r_{ik} is set up in the usual manner. We know approximately its boundary values. From the equation are found the corrections which must be introduced in the previous result.

As the equations from which these corrections r_{ik} are to be found, one can obviously adopt the homogeneous equations corresponding to the finite-difference equations under study, (26), (28), (32), (34) and (36), and which are obtained from the latter by discarding the free members.

The case is more complex when we have to do with a Neumann problem, i.e., when the values assigned on the boundary are not those of the function itself but of its normal derivative:

$$\left.\frac{\partial u}{\partial n}\right|_{\Gamma} = \varphi,$$

or when we have to do with a more general boundary-value problem with boundary conditions of the form

$$\frac{\partial u}{\partial n} + \psi u|_{\Gamma} = \varphi.$$

In such a case not only the values of u are to be shifted from Γ to C, but also the values of the normal derivative.

Let us consider one of the boundary points, identified in Fig. 11 by the number zero, and two of its neighboring points of the net, 1 and 2. Whether the latter are boundary points or not is immaterial to us; we shall require only that the directions of l_1 and l_2 not be parallel. Let us call the angles that (l_1), (l_2) and (n) form with the x axis, φ_1, φ_2 and φ respectively, and their direction cosines α_1, β_1, α_2, β_2, and α, β respectively, so that

$$\alpha_1 = \cos\varphi_1,$$
$$\beta_1 = \sin\varphi_1, \ \ldots$$

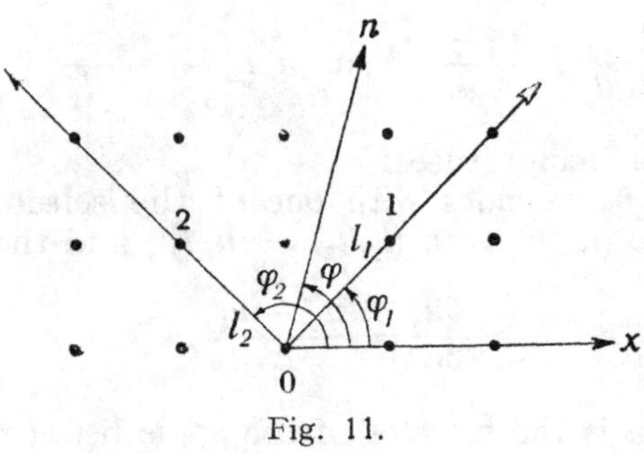

Fig. 11.

We denote by l_1 and l_2 the distances from the point 0 to points 1 and 2 respectively. Using Taylor's formula, we can write

$$\frac{\Delta_1 u}{l_1} = \frac{u_1 - u_0}{l_1} = \frac{u(x + l_1\alpha_1, y + l_1\beta_1) - u(x, y)}{l_1} = \alpha_1 u_x + \beta_1 u_y + R_1,$$

$$\frac{\Delta_2 u}{l_2} = \frac{u_2 - u_0}{l_2} = \frac{u(x + l_2\alpha_2, y + l_2\beta_2) - u(x, y)}{l_2} = \alpha_2 u_x + \beta_2 u_y + R_2,$$

where R_1 and R_2 will be infinitesimals of at least the first order with respect to l_1 and l_2. Hence we find u_x and u_y:

$$u_x = \frac{\dfrac{\Delta_1 u - R_1 l_1}{l_1}\beta_2 - \dfrac{\Delta_2 u - R_2 l_2}{l_2}\beta_1}{\alpha_1\beta_2 - \alpha_2\beta_1},$$

$$u_y = \frac{\dfrac{\Delta_2 u - R_2 l_2}{l_2}\alpha_1 - \dfrac{\Delta_1 u - R_1 l_1}{l_1}\alpha_2}{\alpha_1\beta_2 - \alpha_2\beta_1}.$$

Therefore for $\dfrac{\partial u}{\partial n} = \alpha u_x + \beta u_y$, if we observe that

$$\begin{aligned}
\alpha_1\beta_2 - \alpha_2\beta_1 &= \sin\varphi_2\cos\varphi_1 - \cos\varphi_2\sin\varphi_1 = \sin(\varphi_2 - \varphi_1) = \sin(l_1, l_2),\\
\alpha\beta_2 - \alpha_2\beta &= \sin\varphi_2\cos\varphi - \cos\varphi_2\sin\varphi = \sin(\varphi_2 - \varphi) = \sin(n, l_2),\\
\alpha\beta_1 - \alpha_1\beta &= \sin\varphi_1\cos\varphi - \cos\varphi_1\sin\varphi = -\sin(\varphi - \varphi_1) = -\sin(l_1, n),
\end{aligned}$$

we find

$$\frac{\partial u}{\partial n} = \frac{1}{\sin(l_1, l_2)}\left\{\frac{\Delta_1 u - R_1 l_1}{l_1}\sin(n, l_2) + \frac{\Delta_2 u - R_2 l_2}{l_2}\sin(l_1, n)\right\}.$$

If we discard R_1 and R_2 in this equation, we obtain the following approximate expression for $\dfrac{\partial u}{\partial n}$:

$$\frac{\partial u}{\partial n} = \frac{1}{\sin(l_1, l_2)}\left\{\frac{u_1 - u_0}{l_1}\sin(n, l_2) + \frac{u_2 - u_0}{l_2}\sin(l_1, n)\right\}.$$

We note some particular cases.

1. The normal n coincides with one of the selected directions, for example (l_1). Here $(l_1, n) = 0$, $(l_1, l_2) = (n, l_2)$, and therefore

$$\frac{\partial u}{\partial n} = \frac{u_1 - u_0}{l_1}.$$

2. The normal n is the bisector of the angle between l_1 and l_2. Then

$(l_1, n) = (n, l_2) = \dfrac{(l_1, l_2)}{2}$, and therefore

$$\frac{\partial u}{\partial n} = \frac{1}{2 \cos \frac{(l_1, l_2)}{2}} \left\{ \frac{u_1 - u_0}{l_1} + \frac{u_2 - u_0}{l_2} \right\}.$$

3. If the angle (l_1, l_2) is a right angle, $(n, l_2) = \dfrac{\pi}{2} - (l_1, n)$ and therefore

$$\frac{\partial u}{\partial n} = \frac{u_1 - u_0}{l_1} \cos (l_1, n) + \frac{u_2 - u_0}{l_2} \sin (l_1, n).$$

4. The equation $\Delta^2 u = f(x, y)$ [1]. We shall in addition fix the attention of the reader upon the application of the method of nets to the solution of the partial differential equation of the fourth order:

$$\Delta^2 u = \frac{\partial^4 u}{\partial x^4} + 2 \frac{\partial^4 u}{\partial x^2 \partial y^2} + \frac{\partial^4 u}{\partial y^4} = f(x, y). \tag{37}$$

Let it be required to find in the region G a solution of this equation satisfying certain conditions on the boundary of the region. For our reasoning the form of these conditions is not now of significance, and we shall therefore make no assumptions regarding them.

In constructing the finite-difference equation to replace the given equation approximately, we shall limit ourselves to the simplest case. Let us consider in the plane xy a square net of step h, formed by lines parallel to the axes of coordinates.

In § 1 we saw that if the function $u(x, y)$ has continuous derivatives of up to the sixth order, equation (39) § 1 is fulfilled for it at each node of the net.

Hence it follows that if u satisfies equation (37), then

$$\frac{20u_{0,0} - 8(u_{1,0} + u_{0,1} + u_{-1,0} + u_{0,-1}) + 2(u_{1,1} + u_{-1,1} + u_{-1,-1} + u_{1,-1})}{h^4} + \frac{(u_{2,0} + u_{0,2} + u_{-2,0} + u_{0,-2})}{h^4} = f_{0,0} + R_{0,0}(u),$$

where $R_{0,0}(u)$ has the estimate (40) § 1.

Discarding the remainder term $R_{0,0}(u)$, we obtain a finite-difference equation that approximately replaces differential equation (37):

$$\frac{20u_{0,0} - 8(u_{0,0} + u_{0,1} + u_{-1,0} + u_{0,-1}) + 2(u_{1,1} + u_{-1,1} + u_{1,-1} + u_{-1,-1})}{h^4} + \frac{u_{2,0} + u_{0,2} + u_{-2,0} + u_{0,-2}}{h^4} = f_{0,0}. \tag{38}$$

In this equation, together with the value $u_{0,0}$ of the function at the

[1] See D. IU. Panov, [2].

node (0, 0) in question, there also appear its values at the twelve nodes which stand at distances not greater than $2h$ from (0, 0). The aggregate of such nodes we shall call briefly the $2h$-vicinity of the node (0, 0).

We shall explain the basis of the application of (38) to the approximate solution of boundary-value problems for equation (37). Let us construct the line C which fits as closely as possible the contour Γ of region G and passes along the sides and diagonals of the squares of our net (Fig. 12).

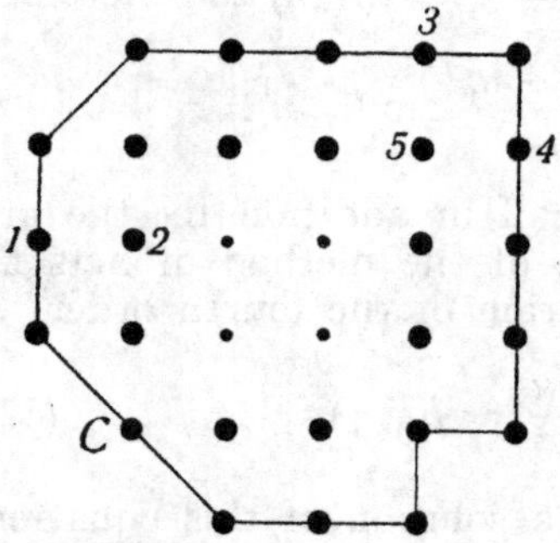

Fig. 12.

Let us take all the points of the net lying on C and add to them the points of the net lying within C and at a distance not greater than the step of the net, h, from any of the nodes on C. In the Figure these are identified by bold-face points. All of them form a certain set of nodes which we shall call the boundary strip. The points of the net which lie within C and do not belong to the boundary strip have the property that the nodes of the $2h$-vicinity of any of them will be either within or on C.

At all points of this kind equation (38) must be fulfilled.

As regards the nodes of the net belonging to the boundary strip, the values of the function u are either assigned at them, or equations are formulated for them which are different from (38) and are set up on the basis of the boundary conditions. For example, on the boundary Γ of the region G let there be given the values of the sought function u itself and of its derivative along the internal normal:

$$u\,|_\Gamma = \varphi, \left(\frac{\partial u}{\partial n}\right)_\Gamma = \psi.$$

We shall transform these boundary conditions into form convenient to us. Let us denote by s the length of the arc Γ, reckoned from an arbitrary point of origin in any definite direction. By the known values of u on Γ we find the value of its derivative with respect to the length of arc, $\frac{\partial u}{\partial s}$. Next by $\frac{\partial u}{\partial s}$ and $\frac{\partial u}{\partial n}$ we determine the derivatives of u with respect to the axes of coordinates, $\frac{\partial u}{\partial x}, \frac{\partial u}{\partial y}$.

The given boundary conditions we can replace by the following equivalent ones:

$$u\,|_\Gamma = \varphi, \left.\frac{\partial u}{\partial x}\right|_\Gamma = \alpha, \quad \left.\frac{\partial u}{\partial y}\right|_\Gamma = \beta,$$

where φ, α and β are known functions of the position of the point on Γ. Let us shift, by whatever method, the values of φ, α and β from Γ to nodes lying on C, and set up analogous boundary conditions for the nodes belonging to C.

Let us now take one of the nodes lying within C and belonging to the boundary strip. We shall consider all the boundary nodes at a distance h from it. Let us assume at the beginning that there is only one such node and that it, together with the node adopted by us, is disposed as are 1 and 2 in the Figure. The value of u at 1 is known and is equal to φ_1. The boundary condition $\left(\frac{\partial u}{\partial x}\right)_1 = \alpha_1$, however, we replace by the approximate condition $\frac{u_2 - u_1}{h} = \alpha_1$, whence we find the value of u at node 2:

$$u_2 = u_1 + h\alpha_1.$$

When it turns out, though, that several boundary nodes lie at a distance h from the adopted node, the value at this node can be computed by several methods.

In view of the approximate character of the equations from which they are computed, the values found need not necessarily coincide. Then the value of u at the node in question is ordinarily taken to be the arithmetic mean of all the values found. Such, for instance, will be node 5 in our Figure.

On considering the pair of points 3 and 5, we shall find the value $u_5 = u_3 - h\beta_3$. If, however, we take the pair of points 4 and 5, we shall find $u_5 = u_4 - h\alpha_4$. As u_5 we adopt the arithmetic mean of the two values found, putting

$$u_5 = \frac{u_3 + u_4 - h(\beta_3 + \alpha_4)}{2}.$$

Proceeding in this fashion, we find the values of u at all the points of the boundary strip.

§ 3. THE SOLUTION OF FINITE-DIFFERENCE EQUATIONS

1. The existence and uniqueness of the solution. In the preceding section we have constructed finite-difference equations approximately replacing the differential equations and the boundary conditions. The object of the present section is the investigation of the solubility of such a system and the exposition of some of the methods most used in solving it.

By a familiar theorem from determinant theory, the existence and uniqueness of the solution of the system will have been established if we show that the corresponding homogeneous system, obtained from that in question by discarding all free terms, has only a null solution. In order to do this one usually has recourse to the following fruitful device, which is the analogue of the use of Green's formulas for proof of the uniqueness of the solution of boundary-value problems for differential equations. For the homogeneous system one sets up a quadratic form in the sought values y_i or $u_{i,k}$ of the functions y and u at the nodes of the net, the form being such that the equations of the system are the conditions for an extremum of this quadratic form. Then, by the quadratic form, one judges for what values of y_i or $u_{i,k}$ it will have an extremum. It often turns out that the quadratic form is positive- or negative-definite, and has, accordingly, a single extremum, attained when all its arguments equal zero. Hence it at once follows that the homogeneous system under consideration has only a null solution.

Such a sequence of ideas is not always applicable to our reasoning. To construct a quadratic form, it is essential that the matrix of the coefficients of the system of equations be symmetric. Of the equations we have found, this, unfortunately, will not always be true. In constructing them we were guided not by considerations of the symmetry of the coefficients, proceeding instead upon what was the more natural requirement for the method of nets, namely that the error of all the equations have the same order of smallness with respect to h.

To preserve the uniformity of the method, we shall therefore proceed everywhere in our discussion from the equations themselves, not resorting to the construction of quadratic forms [1]).

Let us first consider an ordinary differential equation of the second order

$$y'' - q(x)y = f(x) \tag{1}$$

under either the boundary conditions

$$y(a) = \alpha, \quad y(b) = \beta, \tag{2_1}$$

or

$$y'(a) - h_1 y(a) = \alpha, \quad y'(b) + h_2 y(b) = \beta \text{ [2])}. \tag{2_2}$$

---

[1]) For acquaintance with the analogue of Green's formulas see, for instance, R. Courant, K. Friedrichs and H. Lewy, [1].

[2]) The equation we are considering, (1), is a particular case of the more general equation (1) § 2. This more general equation $y'' + p(x)y' + q(x)y = f(x)$ can by simple transformations, however, be reduced to form (1). Indeed, let us replace the unknown function y in it by setting $y = e^{-\int p(x)dx}u$. The equation then takes the form

$$(P(x)u')' - Q(x)u = f(x),$$

where

$$P(x) = e^{-\int p(x)dx}, \; Q(x) = [p'(x) - q(x)]e^{-\int p(x)dx}.$$

We next replace the argument x, putting

$$t = \int_a^x \frac{dx}{P(x)}.$$

The preceding equation is then transformed into the following:

$$\frac{d^2u}{dt^2} - Q_1(t)u = F(t),$$

where

$$Q_1(t) = Q(x)P(x), \; F(t) = f(x)P(x),$$

and this is an equation of the form we require.

The finite-difference equation of the first approximation (4) § 2 has for our case the form:

$$l_i^{(1)}(y) = \frac{y_{i+1} - 2y_i + y_{i-1}}{h^2} - q_i y_i = f_i \qquad (i = 1, 2, \ldots, N-1).$$

The boundary conditions will be either

$$y_0 = \alpha, \quad y_N = \beta,$$

or, in accordance with (6_1) and (6_2) § 2,

$$\frac{-y_2 + 4y_1 - 3y_0}{2h} - h_1 y_0 = \alpha, \quad \frac{3y_N - 4y_{N-1} + y_{N-2}}{2h} + h_2 y_N = \beta.$$

Let us discard the free terms f_i, α, β and take the corresponding homogeneous system:

$$\frac{y_{i+1} - 2y_i + y_{i-1}}{h^2} - q_i y_i = 0 \quad (i = 1, 2, \ldots, N-1) \tag{3}$$

under the boundary conditions

$$y_0 = 0, \quad y_N = 0, \tag{4_1}$$

or

$$\frac{-y_2 + 4y_1 - 3y_0}{2h} - h_1 y_0 = 0, \quad \frac{3y_N - 4y_N + y_{N-2}}{2h} + h_2 y_N = 0. \tag{4_2}$$

We shall now show that if $q_i \geqslant 0$, systems (3) and (4_1) have only a null solution.

Let us construct in the plane xy a broken line with vertices at the points (a, y_0), $(a + h, y_1)$, $\ldots$, $(a + ih, y_i)$, $\ldots$, (b, y_N) and elucidate the geometric meaning of equation (3). For this we shall take three successive vertices of the broken line (Fig. 13): $P_{i-1}(a + (i-1)h, y_{i-1})$, $P_i(a + ih, y_i)$, $P_{i+1}(a + (i+1)h, y_{i+1})$, construct the chord $P_{i-1}P_{i+1}$ and mark its middle point P'_i. Its ordinate is

$$\frac{y_{i+1} + y_{i-1}}{2} = Q_i P'_i.$$

Let us represent equation (3) in the form

$$y_i = \frac{1}{1 + \frac{1}{2}q_i h^2} \cdot \frac{y_{i+1} + y_{i-1}}{2}.$$

Hence it is clear that the ordinate y_i of the point P_i differs from the ordinate $Q_i P'_i$ by the positive factor $\frac{1}{1 + \frac{1}{2}q_i h^2}$, which does not exceed unity in magnitude. Equation (3) has, accordingly, this geometrical

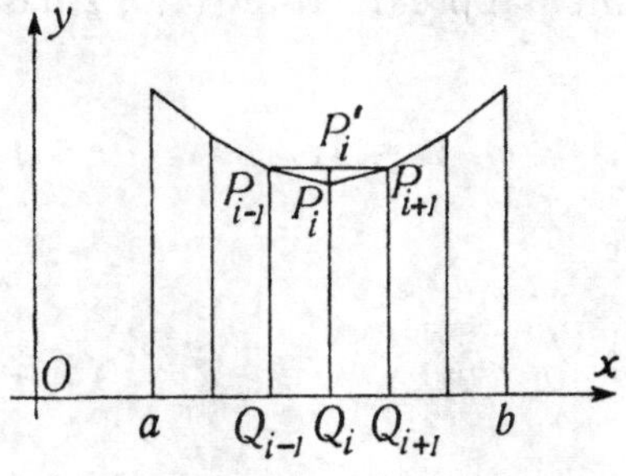

Fig. 13.

meaning, that each vertex P_i of the broken line either lies on the chord connecting the two neighboring vertices P_{i-1} and P_{i+1}, or closer to the axis OX. The part of the broken line lying above OX is turned with its concavity towards the side of positive ordinates, and the part of it lying below OX is turned with its concavity towards the side of negative ordinates.

Hence it is clear that:

1) The ordinate of the broken line can attain its greatest positive value, and least negative value, only at the ends of the interval.

2) The broken line can intersect the axis of abscissas at one point only.

Let us now assume that, together with (3), equations (4_1) are also fulfilled. If of the numbers $y_1, y_2, \ldots, y_N$ there are some different from zero, the ordinate of the broken line would have, within the interval, either a positive maximum or a negative minimum, which cannot be, in view of the first of the indicated properties of the broken line. Therefore we must have $y_0 = y_1 = \ldots = y_N = 0$.

Let us now pass on to the boundary conditions (4_2) and prove that if $q_i \geqslant 0$, $h_1 \geqslant 0$, $h_2 \geqslant 0$, and at least one of these numbers is different from zero, systems (3) and (4_2) have only a null solution.

Let us take the equation for $i = 1$ and the boundary condition for the left end of the interval (a, b):

$$\frac{y_2 - 2y_1 + y_0}{h^2} - q_1 y_1 = 0, \qquad \frac{-y_2 + 4y_1 - 3y_0}{2h} - h_1 y_0 = 0.$$

If we eliminate y_2 from this, we find the following relation between y_0 and y_1:

$$y_1 = \frac{1 + h_1 h}{1 - \frac{1}{2} q_1 h^2} y_0. \tag{*}$$

In exactly the same way we find for the right end of the interval

$$y_{N-1} = \frac{1 + h_2 h}{1 - \frac{1}{2} q_{N-1} h^2} y_N. \tag{**}$$

Let us now assume that not all the numbers $y_0, y_1, \ldots, y_N$ are equal to zero. Then among them there must be either a positive greatest, or a negative least number, where each of these can coincide only with y_0 or y_N. Let us assume, for example, that y_0 is positive and the greatest of the numbers y_i.

Compare the inequality $y_1 \leqslant y_0$ with (*). Since $h_1 \geqslant 0$, $q_1 \geqslant 0$, they can be compatible only in two cases: either a) when $h_1 = q_1 = 0$ and accordingly $y_1 = y_0$; or b) when $\frac{1}{2}q_1h^2 > 1$ and accordingly y_0 and y_1 are of different signs.

Case (a) signifies geometrically that the first link of the broken line runs parallel to the axis OX. Above we spoke of the fact that above the axis OX the broken line must be concave towards the positive ordinates. This means that after the first link it must run either parallel to the axis OX or turn upwards from OX. Since $y_0 \geqslant y_i$, the second possibility is ruled out, and we therefore must have $y_0 = y_1 = \ldots = y_N$. But then it follows from equations (3) and (4_2) that we must have $h_1 = 0$, $q_i = 0$, $h_2 = 0$. If, however, only one of these numbers is different from zero, all of the y_i must be equal to zero, and the homogeneous system, it has turned out, has only a null solution.

Case (b) is of less interest practically, for it holds only for h sufficiently large. Since y_1 is negative here, and $y_0 > 0$, the broken line between a and $a + h$ intersects the axis OX. In view of the second property, to the right of the point $a + h$ it must lie wholly beneath the axis OX: $y_i < 0$ $(i = 1, 2, \ldots, N)$. In particular, it follows from this that the numbers y_{N-1} and y_N are negative. In addition y_N must be the lesser of them. But this, as is at once evident, is incompatible with (**). The assumption $y_0 > 0$ leads to a contradiction here. Among the y_i there cannot be a positive greatest. For exactly the same cause there cannot be a negative least among them. Hence it follows that all y_i must be equal to zero.

Let us now examine the partial differential equation (17) § 2, considering for definiteness that the Dirichlet problem is being solved for it under the boundary condition

$$u|_\Gamma = \varphi(x, y).$$

In No. 2 § 2 we constructed for it the equation in finite differences (26_1) as a first approximation.

The boundary values of u are given on Γ. Shifting them from Γ to the nodes of the contour C of the net region, by any method whatever, we obtain for $u_{i,k}$ the boundary values:

$$u_{i,k}|_C = \varphi|_C.$$

To prove the existence and uniqueness of the solution of equation (26_1) § 2 under this boundary condition, we must show that the corresponding homogeneous system.

$$l_{i,k}(u) = b^{(1)}_{i,k}u_{i+1,k} + b^{(2)}_{i,k}u_{i,k+1} + b^{(3)}_{i,k}u_{i-1,k} + \\ + b^{(4)}_{i,k}u_{i,k-1} - a_{i,k}u_{i,k} = 0, \quad (4)$$

$$u_{i,k}|_C = 0,$$

has only a null solution. The values of the coefficients $b^{(s)}_{i,k}$ and $a_{i,k}$ are given by equations (24) § 2.

We shall first derive one corollary from the equation $l_{i,k}\ (u) = 0$ for the quantities $u_{i,k}$. It will be true for any boundary conditions, and is useful in investigating the uniqueness of the solution not only in the Dirichlet problem, but also in problems with other boundary conditions.

If $q_{i,k} \leqslant 0$ and the numbers Δx_i, Δy_i (the distances between the parallel lines of the net) are so small that for any i and k the inequalities

$$\left.\begin{aligned} |\Delta x_i D_{i,k}| \leqslant 2A_{i,k}, \quad |\Delta x_{i-1} D_{i,k}| \leqslant 2A_{i,k}, \\ |\Delta y_k E_{i,k}| \leqslant 2C_{i,k}, \quad |\Delta y_{k-1} E_{i,k}| \leqslant 2C_{i,k}, \end{aligned}\right\} \tag{5}$$

are fulfilled, the solution of $l_{i,k}(u) = 0$ cannot have either a positive maximum or a negative minimum within the net region. An exception is the trivial case when $u_{i,k}$ are equal to a constant.

The case of a negative minimum reduces to the case of a positive maximum if both sides of the equation $l_{i,k}(u) = 0$ are multiplied by -1, and we shall not examine it separately.

Let $u_{i,k}$ not be constants, and let them attain a positive maximum within the region; call it M. One can indicate an interior point of the net for which $u_{i,k} = M$, whereas at one, at least, of the neighboring points the value of $u_{i,k}$ will be less than M. We shall show that at such a point the equation $l_{i,k}(u) = 0$ cannot be fulfilled.

Given assumptions (5), all the coefficients $b_{i,k}^{(j)}$ will be non-negative, which is evident from their expressions (24) § 2. By adding these equations one can verify that $b_{i,k}^{(1)} + b_{i,k}^{(2)} + b_{i,k}^{(3)} + b_{i,k}^{(4)} = a_{i,k} + q_{i,k}$.

In the expression $b_{i,k}^{(1)}u_{i+1,k} + b_{i,k}^{(2)}u_{i,k+1} + b_{i,k}^{(3)}u_{i-1,k} + b_{i,k}^{(4)}u_{i,k-1}$ let us replace the values of $u_{i+1,k}$, $u_{i,k+1}$, $u_{i-1,k}$, $u_{i,k-1}$ by M. Since at least one of these values is less than M, we obtain as the result of the replacement the expression $M(b_{i,k}^{(1)} + b_{i,k}^{(2)} + b_{i,k}^{(3)} + b_{i,k}^{(4)}) = M(a_{i,k} + q_{i,k})$, which is more than the original. Hence, since $q_{i,k} \leqslant 0$ it follows that

$$b_{i,k}^{(1)}u_{i+1,k} + b_{i,k}^{(2)}u_{i,k+1} + b_{i,k}^{(3)}u_{i-1,k} + b_{i,k}^{(4)}u_{i,k-1} < M(a_{i,k} + q_{i,k}) \leqslant Ma_{i,k},$$

but this contradicts the equation $l_{i,k}(u) = 0$.

After the establishment of this result, that system (4) has just the one null solution is proved without difficulty. Indeed, if among the $u_{i,k}$ at least one were different from zero, $u_{i,k}$ would have attained on the boundary C of the region either a positive greatest or a negative least value, which cannot be, since the values of $u_{i,k}$ on the boundary C are equal to zero.

By just such reasoning can be proved the existence and uniqueness of the solution of the problem, with given boundary values of u, for equations (30), (32), (34), (36) § 2.

2. Two methods of solving finite-difference equations. Examples. Any of the familiar methods of solving systems of linear equations may be applied to the solution of the finite-difference equations by which we have approximated the differential equations. We shall dwell on but two of them, which are the most frequently employed [1]).

A. *The method of successive elimination* [2]). The idea of it can be fully elucidated by the following particular case, to which we shall limit ourselves for simplicity. Let us take equation (26_1) § 2; assume that the values of u on the boundary C of the net region are given, and that C is a rectangle with sides parallel to the coordinate axes (Fig. 14). Inside it a net is constructed by means of the lines $x = x_i$, $i = 0, 1, 2, \ldots, 6$; $y = y_k$, $k = 0, 1, \ldots, 4$. Let us take the inner points on the extreme left, which in our case lie on the line $x = x_1$; these will be the points (x_1, y_1), (x_1, y_2), (x_1, y_3). The values of $u_{i,k}$ corresponding to these points we shall leave undetermined and shall try to express in terms of them the remaining unknown values of u.

The values closest to them on the right can be expressed very simply in terms of them. Let us consider the equations that must be observed at these points:

$$b^{(1)}_{1,k}\, u_{2,k} + b^{(2)}_{1,k}\, u_{1,k+1} + b^{(3)}_{1,k}\, u_{0,k} + b^{(4)}_{1,k}\, u_{1,k-1} - a_{1,k} u_{1,k} = f_{1,k}, \quad (k = 1, 2, 3).$$

From them all $u_{2,k}$ can be found in terms of $u_{1,k}$ and the known values $u_{0,k}$:

$$u_{2,k} = \frac{1}{b^{(1)}_{1,k}} [f_{1,k} + a_{1,k} u_{1,k} - b^{(2)}_{1,k}\, u_{1,k+1} - b^{(3)}_{1,k}\, u_{0,k} - b^{(4)}_{1,k}\, u_{1,k-1}].$$

We next write the equations which must be fulfilled for nodes lying on the line $x = x_2$; from them we find expressions for $u_{3,k}$ in terms of $u_{1,k}$ and the given values, etc. Proceeding in this fashion, we can successively express in terms of $u_{1,k}$ all $u_{i,k}$ for $i \geqslant 2$. The values of $u_{5,k}$ corresponding to the last line $x = x_5$ of the region, we will find from equations relating to the nearest nodes on the left, for the line $x = x_4$. It still remains to satisfy equations for the nodes lying on the last line of the region, $x = x_5$. The number of them will obviously be equal to the number

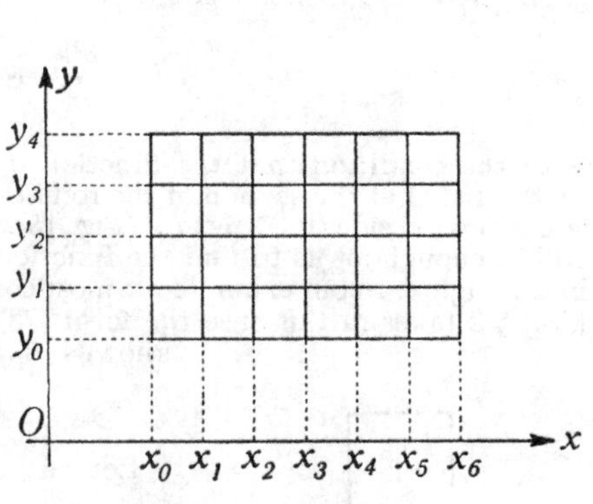

Fig. 14.

[1]) In recent years, in connection with the development of machine mathematics, questions of the employment of digital computing machines in solving boundary-value problems have been studied. They have not found their reflection in a book, and to familiarize him with them we refer the reader to special papers. See Akushskiĭ, I. IA., [1], [2], [3]; Kovner, S. S. and Zhak, D. K., [1].

[2]) See, for example, C. Runge, [1].

of remaining unknowns $u_{1,k}$. By solving them for $u_{1,k}$, we shall have found all the remaining values u_{ik} in terms of them.

In order to simplify all the necessary computations, one can resort to the following device, which permits one to reduce the process here described to a series of simpler processes, each of which can be performed numerically. The sought values $u_{5,k}$ are linearly expressed in terms of $u_{1,k}$. We shall seek separately the free terms and coefficients of $u_{1,k}$ in these expressions. We shall first carry through all the computations for $u_{1,1} = 0$, $u_{1,2} = 0$, $u_{1,3} = 0$. The result to which we come at the nodes of the line $x = x_5$ will give a term in $u_{5,k}$ not depending on $u_{1,k}$. The coefficients of $u_{1,k}$, then, are computed thus. For example, if it is necessary to find the coefficient of $u_{1,1}$, we set $u_{1,1} = 1$, $u_{1,2} = u_{1,3} = 0$; next we discard all the free terms of equation (25_1) § 2 and set all the values of $u_{i,k}$ on the contour equal to zero. In our example the equations for the nodes relating to the line $x = x_i$ will after this take the form:

$$\begin{aligned} b_{i,1}^{(1)}u_{i+1,1} + b_{i,1}^{(2)}u_{i,2} + b_{i,1}^{(3)}u_{i-1,1} + 0 - a_{i,1}u_{i,1} &= 0, \\ b_{i,2}^{(1)}u_{i+1,2} + b_{i,2}^{(2)}u_{i,3} + b_{i,2}^{(3)}u_{i-1,2} + b_{i,2}^{(4)}u_{i,1} - a_{i,2}u_{i,2} &= 0, \\ b_{i,3}^{(1)}u_{i+1,3} + 0 + b_{i,3}^{(3)}u_{i-1,3} + b_{i,3}^{(4)}u_{i,2} - a_{i,3}u_{i,3} &= 0. \end{aligned}$$

After this we conduct the computations in the manner already indicated.

We shall illustrate this argument in an example, which we borrow from the indicated memoir of C. Runge.

Example. The torsion of a homogeneous cylindrical rod is reducible to the solution of the Poisson equation:

$$\frac{\partial^2 u}{\partial x^2} + \frac{\partial^2 u}{\partial y^2} = C$$

with the condition that the function u vanish at the boundary of a cross-section of the rod. Let the section of the rod have the form of a cross formed by five equal squares with sides of length l (Fig. 15). Owing to considerations of symmetry, it will be enough for us to find the function for the eighth part of the cross indicated in the Figure. In our example the most convenient net will be a square one. Equation (26_1) § 2 takes in this case the form (28) § 2, or, if for the unknowns $u_{i,k}$ new unknowns $u_{ik} = -Ch^2 v_{ik}$ be introduced,

$$v_{i+1,k} + v_{i,k+1} + v_{i-1,k} + v_{i,k-1} - 4v_{ik} = -1.$$

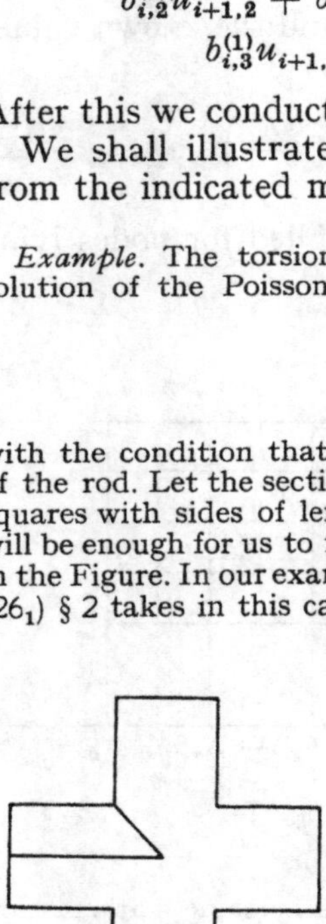

Fig. 15.

Let us now, for example, divide the sides of the squares into eight parts, setting $h = \frac{l}{8}$. The corresponding net for the eighth part of the figure is depicted in Fig. 16. Inside the cross there will be arrayed 273 nodes in all, 42 of them in the part we have under consideration. We must thus solve a system of 42 equations.

Leaving arbitrary the values $v_{11} = \alpha$, $v_{12} = \beta$, $v_{13} = \gamma$, $v_{14} = \delta$, at points A, B, C, D of the drawing respectively, we shall denote by α', β', γ', δ' the values of

v at the nodes of the vertical column adjoining on the right, and shall write the equations that must be observed at the points A, B, C, D. In the equation for the point A there must figure the value of v at the point of the net which lies directly beneath point A along the vertical. In view of the symmetry, it will coincide with β, and we will therefore obtain:

$$
\begin{array}{ll}
\text{at the point } A: & \alpha' + \beta + 0 + \beta - 4\alpha = -1 \\
\quad\text{,,} \qquad B: & \beta' + \gamma + 0 + \alpha - 4\beta = -1 \\
\quad\text{,,} \qquad C: & \gamma' + \delta + 0 + \beta - 4\gamma = -1 \\
\quad\text{,,} \qquad D: & \delta' + 0 + 0 + \gamma - 4\delta = -1.
\end{array}
$$

From this we find the expressions for α', β', γ', δ' in terms of α, β, γ, δ; analogously for the following column; and so on until α''', β''', γ''', δ''' are found for the extreme points on the right; they will be determined from the equations at points A'', B'', C'', D''. It remains for us to write down the equations for the quantities v at the border points. Utilizing the conditions of symmetry, we can write these equations in the following form:

$$
\left.
\begin{aligned}
4\alpha'' - 4\alpha''' &= -1, \\
2\beta'' + 2\alpha'' - 4\beta''' &= -1, \\
2\gamma'' + 2\beta'' - 4\gamma''' &= -1, \\
2\delta'' + 2\gamma'' - 4\delta''' &= -1,
\end{aligned}
\right\} \qquad (*)
$$

and from these equations α, β, γ, δ must be determined.

In conformity with what was said above, we shall not directly determine the equations themselves, but shall determine separately the free terms and the coefficients of the unknowns in the expressions for α'', α''', ...

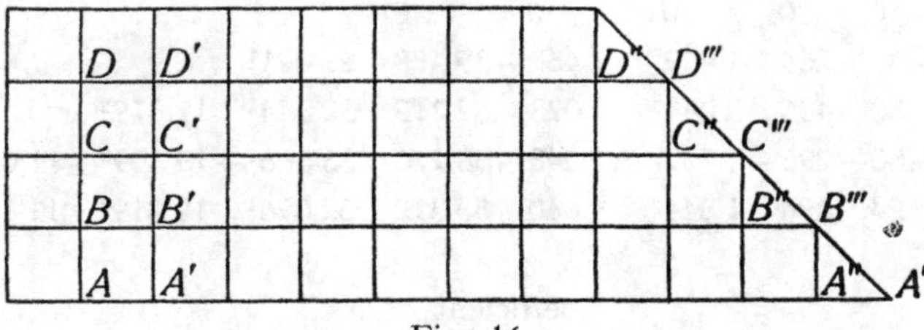

Fig. 16.

For determining the free term we set $\alpha = \beta = \gamma = \delta = 0$ and carry through all the necessary computations. As the result we obtain the following table, which we cite to an accuracy of five figures [1]):

0	0	0	0	0	0	0	0	0				
0	0	− 1	− 4	− 13	− 44	− 165	− 680	− 3 001	− 13 881			
0	0	− 1	− 3	− 5	1	63	445	2 556	13 852	73 821		
0	0	− 1	− 3	− 6	− 11	− 30	− 160	− 1 073	− 7 089	− 44 864	$27\,488^{1}$	
0	0	− 1	− 3	− 6	− 10	− 13	17	400	3 728	28 689	$20\,076^{1}$	$20\,076^{1}$

[1]) Some of the numbers of the table are the product of a five-figure number by some power of ten; the exponent of the power of ten we have affixed to the number as a superscript, so that $13\,241^{2} = 13\,241 \cdot 10^{2}$.

For determining the coefficients of α we put $\alpha = 1$, $\beta = \gamma = \delta = 0$ and discard the free terms. After the computations we obtain the table

0	0	0	0	0	0	0	0	0				
0	0	0	0	−1	−16	−160	−1 296	−9 344	−62 736			
0	0	0	1	12	97	672	4 320	26 656	$16\,048^1$	$95\,094^1$		
0	0	−1	−8	−49	−280	−1 569	−8 752	−48 832	$-27\,294^1$	$-15\,289^2$	$-85\,814^2$	
0	1	4	17	80	401	2 084	11 073	59 712	$32\,544^1$	$17\,879^2$	$98\,840^2$	$54\,911^3$

By the same or a somewhat simpler method, which was proposed by Runge in the article we have cited, and for the explanation of which we shall not tarry here, there may also be found the coefficients of the remaining unknowns. The tables of the computations are subjoined.

Coefficient of β

0	0	0	0	0	0	0	0	0				
0	0	0	1	12	97	672	4 320	26 656	$16\,048^1$			
0	0	−1	−8	−50	−296	−1 729	−10 048	−58 176	$-33\,568^1$	$-19\,309^2$		
0	1	4	18	92	498	2 756	15 393	86 368	$48\,592^1$	$27\,389^2$	$15\,458^3$	
0	0	−2	−16	−98	−560	−3 138	−17 504	−97 664	$-54\,589^1$	$-30\,577^2$	$-17\,163^3$	$-96\,510^3$

Coefficient of γ

0	0	0	0	0	0	0	0	0				
0	0	−1	−8	−48	−264	−1 409	−7 456	−39 488	$-21\,021^1$			
0	1	4	17	80	401	2 084	11 073	59 712	$32\,544^1$	$17\,879^2$		
0	0	−1	−8	−50	−296	−1 729	−10 048	−58 176	$-33\,568^1$	$-19\,309^2$	$-11\,078^3$	
0	0	0	2	24	194	1 344	8 640	53 312	$32\,096^1$	$19\,019^2$	$11\,148^3$	$64\,848^3$

Coefficient of δ

0	0	0	0	0	0	0	0	0				
0	1	4	16	68	304	1 412	6 753	33 056	$16\,496^1$			
0	0	−1	−8	−48	−264	−1 409	−7 456	−39 488	$-21\,021^1$	$-11\,268^2$		
0	0	0	1	12	97	672	4 320	26 656	$16\,048^1$	$95\,094^1$	$55\,742^2$	
0	0	0	0	−2	−32	−320	−2 592	−18 688	$-12\,547^1$	$-80\,416^1$	$-49\,931^2$	$-30\,317^3$

By using these Tables we can form all the quantities α'', β'', ... that figure in the system, for example

$$\delta'' = -3\,001 - 9\,344\alpha + 26\,656\beta - 39\,488\gamma + 33\,056\delta.$$

After solving the system, if we preserve two figures after the decimal, we obtain:

$$\alpha = 2.63, \ \beta = 2.50, \ \gamma = 2.07, \ \delta = 1.30.$$

The table of v_{ik} formed by means of the values found has the form:

0	0	0	0	0	0	0	0	0				
0	1.30	2.12	2.70	3.14	3.53	3.95	4.55	5.65	8.29			
0	2.07	3.50	4.53	5.32	6.02	6.73	7.59	8.78	10.42	11.89		
0	2.50	4.28	5.59	6.61	7.49	8.36	9.32	10.45	11.72	12.86	13.63	
0	2.63	4.53	5.93	7.03	7.98	8.89	9.88	10.98	12.15	13.19	13.90	14.15

B. *The method of iteration.* In the general case it has the following form. Let there be given a system of n equations in the form

$$\begin{aligned} x_1 &= a_{11}x_1 + a_{12}x_2 + \dots + a_{1n}x_n \\ x_2 &= a_{21}x_1 + a_{22}x_2 + \dots + a_{2n}x_n \\ &\dots\dots\dots\dots\dots\dots \\ x_n &= a_{n1}x_1 + a_{n2}x_2 + \dots + a_{nn}x_n. \end{aligned}$$

By some means let us choose first approximations $x_1^{(1)}, x_2^{(1)}, \dots, x_n^{(1)}$ to the unknowns x_i and substitute them in the right members of the equations. We adopt as the second approximations to x_i the values obtained:

$$\begin{aligned} x_1^{(2)} &= a_{11}x_1^{(1)} + a_{12}x_2^{(1)} + \dots + a_{1n}x_n^{(1)} \\ x_2^{(2)} &= a_{21}x_1^{(1)} + a_{22}x_2^{(1)} + \dots + a_{2n}x_n^{(1)} \\ &\dots\dots\dots\dots\dots\dots \\ x_n^{(2)} &= a_{n1}x_1^{(1)} + a_{n2}x_2^{(1)} + \dots + a_{nn}x_n^{(1)}. \end{aligned}$$

We again substitute them in the right sides of the system, taking the results as the third approximation to x_i, and so forth. Approximation number $k + 1$ is computed in terms of the preceding one in accordance with the equations

$$\begin{aligned} x_1^{(k+1)} &= a_{11}x_1^{(k)} + a_{12}x_2^{(k)} + \dots + a_{1n}x_n^{(k)} \\ x_2^{(k+1)} &= a_{21}x_1^{(k)} + a_{22}x_2^{(k)} + \dots + a_{2n}x_n^{(k)} \\ &\dots\dots\dots\dots\dots\dots \\ x_n^{(k+1)} &= a_{n1}x_1^{(k)} + a_{n2}x_2^{(k)} + \dots + a_{nn}x_n^{(k)}. \end{aligned}$$

The computations are repeated until an approximation coincides with the preceding, within the adopted limits of accuracy.

It goes without saying that in practical computations there is no need of finding the approximations $x_i^{(k)}$ themselves, it being sufficient to find only the corrections that have to be added to the values $x_i^{(k)}$ of the kth approximation in order to obtain the $(k + 1)$th approximation. If we designate these corrections as

$$\delta_1^{(k)} = x_1^{(k+1)} - x_1^{(k)}, \dots, \delta_n^{(k)} = x_n^{(k+1)} - x_n^{(k)},$$

for the successive computation of them we shall have the system

$$\delta_1^{(k+1)} = a_{11}\delta_1^{(k)} + a_{12}\delta_2^{(k)} + \ldots + a_{1n}\delta_n^{(k)}$$
$$\cdots\cdots\cdots\cdots\cdots\cdots$$
$$\delta_n^{(k+1)} = a_{n1}\delta_1^{(k)} + a_{n2}\delta_2^{(k)} + \ldots + a_{nn}\delta_n^{(k)}.$$

Moreover

$$x_1^{(k+1)} = x_1^{(1)} + \delta_1^{(1)} + \delta_1^{(2)} + \ldots + \delta_1^{(k)}$$
$$\cdots\cdots\cdots\cdots\cdots\cdots$$
$$x_n^{(k+1)} = x_n^{(1)} + \delta_n^{(1)} + \delta_n^{(2)} + \ldots + \delta_n^{(k)}.$$

We shall make one more observation on the subject of the computations. In finding the $(k+1)$th approximation $x_i^{(k+1)}$ we have used the kth approximation $x_i^{(k)}$. It can be expected that each succeeding approximation will be closer to the true value than the preceding. Therefore for accelerating the convergence one can, in finding x_i in the $(k+1)$th approximation, use the values found without delay, arranging the computations by the scheme

$$x_1^{(k+1)} = a_{11}x_1^{(k)} + a_{12}x_2^{(k)} + \ldots + a_{1n}x_n^{(k)}$$
$$x_2^{(k+1)} = a_{21}x_1^{(k+1)} + a_{22}x_2^{(k)} + \ldots + a_{1n}x_n^{(k)}$$
$$x_3^{(k+1)} = a_{31}x_1^{(k+1)} + a_{32}x_2^{(k+1)} + \ldots + a_{1n}x_n^{(k)}$$
$$\cdots\cdots\cdots\cdots\cdots\cdots$$

As the result of the computations, for each x_i there is obtained the sequence of approximations

$$x_i^{(1)},\ x_i^{(2)},\ x_i^{(3)},\ \ldots \qquad (*)$$

To acquaint the reader with discussions of the convergence of these sequences we refer to special works [1]) and cite without proof the two following very simple tests.

If the coefficients of the system

$$x_i = \sum_{k=1}^{n} a_{i,k}x_k + b_i$$

satisfy the conditions

$$\sum_{k=1}^{n} |a_{i,k}| < 1, \qquad (i = 1, 2, \ldots, n)$$

sequence (*) *converges for any first approximations* $x_1^{(1)}, \ldots, x_n^{(1)}$ *and the limiting values* $x_i = \lim_{k\to\infty} x_i^{(k)}$ *satisfy the given system.*

We note also that the convergence will be the more rapid, the less the numerical value of the sum $\sum_{k=1}^{n} |a_{i,k}|$. It is therefore profitable to transform the system before solution so as to make the coefficients $a_{i,k}$ in it as small as possible.

---

[1]) Mises, R. v. and Pollaczek-Geiringer, [1]. See also D. IU. Panov, [3] or [4]. Cf. Faddeeva, V.N., [1].

The theorem we have formulated is a particular case of an analogous theorem for fully regular infinite systems proved in No. 2 § 2 Chap. I.

The criterion of convergence provided by this theorem is very simple, but relates to a comparatively narrow class. A wider class of the equations encountered in mathematical physics and technology is embraced by a second test, although it does not subsume the first.

If the matrix of the system of equations

$$\sum_{k=1}^{n} a_{i,k}x_k = b_i \qquad (i = 1, 2, \ldots, n)$$

is symmetric, i.e., $a_{i,k} = a_{k,i}$, *and if the quadratic form*

$$F(x_1, x_2, \ldots, x_n) = \sum_{i,k=1}^{n} a_{i,k}x_i x_k,$$

is definite and positive, sequences (*), *constructed by the method of iteration in which the i-th equation is used for the correction of the i-th unknown, will converge for all initial values.*

This second criterion guarantees, for example, the convergence of the method of iteration for equations (28), (32), (34) and (36) § 2 in solving problems with known boundary values of the sought functions.

Example. Let us seek, in the interval $0 \leqslant x < 1$, the solution of the equation

$$y'' - xy = 1$$

under the boundary conditions $y(0) = 0$, $y(1) = 0$. Let us divide the interval (0, 1) into ten equal parts with the points 0, 0.1, 0.2, ..., 1; $h = 0.1$. On replacing the second derivative y'' in the equation by its expression in terms of the second difference

$$y''(ih) = \frac{y_{i+1} - 2y_i + y_{i-1}}{h^2} = \frac{y_{i+1} - 2y_i + y_{i-1}}{0.01},$$

we obtain the following system of equations for the determination of $y_1, y_2, \ldots, y_9$:

$$\frac{y_{i+1} - 2y_i + y_{i-1}}{0.01} - i \cdot 0.1 y_i = 1, \qquad (i = 1, 2, \ldots, 9)$$

$$y_0 = 0, \qquad y_{10} = 0,$$

or

$$y_1 = \frac{y_2 + 0.01}{2.001},$$

$$y_2 = \frac{y_1 + y_3 + 0.01}{2.002},$$

$$\cdots\cdots\cdots\cdots \qquad (**)$$

$$y_i = \frac{y_{i+1} + y_{i-1} + 0.01}{2 + 0.001i},$$

$$\cdots\cdots\cdots\cdots$$

$$y_9 = \frac{y_8 + 0.01}{2.009}.$$

We shall employ the method of iteration for the solution. In order to determine the first approximation to y_i, we shall first solve the problem with a step twice as large: $h = 0.2$; we divide the interval into 5 parts by the points 0, 0.2, 0.4, . . ., 1. To avoid renumbering the unknowns, we shall call the values of y at these points $y_0, y_2, y_4, y_6, y_8, y_{10}$. In the equation we replace the second derivative by its approximate expression

$$y''(ih) = \frac{y_{i+2} - 2y_i + y_{i-2}}{0.04} \qquad (i = 2, 4, \ldots, 8)$$

and obtain the system

$$\frac{y_{i+2} - 2y_i + y_{i-2}}{0.04} - i \cdot 0.1 y_i = 1, \qquad (i = 2, 4, 6, 8)$$

$$y_0 = 0, \qquad y_{10} = 0,$$

or

$$y_2 = \frac{y_4 + 0.04}{2.008},$$

$$y_4 = \frac{y_2 + y_6 + 0.04}{2.016},$$

$$y_6 = \frac{y_4 + y_8 + 0.04}{2.024},$$

$$y_8 = \frac{y_6 + 0.04}{2.032}.$$

We shall eliminate y_2 and y_8, by determining them from the first and last of the equations and substituting in the second and third. The system is then reduced to the following one:

$$2.016 y_4 = y_6 + \frac{y_4 + 0.04}{2.008} + 0.04, \quad 2.024 y_6 = y_4 + \frac{y_6 + 0.04}{2.032} + 0.04.$$

Hence we find these values:

$$y_2 = 0.0768, \; y_4 = 0.1143, \; y_6 = 0.1136, \; y_8 = 0.0756.$$

We shall adopt them for the initial values of our unknowns. To determine y_1, y_3, y_5, y_7, y_9 we make an interpolation by Bessel's formula (14) § 1 (for y_3, y_5, y_7) and by Newton's formula (10) § 1 (for y_1 and y_9). This leads us to the following quantities:

$$y_1 = 0.0434, \; y_3 = 0.1004, \; y_5 = 0.1186, \; y_7 = 0.0993, \; y_9 = 0.0425.$$

We summarize the iterations for system (**) in the following table:

		1st Approx.	2nd Approx.	3rd Approx.
0.1	y_1	0.0434	0.0434	0.0434
0.2	y_2	0.0768	0.0768	0.0768
0.3	y_3	0.1004	0.1004	0.1004
0.4	y_4	0.1143	0.1143	0.1143
0.5	y_5	0.1186	0.1187	0.1187
0.6	y_6	0.1136	0.1137	0.1137
0.7	y_7	0.0993	0.0993	0.0994
0.8	y_8	0.0756	0.0756	0.0756
0.9	y_9	0.0425	0.0426	0.0426

It is clear from it that for the correction of the initial values to the necessary accuracy, two cycles of computations are required altogether.

The exact solution of the equation can be found by means of the power series

$$y = 0.48408\left(x + \frac{x^4}{3\cdot 4} + \frac{x^7}{3\cdot 4\cdot 6\cdot 7} + \frac{x^{10}}{3\cdot 4\cdot 6\cdot 7\cdot 9\cdot 10} + \ldots\right) -$$

$$-\left(\frac{x^2}{1\cdot 2} + \frac{x^5}{1\cdot 2\cdot 4\cdot 5} + \frac{x^8}{1\cdot 2\cdot 4\cdot 5\cdot 7\cdot 8} + \frac{x^{11}}{1\cdot 2\cdot 4\cdot 5\cdot 7\cdot 8\cdot 9\cdot 10\cdot 11} + \ldots\right).$$

The values of y computed from this are

$$\begin{aligned} y(0.1) &= 0.0434, & y(0.6) &= 0.1138,\\ y(0.2) &= 0.0768, & y(0.7) &= 0.0994,\\ y(0.3) &= 0.1004, & y(0.8) &= 0.0757,\\ y(0.4) &= 0.1144, & y(0.9) &= 0.0426.\\ y(0.5) &= 0.1188, \end{aligned}$$

The approximate values which we found differ from these only by a unit of the fourth figure after the decimal.

3. The estimate of error. The convergence of the process. The error occasioned by the application of the method of nets to the solution of differential equations is conditioned by two circumstances: first, by the fact that in shifting the boundary conditions that must be fulfilled on the contour Γ of the given region G to the contour C of the net region, we admit a certain error; second, by the fact that in replacing the differential equation by an equation in finite differences, we introduce an approximation.

In connection with the application of the method of nets, there arise the two fundamental problems which follow. The conditions of convergence of the method must be determined. Putting it otherwise, it must be determined what conditions the coefficients of the differential equation and the region G must satisfy in order that, for $\Delta x_i \to 0$ and $\Delta y_k \to 0$, the solution of the finite-difference equation shall converge to the function solving the posed problem for the differential equation. The second fundamental problem consists in estimating the error of the method of nets, i.e., in estimating the discrepancy between the solution of the finite-difference equation and the corresponding solution of the differential equation.

The convergence of the method of nets has, for partial differential equations, been subjected to investigations in works of L. A. Liusternik and I. G. Petrovskiĭ, [1], and Courant, Friedrichs and Lewy, [2]. Pursuing the aim — one concerned with principles — of elucidating the broadest conditions under which the sequence of functions determined over the net will converge to the solution of the differential equation, they have considered a difference equation of the simplest form, and the investigations themselves were carried through for the Dirichlet problem for the Laplace equation. They have established the convergence under extremely general assumptions concerning the form of the region. The most powerful result in this direction belongs to I. G. Petrovskiĭ, who has shown that if at all points of the boundary of the region there exists a superharmonic barrier, the sequence of "net functions" converges uniformly to the solution of the Dirichlet problem.

The second of the two fundamental problems is the more profound, and as a rule its solution not only permits the establishment of the fact of convergence, but also the drawing of a conclusion concerning its character. It stands to reason that this second problem is solved under more restricted assumptions than is the first.

We shall dwell on this second problem, considering it under the simplest of assumptions, which are nevertheless sufficient for many practical applications. In addition we shall limit ourselves to the consideration of problems with known boundary values of the sought functions and shall not broach the problem of boundary conditions of another type. Let us again begin with the ordinary differential equation $y'' - q(x)y = f(x)$ under the boundary conditions $y(a) = \alpha$, $y(b) = \beta$. Above [see § 2, (3) and (4)] we showed that a function satisfying our differential equation and having continuous derivatives of up to the fourth order will also satisfy the equation

$$l_i^{(1)}(y) = \frac{y(x_{i+1}) - 2y(x_i) + y(x_{i-1})}{h^2} - q_i y(x_i) = f_i + R_i^{(1)*}(y), \quad (1)$$

where $R_i^{(1)*}(y)$ has the estimate:

$$|R_i^{(1)*}(y)| \leqslant \frac{h^2}{12} M_4. \quad (2)$$

Discarding $R_i^{(1)*}(y)$, we obtained an equation in finite differences approximately replacing differential equation (1):

$$l_i^{(1)}(y) = \frac{y_{i+1} - 2y_i + y_{i-1}}{h^2} - q_i y_i = f_i \qquad (i = 1, 2, \ldots, N-1) \quad (3)$$

with the boundary requirements

$$y_0 = \alpha, \ y_N = \beta.$$

We shall have to estimate at the nodes of the net the difference $y(x_i) - y_i$ between solutions of equations (1) and (3) having the same boundary values.

We shall base this estimate on the following simple ideas. Let us take at the nodes of the net an arbitrary system of values $y_0, y_1, y_2, \ldots, y_N$ and show the validity of the following statement.

THEOREM I. *If* $q_1 \geqslant 0$ *and* $l_i^{(1)}(y) \leqslant 0$, y_i *cannot have a negative minimum at internal points of the net. The case when* y_i *is constant is the exception.*

Indeed, let us assume that y_i is not constant and attains its least negative value $(-m)$ at some interior point i of the net. We shall consider this minimum to be proper in the sense that of the values y_{i+1} and y_{i-1} at the neighboring points there is at least one greater than $(-m)$. Let us replace, in $l_i^{(1)}(y)$, y_{i+1} and y_{i-1} by the smaller quantity $(-m)$, owing to which $l_i^{(1)}(y)$ is diminished:

$$l_i^{(1)}(y) = \frac{y_{i+1} - 2y_i + y_{i-1}}{h^2} - q_i y_i >$$

$$> \frac{-m + 2m - m}{h^2} + q_i m = q_i m \geqslant 0 \quad (4)$$

The latter, however, is incompatible with the inequality $l_i^{(1)}(y) \leqslant 0$. As simple corollaries there follow from this the theorems below.

THEOREM II. *If* $q_i \geqslant 0$, $l_i^{(1)}(y) \leqslant 0$, *and if at the boundary points* $y_0 \geqslant 0$ *and* $y_N \geqslant 0$, *then* $y_i \geqslant 0$ *at all interior points.*

Had we assumed that at one of the nodes at least, $y_i < 0$, then, since on the boundary of the interval the values of y_i are non-negative, we should have had to admit that at an interior point of the net y_i has a negative minimum, which cannot be, by Theorem I.

THEOREM III. *If* $q_i \geqslant 0$ *and at the nodes of the net there are given two systems of values* $y_0, y_1, \ldots$ *and* $Y_0, Y_1, \ldots$ *such that* $l_i^{(1)}(Y) \leqslant -|l_i^{(1)}(y)|$ *and if at the boundaries of the interval* $Y_0 \geqslant |y_0|$, $Y_N \geqslant |y_N|$, *hen at all the points of the interval*

$$Y_i \geqslant |y_i|.$$

Actually the inequality $l_i^{(1)}(Y) \leqslant -|l_i^{(1)}(y)|$ is equivalent to the following two:

$$l_i^{(1)}(Y) \leqslant l_i^{(1)}(y), \quad l_i^{(1)}(Y) \leqslant -l_i^{(1)}(y),$$

or, what is the same thing,

$$l_i^{(1)}(Y - y) \leqslant 0, \quad l_i^{(1)}(Y + y) \leqslant 0.$$

On the assumption, at the boundary points $Y_0 - y_0 \geqslant 0$, $Y_N - y_N \geqslant 0$ and $Y_0 + y_0 \geqslant 0$, $Y_N + y_N \geqslant 0$. On the strength of the second theorem, from this and from the foregoing inequalities it follows that at all interior points of the net

$$Y_i - y_i \geqslant 0 \text{ and } Y_i + y_i \geqslant 0,$$

which indeed means that

$$Y_i \geqslant |y_i|.$$

Let us pass now to the problem of estimating the error of the method of nets. We shall again consider $q \geqslant 0$. We saw above that if $y(x)$ satisfies the given differential equation, then it fulfills equation (1). Let us subtract from it, term by term, equation (3) in finite differences. For the sought error $r_i = y(x_i) - y_i$ we obtain the following equation in finite differences:

$$l_i^{(1)}(r) = \frac{r_{i+1} - 2r_i + r_{i-1}}{h^2} - q_i r_i = R_i^{(1)*}(y). \tag{5}$$

Moreover, since $y(x)$ and y_i take identical values at the ends of the interval, the following boundary conditions must be fulfilled for r_i:

$$r_0 = 0, \quad r_N = 0. \tag{6}$$

The error r_i of the method of nets must satisfy an equation of the same form as does y_i, but with another free term. Namely, the free term of the equation for r_i is equal to the error $R_i^{(1)*}(y)$ which occurs owing to the replacement of the differential equation by the equation in finite differences. This relates to any boundary conditions. If, however, we solve the problem with known boundary conditions, the values for the error r_i at the ends of the interval are equal to zero.

$R_i^{(1)*}(y)$ depends on the choice of the solution $y(x)$ of the differential equation and is by the nature of the case an unknown quantity, since $y(x)$ is unknown to us. To make an estimate of $R_i^{(1)*}(y)$, and accordingly of r_i too, is therefore possible only in the sense of its order of smallness relative to the step of the net h.

An estimate for $R_i^{(1)*}(y)$ has been established by us previously, being given by inequality (2). Using it, we shall construct an equation in finite differences which is majorant for (5) in this sense, that its solution, which vanishes at the ends of the interval, will be an upper bound for the error r_i. Indeed, let us consider the equation

$$l_i^{(1)}(\varrho') = \frac{\varrho'_{i+1} - 2\varrho'_i + \varrho'_{i-1}}{h^2} - q_i\varrho'_i = -\frac{h^2}{12} M_4 \tag{7}$$

under the conditions $\varrho'_0 = 0$, $\varrho'_N = 0$.

If r_i and ϱ'_i satisfy equations (5) and (7) respectively, then $l_i^{(1)}(\varrho') \leqslant -|l_i^{(1)}(r)|$ and since ϱ'_i and r_i vanish at the boundary of the interval for $r_0, r_1, \ldots$ and $\varrho'_0, \varrho'_1, \ldots$, the conditions of Theorem III are fulfilled. Therefore

$$|r_i| \leqslant \varrho'_i. \tag{8}$$

To simplify this estimate we shall consider a new equation

$$\lambda_i(\varrho) = \frac{\varrho_{i+1} - 2\varrho_i + \varrho_{i-1}}{h^2} = -\frac{h^2}{12} M_4, \tag{9}$$

$$\varrho_0 = 0, \quad \varrho_N = 0,$$

which is obtained from (7) by putting $q_i = 0$. For the difference $\varrho_i - \varrho$, the following relations are obtainable by differencing (9) and (7):

$$\lambda_i(\varrho - \varrho') = -q_i\varrho'_i \leqslant 0, \quad \varrho_0 - \varrho'_0 = 0, \quad \varrho_N - \varrho'_N = 0,$$

from which, in view of Theorem II, it follows that at all the points of the net we must have $\varrho_i - \varrho_i' \geqslant 0$ and consequently

$$|r_i| \leqslant \varrho_i.$$

The solution of (9), however, is found simply, and is equal to the values which are taken at the nodes of the net by the function

$$\varrho = \frac{h^2}{12} M_4 \frac{(x-a)(b-x)}{2}.$$

This function attains its maximum value at the middle of the interval, for $x = \frac{a+b}{2}$, and $\max \varrho = \frac{h^2(b-a)^2}{96} M_4$.

$$|r_i| \leqslant \frac{h^2(b-a)^2}{96} M_4. \tag{10}$$

Let us assume that the sought solution of the differential equation has a bounded fourth derivative. Such will certainly be the case when $q(x)$ and the free term $f(x)$ have bounded second derivatives. It follows, then, from estimate (10) that the error r_i in the method of nets will tend to zero as $h \to 0$. This establishes the convergence of the successive approximations constructed by the method of nets, to the solution of the differential equation in the case of a finite M_4.

The estimate for partial differential equations is more complicated. Let us consider, for example, differential equation (17) § 2, and make an estimate of the error for the Dirichlet problem. Let us construct, as we did at the beginning of No. 2 § 2, an arbitrary rectangular net with nodes (x_i, y_k), form a closed broken line C from the sides and diagonals of the rectangles of the net, as close as possible to the contour Γ of the given region G, shift from Γ to C the assigned values and replace, at the nodes of the net which lie within C, the differential equation by equation (26_1) § 2. We recall that the error in the equation obtained by such a replacement has the estimate (26_2) § 2. We shall consider below that Δx_i, Δy_k are so small that the coefficients $b_{i,k}^{(j)}$ in equation (26_1) § 2 are positive. For this it is sufficient that inequality (5) § 3 be fulfilled.

For the expression $l_{i,k}(u) = b_{i,k}^{(1)} u_{i+1,k} + b_{i,k}^{(2)} u_{i,k+1} + b_{i,k}^{(3)} u_{i-1,k} + b_{i,k}^{(4)} u_{i,k-1} - a_{i,k} u_{i,k}$ which stands on the left side of (26_1) § 2, theorems can be proved analogous to Theorems I—III. The reasoning involved in the proof of these theorems would be a simple repetition of that we have already carried through for Theorems I—III and we therefore just formulate the theorems and leave them without proof.

Let us assume that at the nodes of the net there is given a certain system of values $u_{i,k}$.

THEOREM IV. *If* $q_{i,k} \leqslant 0$ *and* $l_{i,k}(u) \leqslant 0$, $u_{i,k}$ *cannot have a negative minimum at the interior nodes of the net. Only the case when the* $u_{i,k}$ *are constant is an exception.*

THEOREM V. *If* $q_{i,k} \leqslant 0$, $l_{i,k}(u) \leqslant 0$, *and at the points of the net on the boundary C of the net region* $u_{i,k} \geqslant 0$, *then* $u_{i,k} \geqslant 0$ *also at all nodes of the net within C.*

THEOREM VI. *If* $q_{i,k} \leqslant 0$ *and at the nodes of the net there are given two systems of quantities* $u_{i,k}$ *and* $U_{i,k}$ *such that* $l_{i,k}(U) \leqslant -|l_{i,k}(u)|$, *and at the boundary nodes* $U_{i,k} \geqslant |u_{i,k}|$, *it can then be affirmed that everywhere in the net region*

$$U_{i,k} \geqslant |u_{i,k}|.$$

Let us designate as $\varphi_{i,k}$ the values of $u_{i,k}$ on C. Generally speaking they will not coincide, as we have said above, with the actual values of $u(x_i, y_k)$ which the sought function u takes at the corresponding points.

Let us call the difference between them $\bar{r}_{i,k} = u(x_i, y_k) - \varphi_{i,k}$. An estimate of this error can easily be given. If it is agreed to transfer to C the values of u from the nearest points of the contour Γ, then $|\bar{r}_{i,k}| \leqslant \leqslant M_1 \frac{\delta}{2}$, where δ is the greatest of the numbers Δx_i, Δy_k, and M_1 is the maximum value of the partial derivatives of u with respect to x and y in some region containing both G and the net region.

The function $u(x, y)$ satisfying the differential equation, satisfies also equation (26) § 2. If we subtract from that, term by term, the equation $l_{i,k}(u) = f_{i,k}$ for $u_{i,k}$, we see that the error $r_{i,k} = u(x_i, y_k) - u_{i,k}$ which we obtain by replacing the sought function by the solution of the equation in finite differences, satisfies the equation:

$$l_{i,k}(r) = R_{i,k}(u). \tag{11}$$

In addition, at the nodes of the boundary C of the net region it takes the values

$$r_{i,k} = \bar{r}_{i,k}. \tag{12}$$

We shall separate the influence of the boundary errors $\bar{r}_{ik}$ and the errors R_{ik} in the satisfaction of the differential equation, on the magnitude of the common error r_{ik}. We shall seek the solution of equation (11) in the form of the sum $r_{ik} = v_{ik} + w_{ik}$ of two systems of quantities, one of which satisfies the homogeneous equation

$$l_{ik}(v) = 0$$

and acquires on the contour of the region the given values $v_{ik} = \bar{r}_{ik}$, and the second of which acquires null values on the contour and inside of C satisfies the equation:

$$l_{ik}(w) = R_{ik}.$$

To estimate the magnitude of the first system of quantities presents no difficulty. In fact, in No. 1 § 3 we saw that v_{ik} attains its greatest value on the contour C, and we must have everywhere, therefore,

$$|v_{ik}| \leqslant \max |v_{ik}| \leqslant \frac{\delta}{2} M_1.$$

Hence it is clear that v_{ik} will, generally speaking, be a small quantity of the first order with respect to Δx_i and Δy_k.

It is somewhat more complicated to evaluate w_{ik}. Let us consider $q \leqslant 0$. In accordance with Theorem VI of the present section, to estimate it, it is sufficient to find a function which will acquire non-negative values on the boundary; within the contour the operator l_{ik}, on being applied to it, should give a value not greater than $-|R_{ik}|$. Such a function can be constructed in many ways. We here use an idea worked out by S. A. Gershgorin, [2].

We shall seek the function we require in the form of the product of a constant α by a function W which satisfies the equation

$$L(W) = -1.$$

It is readily seen that W cannot take negative values, since in the contrary case W would have a negative minimum within C, where the relations

$$W < 0, \quad \frac{\partial W}{\partial x} = 0, \quad \frac{\partial W}{\partial y} = 0, \quad \frac{\partial^2 W}{\partial x^2} \geqslant 0, \quad \frac{\partial^2 W}{\partial y^2} \geqslant 0,$$

would have to be fulfilled, and we would therefore have to have $L(W) \geqslant 0$, which contradicts the equation for W.

Substituting αW in equation (26) § 2 in place of u, we obtain

$$l_{ik}(\alpha W) = [-1 + R_{ik}(W)]\alpha.$$

Here $R_{ik}(W)$ has an estimate of form (26) § 2 with this difference, that instead of the maximum values M_3 and M_4 of the third and fourth partial derivatives of u, one must substitute in it N_3 and N_4, the maximum values of the corresponding derivatives of W. In order that the right side of the preceding equation be less than $-|R_{ik}(w)|$, it is sufficient to choose the number α so that

$$\left\{-1 + \left[\left(\frac{\varepsilon}{3} N_3 + \frac{\eta}{6} N_4\right)(A_{ik} + C_{ik}) + \frac{\eta}{3} N_3(|D_{ik}| + |E_{ik}|)\right]\right\}\alpha \leqslant$$
$$\leqslant -\left[\left(\frac{\varepsilon}{3} M_3 + \frac{\eta}{6} M_4\right)(A_{ik} + C_{ik}) + \frac{\eta}{3} M_3(|D_{ik}| + |E_{ik}|)\right].$$

We can always contrive that the expression in braces, the multiplier of α on the left side of the inequality, be negative, by the choice of Δx_i, Δy_k. After this it is sufficient to put

$$\alpha = \frac{\left[\left(\frac{\varepsilon}{3} M_3 + \frac{\eta}{6} M_4\right)(A + C) + \frac{\eta}{3} M_3(|D| + |E|)\right]_{\max}}{1 - \left[\left(\frac{\varepsilon}{3} N_3 + \frac{\eta}{6} N_4\right)(A + C) + \frac{\eta}{3} N_3(|D| + |E|)\right]_{\max}}.$$

The quantities w_{ik} will have the estimate

$$|w_{ik}| \leqslant \alpha N, \quad N = \max W.$$

For r_{ik} we consequently obtain

$$|r_{ik}| \leqslant \frac{\delta}{2} M_1 + \alpha N.$$

When the intervals along each of the coordinate axes are equal, $\Delta x_i = h$, $\Delta y_k = l$, the estimate will take the form:

$$|r_{ik}| \leqslant \frac{\delta}{2} M_1 + N \frac{[M_4(h^2 A + l^2 C) + 2M_3(h^2 |D| + l^2 |E|)]_{\max}}{12 - [N_4(h^2 A + l^2 C) + 2N_3(h^2 |D| + l^2 |E|)]_{\max}}.$$

Hence it is clear that as $\Delta x_i \to 0$, $\Delta y_k \to 0$, the right side too will tend to zero, and therefore $r_{ik} \to 0$.

One could try to construct a simpler majorant for w_{ik} by using with this object in view polynomials whose coefficients were so chosen that this polynomial would have, for the points within C, positive values, and the operator $l_{ik}(u)$ on being applied to it would give a value not greater than $-|R_{ik}|$. We shall dwell only on the simplest case, of a polynomial of the second degree. Let us describe, about the region bounded by the curve C, an ellipse with axes parallel to the axes of coordinates; its semi-axes we shall call s, t, and the coordinates of the center x_c, y_c. Assume that the coefficients of the given differential equation satisfy the inequality:

$$\frac{A}{s^2} + \frac{C}{t^2} - \frac{|D|}{s} - \frac{|E|}{t} + \frac{q}{2} > 0.$$

We shall seek the majorant in the form

$$W = \alpha\left(1 - \frac{(x - x_c)^2}{s^2} - \frac{(y - y_c)^2}{t^2}\right).$$

Let us substitute W in the operator l_{ik}. In view of the fact that R_{ik} depends only on derivatives of order higher than the second, it follows from (26) § 2 that

$$l_{ik}(W) = L(W) =$$

$$= 2\alpha\left\{-\frac{A_{ik}}{s^2} - \frac{C_{ik}}{t^2} - \frac{D_{ik}}{s^2}(x - x_c) - \frac{E_{ik}}{t^2}(y - y_c) + \right.$$

$$\left. + \frac{q}{2}\left(1 - \frac{(x - x_c)^2}{s^2} - \frac{(y - y_c)^2}{t^2}\right)\right\}.$$

Since $|x - x_c| \leqslant s$, $|y - y_c| \leqslant t$ and the expression in parentheses that is the multiplier of q is not greater than unity, for $l_{ik}(W)$ there holds the inequality

$$l_{ik}(W) \leqslant -2\alpha\left(\frac{A_{ik}}{s^2} + \frac{C_{ik}}{t^2} - \frac{|D_{ik}|}{s} - \frac{|E_{ik}|}{t} + \frac{q}{2}\right),$$

and in order that $l_{ik}(W) \leqslant -|R_{ik}|$ it is sufficient to put

$$\alpha = \frac{\left[\left(\frac{\varepsilon}{3}M_3 + \frac{\eta}{6}M_4\right)(A + C) + \frac{\eta}{3}M_4(|D| + |E|)\right]_{\max}}{2\left(\frac{A}{s^2} + \frac{C}{t^2} - \frac{|D|}{s} - \frac{|E|}{t} + \frac{q}{2}\right)_{\min}}.$$

Therefore for r_{ik} we shall obtain the estimate

$$|r_{ik}| \leqslant \frac{\delta}{2}M_1 + \alpha\left(1 - \frac{(x - x_c)^2}{s^2} - \frac{(y - y_c)^2}{t^2}\right).$$

For equal intervals along the coordinate axes

$$|r_{ik}| \leqslant \frac{\delta}{2}M_1 + \frac{[M_4(h^2A + l^2C) + 2M_3(h^2|D| + l^2|E|)]_{\max}}{24\left(\frac{A}{s^2} + \frac{C}{t^2} - \frac{|D|}{s} - \frac{|E|}{t} + \frac{q}{2}\right)_{\min}} \times$$

$$\times\left(1 - \frac{(x - x_c)^2}{s^2} - \frac{(y - y_c)^2}{t^2}\right).$$

For example, for the case of the Poisson equation,

$$\frac{\partial^2 u}{\partial x^2} + \frac{\partial^2 u}{\partial y^2} = f(x, y),$$

since $D = E = q = 0$, $A = C = 1$, we obtain the following estimate:

$$|r_{ik}| \leqslant \frac{\delta}{2}M_1 + \frac{(\eta M_4 + 2\varepsilon M_3)s^2t^2}{6(s^2 + t^2)},$$

or for $\Delta x_i = h$, $\Delta y_k = l$,

$$|r_{ik}| \leqslant \frac{\delta}{2}M_1 + \frac{M_4(h^2 + l^2)s^2t^2}{24(s^2 + t^2)}.$$

In like manner estimates can be made of the errors which are made by replacing the Poisson equation approximately by equations (32), (34) or (36) § 2.

All the estimates of error that we have obtained, despite that they prove the convergence of the successive approximations in the method of nets, given certain assumptions regarding the solution being sought, are inadequate in this respect, that in them appear the maximum values of the moduli of the derivatives of some order of the function being found. These values must be determined on the basis of additional considerations, or computed approximately in terms of approximate values of the functions, by replacing these derivatives by their expressions in terms of differences or difference ratios.

CHAPTER IV

VARIATIONAL METHODS

The purpose of this chapter is to expound the methods for the approximate solution of differential equations that are connected with the calculus of variations. The fundamental problems of mechanics are known to be governed both by differential equations and by so-called minimum principles. Thus, for instance, the equilibrium position of a mechanical system is the position corresponding to the minimum of its potential energy. In view of this, the problem of solving the boundary-value problem for the differential equation of the given mechanical system generally turns out to be equivalent to the problem of finding the function giving the minimum of the integral by which the potential energy of the system is expressed. Mathematically, the problem of solving a boundary-value problem for a differential equation is equivalent to that in the variational calculus of minimizing the integral for which the given differential equation is the Euler-Lagrange equation. Side by side with the Euler equation, direct methods may be employed for the solution of problems of the variational calculus; thanks to the equivalence mentioned above, these methods are simultaneously also methods for the solution of boundary-value problems for the differential equations. The exposition of such methods, the most familiar of which are the methods of Ritz and of B. G. Galerkin, will, then, be the object of the present chapter. In our exposition we shall consider the application of these methods not only to the approximate solution of partial differential equations, but also to ordinary differential equations, because for this case the theory of the method is simpler and serves as a good introduction to the more complex case of partial differential equations, and because the theory cannot, moreover, be considered to be generally known. Finally, the approximate solution of ordinary equations by variational methods (the determination of proper numbers and functions) is frequently utilized in solving partial differential equations (Fourier's method).

Before proceeding to the exposition of the variational methods, we shall recall in the next section certain concepts of the variational calculus and shall set forth for the most important cases the formulas establishing the connection between variational problems and differential equations.

§ 1. VARIATIONAL PROBLEMS CONNECTED WITH THE MOST IMPORTANT DIFFERENTIAL EQUATIONS

Here we wish to consider, at the outset, variational problems connected with the simplest, most important differential equations. From these examples a sufficiently clear idea of the fundamental facts and methods of the variational calculus can be obtained. At the end of each Number are indicated, without detailed consideration, variational problems connected with certain other boundary-value problems for the differential equations.

This section thus has as its purpose the familiarizing of the reader with those ideas of the variational calculus that are particularly important for the use of the direct methods, and also to give formulas that we shall subsequently use, or which the reader may encounter in making independent application of the direct methods to the solution of concrete problems [1]).

1. Problems leading to an ordinary differential equation. Let us consider the integral

$$I = \int_{x_0}^{x_1} [p(x)y'^2 + q(x)y^2 + 2f(x)y]dx. \tag{1}$$

It takes on a definite value for any function (curve) $y=y(x)$ given for $x_0 \leqslant x \leqslant x_1$. Thus the magnitude of the integral I depends on the chosen curve, or, as is often said, it is a *functional*.

Let us pose the following problem: to find the curve $y = y(x)$ passing through the given points (x_0, y_0) and (x_1, y_1), and giving a minimum value [2]) for integral I.

To solve this problem we reason as follows. Assume that $y(x)$ is a function giving the integral I a minimum value. Now let $\eta(x)$ be any function, continuous together with its derivative and vanishing at the ends: $\eta(x_0) = \eta(x_1) = 0$. Then the function $y(x) + \alpha\eta(x)$ satisfies the same boundary conditions at the ends as does $y(x)$, and, for α sufficiently small, is as close as one pleases to the function $y(x)$. Therefore, since $y(x)$ gives the integral I a minimum, for $\alpha \neq 0$ (and small) I must acquire greater values, i.e.,

$$I(y + \alpha\eta) \geqslant I(y).$$

Thus the integral $I(y + \alpha\eta)$, as a function of α, has for $\alpha = 0$ a mini-

---

[1]) For a more detailed acquaintance with the variational calculus we recommend that the reader refer to the following books: D. F. Egorov, [1]; V. I. Smirnov, L. V. Kantorovich, V. I. Krylov,[1]; Courant and Hilbert, [1], Chap. IV; Liusternik and Lavrent'ev, [1]; Giunter, N. M., [1]; Smirnov, V. I., [1], vol. IV.

[2]) We have in view here a relative minimum of the integral, i.e., we require that the integral I have for the curve $y(x)$ a value less than that for neighboring curves, but not less than for all.

mum, and therefore

$$\left.\frac{dI(y + \alpha\eta)}{d\alpha}\right|_{\alpha=0} = 0. \tag{2}$$

Calculating the latter expression, we obtain:

$$\left[\frac{d}{d\alpha} I(y + \alpha\eta)\right]_{\alpha=0} =$$

$$= \left[\frac{d}{d\alpha}\int_{x_0}^{x_1} [p(y' + \alpha\eta')^2 + q(y + \alpha\eta)^2 + 2f(y + \alpha\eta)dx\right]_{\alpha=0} =$$

$$= [\int_{x_0}^{x_1} [2p(y' + \alpha\eta')\eta' + 2q(y + \alpha\eta)\eta + 2f\eta]dx]_{\alpha=0} =$$

$$= \int_{x_0}^{x_1} (2py'\eta' + 2qy\eta + 2f\eta)dx.$$

Thus we find that if $y(x)$ gives the integral a minimum, the expression just obtained must equal zero, whatever be the function $\eta(x)$ satisfying the conditions stipulated above. This condition is often given another formulation. The expression $\left[\frac{d}{d\alpha} I(y + \alpha\eta)\right]_{\alpha=0} \cdot \alpha$, which can be considered even in the case when y is not the function giving the integral an extremum, is, for small α, the principal part of the increment of the integral I in passing from the curve y to the curve $y + \alpha\eta$; if the second factor α be replaced by 1, this expression is called the *variation* of the integral I and is denoted by δI. We found above that

$$\delta I = \left[\frac{d}{d\alpha} I(y + \alpha\eta)\right]_{\alpha=0} \cdot 1 = \int_{x_0}^{x_1} (2py'\eta' + 2qy\eta + 2f\eta)dx. \tag{3}$$

The condition obtained by us above can now be stated in this form: in order that the function y give the integral I a minimum, it is necessary that the variation of the latter equal zero: $\delta I = 0$, whatever $\eta(x)$ may be.

In order to extract further implications, let us transform the expression for δI. For this purpose we shall integrate the first term of integral (3) by parts:

$$\int_{x_0}^{x_1} 2py'\eta' dx = [2py'\eta]_{x=x_0}^{x=x_1} - 2\int_{x_0}^{x_1} \frac{d}{dx}(py')\eta(x)dx =$$

$$= -2\int_{x_0}^{x_1} \frac{d}{dx}(py')\eta(x)dx.$$

Then δI takes the form

$$\delta I = -2 \int_{x_0}^{x_1} \left[\frac{d}{dx}(p(x)y') - qy - f \right] \eta(x) dx. \tag{4}$$

This expression must equal zero, whatever be the function $\eta(x)$ satisfying the conditions indicated above. This is possible only if the function $y(x)$ satisfies the differential equation

$$\frac{d}{dx}(py') - qy - f = 0. \tag{5}$$

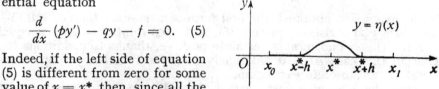

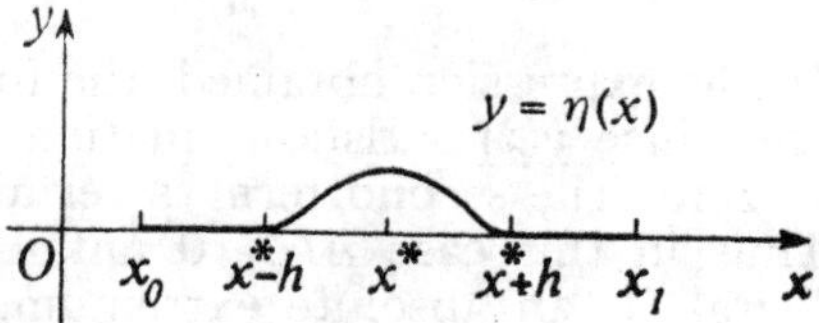

Fig. 17

Indeed, if the left side of equation (5) is different from zero for some value of $x = x^*$, then, since all the necessary functions are assumed to be continuous, it will preserve its sign in a certain vicinity of this point, $(x^* - h, x^* + h)$. Then on constructing a function $\eta(x)$ which is positive in this interval and equal to zero outside of it, and which moreover is continuous together with its derivative (Fig. 17), we would obviously obtain $\delta I \neq 0$, which contradicts the necessary condition obtained above.

Thus we have established that the function $y(x)$ giving a minimum for integral I must necessarily satisfy equation (5) [1]).

The equation that the extremal curves satisfy is called the Euler equation for the given variational problem. In the given case, for the problem of the minimum of integral (1), the role of the Euler equation is played by (5), i.e., by a so-called self-adjoint differential equation of the second order.

The general solution of equation (5) contains two arbitrary constants; thus through the two given points (x_0, y_0) and (x_1, y_1) one can, generally speaking, pass one curve satisfying the equation. It would be expected that this curve is indeed the solution of the given extremal problem. However the last assertions cannot be proved without further limitations, and are not even always valid.

Let us now consider the particular case when it is known that in the given interval $p(x)$ is positive and $q(x)$ is non-negative:

$$p(x) > 0, \quad q(x) \geqslant 0 \quad \text{for} \quad x_0 \leqslant x \leqslant x_1. \tag{6}$$

It can in this case be affirmed that the solution of equation (5) which passes through the given points (x_0, y_0) and (x_1, y_1) gives the absolute extremum for integral (1) without fail. Indeed, let $y(x)$ be this solution, and $\bar{y}(x)$ another function, satisfying the conditions $\bar{y}(x_0) = y_0$, $\bar{y}(x_1) = y_1$. Then setting $\bar{y}(x) - y(x) = \eta(x)$, we obtain $\eta(x_0) = \eta(x_1) = 0$.

[1]) Our reasoning required that the function $y(x)$ have a second derivative. By a modification of it, it can be established that the same conclusion is also true if $y(x)$ be assumed simply to be differentiable.

We have now

$$\Delta I = I(\bar{y}) - I(y) = I(y + \eta) - I(y) = \int_{x_0}^{x_1} [p(y' + \eta')^2 + q(y + \eta)^2 +$$

$$+ 2f(y + \eta)]dx - \int_{x_0}^{x_1} [py'^2 + qy^2 + 2fy]dx =$$

$$= 2\int_{x_0}^{x_1} [py'\eta' + qy\eta + f\eta]dx + \int_{x_0}^{x_1} [p\eta'^2 + q\eta^2]dx. \quad (7)$$

In the expression obtained, the first term is none other than δI [cf. (3)], and since $y(x)$ satisfies equation (5), and $\eta(x_0) = \eta(x_1) = 0$, it is equal to zero. The second term is certainly positive, thanks to conditions (6). Thus in this case $\Delta I > 0$ unfailingly, i.e., the function $y(x)$ gives integral (1) an absolute extremum.

Using the reasoning just carried through, another important fact can be established for the given case — through every two points (x_0, y_0) and (x_1, y_1), one and only one solution of equation (5) can be passed. This fact, that there cannot be two solutions, is clear immediately, for if $\bar{y}$ were another solution passing through the same points as does y, on the basis of (7) we obtain $I(\bar{y}) - I(y) > 0$ simultaneously with $I(y) - I(\bar{y}) > 0$, which is absurd.

We shall show that the solution always exists. Indeed, the general solution of linear equation (5) has the form:

$$y = C_1u_1(x) + C_2u_2(x) + \varphi(x).$$

The constants C_1 and C_2 must be determined on the basis of the initial conditions from the equations

$$C_1u_1(x_0) + C_2u_2(x_0) + \varphi(x_0) = y_0,$$
$$C_1u_1(x_1) + C_2u_2(x_1) + \varphi(x_1) = y_1.$$

Since, by what has been proved, this system can never have two solutions, it has a solution for any free terms. We arrive finally at the conclusion that if condition (6) is met, the problem of the minimum of integral (1) has, for any y_0 and y_1, a unique solution, which is given by the integral curve of equation (5) that passes through the points (x_0, y_0) and (x_1, y_1). Moreover, it is clear from what has been said that the problems: 1) of determining the function $y(x)$ giving integral (1) a minimum and satisfying the conditions $y(x_0) = y_0$ and $y(x_1) = y_1$, and 2) of finding the solution of equation (5) satisfying the same conditions, are completely equivalent.

Thus, for example, the potential energy of a string which is under the action of an external load of intensity $f(x)$ is defined by the formula

$$U = \tfrac{1}{2}\int_0^l \left(\frac{\mu}{2} y'^2 - fy\right) dx, \quad (8)$$

where $\mu > 0$ is a constant. Thus the problem of determining the equilibrium position of such a string with fixed ends consists in determining the minimum of the integral standing on the right, under the conditions $y(0) = y(l) = 0$. Another way of solving the problem — equivalent, by what has been said above — is by solving the equation of the string [the Euler equation for integral (8)]:

$$\mu y'' + f(x) = 0. \tag{9}$$

It is useful to note that any linear differential equation of the second order is the Euler equation for some integral of type (1). For this it is sufficient to show that such an equation can always be reduced to form (5). To establish this, let the given equation

$$py'' + ry' - qy = f \tag{10}$$

be first multiplied by $e^{\int \frac{r-p'}{p} dx}$; it takes the form:

$$\frac{d}{dx}\left(pe^{\int \frac{r-p'}{p} dx} y'\right) - qe^{\int \frac{r-p'}{p} dx} y = fe^{\int \frac{r-p'}{p} dx}, \tag{11}$$

i.e., proves to have been reduced to form (5).

The method by means of which we obtained the Euler equation for integral (1) is general, and with insignificant modifications is applicable to various other problems. Thus, for the so-called general (simplest) problem of the variational calculus, that of the minimum of the integral

$$I = \int_{x_0}^{x_1} F(x, y, y')dx, \tag{1a}$$

we obtain the Euler equation in the form

$$\frac{d}{dx} F'_{y'} - F'_y = 0. \tag{5a}$$

In case derivatives of up to the nth order appear under the integral sign, i.e., in case we have to do with the minimum of an integral of the form

$$I = \int_{x_0}^{x_1} F(x, y, y', y'', \ldots y^{(n)})dx \tag{1b}$$

under the conditions

$$y(x_0) = y_0,\ y'(x_0) = y'_0,\ \ldots\ y^{(n-1)}(x_0) = y_0^{(n-1)},$$
$$y(x_1) = y_1,\ y'(x_1) = y'_1,\ \ldots\ y^{(n-1)}(x_1) = y_1^{(n-1)},$$

then the Euler equation will be of order $2n$, having the form

$$F'_y - \frac{d}{dx} F'_{y'} + \frac{d}{dx^2} F'_{y''} - \ldots + (-1)^n \frac{d}{dx^n} F'_{y^{(n)}} = 0. \tag{5b}$$

If several unknown functions appear beneath the integral sign, i.e., if we have to do with the minimum of an integral of the form

$$I = \int_{x_0}^{x_1} F(x, y_1, y_2, \ldots y_m;\ y_1', y_2', \ldots y_m')dx, \tag{1c}$$

then the Euler equation is replaced by a system of equations, of the form

$$\left.\begin{array}{l} \dfrac{d}{dx} F'_{y_1'} - F'_{y_1} = 0, \\ \cdot\ \cdot\ \cdot\ \cdot\ \cdot\ \cdot\ \cdot\ \cdot\ \cdot \\ \dfrac{d}{dx} F'_{y_m'} - F'_{y_m} = 0. \end{array}\right\} \tag{5c}$$

The reader may endeavor to carry out the derivations of these equations himself, or can acquaint himself with them in textbooks of the calculus of variations.

It is also useful to note that for problems arising from concrete problems of mechanics, not only is the variational problem connected with the given problem usually known to us — on the basis of the variational principles of mechanics (the minimum of the potential energy, the minimum of virtual work, and others) — so is its differential equation as well; this equation is indeed the Euler equation for the given variational problem. In such cases there is, properly speaking, no need to set up the Euler equation for the variational problem; however as this is very simply done and is useful as a check, Euler's equation is often set up regardless, using the general formulas.

2. Variational problems leading to the Laplace and Poisson equations. As a characteristic example of a variational problem relating to a function of two variables, we shall consider a problem that yields, as the Euler equations, the familiar equations of Laplace and Poisson.

Let us consider, namely, the problem of the minimum of the integral

$$I = \iint_D \left[\left(\frac{\partial u}{\partial x}\right)^2 + \left(\frac{\partial u}{\partial y}\right)^2 + 2f(x, y)u\right] dxdy, \tag{12}$$

where the integral extends over the plane region D bounded by the contour Γ. More precisely speaking, the problem is formulated thus: to find, a function $u(x, y)$, continuous in the region D together with its partial derivatives of the first and second orders, acquiring assigned values on the contour Γ:

$$u = \varphi(s) \text{ on } \Gamma \tag{13}$$

and giving integral I, (12), a minimum value.

Let us again find the necessary conditions. Let $u(x, y)$ minimize this integral; let us consider the value of the integral for the function

$u(x, y) + \alpha\eta(x, y)$, where $\eta(x, y)$ is also a function continuous together with its derivatives of the first and second orders, and equalling zero on Γ. As in No. 1, we satisfy ourselves that in this case the variation of the integral must be equal to zero:

$$\delta I = \left[\frac{d}{d\alpha} I(u + \alpha\eta)\right]_{\alpha=0} = 0$$

or as well

$$\delta I = \iint\limits_D \left[2\frac{\partial u}{\partial x}\frac{\partial \eta}{\partial x} + 2\frac{\partial u}{\partial y}\frac{\partial \eta}{\partial y} + 2f\eta\right] dx\,dy = 0. \tag{14}$$

In the expression for δI we transform the first two terms by means of Green's formula:

$$\iint\limits_D \left[\frac{\partial u}{\partial x}\frac{\partial \eta}{\partial x} + \frac{\partial u}{\partial y}\frac{\partial \eta}{\partial y}\right] dx\,dy = \iint\limits_D \left[\frac{\partial}{\partial x}\left(\eta\frac{\partial u}{\partial x}\right) + \right.$$

$$\left. + \frac{\partial}{\partial y}\eta\left(\frac{\partial u}{\partial y}\right)\right] dxdy - \iint\limits_D \left(\frac{\partial^2 u}{\partial x^2} + \frac{\partial^2 u}{\partial y^2}\right)\eta dx\,dy =$$

$$= \int\limits_\Gamma \left(\frac{\partial u}{\partial x}\eta dy - \frac{\partial u}{\partial y}\eta dx\right) - \iint\limits_D \Delta u\eta\, dx\,dy. \tag{15}$$

In the expression obtained, the first curvilinear integral is equal to zero, since $\eta(x, y) = 0$ on Γ by assumption, and only the second integral remains. Replacing by it the first two terms in expression (14) for δI, we obtain a new expression for the variation:

$$\delta I = -2\iint\limits_D (\Delta u - f)\eta dx\,dy. \tag{16}$$

Since δI must be equal to zero, whatever the function η may be, the following equation always must be satisfied:

$$\Delta u = \frac{\partial^2 u}{\partial x^2} + \frac{\partial^2 u}{\partial y^2} = f(x, y). \tag{17}$$

Thus in the given case the Poisson equation has turned up as the Euler equation. If we turn our attention to condition (13), we see that the solution of the problem of the minimum of integral I (12) [under condition (13)] is the solution of the Dirichlet problem for the Poisson equation. If $f = 0$, equation (17) reduces to the Laplace equation:

$$\Delta u = \frac{\partial^2 u}{\partial x^2} + \frac{\partial^2 u}{\partial y^2} = 0. \tag{17'}$$

Integral (12), which for the given case takes the form

$$I(u) = \iint\limits_D \left[\left(\frac{\partial u}{\partial x}\right)^2 + \left(\frac{\partial u}{\partial y}\right)^2\right] dx\, dy, \tag{12'}$$

bears the appellation: the Dirichlet integral. The problem of finding its minimum is connected with the Dirichlet problem for the Laplace equation.

In the given case, just as in No. 1, it can be shown that the conditions obtained are also sufficient. Explicitly, if the function $u(x, y)$ is the solution of the Dirichlet problem for the Poisson equation (17), under condition (13), and integral (12) has for it a finite value, then it provides the absolute minimum for this integral. More exactly, for any function satisfying condition (13) and continuous together with its derivatives of the first order [1]), integral (12) always has a greater value. Indeed, let $\bar{u}(x, y)$ be another function satisfying condition (13). Then $\bar{u}(x, y) = u(x, y) + \eta(x, y)$, where $\eta(x, y) = 0$ on Γ. Therefore we have

$$\begin{aligned} \Delta I = I(\bar{u}) - I(u) = \iint\limits_D \Big[\left(\frac{\partial u}{\partial x} + \frac{\partial \eta}{\partial x}\right)^2 + \left(\frac{\partial u}{\partial y} + \frac{\partial \eta}{\partial y}\right)^2 + \\ + 2f(u + \eta)\Big] dx\, dy - \iint\limits_D \left[\left(\frac{\partial u}{\partial x}\right)^2 + \left(\frac{\partial u}{\partial y}\right)^2 + 2fu\right] dx\, dy = \\ = 2 \iint\limits_D \left(\frac{\partial u}{\partial x}\frac{\partial \eta}{\partial x} + \frac{\partial u}{\partial y}\frac{\partial \eta}{\partial y} + f\eta\right) dx\, dy + \\ + \iint\limits_D \left[\left(\frac{\partial \eta}{\partial x}\right)^2 + \left(\frac{\partial \eta}{\partial y}\right)^2\right] dx\, dy. \end{aligned} \tag{18}$$

The first term differs not at all from the variation δI, and, since u is the solution, it is equal to zero [cf. (14) and (16)]; the second term is obviously positive. So $\Delta I > 0$ and u does give the absolute minimum.

An example of a mechanical problem leading to integral (12) is given by the problem of the equilibrium of a stretched elastic membrane. The potential energy of such a membrane (for small deflections) under a load intensity defined by the function $f(x, y)$, is

$$U = \iint\limits_D \left\{\frac{\mu}{2}\left[\left(\frac{\partial u}{\partial x}\right)^2 + \left(\frac{\partial u}{\partial y}\right)^2\right] + f(x, y)u\right\} dx\, dy,$$

where μ is a constant quantity in the case of a homogeneous membrane. The integral obtained differs immaterially from integral (12), and the

[1]) It can even be assumed that these derivatives do not exist, or experience discontinuities on certain lines.

Euler equation for it is the Poisson equation,

$$\Delta u = \frac{1}{\mu} f(x, y),$$

and since the equilibrium position of the membrane corresponds to the minimum of the potential energy, this equation is also the equation of the surface of the membrane.

The considerations that we utilized in deriving the necessary conditions (the Euler equation) for integral (12) are also suitable for a number of other problems. For example, for the general problem of the minimum of the double integral

$$I = \iint_D F(x, y, u, u'_x, u'_y)\,dx\,dy \tag{12a}$$

we obtain Euler's equation in the form

$$\frac{\partial}{\partial x}\left(\frac{\partial F}{\partial u'_x}\right) + \frac{\partial}{\partial y}\left(\frac{\partial F}{\partial u'_y}\right) - \frac{\partial F}{\partial u} = 0. \tag{17a}$$

The problem of the minimum of the integral

$$I = \iint_D \left[a\left(\frac{\partial u}{\partial x}\right)^2 + b\left(\frac{\partial u}{\partial y}\right)^2 + cu^2 + 2fu\right] dx\,dy \tag{12b}$$

has as its Euler equation the self-adjoint equation of the elliptic type

$$\frac{\partial}{\partial x}\left(a\frac{\partial u}{\partial x}\right) + \frac{\partial}{\partial y}\left(b\frac{\partial u}{\partial y}\right) - cu = f, \tag{17b}$$

which subsumes the Poisson equation.

In the problem of the minimum of a triple integral, the Euler equation turns out to be a partial differential equation in three variables. If derivatives of order higher than the first figure in the integral, then the Euler equation turns out to be of the fourth order or higher.

The most important example of this type of integral is that giving the potential energy of a bent plate, which we shall consider below in No. 4.

3. Other forms of boundary conditions. We considered the problem of the minimum of integral (1) on the assumption that the function $y(x)$ satisfied the conditions: $y(x_0) = y_0$, $y(x_1) = y_1$. This problem led to a boundary-value problem for differential equation (5) with the same conditions. Frequently, however, boundary-value problems with conditions of other types prove to be interesting for differential equations. Let us answer the question: what kind of variational problems lead to them? With this aim, we shall first of all consider the problem of the minimum of integral (1) without imposing any sort of conditions on the function $y(x)$, or under natural conditions, so-called. In this case the course of the reasoning remains the former one, but for $\eta(x)$ the conditions $\eta(x_0) = \eta(x_1) = 0$ can no longer be used. Expression (3) for δI here preserves its force, but on performing the integration by parts for the first term, we can no longer discard the term outside the integral, obtaining, therefore,

$$\int_{x_0}^{x_1} 2py'\eta' dx = [2py'\eta]_{x_0}^{x_1} - 2\int_{x_0}^{x_1} \frac{d}{dx}(py')\eta(x)dx =$$

$$= 2py'\eta\,|_{x=x_1} - 2py'\eta|_{x=x_0} - 2\int_{x_0}^{x_1} \frac{d}{dx}(py')\eta(x)dx.$$

In consequence of this, the expression for the variation takes, instead of (4), the form

$$\delta I = -2\left\{ \int_{x_0}^{x_1} \left[\frac{d}{dx}(py') - qy - f\right]\eta\, dx + py'\eta|_{x=x_0} - py'\eta|_{x=x_1}\right\}. \quad (19)$$

Since δI must be equal to zero for any function $\eta(x)$, on taking $\eta(x)$ equal, in particular, to zero at the ends, we become convinced that for any such function the integral term must equal zero, and consequently, as above, $y(x)$ must satisfy equation (5). Once this is so, then in the case of any $\eta(x)$ the integral term in (19) may be discarded, too. Now choosing $\eta(x)$ such that first $\eta(x_0) = 1$, $\eta(x_1) = 0$, and next such that $\eta(x_0) = 0$, $\eta(x_1) = 1$, we satisfy ourselves that py' must be zero at the ends, and since we assume $p(x) \neq 0$, we arrive at the conditions

$$y'|_{x_0} = 0, \quad y'|_{x_1} = 0. \quad (20)$$

Thus the solution of the problem of the extremum of integral (1), under natural boundary conditions, is the solution of equation (5) under boundary conditions (20). To obtain other forms of boundary conditions, we shall have to complicate somewhat the functional I. We shall add to integral (1) complementary terms, connected with the initial and terminal points, and shall consider the problem of the minimum of the functional that is thus obtained:

$$I = \int_{x_0}^{x_1} [py'^2 + qy^2 + 2fy]dx + [h_1 y^2]_{x=x_1} - [h_0 y^2]_{x=x_0}. \quad (21)$$

In calculating the variation δI in the given case, we shall have to attach to expression (19) summands corresponding to the complementary terms. These terms will obviously be:

$$\frac{d}{d\alpha}\{[h_1(y+\alpha\eta)^2]_{x=x_1} - [h_0(y+\alpha\eta)^2]_{x=x_0}\}|_{\alpha=0} = 2h_1 y\eta|_{x=x_1} - 2h_0 y\eta|_{x=x_0},$$

and the expression for δI will finally take the form:

$$\delta I = -2\Big\{ \int_{x_0}^{x_1} \Big[\frac{d}{dx}(py') - qy - f\Big]\eta\, dx + [py'\eta + h_0 y\eta]_{x=x_0} - \\ - [py'\eta + h_1 y\eta]_{x=x_1}\Big\}. \quad (22)$$

Reasoning as above, we become convinced that the extremal must be the solution of equation (5) under the conditions:

$$\left.\begin{aligned} p(x_0)y'(x_0) + h_0 y(x_0) = 0, \\ p(x_1)y'(x_1) + h_1 y(x_1) = 0. \end{aligned}\right\} \quad (23)$$

The result obtained shows that for any linear homogeneous boundary conditions, the corresponding variational problem can be given; in particular, if $h_1 = h_0 = 0$, conditions (23) reduce to the natural conditions (20).

Analogous results can also be obtained with respect to partial differential equations. In conformity with what we had for ordinary equations, we add to double integral (12) a complementary term in the form of a curvilinear integral and consider the question of the minimum of the functional obtained:

$$I = \iint_D \Big[\Big(\frac{\partial u}{\partial x}\Big)^2 + \Big(\frac{\partial u}{\partial y}\Big)^2 + 2fu\Big]\, dx\, dy + \int_\Gamma [\sigma(s)u^2 - 2\varphi(s)u]\, ds, \quad (24)$$

where the function $u(x, y)$ is not restricted by any kind of boundary condition. In calculating the variation δI for the first term we can utilize equations (14) and (15), but in (15) the curvilinear integral on the right can no longer be discarded, since η is not necessarily equal to zero on Γ. The term corresponding to the second part is found without difficulty:

$$\Big\{\frac{d}{d\alpha}\int_\Gamma [\sigma(s)(u + \alpha\eta)^2 - 2(u + \alpha\eta)\varphi(s)]\, ds\Big\}_{\alpha=0} = 2\int_\Gamma [\sigma(s)u - \varphi(s)]\eta\, ds,$$

and we obtain finally the following value for the variation:

$$\delta I = -2\Big\{\iint_D (\Delta u - f)\eta\, dx\, dy + \\ + \int_\Gamma \Big[-\frac{\partial u}{\partial x}\frac{dy}{ds} + \frac{\partial u}{\partial y}\frac{dx}{ds} - \sigma u + \varphi(s)\Big]\eta\, ds\Big\}. \quad (25)$$

By means of the reasoning already employed above, we can convince ourselves that the multiplier of η, both in the double and in the curvilinear integrals, must be zero, i.e., the function u must satisfy equation

(17) and the boundary condition

$$\frac{\partial u}{\partial x}\frac{dy}{ds} - \frac{\partial u}{\partial y}\frac{dx}{ds} + \sigma u - \varphi(s) = 0.$$

Observing that

$$\frac{dy}{ds} = \cos(s, y) = \cos(n, x)$$

and

$$\frac{dx}{ds} = \cos(s, x) = -\cos(n, y),$$

this condition may be given the form

$$\frac{\partial u}{\partial n} + \sigma(s)u = \varphi(s). \tag{26}$$

In particular, if $\sigma = 0$, we find that the natural condition for the original integral (12) is $\frac{\partial u}{\partial n} = 0$.

We shall make the following observation, which will subsequently be used by us. The problem of the minimum of integral (24) has been considered by us on the assumption that any functions may be regarded as admissible to the equation which do not satisfy any kind of boundary conditions. Nevertheless after having established that condition (26) is fulfilled for the sought extremal, in determining the extremal of integral (24) one can subject all functions admissible for consideration to condition (26) beforehand, as obviously the minimum will be the same as before.

We stated above that integral (12), apart from a factor, is the potential energy of a stretched membrane; the line integral by the addition of which we passed from integral (12) to integral (24), also has a definite mechanical signification, namely: this term is added to the potential energy of the membrane if its boundary is movable and subject to an external force of linear density $\varphi(s)$ and to elastic forces that tend to hold the boundary in equilibrium position, the modulus of elasticity of the boundary, calculated per unit length, equalling $\sigma(s)$.

We shall observe, lastly, that in the same way we have here constructed, with respect to equations (5) and (17), variational problems leading to other boundary conditions, the variational problems can be modified for other equations, too, so as to obtain other boundary conditions, which may prove to be interesting.

4. Variational problems connected with the biharmonic equation. A number of questions of the theory of elasticity (the bending of plates, the plane problem) lead to different boundary-value problems for the biharmonic equation [1])

$$\Delta\Delta u = \frac{\partial^4 u}{\partial x^4} + 2\frac{\partial^4 u}{\partial x^2 \partial y^2} + \frac{\partial^4 u}{\partial y^4} = f(x, y). \tag{27}$$

With this equation is connected the problem of the minimum of the integral

$$U = \iint_D \Big[(\Delta u)^2 - 2(1-\nu)\Big(\frac{\partial^2 u}{\partial x^2}\frac{\partial^2 u}{\partial y^2} - \Big(\frac{\partial^2 u}{\partial x \partial y}\Big)^2\Big) - 2fu\Big]\, dx\, dy - \\ - 2\int_\Gamma p(s) u\, ds + 2\int_\Gamma m(s)\frac{\partial u}{\partial n}\, ds. \tag{28}$$

To obtain the necessary condition, we must again substitute $u + \alpha\eta$ for u and equate to zero the derivative with respect to α (for $\alpha = 0$); however the same thing is obtained if in (28) u is replaced by $u + \eta$ and the terms of the first degree in η and its derivatives are preserved (η is an arbitrary twice-differentiable function). Thus we obtain:

$$\iint_D \Big[2\Delta u \Delta \eta - 2(1-\nu)\Big(\frac{\partial^2 u}{\partial x^2}\frac{\partial^2 \eta}{\partial y^2} + \frac{\partial^2 u}{\partial y^2}\frac{\partial^2 \eta}{\partial x^2} - 2\frac{\partial^2 u}{\partial x \partial y}\frac{\partial^2 \eta}{\partial x \partial y}\Big) - \\ - 2f\eta\Big]\, dx\, dy - 2\int_\Gamma p(s)\eta\, ds + 2\int_\Gamma m(s)\frac{\partial \eta}{\partial n}\, ds = 0. \tag{29}$$

We transform the expressions under the double integral sign by means of Green's formula; we obtain:

$$\begin{aligned}\iint_D \Delta u\, \Delta\eta\, dx\, dy &= \iint_D \Big[\frac{\partial}{\partial x}\Big(\Delta u \frac{\partial \eta}{\partial x}\Big) + \frac{\partial}{\partial y}\Big(\Delta u \frac{\partial \eta}{\partial y}\Big)\Big]\, dx\, dy - \\ &- \iint_D \Big[\frac{\partial}{\partial x}\Big(\eta \frac{\partial}{\partial x}\Delta u\Big) + \frac{\partial}{\partial y}\Big(\eta \frac{\partial}{\partial y}\Delta u\Big)\Big]\, dx\, dy + \\ &+ \iint_D \eta\Big(\frac{\partial^2}{\partial x^2}\Delta u + \frac{\partial^2}{\partial y^2}\Delta u\Big)\, dx\, dy = \\ &= \iint_D (\Delta\Delta u)\eta\, dx\, dy + \int_\Gamma \Delta u\Big(\frac{\partial \eta}{\partial x} dy - \frac{\partial \eta}{\partial y} dx\Big) - \\ &- \int_\Gamma \eta\Big(\frac{\partial(\Delta u)}{\partial x} dy - \frac{\partial(\Delta u)}{\partial y} dx\Big) = \\ &= \iint_D \Delta\Delta u\, \eta\, dx\, dy + \int_\Gamma \Delta u \frac{\partial \eta}{\partial n}\, ds - \int_\Gamma \frac{\partial}{\partial n}\Delta u \eta\, ds. \end{aligned} \tag{30}$$

---

[1]) We note that only the homogeneous equation $\Delta\Delta u = 0$ is ordinarily called biharmonic.

Performing analogous transformations for the second half of the integral also, and substituting the results of the transformation in (29), we bring it to the form:

$$\iint_D (\Delta\Delta u - f)\eta\, dx\, dy + \int_\Gamma [M(u) + m(s)]\frac{\partial\eta}{\partial n}ds - \int_\Gamma (P(u) + p)\eta ds = 0; \qquad (31)$$

for brevity we have introduced the symbols:

$$\left.\begin{aligned} M(u) &= \nu\Delta u + (1-\nu)\left(\frac{\partial^2 u}{\partial x^2}x_n^2 + 2\frac{\partial^2 u}{\partial x\partial y}x_n y_n + \frac{\partial^2 u}{\partial y^2}y_n^2\right), \\ P(u) &= \frac{\partial}{\partial n}\Delta u + (1-\nu)\frac{\partial}{\partial s}\left[\frac{\partial^2 u}{\partial x^2}x_n x_s + \frac{\partial^2 u}{\partial x\partial y}(x_n y_s + x_s y_n) + \frac{\partial^2 u}{\partial y^2}y_n y_s\right], \end{aligned}\right\} \qquad (32)$$

where x_n, y_n and x_s, y_s denote the direction cosines of the normal and the tangent. Since the function η is arbitrary inside D, we conclude from (31), on examining the double integral, that for the function u equation (27) must be fulfilled. Furthermore, since η, as also $\frac{\partial\eta}{\partial n}$, is arbitrary on the contour, for u the following boundary conditions must be fulfilled:

$$M(u) + m = 0, \quad P(u) + p = 0 \text{ on } \Gamma, \qquad (33)$$

which are, indeed, the natural conditions for the given case.

Let us note first of all the case of boundary conditions corresponding to a built-in plate, which is the simplest case:

$$u = 0, \quad \frac{\partial u}{\partial n} = 0 \text{ on } \Gamma. \qquad (34)$$

In this case η too must be taken as satisfying the same conditions; therefore the curvilinear integrals in condition (31) vanish, and only the double integral remains. We thus again find that for u equation (27) must be fulfilled. Functional (28) itself, which is subject to minimization, can in this case be considerably simplified. In the first place, the curvilinear integrals obviously vanish; in the second place, since ν did not appear in the equation, but only in the boundary conditions (33), in place of which we now use conditions (34), ν can be assigned any value. Adopting $\nu = 1$, we find that the problem of solving equation (27) under conditions (34) corresponds to the problem of finding the minimum of the integral

$$I = \iint_D [(\Delta u)^2 - 2fu]dx\, dy. \qquad (35)$$

We could have taken $\nu = 0$; then we would have arrived at the integral

$$\iint_D \left[\left(\frac{\partial^2 u}{\partial x^2}\right)^2 + \left(\frac{\partial^2 u}{\partial y^2}\right)^2 + 2\left(\frac{\partial^2 u}{\partial x \partial y}\right)^2 - 2fu\right] dx\, dy. \tag{35'}$$

We shall also note the case when the plate is simply supported, i.e., it is given that $u = 0$ on the contour. Then only those η are admissible which are equal to zero on the contour, so that there remains only the first of conditions (33), to which we add the condition $u = 0$. The boundary conditions for this case thus have the form:

$$u = 0, \quad M(u) + m = 0 \text{ on } \Gamma. \tag{36}$$

We shall note that conditions (33) are considerably simplified if part of the contour is a straight line; thus on a line of the form $x = a$ they take the form:

$$\frac{\partial^2 u}{\partial x^2} + \nu \frac{\partial^2 u}{\partial y^2} + m = 0, \quad \frac{\partial^3 u}{\partial x^3} + (2 - \nu)\frac{\partial^2 u}{\partial x \partial y^2} + p = 0. \tag{33'}$$

5. Variational problems connected with the determination of proper numbers and proper functions. For definiteness we shall consider this subject for the Laplace equation in the case of null boundary conditions. A proper number here will be a number λ such that the equation

$$\Delta u + \lambda u = 0, \tag{37}$$

under the condition

$$u = 0 \text{ on } \Gamma, \tag{38}$$

admits of a solution not identically equal to zero in D; and this solution will be the proper function. We shall show that the problem of determining the proper numbers and functions is connected with the variational problem of the minimum of the Dirichlet integral:

$$I = \iint_D \left[\left(\frac{\partial u}{\partial x}\right)^2 + \left(\frac{\partial u}{\partial y}\right)^2\right] dx\, dy, \tag{12'}$$

under the condition

$$H = \iint_D u^2\, dx\, dy = 1. \tag{39}$$

This minimum I, under conditions (38) and (39), is equal to λ_1 — the first (least) proper number, and the function $u_1(x, y)$ which furnishes this minimum is the proper function corresponding to λ_1. Indeed, let λ_1 be this minimum, and u_1 the corresponding function. It is obvious that for any function $u(x, y) \neq 0$, one can select a factor c such that $H(cu) = 1$, and then

$$\frac{I(u)}{H(u)} = \frac{I(cu)}{H(cu)} \geqslant \lambda_1,$$

$I(u) - \lambda_1 H(u) \geqslant 0$, i.e., $$\iint_D \left[\left(\frac{\partial u}{\partial x}\right)^2 + \left(\frac{\partial u}{\partial y}\right)^2 - \lambda_1 u^2\right] dx\,dy \geqslant 0.$$

In this inequality let us replace u by $u_1 + \alpha\eta$, where $\eta = 0$ on Γ. Then for $\alpha \neq 0$ the left side will be non-negative; for $\alpha = 0$ it vanishes and attains the minimum. Therefore its derivative with respect to α must, for $\alpha = 0$, be zero:

$$\left\{\frac{d}{d\alpha}\iint_D \left[\left(\frac{\partial u_1}{\partial x} + \alpha\frac{\partial \eta}{\partial x}\right)^2 + \left(\frac{\partial u_1}{\partial y} + \alpha\frac{\partial \eta}{\partial y}\right)^2 - \lambda_1(u_1 + \alpha\eta)^2\right] dxdy\right\}_{\alpha=0} =$$

$$= 2\iint_D \left[\frac{\partial u_1}{\partial x}\,\frac{\partial \eta}{\partial x} + \frac{\partial u_1}{\partial y}\,\frac{\partial \eta}{\partial y} - \lambda_1 u_1 \eta\right] dx\,dy = 0. \quad (40)$$

Transforming the first two terms by means of formula (15), we shall have given the last equation the form:

$$\iint_D (\Delta u_1 + \lambda_1 u_1)\eta\, dx\,dy = 0.$$

Since this equality must be fulfilled for any function η (equal to 0 on Γ), we have

$$\Delta u_1 + \lambda_1 u_1 = 0,$$

that is, λ_1 is indeed a proper number, and u_1 a proper function. Of the fact that λ_1 is the least proper number we convince ourselves directly, since if λ_2 is another proper number and u_2 the respective proper function ($\Delta u_2 + \lambda_2 u_2 = 0$, $u_2 = 0$ on Γ), being normalized so that $\iint_D u_2^2 dx\,dy = 1$, then we find for it, using (15) with $u = \eta = u_2$:

$$\iint_D \left[\left(\frac{\partial u_2}{\partial x}\right)^2 + \left(\frac{\partial u_2}{\partial y}\right)^2\right] dx\,dy = -\iint_D \Delta u_2 u_2\, dx\,dy = \lambda_2 \iint_D u_2^2 dx\,dy = \lambda_2,$$

and since λ_1 is the minimum of integral (12′), $\lambda_2 \geqslant \lambda_1$. The last equation, since its left side is positive, shows, by the way, that equation (37) does not have negative proper values.

The second proper value and function are obtained by considering the problem of the minimum of the same integral (12′), if the function u is subjected, in addition to the former conditions ($u = 0$ on Γ, $\iint_D u^2 dxdy = 1$), to the condition of orthogonality to u_1: $\iint_D u u_1 dx\,dy = 0$. Indeed, let u_2 be the function giving the minimum for integral (12′) under these conditions, and λ_2 the value of this minimum. As in the preceding case, we find that the integral

$$\iint_D \left[\left(\frac{\partial u}{\partial x}\right)^2 + \left(\frac{\partial u}{\partial y}\right)^2 - \lambda_2 u^2\right] dx\,dy \tag{41}$$

has its value $\geqslant 0$ for all functions u satisfying the conditions: $u = 0$ on Γ and $\iint_D uu_1 dx\,dy = 0$; here 0 is attained for $u = u_2$. Now let $\eta(x, y)$ be any function equal to 0 on Γ; let us set $\zeta = \eta + ku_1$, where k is so selected that ζ is orthogonal to u_1. Let us now replace u by $u_2 + \alpha\zeta$ in (41); the function α obtained will, in accordance with what was said above, attain a minimum equal to zero for $\alpha = 0$. Equating to zero the derivative with respect to α for $\alpha = 0$, we shall obtain, as in (40),

$$\iint_D \left[\frac{\partial u_2}{\partial x}\frac{\partial \zeta}{\partial x} + \frac{\partial u_2}{\partial y}\frac{\partial \zeta}{\partial y} - \lambda_2 u_2 \zeta\right] dx\,dy = 0,$$

or, replacing ζ by $\eta + ku_1$,

$$k \iint_D \left[\frac{\partial u_2}{\partial x}\frac{\partial u_1}{\partial x} + \frac{\partial u_2}{\partial y}\frac{\partial u_1}{\partial y} - \lambda_2 u_2 u_1\right] dx\,dy +$$

$$+ \iint_D \left[\frac{\partial u_2}{\partial x}\frac{\partial \eta}{\partial x} + \frac{\partial u_2}{\partial y}\frac{\partial \eta}{\partial y} - \lambda_2 u_2 \eta\right] dx\,dy = 0.$$

The first term on the left side is equal to zero, since on adding to it $k\iint_D (\lambda_2 - \lambda_1)u_2 u_1 dx dy = 0$ (in consequence of the orthogonality of u_2 to u_1), we can give it form (40), but with η replaced by u_2; transforming the second term as we transformed integral (40), the preceding equation can be brought into the form

$$\iint_D [\Delta u_2 + \lambda_2 u_2]\eta\, dx\,dy = 0.$$

From the fact that the expression obtained is equal to zero for any function (equal to 0 on Γ), we conclude that

$$\Delta u_2 + \lambda_2 u_2 = 0,$$

i.e., that λ_2 is indeed a proper number, and u_2 the corresponding proper function. Reasoning as in the proof that λ_1 is the least of the proper values, we can satisfy ourselves that λ_2 is the next larger proper value. It can be established analogously that the subsequent proper numbers represent the values of the minimum of integral (12′) on condition that the function u be subjected, in addition to conditions (38) and (39), to the condition of orthogonality to the preceding proper functions.

In the same way as variational problems connected with the Dirichlet integral lead to the proper values of the equation $\Delta u + \lambda u = 0$, variational problems can also be indicated which lead to the proper values and proper functions for other equations.

§ 2. RITZ'S METHOD AND THE METHOD OF B. G. GALERKIN

In the preceding section we saw that the solution of boundary-value problems for various types of ordinary and partial differential equations will be found if the problem of minimizing one or another integral be solved. For an approximate solution of the last problem a simple and efficient method was proposed in 1908 by Ritz [1]. This method obtained a particularly wide dissemination in applied mechanics thanks to the activity of a number of Russian scientists working in the domain of the theory of elasticity. Its theoretical investigation has been conducted chiefly in works of Soviet mathematicians, particularly N. M. Krylov and N. N. Bogoliubov.

In 1915 there was proposed by B. G. Galerkin [2] a method of solving boundary-value problems which was still simpler in application, and at the same time more general and universal, and which has recently attained a still wider application and dissemination than Ritz's method. The problem of its convergence has been solved in a relatively recent work of M. V. Keldysh.

Although the method of B. G. Galerkin relates not only to problems connected with variational problems, it is this chapter in which we give an exposition of it, in view of its close connection with variational methods and in particular with Ritz's method.

1. The fundamental idea of Ritz's method and the method of B. G. Galerkin. We shall expound the fundamental idea of Ritz's method.

Let us for the time being consider the general problem of the minimum of the double integral [see § 1 (12a)]

$$I(u) = \iint_D F(x, y, u, u'_x, u'_y)\,dx\,dy \tag{1}$$

under the condition

$$u = \varphi(s) \text{ on } \Gamma, \tag{2}$$

where Γ is the contour bounding the region D.

Let $u^*(x, y)$ be the exact solution of this problem, and $I(u^*) = m$ the value of the minimum. It is clear that if we can succeed in constructing

---

[1]) Ritz, W. [1].

[2]) B. G. Galerkin, [3]. Perel'man, M. IA., [1]; in this article is also given a further bibliography. Numerous applications of both methods are given in a book of L. S. Leĭbenzon, [2].

a function $\bar{u}(x, y)$ which satisfies condition (2) and for which the value of the integral $I(u)$ is very close to m, one could expect $\bar{u}(x, y)$ to be a good approximation to the true solution of the problem. If, moreover, we could succeed in finding a minimizing sequence $\bar{u}_n$, i.e., a sequence of functions satisfying conditions (2), and for which $I(\bar{u}_n) \to m$, there would be grounds for expecting such a sequence to converge, in one sense or another, to the solution.

For the actual determination of the function $\bar{u}(x, y)$ giving a value of the integral I close to the minimal one, Ritz proposed this device. Let us consider a family of functions depending on several parameters

$$u = \Phi(x, y, a_1, a_2, \ldots, a_n) \tag{3}$$

and such that for all values of the parameters condition (2) is satisfied. Now we shall limit the class of admissible functions to the functions of family (3), and shall find among them that which gives integral (1) the least value. This problem already represents an incomparably easier one than does the original problem. Indeed, substituting in integral (1), for u, expression (3) and performing the necessary differentiations and integrations, we see that I is converted into a function of the n variables $a_1, a_2, \ldots, a_n$: $I = I(a_1, a_2, \ldots, a_n)$. Since we wish to obtain the minimum of this function, the numbers a_i must satisfy the system of equations

$$\frac{\partial I}{\partial a_k} = 0 \qquad (k = 1, 2, \ldots, n). \tag{4}$$

On solving this system we obtain, generally speaking, definite values of the parameters $\bar{a}_1, \bar{a}_2, \ldots, \bar{a}_n$ giving the function $I(a_1, a_2, \ldots, a_n)$ its absolute minimum; on choosing the function in family (3) corresponding to these values of the parameters, we shall then have obtained the required approximate solution:

$$\bar{u}(x, y) = \Phi(x, y, \bar{a}_1, \bar{a}_2, \ldots, \bar{a}_n). \tag{5}$$

It should still be remarked that in concrete cases the actual process of determining this approximate solution turns out — as we shall see below — to be very simple. The fact is that in the cases most important practically, the integrand of integral I is a polynomial of the second degree in u, u'_x, u'_y, and family (3) is taken to depend linearly on the parameters $a_1, a_2, \ldots, a_n$; in consequence of this, equations (4) prove to be of the first degree, and the number of them is very small, a sufficient approximation being in practice obtained with $n = 2, 3, 4, 5$, and frequently even with $n = 1$.

Let us now discuss the question: under what circumstances may one expect to obtain in this way an arbitrarily close approximation to the actual minimum? By means of Ritz's method a sequence of successively more exact approximations can be obtained. For this purpose a number of families of functions must be considered:

$$u(x, y) = \Phi_n(x, y, a_1, a_2, \ldots, a_n) \qquad (n = 1, 2, \ldots), \tag{6}$$

each of which is broader than the preceding, as the result of the introduction of the additional parameter. Let $\bar{u}_n$ be the nth approximation the function giving the least value for integral I in comparison with all the functions of the nth family. Since each successive family contains all the functions of the preceding, i.e., for each successive problem the class of admissible functions is broader, it is clear that the successive minimums are non-increasing:

$$I(\bar{u}_1) \geqslant I(\bar{u}_2) \geqslant \ldots \tag{7}$$

Let us now ask the question: in what case can it be affirmed that the sequence of functions $\bar{u}_1, \bar{u}_2, \ldots$ is minimal, i.e., that the sequence of integrals $I(\bar{u}_1), I(\bar{u}_2), \ldots$ tends to the true minimum:

$$\lim_{n\to\infty} I\,(\bar{u}_n) = I(u^*) = m? \tag{8}$$

We shall show that the sufficient condition for this is what is known as the relative completeness of the system of families (6), which consists in the following: whatever be the function u, continuous together with $\frac{\partial u}{\partial x}$ and $\frac{\partial u}{\partial y}$ and satisfying boundary condition (2), and whatever be the positive number $\varepsilon > 0$, one can indicate an n, and a function of the nth family (6),

$$u_n^*(x, y) = \Phi_n(x, y, a_1^*, \ldots, a_n^*),$$

such that the following inequality will be valid everywhere in the region D:

$$|u_n^* - u| < \varepsilon, \quad \left|\frac{\partial u_n^*}{\partial x} - \frac{\partial u}{\partial x}\right| < \varepsilon, \quad \left|\frac{\partial u_n^*}{\partial y} - \frac{\partial u}{\partial y}\right| < \varepsilon, \tag{8'}$$

i.e., putting it briefly, any admissible function, together with its partial derivatives, may be approximated as closely as one pleases by means of functions of the given families.

We shall prove that this condition is sufficient for the sequence to be minimal. Indeed, if the condition of completeness is fulfilled, then on applying it to the solution of our problem u^*, we shall be able to select such a function $u_n^*(x, y)$ from a certain, nth, family,

$$u_n^*(x, y) = \Phi_n(x, y, a_1^*, a_2^*, \ldots, a_n^*),$$

that there will be satisfied the inequalities

$$|u_n^* - u^*| < \varepsilon, \quad \left|\frac{\partial u_n^*}{\partial x} - \frac{\partial u^*}{\partial x}\right| < \varepsilon, \quad \left|\frac{\partial u_n^*}{\partial y} - \frac{\partial u^*}{\partial y}\right| < \varepsilon.$$

Then, in consequence of the continuity of F, the difference

$$F\left(x, y, u_n^*, \frac{\partial u_n^*}{\partial x}, \frac{\partial u_n^*}{\partial y}\right) - F\left(x, y, u^*, \frac{\partial u^*}{\partial x}, \frac{\partial u^*}{\partial y}\right)$$

will be arbitrarily small in the region D, and then the difference of the integrals of these functions, too, will be arbitrarily small, i.e.,

$$I(u_n^*) - I(u^*) = \iint_D \left[F\left(x, y, u_n^*, \frac{\partial u_n^*}{\partial x}, \frac{\partial u_n^*}{\partial y}\right) - \right.$$

$$\left. - F\left(x, y, u^*, \frac{\partial u^*}{\partial x}, \frac{\partial u^*}{\partial y}\right)\right] dx\, dy < \varepsilon',$$

where ε' is an arbitrarily small positive number. Moreover, u_n^* is one of the functions of the nth family (6); the function $\bar{u}_n$, however, gives the integral its least value in comparison with all the functions of this family. We must therefore have $I(\bar{u}_n) \leqslant I(u_n^*)$, or

$$I(u^*) \leqslant I(\bar{u}_n) \leqslant I(u_n^*) < I(u^*) + \varepsilon',$$

and since ε' is arbitrarily small, relation (8) is established.

In case the question of the minimum of the integral is being discussed under free boundary conditions, the completeness requirement formulated above is modified in this respect, that there must exist the possibility of approximating, by means of functions of family (6), any functions which are continuous together with their partial derivatives. However if we take into consideration the fact that the sought function, as we know, is such as to satisfy definite boundary conditions (the natural ones), we are led to the conclusion that the possibility of approximating just such functions is sufficient.

We shall promptly turn to the question of the completeness; since it is of fundamental importance — violation of the requirement of completeness can lead to a gross error — in determining how far approximations by Ritz's method are from the correct result.

Further general discussions connected with Ritz's method — the questions: in what case will the minimal sequence converge to the solution of the problem, what is the order of the approximation, what is the error in the nth approximation, etc. — we postpone until § 4.

We shall now also give the general idea of another important direct method, the method of B. G. Galerkin.

The basic idea of the method of B. G. Galerkin can be expounded quite briefly. Let it be required to determine the solution of the equation

$$L(u) = 0,$$

where L is some differential operator in two variables, which solution satisfies homogeneous boundary conditions. We shall seek an approximate solution of the problem in the form

$$\bar{u}(x, y) = \sum_{i=1}^{n} c_i \varphi_i(x, y),$$

where $\varphi_i(x, y)$ $(i = 1, 2, \ldots, n)$ is a certain system of functions, chosen beforehand, and satisfying the same boundary conditions, and the c_i

are undetermined coefficients. We can always consider the functions $\varphi_i(x, y)$ to be linearly independent, and to represent the first n functions of some system of functions $\{\varphi_i(x, y)\}$ $(i = 1, 2, \ldots, n, \ldots)$ which is complete in the given region. In order that $\bar{u}(x, y)$ be the exact solution of the given equation, it is necessary that $L(\bar{u})$ be identically equal to zero; and this requirement, if $L(\bar{u})$ is considered to be continuous, is equivalent to the requirement of the orthogonality of the expression $L(\bar{u})$ to all the functions of the system $\varphi_i(x, y)$ $(i = 1, 2, \ldots, n, \ldots)$. However, having at our disposal only n constants $c_1, c_2, \ldots, c_n$, we can, generally speaking, satisfy only n conditions of orthogonality. Stating these conditions, we arrive at the system of equations

$$\iint_D L(\bar{u}(x, y))\varphi_i(x, y)\, dx\, dy = \iint_D L\left(\sum_{j=1}^{n} c_j\varphi_j(x, y)\right)\varphi_i(x, y)\, dx\, dy = 0$$

$$(i = 1, 2, \ldots, n),$$

which serves for the determination of the coefficients c_i. On finding the c_i from this system — a linear one, in the case of a linear operator L — and substituting them in the expression for $\bar{u}$, we arrive at the required approximate solution.

The equations of the method of B. G. Galerkin can also be obtained from physical considerations, on the basis of the principle of virtual displacements. We shall not tarry on this, referring the reader to the article by B. G. Galerkin himself.

The character of the considerations from which we started in setting up the equations of B. G. Galerkin shows that no connection of the given equation with variational problems was used; this is, therefore, a perfectly universal method. It can be applied with success to equations of diverse types: elliptic, hyperbolic, parabolic, even though they are utterly unconnected with variational problems; it is superior in this respect to Ritz's method. However for problems connected with variational problems, it and Ritz's method are found to be closely related, and in a number of cases it is equivalent to the latter in this sense, that it leads, usually with the simpler computations, to the same approximate solution. Henceforth, therefore, when considering separate classes of problems, it will be expedient to consider both methods in parallel.

2. Application of Ritz's method and that of B. G. Galerkin to ordinary differential equations. Let us consider the application of Ritz's method to the solution of a self-adjoint differential equation of the second order [see § 1 (5)]:

$$L(y) = \frac{d}{dx}(py') - qy - f = 0 \tag{9}$$

under the conditions (rather than x_0 and x_1 we now take 0 and l, which does not lessen the generality)

$$y(0) = y_0, \ y(l) = y_1. \tag{10}$$

This problem, as we saw, may be replaced by the problem of the minimum of the integral

$$I(y) = \int_0^l [py'^2 + qy^2 + 2fy]dx, \tag{11}$$

under the same conditions (10).

In what is to follow, it will be more convenient for us to consider that the end conditions are homogeneous, i.e., have the form

$$y(0) = y(l) = 0; \tag{10'}$$

if this is not the case, it can be contrived by introducing for y in equation (9) a new unknown z defined by the equation

$$y = z + \frac{x}{l}y_1 + \frac{l-x}{l}y_0.$$

We shall in addition assume that in the interval in question the following inequalities are fulfilled:

$$p(x) > 0, \quad q(x) \geqslant 0, \quad (0 \leqslant x \leqslant l). \tag{11'}$$

Let us now be given a system of concrete functions $\varphi_k(x)$ ($k = 1, 2, \ldots, n$), continuous in $[0, l]$ together with their first derivatives, linearly independent and satisfying conditions (10'); as such functions one can adopt, for example,

$$\varphi_k = \sin\frac{k\pi x}{l} \text{ or } \varphi_k = (l-x)x^k \quad (k = 1, 2, \ldots, n). \tag{12}$$

We shall now seek an approximate solution by Ritz's method in the family of linear combinations of the functions φ_k, i.e., among the functions of the form

$$y_n = \sum_{k=1}^n a_k\varphi_k. \tag{13}$$

Substituting the expression for y_n in integral (11), we obtain:

$$I(y_n) = \int_0^l (py_n'^2 + qy_n^2 + 2fy_n)dx = \int_0^l [p(\sum a_k\varphi_k')^2 +$$

$$+ q(\sum a_k\varphi_k)^2 + 2f\sum a_k\varphi_k]dx = \sum_{k,s=1}^n \alpha_{k,s}a_ka_s + 2\sum_{k=1}^n \beta_k a_k, \tag{14}$$

where

$$\alpha_{k,s} = \alpha_{s,k} = \int_0^l (p\varphi_k'\varphi_s' + q\varphi_k\varphi_s)dx, \quad \beta_k = \int_0^l f\varphi_k dx. \tag{15}$$

Differentiating expression (14) with respect to a_s, we obtain the system of equations for the determination of the constants a_k, in two forms

$$\frac{1}{2}\frac{\partial I(y_n)}{\partial a_s} = \int_0^l (py'_n\varphi'_s + qy_n\varphi_s + f\varphi_s)dx = 0 \quad (s = 1, 2, \ldots, n) \qquad \text{(16a)}$$

or

$$\sum_{k=1}^{n} \alpha_{k,s}a_k + \beta_s = 0 \qquad (s = 1, 2, \ldots, n). \qquad (16)$$

We shall designate as $a_k^{(n)}$ $(k = 1, 2, \ldots, n)$ the solution of system of equations (16); then the approximate solution by Ritz's method will be, in the nth approximation,

$$y_n = \sum_{k=1}^{n} a_k^{(n)}\varphi_k. \qquad (17)$$

We shall show that this solution is always determinate and is, moreover, uniquely so, i.e., corresponding circumstances obtain for the system of equations (16). In order to establish this last, it must be shown, as we know, that the corresponding homogeneous system of equations has no solution different from zero. Let us assume the contrary, i.e., that there exist such $\bar{a}_1, \bar{a}_2, \ldots, \bar{a}_n$, not all equal to zero, that these equations are satisfied:

$$\sum_{k=1}^{n} \alpha_{k,s}\bar{a}_k = 0 \qquad (s = 1, 2, \ldots, n)$$

or

$$\int_0^l (p\bar{y}'_n\varphi'_s + q\bar{y}_n\varphi_s)dx = 0, \qquad (s = 1, 2, \ldots, n)$$

where $\bar{y}_n = \sum_{k=1}^{n} \bar{a}_k\varphi_k$. Multiplying the last equations by $\bar{a}_s$ and adding, we obviously obtain

$$\int_0^l (p\bar{y}'^2_n + q\bar{y}^2_n)dx = 0;$$

but this equality, in view of conditions (11′), is possible only if $\bar{y}_n = 0$, i.e., $\sum \bar{a}_k\varphi_k = 0$. This last relation, however, contradicts the assumed linear independence of the functions φ_k.

We shall observe that system of equations (16) can be given still another form, which in some cases renders its formation easier. Namely, by integrating the first term of equation (16a) by parts, using the fact that φ_k and y_n are equal to zero at the ends, we find:

$$\int_0^l py'_n\varphi'_s\,dx = [py'_n\varphi_s]_0^l - \int_0^l \frac{d}{dx}(py'_n)\varphi_s\,dx = -\int_0^l \frac{d}{dx}(py'_n)\varphi_s\,dx.$$

Replacing by the last expression the first term in (16a), changing the sign and utilizing the brief notation (9), we shall have given system (16) this simple form:

$$\int_0^l \left[\frac{d}{dx}(py_n') - qy_n - f\right]\varphi_s dx = \int_0^l L(y_n)\varphi_s dx = 0$$

$$(s = 1, 2, \ldots, n). \quad (16b)$$

After the equations are reduced to such a form, it is evident that they are none other than the equations of the method of B. G. Galerkin, written for the given problem in accordance with the general rule indicated in the preceding No.

The transformation of system (16a) into (16b) given above shows that in application to the given problem the methods of Ritz and of B. G. Galerkin lead to one and the same approximate solution, although the method of B. G. Galerkin makes possible the simpler and more direct setting-up of the respective system. However in the given problem the fundamental advantage of the method of B. G. Galerkin is that in applying it one does not use the connection between the given boundary problem and the variational problem, and it can therefore be employed in the case of any equation of the second order, not requiring the preliminary reduction of the equation to self-adjoint form nor necessarily the satisfaction of conditions (11′) either.

We know (§ 1, No. 1) that under conditions (11′) our variational problem has a definite solution $y^*(x)$; this is indeed the unique solution of equation (9) under conditions (10′). This function (on the assumption of the continuity of p, q and f) will be continuous, together with two derivatives. Let us now pose the question: in what case will the approximations y_n (17) by Ritz' and Galerkin's methods converge to $y(x)$? To answer this, it is first necessary to establish that the sequence y_n is minimal, for which, in turn, as we saw in No. 1, the relative completeness of system of functions (13) must be established. Of course this completeness cannot be established for any choice of φ_k and may even fail to obtain; however, if the φ_k are chosen in one of the two concrete ways which we have indicated above, (12), the condition of completeness is met; i.e., for any function $y(x)$, continuous together with its derivative and satisfying conditions (10′), a linear combination $\sum_{k=1}^{m} a_k\varphi_k$ can be found such that

$$|y(x) - \sum_{k=1}^{m} a_k\varphi_k(x)| < \varepsilon \text{ and } |y'(x) - \sum_{k=1}^{m} a_k\varphi_k'(x)| < \varepsilon. \quad (18)$$

This proposition is obtained almost at once from Weierstrass's theorem concerning the possibility of approximating a continuous function either by a trigonometric or an algebraic polynomial [1]).

[1]) See Smirnov, V. I., [1], vol. II, 1948, p. 441; Vallée-Poussin, [1], 1933, p. 115; L. V. Kantorovich, [1], p. 141.

First let $\varphi_k = \sin \frac{k\pi x}{l}$. Let us take any $y(x)$ satisfying conditions (10'). For the function $y'(x)$ we can find a trigonometric polynomial $T(x) = b_0 + \sum_{k=1}^{n} b_k \cos \frac{k\pi x}{l}$ such that

$$|y'(x) - T(x)| < \frac{\varepsilon}{2(l+1)}.$$

We observe that

$$|b_0| = \left| \frac{1}{l} \int_0^l T(x)dx \right| < \frac{\varepsilon}{2(l+1)} + \left| \frac{1}{l} \int_0^l y'(x)dx \right| = \frac{\varepsilon}{2(l+1)}.$$

Hence

$$\left| y'(x) - \sum_{k=1}^{n} b_k \cos \frac{k\pi x}{l} \right| < |y'(x) - T(x)| + |b_0| < \frac{\varepsilon}{l+1} < \varepsilon.$$

We now put

$$\tau(x) = \int_0^x \sum_{k=1}^{n} b_k \cos \frac{k\pi x}{l}\, dx = \sum_{k=1}^{n} \frac{b_k l}{k\pi} \sin \frac{k\pi x}{l},$$

upon which we have:

$$|y(x) - \tau(x)| = \left| \int_0^x \left[y'(x) - \sum_{k=1}^{n} b_k \cos \frac{k\pi x}{l} \right] dx \right| < \frac{l\varepsilon}{l+1} < \varepsilon,$$

$$|y'(x) - \tau'(x)| = \left| y'(x) - \sum_{k=1}^{n} b_k \cos \frac{k\pi x}{l} \right| < \varepsilon.$$

Since $\tau(x)$ has the form $\sum_{k=1}^{n} a_k \varphi_k$, the completeness is proved for the given case.

Now let $\varphi_k(x) = x^k(l - x)$. By Weierstrass's theorem we can find a polynomial $P_1(x)$ of degree n, such that $|y'(x) - P_1(x)| < \frac{\varepsilon}{2(l+1)}$.

Now put

$$P(x) = \int_0^x P_1(x)dx - \frac{x}{l} \int_0^l P_1(x)dx.$$

Then, since

$$\left| \int_0^l P_1(x)dx \right| = \left| \int_0^l [P_1(x) - y'(x)]dx \right| < \frac{\varepsilon l}{2(l+1)},$$

we have:

$$|y(x) - P(x)| \leqslant \int_0^x |y'(x) - P_1(x)|\, dx + \frac{x}{l}\left|\int_0^l P_1(x)dx\right| <$$

$$< x\frac{\varepsilon}{2(l+1)} + \frac{x}{l}\,\frac{\varepsilon l}{2(l+1)} < \varepsilon,$$

$$|y'(x) - P'(x)| \leqslant |y'(x) - P_1(x)| + \left|\frac{1}{l}\int_0^l P_1(x)\,dx\right| <$$

$$< \frac{\varepsilon}{2(l+1)} + \frac{\varepsilon}{2(l+1)} < \varepsilon.$$

Moreover, from the expression for $P(x)$ it is clear that this is a polynomial of degree $(n + 1)$ and that $P(0) = P(l) = 0$; it therefore has the form

$$P(x) = x(l - x)(a_1 + a_2 x + \ldots + a_n x^{n-1}) = \sum_{k=1}^{n} a_k \varphi_k.$$

Thus for this case too the possibility of satisfying both inequalities (18) is established. In conformity with No. 1, we can now conclude that the sequence of functions (17) will be minimal,

$$\lim_{n\to\infty} I(y_n) = I(y^*). \tag{19}$$

We shall now prove that for the problem of the minimum of integral (11), from the fact that the sequence is minimal [(19) is fulfilled] follows its uniform convergence to the solution:

$$\lim_{n\to\infty} y_n(x) = y^*(x). \tag{20}$$

For proof we shall utilize formula (7) § 1; applying it for $\bar{y} = y_n$ and $y = y^*$, we find

$$I(y_n) - I(y^*) = \int_0^l [p(y_n' - y^{*\prime})^2 + q(y_n - y^*)^2]dx.$$

Hence, since $p(x) > 0$ [see (11')], we find

$$\int_0^l \left[\frac{d}{dx}(y_n - y_n^*)\right]^2 dx \leqslant [I(y_n) - I(y^*)]\frac{1}{\min p}.$$

Then

$$|y_n(x) - y^*(x)| = \left| \int_0^x \frac{d}{dx}(y_n - y^*)dx \right| \leqslant$$

$$\leqslant \left\{ \int_0^x \left[\frac{d}{dx}(y_n - y^*) \right]^2 dx \int_0^x 1^2 \, dx \right\}^{\frac{1}{2}\,1)} \leqslant \frac{l^{\frac{1}{2}}}{\sqrt{\min p}} \sqrt{I(y_n) - I(y^*)}. \quad (21)$$

And, in view of (19), this inequality establishes the uniform convergence of $y_n(x)$ to $y^*(x)$. Summarizing the results obtained, we are led to the conclusion that if the functions defined by equalities (12) be chosen as φ_k, the process of Ritz and of Galerkin gives a sequence of functions $y_n(x)$ which is uniformly convergent to the solution of the problem of the minimum of integral (11) under conditions (10′).

We shall now linger somewhat on the case of other boundary conditions.

We saw above (see § 1, No. 3) that the solution of equation (9) under other boundary conditions leads to the solution of the problem of the minimum of integral (11) or of the complementary integral (21) of § 1 under free boundary conditions. In such a case, therefore, in order that the system be complete, it is necessary that there exist the possibility of approximating any function by combinations of the functions of the system, so that instead of functions (12) we could in this case choose the φ_k in the following manner

$$\varphi_0 = 1, \quad \varphi_k = \cos \frac{k\pi x}{l} \qquad (k = 1, 2, \ldots), \quad (22)$$

or, respectively,

$$\varphi_0 = 1, \quad \varphi_k = x^k \qquad (k = 1, 2, \ldots). \quad (23)$$

It should be remarked that since it is known beforehand what boundary conditions the solution satisfies, it is expedient, in order to reduce the number n of unknown parameters, to subject the combination under consideration, $\sum a_k \varphi_k$, to those conditions too; in this way one can succeed in expressing two of the constants a_k in terms of the remaining ones. Let us consider, for example, the problem of the minimum of integral (11) under free conditions. In this case [see § 1 (20)] the solution is the integral of equation (9), which satisfies the conditions:

$$y'(0) = y'(l) = 0. \quad (24)$$

In this case, instead of the system of functions (23) one can consider this system:

$$\varphi_0 = 1, \quad \varphi_1 = 2x^3 - 3lx^2, \quad \varphi_k = (l - x)^2 x^k \quad (k = 2, 3, \ldots). \quad (25)$$

It is immediately clear that the functions of system (25) satisfy conditions (24); it is also easy to convince oneself that any function that is

---

1) We have utilized here Buniakovskii's inequality:

$|\int_a^b f\varphi dx| \leqslant [\int_a^b f^2 dx \cdot \int_a^b \varphi^2 dx]^{\frac{1}{2}}$ for the case $f(x) = \frac{d}{dx}(y_n - y^*)$, $\varphi(x) = 1$.

continuous together with its derivative and that satisfies conditions (24) may be approximated by means of some linear combination of functions (25).

Analogously, in the case of the conditions

$$y'(0) + k_0 y(0) = 0, \quad y'(l) + k_1 y(l) = 0, \tag{26}$$

it is possible to choose this system of functions:

$$\varphi_0 = x^2\left(x - l - \frac{l}{2 + k_1 l}\right), \quad \varphi_1 = (l - x)^2\left(x + \frac{1}{2 - k_0 l}\right),$$

$$\varphi_k = x^k(l - x)^2 \qquad (k = 2, 3, \ldots). \tag{27}$$

Finally, analogous considerations can be advanced with respect to equations of the fourth order, too. Without dwelling on this, we shall now consider three concrete examples of the application of the methods of Ritz and of Galerkin.

Example 1. Let us consider the problem of finding the solution of the equation

$$y'' + y + x = 0, \quad y(0) = y(1) = 0,$$

the exact solution of which, as one can readily satisfy oneself, is $y = \dfrac{\sin x}{\sin 1} - x$.

The corresponding variational problem [cf. (9) and (11)] is the problem of the minimum of the integral

$$I = \int_0^1 (y'^2 - y^2 - 2xy)dx.$$

In accordance with the methods of Ritz and Galerkin [see (12), (13)], the approximate solution should be sought in the form

$$y_n = x(1 - x)(a_1 + a_2 x + \ldots + a_n x^{n-1}).$$

Let us first take $n = 1$. Substituting $y_1 = a_1\varphi_1 = a_1 x(1 - x)$ in system (16b) for y_n, we obtain for the determination of a_1 the equation

$$\int_0^1 L(y_1)\varphi_1\, dx = \int_0^1 [-2a_1 + a_1 x\, x(1 - x) + x]x(1 - x)dx = 0,$$

or [1])

$$-\tfrac{3}{10}a_1 + \tfrac{1}{12} = 0,$$

[1]) In performing the computations in this and in other cases it is convenient to use the formula

$$\int_0^l x^k(l - x)^m dx = \frac{k!\, m!}{(k + m + 1)!}\, l^{k+m+1},$$

and also

$$\int_0^l x^{2k}(l^2 - x^2)^m dx = \frac{1}{2}\, \frac{\dfrac{2k-1}{2}\cdot\dfrac{2k-3}{2}\cdots\dfrac{1}{2}m(m-1)\ldots 1}{\left(m + \dfrac{2k+1}{2}\right)\left(m + \dfrac{2k-1}{2}\right)\cdots\dfrac{1}{2}}\, l^{2k+2m+1}.$$

whence $a_1 = \frac{5}{18}$, and the required approximate solution will be

$$y_1 = \tfrac{5}{18}x(1 - x).$$

Now let $n = 2$. Then $\varphi_1 = x(1 - x)$, $\varphi_2 = x^2(1 - x)$, and y_2 is sought in the form

$$y_2 = x(1 - x)(a_1 + a_2x).$$

Substituting the expression for y_2 into system (16b) $[L(y) = y'' + y + x]$, we arrive at the equation

$$\int_0^1 L(y_2)\varphi_1 dx = \int_0^1 [-2a_1 + a_2(2 - 6x) + x(1 - x)(a_1 + a_2x) + x]x(1 - x)dx = 0,$$

$$\int_0^1 L(y_2)\varphi_2 dx = \int_0^1 [-2a_1 + a_2(2 - 6x) + x(1 - x)(a_1 + a_2x) + x]x^2(1 - x)dx = 0,$$

or, after computing the integrals:

$$\tfrac{3}{10}a_1 + \tfrac{3}{20}a_2 = \tfrac{1}{12}, \quad \tfrac{3}{20}a_1 + \tfrac{13}{105}a_2 = \tfrac{1}{20},$$

whence

$$a_1 = \tfrac{71}{369}, \; a_2 = \tfrac{7}{41} \text{ and } y_2 = x(1 - x)(\tfrac{71}{369} + \tfrac{7}{41}x).$$

Comparing the values of the exact solution $y = \dfrac{\sin x}{\sin 1} - x$ with y_1 and y_2 for $x = \frac{1}{4}, \frac{1}{2}, \frac{3}{4}$, we find

	y	y_1	y_2
$x = \frac{1}{4}$	0.044	0.052	0.044
$x = \frac{1}{2}$	0.070	0.069	0.069
$x = \frac{3}{4}$	0.060	0.052	0.060

From this it is seen that the error of the first approximation is of the order of 0.01; of the second, 0.001.

Example 2. The integration of the Bessel equation

$$x^2y'' + xy' + (x^2 - 1)y = 0$$

in the interval (1, 2), under the conditions $y(1) = 1$, $y(2) = 2$.

By means of the substitution $y = z + x$ we shall reduce the initial equation to the equation

$$x^2z'' + xz' + (x^2 - 1)z + x^3 = 0.$$

The new boundary conditions are $z(1) = z(2) = 0$.

Let us transform the equation to self-adjoint form:

$$xz'' + z' + \frac{x^2 - 1}{x} z + x^2 = 0.$$

(On the other hand, by employing the method of B. G. Galerkin, it would be possible to avoid this transformation.)

Let us put $\varphi_1 = (x - 1)(2 - x)$, $z_1 = a_1\varphi_1$. For determining a_1 we have the equation

$$\int_1^2 \left[-2a_1x + (3 - 2x)a_1 + \frac{x^2 - 1}{x}(x - 1)(2 - x)a_1 + x^2\right](x - 1)(2 - x)dx = 0.$$

On solving it, we obtain $a_1 = 0.8110$. Then

$$y_1 = 0.8110(x - 1)(2 - x) + x.$$

The exact solution is expressible in terms of the Bessel functions, having the form

$$y = 3.6072 I_1(x) + 0.75195 Y_1(x).$$

Let us compare y with y_1 for particular values of x:

x	y	y_1
1.3	1.4706	1.4703
1.5	1.7026	1.7027
1.8	1.9294	1.9297.

Thus in this example a very good result is obtained even in the first approximation.

Example 3. Let us consider the solution of the equation of the fourth order

$$[(x + 2l)y'']'' + qy - kx = 0$$

under the conditions

$$y = 0, \; y' = 0 \text{ for } x = l,$$

$$(x + 2l)y'' = 0, \; \frac{d}{dx}[(x + 2l)y''] = 0 \text{ for } x = 0.$$

For the constants k, l, q we shall have in view below the values $l = q = 1$, $k = 3$
The fundamental functions φ_1, φ_2 we choose in the form of polynomials, and such as to satisfy the boundary conditions given above (see p. 280). One can adopt these:

$$\varphi_1(x) = (x - l)^2(x^2 + 2lx + 3l^2), \quad \varphi_2(x) = (x - l)^3(3x^2 + 4lx + 3l^2).$$

First approximation. Let us seek u_1 in the form $u_1 = a_1^{(1)}\varphi_1$. We find $a_1^{(1)}$. Equation (16b) takes the form

$$\int_0^l L(u_1)\varphi_1 dx = \int_0^l \{a_1^{(1)}[(x + 2l)\varphi_1'']'' + q\varphi_1 - kx\}\varphi_1 dx = 0,$$

or, if we carry out the computation of the necessary integrals,

$$24a_1^{(1)}l + 57.6a_1^{(1)} + \tfrac{161}{315}qa_1^{(1)}l^4 + \tfrac{9}{5}qa_1^{(1)}l^2 + \frac{k}{3}l = 0,$$

and, if we take into consideration the numerical values of the constants l, q, k, we find

$$a_1^{(1)} = 0.011917, \; u_1 = a_1^{(1)}\varphi_1 = 0.011917(x - 1)^2(x^2 + 2x + 3).$$

Second approximation. We now seek the solution in the form

$$u_2 = a_1^{(2)}\varphi_1 + a_2^{(2)}\varphi_2.$$

For determining $a_1^{(2)}$ and $a_2^{(2)}$ we have the system of equations

$$\int_0^l L(u_2)\varphi_1 dx = 0, \quad \int_0^l L(u_2)\varphi_2 dx = 0,$$

which in developed form, if we take the numerical values of the constants into consideration, becomes

$$83.911 a_1^{(2)} - 67.213 a_2^{(2)} = 1$$

$$67.213 a_1^{(2)} - 91.882 a_2^{(2)} = 0.71430,$$

whence

$$a_1^{(2)} = 0.013743, \quad a_2^{(2)} = 0.002279,$$

$$u_2 = 0.013743(x-1)^2(x^2+2x+3) + 0.002279(x-1)^3(3x^2+4x+3).$$

Not having the possibility of comparing the approximations with the exact solution, to get an idea of the magnitude of the error we shall compare them with each other; for $x = \frac{1}{2}$, for instance, we have:

$$u_1(\tfrac{1}{2}) = 0.012662; \quad u_2(\tfrac{1}{2}) = 0.012964$$ [1].

3. The application of the methods of Ritz and B. G. Galerkin to the solution of partial differential equations of the second order. Let us consider the problem of solving the Poisson equation

$$L(u) = \frac{\partial^2 u}{\partial x^2} + \frac{\partial^2 u}{\partial y^2} - f = 0 \tag{28}$$

under the condition

$$u = \varphi(s) \text{ on } \Gamma. \tag{29}$$

One can always easily manage to replace condition (29) by the simpler one

$$u = 0 \text{ on } \Gamma; \tag{29'}$$

for this it is sufficient to construct any function $h(x, y)$ [2] that reduces to $\varphi(s)$ on Γ; then on introducing for u a new unknown function z defined by the formula $u = z + h(x, y)$, we shall have obtained for it a condition of form (29'); this function will also satisfy the Poisson equation, but with another free term. We shall therefore consider below the problem with just the condition (29').

This problem, as we have seen (§ 1 No. 2), is equivalent to the problem

---

[1] Numerous examples of the application of the method of B. G. Galerkin are given in a book of IA. A. Protusevich, [1].

[2] We shall remark that sometimes the function $h(x, y)$ has to be constructed so that it is given differently in different parts of the region.

of the minimum of the integral

$$I = \iint\limits_D \left[\left(\frac{\partial u}{\partial x}\right)^2 + \left(\frac{\partial u}{\partial y}\right)^2 + 2fu\right] dx\,dy \tag{30}$$

under the same condition (29′). In order to employ the methods of Ritz and B. G. Galerkin, one must have a system of functions $\varphi_k(x, y)$ that vanish on Γ, and which is complete in the region D, i.e., is such that by linear combinations of them one can approximate any function $u(x, y)$ that is continuous together with its partial derivatives and which satisfies condition (29′); more exactly speaking, for an $\varepsilon > 0$ and u one can indicate such a_k ($k = 1, 2, \ldots, n$) that the following inequalities are fulfilled:

$$\left.\begin{aligned} &\left|u - \sum_1^n a_k \varphi_k\right| < \varepsilon, \\ &\left|\frac{\partial u}{\partial x} - \sum_1^n a_k \frac{\partial \varphi_k}{\partial x}\right| < \varepsilon, \\ &\left|\frac{\partial u}{\partial y} - \sum_1^n a_k \frac{\partial \varphi_k}{\partial y}\right| < \varepsilon. \end{aligned}\right\} \tag{31}$$

These conditions can, however, be somewhat weakened; namely, the last two inequalities may be replaced by the following ones:

$$\left.\begin{aligned} &\iint\limits_D \left(\frac{\partial u}{\partial x} - \sum_1^n a_k \frac{\partial \varphi_k}{\partial x}\right)^2 dx\,dy < \varepsilon \\ \text{and}\quad &\iint\limits_D \left(\frac{\partial u}{\partial y} - \sum_1^n a_k \frac{\partial \varphi_k}{\partial y}\right)^2 dx\,dy < \varepsilon. \end{aligned}\right\} \tag{31′}$$

If such a system has been chosen, the nth approximation by Ritz's method is sought in the form

$$u_n = \sum_{k=1}^n a_k \varphi_k. \tag{32}$$

Substituting this expression for u in (30), we obtain:

$$\begin{aligned} I(u_n) &= \iint\limits_D \left[\left(\frac{\partial u_n}{\partial x}\right)^2 + \left(\frac{\partial u_n}{\partial y}\right)^2 + 2fu_n\right] dx\,dy = \\ &= \iint\limits_D \left[\left(\sum_1^n a_k \frac{\partial \varphi_k}{\partial x}\right)^2 + \left(\sum_1^n a_k \frac{\partial \varphi_k}{\partial y}\right)^2 + 2f \sum_1^n a_k \varphi_k\right] dx\,dy = \\ &= \sum_{k,s=1}^n \alpha_{k,s} a_k a_s + 2 \sum_{k=1}^n \beta_k a_k, \end{aligned} \tag{33}$$

where

$$\alpha_{k,s} = \alpha_{s,k} = \iint\limits_D \left(\frac{\partial \varphi_k}{\partial x}\frac{\partial \varphi_s}{\partial x} + \frac{\partial \varphi_k}{\partial y}\frac{\partial \varphi_s}{\partial y}\right) dx\,dy,$$

$$\beta_k = \iint\limits_D f\varphi_k\,dx\,dy. \tag{34}$$

Hence the system defining the a_k can be written in either of two forms:

$$\frac{1}{2}\frac{\partial I(u_n)}{\partial a_s} = \iint\limits_D \left(\frac{\partial u_n}{\partial x}\frac{\partial \varphi_s}{\partial x} + \frac{\partial u_n}{\partial y}\frac{\partial \varphi_s}{\partial y} + f\varphi_s\right) dx\,dy = 0$$

$$(s = 1, 2, \ldots, n), \tag{35a}$$

or

$$\sum_{k=1}^{n} \alpha_{k,s} a_k + \beta_s = 0 \qquad (s = 1, 2, \ldots, n). \tag{35}$$

Let us denote by $a_k^{(n)}$ the solution of system of equations (35) [the fact that its solution exists and is unique could be established just as was done above with respect to system (16) No. 2]. Then the nth approximation by Ritz's method will be

$$u_n(x, y) = \sum_{k=1}^{n} a_k^{(n)} \varphi_k(x, y). \tag{36}$$

Lastly, to system of equations (35) there may again be given yet another form, to wit: using formula (15) from § 1 for $u = u_n$ and $\eta = \varphi_s$, we find that

$$\iint\limits_D \left(\frac{\partial u_n}{\partial x}\cdot\frac{\partial \varphi_s}{\partial x} + \frac{\partial u_n}{\partial y}\cdot\frac{\partial \varphi_s}{\partial y}\right) dx\,dy = -\iint\limits_D \left(\frac{\partial^2 u_n}{\partial x^2} + \frac{\partial^2 u_n}{\partial y^2}\right)\varphi_s\,dx\,dy.$$

Applying this transformation to the left side of equation (35a) and changing the sign, we obtain the system of equations defining a_k in the form of the system of equations of the method of B. G. Galerkin:

$$\iint\limits_D \left(\frac{\partial^2 u_n}{\partial x^2} + \frac{\partial^2 u_n}{\partial y^2} - f\right)\varphi_s\,dx\,dy =$$

$$= \iint\limits_D L(u_n)\varphi_s\,dx\,dy = 0 \qquad (s = 1, 2, \ldots, n). \tag{35b}$$

Quite the same reasoning is also applicable to the problem of solving the general self-adjoint equation of the elliptic type [see § 1 (17b), p. 249],

$$L(u) = \frac{\partial}{\partial x}\left(a\frac{\partial u}{\partial y}\right) + \frac{\partial}{\partial y}\left(b\frac{\partial u}{\partial y}\right) - cu - f = 0 \tag{37}$$

under boundary conditions (29′). This problem is equivalent to the problem of the minimum of the integral

$$\iint_D \left[a\left(\frac{\partial u}{\partial x}\right)^2 + b\left(\frac{\partial u}{\partial y}\right)^2 + cu^2 + 2fu\right] dx\,dy. \tag{38}$$

In finding the approximate solution in form (32), for the determination of the constants a_k we are again led to system (35), where this time, however,

$$\left.\begin{aligned} \alpha_{k,s} = \alpha_{s,k} &= \iint_D \left[a\frac{\partial \varphi_k}{\partial x}\cdot\frac{\partial \varphi_s}{\partial x} + b\frac{\partial \varphi_k}{\partial y}\cdot\frac{\partial \varphi_s}{\partial y} + c\varphi_k\varphi_s\right] dx\,dy, \\ \beta_k &= \iint_D f\varphi_k\,dx\,dy. \end{aligned}\right\} \tag{39}$$

And in the given case system (35) has a definite solution provided the following inequalities are fulfilled:

$$a(x, y) > 0, \quad b(x, y) > 0, \quad c(x, y) \geqslant 0. \tag{40}$$

Lastly, we shall remark that by means of a suitable transformation, the system defining the a_k can again be brought into the form of the system of equations of B. G. Galerkin:

$$\iint_D \left[\frac{\partial}{\partial x}\left(a\frac{\partial u_n}{\partial x}\right) + \frac{\partial}{\partial y}\left(b\frac{\partial u_n}{\partial y}\right) - cu_n - f\right]\varphi_s\,dx\,dy = 0.$$

$$(s = 1, 2, \ldots, n). \tag{41}$$

As the functions $\varphi_k(x, y)$ for problems with two independent variables, there are usually chosen different combinations of trigonometric functions, or polynomials, satisfying the boundary conditions. The completeness of the system can usually be verified if one uses the following generalized Weierstrass theorem:

If the function $v(x, y)$ is continuous in a closed bounded region D, together with its partial derivatives $\frac{\partial v}{\partial x}$ and $\frac{\partial v}{\partial y}$, then, for an $\varepsilon > 0$, one can indicate a polynomial $P(x, y)$ such that in the region D there will be fulfilled the inequalities [1])

$$|v(x, y) - P(x, y)| < \varepsilon, \quad \left|\frac{\partial v}{\partial x} - \frac{\partial P}{\partial x}\right| < \varepsilon, \quad \left|\frac{\partial v}{\partial y} - \frac{\partial P}{\partial y}\right| < \varepsilon. \tag{42}$$

---

[1]) Vallée-Poussin, [1], vol. II, p. 120. Granted, there this theorem is proved for the case when the region D is a square. But for the regions ordinarily considered the function $v(x, y)$ can, while preserving its continuity and that of $\frac{\partial v}{\partial x}$ and $\frac{\partial v}{\partial y}$, be extended to some square containing the given region. See Fikhtengol'ts, G. M., [1], vol. I, Appendix.

In particular, by means of the Theorem the completeness of the following system of functions can be established. Let us assume that we have succeeded in defining, in the region, a function $\omega(x, y)$, continuous, having within D bounded and continuous derivatives $\frac{\partial \omega}{\partial x}$ and $\frac{\partial \omega}{\partial y}$, and satisfying the conditions

$$\omega(x, y) > 0 \text{ within } D, \quad \omega(x, y) = 0 \text{ on } \Gamma. \tag{43}$$

Then as the fundamental system of functions one can adopt a system consisting of the products of ω and various powers of x and y:

$$\varphi_0 = \omega, \quad \varphi_1 = \omega x, \quad \varphi_2 = \omega y, \quad \varphi_3 = \omega x^2, \quad \varphi_4 = \omega xy. \ \ldots \tag{44}$$

It is of course obvious that the functions $\varphi_k = 0$ on Γ.

Let us pass to the proof of the completeness of system of functions (44).

Let $u(x, y)$ be a function continuous in the region D, equal to 0 on Γ, whose derivatives are continuous within D, and such that the integral

$$\iint\limits_D \left[\left(\frac{\partial u}{\partial x}\right)^2 + \left(\frac{\partial u}{\partial y}\right)^2\right] dx\, dy$$

is finite. It is not hard to satisfy oneself, on the basis of the existence of the last integral, that the contour Γ and the lines within D on which $u(x, y) = 0$ can be included in an open region D_1 such that

$$\iint\limits_{D_1} \left[\left(\frac{\partial u}{\partial x}\right)^2 + \left(\frac{\partial u}{\partial y}\right)^2\right] dx\, dy < \varepsilon.$$

Denote by $\delta > 0$ the minimum of $|u(x, y)|$ in $D_2 = D - D_1$. It is easily shown that if $\delta_2 > 0$ be some number less than δ and ε, and δ_1 $(0 < \delta_1 < \delta_2)$ be chosen sufficiently small, then a function $M(u)$, continuous together with its derivative, can be constructed which satisfies the conditions: 1) $M(u) = 0$ for $|u| \leqslant \delta_1$, 2) $M(u) = u$ for $|u| \geqslant \delta_2$, 3) $|M'(u)| < 1 + \varepsilon$. Now set $\bar{u}(x, y) = M(u(x, y))$. Obviously by 2) in D_2 we shall have $\bar{u}(x, y) = u(x, y)$. Therefore:

$$\iint\limits_D \left[\frac{\partial u}{\partial x} - \frac{\partial \bar{u}}{\partial x}\right]^2 dx\, dy =$$

$$= \iint\limits_{D_1} (M'(u) - 1)^2 \left(\frac{\partial u}{\partial x}\right)^2 dx\, dy < (2 + \varepsilon)^2 \iint\limits_{D_1} \left(\frac{\partial u}{\partial x}\right)^2 dx\, dy < (2 + \varepsilon)^2 \varepsilon.$$

An analogous inequality is fulfilled for the derivatives with respect to the variable y, too. The function $\bar{u}(x, y)$, by 1), is equal to zero near the contour Γ; therefore, since $\omega(x, y) > 0$ within D, $v(x, y) = \dfrac{\bar{u}(x, y)}{\omega(x, y)}$ is continuous together with its partial derivatives in the region D. If we enclose the region D in a rectangle and put $v(x, y) = 0$ in the remaining part of it, these circumstances are not violated. In such a case, by the Theorem given above, we shall be able to choose a polynomial $P(x, y)$ such that inequalities (42) will be satisfied.

For the function $\pi(x, y) = \omega(x, y)P(x, y)$, we shall then have:

$$|\bar{u}-\pi| = |\omega|\,|v-P| < K\varepsilon,$$

$$\left|\frac{\partial \bar{u}}{\partial x} - \frac{\partial \pi}{\partial x}\right| = \left|\omega\frac{\partial v}{\partial x} + v\frac{\partial \omega}{\partial x} - \omega\frac{\partial P}{\partial x} - P\frac{\partial \omega}{\partial x}\right| < K_1\varepsilon,$$

$$\left|\frac{\partial u}{\partial y} - \frac{\partial \pi}{\partial y}\right| = \left|\omega\frac{\partial v}{\partial y} + v\frac{\partial \omega}{\partial y} - \omega\frac{\partial P}{\partial y} - P\frac{\partial \omega}{\partial y}\right| < K_2\varepsilon.$$

Hence

$$|u - \pi| \leqslant |u - \bar{u}| + |\bar{u} - \pi| < (2+\varepsilon)\delta_2 + K\varepsilon < K'\varepsilon,$$

$$\iint\limits_D \left(\frac{\partial u}{\partial x} - \frac{\partial \pi}{\partial x}\right)^2 dx\,dy \leqslant 2\iint\limits_D \left(\frac{\partial u}{\partial x} - \frac{\partial \bar{u}}{\partial x}\right)^2 dx\,dy +$$

$$+ 2\iint\limits_D \left(\frac{\partial \bar{u}}{\partial x} - \frac{\partial \pi}{\partial x}\right)^2 dx\,dy < \varepsilon(2+\varepsilon)^2 + 2K_1^2\varepsilon^2 \iint\limits_D dx\,dy < K''\varepsilon,$$

$$\iint\limits_D \left(\frac{\partial u}{\partial y} - \frac{\partial \pi}{\partial y}\right)^2 dx\,dy < K'''\varepsilon.$$

Thus an inequality of type (31′) for some combination of the functions of the system $\pi = \omega P$ has been established, and the completeness of the system, which secures the minimality of the sequence $u_n(x, y)$, has been proved [1]).

We note finally that until now we have considered only the possibility of approximating by algebraic polynomials; but by this same Theorem of Weierstrass, with the aid of a suitable change of the independent variables (for instance $x = \sin\varphi$, $y = \sin\theta$ or $x = r\cos\theta$, $y = r\sin\theta$), there can also be established the possibility of approximating by, as well as the completeness of, various other systems of functions.

We shall yet dwell a bit on the method of constructing the function $\omega(x, y)$. In some cases the function $\omega(x, y)$ is very simply defined. Thus, for example, for the rectangle $[-a, a; -b, b]$ one can adopt $\omega(x, y) = (x^2 - a^2)(y^2 - b^2)$; for a circle of radius R with center at the origin, $\omega(x, y) = R^2 - x^2 - y^2$. We shall indicate some general rules for the construction of the function $\omega(x, y)$.

1) If the contour bounding the region D admits of an equation of the form $F(x, y) = 0$, where F is a function continuous together with its partial derivatives, then one can adopt $\omega(x, y) = \pm F(x, y)$.

An example of this kind is the definition given above of the function ω for the circle.

2) For the case of a convex polygon, the equations of whose sides are

---

1) See L. V. Kantorovich, [12].

$a_1x + b_1y + c_1 = 0, \ldots, a_mx + b_my + c_m = 0$, one can adopt

$$\omega(x, y) = \pm (a_1x + b_1y + c_1) \ldots (a_mx + b_my + c_m).$$

An example of this kind is the case of the rectangle mentioned above. This same example of the combining of equations is suitable also in different cases of regions bounded by curved lines. So, for example, for a sector formed by circles of radii a and $\frac{a}{2}$ (see Fig. 18), one can adopt

$$\omega(x, y) = (a^2 - x^2 - y^2)(x^2 - ax + y^2).$$

3) The matter is considerably more complicated in the case of non-convex polygons. Here, when constructing $\omega(x, y)$, in order to take into consideration a re-entrant angle one must introduce moduli, and in

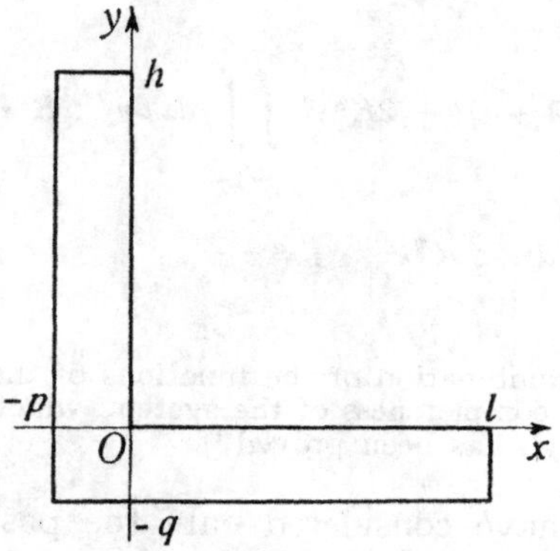

Fig. 18.

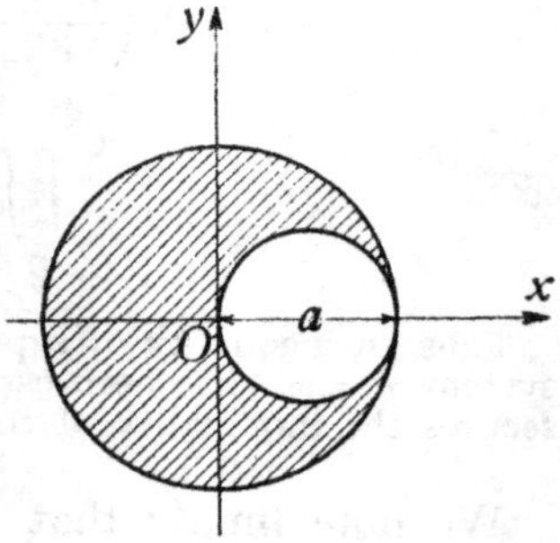

Fig. 19.

any case the function $\omega(x, y)$ has to be assigned piecewise in different parts of the region. We shall show a method of defining $\omega(x, y)$ in the example of the corner depicted in Fig. 19. There $\omega(x, y)$ can be defined thus:

$$\omega(x, y) = (|x| + |y| - x - y)(x + p)(y + q)(l - x)(h - y) =$$
$$= \begin{cases} -2y(x + p)(y + q)(l - x)(h - y) \text{ in the rectangle } [0, l; -q, 0] \\ -2(x + y)(x + p)(y + q)(l - x)(h - y) \text{ in the rectangle } [-p, 0; -q, 0] \\ -2x(x + p)(y + q)(l - x)(h - y) \text{ in the rectangle } [-p, 0; 0, h]. \end{cases}$$

We could also have adopted

$$\omega(x, y) = (x^2 + y^2 - x|x| - y|y|)(x + p)(y + q)(l - x)(h - y);$$

not only the continuity of ω, but also of its partial derivatives of the first order, would then have been secured. Here one can also proceed somewhat differently: assign the function $u_n(x, y)$ directly in the three parts of the region by different expressions:

$$u_n(x, y) = \begin{cases} (x+p)x(h-y)(a_1 + a_2x + a_3y + \ldots + a_ny^m) \text{ in } [-p, 0; 0, h] \\ (x+p)(y+q)(b_1 + b_2x + b_3y + \ldots + b_ny^m) \text{ in } [-p, 0; -q, 0] \\ (y+q)y(l-x)(c_1 + c_2x + c_3y + \ldots + c_ny^m) \text{ in } [0, l; -q, 0], \end{cases}$$

where $a_1, a_2, \ldots, a_n$; $b_1, b_2, \ldots, b_n$; $c_1, c_2, \ldots, c_n$ are parameters which must be connected by a condition of conjugacy [of the identical equality of two expressions $u_n(x, y)$] on the lines $x = 0$ and $y = 0$, viz.:

$$(x+p)xh(a_1 + a_2x + a_4x^2 + \ldots) = (x+p)q(b_1 + b_2x + b_4x^2 + \ldots)$$

$$p(y+q)(b_1 + b_3y + \ldots + b_ny^m) = (y+q)yl\,(c_1 + c_3y + \ldots + c_ny^m).$$

In the first approximations one can limit oneself to the conditions we have set forth; in further approximations it is expedient to add conditions set up analogously for the first and second partial derivatives [1]).

Finally, we shall yet mention a possibility of solving problems for this kind of contours by the methods of Ritz and Galerkin. Let us assume that the given regions can be separated into several parts (for example rectangles), for each of which we can solve the Dirichlet problem. Let there be given on the lines separating the regions an expression of definite form for the unknown function, containing only some undetermined parameters. Then for each partial region we can set up an expression for the unknown function. After this we can compute the value of the integral I for the entire region, but it will of course depend on the undetermined parameters introduced above. Equating to zero the derivatives of the last expression with respect to these parameters, we shall obtain the equations for their determination. After they have been found, the determination of the approximate solution will have been completed.

Until now we have concentrated on the Dirichlet problem. Let us now dwell on other forms of boundary conditions. Let us consider the Poisson equation (28) under homogeneous boundary conditions:

$$\frac{\partial u}{\partial n} + \sigma(s)u = 0 \text{ on } \Gamma. \tag{45}$$

This problem, as we have seen (§ 1, No. 3), reduces to the question of the minimum of

$$I = \iint\limits_D \left[\left(\frac{\partial u}{\partial x}\right)^2 + \left(\frac{\partial u}{\partial y}\right)^2 + 2fu\right] dx\,dy + \int\limits_\Gamma u^2\sigma(s)ds \tag{46}$$

under free boundary conditions.

---

1) On the subject of the solution of boundary-value problems for this kind of region, see L. V. Kantorovich, [12]. A solution of a numerical example is given in a note of S. M. Lekhtik, [1], found in the same issue of the journal.

For the application of Ritz's method in the given case one must have a system of functions $\{\varphi_k(x, y)\}$ satisfying only the condition of completeness, i.e., such that any function continuous together with its partial derivatives in the region D can be approximated, together with its derivatives, by a linear combination of the functions of the system. If such a system exists, then we again seek the nth approximation in the form

$$u_n(x, y) = \sum_{k=1}^{n} a_k \varphi_k(x, y). \tag{32}$$

Substituting expression (32) in integral (46) and equating to zero the derivatives with respect to the a_s, we shall obtain for determining the a_k, as above, the system of equations

$$\iint\limits_D \left[\frac{\partial u_n}{\partial x} \cdot \frac{\partial \varphi_s}{\partial x} + \frac{\partial u_n}{\partial y} \frac{\partial \varphi_s}{\partial y} + f\varphi_s\right] dx\, dy + \int\limits_\Gamma \sigma(s) u_n \varphi_s \, ds = 0. \tag{47a}$$

Transforming the first two summands by means of Green's formula [cf. § 1 (15) and also p. 251], we find [1])

$$\iint\limits_D \left[\frac{\partial u_n}{\partial x} \cdot \frac{\partial \varphi_s}{\partial x} + \frac{\partial u_n}{\partial y} \cdot \frac{\partial \varphi_s}{\partial y}\right] dx\, dy =$$

$$= - \iint\limits_D \left(\frac{\partial^2 u_n}{\partial x^2} + \frac{\partial^2 u_n}{\partial y^2}\right) \varphi_s \, dx\, dy + \int\limits_\Gamma \left(\frac{\partial u_n}{\partial x} \varphi_s \, dy - \frac{\partial u_n}{\partial y} \varphi_s \, dx\right) =$$

$$= - \iint\limits_D \Delta u_n \varphi_s \, dx\, dy + \int\limits_\Gamma \frac{\partial u_n}{\partial n} \varphi_s \, ds.$$

Substituting in (47a) the expression obtained, and changing the sign, we impart to the system for determining the a_k the following form:

$$\iint\limits_D L(u_n) \varphi_s \, dx\, dy - \int\limits_\Gamma \left(\frac{\partial u_n}{\partial n} + \sigma(s) u_n\right) \varphi_s \, ds =$$

$$= \iint\limits_D \left(\frac{\partial^2 u_n}{\partial x^2} + \frac{\partial^2 u_n}{\partial y^2} - f\right) \varphi_s \, dx\, dy - \int\limits_\Gamma \left(\frac{\partial u_n}{\partial n} + \sigma(s) u_n\right) \varphi_s \, ds = 0. \tag{47b}$$

In case $\sigma(s) > 0$, the system for the determination of the a_k will be determinate; if $\sigma(s) \equiv 0$, the function $u_n(x, y)$ will be defined only accurate to a constant.

We shall remark that, as we have mentioned in No. 1, although we

---

[1]) We call attention to the fact that the letters n and s, when standing as indices, denote the ordinal number; in other cases the same letters denote the length of the arc and the normal.

are bound by no conditions, it is expedient to choose the functions $\varphi_k(x, y)$ beforehand in such a way that boundary condition (45) will have been satisfied, since in this case, in the first place, the determination of the approximate solution is facilitated, as the same accuracy can be obtained with a smaller number of independent parameters; in the second place, thanks to the fact that for φ_k and consequently for u_n condition (45) is met, system (47b) for the determination of the a_k is simplified, taking the form:

$$\iint_D L(u_n)\varphi_s dx\,dy = 0 \qquad (s = 1, 2, \ldots, n). \tag{48}$$

In a free variation, products of powers of x and y may be taken as the functions φ_k, so that $u_n(x, y)$ will be a polynomial:

$$u_n(x, y) = a_1 + a_2x + a_3y + a_4x^2 + a_5xy + \ldots + a_ny^m.$$

This system will be complete in any region. As we indicated above, in case this is easily done, it is desirable to contrive that on the entire contour, or part of it, the function $u_n(x, y)$ satisfy condition (45). If we subject $u_n(x, y)$ to such a condition, part of the parameters a_k are thereby eliminated. We note, however, that in case this is hard to accomplish (non-convex polygons), we can leave the variation free. Thus for such cases the boundary problem (45) is perhaps simpler than the Dirichlet problem, and in such cases the solution of the Dirichlet problem may be replaced by the solution of the problem under condition (45) with constant σ, which afterwards should go to ∞.

We mention lastly that for some particular cases, by introducing immediately an infinite number of parameters, one can succeed in determining them by Ritz's method and in obtaining the exact solution of the problem in this manner.

We shall now present several examples of the application of the methods of Ritz and of B. G. Galerkin.

Example 1. Let us consider the problem of the torsion of a prismatic rod of rectangular section (see Chap. I, § 1, No. 1). This problem reduces to the solution of the Poisson equation $\Delta u = -2$ under the condition that $u = 0$ on the contour. We shall consider two ways of solving this problem by the method of B. G. Galerkin.

A. Let us take the sides of the rectangle as equal to $2a$ and $2b$, the origin being taken at the center of the rectangle. Choosing $\omega(x, y) = (a^2 - x^2)(b^2 - y^2)$, we seek the solution, in view of its obvious symmetry, in the following form:

$$u_0(x, y) = (a^2 - x^2)(b^2 - y^2)(A_1 + A_2x^2 + A_3y^2 + \ldots + A_nx^{2i}y^{2j}).$$

First let $n = 1$. We set up the equation for the determination of A_1 [cf. (35b): $f(x, y) = -2$], obtaining:

$$\int_{-a}^{a}\int_{-b}^{b} [-2(b^2 - y^2)A_1 - 2(a^2 - x^2)A_1 + 2](a^2 - x^2)(b^2 - y^2)dxdy = 0,$$

or

$$-\tfrac{128}{45}a^3b^3(a^2 + b^2)A_1 + \tfrac{32}{9}a^3b^3 = 0,$$

whence

$$A_1 = \frac{5}{4(a^2+b^2)}, \quad u_1(x, y) = \tfrac{5}{4}\,\frac{(a^2-x^2)(b^2-y^2)}{a^2+b^2}.$$

The approximate value for the moment proves to be

$$M = 2G\vartheta \int_{-a}^{a}\int_{-b}^{b} u_1(x, y)\,dx\,dy = 2G\vartheta\,\tfrac{5}{4}\cdot\tfrac{16}{9}\,\frac{a^3b^3}{a^2+b^2} = \tfrac{40}{9}G\vartheta\,\frac{a^3b^3}{a^2+b^2}.$$

In particular, in the case of a square section ($b = a$), we obtain $M = \frac{20}{9}G\vartheta a^4 = 0.1388(2a)^4G\vartheta$ rather than the exact $0.1406(2a)^4G\vartheta$.

In finding the second approximation let us limit ourselves to the case of a square section; then, on the strength of the symmetry with respect to the diagonal, we must obviously have $A_2 = A_3$, and consequently this approximation must be sought in the form

$$u_2(x, y) = (a^2 - x^2)(a^2 - y^2)(A_1 + A_2(x^2 + y^2)).$$

Without performing the computations in detail here, we shall indicate that for A_1 and A_2 there are obtained the values $A_1 = \frac{5}{8}\cdot\frac{259}{177}\cdot\frac{1}{a^2}$, $A_2 = \frac{5}{8}\cdot\frac{3}{2}\cdot\frac{35}{277}\cdot\frac{1}{a^4}$, and accordingly

$$u_2(x, y) = \frac{35}{16\cdot 277}\,\frac{1}{a^2}\,(a^2 - x^2)(a^2 - y^2)\left(74 + 15\,\frac{x^2+y^2}{a^2}\right).$$

From this we find for the moment the value $M_2 = 0.1404(2a)^4G\vartheta$. The error is about 0.15%.

For the values at the center we obtain $u_1(0, 0) = 0.625a^2$, $u_2(0, 0) = 0.584a^2$; exactly, however, $u(0, 0) = 0.586a^2$. A considerably greater error is obtained in the stresses that are expressible in terms of $\frac{\partial u}{\partial x}$ and $\frac{\partial u}{\partial y}$. Thus in the second approximation this error attains 4%.

B. Now we shall seek the solution in the form of an expression composed of trigonometric functions, and shall at once introduce an infinite set of parameters

$$u(x, y) = \sum_{n=1,3,5,\ldots}\ \sum_{m=1,3,5,\ldots} a_{m,n}\cos\frac{m\pi x}{2a}\cos\frac{n\pi y}{2b}.$$

The fact that the functions $\cos\frac{m\pi x}{2a}\cos\frac{n\pi y}{2b}$ vanish at the borders of the rectangle, and form a complete system relative to the functions satisfying the conditions of symmetry, is obvious. In view of the orthogonality of these functions to one another, it is easy to convince oneself that equation (35b), serving for the determination of the $a_{m,n}$, in the given case acquires the form

$$\int_{-a}^{a}\int_{-b}^{b}\Big[-\sum_k\sum_l a_{k,l}\frac{k^2\pi^2}{4a^2}\cos\frac{k\pi x}{2a}\cos\frac{l\pi y}{2b} - \sum_k\sum_l a_{k,l}\frac{l^2\pi^2}{4b^2}\cos\frac{k\pi x}{2a}\times$$
$$\times\cos\frac{l\pi y}{2b} + 2\Big]\cos\frac{m\pi x}{2a}\cos\frac{n\pi y}{2b}\,dx\,dy = 0,$$

or

$$-ab\left(\frac{m^2\pi^2}{4a^2} + \frac{n^2\pi^2}{4b^2}\right)a_{m,n} + 2\,\frac{16ab}{\pi^2 mn}(-1)^{\frac{m+n}{2}-1} = 0,$$

whence, definitively,

$$a_{m,n} = \frac{128a^2b^2(-1)^{\frac{m+n}{2}-1}}{\pi^4 mn(b^2m^2+a^2n^2)},$$

$$u(x, y) = \frac{128a^2b^2}{\pi^4} \sum_{n,m=1,3,5,\ldots}\sum (-1)^{\frac{m+n}{2}-1} \frac{\cos\frac{m\pi x}{2a}\cos\frac{n\pi y}{2b}}{mn(b^2m^2+a^2n^2)}.$$

The solution found coincides with that obtained by means of double series (see Chap. I, § 4 No. 2) for $f(x, y) = -2$.

Example 2. Let us find, for this same rectangle, the solution of the Neumann problem: $\frac{\partial u}{\partial n} = 0$ on the contour, for the Poisson equation

$$\Delta u = y.$$

We shall seek the solution in the form of a polynomial; from considerations of symmetry it is clear that it must be an odd function of y and an even function of x; we seek it, accordingly, in the form

$$u_n(x, y) = A_1 + A_2y + A_3x^2 + A_4y^3 + A_5x^2y + A_6x^4 + A_7x^4y + A_8x^2y^3 + A_9y^5 + \ldots$$

As we indicated, it is expedient to subject this solution to the boundary conditions (we limit ourselves to the coefficients already introduced):

$$\left.\frac{\partial u_n}{\partial x}\right|_{x=\pm a} = \pm (A_3 2a + A_5 2ay + A_6 4a^3 + A_7 4a^3y + A_8 2ay^3) = 0$$

$$\left.\frac{\partial u_n}{\partial y}\right|_{y=\pm b} = \pm (A_2 + A_4 3b^2 + A_5 x^2 + A_7 x^4 + A_8 x^2 3b^2 + A_9 5b^4) = 0.$$

Equating to zero the coefficients of different powers of x and y, we obtain a number of equations connecting the constants, which give: $A_5 = A_7 = A_8 = 0$, $A_3 = -2a^2A_6$, $A_2 = -3b^2A_4 - 5b^4A_9$. In addition, since the solution is defined accurate to a constant, we can adopt $A_1 = 0$. Accordingly we must seek the approximate solution in the form

$$\bar{u}(x, y) = A_6(x^4 - 2a^2x^2) + A_4(y^3 - 3b^2y) + A_9(y^5 - 5b^4y).$$

Since the functions we have chosen, $\varphi_1 = x^4 - 2a^2x^2$, $\varphi_2 = y^3 - 3b^2y$, $\varphi_3 = y^5 - 5b^4y$, satisfy the boundary conditions, for determining the constants A we can utilize equations (48) (we should adopt $f(x, y) = y$). As the result of the computations it turns out that $A_6 = A_9 = 0$, $A_4 = \frac{1}{6}$, which in the given case leads to the exact solution of the problem [1] $u = \frac{1}{6}(y^3 - 3b^2y)$.

4. Application to the biharmonic equation.

As we have seen, (§ 1, No. 4), to the problem of solving the biharmonic equation

$$L(u) = \Delta\Delta u - f(x, y) = 0 \tag{49}$$

---

[1]) With respect to other applications of the methods of Ritz and B. G. Galerkin, besides the monographs of L. S. Leĭbenzon and IA͡. A. Protusevich cited above, we refer the reader to the articles: Sokolov, B. A., [1], Stepanianţs, L. G. [1], Panov, D. IU͡., [5], where there is given an example of the application of the method of B. G. Galerkin to a non-linear problem.

under the conditions

$$u = 0, \quad \frac{\partial u}{\partial n} = 0 \text{ on } \Gamma, \tag{50}$$

there corresponds the variational problem of the minimum of the integral

$$I = \iint_D [(\Delta u)^2 - 2fu]dx\,dy = 0 \tag{51}$$

under these same boundary conditions. Ritz's method may also be employed for the solution of this problem. Let us have succeeded in finding a system of functions $\varphi_k(x, y)$ satisfying conditions (50) and such that any function $u(x, y)$ continuous together with its partial derivatives of the first and second order and satisfying conditions (50) can be approximated by some linear combination of the functions of the system, so that these inequalities [1]) are fulfilled:

$$\left.\begin{aligned} \left|u - \sum a_k\varphi_k\right| < \varepsilon, \quad \left|\frac{\partial u}{\partial x} - \sum a_k \frac{\partial \varphi_k}{\partial x}\right| < \varepsilon, \\ \left|\frac{\partial u}{\partial y} - \sum a_k \frac{\partial \varphi_k}{\partial y}\right| < \varepsilon, \quad \left|\frac{\partial^2 u}{\partial x^2} - \sum a_k \frac{\partial^2 \varphi_k}{\partial x^2}\right| < \varepsilon, \\ \left|\frac{\partial^2 u}{\partial x\,\partial y} - \sum a_k \frac{\partial^2 \varphi_k}{\partial x\,\partial y}\right| < \varepsilon, \quad \left|\frac{\partial^2 u}{\partial y^2} - \sum a_k \frac{\partial^2 \varphi_k}{\partial y^2}\right| < \varepsilon. \end{aligned}\right\} \tag{52}$$

We seek the approximate solution in the form $u_n = \sum_1^n a_k\varphi_k$; for the determination of the constants a_k, substituting the expression for u_n in place of u in (51) and equating to zero the derivative with respect to a_s, we obtain the system of equations:

$$\iint_D [\Delta u_n \Delta\varphi_s - f\varphi_s]dx\,dy =$$

$$= \iint_D \left[\sum_{k=1}^n a_k\left(\frac{\partial^2\varphi_k}{\partial x^2} + \frac{\partial^2\varphi_k}{\partial y^2}\right)\left(\frac{\partial^2\varphi_s}{\partial x^2} + \frac{\partial^2\varphi_s}{\partial y^2}\right) - f\varphi_s\right] dx\,dy = 0. \tag{53}$$

Applying to the integral here the transformation used earlier [§ 1, (30)], we can give system (53) another form:

$$\iint_D L(u_n)\varphi_s dx\,dy = \iint_D (\Delta\Delta u_n - f)\varphi_s dx\,dy = 0$$
$$(s = 1, 2, \ldots, n). \tag{53a}$$

And in the given case the system of equations is determinate; let its

---

[1]) As above [cf. (31′)], the last three conditions may be replaced by the weaker ones:

$$\iint_D \left[\frac{\partial^2 u}{\partial x^2} - \sum a_k \frac{\partial^2\varphi_k}{\partial x^2}\right]^2 dx\,dy < \varepsilon, \quad \iint_D \left[\frac{\partial^2 u}{\partial y^2} - \sum a_k \frac{\partial^2\varphi_k}{\partial y^2}\right]^2 dx\,dy < \varepsilon, \text{ etc.}$$

solution be $a_k^{(n)}$, then the nth approximation by Ritz's method is

$$u_n(x, y) = \sum_{k=1}^{n} a_k^{(n)} \varphi_k(x, y). \tag{54}$$

Since the system $\{\varphi_k\}$ has been assumed to be complete [see (52)] and the reasoning of No. 1 preserves its force in the given case, it can be affirmed that the sequence of functions (54) will be minimal, i.e., $I(u_n) \to I(u)$. Here, however, it can be affirmed that from the fact of minimality follows the uniform convergence of the sequence $u_n(x, y)$ to the exact solution of the problem; we shall not stop for a proof of this [1]).

Let us now consider the case of other boundary conditions. As we saw (§ 1, No. 4), to the solution of equation (49) under the (homogeneous) conditions

$$\left.\begin{aligned} M(u) &= \nu \Delta u + (1-\nu)\left(\frac{\partial^2 u}{\partial x^2} x_n^2 + 2\frac{\partial^2 u}{\partial x \partial y} x_n y_n + \frac{\partial^2 u}{\partial y^2} y_n^2\right) = 0 \text{ on } \Gamma, \\ P(u) &= \frac{\partial}{\partial n}\Delta u + (1-\nu)\frac{\partial}{\partial s}\left(\frac{\partial^2 u}{\partial x^2} x_n x_s + \frac{\partial^2 u}{\partial x \partial y}(x_n y_s + x_s y_n) + \right. \\ &\qquad\qquad \left. + \frac{\partial^2 u}{\partial y^2} y_n y_s\right) = 0 \text{ on } \Gamma \end{aligned}\right\} \tag{55}$$

there corresponds the problem of the minimum of the integral

$$I = \iint_D \left[(\Delta u)^2 - 2(1-\nu)\left(\frac{\partial^2 u}{\partial x^2} \cdot \frac{\partial^2 u}{\partial y^2} - \left(\frac{\partial^2 u}{\partial x \partial y}\right)^2\right) - 2fu\right] dx\, dy \tag{56}$$

under free boundary conditions. In this case one can proceed from any system of functions $\{\varphi_k(x, y)\}$ by means of which the approximation of an arbitrary function twice differentiable in D is realizable, (52). Seeking the solution in the form $u_n = \sum a_k \varphi_k$, we obtain for the determination of the constants a_k a system of equations of the following form:

$$\iint_D \left[\Delta u_n \Delta \varphi_s - (1-\nu)\left(\frac{\partial^2 u_n}{\partial x^2}\frac{\partial^2 \varphi_s}{\partial y^2} + \frac{\partial^2 u_n}{\partial y^2}\frac{\partial^2 \varphi_s}{\partial x^2} - 2\frac{\partial^2 u_n}{\partial x \partial y}\frac{\partial^2 \varphi_s}{\partial x \partial y}\right) - \right.$$
$$\left. - f\varphi_s\right] dx dy = 0 \qquad (s = 1, 2, \ldots, n). \tag{57}$$

On transforming these expressions by integrating twice by parts, we can bring this system of equations into the form [cf. § 1 (30) and (31)]:

$$\iint_D L(u_n)\varphi_s dx\, dy + \int_\Gamma M(u_n)\frac{\partial \varphi_s}{\partial n} ds -$$
$$- \int_\Gamma P(u_n)\varphi_s ds = 0 \qquad (s = 1, 2, \ldots, n). \tag{57a}$$

---

1) See, e.g., Trefftz, [1], pp. 149—164 or Shevchenko, K. N., [1].

It is expedient to choose the functions φ_k so that boundary conditions (55) are fulfilled for them and the possibility of the approximation by functions satisfying these conditions is preserved; in this case the number of parameters subject to determination is reduced, in the first place, and in the second place, system of equations (57a) is considerably simplified, thanks to the fact that for the φ_k, and consequently for the u_n too, conditions (55) are met, the curvilinear integrals vanish, and system (57a) takes the form of the system of equations of B. G. Galerkin's method:

$$\iint_D L(u_n)\varphi_s dx\,dy = 0 \qquad (s = 1, 2, \ldots, n). \tag{57b}$$

As regards the form in which the approximate solution should be sought (how to choose the φ_k), the suggestions given in the preceding No. still apply. In particular, in the case of conditions (50), one should seek $u_n(x, y)$ in the form

$$u_n(x, y) = \omega(x, y)(a_1 + a_2x + a_3y + \ldots + a_nx^iy^j),$$

where $\omega(x, y)$ is a positive function in D and satisfies conditions (50). Moreover, if the contour has the equation $F(x, y) = 0$, where F is a continuous function together with its partial derivatives, one can adopt

$$\omega(x, y) = F^2(x, y).$$

We now give examples of the solution of biharmonic problems by the methods of Ritz and of B. G. Galerkin [1]).

Example 1. Let us consider the problem of solving the equation

$$\Delta\Delta u = 0$$

for the rectangle $[-a, a; -b, b]$, under the boundary conditions

$$\frac{\partial^2 u}{\partial x\,\partial y} = 0, \quad \frac{\partial^2 u}{\partial y^2} = S\left(1 - \frac{y^2}{b^2}\right), \qquad \text{for } x = \pm a$$

$$\frac{\partial^2 u}{\partial x\,\partial y} = 0, \quad \frac{\partial^2 u}{\partial x^2} = 0, \qquad \text{for } y = \pm b.$$

To this problem reduces that of the extension of a rectangular plate under defined tensile forces [2]).

Let us reduce the boundary conditions to homogeneous ones at the outset. The function $u_0 = \frac{1}{2}Sy^2\left(1 - \frac{1}{6}\frac{y^2}{b^2}\right)$ obviously satisfies the given boundary conditions (it would be easily found, taken in the form of a polynomial in y with undetermined coefficients). Then setting $u = u_0 + \tilde{u}$, we shall obtain for $\tilde{u}$ the

---

[1]) A valuable variational method of reducing the biharmonic problem to the solution of Poisson equations is given in a recent work of Z. K͡H. Rafal'son, [1].

[2]) See S. Timoshenko, [1], p. 171.

equation

$$\Delta\Delta\bar{u} = \frac{2S}{b^2},$$

under the conditions

$$\frac{\partial^2\bar{u}}{\partial x\,\partial y} = 0, \quad \frac{\partial^2\bar{u}}{\partial y^2} = 0, \qquad \text{for } x = \pm a$$

$$\frac{\partial^2\bar{u}}{\partial x\,\partial y} = 0, \quad \frac{\partial^2\bar{u}}{\partial x^2} = 0, \qquad \text{for } y = \pm b.$$

Those conditions, as one is readily convinced, will be satisfied if these conditions are met:

$$\bar{u} = 0, \quad \frac{\partial\bar{u}}{\partial x} = 0 \text{ for } x = \pm a, \quad \bar{u} = 0, \quad \frac{\partial\bar{u}}{\partial y} = 0 \text{ for } y = \pm b,$$

and the last conditions are none other than conditions (50), for the particular case of the rectangle [1]). In accordance with what has been said, the problem then reduces to the question of the minimum of the integral [2])

$$I = \iint\limits_D \left[(\Delta\bar{u})^2 - \frac{4S}{b^2}\,\bar{u}\right] dx\,dy.$$

Taking the symmetry into account, we seek the approximate solution in the form

$$\bar{u}_n(x, y) = (x^2 - a^2)^2(y^2 - b^2)^2[a_1 + a_2x^2 + a_3y^2 + \ldots].$$

Limiting ourselves to the single coefficient a_1, we obtain for its determination the equation

$$a_1\left(\tfrac{64}{7} + \tfrac{256}{49}\frac{b^2}{a^2} + \tfrac{64}{7}\frac{b^4}{a^4}\right) = \frac{S}{a^4b^2}.$$

In particular, for a square plate this gives

$$a_1 = 0.04253\frac{S}{a^6}, \quad \bar{u}_1(x, y) = 0.04253\frac{S}{a^6}(x^2 - a^2)^2(y^2 - b^2)^2.$$

With $n = 3$, for the determination of the coefficients we obtain the equations:

$$a_1\left(\tfrac{64}{7} + \tfrac{256}{49}\frac{b^2}{a^2} + \tfrac{64}{7}\frac{b^4}{a^4}\right) + a_2a^2\left(\tfrac{64}{77} + \tfrac{64}{49}\frac{b^4}{a^4}\right) + a_3a^2\left(\tfrac{64}{49}\frac{b^2}{a^2} + \tfrac{64}{77}\frac{b^6}{a^6}\right) = \frac{S}{a^4b^2},$$

$$a_1\left(\tfrac{64}{11} + \tfrac{64}{7}\frac{b^4}{a^4}\right) + a_2a^2\left(\tfrac{192}{143} + \tfrac{256}{77}\frac{b^2}{a^2} + \tfrac{192}{7}\frac{b^4}{a^4}\right) + a_3a^2\left(\tfrac{64}{77}\frac{b^2}{a^2} + \tfrac{64}{77}\frac{b^6}{a^6}\right) = \frac{S}{a^4b^2},$$

$$a_1\left(\tfrac{64}{7} + \tfrac{64}{11}\frac{b^4}{a^4}\right) + a_2a^2\left(\tfrac{64}{77} + \tfrac{64}{77}\frac{b^4}{a^4}\right) + a_3a^2\left(\tfrac{192}{7}\frac{b^2}{a^2} + \tfrac{256}{77}\frac{b^4}{a^4} + \tfrac{192}{143}\frac{b^6}{a^6}\right) = \frac{S}{a^4b^2}.$$

---

[1]) We shall remark that by altering the boundary conditions in this manner, we are discarding a number of solutions, but since it is sufficient to find only one solution, this is admissible.

[2]) One can, however — as Timoshenko does — utilize integral (35′) of § 1; the equations obtained for the determination of the a_k, as their form (57b) shows, would be the same.

For a square plate ($b = a$) these equations give

$$a_1 = 0.04040\frac{S}{a^6}, \quad a_2 = a_3 = 0.01174\frac{S}{a^8},$$

and finally (for $b = a$):

$$u_3 = u_0 + \tilde{u}_3 = \tfrac{1}{2}Sy^2\left(1 - \tfrac{1}{6}\frac{y^2}{a^2}\right) +$$

$$+ \frac{S}{a^6}(x^2 - a^2)^2(y^2 - a^2)^2\left[0.04040 + 0.01174\frac{x^2 + y^2}{a^2}\right].$$

Example 2. We borrow our second example from the fundamental work of W. Ritz which we mentioned earlier. Let the discussion concern the determination of the equilibrium position of a homogeneous isotropic plate having the form of a rectangle with sides a and b and with built-in edges in the horizontal direction, under the action of external transverse forces.

Let us denote by $f(x, y)$ the density of the external forces, reckoned per unit area of the plate.

We shall arrange the axes of coordinates as is indicated in Fig. 20. The problem reduces to finding the solution of the equation

$$\Delta\Delta u = f(x, y),$$

satisfying the boundary conditions

$$u = 0 \text{ and } \frac{\partial u}{\partial n} = 0 \text{ on the contour } \Gamma, \tag{*}$$

or, if we take the form of the plate into account,

$$\left.\begin{aligned} u = 0, \quad \frac{\partial u}{\partial x} = 0 \text{ for } x = 0, \ y = a, \\ u = 0, \quad \frac{\partial u}{\partial y} = 0 \text{ for } x = 0, \ y = b. \end{aligned}\right\} \tag{**}$$

The corresponding variational problem has the form:

$$I(u) = \int_0^a\int_0^b [(\Delta u)^2 - 2f(x, y)u]dxdy = \min$$

under conditions (**).

Let us now select the fundamental functions φ. The form of the plate, and the physical analogy between the deformation of a plate with built-in edges and the deformation of a built-in beam, naturally lead us to the idea of using for the construction of such functions the proper functions of the equation of a built-in beam:

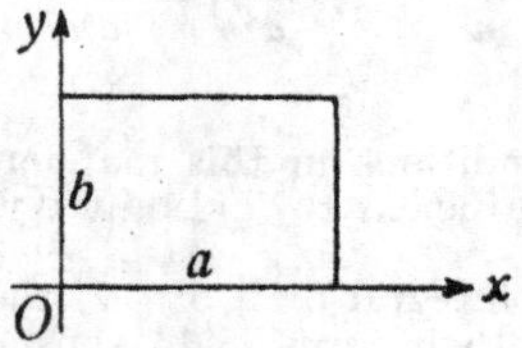

Fig. 20.

$$\frac{d^4u}{dx^4} - \lambda u = 0.$$

Let the beam have the length a and be built-in

at both ends:

$$u = 0 \text{ and } \frac{du}{dx} = 0 \text{ for } x = 0 \text{ and } x = a.$$

The proper numbers are equal to λ_n^4, where λ_n is a solution of the following equation:

$$\cos \lambda_n a \operatorname{ch} \lambda_n a = 1.$$

Rather than λ_n, we shall henceforth often use the quantity $\mu_n = \lambda_n a$, which is independent of the length of the beam. For large n, μ_n differs very little from $(n+\frac{1}{2})\pi$. We shall cite here several of the first values of μ, which will be useful to us in the future; they have been found accurate to four places of decimals.

$$\mu_1 = \frac{3\pi}{2} + 0.01765 = 4.7300, \qquad \mu_2 = \frac{5\pi}{2} - 0.00078 = 7.8532,$$

$$\mu_3 = \frac{7\pi}{2} + 0.00003 = 10.9956, \quad \mu_4 = \frac{9\pi}{2} = 14.1372,$$

$$\mu_5 = \frac{11\pi}{2} = 17.2787, \qquad \mu_6 = \frac{13\pi}{2} = 20.4203.$$

The proper functions satisfy the equation

$$u^{\mathrm{IV}} = \lambda_n^4 u.$$

If we normalize them in the usual manner:

$$\int_0^a u^2 dx = 1,$$

they will have the following form:

$$\xi_n(x) = \frac{(\cos \lambda_n x - \operatorname{ch} \lambda_n x)(\sin \mu_n - \operatorname{sh} \mu_n) - (\sin \lambda_n x - \operatorname{sh} \lambda_n x)(\cos \mu_n - \operatorname{ch} \mu_n)}{\sqrt{a}(\sin \mu_n - \operatorname{sh} \mu_n)}$$ [1].

Let us denote by $\eta_n(x)$ the proper functions of a beam of length b, which are obtained from ξ_n by replacing a by b and by replacing the proper numbers λ_n^4 by $\left(\frac{b}{a}\right)^4 \lambda_n^4 = \lambda_n'^4$. We put:

$$\varphi_{kj} = \xi_k(x)\eta_j(y),$$

$$U_{n,m} = \sum_{k=1}^{n} \sum_{j=1}^{m} a_{kj}\xi_k(x)\eta_j(y).$$

Equations (53a) will in the given case take the form

$$\int_0^a \int_0^a [\Delta\Delta U_{n,m} - f]\xi_r\eta_s dx\, dy = 0. \qquad (***)$$

But since $\xi_k^{\mathrm{IV}} = \lambda_k^4 \xi_k$, $\eta_j^{\mathrm{IV}} = \lambda_j'^4 \eta_j$, we have

$$\Delta\Delta U_{n,m} = \sum_{k=1}^{n} \sum_{j=1}^{m} a_{kj}[(\lambda_k^4 + \lambda_j'^4)\xi_k\eta_j + 2\xi_k''\eta_j'']. \qquad (****)$$

[1]) Another, more convenient form for these functions is given in a note by V. N. Faddeeva, [2].

We shall use the symbols

$$\alpha_k = \frac{\cos \mu_k - \text{ch } \mu_k}{\sin \mu_k - \text{sh } \mu_k}, \quad \sigma_{kj} = \int_0^a \xi_k \xi_j'' dx.$$

We cite the values of α_k:

$$\alpha_1 = 0.98250, \ \alpha_2 = 1.00078, \ \alpha_3 = 0.99997, \ \alpha_4 = 1.00000.$$

On taking into consideration the expression for ξ_k and integrating twice by parts, we have

$$\sigma_{kj} = \sigma_{jk} = \frac{1}{\lambda_k^4 - \lambda_j^4} [\xi_k''' \xi_j'' - \xi_k'' \xi_j''']_0^a = \frac{1}{a} \frac{4\lambda_k^2\lambda_j^2}{\lambda_k^4 - \lambda_j^4} [(-1)^{k+j} + 1](\lambda_k \alpha_k - \lambda_j \alpha_j) =$$

$$= \frac{1}{a^2} \frac{4\mu_k^2\mu_j^2}{\mu_k^4 - \mu_j^4} [(-1)^{k+j} + 1](\mu_k \alpha_k - \mu_j \alpha_j),$$

$$\sigma_{kk} = \int_0^a \xi_k \xi_k'' \, dx = -\lambda_k^2 \alpha_k^2 + 2\lambda_k \frac{\alpha_k}{a} = \frac{1}{a^2} [2\mu_k \alpha_k - \mu_k^2 \alpha_k^2].$$

For $k > 2$, $j > 2$, accurate to 0.0001:

$$\sigma_{kj} = \frac{1}{a^2} \frac{4\pi(k + \frac{1}{2})^2 (j + \frac{1}{2})^2 (k - j)[(-1)^{k+j} + 1]}{(k + \frac{1}{2})^2 - (j + \frac{1}{2})^2},$$

$$\sigma_{kk} = \frac{1}{a^2} [(2k + 1)\pi - (k + \tfrac{1}{2})^2 \pi^2].$$

System (****), if we use the orthogonality of the proper functions, may be rewritten in the form:

$$a_{rs}(\lambda_r^4 + \lambda_s'^4 + 2\sigma_{rr}\sigma_{ss}) + 2 \sum_{k=1}^{n}{}' \sum_{j=1}^{m}{}' a_{kj} \sigma_{kr} \sigma_{js} = f_{rs}, \qquad (\text{*****})$$

$$f_{rs} = \int_0^a \int_0^b f \xi_r \eta_s dx dy, \qquad (r = 1, 2, \ldots, n; \ s = 1, 2, \ldots, m).$$

The prime sign $'$ on the sum shows that in making the summation the summand containing a_{rs} must be omitted.

The further computations will depend on the form of the function f; below we shall assume the plate to be square, with sides a, and the external force acting to be distributed uniformly over the entire surface of the plate: $f(x, y) = C$. Since

$$\int_0^a \xi_k \, dx = \begin{cases} -\dfrac{4\alpha_k}{\sqrt{a\lambda_k}} & \text{for } k \text{ odd,} \\ 0 & \text{for } k \text{ even,} \end{cases}$$

we have, in view of the fact that $\eta_j(y) = \xi_j(y)$,

$$f_{rs} = \int_0^a \int_0^a C \xi_r(x) \xi_s(y) dx \, dy =$$

$$= \begin{cases} C \dfrac{16\alpha_r\alpha_s}{a\lambda_r\lambda_s} = C \dfrac{16a\alpha_r\alpha_s}{\mu_r\mu_s} & \text{if } r \text{ and } s \text{ are odd;} \\ 0, \text{ if one or both of the numbers } r \text{ or } s \text{ is even.} \end{cases}$$

On the strength of the symmetry, the axes x and y must be of equal importance; therefore $a_{kj} = a_{jk}$ and system (${*}^{**}_{**}$) may be simplified; one can, namely, discard all the equations for which $r > s$. Furthermore, the first proper function of the beam does not have a zero in the interval and is symmetric with respect to the middle of the beam; the second will be antisymmetric with respect to the middle, etc. On the strength of the symmetry of our problem, all proper functions ξ and η with even indices must obviously be absent from the solution, and therefore only those such a_{kj} can be different from zero for which both indices are odd.

For notational simplicity let $l = 8Ca^5 10^{-4}$. Let us first put $n = m = 1$. Then the system for the determination of the a_{kj} degenerates into one equation:

$$651.8a_{11} = 431.5l,$$

$$a_{11} = 0.6620l, \quad u_{11} = 0.6620l\ \xi_1(x)\xi_1(y).$$

For $n = m = 3$ we obtain the system

$$\begin{aligned} 651.8a_{11} - 239.2a_{13} + 94.7a_{33} &= 431.5l, \\ -119.6a_{11} + 8867a_{13} - 961a_{33} &= 188.5l, \\ 94.6a_{11} - 1922a_{13} + 24390a_{33} &= 82.7l. \end{aligned}$$

Rather than solving this system by means of determinants, we prefer to find its solution by the method of successive approximations, in view of the fact that the coefficients of its diagonal terms considerably exceed the other coefficients.

Solving the system for a_{11}, a_{13}, a_{33}, we have:

$$\begin{aligned} a_{11} &= 0.6620l + 0.3670a_{13} - 0.1453a_{33} \\ a_{13} &= 0.0213l + 0.0135a_{11} + 0.1084a_{33} \\ a_{33} &= 0.0034l - 0.0039a_{11} + 0.0792a_{13} \end{aligned}$$

As regards the initial values we could, for instance, adopt the free terms on the right sides for them. But since we know a first approximation for a_{11} it will be preferable to begin the computations with $a_{11} = 0.6620l$, $a_{13} = a_{33} = 0$. As the result we obtain the table

Coefficient	Initial Values	First Approx.	Second Approx.	Third Approx.
a_{11}	$0.6620\ l$	$0.6620\ l$	$0.6730\ l$	$0.6727\ l$
a_{13}	0	$0.0302\ l$	$0.0303\ l$	$0.0307\ l$
a_{33}	0	$0.0008\ l$	$0.0031\ l$	$0.0031\ l$

$$U_{33} = 0.6727l\xi_1(x)\xi_1(y) + 0.0307l[\xi_1(x)\xi_3(y) + \xi_3(x)\xi_1(y)] + 0.0031l\xi_3(x)\xi_3(y).$$

The advantage of the method of successive approximations becomes still clearer if we turn to the subsequent approximations. Thus for $n = m = 5$ we have:

$$\begin{aligned} 651.8a_{11} - 239.2a_{13} + 94.7a_{33} - 188a_{15} + 148a_{35} + 58a_{55} &= 431.5l, \\ -119.6a_{11} + 8867a_{13} - 961a_{33} - 226a_{15} - 515a_{35} + 186a_{55} &= 188.5l, \\ 94.6a_{11} - 1922a_{13} + 24390a_{33} + 474a_{15} - 4820a_{35} + 592a_{55} &= 82.7l, \\ -93.8a_{11} - 226a_{13} + 237a_{33} + 48100a_{15} - 2370a_{35} - 2015a_{55} &= 120.0l, \\ 74.3a_{11} - 515a_{13} - 2410a_{33} - 2370a_{15} + 78600a_{35} - 6420a_{55} &= 52.5l, \\ 58a_{11} + 372a_{13} + 592a_{33} - 4030a_{15} - 12840a_{35} + 158500a_{55} &= 33.5l. \end{aligned}$$

Since the system contains many terms, we prefer to apply the successive approximations directly to it without solving it with respect to the diagonal terms. As the initial values for a_{11}, a_{13}, a_{33} we shall adopt the values found for $n = m = 3$, and as a_{15}, a_{35}, a_{55}, zeros. Substituting in the last three equations of the system the first approximations found for the coefficients:

$$a_{11} = 0.6727l, \; a_{13} = 0.0307l, \; a_{33} = 0.0031l,$$

we obtain:

$$a_{15} = 0.0039l, \; a_{35} = 0.0003l, \; a_{55} = 0.0001l.$$

Substituting the values found for a_{ik} in all the terms of the system except the diagonal ones, we obtain second approximations for the a_{ik}, which will already be sufficiently exact, further computations yielding no substantial modification of them:

$$a_{11} = 0.6740l, \; a_{13} = 0.0308l, \; a_{33} = 0.0032l,$$

$$a_{15} = 0.0040l, \; a_{35} = 0.0004l, \; a_{55} = 0.0001l.$$

It goes without saying that in computing the successive approximations there is no need of finding the approximations themselves; it is enough to limit oneself to the simpler problem of finding the successive corrections to them.

5. Application to the determination of the proper values and functions. The reasoning of B. G. Galerkin, which has been set forth above (No. 1), leads most simply to a way of solving this problem. For a beginning let us consider the problem of the determination of the proper numbers and functions of the ordinary equation:

$$L(y) = \frac{d}{dx}(p(x)y') - qy + \lambda y = 0, \tag{58}$$

under the boundary conditions

$$y(0) = y(l) = 0. \tag{59}$$

Seeking the solution in the form of a sum

$$y_n = \sum_{k=1}^{n} a_k \varphi_k(x), \tag{60}$$

where φ is a complete system of functions that satisfy conditions (59), we need contrive, instead of the identical satisfaction of equation (58), only the orthogonality of the left side of it to the functions $\varphi_1, \ldots, \varphi_n$. This leads us to the system of equations

$$\int_0^l L(y_n)\varphi_j dx = \int_0^l \left[\frac{d}{dx}(p(x)y_n') - qy_n + \lambda y_n\right]\varphi_j dx = 0$$

$$(j = 1, 2, \ldots, n). \tag{61}$$

This system can be written in the form

$$\sum_{i=1}^{n} (\alpha_{i,j} + \lambda\gamma_{i,j})a_i = 0 \qquad (j = 1, 2, \ldots, n), \tag{61a}$$

where

$$\alpha_{i,j} = \int_0^l \left[\frac{d}{dx}(p(x)\varphi_i') - q\varphi_i\right]\varphi_j\,dx, \qquad \gamma_{i,j} = \int_0^l \varphi_i\varphi_j\,dx. \tag{62}$$

The system of equations obtained is a homogeneous system of n equations in n unknowns; it has a non-trivial solution only in case the determinant of this system equals zero:

$$\Delta = \begin{vmatrix} \alpha_{1,1} + \lambda\gamma_{1,1}; & \ldots; & \alpha_{n,1} + \lambda\gamma_{n,1} \\ \cdot & \cdot\;\cdot\;\cdot & \cdot \\ \cdot & \cdot\;\cdot\;\cdot & \cdot \\ \alpha_{1,n} + \lambda\gamma_{1,n}; & \ldots; & \alpha_{n,n} + \lambda\gamma_{n,n} \end{vmatrix} = 0. \tag{63}$$

Equation (63) is of the nth degree in λ; it will give, generally speaking[1]), n roots $\lambda_1^{(n)}, \lambda_2^{(n)}, \ldots, \lambda_n^{(n)}$. For each $\lambda = \lambda_m^{(n)}$ system of equations (61) will have a non-zero solution $a_i^{(m)}$, which will give us the function which corresponds to $\lambda_m^{(n)}$, $\varphi_m^{(n)}(x) = \sum a_i^{(m)}\varphi_i(x)$ (it is understood that these functions are determinate only accurate to a factor).

The problem of determining the proper values of equation (58) consists in finding those values of λ for which a non-zero solution of the boundary-value problem (59) exists. Since system (61) has arisen as an approximation to equation (58), the values $\lambda_1^{(n)}, \lambda_2^{(n)}, \ldots$, found above, for which it has a non-trivial solution, will be approximations to the first proper values $\lambda_1, \lambda_2, \ldots$ of equation (58), and the functions $\varphi_1^{(n)}(x)$, $\varphi_2^{(n)}(x), \ldots$ will be approximations to the respective proper functions.

In exactly the same fashion the corresponding equations for other problems can be set up. For instance, for the equation

$$\frac{\partial^2 u}{\partial x^2} + \frac{\partial^2 u}{\partial y^2} + \lambda u = 0 \tag{64}$$

under the boundary condition

$$u = 0 \text{ on } \Gamma, \tag{65}$$

the corresponding system will be

$$\iint_D \left[\frac{\partial^2 u_n}{\partial x^2} + \frac{\partial^2 u_n}{\partial y^2} + \lambda u_n\right]\varphi_j\,dx\,dy = 0 \qquad (j = 1, 2, \ldots, n), \tag{66}$$

where $u_n = \sum_1^n a_k\varphi_k$, $\{\varphi_k\}$ being a complete system of functions satisfying boundary condition (65). The proper values are found approximately by equating to zero the determinant of system (66). System (66) also

---

[1]) On the topic of methods for the numerical solution of equation (63), see the footnote on page 103.

preserves its form for the case when condition (65) is replaced by the condition

$$\frac{\partial u}{\partial n} + \sigma(s)u = 0, \tag{67}$$

but it must again be assumed that this same condition is satisfied for the φ_k too. An analogous system, finally, may be set up even for the determination of the proper numbers and functions in the case of the biharmonic problem, to wit, considering the equation

$$\Delta\Delta u - \lambda u = 0, \tag{68}$$

under the conditions

$$u = 0, \quad \frac{\partial u}{\partial n} = 0 \text{ on } \Gamma, \tag{69}$$

then choosing φ_k satisfying conditions (69) and setting $u_n = \sum a_k\varphi_k$, we obtain a system of the form

$$\iint_D (\Delta\Delta u_n - \lambda u_n)\varphi_j dxdy = 0 \qquad (j = 1, 2, \ldots, n). \tag{70}$$

One can also arrive at these same systems of equations, (61), (66) and (70), from the consideration of variational problems. We shall do this for the case of equation (64) under condition (65), which we considered in § 1 No. 5. As we saw there, the least proper number of the given problem is the minimum of the integral

$$\iint_D \left[\left(\frac{\partial u}{\partial x}\right)^2 + \left(\frac{\partial u}{\partial y}\right)^2\right] dx\,dy, \tag{71}$$

under condition (65) and

$$\iint_D u^2\,dx\,dy = 1, \tag{72}$$

and the function furnishing this extremum is the corresponding proper function. Now let $\{\varphi_k\}$ be a complete system of functions that satisfy condition (65); we shall seek the approximate solution of this extremal problem in the form $u_n = \sum a_k\varphi_k$. Replacing u by u_n in (71) and (72), we find that the problem reduces to the determination of the minimum of the expression

$$\iint_D \left[\left(\frac{\partial u_n}{\partial x}\right)^2 + \left(\frac{\partial u_n}{\partial y}\right)^2\right] dx\,dy, \tag{73}$$

which is a quadratic form in a_k under the condition

$$\iint_D u_n^2\,dx\,dy = 1. \tag{74}$$

By Lagrange's rule for the determination of relative extrema, the problem may be replaced by the determination of the minimum of the expression

$$\iint\limits_D \left[\left(\frac{\partial u_n}{\partial x}\right)^2 + \left(\frac{\partial u_n}{\partial y}\right)^2\right] dx\,dy - \lambda \iint\limits_D u_n^2\,dx\,dy, \tag{75}$$

where λ is an undetermined factor. Equating to zero the derivatives of this expression with respect to a_j, we obtain:

$$\iint\limits_D \left[\frac{\partial u_n}{\partial x}\frac{\partial \varphi_j}{\partial x} + \frac{\partial u_n}{\partial y}\frac{\partial \varphi_j}{\partial y}\right] dx\,dy - \lambda \iint\limits_D u_n \varphi_j\,dx\,dy = 0, \tag{76}$$

or, transforming the first two terms by means of Greens' formula [cf. § 1, (15)], we can obviously bring system (76) into the form given earlier, (66). However considerations connected with the extremal properties of proper values not only permit us to arrive at system (66) again and justify in another way the proposed approximate method of determining the proper numbers and functions, but also to obtain additional facts about these approximations.

Let $\lambda_1^{(n)}$ be the least root of the determinant of system (66) [or (76)]. It is not difficult to show, reasoning as in § 1 No. 5, that $\lambda_1^{(n)}$ is the minimum of (73) under condition (74). In other words, $\lambda_1^{(n)}$ is the minimum of the integral (71) under condition (72), in case we limit ourselves to the family of functions of the form $\sum a_k\varphi_k$. It is evident that such a minimum will be $\geqslant$ the minimum of the same problem without the last limiting condition, i.e., $\lambda_1^{(n)} \geqslant \lambda_1$, the true first proper value. Thus Ritz's method gives for the proper value an excessive approximation. We have satisfied ourselves of this with regard to the first proper value, but analogous considerations show that the same thing remains true for the rest of the proper values too.

Moreover, it is clear that, thanks to the assumed completeness of the system of functions $\{\varphi_k\}$, by taking n sufficiently large, one can approximate the first proper function $\varphi_1(x)$, which gives the minimum of integral (71) [under condition (72)], as well as one pleases by means of the functions $\sum a_k\varphi_k$. Thus the minimum of integral (73) [under condition (74)] will differ by as little as one pleases from the minimum of integral (71) [under condition (72)]. The last signifies that the nth approximation to the first proper value as $n \to \infty$ converges to its true magnitude. The same thing is true also with respect to the further proper numbers. Thus Ritz's method always gives a convergent process for the computation of the proper numbers [1]).

We shall yet say a few words about what is known as Rayleigh's method. The English physicist Rayleigh suggested, long before Ritz's

[1]) On the subject of further applications of the extremal properties of proper numbers, in particular for an investigation of the asymptotic character of the latter, we refer the reader to the book of Courant and Hilbert [1].

works, a method of determining the principal proper frequency of vibration of a mechanical system. Without setting forth here the purely mechanical considerations on the basis of which it was proposed [1]), we shall say briefly that it consists essentially in choosing a definite function φ satisfying the boundary conditions for the function u, which is assigned the form $u = A\varphi$, the constant A being found from the condition of minimum potential energy. Thanks to this it turns out that Rayleigh's method for the determination of the first proper number gives precisely the same result as does the first approximation of Ritz's method.

We shall now give examples of the determination of proper values by the methods of Ritz and of Galerkin [2]).

Example 1 [3]). Study of the vibrations of a homogeneous string by Fourier's method leads, as we know, to the integration of a differential equation of the second order of the form:

$$y'' + k^2 y = 0.$$

Let us assume that our string has the length 2 and is fastened at the points -1 and $+1$:

$$y(-1) = y(1) = 0.$$

The general solution of the equation is

$$y = A \sin (kx + \varphi).$$

From the end conditions it is clear that the fundamental tone of the string is given exactly by the solution

$$y = \cos \frac{\pi x}{2}, \quad k = \frac{\pi}{2}$$

$$\left[\text{the first proper number } \lambda_1 = k^2 = \left(\frac{\pi}{2}\right)^2\right].$$

---

[1]) On this see, for example, S. Timoshenko, [2], p. 60.

[2]) Examples of the determination of proper values, and estimates of the error of their determination, by the methods of Ritz and of B. G. Galerkin can be found in other than the monographs of L. S. Leĭbenzon and IA. A. Protusevich and the various courses in the theory of elasticity already mentioned: the book of P. F. Papkovich, very rich in material, [2], vol. 2; see also the survey article of L. Collatz, [3], and also his monograph [4].

In these monographs, as well as in the authors' survey article in the collection "Thirty Years of Mathematics in the USSR", one can find the literature on approximate methods for the solution of these same problems which differ from the methods of Ritz and Galerkin and which have not obtained reflection in this book: the method of least squares, the method of moments (developed in articles of N. M. Krylov and M. F. Kravchuk), Trefftz' method, the method of successive approximations, interpolation methods.

On the subject of the investigation of the first of the above-named methods, see the recent work of S. G. Mikhlin, [2], and also [5].

[3]) The exhibited problem is borrowed from Ritz's memoir, [1].

the first overtone by

$$y = \sin \pi x, \ k = \pi,$$

the second overtone by

$$y = \cos \frac{3\pi x}{2}, \quad k = \frac{3\pi}{2}, \text{ and so forth.}$$

We seek approximately, for comparison, the even solutions (the even tones of the string) in the form of a polynomial arrayed in even powers of x. The general form of the polynomial satisfying the end conditions will be:

$$y_n(x) = (1 - x^2)(a_1 + a_2x^2 + a_3x^4 + \ldots + a_nx^{2n-2}).$$

We shall first limit ourselves to two terms only; we have:

$$y_2 = (1 - x^2)(a_1 + a_2x^2),$$

$$y_2' = 2(a_2 - a_1)x - 4a_2x^3,$$

$$y_2'' = 2(a_2 - a_1) - 12a_2x^2.$$

System (61) in the given case will be:

$$\int_{-1}^{1} [2(a_2 - a_1) - 12a_2x^2 + k^2(1 - x^2)(a_1 + a_2x^2)](1 - x^2)dx = 0,$$

$$\int_{-1}^{1} [2(a_2 - a_1) - 12a_2x^2 + k^2(1 - x^2)(a_1 + a_2x^2)]x^2(1 - x^2)dx = 0.$$

After several cancellations it takes the form:

$$(35 - 14k^2)a_1 + (7 - 2k^2)a_2 = 0,$$

$$(21 - 6k^2)a_1 + (33 - 2k^2)a_2 = 0.$$

Equating to zero the determinant, we have the equation for the determination of k,

$$k^4 - 28k^2 + 63 = 0,$$

the roots of which will be

$$k_1^2 = 2.46744, \ k_2^2 = 25.6.$$

From the exact solutions, however, there are obtained

$$k_1^2 = 2.467401100, \ k_2^2 = 22.207.$$

From this it is seen that with the second approximation the characteristic number corresponding to the fundamental tone of the string has already been determined with an error less than 0.00004, and the second overtone with an accuracy of 15%.

In the third approximation

$$y_3 = (1 - x^2)(a_1 + a_2x^2 + a_3x^4)$$

for the determination of k one finds the equation

$$4k^6 - 450k^4 + 8910k^2 - 19305 = 0,$$

from which we find:

$$k_1^2 = 2.467401108, \quad k_2^2 = 22.301.$$

The error of the third approximation to the characteristic number of the fundamental tone does not exceed 0.000000008, and to the second, 0.5%.

Next let us substitute the third approximate value for k_1^2 in the equations defining a_1, a_2, a_3. From the system obtained, the coefficients a_1, a_2, a_3 are found accurate to a factor. If we then find this undetermined factor from the condition of normality,

$$\int_{-1}^{1} [\psi_1^{(3)}(x)]^2 dx = 1,$$

which the exact solution $y = \cos \frac{\pi x}{2}$ satisfies, we obtain

$$\psi_1^{(3)}(x) = (1 - x^2)(1 - 0.233430x^2 + 0.018962x^4).$$

For $x = -1$ and $x = 1$ the first fundamental function and its second approximation coincide, by the very construction. How little $\cos \frac{\pi x}{2}$ differs from $\psi_1^{(3)}(x)$ in the whole interval is shown in the following table, in which the mantissas of the logarithms of these functions are exhibited:

x	0.1	0.2	0.3	0.4	0.5	0.6	0.7	0.8	0.9
$\lg \cos \frac{\pi x}{2}$	994620	978206	949881	907958	849485	769219	657047	489982	194332
$\lg \psi_1^{(3)}(x)$	994621	978212	949889	907952	849493	769221	657043	489978	194345

In only one place does the difference in the mantissas exceed a unit in the fifth position.

Finally, just such computations for the third approximation to the characteristic function $\cos \frac{3\pi x}{2}$, corresponding to the second overtone, yield:

$$\psi_2^{(3)}(x) = (1 - x^2)(-1 + 9.3335x^2 - 6.219x^4).$$

The nodes of this overtone, found theoretically from $\cos \frac{3\pi x}{2}$, are situated at the points ± 1, $\pm \frac{1}{3}$.

The values for the nodes found on the basis of $\psi_2^{(3)}(x)$, however, will be:

$$x = \pm 1, \quad x = \pm 0.3408.$$

The error is less than 0.01.

Example 2. Let us consider the proper functions of a homogeneous membrane having the form of a circle of radius a. We shall have in view the simplest case, where the deformed surface of the membrane is symmetric with respect to the center of the circle and its edge is fixed. In this case the deflection depends only on the radial distance r from the center.

The deflection of the homogeneous membrane satisfies the Poisson equation, and the equation defining the proper numbers and functions is the equation

$$\Delta u + \lambda u = 0, \tag{64}$$

which we have considered above. If it be transformed to polar coordinates [cf. Chap. I, § 1 (35)], and if the fact that u does not depend on θ be utilized, we obtain

$$\Delta u + \lambda u = \frac{1}{r}\frac{d}{dr}(ru_r') + \lambda u = 0.$$

Let us take as the fundamental functions the following:

$$\varphi_1 = \cos\frac{\pi}{2a}r,\ \varphi_2 = \cos\frac{3\pi}{2a}r,\ \varphi_3 = \cos\frac{5\pi}{2a}r,\ \ldots.$$

We shall first consider the first approximation:

$$u_1 = a_1 \cos\frac{\pi}{2a}r.$$

Equation (66), if it brought into polar coordinates, acquires the form

$$\int\int (\Delta u_1 + \lambda u_1)\varphi_1\, dx\, dy =$$

$$= 2\pi \int_0^a \left\{\frac{1}{r}\frac{d}{dr}\left[r\frac{\pi}{2a}\left(-\sin\frac{\pi}{2a}r\right)\right]a_1 + \lambda a_1 \cos\frac{\pi}{2a}r\right\}\cos\frac{\pi}{2a}r \cdot r dr = 0,$$

or, if the factor πa_1 be discarded, the equation for the determination of λ will be

$$\frac{\pi^2}{4}\left(\frac{1}{2} + \frac{2}{\pi^2}\right) - \lambda a^2\left(\frac{1}{2} - \frac{2}{\pi^2}\right) = 0,$$

whence it follows that

$$\lambda_1 = \frac{5.832}{a^2}.$$

The exact value is $\lambda = \dfrac{5.779}{a^2}$ and the error of the first approximation is less than 1%.

Now let us take the second approximation:

$$u_2 = a_1 \cos\frac{\pi}{2a}r + a_2 \cos\frac{3\pi}{2a}r.$$

Equations (66) for the coefficients a_1 and a_2 and the equation for the number λ will be:

$$(1.7337 - 0.29736\lambda a^2)a_1 + (0.20264\lambda a^2 - 1.5)a_2 = 0,$$
$$(0.20264\lambda a^2 - 1.5)a_1 + (11.603 - 0.47748\lambda a^2)a_2 = 0,$$

$$\begin{vmatrix} 1.7337 - 0.29736\lambda a^2, & 0.20264\lambda a^2 - 1.5 \\ 0.20264\lambda a^2 - 1.5, & 11.603 - 0.47748\lambda a^2 \end{vmatrix} = 0$$

or

$$0.10092\lambda^2 a^4 - 3.6701\lambda a^2 + 17.866 = 0.$$

The lesser root of the equation is

$$\lambda_1 = \frac{5.792}{a^2}.$$

The error of the second approximation to the first proper number will, accordingly, be less than 0.23%.

The substitution in the first of the equations for λ_1 gives

$$0.0114a_1 - 0.3263a_2 = 0.$$

We can therefore put

$$a_1 = 0.3263C, \ a_2 = 0.0114C,$$

where C must be determined from the normalizing condition.

Example 3. Let us find the fundamental tone of the vibration of a homogeneous plate having the form of a circle of radius a with center at the origin of coordinates, built-in horizontally along the edge. This is equivalent to finding the first proper function and the first proper number of the equation

$$\Delta\Delta u - \lambda u = 0, \tag{68}$$

under the end conditions

$$u\Big|_{r=a} = \frac{\partial u}{\partial r}\Big|_{r=a} = 0.$$

The corresponding variational problem is

$$I(u) = \int\int_D (\Delta u)^2 \, dx \, dy = \min,$$

under the conditions

$$\int\int_D u^2 \, dx \, dy = 1, \ u\Big|_{r=a} = \frac{\partial u}{\partial r}\Big|_{r=a} = 0.$$

In the case of the lowest tone, the form of the vibrating plate is symmetric with respect to the origin of coordinates, and u will be a function of r only.

For the application of Ritz's method we make a preliminary transformation of equation (68) into polar coordinates. If we take into consideration the fact that u does not depend on θ, it will take the form

$$\left(\frac{d}{dr^2} + \frac{1}{r}\frac{d}{dr}\right)\left(\frac{d^2u}{dr^2} + \frac{1}{r}\frac{du}{dr}\right) - \lambda u = 0.$$

Let us adopt u_n in the form:

$$u_n = a_1\left(1 - \frac{r^2}{a^2}\right)^2 + a_2\left(1 - \frac{r^2}{a^2}\right)^3 + \ldots + a_n\left(1 - \frac{r^2}{a^2}\right)^{n+1}$$

and limit ourselves to the two first terms only in our computations. Equations (70) will take the form:

$$a_1\left(\frac{192}{9} - \frac{\lambda a^4}{5}\right) + a_2\left(\frac{144}{9} - \frac{\lambda a^4}{6}\right) = 0,$$

$$a_1\left(\frac{144}{9} - \frac{\lambda a^4}{6}\right) + a_2\left(\frac{96}{5} - \frac{\lambda a^4}{7}\right) = 0.$$

Hence for λ there follows the equation:

$$(\lambda a^4)^2 - \frac{204.48}{5}\lambda a^4 + 768 \cdot 36 \cdot 7 = 0,$$

the lesser root of which will be

$$\lambda = \frac{104.39}{a^4}.$$

Substituting the value found for λ in the first equation of the system, we find

$$0.455a_1 - 1.398a_2 = 0, \quad a_2 = 0.325a_1,$$

$$u_2 = a_1\left[\left(1 - \frac{r^2}{a^2}\right)^2 + 0.325\left(1 - \frac{r^2}{a^2}\right)^3\right].$$

Example 4. Soon after the appearance of his fundamental memoir, Ritz published an example of the application of his method to the solution of a difficult problem concerning the determination of the proper numbers and functions of the vibration of a homogeneous plate with free boundaries [1]. We shall not present *in toto* his detailed investigations, but shall limit ourselves to that part of them that is of significance for our trend of thought on the approximate solution of equations.

Let us have to do with a square plate with side 2. We shall choose the axes as is shown in Fig. 21. Analytically the problem posed reduces, if the constant factors which are immaterial to us be discarded, to the determination of the function which gives the least value for the integral

$$I(u) = \int_{-1}^{+1}\int_{-1}^{+1}\{(\Delta u)^2 - 2(1 - \nu)(u_{xx}u_{yy} - u_{xy}^2)\}dx\,dy$$

under the condition

$$\int_{-1}^{+1}\int_{-1}^{+1} u^2\,dx\,dy = 1.$$

The boundary values of the admissible functions and their normal derivatives are free.

As for the case, analysed in No. 4, of a plate with built-in boundaries, we shall construct a system of fundamental functions $\varphi_n(x, y)$ using the proper functions of a free beam. As is known, they satisfy the equation

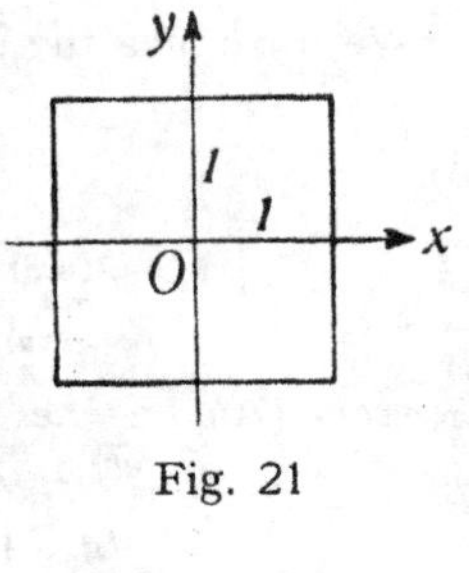

Fig. 21

$$\xi_n^{IV} - k_n\xi_n = 0$$

and the conditions

$$\xi_n'' = 0, \quad \xi_n''' = 0$$

at the ends of the beam. In addition, they satisfy conditions of orthogonality and must always be normalized:

$$\int u_m u_n\,dx = 0, \quad \int u_n^2\,dx = 1.$$

---

[1]) W. Ritz, [3].

The integrals are taken along the entire length of the beam. To make the computations more symmetric, we shall have the ends of the beam lying at the points $x = \pm 1$.

Then for the proper functions there are obtained the following expressions, the validity of which is easily verifiable.

For even m:

$$\xi_m = \frac{\operatorname{ch} k_m \cos k_m x + \cos k_m \operatorname{ch} k_m x}{\sqrt{\operatorname{ch}^2 k_m + \cos^2 k_m}}, \qquad (m = 2, 4, \ldots)$$

where k_m is a root of the equation

$$\operatorname{tg} k + \operatorname{th} k = 0.$$

For odd m:

$$\xi_m = \frac{\operatorname{sh} k_m \sin k_m x + \sin k_m \operatorname{sh} k_m x}{\sqrt{\operatorname{sh}^2 k_m - \sin^2 k_m}}, \qquad (m = 3, 5, \ldots)$$

where k_m is a root of the equation

$$\operatorname{tg} k - \operatorname{th} k = 0.$$

To this must still be added two proper functions corresponding to the case $k_0 = k_1 = 0$

$$\xi_0(x) = \frac{1}{\sqrt{2}}, \quad \xi_1(x) = \sqrt{\frac{3}{2}}\, x.$$

They correspond to the movement of the beam as a whole without deformation.

For $m = 2$ we shall have the fundamental tone with two nodes; for $m = 3$, the first overtone with three nodes, etc.

The numbers k_m for m even differ very little from $\frac{m\pi}{2} - \frac{\pi}{4}$: $k_2 = 2.3650$, and k_4 is already different from $(2 - \frac{1}{4})\pi$ only in the sixth decimal place, and so on.

The numbers k_m for m odd are also very little different from $\frac{m\pi}{2} - \frac{\pi}{4}$, so, accurate to five figures of decimals, one can consider:

$$k_3 = 3.92660, \quad k_5 = \left(\frac{5}{2} - \frac{1}{4}\right)\pi, \; \ldots .$$

Let us now form the functions φ:

$$\varphi_{kj} = \xi_k(x)\xi_j(y), \quad u_s = \sum_{n,m=0}^{s} a_{m,n}\xi_m(x)\xi_n(y).$$

We introduce the symbols

$$\omega_{mn} = \int_{-1}^{1} \xi''_m \xi_n dx, \quad \Delta_{mn} = \Delta_{nm} = \int_{-1}^{1} \xi'_m \xi'_n dx,$$

$$\alpha_{mn}^{(mn)} = k_m^4 + k_n^4 + 2\nu\omega_{mm}\omega_{nn} + 2(1 - \nu)\Delta_{mm}\Delta_{nn}$$

$$\alpha_{mn}^{(pq)} = \nu[\omega_{mp}\omega_{qn} + \omega_{pm}\omega_{nq}] + 2(1 - \nu)\Delta_{mp}\Delta_{nq}.$$

System (70) for the determination of the numbers a_{mn} will take the form:

$$[\alpha_{00}^{(00)} - \lambda]a_{00} + \alpha_{01}^{(00)}a_{01} + \alpha_{10}^{(00)}a_{10} + \ldots + \alpha_{ss}^{(00)}a_{ss} = 0,$$

$$\alpha_{00}^{(01)}a_{00} + [\alpha_{01}^{(01)} - \lambda]a_{01} + \alpha_{10}^{(01)}a_{10} + \ldots + \alpha_{ss}^{(01)}a_{ss} = 0,$$

$$\cdots\cdots\cdots\cdots\cdots\cdots\cdots\cdots$$

$$\alpha_{00}^{(ss)}a_{00} + \alpha_{01}^{(ss)}a_{01} + \alpha_{10}^{(ss)}a_{10} + \ldots + [\alpha_{ss}^{(ss)} - \lambda]a_{ss} = 0.$$

It can be considerably simplified if one utilizes the symmetry of the problem.

We shall, for example, seek solutions which are symmetric with respect to the diagonals of the coordinate angles, i.e., which remain invariant when x is replaced by y, and antisymmetric with respect to the coordinate axes, i.e., are odd with respect to x and y. Among them will be found the solution corresponding to the fundamental tone.

In conformity with this, we drop from u_s all terms containing $\xi_m(x)\xi_n(y)$ with even indices, and put the a_{mn} equal to a_{nm}.

Let us limit ourselves to terms of up to ξ_5, setting $s = 5$, and let us choose for ν the value $\nu = 0.225$ (glass). Designating $\xi_n(y) = \eta_n$ for the sake of simplicity, we have:

$$u_5 = A_0\xi_1\eta_1 + A_1(\xi_1\eta_3 + \xi_3\eta_1) + A_2\xi_3\eta_3 + A_3(\xi_1\eta_5 + \xi_5\eta_1) + A_4(\xi_3\eta_5 + \xi_5\eta_3) + A_5\xi_5\eta_5.$$

The system of equations for A_k, if we introduce the numerical values of $\alpha_{mn}^{(pq)}$ and ν, will be:

$$\begin{aligned}
(13.95 - \lambda)A_0 - 32.08A_1 - 18.6A_2 + 32.08A_3 - 37.20A_4 + 18.60A_5 &= 0,\\
-16.04A_0 + (411.8 - \lambda)A_1 - 120.0A_2 - 133.6A_3 + 166.8A_4 + 140A_5 &= 0,\\
18.60A_0 - 240.0A_1 + (1686 - \lambda)A_2 - 218.0A_3 - 1134A_4 + 330A_5 &= 0,\\
16.04A_0 - 133.6A_1 + 109.0A_2 + (2945 - \lambda)A_3 - 424A_4 + 179A_5 &= 0,\\
-18.6A_0 + 166.8A_1 - 567A_2 - 424A_3 + (6308 - \lambda)A_4 - 1437A_5 &= 0,\\
18.6A_0 + 280A_1 - 330A_2 + 358A_3 - 2874A_4 + (13674 - \lambda)A_5 &= 0.
\end{aligned}$$

To proceed in the usual way in solving this system would be extremely difficult, for the equation for λ obtained by equating the determinant of the system to zero is now a secular equation of the sixth degree, and the actual development of it would occupy a great deal of time. One peculiarity of the system, however, permits one to simplify greatly the computations. As in the example given in the preceding No., the coefficients of the diagonal terms here will, generally speaking, be dominant. If the system were really reducible to the diagonal terms, we would obtain for λ the following roots: $\lambda_1 = 13.95$; $\lambda_2 = 411.8$; $\lambda_3 = 1686$, ... If we stop with the first of them, A_0 remains arbitrary, and the remaining A_k should be adopted as equalling zero; if we stop with the second, A_1 remains arbitrary, etc. Since the constant factor has for us no significance at present, in determining the fundamental tone we can consider that $A_0 = 1$.

Thus let us take for the initial values $\lambda_0 = 13.95$, $A_1 = A_2 = \ldots = 0$, and substitute these values in all terms except the diagonals of the last five equations. From them we find the first approximations for the A_k and from the first, for λ; by means of the values found we again correct all the values of the unknowns.

After several steps we shall have the following values of the sought quantities:

$$A_1 = 0.0394, \quad A_2 = -0.0040, \quad A_3 = -0.0034, \quad A_4 = 0.0011, \quad A_5 = -0.0019,$$
$$\lambda = 12.43.$$

Consequently the proper function corresponding to the fundamental tone will be approximately equal to

$$u = \xi_1\eta_1 + 0.0394(\xi_1\eta_3 + \eta_3\xi_1) - 0.0040\xi_2\eta_3 - 0.0034(\xi_1\eta_5 + \xi_5\eta_1) + \\ + 0.0011(\xi_3\eta_5 + \xi_5\eta_3) - 0.0019\xi_5\eta_5.$$

All the subsequent proper numbers and functions are computed analogously.

§ 3. REDUCTION TO ORDINARY DIFFERENTIAL EQUATIONS [1]

This method occupies a position intermediate between the exact solution of the problem and the methods of Ritz and Galerkin. In Ritz's method the problem of the minimum of a double integral is reduced to the problem of the minimum of a function of several variables. This is accomplished by choosing the form of the solution *a priori* and only afterwards selecting the best values of the constants figuring in it.

In the reduction to ordinary equations the solution is sought in such a form that there figure in it undetermined functions of one variable. Thus the problem of the minimum of a double integral is reduced to the problem of the minimum of a simple integral. The advantage of this method, apart from its greater accuracy, consists in that only part of the expression giving the solution is chosen *a priori*, part of the functions being determined in accordance with the character of the problem.

In order to explain the gist of the method, we present the following simple example.

Example. To be found, the approximate solution of the Poisson equation $\Delta u = -1$ for the rectangle $[-a, a; -b, b]$ on condition that $u = 0$ on the contour.

We seek the solution in the form

$$\bar{u} = (b^2 - y^2)f(x); \tag{1}$$

it satisfies the boundary condition on the lines $y = \pm b$. Substituting this expression in the integral corresponding to the given problem, we find:

$$I(\bar{u}) = \int_D\int \left[\left(\frac{\partial \bar{u}}{\partial x}\right)^2 + \left(\frac{\partial \bar{u}}{\partial y}\right)^2 - 2\bar{u}\right] dx\, dy =$$

$$= \int_{-a}^{a} dx \int_{-b}^{b} [(b^2 - y^2)^2 f'^2 + 4y^2 f^2 - 2f(b^2 - y^2)]dy.$$

Performing the integration with respect to y, we shall have reduced the problem to that of the minimum of a simple integral:

$$I(\bar{u}) = \int_{-a}^{a} [\tfrac{16}{15}b^5 f'^2 + \tfrac{8}{3}b^3 f^2 - \tfrac{8}{3}b^3 f]dx.$$

The corresponding variational equation [cf. § 1 (5a)] will be

$$\frac{d}{dx}[\tfrac{32}{15}b^5 f'] - \tfrac{16}{3}b^3 f + \tfrac{8}{3}b^3 = 0 \text{ or } f'' - \frac{5}{2b^2}f = -\frac{5}{4b^2}.$$

Solving this equation with constant coefficients, we find

$$f(x) = C_1 \operatorname{ch} \sqrt{\tfrac{5}{2}}\frac{x}{b} + C_2 \operatorname{sh} \sqrt{\tfrac{5}{2}}\frac{x}{b} + \tfrac{1}{2}.$$

[1] This method was proposed in a work of L. V. Kantorovich, [13].

The constants C_1 and C_2 are determined from the conditions $f(a) = f(-a) = 0$, which secure the fulfillment of the condition $u = 0$ on the sides $x = \pm a$. Since the solution is obviously an even function of x, $C_2 = 0$; we have for the determination of C_1

$$C_1 \operatorname{ch} \sqrt{\tfrac{5}{2}}\frac{a}{b} + \tfrac{1}{2} = 0.$$

Substituting the value of C_1 found from this, we obtain finally

$$f = \tfrac{1}{2}\left(1 - \frac{\operatorname{ch}\sqrt{\tfrac{5}{2}}\dfrac{x}{b}}{\operatorname{ch}\sqrt{\tfrac{5}{2}}\dfrac{a}{b}}\right), \quad \bar{u}(x, y) = \tfrac{1}{2}(b^2 - y^2)\left(1 - \frac{\operatorname{ch}\sqrt{\tfrac{5}{2}}\dfrac{x}{b}}{\operatorname{ch}\sqrt{\tfrac{5}{2}}\dfrac{a}{b}}\right). \quad (2)$$

Hence, for example, for the moment we obtain [1])

$$M = 4G\theta \iint_D \bar{u}(x, y)\,dx\,dy =$$

$$= 2G\theta \int_{-a}^{a}\left(1 - \frac{\operatorname{ch}\sqrt{\tfrac{5}{2}}\dfrac{x}{b}}{\operatorname{ch}\sqrt{\tfrac{5}{2}}\dfrac{a}{b}}\right)dx \int_{-b}^{b}(b^2 - y^2)\,dy =$$

$$= \tfrac{16}{3}b^3 a\left[1 - \sqrt{\tfrac{2}{5}}\frac{b}{a}\cdot \operatorname{th}\sqrt{\tfrac{5}{2}}\frac{a}{b}\right]G\theta.$$

Let us compare this value of M with the exact value (Chap. I, § 1, p. 24) and with the approximate value (the first approximation) by Ritz's method [2]) (§ 2, p. 269):

$\dfrac{a}{b} =$	1	2	3	∞
Exact M	2.244	3.659	4.213	5.333
Method of ordinary equations	2.234	3.653	4.209	5.333
Ritz's Method	2.222	3.555	4.000	4.444

As we see, the greatest error, of about 0.5%, is obtained for a square section; if $\dfrac{a}{b} \geqslant 2$, the agreement with the exact solution is almost complete. From the comparison with Ritz's method it is clear that in the most unfavorable case, $\dfrac{a}{b} = 1$, the above method gives an error

---

[1]) Cf. Chap. I, p. 10.

[2]) Exhibited here are not the values of the moment themselves, but only the numerical factors that stand with $ab^3G\theta$.

half as large as does Ritz's method. An error little greater than that in the moments is obtained in the values of the functions themselves; so, for example, for $a = 1$, $b = 0.5$, we find $u(0, 0) = 0.11443$ instead of the exact 0.11387 — an error of 0.5%.

After this introductory example, we pass on to the exposition of the method in general form.

1. Fundamental equations. Let us again consider the Poisson equation

$$L(u) = \frac{\partial^2 u}{\partial x^2} + \frac{\partial^2 u}{\partial y^2} - p(x, y) = 0, \tag{3}$$

under the condition

$$u(x, y) = 0 \text{ on } \Gamma, \tag{4}$$

where Γ is the boundary of the region D (Fig. 22). For definiteness we shall consider that the region D is bounded by the lines $x = a$ and $x = b$ and by the curves $y = g(x)$, $y = h(x)$; this condition is not essential, however.

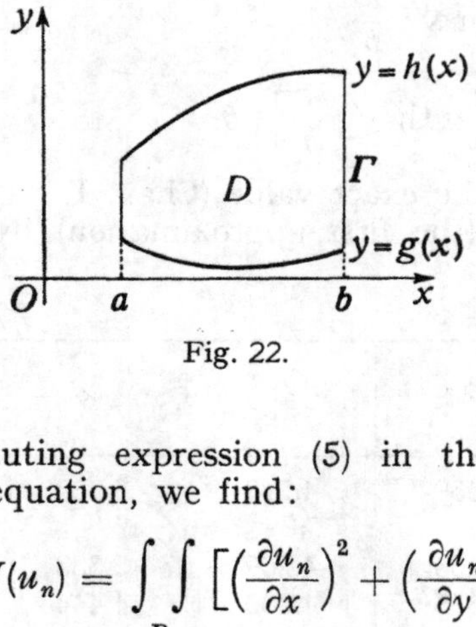

Fig. 22.

We seek the solution in the form

$$u_n(x, y) = \sum_{k=1}^{n} \chi_k(x, y) f_k(x) + \chi_0(x, y), \tag{5}$$

where the functions $\chi_k(x, y)$ $(k = 1, 2, \ldots, n)$ are equal to zero on the contour Γ, with the possible exception of the lines $x = a$ and $x = b$, but $\chi_0(x, y) = 0$ everywhere on Γ. Substituting expression (5) in the integral corresponding to the Poisson equation, we find:

$$I(u_n) = \iint_D \left[\left(\frac{\partial u_n}{\partial x}\right)^2 + \left(\frac{\partial u_n}{\partial y}\right)^2 + 2pu_n\right] dx\, dy =$$

$$= \int_a^b dx \int_{g(x)}^{h(x)} \left\{\left[\sum_{k=1}^{n} \frac{\partial \chi_k}{\partial x} f_k + \sum_{k=1}^{n} \chi_k f_k' + \frac{\partial \chi_0}{\partial x}\right]^2 + \right.$$

$$\left. + \left[\sum_{k=1}^{n} \frac{\partial \chi_k}{\partial y} f_k + \frac{\partial \chi_0}{\partial y}\right]^2 + 2p\left[\sum_{k=1}^{n} \chi_k f_k + \chi_0\right]\right\} dy =$$

$$= \int_a^b \Phi(x, f_k, f_k')\, dx, \tag{6}$$

where by Φ we have briefly designated the integrand of the outer integral of (6); in it f_k and f_k' will figure, but the rest of it is a known function of x (the integration with respect to y is assumed to have been performed). Thus the problem is reduced to the minimum of a simple integral; the corresponding system of ordinary differential equations has the form [see § 1 (5c), p. 246]

$$\frac{d}{dx}\frac{\partial \Phi}{\partial f_k'} - \frac{\partial \Phi}{\partial f_k} = 0 \qquad (k = 1, 2, \ldots, n) \qquad (7)$$

or, *in extenso*, after cancelling 2:

$$\frac{d}{dx}\int_{g(x)}^{h(x)} \frac{\partial u_n}{\partial x}\chi_k\,dy - \int_{g(x)}^{h(x)}\left[\frac{\partial u_n}{\partial x}\frac{\partial \chi_k}{\partial x} + \frac{\partial u_n}{\partial y}\frac{\partial \chi_k}{\partial y} + p\chi_k\right]dy = 0 \quad (7a)$$

$$(k = 1, 2, \ldots, n).$$

Let us carry out the differentiation in the first term, in accordance with the rule for the differentiation of an integral with respect to a parameter, taking into consideration that $\chi_k(x, y)$ is equal to zero on the lines $y = g(x)$ and $y = h(x)$; we find

$$\frac{d}{dx}\int_{g(x)}^{h(x)} \frac{\partial u_n}{\partial x}\chi_k\,dy = \int_{g(x)}^{h(x)} \frac{d}{dx}\left[\frac{\partial u_n}{\partial x}\chi_k\right]dy + \left[\frac{\partial u_n}{\partial x}\chi_k h'(x)\right]_{y=h(x)} -$$

$$-\left[\frac{\partial u_n}{\partial x}\chi_k g'(x)\right]_{y=g(x)} = \int_{g(x)}^{h(x)}\left[\frac{\partial^2 u_n}{\partial x^2}\chi_k + \frac{\partial u_n}{\partial x}\frac{\partial \chi_k}{\partial x}\right]dy.$$

Next we transform the second summand of the second integral in (7a) by an integration by parts; we obtain

$$\int_{g(x)}^{h(x)} \frac{\partial u_n}{\partial y}\frac{\partial \chi_k}{\partial y}\,dy = \left[\frac{\partial u_n}{\partial y}\chi_k\right]_{g(x)}^{h(x)} - \int_{g(x)}^{h(x)} \frac{\partial^2 u_n}{\partial y^2}\chi_k\,dy = -\int_{g(x)}^{h(x)} \frac{\partial^2 u_n}{\partial y^2}\chi_k\,dy.$$

Substituting the results of these transformations in (7a), we can give this system of equations the following form, analogous to the equations of Galerkin's method:

$$\int_{g(x)}^{h(x)} L(u_n)\chi_k\,dy = \int_{g(x)}^{h(x)}\left[\frac{\partial^2 u_n}{\partial x^2} + \frac{\partial^2 u_n}{\partial y^2} - p\right]\chi_k\,dy = 0$$

$$(k = 1, 2, \ldots, n). \quad (7b)$$

This form of the equations considerably shortens the task of setting them up [1]). The system of ordinary equations for the functions f_k which

---

[1]) Another derivation of this system of equations is given in an article of L. V. Kantorovich, [10].

we have obtained should be solved under the boundary conditions $f_k(a) = f_k(b) = 0$ $(k = 1, 2, \ldots, n)$. Then for the function u_n, (5), condition (4) will be satisfied. In case the region has a more complicated form than that depicted in Fig. 22, the equations for the determination of the functions f_k will take, instead of (7b), the following form:

$$\int_{D_x} \left[\frac{\partial^2 u_n}{\partial x^2} + \frac{\partial^2 u_n}{\partial y^2} - p\right] \chi_k dy = 0 \qquad (k = 1, 2, \ldots, n),$$

where D_x denotes the section of the region D by the line $x = \text{const.}$

Like the methods of Ritz and of Galerkin, the method under discussion may be applied not only in the case of the Dirichlet problem, but also for the solution of other boundary-value problems. We shall not dwell on this. We shall now indicate possible rules for the choice of the functions $\chi_k(x, y)$.

Let $\varphi_k(y)$ be a system of linearly independent functions, relatively complete in [0, 1] (cf. p. 265). Then the system of functions $\left\{\varphi_k\left(\frac{y-g}{h-g}\right)\right\}$ is relatively complete in $[g, h]$. Therefore for the region bounded by the lnes $x = a$, $x = b$ and the curves $y = g(x)$, $y = h(x)$, one can adopt is the $\chi_k(x, y)$ the system of functions

$$\chi_k(x, y) = \varphi_k\left(\frac{y - g(x)}{h(x) - g(x)}\right) \qquad (k = 1, 2, \ldots). \quad (8)$$

It is obvious that family (5), and together with it the solution too, will not change if for χ_k we introduce χ_k^* of a kind such that each function χ_k is a linear combination of the χ_k^* $(k = 1, 2, \ldots, n)$ and, conversely, each χ_k^* is a linear combination of the χ_k. In particular, the functions χ_k may be chosen as follows [cf. § 2, (12), p. 263]:

$$\chi_k(x, y) = (y - g(x))(h(x) - y)y^{k-1} \qquad (k = 1, 2, \ldots), \quad (8a)$$

$$\chi_k(x, y) = \sin \frac{k\pi(y - g(x))}{h(x) - g(x)} \qquad (k = 1, 2, \ldots). \quad (8b)$$

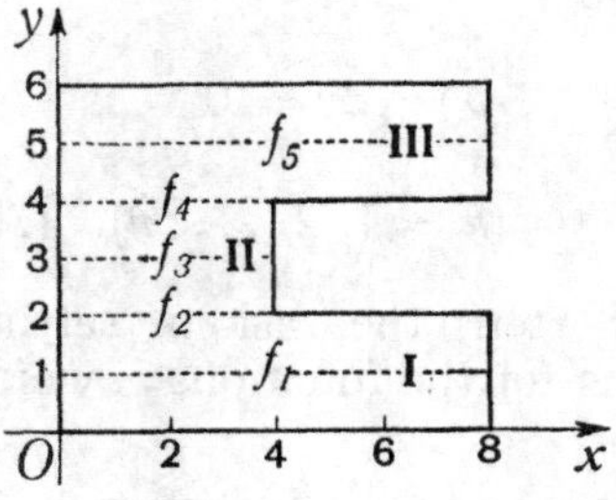

Fig. 23.

If the region has a more complicated form, in the various parts of the region it will often be necessary to resort to assignments piecewise of the χ_k and u_n, in constructing which it is expedient to use interpolation formulas. Explicitly, the region should be divided into a finite number of parts of the form indicated in Fig. 22, and in each part the expression for u_n should be set up as indicated above. Thus, for the region depicted in Fig. 23, it is convenient to introduce five unknown functions f_1, f_2, f_3, f_4, f_5, corresponding to the lines $y = 1$, $y = 2$, $y = 3$, $y = 4$, $y = 5$ and, basing oneself on them, to assign the u_n by interpolation formulas.

If, moreover, the functions f_2, f_3, f_4 be considered as equal to zero in the interval (4, 8), then the expression u_5 corresponding to the condition $u = 0$ on the contour will be the following:

$$u_5(x,y) = \begin{cases} \dfrac{y(y-2)}{-1} f_1 + \dfrac{y(y-1)}{2} f_2 \text{ in region I: } [0, 8;\ 0, 2], \\ \dfrac{(y-3)(y-4)}{2} f_2 + \dfrac{(y-2)(y-4)}{-1} f_3 + \dfrac{(y-2)(y-3)}{2} f_4 \\ \qquad \text{in region II: } [0, 4;\ 2, 4], \\ \dfrac{(y-5)(y-6)}{2} f_4 + \dfrac{(y-4)(y-6)}{-1} f_5 \\ \qquad \text{in region III: } [0, 8;\ 4, 6]. \end{cases}$$

Other methods of linking the different definitions are also possible, of course. We note in addition that in constructing the expressions for u_n by means of interpolation there is no need to reduce the conditions to null ones beforehand. Explicitly, if the contour function on the vertical portions of the contour is a polynomial of degree not higher than that which is used for the interpolation, and on the remaining portions of the contour is arbitrary, then the expression constructed by means of interpolation will exactly satisfy the boundary conditions. Here, of course, in solving the differential system defining the functions f_k, the boundary conditions for the functions f_k must be chosen in accordance with the assigned values at the points of intersection of the line corresponding to the function f_k and the contour of the region.

We call attention to the fact, finally, that in a number of cases it is most expedient to simplify the form of the region by means of another choice of the coordinate system.

2. Examples of the determination of the first approximation.

Example 1. Let us consider the Poisson equation $\Delta u = p(x, y)$ for the narrow region bounded by the curves $y = \pm \varphi(x)$ and the lines $x = a$ and $x = b$, under the condition $u = 0$ on the contour [1]. For the first approximation we seek the solution in the form of a polynomial of the second degree in y, and since it must vanish for $y = \pm \varphi(x)$, we seek it in the form

$$\bar{u} = (y^2 - \varphi^2)f. \tag{9}$$

[1] See L. V. Kantorovich and P. V. Frumkin, [1].

Equation (7b) acquires in the given case the form

$$\int\limits_{D_*}\left[\frac{\partial^2\bar{u}}{\partial x^2}+\frac{\partial^2\bar{u}}{dy^2}-p\right](y^2-\varphi^2)dy=$$

$$=\int\limits_{-\varphi}^{\varphi}[(y^2-\varphi^2)f''-4\varphi\varphi'f-2(\varphi'^2+\varphi\varphi'')f+2f-p](y^2-\varphi^2)dy=0,$$

or, if the integration be performed and $\frac{8}{3}\varphi^3$ cancelled:

$$\tfrac{2}{5}\varphi^2f''+2\varphi\varphi'f+(\varphi\varphi''-\varphi'^2-1)f=\frac{3}{8\varphi^3}\Phi, \tag{10}$$

where

$$\Phi(x)=\int\limits_{-\varphi}^{\varphi}p(x,y)(y^2-\varphi^2)dy.$$

In particular, in case $p(x, y) = -1$ we obtain

$$\tfrac{2}{5}\varphi^2f''+2\varphi\varphi'f+(\varphi\varphi''+\varphi'^2-1)f=\tfrac{1}{2}. \tag{11}$$

The example presented in the introduction, where the region is a rectangle, is obtained from this anew if one adopts $\varphi(x) = b$. Another particular case where equation (11) is solved in finite form is when the function $\varphi(x)$ is linear: $\varphi(x) = kx$. Equation (11) then takes the form

$$x^2f''+5xf'+\frac{5(k^2-1)}{2k^2}f=\frac{5}{4k^2}, \tag{12}$$

i.e., turns out to be an equation of the Euler type [1]). Seeking the solution of the homogeneous equation in the form x^ν, for the determination of ν we obtain the equation

$$\nu(\nu-1)+5\nu+\frac{5(k^2-1)}{2k^2}=0,$$

$$\text{whence } \nu_{1,2}=\frac{-4\pm\sqrt{6+\dfrac{10}{k^2}}}{2}. \tag{13}$$

Thus the general solution of equation (12) is

$$f=C_1x^{\nu_1}+C_2x^{\nu_2}+\lambda, \tag{14}$$

where

$$\nu_1=\frac{-4+\sqrt{6+\dfrac{10}{k^2}}}{2},\quad \nu_2=\frac{-4-\sqrt{6+\dfrac{10}{k^2}}}{2},\quad \lambda=\frac{1}{2(k^2-1)}.$$

---

[1]) When $\varphi(x) = ax^\nu$, equation (11) is an equation of the Bessel type. See A. I. Lur'e, [1].

The constants C_1 and C_2 are determined from the boundary conditions. The sought approximate solution will accordingly be

$$\bar{u}(x, y) = (y^2 - k^2x^2)(C_1x^{\nu_1} + C_2x^{\nu_2} + \lambda). \tag{15}$$

Let us consider the determination of the constants C_1 and C_2 for particular cases:

The equilateral trapezoid, bounded by the lines $y = \pm kx$, $x = a$, $x = b$. In this case the constants C_1 and C_2 are determined from the equations

$$f(a) = C_1a^{\nu_1} + C_2a^{\nu_2} + \lambda = 0, \quad f(b) = C_2b^{\nu_1} + C_2b^{\nu_2} + \lambda = 0. \tag{16}$$

The triangle bounded by the lines $y = \pm kx$, $x = b$. In this case $a = 0$, and the first of equations (16) loses meaning. But then, since $\nu_2 < -2$, one should adopt $C_2 = 0$ in order that $\bar{u}(x, y)$ remain bounded near $x = 0$. After this C_1 is determined from the second of conditions (16), namely $C_1 = -\lambda b^{-\nu_1}$ and therefore

$$\bar{u}(x, y) = (y^2 - k^2x^2)\frac{1 - \left(\frac{x}{b}\right)^{\nu_1}}{2(k^2 - 1)}. \tag{17}$$

In particular, for the equilateral triangle, $k = \operatorname{tg} 30° = \dfrac{1}{\sqrt{3}}$, $\nu_1 = 1$, and the solution takes the especially simple form

$$\bar{u}(x, y) = -\tfrac{3}{4}\left(1 - \frac{x}{b}\right)\left(y^2 - \tfrac{1}{3}x^2\right), \tag{18}$$

coinciding with the known exact solution for this case.

Example 2 [1]). Let us consider the solution of the same problem for the parallelogram. Here it is expedient to utilize oblique coordinates. For brevity we limit ourselves to the case when the acute angle θ of the parallelogram is 45°.

Then the transformation formulas will be

$$\left.\begin{aligned} x &= \xi + \frac{\sqrt{2}}{2}\eta, \\ y &= \frac{\sqrt{2}}{2}\eta. \end{aligned}\right\} \tag{19}$$

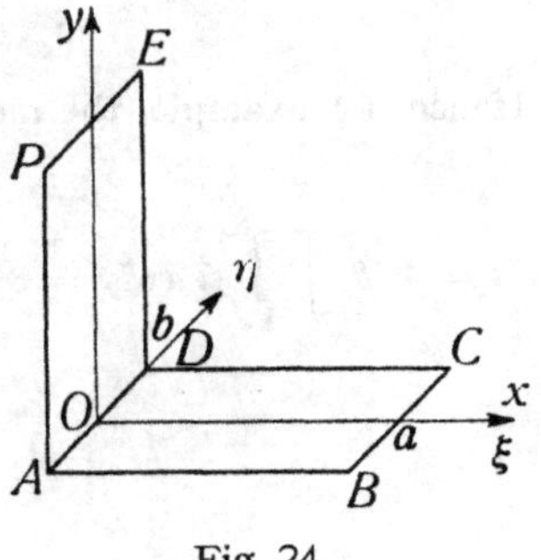

Fig. 24.

We seek the solution for the parallelogram $ABCD$ (Fig. 24) in the form

$$\bar{u}(x, y) = f(\xi)(b^2 - \eta^2). \tag{20}$$

In the given case it is more convenient to apply the method in direct form, without

[1]) On the subject of Examples 2 and 3 see T. K. Chepov, [1].

utilizing equations (7b), in order to avoid transforming the Laplace operator into the new coordinates, effecting this transformation in the integral $I(u)$. We find:

$$I(\bar{u}) = \iint\limits_{ABCD} \left[\left(\frac{\partial \bar{u}}{\partial x}\right)^2 + \left(\frac{\partial \bar{u}}{\partial y}\right)^2 - 2\bar{u}\right] dx\,dy = 2\sqrt{2}\int\limits_0^a d\xi \int\limits_{-b}^{b} \left[\tfrac{1}{2}\left(\frac{\partial \bar{u}}{\partial \xi}\right)^2 -\right.$$

$$\left. - \frac{\sqrt{2}}{2}\,\frac{\partial \bar{u}}{\partial \xi}\,\frac{\partial \bar{u}}{\partial \eta} + \tfrac{1}{2}\left(\frac{\partial \bar{u}}{\partial \eta}\right)^2 - \tfrac{1}{2}\bar{u}\right] d\eta = \sqrt{2}\int\limits_0^a d\xi \int\limits_{-b}^{b} [(b^2 - \eta^2)^2 f'^2 +$$

$$+ 2\sqrt{2}\eta(b^2 - \eta^2)ff' + 4\eta^2 f^2 - (b^2 - \eta^2)f]d\eta =$$

$$= \sqrt{2}\int^a (\tfrac{16}{15}b^5 f'^2 + \tfrac{8}{3}b^3 f^2 - \tfrac{4}{3}b^3 f)d\xi. \tag{21}$$

The Euler equation will be

$$4b^2 f'' - 10f + \tfrac{5}{2} = 0. \tag{22}$$

Its solution satisfying the conditions $f(0) = f(a) = 0$ has the form

$$f(\xi) = \tfrac{1}{4}\left(1 - \operatorname{ch}\sqrt{\frac{5}{2}}\frac{\xi}{b}\right) - \tfrac{1}{4}\left(1 - \operatorname{ch}\sqrt{\frac{5}{2}}\frac{a}{b}\right)\frac{\operatorname{sh}\sqrt{\tfrac{5}{2}}\dfrac{\xi}{b}}{\operatorname{sh}\sqrt{\tfrac{5}{2}}\dfrac{a}{b}}.$$

Using the same coordinates, one can find the solution for the oblique corner $ABCDEF$ depicted in the same Fig. 24. The solution for part of it $ABCD$ may be sought. The problem obviously reduces to the minimum of integral (21) with a free variation on the side AD. Therefore in determining the function f from differential equation (22) we must start from the conditions $f(a) = 0$, $f'(0) = 0$ [cf. § 1 (20)]. On determining the constants from this and substituting in (20) the expression found, we have:

$$\bar{u}(\xi, \eta) = \tfrac{1}{4}(b^2 - \eta^2)\left(1 - \frac{\operatorname{ch}\dfrac{\sqrt{10}}{2b}\xi}{\operatorname{ch}\dfrac{\sqrt{10}}{2b}a}\right). \tag{23}$$

Hence, for example, the moment of torsion of this corner:

$$M = 4G\vartheta \iint \bar{u}\,dx dy = 8G\vartheta \int\limits_0^a \int\limits_{-b}^{b} \tfrac{1}{4}(b^2 - \eta^2)\left(1 - \frac{\operatorname{ch}\dfrac{\sqrt{10}}{2b}\xi}{\operatorname{ch}\dfrac{\sqrt{10}}{2b}a}\right)\frac{\sqrt{2}}{2}\,d\xi\,d\eta =$$

$$= \frac{4\sqrt{2}}{2}\,b^3\left(a - \frac{2b}{\sqrt{10}}\operatorname{th}\frac{a\sqrt{10}}{2b}\right)G\vartheta. \tag{24}$$

Example 3. Let us consider the solution of the Poisson equation $\Delta u = -1$ under the condition $u = 0$ on the contour for a trapezoid having close parallel sides. The dimensions of the trapezoid $ABCD$ are given in Fig. 25.

It is expedient to introduce, instead of x and y, new coordinates $\xi = \operatorname{tg}\theta$ and η, connected with the previous ones by the formulas

$$x = \eta\xi, \; y = -\eta + a - \frac{b}{2}. \tag{25}$$

Expression $I(u)$ is transformed as follows:

$$I = \int\int\left[\left(\frac{\partial u}{\partial x}\right)^2 + \left(\frac{\partial u}{\partial y}\right)^2 - 2u\right]dx\,dy = \int\limits_{-\operatorname{tg}\alpha}^{\operatorname{tg}\beta}\int\limits_{a-b}^{a}\left[\frac{1+\xi^2}{\eta^2}\left(\frac{\partial u}{\partial \xi}\right)^2 + \right.$$

$$\left. + \left(\frac{\partial u}{\partial \eta}\right)^2 - 2\frac{\xi}{\eta}\frac{\partial u}{\partial \xi}\frac{\partial u}{\partial \eta} - 2u\right]\eta\,d\xi\,d\eta. \tag{26}$$

In the coordinates ξ, η, the trapezoid is defined by the inequalities $-\operatorname{tg}\alpha \leqslant \xi \leqslant \operatorname{tg}\beta$, $a - b \leqslant \eta \leqslant a$; we therefore seek the solution in the form

$$\bar{u}(\xi, \eta) = (\eta - a + b)(\eta - a)f(\xi). \tag{27}$$

Substituting $\bar{u}$ in the expression for I, (26), and performing the integration with respect to η, we see that, apart from a constant factor, $I(\bar{u})$ turns out to be equal to

$$\int\limits_{-\operatorname{tg}\alpha}^{\operatorname{tg}\beta}[(1+\xi^2)f'^2(\xi) + c(f^2(\xi) + f(\xi))]d\xi, \tag{28}$$

where

$$c = \frac{b^3(2a-b)}{-6a^3b + 9a^2b^2 - 2ab^3 - \frac{1}{2}b^4 + 6a^2(a-b)^2 \lg\dfrac{a}{a-b}}.$$

The Euler equation corresponding to this problem will be

$$(1+\xi^2)f''(\xi) + 2\xi f'(\xi) - cf(\xi) - \frac{c}{2} = 0, \tag{29}$$

and the boundary conditions for the determination of f are $f(-\operatorname{tg}\alpha) = f(\operatorname{tg}\beta) = 0$.

Equation (29) is the Gauss equation, and is not soluble in elementary functions. Its solution can be found in the form of a series — numerically, by means, for example, of Störmer's method and finally by Ritz's method, using the fact that it is connected with the problem of the minimum of integral (28).

We shall yet show that, starting from the same equations, one can obtain an approximate solution of the problem for a rectangular corner — half of such a corner being the trapezoid $ABCD$ of the form indicated in Fig. 25, corresponding to $\alpha = 0$ and $\beta = 45°$. The solution for this half of the corner may therefore again be sought in form (27), but this time the function f will be the solution of equation (29) corre-

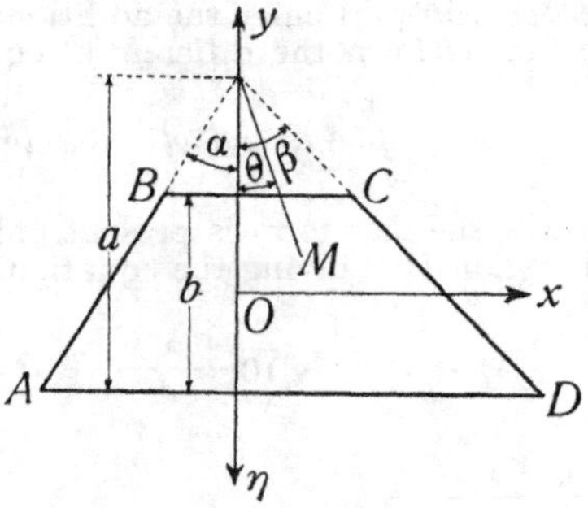

Fig. 25.

sponding to the boundary conditions

$$f(0) = 0,$$

$$f'(\operatorname{tg} 45°) = f'(1) = 0.$$

Example 4. Lastly, in a single example — in a problem of the torsion of a symmetric corner — we shall show the method of "abutment"—of joining solutions which are different for different parts of the region [1]).

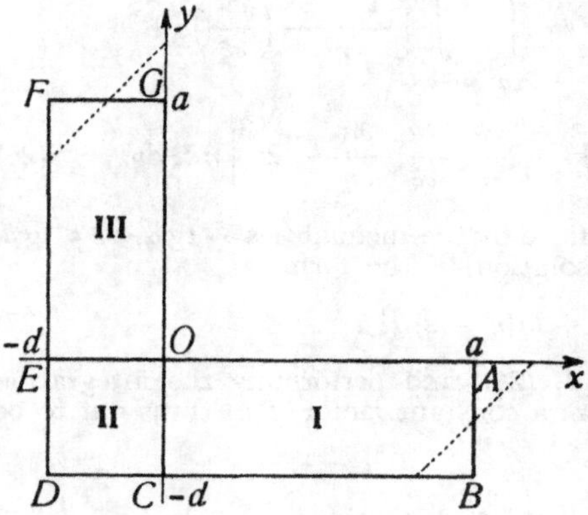

Fig. 26.

Let us consider the problem of the solution of the Poisson equation $\Delta u = -1$ for the symmetric corner whose dimensions are given in Fig. 26, under null boundary conditions. Let us first single out the rectangle $ABDE$. The boundary values of u on all of its sides are null; they are undefined only on the portion EO. It is convenient to adopt $u_1(x, y)$ as having here, as a function of x, the form of a parabolic function with an undetermined factor:

$$u_1(x, 0) = Cx(x + d) \qquad (-d \leqslant x \leqslant 0).$$

Next we seek the solution for the rectangle $ABDE$ in the form

$$u_1(x, y) = f(x)y(y + d) + u_0(x, y),$$

where

$$u_0(x, y) = \begin{cases} 0 & \text{in region I}, \\ \dfrac{C}{d}\, x(x + d)(y + d) & \text{in region II}. \end{cases}$$

In this case, if in determining f we subject it to the conditions

$$f(-d) = f(a) = 0,$$

for $u_1(x, y)$ all the boundary conditions will be satisfied on the sides of the rectangle $ABDE$, in particular the additionally imposed condition on the segment EO.

Let us form the differential equation [see (7b)] for the determination of $f(x)$:

$$\int_{-d}^{0} [\Delta u_1 + 1]y(y + d)dy = \tfrac{1}{30}d^5 f'' - \tfrac{1}{3}d^3 f - \tfrac{1}{6}d^3 - \tfrac{1}{6}Cd^3 = 0 \qquad (x < 0),$$

where the last term is present only if $x < 0$, i.e., we are located within the limits of region II. Solving the equation obtained, we can write its solution in the form

$$f(x) = A \operatorname{ch} \sqrt{10}\,\frac{x + d}{d} + B \operatorname{sh} \sqrt{10}\,\frac{x + d}{d} - \tfrac{1}{2} - \frac{C}{2}\left(1 - \operatorname{ch} \sqrt{10}\,\frac{x}{d}\right)$$

$$(x < 0).$$

---

[1]) See L. V. Kantorovich, [10].

Determining the constants A and B from the conditions $f(-d) = f(a) = 0$, we find

$$A = \tfrac{1}{2} + \frac{C}{2}(1 - \operatorname{ch}\sqrt{10}),$$

$$B = \tfrac{1}{2}\frac{1}{\operatorname{sh}\sqrt{10}\left(1 + \frac{a}{d}\right)} - \tfrac{1}{2}\operatorname{cth}\sqrt{10}\left(1 + \frac{a}{d}\right)\cdot\left[1 + C\left(1 - \operatorname{ch}\sqrt{10}\right)\right].$$

Furthermore, thanks to the symmetry, it is clear that on the line CO the function must have the same form as on EO, and since $u(0, y) = f(0)\,(y + d)y \quad (-d \leqslant y \leqslant 0)$, we must have $f(0) = C$, or

$$A \operatorname{ch}\sqrt{10} + B \operatorname{sh}\sqrt{10} - \tfrac{1}{2} = C.$$

Substituting in this equation the values for A and B, we make it possible to determine C. We find, to wit, that

$$C = \tfrac{1}{2}\,\frac{-1 + \operatorname{ch}\sqrt{10} + \dfrac{\operatorname{sh}\sqrt{10}}{\operatorname{sh}\sqrt{10}\left(1 + \frac{a}{d}\right)} - \operatorname{sh}\sqrt{10}\operatorname{cth}\sqrt{10}\left(1 + \frac{a}{d}\right)}{1 + \tfrac{1}{2}\operatorname{ch}\sqrt{10}\,(\operatorname{ch}\sqrt{10} - 1) - \tfrac{1}{2}\operatorname{sh}\sqrt{10}\,(\operatorname{ch}\sqrt{10} - 1)\operatorname{cth}\sqrt{10}\left(1 + \frac{a}{d}\right)}.$$

The expressions for A, B and C simplify greatly if we consider the ratio $\frac{a}{d}$ to be so large that the quantity $e^{-\sqrt{10}\left(\frac{a}{d}+1\right)}$ can be neglected (already for $\frac{a}{d} = 1$ this quantity < 0.002). We then obtain:

$$C = -\tfrac{1}{2}\,\frac{1 - \operatorname{ch}\sqrt{10} + \operatorname{sh}\sqrt{10}}{1 + \tfrac{1}{2}(\operatorname{ch}\sqrt{10} - \operatorname{sh}\sqrt{10})(\operatorname{ch}\sqrt{10} - 1)} = -0.3895,$$

$$A = \tfrac{1}{2} + \frac{C}{2}(1 - \operatorname{ch}\sqrt{10}) \cong 2.610,$$

$$B \cong \tfrac{1}{2}\cdot\frac{1}{\operatorname{sh}\sqrt{10}\left(1 + \frac{a}{d}\right)} - A \cong -A = -2.610.$$

For this case we also calculate the quantity $\iint u_1 dxdy$, which is required for the determination of the moment of torsion. We have:

$$\iint\limits_{II} u_1\,dx\,dy = \int\limits_{-d}^{0} dx \int\limits_{-d}^{0}\left\{\left[A \operatorname{ch}\sqrt{10}\frac{x+d}{d} + B \operatorname{sh}\sqrt{10}\frac{x+d}{d} - \tfrac{1}{2} - \right.\right.$$

$$\left.\left. - \frac{C}{2}\left(1 - \operatorname{ch}\sqrt{10}\frac{x}{d}\right)\right]\cdot y(y+d) + \frac{C}{2}x(x+d)(y+d)\right\} dy =$$

$$= \frac{d}{\sqrt{10}}\left[A \operatorname{sh} \sqrt{10} + B(\operatorname{ch} \sqrt{10} - 1) - \frac{\sqrt{10}}{2} + \frac{C}{2}(\operatorname{sh} \sqrt{10} - \sqrt{10})\right]\left(-\frac{d^3}{6}\right) +$$

$$+ \frac{C}{d}\left(-\frac{d^3}{6}\right)\left(\frac{d^2}{2}\right) \cong 0.0726 d^4,$$

$$\iint\limits_{III} u_1\, dx\, dy = \iint\limits_{I} u_1\, dx\, dy = \int_0^a dx \int_{-d}^0 \left[A \operatorname{ch} \sqrt{10}\frac{x+d}{d} + \right.$$

$$\left. + B \operatorname{sh} \sqrt{10}\frac{x+d}{d} - \tfrac{1}{2}\right] y \cdot (y+d) dy = \left[\frac{d}{\sqrt{10}} A \operatorname{sh} \sqrt{10}\left(1 + \frac{a}{d}\right) + \right.$$

$$\left. + \frac{d}{\sqrt{10}} B \operatorname{ch} \sqrt{10}\left(1 + \frac{a}{d}\right) - \tfrac{1}{2}a - \frac{d}{\sqrt{10}}(A \operatorname{sh} \sqrt{10} + B \operatorname{ch} \sqrt{10})\right]\left(-\frac{d^3}{6}\right).$$

If we now utilize the expression for B and by means of it eliminate A from this, we find:

$$\iint\limits_{I} u_1\, dx\, dy \cong \left[\frac{d}{\sqrt{10}} \tfrac{1}{2} \operatorname{cth} \sqrt{10}\left(1 + \frac{a}{d}\right) - \right.$$

$$\left. - \tfrac{1}{2}a - \frac{d}{\sqrt{10}} B(\operatorname{ch} \sqrt{10} - \operatorname{sh} \sqrt{10})\right]\left(-\frac{d^3}{6}\right) \cong$$

$$\cong \tfrac{1}{2}\left(a + \frac{d}{\sqrt{10}}\left[2B(\operatorname{ch} \sqrt{10} - \operatorname{sh} \sqrt{10}) - 1\right]\right)\frac{d^3}{6}.$$

For the moment of torsion we find at last

$$M = 4C\vartheta \iint u_1\, dx\, dy = 4G\vartheta \left[\iint\limits_{I} u_1\, dx\, dy + \iint\limits_{II} u_1\, dx\, dy + \iint\limits_{III} u_1\, dx\, dy\right] \cong$$

$$\cong \left[1 + 0.050\frac{d}{a}\right] \tfrac{2}{3} a d^3 G\vartheta\ ^{1)}.$$

Let us compare the value obtained with the moment of torsion of the oblique corner depicted by the dotted lines in Fig. 26. The latter, if formula (24) is utilized, turns out to be equal to

$$M = \left[1 - 0.3942\frac{d}{a}\right] \tfrac{2}{3} a d^3 G\vartheta.$$

As we see, the difference is quite considerable [2]).

3. Examples of the refinement of the solution. Here, in a single example — the solution of the Poisson equation for a rectangle — we shall show methods of obtaining a more exact solution than that given by the first approximation.

---

1) The method of reduction to ordinary differential equations is applied in a different manner to a problem of the torsion of rolled profiles in a work of N. K͡H. Arutiu͡nian, [2].

2) An application of the method under consideration to a problem with three variables is given in a work of N. Bystrov, [1].

1) *The construction of a second approximation.* We shall now seek the solution of the equation $\Delta u = -1$ for the rectangle $[-a, a; -b, b]$ in the form of a polynomial of the fourth degree in y, introducing two unknown functions f_1 and f_2

$$u = (y^2 - b^2)(f_1(x) + f_2(x)y^2). \tag{30}$$

Equations (7b) here take the form:

$$\left.\begin{aligned} &\int_{-b}^{b} [(y^2 - b^2)(f_1'' + f_2''y^2) + 2f_1 + f_2(12y^2 - 2b^2) + 1](y^2 - b^2)\,dy = 0, \\ &\int_{-b}^{b} [(y^2 - b^2)(f_1'' + f_2''y^2) + 2f_1 + f_2(12y^2 - 2b^2) + 1](y^2 - b^2)y^2\,dy = 0, \end{aligned}\right\} \tag{31}$$

or after integration with respect to y:

$$\left.\begin{aligned} \tfrac{8}{15}b^5 f_1'' + \tfrac{8}{105}b^7 f_2'' - \tfrac{4}{3}b^3 f_1 - \tfrac{4}{15}b^5 f_2 - \tfrac{2}{3}b^3 = 0, \\ \tfrac{8}{105}b^7 f_1'' + \tfrac{8}{315}b^9 f_2'' - \tfrac{4}{15}b^5 f_1 - \tfrac{44}{105}b^7 f_2 - \tfrac{2}{15}b^5 = 0. \end{aligned}\right\} \tag{31'}$$

For the solution of the corresponding homogeneous system we introduce an auxiliary function V, setting [1])

$$f_1 = -\left(\tfrac{8}{105}b^4 D^2 - \tfrac{4}{15}b^2\right)V, \quad f_2 = \left(\tfrac{8}{15}b^2 D^2 - \tfrac{4}{3}\right)V. \tag{32}$$

Then for the determination of V we shall obtain the equation:

$$\left[\left(\tfrac{8}{15}b^2 D^2 - \tfrac{4}{3}\right)\left(\tfrac{8}{315}b^2 D^2 - \tfrac{44}{105}\right) - \left(\tfrac{8}{105}b^2 D^2 - \tfrac{4}{15}\right)^2\right]V = 0.$$

The characteristic equation, after collection of terms, is

$$b^4 D^4 - 28 b^2 D^2 + 63 = 0. \tag{33}$$

Its roots are $\pm \alpha \frac{1}{b}$ and $\pm \gamma \frac{1}{b}$, where

$$\left.\begin{aligned} \alpha &= \sqrt{14 - \sqrt{133}} = \sqrt{2.467437} = 1.570815, \\ \gamma &= \sqrt{14 + \sqrt{133}} = \sqrt{25.5326} = 5.05297. \end{aligned}\right\} \tag{34}$$

Hence the general solution for V will be

$$V = A \operatorname{ch} \alpha \frac{x}{b} + A_1 \operatorname{sh} \alpha \frac{x}{b} + C \operatorname{ch} \gamma \frac{x}{b} + C_1 \operatorname{sh} \gamma \frac{x}{b}.$$

Here, since the solution must be an even function of x, it is possible to at once adopt $A_1 = C_1 = 0$. Hence for f_1 and f_2, if we yet take into consideration the particular solutions $f_1 = -\frac{1}{2}$, $f_2 = 0$, we obtain the

---

1) D is the operator for differentiation. See Smirnov, [1], vol. II, 1948, p. 128.

following expressions:

$$f_1 = -b^2(\tfrac{8}{105}\alpha^2 - \tfrac{4}{15})A \operatorname{ch} \alpha \frac{x}{b} - b^2(\tfrac{8}{105}\gamma^2 - \tfrac{4}{15})C \operatorname{ch} \gamma \frac{x}{b} - \tfrac{1}{2} =$$

$$= 0.0786b^2 A \operatorname{ch} \alpha \frac{x}{b} - 1.6787 b^2 C \operatorname{ch} \gamma \frac{x}{b} - \tfrac{1}{2},$$

$$f_2 = (\tfrac{8}{15}\alpha^2 - \tfrac{4}{3})A \operatorname{ch} \alpha \frac{x}{b} + (\tfrac{8}{15}\gamma^2 - \tfrac{4}{3})C \operatorname{ch} \gamma \frac{x}{b} =$$

$$= -0.01738 A \operatorname{ch} \alpha \frac{x}{b} + 12.284 C \operatorname{ch} \gamma \frac{x}{b}.$$

Now from the conditions $f_1(a) = f_2(a) = 0$ we determine the constants A and C, and find at last:

$$f_1(x) = 0.51558 \frac{\operatorname{ch} \alpha \frac{x}{b}}{\operatorname{ch} \alpha \frac{a}{b}} - 0.01558 \frac{\operatorname{ch} \gamma \frac{x}{b}}{\operatorname{ch} \gamma \frac{a}{b}} - \tfrac{1}{2},$$

$$f_2(x) = -0.11400 \frac{\operatorname{ch} \alpha \frac{x}{b}}{b^2 \operatorname{ch} \alpha \frac{a}{b}} + 0.11400 \frac{\operatorname{ch} \gamma \frac{x}{b}}{b^2 \operatorname{ch} \gamma \frac{a}{b}},$$

$$\bar{u}(x, y) = (y^2 - b^2) \left[0.51558 \frac{\operatorname{ch} \alpha \frac{x}{b}}{\operatorname{ch} \alpha \frac{a}{b}} - 0.01558 \frac{\operatorname{ch} \gamma \frac{x}{b}}{\operatorname{ch} \gamma \frac{a}{b}} - \tfrac{1}{2} + \right.$$

$$\left. + \left(-0.11400 \frac{\operatorname{ch} \alpha \frac{x}{b}}{b^2 \operatorname{ch} \alpha \frac{a}{b}} + 0.11400 \frac{\operatorname{ch} \gamma \frac{x}{b}}{b^2 \operatorname{ch} \gamma \frac{a}{b}} \right) y^2 \right]. \tag{35}$$

For verifying the accuracy of the solution we compute its values for $a = 2$, $b = 1$, at two points: we obtain $\bar{u}(0, 0) = 0.45552$, $\bar{u}(1, \tfrac{1}{2}) = 0.29586$. The exact values are respectively equal to 0.45549 and 0.29501.

2) *The satisfaction of the equation on the boundaries.* A solution in the same form as that of the preceding case, somewhat less exact, but with considerably simpler calculations, may be obtained in another way.

We shall seek the solution in the same form (30), but shall introduce beforehand the requirement that on the lines $y = \pm b$ the equation $\Delta u = -1$ be satisfied [1]).

[1]) The satisfaction of the equation on certain lines, like the method of the approximate replacement of a partial differential equation by one or a system of ordinary equations, was proposed in a note of L. V. Kantorovich, [11]. Here we combine this device with the variational method.

This gives us

$$[(y^2 - b^2)(f_1'' + f_2''y^2) + 2f_1 + f_2(12y^2 - 2b^2) + 1]_{y=b} = 0. \tag{36}$$

Hence $f_2 = -\frac{1}{10b^2}(2f_1 + 1)$. Substituting this expression in (30), we must now seek u in the form

$$u(x, y) = f_1\left(-\frac{1}{5b^2}y^4 + \tfrac{6}{5}y^2 - b^2\right) - \frac{1}{10b^2}(y^4 - b^2y^2). \tag{37}$$

Then equation (7b) will take the form

$$\int_{-b}^{b}\Big[f_1''\left(-\frac{1}{5b^2}y^4 + \tfrac{6}{5}y^2 - b^2\right) + f_1\left(-\frac{12}{5b^2}y^2 + \tfrac{12}{5}\right) -$$

$$- \frac{1}{10b^2}(12y^2 - 2b^2) + 1\Big] \times \left(-\frac{1}{5b^2}y^4 + \tfrac{6}{5}y^2 - b^2\right)dy = 0,$$

or, after integration with respect to y,

$$25.19365b^2f_1'' - 62.17143f_1 - 31.08572 = 0. \tag{38}$$

Solving this equation under the condition $f_1(a) = 0$, we find

$$f_1 = C \operatorname{ch} \alpha \frac{x}{b} - \tfrac{1}{2},$$

where $\alpha = 1.57092$, or finally:

$$f_1(x) = \tfrac{1}{2}\left(\frac{\operatorname{ch} \alpha \frac{x}{b}}{\operatorname{ch} \alpha \frac{a}{b}} - 1\right), \tag{39}$$

$$\bar{u}(x, y) = \tfrac{1}{2}\left(-\frac{1}{5b^2}y^4 + \tfrac{6}{5}y^2 - b^2\right)\left(\frac{\operatorname{ch} \alpha \frac{x}{b}}{\operatorname{ch} \alpha \frac{a}{b}} - 1\right) - \frac{1}{10b^2}(y^4 - b^2y^2).$$

Again computing $\bar{u}(0, 0)$ and $\bar{u}(1, \tfrac{1}{2})$ for $a = 2$, $b = 1$, we find

$$\bar{u}(0, 0) = 0.456865, \quad \bar{u}(1, \tfrac{1}{2}) = 0.297886.$$

It should be remarked, granted, that the approximate solution obtained here, as distinct from the solution obtained in 1) and in the introduction to this section, does not exactly satisfy the boundary condition $u = 0$ on the lines $x = \pm a$, satisfying it only approximately.

3) *The exact solution of the problem.* We shall now show that with another choice of the form of the solution as a function of y, one can arrive by this method at the exact solution of the problem. We shall,

namely, seek the solution in the form

$$u(x, y) = \cos \frac{\pi}{2} \frac{y}{b} f_1(x) + \cos 3 \frac{\pi}{2} \frac{y}{b} f_2(x) + + \cos 5 \frac{\pi}{2} \frac{y}{b} f_3(x) + \ldots \qquad (40)$$

Then the satisfaction of the boundary condition $u = 0$ is secured on the sides $y = \pm b$. Equations (7b) take the form:

$$\int_{-b}^{b} (\Delta u + 1) \cos\left(s\pi - \frac{\pi}{2}\right)\frac{y}{b}\,dy = \int_{-b}^{b} \Big[\sum_{k=1}^{\infty} \cos\left(k\pi - \frac{\pi}{2}\right)\frac{y}{b} f_k'' - \sum_{k=1}^{\infty} (k - \tfrac{1}{2})^2 \frac{\pi^2}{b^2} \cos\left(k\pi - \frac{\pi}{2}\right)\frac{y}{b} f_k + 1\Big] \times \cos\left(s\pi - \frac{\pi}{2}\right)\frac{y}{b}\,dy = 0 \qquad (s = 1, 2, \ldots), \qquad (41)$$

or after integration with respect to y, taking into account the mutual orthogonality of the functions $\cos\left(k\pi - \frac{\pi}{2}\right)\frac{y}{b}$, we obtain the system in this form:

$$bf_s'' - (s - \tfrac{1}{2})^2 \frac{\pi^2}{b} f_s - (-1)^s \frac{2b}{\pi(s - \frac{1}{2})} = 0 \qquad (s = 1, 2, \ldots). \qquad (41')$$

In the given case the unknown functions have separated; in each equation there figures only one function. Solving the equations under the conditions $f_s(a) = f_s(-a) = 0$, we find

$$f_s(x) = \frac{(-1)^{s-1} 2b^2}{\pi^3 (s - \frac{1}{2})^3} \left[1 - \frac{\operatorname{ch}(s - \frac{1}{2})\pi \frac{x}{b}}{\operatorname{ch}(s - \frac{1}{2})\pi \frac{a}{b}}\right], \qquad (42)$$

whence

$$u(x, y) = \sum_{s=1}^{\infty} \frac{(-1)^{s-1} 16b^2}{\pi^3 (2s - 1)^3} \left[1 - \frac{\operatorname{ch}\,(2s - 1)\frac{\pi}{2}\frac{x}{b}}{\operatorname{ch}\,(2s - 1)\frac{\pi}{2}\frac{a}{b}}\right] \cos\,(2s - 1)\frac{\pi}{2}\frac{y}{b}. \qquad (43)$$

But the first part of the given sum

$$\frac{16b^2}{\pi^3} \sum_{s=1}^{\infty} \frac{(-1)^{s-1}}{(2s - 1)^3} \cos\,(2s - 1)\frac{\pi}{2}\frac{y}{b}, \qquad (44)$$

as one is readily convinced, is the expansion of the function $\frac{1}{2}(b^2 - y^2)$ in the interval $(0, b)$ in a series involving the functions $\cos \frac{\pi}{2} \frac{y}{b}$, $\cos 3 \frac{\pi}{2} \frac{y}{b}, \ldots$. Therefore the solution can finally be given this form:

$$u(x, y) =$$

$$= \tfrac{1}{2}(b^2 - y^2) - \frac{16b^2}{\pi^3} \sum_{n=1, 3, 5 \ldots} \frac{(-1)^{\frac{n-1}{2}} \operatorname{ch} n \frac{\pi}{2} \frac{x}{b}}{\operatorname{ch} n \frac{\pi}{2} \frac{a}{b}} \cos n \frac{\pi}{2} \frac{y}{b}, \quad (45)$$

which differs from the solution of the same problem which was obtained in Chapter I (p. 11) only in the disposition of the coordinate system.

The fact that (45) actually represents the solution of the problem may be proved directly, too; one can verify, namely, that for the series formed for $\Delta u + 1$, a term-by-term integration with respect to y is admissible after a premultiplication by $\cos (2s - 1) \frac{\pi}{2} \cdot \frac{y}{b}$.

Since $f_s(x)$ satisfies equations (41′), equations (41) are fulfilled; these show that for any x the expression for $\Delta u + 1$, as a function of y, is orthogonal to all functions $\cos \frac{\pi}{2} \frac{y}{b}$, $\cos 3 \frac{\pi}{2} \frac{y}{b}, \ldots$, which is possible only if it is identically zero. Thus u satisfies the equation $\Delta u + 1 = 0$; the fact that $u = 0$ on the contour is verifiable directly.

We note in addition that even for the general equation $\Delta u = p(x, y)$ the exact solution can be found for the rectangle by this same method [1]).

4) *Some numerical methods.* In the case of a comparatively complicated contour, the exact solution of the system of differential equations that is obtained for the determination of the functions $f_1, f_2, \ldots$ turns out to be impossible. In such a case the numerical integration of the system, which for ordinary equations has been well worked out, can be recommended.

Another way is also possible. The contour of the region may be approximately replaced by a polygon. Then the equations for the determination of the functions $f_1, f_2, \ldots$ will be equations of the Euler type (see No. 2, Example 1), and can therefore be integrated in finite form [2]).

---

1) See L. V. Kantorovich, [10].

2) The method in which the partial differential equations are replaced by a system of ordinary differential equations has been developed by M. G. Slobodianskii, [1], [2], in another form: the replacement of the derivatives with respect to one of the variables by differences. See also the work of V. N. Faddeeva, [2], where a further elaboration of this method is given, as well as auxiliary tables, and the interrelations of this method and that expounded in L. V. Kantorovich's work [11] are considered.

4. An example of the application of the method to the biharmonic equation. Let us consider the problem of the deflection of a rectangular plate clamped along the sides $y = \pm b$ and with any mode of mounting along the sides $x = \pm a$.

The problem reduces to the solution of the equation

$$\Delta\Delta u = \frac{\partial^4 u}{\partial x^4} + 2\frac{\partial^4 u}{\partial x^2 \partial y^2} + \frac{\partial^4 u}{\partial y^4} = \frac{p(x, y)}{D}, \tag{46}$$

where p is the load function and D is a constant connected with the elastic constants. The boundary conditions will be

$$u = 0, \quad \frac{\partial u}{\partial y} = 0 \text{ for } y = \pm b, \tag{47}$$

and also the conditions on the sides $x = \pm a$ that answer to the manner in which they are mounted. In order to secure the satisfaction of conditions (47), the solution must be sought in the form

$$u = \varphi_1(y)f_1(x) + \varphi_2(y)f_2(x) + \ldots + \varphi_n(y)f_n(x), \tag{48}$$

where $\varphi_1(y), \varphi_2(y), \ldots$ are functions satisfying the conditions

$$\varphi_k(\pm b) = \varphi_k'(\pm b) = 0 \quad (k = 1, 2, \ldots, n). \tag{49}$$

Such functions may be chosen in different fashions, in particular — in case all the conditions of the problem are symmetric with respect to the axis OX, and by that fact u is an even function of y — one may put

$$\varphi_k(y) = (y^2 - b^2)^2 \cdot y^{2k-2}. \tag{50}$$

Stopping on this case and limiting ourselves to the first approximation, we must adopt

$$\bar{u}(x, y) = (y^2 - b^2)^2 f(x). \tag{51}$$

In the given case, just as in the case of the Poisson equation, the equation may be used in Galerkin's form. We shall therefore obtain the differential equation for the determination of f in this form:

$$\int_{-b}^{b} \left(\Delta\Delta u - \frac{p}{D}\right)\varphi_1(y)dy =$$

$$= \int_{-b}^{b} \left[24f + 2(12y^2 - 4b^2)f'' + (y^2 - b^2)f^{IV} - \frac{p}{D}\right] \cdot (y^2 - b^2)^2 dy = 0, \tag{52}$$

or

$$\frac{2 \cdot 128}{315} b^9 f^{IV} - \frac{2 \cdot 256}{315} b^7 f'' + \frac{2 \cdot 64}{5} b^5 f = p_1(x), \tag{52'}$$

where

$$p_1(x) = \int_{-b}^{b} \frac{p(x, y)}{D} (y^2 - b^2)^2 dy.$$

The characteristic equation has the form:

$$\frac{2 \cdot 128}{315} b^9 \lambda^4 - \frac{2 \cdot 256}{315} b^7 \lambda^2 + \frac{2 \cdot 64}{5} b^5 = 0,$$

its roots being

$$\lambda_{1,2,3,4} = (\pm \alpha \pm \beta i) \frac{1}{b}, \quad \alpha = 2.075, \ \beta = 1.143.$$

The general solution of equation (52′) will be

$$f(x) = A_1 \operatorname{ch} \alpha \frac{x}{b} \cos \beta \frac{x}{b} + A_2 \operatorname{ch} \alpha \frac{x}{b} \sin \beta \frac{x}{b} + B_1 \operatorname{sh} \alpha \frac{x}{b} \sin \beta \frac{x}{b} +$$

$$+ B_2 \operatorname{sh} \alpha \frac{x}{b} \cos \beta \frac{x}{b} + f_0(x),$$

where $f_0(x)$ is a particular solution of equation (52′). If the load is constant ($p = \text{const.}$), then $p_1 = \frac{2 \cdot 8}{15} b^5 \frac{p}{D}$, and therefore $f_0 = \frac{1}{24} \frac{p}{D}$. Moreover, if the conditions on the sides $x = + a$ and $x = - a$ are identical, $u(x, y)$ and therefore f too must be even functions of x, and $f(x)$ will take the form

$$f(x) = A \operatorname{ch} \alpha \frac{x}{b} \cos \beta \frac{x}{b} + B \operatorname{sh} \frac{x}{b} \sin \beta \frac{x}{b} + \frac{1}{24} \frac{p}{D}. \tag{53}$$

The constants A and B must be determined in accordance with the mounting conditions of the sides $x = \pm a$. In particular, if these sides are also clamped, they are determined from the equations

$$f(a) = A \operatorname{ch} \alpha \frac{a}{b} \cos \beta \frac{a}{b} + B \operatorname{sh} \alpha \frac{a}{b} \sin \beta \frac{a}{b} + \frac{1}{24} \frac{p}{D} = 0,$$

$$f'(a) = A \left(\frac{\alpha}{b} \operatorname{sh} \alpha \frac{a}{b} \cos \beta \frac{a}{b} - \frac{\beta}{b} \operatorname{ch} \alpha \frac{a}{b} \sin \beta \frac{a}{b} \right) +$$

$$+ B \left(\frac{\alpha}{b} \operatorname{ch} \alpha \frac{a}{b} \cdot \sin \beta \frac{a}{b} + \frac{\beta}{b} \operatorname{sh} \alpha \frac{a}{b} \cos \beta \frac{a}{b} \right) = 0,$$

whence

$$A = \frac{\gamma_1}{\gamma_0} \cdot \frac{1}{24} \frac{p}{D}, \quad B = \frac{\gamma_2}{\gamma_0} \cdot \frac{1}{24} \frac{p}{D},$$

where

$$\gamma_0=\beta \operatorname{sh} \alpha\mu \operatorname{ch} \alpha\mu+\alpha \sin \beta\mu \cos \beta\mu,\ \gamma_1=-(\alpha \operatorname{ch} \alpha\mu \sin \beta\mu+\beta \operatorname{sh} \alpha\mu \cos \beta\mu),$$

$$\gamma_2 = \alpha \operatorname{sh} \alpha\mu \cos \beta\mu - \beta \operatorname{ch} \alpha\mu \sin \beta\mu,\quad \mu = \frac{a}{b}. \tag{54}$$

The final form of the approximate solution will be

$$u(x, y) = (b^2 - y^2)^2 \frac{1}{24\gamma_0} \frac{p}{D} \left(\gamma_1 \operatorname{ch} \alpha \frac{x}{b} \cos \beta \frac{x}{b} + \gamma_2 \operatorname{sh} \alpha \frac{x}{b} \sin \beta \frac{x}{b}\right) + \\ + \tfrac{1}{24} \frac{p}{D} (y^2 - b^2)^2. \tag{55}$$

It is easy to satisfy oneself that if $a = \infty$ and the ratio $\mu = \dfrac{a}{b} = \infty$, then (55) gives $u = \frac{1}{24} \dfrac{p}{D} (y^2 - b^2)^2$, the exact solution for the given case. We shall also exhibit some numerical data for the case of a square plate $\mu = 1$.

For the maximum deflection we find:

$$u(0, 0) = 0.479 \tfrac{1}{24} \frac{p}{D} b^4 = 0.01363 \frac{p(2b)^4}{Eh^3},$$

$$\left(D = \frac{Eh^3}{12(1 - \nu^2)},\ \nu = 0.3\right),$$

instead of the exact 0.0138. For the moment in the middle we obtain [1])

$$\sigma_x = 0.140 \frac{p(2b)^2}{h^2},\quad \sigma_y = 0.138 \frac{p(2b)^2}{h^2},$$

instead of the exact $\sigma_x = \sigma_y = 0.137 \dfrac{p(2b)^2}{h^2}$. As we see, the error in the deflection, and in the moment at the center, does not, even in this worst case, exceed 1 to 2%.

With appropriate modifications this method is applicable even to other cases of the mounting of plates. Thus, for example, if the sides $y = \pm b$ are supported, the functions $\varphi_k(y)$ should be chosen so that $\varphi_k(\pm b) = \varphi_k''(\pm b) = 0$. In particular, one can adopt $\varphi_1(y) = y^4 - 6b^2y^2 + 5b^4$. In case the side $y=-b$ is built-in, but $y=+b$ is free, the first approximation for u can be assigned in the form $u = (y + b)^2 f(x)$.

---

[1]) The numerical data and some computations are taken from the work of Engineer Semenov, [1]. Other applications of this method in the theory of plates are given in works of T. N. Rogov, [1], (triangular plate), I. A. Baslavskii, [1], (plate of variable stiffness). In the theory of shells a similar method has been employed by V. Z. Vlasov, [1].

The integral $I(u)$, however, must in the given case be taken in complete form [§ 1 (28)]. Here, since we are not initially satisfying the boundary conditions, the equation cannot be used in Galerkin's form.

5. Application of the method to the determination of proper values and proper functions. Like the methods of Ritz and of Galerkin, the method of reduction to ordinary equations can be used with success even for the problem of finding the proper values and functions.

For definiteness let us consider the equation

$$\Delta u + \lambda u = 0, \tag{56}$$

under the condition $u = 0$ on Γ. We shall seek the solution in a form

$$u_n = \sum_{k=1}^{n} \chi_k(x, y) f_k(x) + \chi_0(x, y)$$

such that the expression for u_n vanishes everywhere on Γ by virtue of the choice of the functions χ_k and of the boundary conditions for f_k. The task of setting up the necessary equations can be approached most simply from the considerations of Galerkin's method. In this case we have that the functions $f_1, \ldots, f_n$ must satisfy the system of equations

$$\int_{D_x} [\Delta u_n + \lambda u_n] \chi_s dy = 0 \qquad (s = 1, 2, \ldots, n) \tag{57}$$

and must vanish for the terminal values of the argument. As a rule the sole solution of system (57) under such conditions will be the trivial one, $f_s = 0$ $(s = 1, 2, \ldots, n)$; those values of λ for which there is a non-trivial solution will give approximate values for the proper values, and the solutions themselves will give approximations to the proper functions. One could have arrived at these same equations and conditions by starting from the extremal properties of the proper values and functions (§ 1, No. 5); namely, by reasoning as in § 2, No. 5, we would here, for the determination of the functions f_s, have arrived at an isoperimetric problem, which would have again led us to equations (57).

We shall illustrate the application of the method in an example, using the rectangle $[-a, a; -b, b]$. We again seek the solution in the form $u = (y^2 - b^2)f$. Equation (57) will take the form:

$$\int_{-b}^{b} [2f + (y^2 - b^2)f'' + \lambda(y^2 - b^2)f](y^2 - b^2)dy = 0, \tag{58}$$

or

$$\tfrac{16}{15}b^5 f'' + (\tfrac{16}{15}b^5\lambda - \tfrac{8}{3}b^3)f = 0, \tag{58'}$$

under the conditions $f(a) = f(-a) = 0$. The general solution of equation (58′) is

$$f = A \sin \sqrt{\lambda - \frac{5}{2b^2}}\, x + B \cos \sqrt{\lambda - \frac{5}{2b^2}}\, x.$$

If the symmetry and the end condition $f(a) = 0$ be taken into account, we find $A = 0$, $B \cos \sqrt{\lambda - \frac{5}{2b^2}}\, a = 0$, whence it is clear that a non-trivial solution is obtained only if

$$\sqrt{\lambda - \frac{5}{2b^2}}\, a = (2k-1)\frac{\pi}{2}, \quad \lambda = \frac{(2k-1)^2\pi^2}{(2a)^2} + \frac{5}{2b^2}.$$

In particular for $k = 1$, we find $\lambda = \frac{\pi^2}{(2a)^2} + \frac{10}{(2b)^2}$ rather than the exact value $\lambda = \frac{\pi^2}{(2a)^2} + \frac{\pi^2}{(2b)^2}$, the error being less than 1.3%. Next determining the function f for the given λ, we obtain for the first proper function the approximation

$$u = (y^2 - b^2)\cos\frac{\pi}{2}\,\frac{x}{a}.$$

In this problem all the devices for refining the solution which were indicated in No. 3 may be employed. We shall employ the device given in No. 3, 2.

Thus we shall seek the solution of the same problem in the form

$$u = (y^2 - b^2)(f_1 + f_2 y^2).$$

The requirement that u must satisfy the equation $\Delta u + \lambda u = 0$ for $y = \pm b$ gives $f_1 = -5b^2 f_2$. Taking this into consideration, we must seek the solution in the form

$$u = (y^2 - b^2)(y^2 - 5b^2)f.$$

Equation (57) in the given case takes the form

$$\int_{-b}^{b} (\Delta u + \lambda u)\chi_1 dy = \int_{-b}^{b} [(y^2 - b^2)(y^2 - 5b^2)f'' + 12(y^2 - b^2)f +$$
$$+ \lambda(y^2 - b^2)(y^2 - 5b^2)f](y^2 - b^2)(y^2 - 5b^2)dy = 0$$

or, if we perform the integration,

$$62b^2 f'' + (62b^2\lambda - 153)f = 0.$$

Hence the even solution of the equation will be

$$f(x) = C\cos\sqrt{\lambda - \frac{153}{62b^2}}\, x.$$

If we now take into consideration the condition $f(a) = 0$, we see that a non-trivial solution is obtained only when

$$\cos\sqrt{\lambda - \frac{153}{62b^2}}\, a = 0, \quad \sqrt{\lambda - \frac{153}{62b^2}}\, a = (2k-1)\frac{\pi}{2}.$$

In particular for $k = 1$ we find

$$\lambda = \frac{\pi^2}{4a^2} + \frac{153}{62b^2} = \frac{\pi^2}{(2a)^2} + \frac{\frac{306}{31}}{(2b)^2}.$$

In the worst case $b = a$ we obtain $\lambda = \dfrac{9.87028}{2a^2}$ rather than the exact $\lambda = \dfrac{\pi^2}{2a^2} = \dfrac{9.86960}{2a^2}$; the error is equal to 0.007%.

§ 4. ESTIMATE OF THE ERROR IN VARIATIONAL METHODS AND THEIR ORDER OF CONVERGENCE

The object of this section is to expound some investigations into the question of the estimation of the maximum error that is obtained at the nth step in the approximate solution by Ritz's method, and also the order, relative to $\dfrac{1}{n}$, with which this error tends to zero as $n \to \infty$. As regards ordinary equations, there exist very detailed estimates for a number of cases, thanks to the work of Acad. N. M. Krylov. As regards partial differential equations, the question has been much less worked out, and there we must forgo exact estimates of the error and state only a result indicating the rapidity of its approach to zero. We give an investigation of the latter question for Ritz's method and for the method of reduction to ordinary equations as well.

1. The case of ordinary differential equations. We shall here expound a few results only of the numerous investigations of Acad. N. M. Krylov. The fundamental idea of these investigations is to obtain the most exact estimates possible of the error in the nth approximation, i.e., estimates as close as possible to the real error, so that the results obtained may be used for small values of $n = 1, 2, 3, 4, 5, 6$. On the other hand, the condition is laid down that this estimate of the error must contain only quantities that can be determined easily from the data of the problem (the coefficients of the equation, etc.).

Given these conditions, the question can be satisfactorily resolved only in case it is not considered for a too-general class of problems: it must be considered for a definite type of equation, a concrete choice of the fundamental functions, etc.

To show the methods that are applicable to the obtaining of such estimates, we shall set forth the derivations of some of the latter.

Let us consider the problem of finding the solution of the equation

$$\frac{d}{dx}(py') - qy = f \tag{1}$$

under the boundary conditions

$$y(0) = y(1) = 0 \tag{2}$$

by Ritz's method. We assume in addition that these inequalities are fulfilled:

$$p(x) > 0, \ q(x) \geqslant 0 \text{ in } [0, 1]. \tag{3}$$

As a preliminary we shall give an estimate of the solution of this equation, its derivatives, and several quantities connected with them. It will be convenient for us to introduce the notation $\|g\| = [\int_0^1 g^2(x)dx]^{\frac{1}{2}}$ if $g(x)$ is a square-integrable function in $[0, 1]$. If g and h are two such functions, then the Buniakovskiĭ and Cauchy inequalities are valid:

$$|\int_0^1 ghdx| \leqslant \|g\| \, \|h\| \text{ and } \|g + h\| \leqslant \|g\| + \|h\|; \tag{4}$$

we note in addition that $\|g\| \leqslant \max |g|$; $\|g\| \min |h| \leqslant \|hg\| \leqslant \|g\| \max |h|$ and $\|ag\| = |a| \cdot \|g\|$, if a is a constant.

Since the function y gives a minimum for the integral

$$I(y) = \int_0^1 [py'^2 + qy^2 + 2fy]dx, \tag{5}$$

then for any η equal to zero at the ends, the equality $\delta I = 0$ is valid [see § 1 (3)]; setting $\eta = y$ in it, we find

$$\int_0^1 (py'^2 + qy^2 + fy)dx = 0.$$

Hence

$$\int_0^1 py'^2 \, dx \leqslant |\int_0^1 fy \, dx| \leqslant \|f\| \cdot \|y\|,$$

or

$$\|y'\|^2 = \int_0^1 y'^2 \, dx \leqslant \frac{1}{\min p} \|f\| \cdot \|y\|. \tag{6}$$

Next we shall utilize V. A. Steklov's inequality, true for a function satisfying conditions (2) or the condition $\int_0^1 y \, dx = 0$,

$$\|y\| \leqslant \frac{1}{\pi} \|y'\|, \tag{7}$$

which is easily obtained on the basis of the equation of closure. Indeed, if $y = \sum_{k=1}^{\infty} b_k \sin k\pi x$, we have

$$\|y\|^2 = \tfrac{1}{2} \sum_{k=1}^{\infty} b_k^2, \quad \|y'\|^2 = \tfrac{1}{2} \sum_{k=1}^{\infty} (b_k k\pi)^2,$$

whence (7) follows at once[1]). Comparing (6) and (7), we find:

$$\|y'\| \leqslant \frac{1}{\pi \min p} \|f\|. \tag{8}$$

Let us now estimate $|y(x)|$. Since in both parts of the interval $[0, 1]$ the estimate is the same, we shall assume that x lies in the interval $[0, \frac{1}{2}]$. We obtain[2])

$$|y(x)| = |\int_0^x y'(x)dx| \leqslant \sqrt{\int_0^x y'^2 dx} \sqrt{\int_0^x 1^2 dx} \leqslant \frac{1}{\sqrt{2}} \|y'\|, \tag{9}$$

or on the basis of (8)

$$\max |y(x)| \leqslant \frac{1}{\sqrt{2}} \|y'\| \leqslant \frac{1}{\pi\sqrt{2} \min p} \|f\|.$$

Finally, from equation (1) we find

$$py'' = -p'y' + qy + f,$$

whence, on the basis of (8) and (7),

$$\|py''\| \leqslant \|p'y'\| + \|qy\| + \|f\| \leqslant \max |p'| \, \|y'\| + \max q \, \|y\| + \|f\| \leqslant$$
$$\leqslant \left(\frac{\max |p'|}{\pi \min p} + \frac{\max q}{\pi^2 \min p} + 1\right) \|f\|.$$

Hence

$$\|y''\| \leqslant \frac{1}{\pi (\min p)^2} \left(\max |p'| + \frac{1}{\pi} \max q + \pi \min p\right) \|f\|. \tag{10}$$

Let us now consider the application of Ritz's method to the given problem, with the fundamental functions

$$\varphi_k(x) = \sqrt{2} \sin k\pi x \qquad (k = 1, 2, \ldots). \tag{11}$$

By y_n we shall designate the nth approximation for y by Ritz's method; by Y_n, the segment $\sum_{k=1}^{n} b_k \sin k\pi x$ consisting of the first n terms of the Fourier series for $y(x)$.

At the outset we shall note the following inequality:

$$\|y - Y_n\| \leqslant \frac{1}{\pi(n+1)} \|y' - Y_n'\| \leqslant \frac{1}{\pi^2(n+1)^2} \|y'' - Y_n''\|, \tag{12}$$

---

[1]) For more detail see G. M. Fikhtengol'ts, [1], vol. III, p. 716 (1949).

[2]) Somewhat more complicated reasoning permits the replacement here of the factor $\frac{1}{\sqrt{2}}$ by $\frac{1}{2}$.

which is obtained by means of the equation of closure, and from which it follows that

$$\|y - Y_n\|^2 = \tfrac{1}{2}\sum_{n+1}^{\infty} b_k^2, \quad \|y' - Y'_n\|^2 = \tfrac{1}{2}\sum_{n+1}^{\infty} \pi^2 k^2 b_k^2, \quad \|y'' - Y''_n\|^2 = \tfrac{1}{2}\sum_{n+1}^{\infty} \pi^4 k^4 b_k^2. \quad (12)$$

Now on the basis of the fact that of all functions of the form $\sum_1^n a_k \varphi_k$, the function y_n gives the least value for the integral I, (5), we have $I(y_n) - I(y) \leqslant I(Y_n) - I(y)$, whence [cf. § 1 (7)], using (12), we find:

$$\int_0^1 [p(y'_n - y')^2 + q(y_n - y)^2]dx \leqslant \int_0^1 [p(Y'_n - y')^2 + q(Y_n - y)^2]dx \leqslant$$

$$\leqslant \left\{\max p + \frac{\max q}{(n+1)^2\pi^2}\right\} \|Y'_n - y'\|^2.$$

Then

$$\|y'_n - y'\|^2 \leqslant \frac{1}{\min p}\left\{\max p + \frac{\max q}{(n+1)^2\pi^2}\right\} \|Y'_n - y'\|^2. \quad (13)$$

Furthermore on the basis of (12) and (10) we obtain

$$\|Y'_n - y'\| \leqslant \frac{1}{\pi(n+1)} \|Y''_n - y''\| \leqslant \frac{1}{\pi(n+1)} \|y''\| \leqslant$$

$$\leqslant \frac{\max |p'| + \dfrac{1}{\pi} \max q + \pi \min p}{\pi^2(n+1)(\min p)^2} \cdot \|f\|.$$

Replacing y in (9) by $(y_n - y)$, we find

$$\max |y_n(x) - y(x)| \leqslant \frac{1}{\sqrt{2}} \|y'_n - y'\|.$$

As the result, (13) gives us the required estimate of the error:

$$\max |y_n(x) - y(x)| \leqslant \frac{L}{n+1},$$

where

$$L = \left\{\max p + \frac{\max q}{(n+1)^2\pi^2}\right\}^{\frac{1}{2}} \frac{1}{\pi^2\sqrt{2}(\min p)^{5/2}} \Big[\max |p'| +$$

$$+ \frac{1}{\pi} \max q + \pi \min p\Big] \Big[\int_0^1 f^2 dx\Big]^{\frac{1}{2}}. \quad (14)$$

We shall now show another, more refined method, which permits one to obtain more exact estimates. This method utilizes essentially the properties of the so-called Green function. In order not to complicate the exposition unduly, we shall limit ourselves to the case when $p(x) = 1$.

At the outset we shall again give an estimate for $\|y''\|$. We have:

$$y'' = qy + f,\ y''^2 = qyy'' + fy'',$$

whence

$$\int_0^1 \frac{y''^2}{q}\, dx = \int_0^1 yy''\, dx + \int_0^1 \frac{f}{q} y''\, dx \leqslant \left| \int_0^1 \frac{f}{q} y''\, dx \right|, \tag{15}$$

for

$$\int_0^1 yy''\, dx = [yy']_0^1 - \int_0^1 y'^2 dx = -\int_0^1 y'^2 dx \leqslant 0.$$

From the inequality obtained we find

$$\int_0^1 \frac{y''^2}{q}\, dx = \left\| \frac{y''}{\sqrt{q}} \right\|^2 \leqslant \left| \int_0^1 \frac{f}{q} y''\, dx \right| \leqslant \left\| \frac{f}{\sqrt{q}} \right\| \cdot \left\| \frac{y''}{\sqrt{q}} \right\|$$

(we assume the existence of the integral $\int_0^1 \frac{f^2}{q}\, dx$).

Now we have

$$\left\| \frac{y''}{\sqrt{q}} \right\| \leqslant \left\| \frac{f}{\sqrt{q}} \right\|$$

and

$$\|y''\| \leqslant \max \sqrt{q} \left\| \frac{f}{\sqrt{q}} \right\|. \tag{16}$$

We shall next cite the definition of Green's function. This is a continuous function $K(x, \xi)$ which for $x \neq \xi$ satisfies the equation $\frac{d^2}{dx^2} K(x, \xi) = 0$ (in the given case) and which fulfils the conditions

$$K(0, \xi) = K(1, \xi) = 0,\ \left[\frac{dK}{dx}\right]_{x=\xi+0} - \left[\frac{dK}{dx}\right]_{x=\xi-0} = 1. \tag{17}$$

Obviously we shall have such a function in

$$K(x, \xi) = \begin{cases} -(1-\xi)x & \text{for } 0 \leqslant x \leqslant \xi, \\ -\xi(1-x) & \text{,, } \xi \leqslant x \leqslant 1. \end{cases} \tag{18}$$

By means of this function the solution y of the equation $y'' = g(x)$ satisfying the conditions $y(0) = y(1) = 0$ is determined as follows:

$$y(x) = \int_0^1 K(x, \xi) y''(\xi) d\xi = \int_0^1 K(x, \xi) g(\xi) d\xi. \tag{19}$$

Of this one satisfies oneself by twice differentiating expression (19). Indeed:

$$y(x) = \int_0^1 K(x, \xi)g(\xi)d\xi = -\int_0^x \xi(1-x)g(\xi)d\xi - \int_x^1 (1-\xi)xg(\xi)d\xi,$$

$$y'(x) = -x(1-x)g(x) + \int_0^x \xi g(\xi)d\xi + (1-x)xg(x) - \int_x^1 (1-\xi)g(\xi)d\xi,$$

$$y''(x) = xg(x) + (1-x)g(x) = g(x).$$

Next we exhibit the expansion of the function $K(x, \xi)$ in a double Fourier series,

$$K(x, \xi) = -2\sum_{k=1}^{\infty} \frac{\sin k\pi x \sin k\pi\xi}{\pi^2 k^2}. \tag{20}$$

To convince oneself of the validity of this, it is sufficient, for example, to perform the summation of the series on the right, using the series for the function $\sigma_1(x)$ (cf. Chap. I, § 5, p. 81); we have:

$$2\sum_{k=1}^{\infty} \frac{\sin k\pi x \sin k\pi\xi}{\pi^2 k^2} = \sum_{k=1}^{\infty} \frac{\cos k\pi(x-\xi)}{\pi^2 k^2} - \sum_{k=1}^{\infty} \frac{\cos k\pi(x+\xi)}{\pi^2 k^2} =$$

$$= \frac{1}{\pi^2}[\sigma_1(\pi(x+\xi)) - \sigma_1(\pi(x-\xi))] =$$

$$= \begin{cases} \dfrac{1}{\pi^2}\left\{\left[\dfrac{\pi^2}{12} - \dfrac{(\pi-(x+\xi)\pi)^2}{4}\right] - \left[\dfrac{\pi^2}{12} - \dfrac{(\pi-(x-\xi)\pi)^2}{4}\right]\right\} = (1-x)\xi, & \text{if } x-\xi \geqslant 0 \\ \dfrac{1}{\pi^2}\left\{\left[\dfrac{\pi^2}{12} - \dfrac{(\pi-(x+\xi)\pi)^2}{4}\right] - \left[\dfrac{\pi^2}{12} - \dfrac{(\pi+(x-\xi)\pi)^2}{4}\right]\right\} = x(1-\xi), & \text{if } x-\xi \leqslant 0. \end{cases}$$

From expansion (20), by the way, follows the negative definiteness of $K(x, \xi)$, i.e., that for any function $\psi(x)$ integrable in [0, 1],

$$\int_0^1\int_0^1 K(x, \xi)\psi(x)\psi(\xi)dx\,d\xi \leqslant 0. \tag{21}$$

Indeed, if $\beta_k = \int_0^1 \sin k\pi x\psi(x)dx$, on multiplying expansion (20) by $\psi(x)$, integrating it with respect to x, and then multiplying by $\psi(\xi)$ and integrating it with respect to ξ, we satisfy ourselves that this integral is equal to

$$-\sum_{k=1}^{\infty} \frac{2\beta_k^2}{\pi^2 k^2} \leqslant 0.$$

Let us now pass to the estimate of the error itself. As previously, by y_n we designate the nth approximation to the function y by Ritz's method, by Y_n the sum of the first n terms of its expansion in a series involving $\sin k\pi x$. For any function $g(x)$ integrable in $[0, 1]$ we shall generally denote by $[g]_n$ the sum of the first n terms of its expansion in a series involving $\sin k\pi x$. We shall in addition introduce the notations $r_n = y - y_n$, $R_n = y - Y_n$. We shall now utilize equations (16b), [§ 2, p. 265], which for $p = 1$ and $l = 1$ acquire the form

$$\int_0^1 (y_n'' - qy_n - f)\varphi_k dx = 0 \qquad (k = 1, 2, \ldots, n). \tag{22}$$

Multiplying each of equations (22) by φ_k, summing them, and taking into consideration the fact that the function $[y_n'']_n = y_n''$, since y_n'' is a combination of the first n functions φ_k, (11), we find that

$$y_n'' - [qy_n]_n = [f]_n. \tag{23}$$

Subtracting the last equation from (1) (for $p = 1$), however, we obtain

$$y'' - y_n'' - qy + [qy_n]_n = f - [f]_n.$$

Hence, adding $qy_n - [qy_n]_n$ to both sides, we obtain

$$\begin{aligned} r_n'' - qr_n &= f - [f]_n + qy_n - [qy_n]_n = \\ &= f + qy - [f + qy]_n - qr_n + [qr_n]_n = \\ &= y'' - Y_n'' - \sum_{k=n+1}^{\infty} \varphi_k(x) \int_0^1 qr_n\varphi_k\, dx. \end{aligned} \tag{24}$$

Let us multiply both parts of the equation obtained (replacing x in it by ξ) by $K(x, \xi)$, and integrate with respect to ξ from 0 to 1; then, using (19), and also the fact that $\varphi_k = -\dfrac{1}{k^2\pi^2}\varphi_k''$ [cf. (11)], we shall obtain:

$$\begin{aligned} r_n(x) - \int_0^1 K(x, \xi)q(\xi)r_n(\xi)d\xi &= \\ = \{y(x) + Y_n(x)\} &+ \sum_{k=n+1}^{\infty} \frac{\varphi_k(x)}{k^2\pi^2} \int_0^1 qr_n\varphi_k dx. \end{aligned} \tag{25}$$

Now multiplying (25) by $r_n(x)q(x)$ and integrating with respect to x from 0 to 1, we obtain:

$$\begin{aligned} \int_0^1 qr_n^2 dx - \int_0^1\int_0^1 K(x, \xi)r_n(\xi)q(\xi)r_n(x)q(x)dx\,d\xi &= \\ = \int_0^1 (y - Y_n)qr_n dx &+ \sum_{k=n+1}^{\infty} \frac{1}{\pi^2k^2}\left[\int_0^1 qr_n\varphi_k dx\right]^2. \end{aligned}$$

The subtrahend on the left side, on the strength of (21), is negative or zero; therefore by discarding it we shall convert this equality into an inequality; moreover, in the second summand on the right side we will replace k^2 by $(n+1)^2$ and utilize the obvious inequality

$$\sum_{n+1}^{\infty} \Big(\int_0^1 q r_n \varphi_k \, dx\Big)^2 \leqslant \sum_1^{\infty} \Big(\int_0^1 q r_n \varphi_k \, dx\Big)^2 = \int_0^1 q^2 r_n^2 \, dx,$$

whereupon we shall obtain:

$$\int_0^1 q r_n^2 \, dx \leqslant \int_0^1 q r_n (y - Y_n) dx + \frac{1}{\pi^2 (n+1)^2} \int_0^1 q^2 r_n^2 \, dx \leqslant$$

$$\leqslant \Big[\int_0^1 q^2 r_n^2 \, dx \int_0^1 (y - Y_n)^2 dx\Big]^{\frac{1}{2}} + \frac{\max q}{\pi^2 (n+1)^2} \int_0^1 q r_n^2 \, dx \leqslant$$

$$\leqslant \max q^{\frac{1}{2}} \Big[\int_0^1 q r_n^2 \, dx\Big]^{\frac{1}{2}} \|y - Y_n\| + \frac{\max q}{\pi^2 (n+1)^2} \int_0^1 q r_n^2 \, dx,$$

whence

$$\left\{\int_0^1 q r_n^2 \, dx\right\}^{\frac{1}{2}} \leqslant \frac{\max q^{\frac{1}{2}} \|y - Y_n\|}{1 - \dfrac{\max q}{\pi^2 (n+1)^2}} \tag{26}$$

(the denominator being assumed to be positive).

On the other hand, using (18), we have:

$$\Big|\int_0^1 K(x, \xi) q(\xi) r_n(\xi) d\xi\Big| \leqslant \Big[\int_0^1 K^2(x, \xi) d\xi \int_0^1 q^2(\xi) r_n^2(\xi) d\xi\Big]^{\frac{1}{2}} \leqslant$$

$$\leqslant \left\{\tfrac{1}{3} x^2 (1-x)^2 \int_0^1 q^2(\xi) r_n^2(\xi) d\xi\right\}^{\frac{1}{2}} \leqslant \frac{\max \sqrt{q}}{4\sqrt{3}} \left\{\int_0^1 q r_n^2 \, dx\right\}^{\frac{1}{2}}. \tag{27}$$

Now, returning to (25), we find:

$$|r_n| \leqslant |y - Y_n| + \frac{\max \sqrt{q}}{4\sqrt{3}} \left\{\int_0^1 q r_n^2 \, dx\right\}^{\frac{1}{2}} +$$

$$+ \left|\sum_{n+1}^{\infty} \frac{2 \sin k\pi x}{k^2 \pi^2} \int_0^1 q(x) r_n(x) \sin k\pi x \, dx\right|. \tag{28}$$

But for the last sum, using the inequality $|\sum a_i b_i| \leqslant (\sum a_i^2 \sum b_i^2)^{\frac{1}{2}}$, the equation of closure, and the facts that $\varphi_k^2(x) \leqslant 2$ and $\sum\limits_{n+1}^{\infty} \dfrac{1}{k^4} \leqslant \dfrac{1}{3n^3}$, we find that it does not exceed

$$\Big\{\sum_{n+1}^{\infty} \frac{\varphi_k^2(x)}{\pi^4 k^4} \sum_{n+1}^{\infty} \Big[\int_0^1 q(x) r_n(x) \varphi_k(x)\, dx\Big]^2\Big\}^{\frac{1}{2}} \leqslant$$

$$\leqslant \Big\{\frac{2}{\pi^4} \cdot \frac{1}{3} \frac{1}{n^3} \int_0^1 q^2 r_n^2(x)\, dx\Big\}^{\frac{1}{2}} \leqslant \sqrt{\tfrac{2}{3}}\, \frac{\max q^{\frac{1}{2}}}{\pi^2 n^{3/2}} \Big\{\int_0^1 q r_n^2\, dx\Big\}^{\frac{1}{2}}.$$

Using this, and also (26), inequality (28) can be rewritten thus:

$$|y - y_n| = |r_n| \leqslant |y - Y_n| +$$

$$+ \frac{\max q}{4\sqrt{3}\Big(1 - \dfrac{\max q}{\pi^2 (n+1)^2}\Big)} \Big[1 + \frac{4\sqrt{2}}{n^{3/2}\pi^2}\Big] \|y - Y_n\|. \tag{29}$$

From (12) and (16) we find:

$$\|y - Y_n\| \leqslant \frac{1}{(n+1)^2 \pi^2} \|y'' - Y_n''\| < \frac{1}{(n+1)^2\pi^2} \|y''\| < \frac{\max q^{\frac{1}{2}}}{\pi^2 (n+1)^2} \Big\|\frac{f}{\sqrt{q}}\Big\|, \tag{30}$$

$$\|y' - Y_n'\| < \frac{\max q^{\frac{1}{2}}}{(n+1)\pi} \Big\|\frac{f}{\sqrt{q}}\Big\|. \tag{31}$$

Hence, using our notation $R_n = y - Y_n$, we obtain

$$|y(x) - Y_n(x)| = |R_n(x)| = \{2 \int_0^x R_n(\xi) R_n'(\xi) d\xi\}^{\frac{1}{2}} \leqslant$$

$$\leqslant \sqrt{2} \{\int_0^x R_n^2(\xi) d\xi \int_0^x R_n'^2(\xi) d\xi\}^{\frac{1}{4}} \leqslant \sqrt{2}\, \|R_n\|^{\frac{1}{2}}\, \|R_n'\|^{\frac{1}{2}} \leqslant$$

$$\leqslant \frac{\sqrt{2} \max q^{\frac{1}{2}}}{\pi^{3/2} (n+1)^{3/2}} \Big\|\frac{f}{\sqrt{q}}\Big\|, \tag{32}$$

and therefore (29) finally takes the form:

$$|y - y_n| \leqslant \frac{\sqrt{2} \max q^{\frac{1}{2}}}{\pi^{3/2} (n+1)^{3/2}} \Big\{1 + \frac{\max q}{4\sqrt{6}\pi (n+1)} \Big(1 + \frac{4\sqrt{2}}{n^{3/2}\pi^2}\Big) \times$$

$$\times \Big(1 - \frac{\max q}{\pi^2 (n+1)^2}\Big)^{-1}\Big\} \cdot \Big[\int_0^1 \frac{f^2}{q} dx\Big]^{\frac{1}{2}}. \tag{33}$$

We shall now cite some further results of N. M. Krylov, henceforth without proofs, referring to the author's works for them [1]).

Certain modifications in the preceding reasoning permit one to obtain the asymptotic equality

$$\lim_{n\to\infty} \frac{\max|y - y_n|}{\max|y - Y_n|} = 1, \tag{34}$$

showing that the order of the approximation to the solution by Ritz's method coincides with the order of the approximation given by segments of the Fourier series.

Moreover, analogous results are also obtained in the case when the fundamental system is $\varphi_k(x) = x^k(1 - x)$ $(k = 1, 2, \ldots)$. A number of estimates are obtained for the error in determining the proper values by Ritz's method. We shall cite some of them, designating by λ_k the kth proper number of the problem

$$y'' + \lambda q y = 0 \ (q > 0), \ y(0) = y(l) = 0, \tag{35}$$

and by $\lambda_k^{(n)}$ the value of the nth approximation to it; then we have the following estimates:

$$|\lambda_k - \lambda_k^{(n)}| < \frac{2\lambda_k^2 \max q^{3/2}}{\min q^{\frac{1}{2}}(n+1)^2\pi^2 - 2\max q^{3/2}\lambda_k} \quad \left(\lambda_k \leqslant \frac{k^2\pi^2}{\min q}\right), \tag{36}$$

$$\left|\frac{\lambda_k - \lambda_k^{(n)}}{\lambda_k}\right| < \frac{\lambda_k^{(n)}}{\pi^4(n+1)^4}\left[|\lambda_k^{(n)}| \max q^2 + \max\left|\frac{q'^2}{q}\right|\right], \tag{37}$$

$$\left|\frac{\lambda_k - \lambda_k^{(n)}}{\lambda_k}\right| < \frac{|\lambda_k^{(n)}|^3}{\pi^6(n+1)^6}\Bigl\{\max q^{3/2} +$$

$$+ \frac{2\max|q'|}{\sqrt{|\lambda_k^{(n)}|}}\left(1 + \frac{\max q\,|\lambda_k^{(n)}|}{8(n+1)^2\pi^2}\right) + \frac{1}{|\lambda_k^{(n)}|}\max\left|\frac{q''}{\sqrt{q}}\right|\Bigr\}^2.$$

These three estimates relate to the case when $\varphi_l(x) = \sqrt{2}\sin \pi l x$ $(l = 1, 2, \ldots, n)$; for the case when $\varphi_l(x) = x^l(1 - x)$ $(l = 1, 2, \ldots, n)$,

---

[1]) See N. M. Krylov, [3] and [4]. See also the survey article of N. M. Krylov in the book "Jubilee symposium dedicated to the thirtieth anniversary of the great October socialist revolution", vol. 1, 231—241. In it is given also the literature relating to questions of the convergence of other methods, in particular the method of least squares and the method of moments, which were studied particularly in the works of N. M. Krylov and in those of M. F. Kravchuk, [1], [2].

The question of the convergence of the method of B. G. Galerkin for the cases when it does not coincide with Ritz's method has been investigated only recently. For a more particular type of the problems relating to ordinary differential equations, such an investigation has been carried through by G. I. Petrov, [1] and [2]. A very complete resolution of the problem has been obtained in the fundamental work of M. V. Keldysh, [1], where the convergence is proved for problems relating to ordinary equations and to equations of the second order of elliptic type. On this question see also S. G. Mikhlin, [3], [5], and L. V. Kantorovich, [3], Chap. II, § 7.

the following inequality holds:

$$\left| \frac{\lambda_k^{(n)} - \lambda_k}{\lambda_k} \right| < \frac{N\lambda_k^{(n)}}{n^2(n+1)^2(n+2)(n-1)}, \tag{38}$$

where

$$N = \left\{\max\left|\frac{q''}{q^{\frac{1}{2}}}\right| + 2\max|q'| \sqrt{|\lambda_k^{(n)}| \frac{\max q^{\frac{1}{2}}}{\min q^{\frac{1}{2}}}} + \max q^{3/2}\, |\lambda_k^{(n)}|\right\}^2. \tag{39}$$

Estimates of like character are also obtained for the error in the determination of the proper functions and for the error in the determination of the solution of the non-homogeneous equation

$$y'' + \lambda q(x) = f(x).$$

We shall present one of the estimates of the latter form:

$$|y - y_n| < \frac{\sqrt{\frac{2}{3}}\{\|f\| + \lambda \max|q|\, M\, \|f\|\}}{(n+1)^{3/2}\pi^2}\Big[1 + \\ + \sqrt{\tfrac{3}{2}}\, \frac{1}{\pi^2(n+1)^{3/2}}\, \frac{\lambda\nu M}{\left(1 - \dfrac{\lambda M \max|q|}{(n+1)^2\pi^2}\right)}\Big],$$

where

$$M = \max_{i=1,2,\ldots}\left|\frac{\lambda_i}{\lambda_i - \lambda}\right|, \quad \nu = \|q\| + \max|q| + \sqrt{\int_0^1 [\int_0^1 K_x'^2 q'^2 dt] dx}.$$

The cited estimates are sufficiently close. For instance, estimate (39), in application to the determination of the first proper number of the equation $y'' + \lambda y = 0$, under the condition $y(-1) = y(1) = 0$, if the mth approximation be sought in the form

$$y = (1 - x^2)(a_0 + a_1x^2 + \ldots + a_{m-1}x^{2m-2}),$$

gives 0.2% for $m = 2$ and 0.02% for $m = 3$.

However the real error is considerably less yet, namely 0.004% for $m = 2$ and 0.0000008% for $m = 3$ (cf. § 2, No. 5, Example 1).

In other cases the estimate proves to be considerably closer to the real error.

2. The problem of the convergence of the minimal sequences for equations of elliptic type. For ordinary self-adjoint equations of the second order (with the conditions $p > 0$, $q \geqslant 0$) it has been shown [see § 2, No. 2 (21)] that each minimal sequence (i.e., one such that $I(y_n) \to I(y)$) converges uniformly to the solution: $\lim_{n\to\infty} y_n(x) = y(x)$. For partial differential equations this circumstance no longer obtains.

We shall show this in an exampleof the Laplace equation. Let us consider for it the Dirichlet problem for a unit circle under null boundary conditions: $u(r, \theta) = 0$ for $r = 1$. Its obvious solution is $u = 0$, and the value of the Dirichlet integral

$$I(u) = \iint\limits_D \left[\left(\frac{\partial u}{\partial x}\right)^2 + \left(\frac{\partial u}{\partial y}\right)^2\right] dx\,dy = \iint\limits_D \left[\left(\frac{\partial u}{\partial r}\right)^2 + \frac{1}{r^2}\left(\frac{\partial u}{\partial \theta}\right)^2\right] r\,dr\,d\theta,$$

corresponding to it is equal to zero: $I(u) = 0$. Now take any sequence $a_n \to 0$ $(1 > a_n > 0)$ and define the functions $u_n(r, \theta)$ as follows:

$$u_n(r, \theta) = \left\{\begin{array}{ll} 0 & \text{for } a_n \leqslant r \leqslant 1 \\ \dfrac{1}{\ln a_n} \ln \dfrac{r}{a_n} & \quad ,, \quad a_n^2 \leqslant r \leqslant a_n \\ 1 & \quad ,, \quad 0 \leqslant r \leqslant a_n^2 \end{array}\right\}. \tag{40}$$

Then we have

$$I(u_n) = \int_0^{2\pi} d\theta \int_{a_n^2}^{a_n} \left(\frac{1}{\ln a_n} \cdot \frac{1}{r}\right)^2 r\,dr = -\frac{2\pi}{\ln a_n}.$$

Hence it is clear that as $n \to \infty$ we shall have $I(u_n) \to I(u) = 0$, i.e., the sequence u_n is minimal. However $u_n(0, 0) = 1$, and this sequence will not converge to the solution. It would be possible to construct a sequence of an analogous kind which did not converge to the solution at even one point. [1]) However under certain supplementary limitations, the uniform convergence of the minimal sequence to the solution can be proved even for problems with two variables. Hence are obtained some further theorems on the convergence of variational methods.

We shall first of all prove that uniform convergence obtains if the upper bound of one of the partial derivatives of the functions u_n does not increase too rapidly by comparison with the difference [2]) $I(u_n) - I(u)$. We shall carry through the proof for the general equation of elliptic type (see § 1, p. 248). Our proposition will follow from the following lemma.

LEMMA 1. *Let Γ be the contour bounding the region D. Let $\eta(x, y)$ be a function that is continuous in the closed region and which has partial derivatives that are continuous within the region, and let the function satisfy the following conditions*:

1) $\eta(x, y) = 0$ on Γ, (41)

2) $$\iint\limits_D \left[a\left(\frac{\partial \eta}{\partial x}\right)^2 + b\left(\frac{\partial \eta}{\partial y}\right)^2 + c\eta^2\right] dx\,dy \leqslant \varepsilon, \tag{42}$$

where a, b, c are functions of x and y which are continuous in D, and

---

1) On the subject of convergence in the mean see Courant and Hilbert, [1], and S. G. Mikhlin, [5].

2) On the subject of what follows below see L. V. Kantorovich, [14], [15].

$a > 0$, $b > 0$ and $c \geqslant 0$ in the closed region D;

$$3) \int_{\alpha}^{\beta} \left(\frac{\partial \eta}{\partial y}\right)^2 dy \leqslant K, \qquad (43)$$

if the segment joining the points (x, α) and (x, β) lies in the region D; K is a constant.

Then the following estimate holds for the function η in the region D:

$$|\eta(x, y)| < C_2 \sqrt{\varepsilon \ln \frac{dK}{\varepsilon}} + C_3 \sqrt{\varepsilon}, \qquad (44)$$

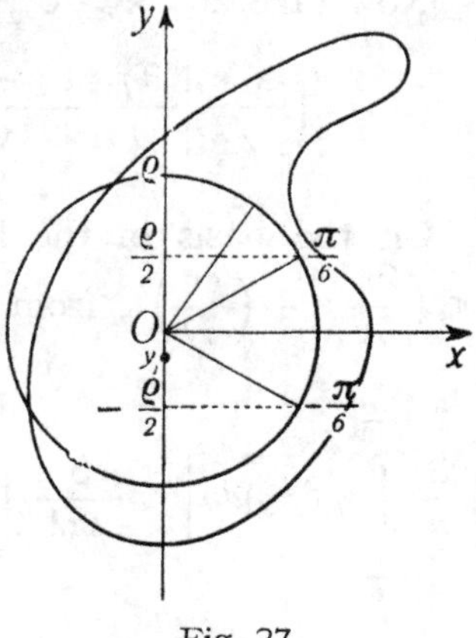

Fig. 27.

where d is the diameter of the region D, C_2 and C_3 are constants dependent only on the coefficients of the equation and the form of the region [1]).

PROOF. Let us extend the definition of $\eta(x, y)$ to have it null outside of the region D. We first of all note that from (42) we obtain the inequality:

$$\iint_D \left[\left(\frac{\partial \eta}{\partial x}\right)^2 + \left(\frac{\partial \eta}{\partial y}\right)^2\right] dx\, dy \leqslant C_1 \varepsilon, \text{ where } C_1 = \max\left(\frac{1}{\min a}, \frac{1}{\min b}\right). \qquad (45)$$

Let us choose any point M within D; it is for it that we shall make the estimate $|\eta(M)|$. Construct Cartesian and polar systems of coordinates with origin at this point. Take any number $\varrho > 0$. Since the point (θ, d) lies on the boundary or outside of D, we have $\eta(\theta, d) = 0$ and therefore

$$\eta(\theta, \varrho) = \int_d^{\varrho} \frac{\partial \eta}{\partial r} dr.$$

Hence, denoting by D_1 the region contained between the circles $r = \varrho$, $r = d$ and the rays $\theta = -\frac{\pi}{6}$, $\theta = \frac{\pi}{6}$ (Fig. 27), we have:

$$\left| \frac{3}{\pi} \int_{-\frac{\pi}{6}}^{\frac{\pi}{6}} \eta(\theta, \varrho) d\theta \right| \leqslant \frac{3}{\pi} \int_{-\frac{\pi}{6}}^{\frac{\pi}{6}} d\theta \int_{\varrho}^{d} \left| \frac{\partial \eta}{\partial r} \right| dr = \frac{3}{\pi} \iint_{D_1} \left| \frac{\partial \eta}{\partial r} \right| dr\, d\theta. \qquad (46)$$

---

[1]) Here and henceforth we denote by the letter C, with various indices, constants of such a kind.

Now take any $\lambda > 0$; we have:

$$\left[\frac{\sqrt{r}}{\sqrt{\lambda}}\left|\frac{\partial\eta}{\partial r}\right| - \frac{\sqrt{\lambda}}{\sqrt{r}}\right]^2 \geqslant 0, \quad \left|\frac{\partial\eta}{\partial r}\right| \leqslant \frac{\lambda}{2r} + \frac{r}{2\lambda}\left(\frac{\partial\eta}{\partial r}\right)^2.$$

On the basis of the last inequality, using the fact that $\left|\frac{\partial\eta}{\partial r}\right|^2 \leqslant \left(\frac{\partial\eta}{\partial x}\right)^2 + \left(\frac{\partial\eta}{\partial y}\right)^2$, from (46) and (45) we find:

$$\left|\frac{3}{\pi}\int_{-\frac{\pi}{6}}^{\frac{\pi}{6}} \eta(\theta, \varrho)d\theta\right| \leqslant \frac{3}{2\pi\lambda}\iint_D \left(\frac{\partial\eta}{\partial r}\right)^2 r\,dr\,d\theta + \frac{3\lambda}{2\pi}\int_{-\frac{\pi}{6}}^{\frac{\pi}{6}} d\theta \int_\varrho^d \frac{dr}{r} \leqslant$$

$$\leqslant \frac{3}{2\pi\lambda}\iint_D \left[\left(\frac{\partial\eta}{\partial x}\right)^2 + \left(\frac{\partial\eta}{\partial y}\right)^2\right] dx\,dy + \frac{3\lambda}{2\pi}\int_{-\frac{\pi}{6}}^{\frac{\pi}{6}} \ln\frac{d}{\varrho}\,d\theta \leqslant$$

$$\leqslant \frac{3C_1\varepsilon}{2\pi\lambda} + \frac{\lambda}{2}\ln\frac{d}{\varrho} = q. \tag{47}$$

Inequality (47) shows that the mean value of η on the arc of the circumference does not exceed the quantity that we have designated briefly by q. From it may be obtained an analogous statement relative to the mean value of η on the vertical segment. Indeed, passing to Cartesian coordinates in (47) ($\varrho \sin\theta = y$), we shall obtain

$$\left|\frac{3}{\pi}\,\frac{1}{\varrho}\int_{-\frac{\varrho}{2}}^{\frac{\varrho}{2}} \eta(\sqrt{\varrho^2 - y^2}, y)\frac{dy}{\sqrt{1 - \left(\frac{y}{\varrho}\right)^2}}\right| \leqslant q.$$

Hence

$$\left|\frac{1}{\varrho}\int_{-\frac{\varrho}{2}}^{\frac{\varrho}{2}} \eta(0, y)\frac{dy}{\sqrt{1 - \left(\frac{y}{\varrho}\right)^2}}\right| =$$

$$= \left|\frac{1}{\varrho}\int_{-\frac{\varrho}{2}}^{\frac{\varrho}{2}} \left[\eta(\sqrt{\varrho^2 - y^2}, y) + \int_{\sqrt{\varrho^2-y^2}}^{0} \frac{\partial\eta}{\partial x}dx\right]\frac{dy}{\sqrt{1 - \left(\frac{y}{\varrho}\right)^2}}\right| \leqslant$$

$$< \frac{\pi}{3} q + \left| \frac{1}{\varrho} \frac{2}{\sqrt{3}} \int_{-\frac{\varrho}{2}}^{\frac{\varrho}{2}} dy \int_0^{\varrho} \left| \frac{\partial \eta}{\partial x} \right| dx \right| \leqslant$$

$$\leqslant \frac{\pi}{3} q + \frac{2}{\sqrt{3}} \frac{1}{\varrho} \sqrt{\int_{-\frac{\varrho}{2}}^{\frac{\varrho}{2}} \int_0^{\varrho} \left(\frac{\partial \eta}{\partial x} \right)^2 dx\, dy} \sqrt{\int_{-\frac{\varrho}{2}}^{\frac{\varrho}{2}} \int_0^{\varrho} 1^2\, dx\, dy} \leqslant$$

$$\leqslant \frac{\pi}{3} q + \frac{2}{\sqrt{3}} (C_1 \varepsilon)^{\frac{1}{2}}. \tag{48}$$

It is clear, therefore, that one can find a y_1 $\left(|y_1| \leqslant \frac{\varrho}{2}\right)$ such that

$$\left| \eta(0, y_1) \frac{1}{\sqrt{1 - \left(\frac{y_1}{\varrho}\right)^2}} \right| \leqslant \frac{\pi}{3} q + \frac{2}{\sqrt{3}} (C_1 \varepsilon)^{\frac{1}{2}},$$

and the more so will

$$|\eta(0, y_1)| < \frac{\pi}{3} q + \frac{2}{\sqrt{3}} (C_1 \varepsilon)^{\frac{1}{2}}. \tag{49}$$

Let us utilize, finally, inequality [1]) (43); we then have

$$|\eta(0, 0)| = \left| \eta(0, y_1) + \int_{y_1}^{0} \frac{\partial \eta}{\partial y} dy \right| \leqslant$$

$$\leqslant |\eta(0, y_1)| + \sqrt{\int_0^{y_1} \left(\frac{\partial \eta}{\partial y} \right)^2 dy} \sqrt{\int_0^{y_1} 1^2 dy} \leqslant$$

$$\leqslant \frac{\pi}{3} q + \frac{2}{\sqrt{3}} (C_1 \varepsilon)^{\frac{1}{2}} + \sqrt{\frac{\varrho}{2}} \sqrt{K} =$$

$$= \frac{C_1 \varepsilon}{2\lambda} + \frac{\pi \lambda}{6} \ln \frac{d}{\varrho} + \frac{2}{\sqrt{3}} (C_1 \varepsilon)^{\frac{1}{2}} + \frac{1}{\sqrt{2}} \sqrt{K \varrho}. \tag{50}$$

The numbers λ and ϱ we can choose arbitrarily; let us adopt

$$\lambda = \sqrt{\frac{\varepsilon}{\ln \frac{dK}{\varepsilon}}}, \quad \varrho = \frac{\varepsilon}{K}$$

---

[1]) Inequality (43) cannot be used, granted, if the segment from the point (0, 0) to $(0, y_1)$ does not lie wholly in the region D; but then the point $(0, y_1)$ can be replaced by the point $(0, y_2)$ of the curve Γ which is closest to the origin along the vertical $(0 < y_2 < y_1)$.

(we consider that $dK > \varepsilon$); then inequality (50) will acquire the form

$$|\eta(0, 0)| \leqslant C_2 \sqrt{\varepsilon \ln \frac{dK}{\varepsilon}} + C_3\sqrt{\varepsilon},$$

which does not differ from (44), since any interior point of the region D could have been chosen as origin [1]).

From this lemma there is almost directly obtained the following theorem.

THEOREM 1. *In the region D bounded by the rectifiable contour Γ, let $u(x, y)$ be the solution of the elliptic equation*

$$\frac{\partial}{\partial x}\left(a\frac{\partial u}{\partial x}\right) + \frac{\partial}{\partial y}\left(b\frac{\partial u}{\partial y}\right) - cu = f \qquad (a, b > 0,\ c \geqslant 0), \tag{51}$$

vanishing on the contour Γ. We shall assume that the integral

$$I_1(u) = \iint\limits_D \left[a\left(\frac{\partial u}{\partial x}\right)^2 + b\left(\frac{\partial u}{\partial y}\right)^2 + cu^2\right] dx\,dy < +\infty \tag{52}$$

is finite and that the following integral is bounded:

$$\int\limits_\alpha^\beta \left(\frac{\partial u}{\partial y}\right)^2 dy \leqslant K_0, \tag{53}$$

α and β being such that the segment joining the points (x, α) and (x, β) is contained in D.

Let $u_n(x, y)$ be a sequence of functions which are also equal to 0 on Γ, are continuous in the closed region and have continuous derivatives within D. Furthermore let it be given that [2])

$$I(u_n) - I(u) =$$

$$= \iint\limits_D \left[a\left(\frac{\partial u_n}{\partial x} - \frac{\partial u}{\partial x}\right)^2 + b\left(\frac{\partial u_n}{\partial y} - \frac{\partial u}{\partial y}\right)^2 + c(u_n - u)^2\right] dx\,dy = \varepsilon_n, \tag{54}$$

$$\int\limits_\alpha^\beta \left(\frac{\partial u_n}{\partial y}\right)^2 dy \leqslant K_n, \tag{55}$$

$$\lim_{n\to\infty} \varepsilon_n = 0, \quad \lim_{n\to\infty} \varepsilon_n \ln K_n = 0. \tag{56}$$

---

[1]) We note that since condition (43) has been used only at the end of the reasoning, if it is fulfilled for some segments, the conclusion of the lemma remains in force for points (x, y) for which there exists such a segment joining the points (x, α), (x, β).

[2]) $I(u)$ here denotes the integral connected with equation (51) [see § 1, (12c)]. This form of the difference $I(u_n) - I(u)$ is obtained by means of transformations like those exhibited in § 1 (18).

Then the sequence of functions u_n converges uniformly to $u(x, y)$ in the region D, and the order of convergence is

$$|u_n - u| = O\left(\varepsilon_n^{\frac{1}{2}} + \left|\varepsilon_n \ln \frac{K_n + K_0}{\varepsilon_n}\right|^{\frac{1}{2}}\right). \tag{57}$$

PROOF. Let us put $\eta_n(x, y) = u_n(x, y) - u(x, y)$. Then for the function $\eta_n(x, y)$ the conditions of the lemma will be fulfilled provided one adopts $\varepsilon = \varepsilon_n$ and $K = 2(K_0 + K_n)$; the latter is clear from the fact that

$$\int_\alpha^\beta \left(\frac{\partial \eta_n}{\partial y}\right)^2 dy \leqslant 2\left[\int_\alpha^\beta \left(\frac{\partial u_n}{\partial y}\right)^2 dy + \int_\alpha^\beta \left(\frac{\partial u}{\partial y}\right)^2 dy\right] = 2(K_0 + K_n).$$

Employing inequality (44), therefore, we find

$$|\eta_n(x, y)| < C_2 \sqrt{\varepsilon_n \ln \frac{2d(K_0 + K_n)}{\varepsilon_n}} + C_3\sqrt{\varepsilon_n},$$

whence, on the strength of (56), it at once follows that $\eta_n(x, y) \to 0$ uniformly in the region D, and the order of convergence is precisely that indicated in (57).

Theorem 1 may be given a form more convenient for use, relying on the fact that when the form of the functions $u_n(x, y)$ is known, one can on the basis of this give an estimate of the rate of growth of the K_n. We shall do this for the case when u_n is a polynomial in one of the variables.

THEOREM 2. *Relative to the region D and the function u we make the same assumptions as were made in Theorem 1. In addition we assume that at points with extremal abscissas, of which there is a finite number, the contour Γ has a finite curvature.*

As regards $u_n(x, y)$, we shall assume that it is equal to 0 on Γ and has the form

$$u_n(x, y) = a_0(x) + a_1(x)y + \ldots + a_n(x)y^n, \tag{58}$$

where $a_0, a_1, \ldots, a_n$ are functions continuous together with their derivatives $a_0', a_1', \ldots, a_n'$, and finally, that

$$\lim_{n\to\infty} \varepsilon_n \ln n = 0. \tag{59}$$

Then the uniform convergence of the sequence $u_n(x, y)$ to $u(x, y)$ obtains, and

$$|u_n - u| = O\left(\sqrt{\varepsilon_n \ln \frac{n}{\varepsilon_n}}\right). \tag{60}$$

PROOF. Let us take a number $h > 0$ which we shall define more exactly below. Divide the region D into two parts D_h and D'_h. To the first of them we assign those points which together with a vertical segment of length $\geqslant h$ belong to D, and to the second the other points of D. One is readily convinced that for an h sufficiently small the region D'_h will consist of several arbitrarily small segments situated at points with extremal abscissas. We shall first of all estimate $|\eta_n(x, y)|$ in the region D_h. Let us put

$$\max |\eta_n(x, y)| = M_n, \quad \max |u(x, y)| = M_0. \tag{61}$$

Then

$$\max |u_n(x, y)| \leqslant M_0 + M_n. \tag{62}$$

We shall now utilize a theorem of A. A. Markov [1]) according to which the maximum of the modulus of the derivative of a polynomial of the nth degree in an interval of length h does not exceed the product of $\frac{2n^2}{h}$ by the maximum of the modulus of the polynomial itself. Then in the region D_h we have

$$\left| \frac{\partial u_n}{\partial y} \right| \leqslant (M_0 + M_n) \frac{2n^2}{h}. \tag{63}$$

Let it now be known to us that the segment whose ends are (x, α) and (x, β) is contained in the region D_h. Then in establishing the inequality which follows below, one can consider without diminishing the generality that its length $\beta - \alpha \geqslant h$, and therefore

$$\int_\alpha^\beta \left(\frac{\partial u_n}{\partial y}\right)^2 dy \leqslant \left[(M_0 + M_n) \frac{2n^2}{\beta - \alpha}\right]^2 (\beta - \alpha) \leqslant (M_0 + M_n)^2 \frac{4n^4}{h}. \tag{64}$$

Hence on the basis of (53)

$$\int_\alpha^\beta \left(\frac{\partial \eta_n}{\partial y}\right)^2 dy \leqslant 2 \int_\alpha^\beta \left(\frac{\partial u_n}{\partial y}\right)^2 dy + 2 \int_\alpha^\beta \left(\frac{\partial u}{\partial y}\right)^2 dy \leqslant$$

$$\leqslant \frac{(M_0 + M_n)^2 8n^4}{\beta - \alpha} + 2K_0 \leqslant (M_0 + M_n)^2 \frac{8n^4}{h} + 2K_0. \tag{65}$$

Thanks to inequality (65), to points of the region D_h estimate (44) [see note [1]) on p. 342] is applicable, but as K the right side of inequality

[1]) Markov, A. A., [2]. See also I. P. Natanson, [1], p. 178.

(65) must be taken. We obtain in the region D_h the estimate

$$|\eta_n(x, y)| < C_2 \sqrt{\varepsilon_n \ln \frac{d(M_0 + M_n)^2 8n^4 + 2hK_0 d}{h\varepsilon_n}} + C_3\sqrt{\varepsilon_n}. \quad (66)$$

Let us now estimate $|\eta_n|$ in the region D_h'. We take some point (x_0, y_0) of this region (Fig. 28). Let the points of intersection of the line $x = x_0$ with the contour Γ that are closest to (x_0, y_0) be (x_0, α) and (x_0, β) $(\alpha < y_0 < \beta)$. Since (x_0, y_0) is in the region D_h', we have $\beta - \alpha < h$. A portion of the contour Γ parted by this segment is located close to one of the points with extremal abscissa, and admits — if h be chosen sufficiently small — of being given explicitly in the form $x = \varphi(y)$ $(\alpha \leqslant y \leqslant \beta)$, where, thanks to the finiteness of curvature at the extremal point, we have

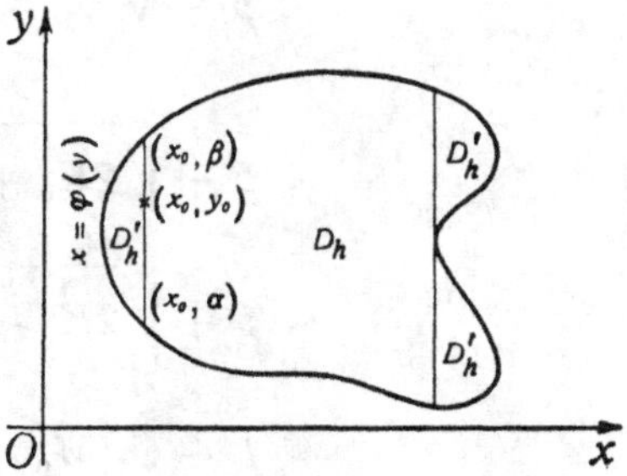

Fig. 28.

$$|\varphi(y) - x_0| < C_4(\beta - \alpha)^2,$$

where C_4 is a constant dependent only on the form of the region. We have now, since $\eta_n(y, \varphi(y)) = 0$ for $\alpha \leqslant y \leqslant \beta$:

$$\int_\alpha^\beta \eta_n^2(x_0, y)dy = \int_\alpha^\beta \Big(\int_{\varphi(y)}^{x_0} \frac{\partial \eta_n}{\partial x} dx\Big)^2 dy \leqslant \int_\alpha^\beta \Big[\int_{\varphi(y)}^{x_0} \Big(\frac{\partial \eta_n}{\partial x}\Big)^2 dx\Big] |\varphi(y) - x_0| \, dy \leqslant$$

$$\leqslant C_4(\beta - \alpha)^2 \iint_D \Big(\frac{\partial \eta_n}{\partial x}\Big)^2 dx\, dy \leqslant C_1 C_4 \varepsilon_n (\beta - \alpha)^2.$$

Let us now take a number $p > 0$ and denote by e the system of those intervals in which $|\eta_n(x_0, y)| > p$, and by me their total length. Then we have

$$p^2 me \leqslant \int_\alpha^\beta \eta_n^2(x_0, y)dy \leqslant C_1 C_4 \varepsilon_n (\beta - \alpha)^2,$$

$$me \leqslant C_1 C_4 \frac{\varepsilon_n(\beta - \alpha)^2}{p^2}.$$

At points not belonging to e we have $|\eta_n(x_0, y)| \leqslant p$; if, however, the point (x_0, y_0) belongs to e, then at a distance $\leqslant \frac{me}{2}$ from (x_0, y_0) there is a point (x_0, y_1) not belonging to e. Therefore, using again the first of inequalities (65), we find:

$$|\eta_n(x_0, y_0)| \leqslant |\eta_n(x_0, y_1)| + \left| \int_{y_1}^{y_0} \frac{\partial \eta_n}{\partial y} \, dy \right| \leqslant$$

$$\leqslant p + \sqrt{|y_1 - y_0|} \sqrt{\int_{y_1}^{y_0} \left(\frac{\partial \eta_n}{\partial y}\right)^2 dy} \leqslant$$

$$\leqslant p + \sqrt{2K_0 + \frac{(M_0 + M_n)^2 8n^4}{\beta - \alpha}} \sqrt{me} \leqslant$$

$$\leqslant p + \sqrt{2K_0 + \frac{(M_0 + M_n)^2 8n^4}{\beta - \alpha}} \sqrt{C_1 C_4} \frac{\sqrt{\varepsilon_n}(\beta - \alpha)}{p} \leqslant$$

$$\leqslant p + C_5(M_0 + M_n) n^2 \sqrt{\varepsilon_n} h \frac{1}{p} = 2p, \tag{67}$$

if we adopt

$$p = C_5^{\frac{1}{2}}(M_0 + M_n)^{\frac{1}{2}} n \varepsilon_n^{\frac{1}{4}} h^{\frac{1}{2}}. \tag{68}$$

Inequality (67) is valid for the region D'_h. Comparing inequalities (66) and (67), we have the inequality

$$|\eta_n(x, y)| \leqslant C_2 \sqrt{\varepsilon_n \ln \frac{d(M_0 + M_n)^2 8n^4 + 2hK_0 d}{h\varepsilon_n}} + C_3\sqrt{\varepsilon_n} +$$

$$+ 2C_5^{\frac{1}{2}}(M_0 + M_n)^{\frac{1}{2}} n \varepsilon_n^{\frac{1}{4}} h^{\frac{1}{2}}, \tag{69}$$

now valid in the whole region D. It is now natural to adopt $h = \varepsilon_n n^{-4}$. We then obtain:

$$M_n \leqslant C_2 \sqrt{\varepsilon_n \ln \left(\frac{8d(M_0 + M_n)^2 n^8}{\varepsilon_n^2} + \frac{2K_0 d}{\varepsilon_n}\right)} +$$

$$+ C_3\sqrt{\varepsilon_n} + 2C_5^{\frac{1}{2}}(M_0 + M_n)^{\frac{1}{2}} \sqrt{\varepsilon_n}. \tag{70}$$

From inequality (70) it at once follows, thanks to condition (59), that $M_n \to 0$. Therefore the quantities M_0, M_n are bounded and $\leqslant M$. Taking this into account, we can considerably simplify inequality (70), giving it the form

$$M_n \leqslant C_6 \sqrt{\varepsilon_n \ln \frac{n}{\varepsilon_n}}, \tag{71}$$

which proves the statement of the Theorem, (60).

Observation 1. In Theorem 2 the function u_n was assumed to be a polynomial in y; all the reasoning of the Theorem preserves its force, however, if, as a function of y, it is of some other definite type — it is important only that a theorem of the type of A. A. Markov's theorem hold. Thus, for example, $u_n(x, y)$ may be a trigonometric polynomial of the nth order in y; the entire difference will consist in this, that rather than A. A. Markov's theorem the theorem of S. N. Bernshteĭn [1]) will have to be used.

Observation 2. By a certain complication of the proof, Theorem 2 could have been strengthened, viz.: it is sufficient to assume that $u_n(x, y)$ has the form

$$u_n(x, y) = \omega(x, y)(a_0(x) + a_1(x)y + \ldots + a_n(x)y^n),$$

where $\omega(x, y)$ is a function satisfying certain conditions. Condition (53) could also have been relaxed essentially.

3. The convergence of Ritz's method and the method of reduction to ordinary equations. In the preceding No. were given the conditions securing the convergence of the minimal sequence to the solution. The essential role in them was, as we saw, played by the rate of diminution of ε_n. We shall now show under what conditions the necessary rate of diminution of ε_n is secured, if we are concerned with Ritz's method and the method of reduction to ordinary equations. It is under these conditions that the convergence of the above-named methods will be secured.

It will be the more convenient for us to begin with the proof of the convergence of the second method. Moreover, to abridge the proof we shall limit ourselves to the case of a region of the simplest form.

LEMMA 2. *Let D be a region bounded by the lines $x = 0$, $x = l$, $y = g(x)$, $y = h(x)$, where the functions $g(x)$ and $h(x)$ are defined together with their derivatives of the first and second order, and the latter are bounded for $0 \leqslant x \leqslant l$; $h(x) > g(x)$ for the indicated values of x. Let $u(x, y)$ be a function that is defined in the region D, equal to 0 on the boundary of the region and such that $\frac{\partial^2 u}{\partial y^2}$ and $\frac{\partial^2 u}{\partial x \partial y}$ are square-integrable in D; $a(x, y)$, $b(x, y)$, $c(x, y)$ are some bounded continuous functions. Then there may be constructed functions $\bar{u}_n(x, y)$ and $\tilde{u}_n(x, y)$ of the form*

$$\bar{u}_n(x, y) = \sum_{k=1}^{n} f_k(x) \sin \frac{k\pi(y - g(x))}{h(x) - g(x)}, \tag{72}$$

$$\tilde{u}_n(x, y) = \sum_{k=1}^{n} f_k(x) y^{k-1}(y - g(x))(y - h(x)), \tag{72'}$$

where $f_k(x)$ are functions of x equal to 0 for $x = 0$ and $x = l$, and such that the quantities

$$\bar{\varepsilon}_n = \iint\limits_D \left[a\left(\frac{\partial u}{\partial x} - \frac{\partial \bar{u}_n}{\partial x}\right)^2 + b\left(\frac{\partial u}{\partial y} - \frac{\partial \bar{u}_n}{\partial y}\right)^2 + c(u - \bar{u}_n)^2\right] dx\, dy, \tag{73}$$

---

[1]) See V. L. Goncharov, [1], p. 303, or I. P. Natanson, [1], p. 123.

$$\tilde{\varepsilon}_n = \iint\limits_D \Big[a\Big(\frac{\partial u}{\partial x} - \frac{\partial \tilde{u}_n}{\partial y}\Big)^2 + b\Big(\frac{\partial u}{\partial y} - \frac{\partial \tilde{u}_n}{\partial y}\Big)^2 + c(u - \tilde{u}_n)^2\Big]\,dx\,dy \quad (73')$$

are of the order $O\Big(\frac{1}{n^2}\Big)$.

PROOF. We shall first carry out the construction of $\bar{u}_n$. Put

$$v(x, z) = u\Big(x, g + \frac{z}{\pi}(h - g)\Big) \quad 0 \leqslant z \leqslant \pi, \quad 0 \leqslant x \leqslant l, \tag{74}$$

$$\sigma(x, z) = \sum_{k=1}^{n} a_k \sin kz, \quad a_k = \frac{2}{\pi}\int_0^\pi v(x, z) \sin kz dz, \tag{75}$$

$$\varrho(x, z) = v(x, z) - \sigma(x, z).$$

Then, performing yet an integration by parts and using the fact that $v = 0$ for $z = 0$ and $z = \pi$, we find:

$$\frac{\partial \varrho}{\partial z} = \frac{\partial v}{\partial z} - \sum_{k=1}^{n} k a_k \cos kz = \frac{\partial v}{\partial z} - \sum_{k=1}^{n}\Big(\frac{2k}{\pi}\int_0^\pi v(x, z) \sin kz\,dz\Big) \cos kz =$$

$$= \frac{\partial v}{\partial z} - \sum_{k=1}^{n}\Big(\frac{2}{\pi}\int_0^\pi \frac{\partial v}{\partial z} \cos kz\,dz\Big) \cos kz.$$

Hence [cf. (12a)]:

$$\frac{2}{\pi}\int_0^\pi \Big(\frac{\partial \varrho}{\partial z}\Big)^2 dz = \sum_{k=n+1}^{\infty}\Big(\frac{2}{\pi}\int_0^\pi \frac{\partial v}{\partial z} \cos kz\,dz\Big)^2 =$$

$$= \sum_{k=n+1}^{\infty}\Big(\frac{2}{k\pi}\int_0^\pi \frac{\partial^2 v}{\partial z^2} \sin kz\,dz\Big)^2 \leqslant \frac{1}{(n+1)^2}\,\frac{2}{\pi}\int_0^\pi \Big(\frac{\partial^2 v}{\partial z^2}\Big)^2 dz,$$

$$\int_0^l dx \int_0^\pi \Big(\frac{\partial \varrho}{\partial z}\Big)^2 dz \leqslant \frac{1}{(n+1)^2}\int_0^l dx \int_0^\pi \Big(\frac{\partial^2 v}{\partial z^2}\Big)^2 dz. \tag{76}$$

The fact that the integral on the right side has a finite value we establish below.

In just the same manner, and using the fact that $\dfrac{\partial v}{\partial x} = 0$ for $z = 0$

and $z = \pi$, we obtain:

$$\frac{\partial \varrho}{\partial x} = \frac{\partial v}{\partial x} - \sum_{k=1}^{n} a_k' \sin kz = \frac{\partial v}{\partial x} - \sum_{k=1}^{n} \left(\frac{2}{\pi} \int_0^\pi \frac{\partial v}{\partial x} \sin kz\, dz\right) \sin kz,$$

$$\frac{2}{\pi} \int_0^\pi \left(\frac{\partial \varrho}{\partial x}\right)^2 dz = \sum_{k=n+1}^{\infty} \left(\frac{2}{\pi} \int_0^\pi \frac{\partial v}{\partial x} \sin kz\, dz\right)^2 =$$

$$= \sum_{k=n+1}^{\infty} \left(\frac{2}{k\pi} \int_0^\pi \frac{\partial^2 v}{\partial x \partial z} \cos kz\, dz\right)^2 \leqslant \frac{1}{(n+1)^2} \cdot \frac{2}{\pi} \int_0^\pi \left(\frac{\partial^2 v}{\partial x \partial z}\right)^2 dz,$$

$$\int_0^l dx \int_0^\pi \left(\frac{\partial \varrho}{\partial x}\right)^2 dz \leqslant \frac{1}{(n+1)^2} \int_0^l dx \int_0^\pi \left(\frac{\partial^2 v}{\partial x \partial z}\right)^2 dz. \tag{77}$$

Finally:

$$\frac{2}{\pi} \int_0^\pi \varrho^2\, dz = \sum_{k=n+1}^{\infty} a_k^2 \leqslant \frac{1}{(n+1)^2} \sum_{k=n+1}^{\infty} (ka_k)^2 =$$

$$= \frac{1}{(n+1)^2} \cdot \frac{2}{\pi} \int_0^\pi \left(\frac{\partial \varrho}{\partial z}\right)^2 dz \leqslant \frac{1}{(n+1)^4} \cdot \frac{2}{\pi} \int_0^\pi \left(\frac{\partial^2 v}{\partial z^2}\right)^2 dz,$$

$$\int_0^l dx \int_0^\pi \varrho^2\, dz \leqslant \frac{1}{(n+1)^4} \int_0^l dx \int_0^\pi \left(\frac{\partial^2 v}{\partial z^2}\right)^2 dz. \tag{78}$$

Comparing these inequalities, we have:

$$\int_0^l \int_0^\pi \left[\left(\frac{\partial \varrho}{\partial x}\right)^2 + \left(\frac{\partial \varrho}{\partial z}\right)^2 + \varrho^2\right] dx\, dz \leqslant \frac{C_7}{(n+1)^2}. \tag{79}$$

Let us pass from the variable z to y, taking $z = \dfrac{\pi(y-g)}{h-g}$. We shall set

$$S(x, y) = \sigma\left(x, \frac{\pi(y-g)}{h-g}\right) = \sum_{k=1}^{n} a_k \sin \frac{k\pi(y-g)}{h-g}, \tag{80}$$

$$R(x, y) = u(x, y) - S(x, y) = \varrho\left(x, \frac{\pi(y-g)}{h-g}\right). \tag{81}$$

Then, forming the expressions for the derivatives of the functions

v and R, for example

$$\frac{\partial v}{\partial x} = \frac{\partial u}{\partial x} + \frac{\partial u}{\partial y}\left[g' + \frac{z}{\pi}(h' - g')\right],$$

$$\frac{\partial R}{\partial x} = \frac{\partial \varrho}{\partial x} + \frac{\partial \varrho}{\partial z}\frac{\pi[-g'(h-g) - (y-g)(h'-g')]}{(h-g)^2}, \qquad (82)$$

one can satisfy oneself that from the assumed boundedness of the integrals for $\dfrac{\partial^2 u}{\partial y \partial x}$ and $\dfrac{\partial^2 u}{\partial y^2}$ and the boundedness of $h'(x)$ and $g'(x)$ follows the finiteness of the integrals standing on the right side of inequalities (76), (77) and (78), which establishes finally the validity of inequality (79). On the other hand, once it has been established, it is not hard to obtain from it, by the transformation of variables, using formulas of form (82), the inequality

$$\iint\limits_{D} \left[\left(\frac{\partial R}{\partial x}\right)^2 + \left(\frac{\partial R}{\partial y}\right)^2 + R^2\right] dx\, dy < \frac{C_8}{(n+1)^2}. \qquad (83)$$

Finally, from this, in view of the boundedness of the functions a, b, c, we obtain

$$\iint\limits_{D} \left[a\left(\frac{\partial R}{\partial x}\right)^2 + b\left(\frac{\partial R}{\partial y}\right)^2 + cR^2\right] dx\, dy < \frac{C_9}{(n+1)^2}. \qquad (84)$$

The last inequality, if we take (80) and (81) into consideration, shows that one can adopt $\bar{u}_n(x, y) = S(x, y)$, and then we will have $\bar{\varepsilon}_n < \dfrac{C_9}{(n+1)^2}$, i.e., $\bar{\varepsilon}_n = O\left(\dfrac{1}{n^2}\right)$. Moreover, from formula (75) defining a_k, it is clear that $a_k(0) = a_k(l) = 0$, and therefore $f_k(0) = f_k(l) = 0$ too.

Let us pass now to the construction of $\tilde{u}_n(x, y)$. We shall put $\cos\zeta = 4\left(\dfrac{z}{\pi}\right)^3 - 6\left(\dfrac{z}{\pi}\right)^2 + 1$ for $0 \leqslant x \leqslant \pi$; in accordance with this for $0 \leqslant \zeta \leqslant \pi$ a function $z = \varkappa(\zeta)$ is defined, with regard to which it is easily verified that it is differentiable any number of times and that $\varkappa'(\zeta) \neq 0$ for $0 \leqslant \zeta \leqslant \pi$. Furthermore, put

$$w(x, \zeta) = \frac{\sin\zeta}{z(\pi - z)}v(x, z), \quad \text{where } z = \varkappa(\zeta), \quad 0 \leqslant \zeta \leqslant \pi. \qquad (85)$$

Again it is easy to satisfy oneself that the multiplier of v on the right side is a differentiable function for $0 \leqslant \zeta \leqslant \pi$. Now put

$$\tau(x, \zeta) = \sum_{k=1}^{n} b_k \sin k\zeta, \quad b_k = \frac{2}{\pi}\int_0^\pi w(x, \zeta)\sin k\zeta\, d\zeta. \qquad (86)$$

Then, as for (79), using the fact that the derivatives of w are simply evaluated in terms of the derivatives of v, one can establish that

$$\int_0^l dx \int_0^\pi \left[\left(\frac{\partial w}{\partial x} - \frac{d\tau}{\partial x}\right)^2 + \left(\frac{\partial w}{\partial \zeta} - \frac{\partial \tau}{\partial \zeta}\right)^2 + (w-\tau)^2\right] d\zeta < \frac{C_{10}}{(n+1)^2}. \quad (87)$$

Furthermore we have

$$\tau(x, \zeta) = \sin \zeta P_{n-1}(\cos \zeta),$$

where P_{n-1} is a polynomial of the indicated degree. Hence, replacing $\cos \zeta$ by its expression in terms of z, we can write:

$$\tau(x, \zeta) = \frac{\sin \zeta}{z(\pi - z)} T(x, z),$$

where

$$T(x, z) = z(\pi - z)P_{n-1}(\cos \zeta) = z(\pi - z)P_{3n-3}(z). \quad (88)$$

Now replacing w in inequality (87) by its expression in terms of v (85) and τ by its expression in terms of T (88), we find that

$$\int_0^l dx \int_0^\pi \left[\left(\frac{\partial v}{\partial x} - \frac{\partial T}{\partial x}\right)^2 + \left(\frac{\partial v}{\partial z} - \frac{\partial T}{\partial z}\right)^2 + (v-T)^2\right] dz < \frac{C_{11}}{(n+1)^2}. \quad (89)$$

Now reverting to the variable y, we can adopt

$$\tilde{u}_{3n-2}(x, y) = \tilde{u}_{3n-1}(x, y) = \tilde{u}_{3n}(x, y) = T\left(x, \pi \frac{y - g(x)}{h(x) - g(x)}\right);$$

then it is easy to satisfy oneself [see (88)] that the $\tilde{u}_n$ have the form indicated in (72′), and using (89) and reasoning as in the derivation of (84), to verify that

$$\tilde{\varepsilon}_n = \iint_D \left[a\left(\frac{\partial u}{\partial x} - \frac{\partial \tilde{u}_n}{\partial x}\right)^2 + b\left(\frac{\partial u}{\partial y} - \frac{\partial \tilde{u}_n}{\partial y}\right)^2 + \right.$$
$$\left. + c(u - \tilde{u}_n)^2\right] dx\, dy < \frac{C_{12}}{n^2}, \quad (90)$$

and the Lemma has been proved in full.

Observation. The conditions of the Lemma could have been relaxed in this respect, that in formulating it, rather than the square-integrability of $\frac{\partial^2 u}{\partial x \partial y}$ and $\frac{\partial^2 u}{\partial y^2}$, there could have been required the square-integrability in the region D

of the derivatives $\frac{\partial^{1+p}u}{\partial x \partial y^p}$, $\frac{\partial^{1+p}u}{\partial y^{1+p}}$ $(0 < p < 1)$. By the derivatives $\frac{\partial^{1+p}}{\partial x \partial y^p}$ and $\frac{\partial^{1+p}}{\partial y^{1+p}}$ there should be understood the derivatives of fractional order which were at one time introduced by Liouville. In conformity with this we would have obtained only $\varepsilon_n = O\left(\frac{1}{n^{2p}}\right)$. The proof also remains unmodified, except that rather than the elementary estimate of $\int_0^\pi \left(\frac{\partial \varrho}{\partial z}\right) dz$ one must use the Hardy-Littlewood theorem according to which, if $f(z)$ is a function for which the derivative of order p is square-summable, then

$$\int_0^\pi (f - S_n)^2 dx = O\left(\frac{1}{n^{2p}}\right),$$

where S_n is the nth Fourier sum for $f(x)$ [1]).

THEOREM 3. *If the region D has the form indicated in Lemma 2, and the function $u(x, y)$ satisfies the conditions of Theorem 2 and Lemma 2, then there obtains the uniform convergence of the sequence of approximations found by the method of reduction to ordinary equations, if they be sought in either of the two forms:*

$$\sum_{k=1}^{n} f_k(x) \sin \frac{k\pi(y-g)}{h-g}, \quad \sum_{k=0}^{n-1} f_k(x)(y-g)(y-h)y^{k-1}, \tag{91}$$

and moreover the order of the error of the n-th approximation is $O\left(\frac{\sqrt{\ln n}}{n}\right)$.

PROOF. For definiteness we shall dwell on the case when the nth approximation is sought in the form

$$u_n(x, y) = \sum_{k=0}^{n-1} f_k(x)(y-g)(y-h)y^{k-1}. \tag{92}$$

Since u_n gives a minimum for the integral [2])

$$I(u) = \iint_D \left[a\left(\frac{\partial u}{\partial x}\right)^2 + b\left(\frac{\partial u}{\partial y}\right)^2 + cu^2 - 2fu\right] dx\, dy,$$

by comparison with all other functions of the same form, we have $I(u_n) \leqslant I(\tilde{u}_n)$, where $\tilde{u}_n$ is the function constructed in Lemma 2. Then

$$\varepsilon_n = I(u_n) - I(u) \leqslant I(\tilde{u}_n) - I(u) = \tilde{\varepsilon}_n = O\left(\frac{1}{n^2}\right),$$

and therefore on the basis of Theorem 2, since obviously $\varepsilon_n \ln n \to 0$, the functions $u_n(x, y)$ converge uniformly to $u(x, y)$. Moreover, as is

---

[1]) Hardy and Littlewood, [1]. See also Quade, [1].

[2]) Attention should be directed to the fact that equation (51) serves as the Euler equation for the integral $I(u)$ [cf. § 1, (17b), p. 249].

there indicated,

$$|u - u_n| = O\left(\sqrt{\varepsilon_n \ln \frac{n}{\varepsilon_n}}\right) = O\left(\frac{\sqrt{\ln n}}{n}\right), \tag{93}$$

which is what was required to be proved.

Observation. We shall note that by imposing conditions on the derivatives of higher order, a higher order of smallness for $|u_n - u|$ could have been obtained. Thus the square-integrability of the derivatives of order $(p + 1)$ secures that $|u_n - u| = O\left(\frac{\sqrt{\ln n}}{n^p}\right)$, if u_n is sought in the second of forms (91). The sole modification that must be introduced in the proof of Lemma 2 in this case consists in replacing the variable z in the function $v(x, z)$ by a ζ such as to secure the fulfilment of the conditions $\left.\frac{\partial^2 v}{\partial \zeta^2}\right|_{\zeta=\pi} = -\left.\frac{\partial^2 v}{\partial \zeta^2}\right|_{\zeta=0}$, etc. This would permit one to employ repeated integration by parts in transforming the Fourier coefficients (p. 348).

Let us now pass on to Ritz's method. Again we begin with a Lemma; in view of the awkwardness of the proof, we shall not give it in all its details.

LEMMA 3. *Let $u(x, y)$ be a function defined in the region D, of the same form as that of Lemma 2, and such that its partial derivatives of the third order are square-summable in D. Then one can select a function $u_n^*(x, y)$ of the form*

$$u_n^*(x, y) = x(l - x)(y - g(x))(y - h(x)) \sum_{i,k \leqslant n} a_{ik} x^i y^k, \tag{94}$$

such that the quantity

$$\varepsilon_n^* = \iint\limits_D \left[a\left(\frac{\partial u}{\partial x} - \frac{\partial u_n^*}{\partial x}\right)^2 + b\left(\frac{\partial u}{\partial y} - \frac{\partial u_n^*}{\partial y}\right)^2 + c(u - u_n^*)^2\right] dx\, dy \tag{95}$$

has the order $O\left(\frac{\ln n}{n^2}\right)$.

PROOF. Let us consider the function $\sigma(x, z)$ [see (75)]:

$$\sigma(x, z) = \sum_{k=1}^{n} a_k \sin kz = 2 \sin z \sum_{k=1}^{n} a_k (\cos (k - 1)z +$$
$$+ \cos (k - 3)z + \ldots)^{1)} = (a_1 + a_3 + \ldots) \sin z +$$
$$+ 2 \sin z \sum_{k=1}^{n-1} (a_{k+1} + a_{k+3} + \ldots) \cos kz = \sin z H(x, z); \tag{96}$$

here in the last sum there should not be included the a with indices $> n$.

[1]) The last summand in the parentheses is $\cos z$ if k is even, and $\frac{1}{2}$ if k is odd.

Let us estimate the integral of H^2; we have

$$\frac{2}{\pi}\int_0^\pi H^2\,dz \leqslant 4\sum_{k=0}^{n-1}(a_{k+1}+a_{k+3}+\ldots)^2 \leqslant$$

$$\leqslant 4\sum_{k=0}^{n-1}(|a_{k+1}|+|a_{k+2}|+\ldots+|a_n|)^2 =$$

$$= 4\sum_{k=0}^{n-1}\left(\frac{1}{k+1}(k+1)\,|a_{k+1}|+\ldots+\frac{1}{n}\,n\,|a_n|\right)^2 \leqslant$$

$$\leqslant 4\sum_{k=0}^{n-1}\left[\frac{1}{(k+1)^2}+\ldots+\frac{1}{n^2}\right]\left[(k+1)^2a_{k+1}^2+\ldots+n^2a_n^2\right] \leqslant$$

$$\leqslant 4\frac{\pi^2}{6}(1^2a_1^2+\ldots+n^2a_n^2)+4\sum_{k=1}^{n-1}\frac{1}{k}[(k+1)^2a_{k+1}^2+\ldots+n^2a_n^2] \leqslant$$

$$\leqslant 4\left(\frac{\pi^2}{6}+1+\tfrac{1}{2}+\ldots+\frac{1}{n-1}\right)\sum_{k=1}^{n}k^2a_k^2 < 4(\ln n+3)\,\frac{2}{\pi}\int_0^\pi\left(\frac{\partial v}{\partial z}\right)^2 dz,$$

whence

$$\int_0^l dx\int_0^\pi H^2dz \leqslant 4\,(\ln n+3)\int_0^l dx\int_0^\pi\left(\frac{\partial v}{\partial z}\right)^2 dz. \tag{97}$$

In just the same fashion one can estimate the integrals of the squares of the first and second derivatives of the function H in terms of just such integrals of the second and third derivatives of the function v, and since the latter integrals are finite, thanks to the corresponding assumption concerning the function u, one is convinced that the integrals of the squares of the derivatives of the function H have the order $O(\ln n)$ [1]).

Reverting to the variable y, we have:

$$S(x,y) = \sigma\left(x,\pi\frac{y-g}{h-g}\right) =$$

$$= (h-y)(y-g)\frac{\sin\pi\dfrac{y-g}{h-g}}{(h-y)(y-g)}\,H\left(x,\pi\frac{y-g}{h-g}\right). \tag{98}$$

---

[1]) It should be said that in respect of the second derivatives this will be true only if one contrives to meet beforehand the conditions $\dfrac{\partial^2 v}{\partial y^2}=0$ for $y=0$ and $y=\pi$. If this were not done, it would be possible to show only that this integral is of the order $O(n)$. This is sufficient, however, for the forthcoming proof of Ritz's method.

Choosing L sufficiently large and executing, if necessary, a translation, we obviously can consider that $0 < g(x) < h(x) < L$. Let us now define the function $G(x, y)$, setting

$$G(x, y) = \frac{\sin \pi \dfrac{y-g}{h-g}}{(h-y)(y-g)} H\left(x, \pi \frac{y-g}{h-g}\right) \text{ for } g(x) \leqslant y \leqslant h(x), \tag{99}$$

and extend its definition for $0 \leqslant y \leqslant L$ so that $G(x, 0) = G(x, L) = 0$ [1]). Thanks to what was said above about the function H, one is readily convinced that the integrals of the squares of G and of its derivatives of the first and second order over the rectangle $[0, l;\ 0, L]$ have the order $\ln n$. We note in addition that, as is easily verified, $G(0, y) = G(l, y) = 0$.

Now, starting with $G(x, y)$ rather than $w(x, z)$ and repeating the reasoning carried through in Lemma 2, we can construct $T(x, y)$, a polynomial in y of degree $< 3n$, so that

$$\int_0^l dx \int_0^L \left[\left(\frac{\partial G}{\partial x} - \frac{\partial T}{\partial x}\right)^2 + \left(\frac{\partial G}{\partial y} - \frac{\partial T}{\partial y}\right)^2 + (G-T)^2\right] dy < \frac{C_{13} \ln n}{n^2}. \tag{100}$$

Here, by the construction, we shall have $T(0, y) = T(l, y) = 0$. It is also easily verified that the integrals of the squares of the derivatives of the function T have the same order as do those of G, i.e., $O(\ln n)$. The same process can be applied anew to the function $T(x, y)$, changing only the role of the variables. Then we shall define a function $P(x, y)$, a polynomial in x of degree $< 3n$; it is easily verified that it will be, as is T, a polynomial in y of degree $< 3n$, and, since $T(0, y) = T(l, y) = 0$, it will have the factors x and $l - x$, i.e.,

$$P(x, y) = x(l-x) \sum_{i,k<3n} a_{i,k} x^i y^k, \tag{101}$$

and, finally, there will be valid the inequality

$$\int_0^L dy \int_0^l \left[\left(\frac{\partial T}{\partial x} - \frac{\partial P}{\partial x}\right)^2 + \left(\frac{\partial T}{\partial y} - \frac{\partial P}{\partial y}\right)^2 + (T-P)^2\right] dx < C_{14} \frac{\ln n}{n^2}. \tag{102}$$

---

[1]) It is most convenient, in determining $G(x, y)$ for values of y lying outside the interval $(g(x), h(x))$, to utilize the same expression (99), but premultiplied (for $h(x) \leqslant y \leqslant L$) by a function smoothly descending from the value 1 for $y = h(x)$ to the value 0 for $y = L$.

Now we can adopt

$$u^*_{3n-1}(x, y) = u^*_{3n}(x, y) = u^*_{3n+1}(x, y) =$$
$$= (h(x) - y)(y - g(x))P(x, y). \qquad (103)$$

Indeed, as is clear from the preceding estimates and (84),

$$\varepsilon^*_n = \iint\limits_D \Big[a\Big(\frac{\partial u^*_n}{\partial x} - \frac{\partial u}{\partial x}\Big)^2 + b\Big(\frac{\partial u^*_n}{\partial y} - \frac{\partial u}{\partial y}\Big)^2 +$$
$$+ c(u^*_n - u)^2\Big]\,dx\,dy < \frac{C_{15}\ln n}{n^2}, \qquad (104)$$

which proves the statement of the Lemma.

THEOREM 4. *If the region D and the function u satisfy the conditions of Lemma 3, then the convergence of Ritz's method obtains if the solution be sought in the form*

$$u_n(x, y) = x(l - x)(h(x) - y)(g(x) - y) \cdot \sum_{i,k\leqslant n} a_{i,k}x^i y^k; \qquad (105)$$

moreover, $|u_n - u|$ has the order $O\Big(\frac{\ln n}{n}\Big)$.

Indeed, by the Lemma, $\varepsilon^*_n = O\Big(\frac{\ln n}{n^2}\Big)$; then the more so, for the nth approximation by Ritz's method, will $\varepsilon_n = O\Big(\frac{\ln n}{n^2}\Big)$, and in this case, by Theorem 2, the convergence of Ritz's method obtains and $|u_n - u|$ has the order

$$O(\sqrt{\varepsilon_n \ln n}) = O\Big(\frac{\ln n}{n}\Big).$$

Observation 1. The question of the convergence of Ritz's method in the case of the Dirichlet problem for an equation of elliptic type with two variables has been studied previously by N. M. Krylov and N. N. Bogoli͡ubov [1]). From these results, however, the convergence of the method follows only on the imposition of conditions on derivatives of higher orders (the boundedness of the seventh derivatives).

Observation 2. The method of reasoning that we executed for Theorems 3 and 4 is also applicable in the case of boundary problems other than the Dirichlet problem, as well as for proof of the convergence of other variational processes. In particular, in the method of reduction to ordinary differential equations the transition to the next approximation can be effected by subdividing the region into ever narrower strips, in each of which the solution is sought in the form of a polynomial of fixed degree. The convergence also holds for this method, and can be investigated by the same methods.

Observation 3. The results of the last two sections have been substantially

---

[1]) Krylov, N. M. and Bogoli͡ubov, N. N. [2].

strengthened and generalized in recent works of V. I. Il'in and of IŪ. Kharrik [1]).

In I. IŪ. Kharrik's works is investigated the question of the possible order of the approximation, by expressions of the form $\omega(x, y)P_n(x, y)$ (P_n being a polynomial of the nth degree) in a region D bounded by the curve $\omega(x, y) = 0$, of a function $u(x, y)$ vanishing on the contour. Also investigated is the analogous expression for spaces of many dimensions.

From these results it follows, for example, under certain assumptions with respect to $\omega(x, y)$, that if u has first partial derivatives satisfying the Lipschitz condition with exponent $\alpha \leqslant \frac{1}{2}$, then the order of simultaneous approximation of u, $\frac{\partial u}{\partial x}$, $\frac{\partial u}{\partial y}$ possible is $n^{-\alpha}$. In particular, it could have been established in Lemma 3 that for such a function $u(x, y)$ we will have $\varepsilon_n = O(n^{-2\alpha})$. Thanks to this also, in Theorem 4 the convergence of Ritz's method could have been obtained under the condition on u cited above, which is considerably weaker than that indicated in the text of the Theorem. For this $|u_n - u| = O(n^{-\alpha}\sqrt{\ln n})$. Under stronger assumptions with respect to u, for $|u_n - u|$ is obtained a correspondingly higher order.

In the works of V. P. Il'in, there are first of all systematically investigated the possible estimates of a function of m variables, given in a certain region, if estimates are known for the integrals of its partial derivatives over the entire region and along its sections of lower dimensions (Lemma 1 represents an example of similar estimates). By means of them are established, in the case of the Dirichlet problem for an equation of elliptic type with m variables, theorems analogous to Theorems 1 and 2. In particular a comparison of these results with the aforementioned ones of I. IŪ. Kharrik leads to the establishment of the uniform convergence of Ritz's method for this problem when the approximate solution is sought in the form $\omega(x, y)P_n(x, y)$, if the exact solution has m continuous derivatives.

It has also been shown by V. P. Il'in that analogous results hold also with respect to the Neumann problem and the mixed problem for an elliptic equation. Finally, from these works are obtained even the conditions of the uniform convergence of Ritz's method for the basic biharmonic problem $\left(u = 0, \frac{\partial u}{\partial n} = 0 \text{ on the contour}\right)$ with m variables, when its solution is sought in the form $\omega^2(x, y)P_n(x, y)$.

[1]) V. P. Il'in, [1]; I. IŪ. Kharrik, [1].

CHAPTER V

THE CONFORMAL TRANSFORMATION OF REGIONS

§ 1. INTRODUCTION

1. Conformal transformation and the Laplace equation. A very large number of the steady-state problems of mathematical physics reduce to solving the simplest of the equations of elliptic type — the Laplace equation

$$\Delta_{xy}u = \frac{\partial^2 u}{\partial x^2} + \frac{\partial^2 u}{\partial y^2} = 0. \tag{1}$$

To these belong, for example, plane problems of the theory of electrostatics, of the theory of heat, problems of the motion of a perfectly incompressible fluid, etc.

Here the sought function solving the problem posed is defined first by the region in which it is sought, being dependent upon the form of the object under study, and second, by the conditions that must be fulfilled on the boundary of this region.

It often happens that for the simplest regions — circle, square, annulus, plane with cuts, etc. — the problem can be solved relatively simply for comparatively complex boundary requirements. If the region has a complex structure, however, the solution of the problem directly for this region sometimes presents very great difficulties even for such a simple problem as the Dirichlet problem, for example.

An effort is therefore made to transform the given region into simplest form beforehand. In transformations of such a kind there will be altered, generally speaking, not only the region for which the solution is sought, and the boundary conditions, but also the differential equation that the sought function must satisfy.

For the Laplace equation that transformation will obviously be of the greatest value for which the Laplace equation itself remains invariant.

Instead of the variables x and y let us have introduced new variables ξ and η, by means of the equations

$$\xi = \xi(x, y), \quad \eta = \eta(x, y). \tag{2}$$

When the variables x and y vary in the region B, the new variables ξ and η vary in some region G. Let us inquire into how the equation is changed by the transformation:

$$\frac{\partial u}{\partial x} = \frac{\partial u}{\partial \xi}\,\frac{\partial \xi}{\partial x} + \frac{\partial u}{\partial \eta}\,\frac{\partial \eta}{\partial x},$$

$$\frac{\partial^2 u}{\partial x^2} = \frac{\partial^2 u}{\partial \xi^2}\left(\frac{\partial \xi}{\partial x}\right)^2 + 2\,\frac{\partial^2 u}{\partial \xi \partial \eta}\,\frac{\partial \xi}{\partial x}\,\frac{\partial \eta}{\partial x} + \frac{\partial^2 u}{\partial \eta^2}\left(\frac{\partial \eta}{\partial x}\right)^2 + \frac{\partial u}{\partial \xi}\,\frac{\partial^2 \xi}{\partial x^2} + \frac{\partial u}{\partial \eta}\,\frac{\partial^2 \eta}{\partial x^2}$$

and analogously

$$\frac{\partial^2 u}{\partial y^2} = \frac{\partial^2 u}{\partial \xi^2}\left(\frac{\partial \xi}{\partial y}\right)^2 + 2\,\frac{\partial^2 u}{\partial \xi \partial \eta}\,\frac{\partial \xi}{\partial y}\,\frac{\partial \eta}{\partial y} + \frac{\partial^2 u}{\partial \eta^2}\left(\frac{\partial \eta}{\partial y}\right)^2 + \frac{\partial u}{\partial \xi}\,\frac{\partial^2 \xi}{\partial y^2} + \frac{\partial u}{\partial \eta}\,\frac{\partial^2 \eta}{\partial y^2}.$$

The new equation for the function u in the variables ξ and η will therefore be

$$\Delta_{xy} u = \frac{\partial^2 u}{\partial \xi^2}\left[\left(\frac{\partial \xi}{\partial x}\right)^2 + \left(\frac{\partial \xi}{\partial y}\right)^2\right] + 2\,\frac{\partial^2 u}{\partial \xi \partial \eta}\left(\frac{\partial \xi}{\partial x}\,\frac{\partial \eta}{\partial x} + \frac{\partial \xi}{\partial y}\,\frac{\partial \eta}{\partial y}\right) +$$

$$+ \frac{\partial^2 u}{\partial \eta^2}\left[\left(\frac{\partial \eta}{\partial x}\right)^2 + \left(\frac{\partial \eta}{\partial y}\right)^2\right] + \frac{\partial u}{\partial \xi}\,\Delta_{xy}\xi + \frac{\partial u}{\partial \eta}\,\Delta_{xy}\eta = 0.$$

In order that it be a Laplace equation, it is obviously necessary that transformation (2) satisfy the following requirements:

$$\Delta_{xy}\xi = 0, \quad \Delta_{xy}\eta = 0,$$

$$\frac{\partial \xi}{\partial x}\,\frac{\partial \eta}{\partial x} + \frac{\partial \xi}{\partial y}\,\frac{\partial \eta}{\partial y} = 0, \quad \left(\frac{\partial \xi}{\partial x}\right)^2 + \left(\frac{\partial \xi}{\partial y}\right)^2 = \left(\frac{\partial \eta}{\partial x}\right)^2 + \left(\frac{\partial \eta}{\partial y}\right)^2.$$

The first two of these equations tell us that ξ and η must be harmonic functions of x and y. The third equation gives

$$\frac{\partial \eta}{\partial x} : \frac{\partial \xi}{\partial y} = -\frac{\partial \eta}{\partial y} : \frac{\partial \xi}{\partial x} = -\mu, \quad \frac{\partial \eta}{\partial x} = -\mu\,\frac{\partial \xi}{\partial y}, \quad \frac{\partial \eta}{\partial y} = \mu\,\frac{\partial \xi}{\partial x}.$$

If we now substitute these values for the derivatives of η in the last equation, we obtain

$$(1 - \mu^2)\left[\left(\frac{\partial \xi}{\partial x}\right)^2 + \left(\frac{\partial \xi}{\partial y}\right)^2\right] = 0.$$

From this it is seen that $\mu = \pm 1$, and consequently either

$$\frac{\partial \xi}{\partial x} = \frac{\partial \eta}{\partial y}, \quad \frac{\partial \xi}{\partial y} = -\frac{\partial \eta}{\partial x}, \tag{3}$$

or

$$\frac{\partial \xi}{\partial x} = -\frac{\partial \eta}{\partial y}, \quad \frac{\partial \xi}{\partial y} = \frac{\partial \eta}{\partial x}. \tag{3'}$$

Equation (3′) differs from (3) only in that ξ and η have exchanged places; we shall dwell therefore only on the first of the conditions. It expresses the circumstance that ξ and η are conjugate harmonic functions. Transformations given by functions of this kind will be conformal — carrying infinitesimal figures of the plane x, y into figures of the plane ξ, η which are similar to them, everywhere where $\left(\frac{\partial \xi}{\partial x}\right)^2 + \left(\frac{\partial \xi}{\partial y}\right)^2 \neq 0$.

Let us form the following complex function: $f(z) = \xi + i\eta$ of the complex variable $z = x + iy$. For it (3) will be the familiar Cauchy-Riemann conditions of monogeneity.

Thus the Laplace equation remains invariant with respect to transformations furnished by monogenic functions of a complex variable.

In the sequel we shall find certain general propositions useful which relate to conformal transformations. We cite them now without proof.

2. The transformation of simply connected regions. The simplest case of conformal transformation will be that where the regions to be transformed are simply connected. The first question to arise in connection with the study of the conformal transformation of them is the question of the existence of a holomorphic function effecting a reciprocally single-valued transformation of them into each other.

The answer is given by the following proposition, known as the Riemann Theorem:

If B and G are simply connected single-sheeted regions of the planes of the complex variables $z = x + iy$ and $\zeta = \xi + i\eta$, the regions being different from the entire plane and the entire plane without one point, there exists a function $\zeta = f(z)$, holomorphic in B and effecting the conformal mapping of B onto G in a one-to-one manner, this mapping being such that a pre-assigned point a in the region B is carried to an arbitrarily chosen point α of G and a given direction at the point a is carried into a given direction at α.

By the requirement of the correspondence of the interior points and of the directions at them, the transforming function is fully determined.

Thus the function transforming B into G will depend on three parameters, in view of the fact that the point α, arbitrarily assigned, is determined by two parameters, and the direction at it, by one. We can arrange the choice of the three parameters differently from that of Riemann's Theorem. For example, for regions subject to certain limitations, which for the cases encountered in practice are certainly fulfilled, one can require that three points, given in advance, of the boundary of region B be carried to three points, given in advance, of the boundary of G.

We shall in addition draw the attention of the reader to certain facts relating to the behavior of the transforming function on the boundary L of the region B that is under transformation.

Let us assume, for the sake of simplicity, that the regions B and G are bounded. For practical applications it will be sufficient for us to consider their boundaries as being piece-wise smooth curves [1]. We shall not, however, assume these curves to be simple: several boundary points may have one and the same complex coordinate z.

Describe about a boundary point with coordinate z a circle of small radius ε; it will intersect B over a certain aggregate of regions. Take those of them that have z for a boundary point. Let these be $B_\varepsilon^{(1)}, B_\varepsilon^{(2)}, \ldots$. The point z is called a simple point of the boundary L of the region B, if for any ε only one of the regions B_ε exists. In the contrary case the point is called multiple. Boundary points with one and the same complex coordinate are called different if different $B_\varepsilon^{(k)}$ correspond to them for a certain ε. As an individual, the boundary point is determined by its coordinate and, for arbitrarily small ε, by the region B_ε contiguous to it; the latter is called the B_ε-neighborhood of the boundary point.

It can be shown that the transforming function $\zeta = f(z)$ will tend to a definite limit if the argument z approaches a boundary point. More than that: if one adjoins to the region all of its boundary points, the transforming function will be continuous everywhere in the closure of the region B obtained.

The problem of the effective construction of a function effecting the transformation of B into G, even in the simple case of the transformation of a region into a circle or a circle into a region, very often presents very great or even insurmountable difficulties. It is sufficient to recall the problem of determining the constants in the Christoffel-Schwarz formula for the function transforming a circle into a polygon to be convinced of this. Even if the means for representing the transforming function is known, its structure will occasionally be so complex that use of it entails a great expenditure of labor.

Therefore, for the inexact solution of problems, the simpler methods for the approximate representation of a transforming function are of very great value. Usually the approximation to the sought function is attempted by forming certain combinations of well-studied functions. The simplest and most frequently employed functions are the polynomials. The possibility of employing them was investigated by C. Runge; it was established by him that it is always possible in the region B to construct a polynomial which differs arbitrarily little in any interior part of B from the function transforming the region into a circle.

If the boundary of the region B has multiple points, polynomials are then obviously insufficient to approximate the mapping function

---

[1]) A curve is said to be piece-wise smooth if it consists of a finite number of rcs with a continuously varying tangent.

uniformly everywhere in B. If, however, the boundary of the region B is a simple curve, i.e., a curve without multiple points, then, as Walsh's investigations have shown, a polynomial can be chosen in B such as to differ arbitrarily little from the transforming function everywhere in B, including even the boundary L.

3. The transformation of multiply connected regions. The conformal transformation of multiply connected regions is subject to greater limitations than is the transformation of simply connected ones. To make possible the reciprocally single-valued transformation of one region into another it is necessary that they be of like connectivity. But the equal connectivity of the regions is still not sufficient to make possible their transformation; it is not any two regions of one and the same connectivity that can be transformed into one another.

We shall first dwell on the case of doubly connected regions, and on the example of them shall explain the idea of constructing the apparatus that permits one to determine the transforming function to any degree of accuracy. In similar fashion the apparatus can be constructed in the case of a region of any connectivity. Of the doubly connected regions, the simplest is the ring between two concentric circles. With respect to the transformation into it of any other region, the following theorem is known:

Any doubly connected region can be transformed, conformally and with reciprocal single-valuedness, into an annulus with the ratio of the radii of its bounding circumferences finite or infinite.

Let us denote the boundary curves of the region by L_1 and L_2, and the radii of the circumferences corresponding to them by ϱ_1 and ϱ_2. If it is indicated which of the boundary curves of the region must pass to the outer, and which to the inner circumference of the annulus the ratio $\frac{\varrho_1}{\varrho_2}$ of the radii ϱ_1 and ϱ_2 will be fully defined. The finding of this ratio is one of the difficult problems of the theory of conformal transformation.

One can simply see that the transformation of one ring into another with the same ratio of radii, without an interchange of inner and outer circumferences, reduces just to a similarity transformation and a rotation about the center. Therefore if the origin of coordinates lies at the center of the ring, the function transforming region B into a ring is defined accurate to an arbitrary complex factor.

For the approximate transformation here, rational functions of a particular form can be used.

Let the infinitely distant point of the plane not belong to the region. We denote the outer contour by L_2 and the inner by L_1. Place the origin of coordinates within or on L_1. The transforming function is, as is known, the sum of two terms, one of which, $\varphi_0(z)$, is holomorphic inside the outer contour, and the other, $\varphi_1(z)$, outside the inner. One can approximate $\varphi_0(z)$ uniformly in any interior part of the region bounded by L_2, by means of a polynomial of the form

$$P_n(z) = a_0 + a_1 z + a_2 z^2 + \ldots + a_n z^n,$$

while $\varphi_1(z)$ can be uniformly approximated as exactly as one pleases at any interior part of the region lying outside of L_1 by means of a rational function having the form:

$$Q_m(z) = a_0 + \frac{a_{-1}}{z} + \frac{a_{-2}}{z^2} + \ldots + \frac{a_{-m}}{z^m}.$$

Thus the function transforming into an annulus the region B between L_1 and L_2 can, in any interior part of B, be uniformly represented with arbitrary accuracy by means of a rational function of the type:

$$\sum_{\nu=-m}^{n} a_\nu z^\nu.$$

If the point at infinity and the origin of coordinates are points exterior to B, and the curves L_1 and L_2 do not have multiple points, then a uniform approximation is possible everywhere in B, including even the contour.

The matter will be somewhat different in the case where the infinitely distant point belongs to the region. B will then be the exterior of two curves L_1 and L_2. Let a_1 and a_2 be two points lying within the curves or upon them.

Here the transforming function will be the sum of two functions, one of which is holomorphic outside of L_1; it can be approximated arbitrarily closely in any interior part of the region exterior to L_1 by means of the rational function:

$$R_1 = \alpha_0 + \frac{\alpha_1}{(z - a_1)} + \frac{\alpha_2}{(z - a_1)^2} + \ldots + \frac{\alpha_n}{(z - a_1)^n};$$

the second, however, will be holomorphic outside of L_2, and one can approximate it in quite the same way by means of the rational function:

$$R_2 = \beta_0 + \frac{\beta_1}{(z - a_2)} + \frac{\beta_2}{(z - a_2)^2} + \ldots + \frac{\beta_n}{(z - a_2)^n}.$$

The transforming function can therefore be represented approximately by a function of the form:

$$\sum_{k=0}^{n} \left\{ \frac{\alpha_k}{(z - a_1)^k} + \frac{\beta_k}{(z - a_2)^k} \right\}.$$

When L_1 and L_2 are simple curves and the points a_1 and a_2 lie within them, then, as above, a uniform approximation to $\varphi(z)$ in B, including even the boundary, is possible.

Let us pass to the transformation of regions of any connectivity. For the sake of simplicity we shall have in view a region of finite connectivity n.

As the canonical region into which is transformed any other region, here there is most frequently chosen the entire plane with cuts parallel to a certain direction, for example the direction of the real axis.

In the transformation we can arbitrarily assign the point a of the region B of the z-plane that must go to the infinitely distant point of the ζ-plane of the image. Moreover, there can be arbitrarily assigned the direction at the point a that must be carried over to the direction of the real axis at the point $\zeta = \infty$. If a lies at a finite distance, the transforming function has near it the expansion:

$$\zeta = \frac{c}{z-a} + c_0 + c_1(z-a) + c_2(z-a)^2 + \dots, \tag{4}$$

but when $a = \infty$, then

$$\zeta = cz + c_0 + \frac{c_1}{z} + \frac{c_2}{z^2} + \dots \tag{4'}$$

In both cases the choice of the indicated direction corresponds to the choice of the argument of the coefficient c of the principal term of the series.

Given the multiplication of ζ by any real positive number, which corresponds to an arbitrary similarity transformation from the origin of coordinates of the ζ-plane, the cuts parallel to the real axis will obviously not change their direction, still remaining parallel to the real axis.

But given the multiplication of ζ by a positive number, the modulus of c is multiplied by this number. This means that in our problem we must consider as being arbitrary not only the argument of c, but also its modulus. Finally, the cuts will not change their direction with any parallel transfer of the ζ-plane, or putting it differently, with any change of the free term c_0.

Thus in the case of the similar type of transformation, in expressions (4) or (4′) the transforming functions remain arbitrary as regards a, c, c_0, or c, c_0, respectively.

It can be shown simply that if these quantities are given, the transforming function is completely defined.

Indeed, let the region B be transformed conformally into a plane with cuts parallel to the real axis by means of — in addition to ζ — a function ζ^* having as its expansion about a either

$$\zeta^* = \frac{c}{z-a} + c_0 + c_1^*(z-a) + c_2^*(z-a)^2 + \dots,$$

or

$$\zeta^* = cz + c_0 + \frac{c_1^*}{z} + \frac{c_2^*}{z^2} + \dots$$

Form the difference between ζ and ζ^*:

$$\zeta - \zeta^* = p + iq.$$

Near a it has either the expansion

$$\zeta - \zeta^* = (c_1 - c_1^*)(z - a) + (c_2 - c_2^*)(z - a)^2 + \dots,$$

or

$$\zeta - \zeta^* = \frac{c_1 - c_1^*}{z} + \frac{c_2 - c_2^*}{z^2} + \dots.$$

Since upon subtraction the principal terms in the expansions of ζ and ζ^* cancel, the difference $\zeta - \zeta^* = \Phi(z)$ will be a function that is regular and bounded everywhere in B [1]). On each boundary curve the imaginary parts of ζ and ζ^* preserve constant values, for each such curve is carried by the functions $\zeta = \zeta(z)$ and $\zeta = \zeta^*(z)$ into a cut parallel to the real axis. Therefore so will the imaginary part of $\Phi(z)$ preserve a constant value on each of the boundary curves of the region.

Moreover, as the free terms c_0 of the expansions also cancel in the subtraction, $\Phi(z)$ will vanish at the point a, or, for case (4′), at the point at infinity.

Let us assume that $\Phi(z)$ is different from a constant quantity. Then it will provide a transformation of the region B into a certain bounded region, the boundary of which consists of straight-line cuts parallel to the real axis. But such a region is obviously impossible, and therefore $\Phi(z) = \zeta - \zeta^* = \text{const}$. On the other hand, from $\Phi(a) = 0$ or $\Phi(\infty) = 0$ it follows that the constant to which $\Phi(z)$ is equal can only be zero, and, accordingly, ζ and ζ^* necessarily coincide.

§ 2. THE PROPERTY OF MINIMUM AREA IN THE TRANSFORMATION OF A REGION INTO A CIRCLE

1. An extremal property of the function transforming a region into a circle. Of the functions conformally transforming a given simply connected, single-sheeted region B into some other region, the function mapping onto a circle has one interesting extremal property that permits one to give a method for the approximate conformal transformation of the region into a circle, a method which is comparatively simple in its idea, and which is in some cases also computationally simple [2]).

---

[1]) The regularity of $\Phi(z)$ everywhere in B is obvious. Its boundedness, however, emerges from the following considerations. Each of the functions ζ and ζ^* is bounded near the contour of the region, since each of them carries the boundary curves into cuts lying at a finite distance. Therefore $\Phi(z) = \zeta - \zeta^*$ will also be bounded near the contour. Thus it turns out that $\Phi(z)$ is regular in B and is bounded near the boundary of B. But then it is bounded everywhere in B.

[2]) L. Bieberbach, [1]. An analogous extremal property for the function conformally transforming a region into a circular ring has been considered by G. IÂ. Khazhaliia, [1], [2].

We shall first establish the minimum principle. Let the function:

$$\zeta = f(z) = a_0 + a_1 z + a_2 z^2 + \ldots, \tag{1}$$

regular in a circle of radius R with center at the origin of coordinates, transform it into some region of the complex variable ζ.

Its area, as we know, is expressed by an integral of the form:

$$I = \int_0^R \int_0^{2\pi} f'(z)\overline{f'(z)}\, r\, dr\, d\varphi, \quad z = re^{i\varphi}. \tag{2}$$

Upon substitution under the integral sign of the value from (1) of the derivative $f'(z)$, we obtain there a sum of terms of the form:

$$nka_n\bar{a}_k r^{n+k-1} e^{i(n-k)\varphi}.$$

In integrating with respect to φ, two cases are conceivable: if $n \neq k$,

$$\int_0^{2\pi} e^{i(n-k)\varphi} d\varphi = 0,$$

if $n = k$, however, integration with respect to φ gives simply 2π, and

$$I = \int_0^R \sum_{n=0}^{\infty} |a_n|^2 n^2 2\pi r^{2n-1} dr = \pi R^2 |a_1|^2 + \pi \sum_{n=2}^{\infty} n\, |a_n|^2 R^{2n}, \tag{3}$$

from which the following result is obtained.

In mapping a circle of radius R by means of function (1), regular within it and having its first derivative at the center equal to a_1, the area of the mapped region is always larger than $\pi R^2 |a_1|^2$; it can coincide with this number only in case the transformation is linear: $\zeta = a_0 + a_1 z$.

We shall be particularly interested below in the case when $a_1 = 1$; here the result of the preceding theorem tells us that with such a transformation the area is increased, if the case of translation $\zeta = a_0 + z$ be excluded.

Let us now assume that we have to find a conformal transformation of a single-sheeted region B into a circle, such that the assigned interior point of it, which we can always adopt as the origin of coordinates, is carried to the center of the circle, and the definite direction at this point, adopted by us as the positive direction of the real axis, coincides with the positive direction of the real axis of the transformed region. These conditions are equivalent to the requirements

$$f(0) = 0 \quad f'(0) = a_1 > 0.$$

We can always contrive that a_1 equal 1; for this it is sufficient to consider the function $\dfrac{f(z)}{a_1}$, which will give the same mapping of B onto the circle as does $f(z)$, but with perhaps another radius, which is immaterial to us. We can therefore always assume:

$$f(0) = 0 \text{ and } f'(0) = 1. \tag{4}$$

Let us now consider all possible conformal transformations of the region B into certain other regions, given by functions that satisfy the same conditions (4). It is seen without difficulty that of all the regions into which B is carried by means of functions of that type, the circle will possess the least area. Indeed, the mapping of B onto a certain region B' can be imagined to have been carried out by means of the following two steps: first B is mapped onto a circle, and afterwards the circle upon B'. Here the function effecting the latter transformation will satisfy requirements (4). Indeed, let the function transforming region B into the circle be $\zeta = \varphi(z)$, and the function transforming B into B' be $t = f(z)$. Both of them satisfy conditions (4). Finally, let us denote the function furnishing the transformation of the circle into B' with the preservation of the origin of coordinates, by $t = \psi(\zeta)$. It will satisfy the first requirement of (4). We must in addition establish the equality $\psi'(0) = 1$. Carrying out the aforementioned transformation of B into B' via the circle, we find $f(z) = \psi(\zeta)$. Differentiating this equation with respect to z and setting $z = 0$, we find $f'(0) = \psi'(0)\varphi'(0)$, from which, indeed, follows our assertion. But as we have just now seen, under such a transformation the area of the circle can only increase, and therefore the area of B' is always greater than the area of the circle. Therefore the sought function gives the minimum for the integral

$$I = \int_B\int f'(z)\overline{f'(z)}\,dx\,dy, \tag{5}$$

expressing the area of the transformed region, by comparison with all the other functions that satisfy requirements (4).

The function giving the required conformal transformation can be sought, accordingly, as the solution of the following variational problem:

Of the functions that are regular in the region B and that are subject to conditions (4), *to find: that which gives the minimum for integral* (5).

We are convinced of the existence and the uniqueness of the solution of the posed extremal problem by Riemann's Theorem.

It goes without saying that we cannot find $f(z)$ directly, but only $f'(z)$, as $f(z)$ does not appear under the integral sign; from the $f'(z)$ found, $f(z)$ is reconstructed.

2. Application of Ritz's method. The property just formulated of the function transforming the region into a circle permits the application to the finding of it of any of the many approximate methods for the solution of variational problems. In the memoir on this question which we have cited above, Bieberbach indicates Ritz's method, in a particular form, finding the solution of the problem approximately in the form of a polynomial. We are not going to limit ourselves from the outset to the class of regions for which polynomials are sufficient for the approximate representation of the sought function to any degree of accuracy, for in practical problems regions are encountered for which polynomials are patently insufficient. Of regions of such a kind we shall have occasion to speak below. On the other hand, it may turn out that computations with polynomials are not always convenient. Finally, the given region may be unbounded, in which case the approximation by polynomials of the function transforming it becomes impossible, if one pursues the computational method that we have indicated here, since the integrals that have to be computed will then be divergent. In the last case, one will naturally resort to other functions than polynomials; the choice of them will of course be connected with the form of the region if we wish to attain the greatest computational simplicity.

Let us therefore take an arbitrary system of linearly independent functions $u_0(z), u_1(z), \ldots$ [1] which are regular in the region B and such that for them the integrals $\iint_B \overline{u_i(z)} u_i(z) dx\, dy$ remain finite. Assume in addition that one of them, for instance $u_0(z)$, is different from zero at the point $z = 0$. We next form, as usual, their linear combination

$$\varphi_n(z) = c_0 u_0(z) + c_1 u_1(z) + \ldots + c_n u_n(z), \tag{6}$$

where the coefficients c_k are some complex numbers, $c_k = \gamma_k + i\delta_k$, and we subject the combination to the requirement:

$$c_0 u_0(0) + c_1 u_1(0) + \ldots + c_n u_n(0) = 1, \tag{7}$$

to satisfy which, under our assumption that $u_0(0) \neq 0$, is always possible. We then introduce $\varphi_n(z)$ under the integral sign in (5) instead of $f'(z)$. After performing all the necessary quadratures we obtain the integral as a function of the quantities $\gamma_0, \ldots, \gamma_n, \delta_0, \ldots, \delta_n$:

$$I(\varphi_n) = \iint_B \varphi_n \overline{\varphi}_n \, dx\, dy.$$

Its numerical value will be equal to the area of the image of the region B.

We now select the numbers c_k so that this value will be the least by comparison with the values of the same integral for any other linear combination ψ_n of the functions $u_0(z), \ldots, u_n(z)$, the value of which combination at the point $z = 0$ is equal to unity.

Let us conceive of it in the form:

$$\psi_n(z) = \varphi_n(z) + \varepsilon \eta_n(z),$$

where $\eta_n(z)$ is a linear combination of the $u_i(z)$ $(i = 0, 1, \ldots, n)$ and ε is some complex number. In order that $\psi_n(0) = 1$ for all ε, it is obviously necessary that $\eta_n(z)$ equal zero at the point $z = 0$. The value of the

---

[1] The linear independence of the system is here understood in the usual way: whatever finite number of the functions $u_0(z), u_1(z), \ldots$ of the system we take, they prove to be linearly independent.

integral computed for ψ_n will be:

$$I(\psi_n) = \iint_B \varphi_n\bar{\varphi}_n\,dx\,dy + \bar{\varepsilon}\iint_B \varphi_n\bar{\eta}_n\,dx\,dy + \\ + \varepsilon\iint_B \bar{\varphi}_n\eta_n\,dx\,dy + \varepsilon\bar{\varepsilon}\iint_B \eta_n\bar{\eta}_n\,dx\,dy.$$

The sign of the difference between it and $I(\varphi_n)$ for small ε will depend on the terms containing ε and $\bar{\varepsilon}$ linearly, for the last term will be of the second order of smallness, and in order that $I(\psi_n) - I(\varphi_n)$ be not less than zero, it is necessary that

$$\iint_B \varphi_n\bar{\eta}_n\,dx\,dy = 0, \quad \iint_B \bar{\varphi}_n\eta_n\,dx\,dy = 0 \tag{8}$$

for any linear combination η_n satisfying the condition $\eta_n(0) = 0$.

In the contrary case, by the choice of ε it is always possible to manage that $I(\psi_n) < I(\varphi_n)$, which would contradict the minimal properties of $\varphi_n(z)$.

One of the equalities (8) written above is a consequence of the other, for on the left sides stand conjugate complex quantities and we can use either of them at will.

Conversely, if $\varphi_n(z)$ satisfies the condition of orthogonality (8) with respect to any linear combination $\eta_n(z)$ with $\eta_n(0) = 0$, then $\varphi_n(z)$ will give the integral $I(\varphi)$ its minimum value by comparison with all the combinations $\psi_n(z)$ of the same number of functions $u_k(z)$ satisfying the requirement $\psi_n(0) = 1$, for then

$$I(\psi_n) - I(\varphi_n) = \varepsilon\bar{\varepsilon}\iint_B \eta_n\bar{\eta}_n\,dx\,dy\ ^{1)}.$$

The observance of equations (8) is a necessary and sufficient condition that the function $\varphi_n(z)$ must satisfy. We shall rewrite them in a somewhat different, equivalent form, choosing $\eta_n(z)$ in a special fashion. Obviously each of the following functions

$$v_1(z) = u_1(z) - \frac{u_1(0)}{u_0(0)}\,u_0(z), \ldots, v_n(z) = u_n(z) - \frac{u_n(0)}{u_0(0)}\,u_0(z), \ldots \tag{9}$$

satisfies all the requirements imposed on $\eta_n(z)$, and therefore from (8) there follow these equations:

$$\sum_{j=0}^{n} \alpha_{kj}c_j = 0, \text{ where } \alpha_{kj} = \iint_B u_j\bar{v}_k\,dx\,dy \quad (k = 1, 2, \ldots, n). \tag{10}$$

---

1) From this it is seen that $\varphi_n(z)$ will be unique if it exists, since the integral on the right side of the equation can be equal to zero only in case $\eta_n = 0$, i.e., $\psi_n = \varphi_n$. As regards the existence of φ_n, one can satisfy oneself of it on the basis of the following simple observation. As is evident from the structure of $I(\varphi_n)$, this is a positive-definite quadratic form in the real γ_k and the imaginary δ_k parts of the coefficients c_k, and we have here the problem of the minimum of a positive quadratic form with linear constraints; but such a problem, as we know, always has a solution.

If we add yet the condition

$$\sum_{k=0}^{n} u_k(0)c_k = 1,$$

we shall have obtained a system of $n + 1$ equations, from which there must be found the numbers $c_0, \ldots, c_n$. As regards the possibility of its solution, we refer the reader to note [1]), p. 369.

In the approximate determination of the function giving the conformal transformation by the method here expounded, the essential role, from the computational point of view, is played by the following two circumstances. First, the determination of the integrals $\iint_B u_j \bar{v}_k \, dx \, dy$, which are the coefficients of the c_j in the equations of system (10), and second, the solution of this system itself. And both, for a region B the least bit complex and for large n, represent no easy task; it is usually important, therefore, to utilize the freedom remaining to the computor to introduce all possible simplifications beforehand.

The system of functions $u_k(z)$ still remains undefined. Apart from their linear independence and, as we shall see below, the familiar completeness requirement, they are subject to no kind of limitation.

The determination of the integrals cited above is at once simplified if the $u_k(z)$ be chosen beforehand so that they will all, with the exception of the first, vanish at the point $z = 0$, but so the first will reduce to unity: $u_0(0) = 1$. First, then, $v_k(z) = u_k(z)$; next the integral $\iint_B u_j \bar{v}_k \, dx \, dy$ would reduce to $\iint_B u_j \bar{u}_k \, dx \, dy$; finally, the equation of connection $\sum_{k=0}^{n} c_k u_k(0) = 1$ would degenerate to $c_0 = 1$, and the number of unknowns would at once be reduced by one.

Furthermore, if the system of functions $u_0(z), u_1(z), \ldots$ is orthogonal in the region B in the sense that

$$\iint_B u_j \bar{u}_k \, dx \, dy = 0 \text{ for } k \neq j,$$

then

$$\iint_B u_j \bar{v}_k \, dx \, dy = 0 \text{ for } j \neq 0 \text{ and } k \neq j,$$

whereas

$$\iint_B u_0 \bar{v}_k \, dx \, dy = -\frac{\overline{u_k(0)}}{\overline{u_0(0)}} \iint_B u_0 \bar{u}_0 \, dx \, dy$$

and

$$\iint_B u_k \bar{v}_k \, dx \, dy = \iint_B u_k \bar{u}_k \, dx \, dy.$$

The equations of the system will have been considerably simplified, having taken the following form:

$$c_k \iint_B u_k \bar{u}_k \, dx\, dy - c_0 \frac{\overline{u_k(0)}}{u_0(0)} \iint_B u_0 \bar{u}_0 \, dx\, dy = 0.$$

It is therefore usually advantageous to start with an orthogonal system, if it is known, or, in certain cases, to reduce the adopted system of functions $u_k(z)$ to an orthogonal one by a known process of orthogonalization. Of this we shall speak in greater detail when expounding the application to conformal transformation of polynomials that are orthogonal in the sense of Bochner.

3. Minimization by means of polynomials. We shall dwell in more detail on the most widely used sequence of functions $u_n(z)$ — the powers of z:

$$u_0(z) = 1,\ u_1(z) = z,\ \ldots,\ u_n(z) = z^n \ldots \tag{11}$$

$$\varphi_n(z) = 1 + c_1 z + c_2 z^2 + \ldots + c_n z^n. \tag{12}$$

System (10) for the determination of the coefficients c_k has here the form:

$$\left.\begin{array}{l} \alpha_{10} + \alpha_{11}c_1 + \alpha_{12}c_2 + \ldots + \alpha_{1n}c_n = 0 \\ \alpha_{20} + \alpha_{21}c_1 + \alpha_{22}c_2 + \ldots + \alpha_{2n}c_n = 0 \\ \ldots\ldots\ldots\ldots\ldots\ldots\ldots\ldots \\ \alpha_{n0} + \alpha_{n1}c_1 + \alpha_{n2}c_2 + \ldots + \alpha_{nn}c_n = 0 \\ \alpha_{kj} = \iint_B \bar{z}^k z^j \, dx\, dy,\ \alpha_{kj} = \bar{\alpha}_{jk}. \end{array}\right\} \tag{13}$$

The exact computation of the coefficients α_{kj} even for comparatively simple cases, presents considerable difficulties. One can indicate, however, a sufficiently broad class of regions for which the approximate computation of these coefficients represents no labor and can be performed by widely known methods of harmonic analysis.

Let us assume that the region B is star-shaped, i.e., that its contour intersects in only one point each ray drawn from the origin. Let the equation of the contour be $r = r(\varphi)$.

Let us transform the integral α_{kj} into polar coordinates, putting $z = re^{i\varphi}$.

$$\alpha_{kj} = \int_0^{2\pi} \int_0^{r(\varphi)} r^{j+k} e^{i(j-k)\varphi} r\, dr\, d\varphi = \frac{1}{k+j+2} \int_0^{2\pi} r^{k+j+2}(\varphi) e^{i(j-k)\varphi}\, d\varphi =$$

$$= \frac{1}{k+j+2} \int_0^{2\pi} r^{k+j+2}(\varphi) \cos (j-k)\, \varphi\, d\varphi +$$

$$+ \frac{i}{k+j+2} \int_0^{2\pi} r^{k+j+2}(\varphi) \sin (j-k) \varphi\, d\varphi. \tag{14}$$

From this it is evident that the real and imaginary parts of α_{kj} will differ from the coefficients of the Fourier expansion of $r^{k+j+2}(\varphi)$ only by the factor $\frac{\pi}{k+j+2}$, and therefore their computation can be performed by the ordinary methods.

4. The convergence of the successive approximations. The completeness of the system of coordinate functions. We have been trying to construct the function $f(z)$ giving the transformation of the region B into a circle, via the minimization of the integral $I(\varphi) = \iint_B \varphi\bar{\varphi}\, dx\, dy$ under the condition $\varphi(0) = 1$, constructing the successive approximations to the sought function by Ritz's method. For our subsequent reasoning, the method of constructing the approximations is immaterial. Let us have succeeded, in one way or another, in constructing a so-called "minimizing sequence" of approximations $\varphi_1(z), \varphi_2(z), \ldots$, i.e., one such that the values of the integral I corresponding to these functions, $I(\varphi_1), I(\varphi_2), \ldots$ converge to the least value of $I(\varphi)$:

$$\lim_{n\to\infty} I(\varphi_n) = I(\varphi).$$

We shall now show that from the fact of the convergence of the values of the integral to the minimum follows the convergence of the minimizing sequence to the function effecting the required transformation.

From the subsequent reasoning it will in addition be seen that the convergence will be uniform in any interior part of the region B.

We shall require as a preliminary a study of some of the properties of the function $\varphi(z)$.

Let us consider the following function, which is admissible in our variational problem:

$$\Phi(z) = \varphi(z) + \varepsilon\omega(z), \quad \omega(0) = 0,$$

ε being an arbitrary parameter.

It is important to note that any admissible function $\psi(z)$ can be represented in that form, for in the preceding equality it is sufficient to put $\varepsilon = 1$ and $\omega(z) = \psi(z) - \varphi(z)$

$$I(\Phi) - I(\varphi) = \iint_B (\varphi + \varepsilon\omega)(\bar{\varphi} + \bar{\varepsilon}\bar{\omega})\, dx\, dy - \iint_B \varphi\bar{\varphi}\, dx\, dy =$$

$$= \bar{\varepsilon}\iint_B \varphi\bar{\omega}\, dx\, dy + \varepsilon\iint_B \bar{\varphi}\omega\, dx\, dy + \varepsilon\bar{\varepsilon}\iint_B \omega\bar{\omega}\, dx\, dy.$$

By means of reasoning perfectly analogous to that by means of which equations (8) were derived, from this we arrive at the conclusion that the integrals that stand as multipliers of ε and $\bar{\varepsilon}$ must be equal to zero.

Thus the function $\varphi(z)$ solving the variational problem possesses the property that for all $\omega(z)$ that are regular in the region B and vanish for $z = 0$, the integral

$$\iint_B \varphi(z)\overline{\omega(z)}\, dx\, dy = 0.$$

If $\varphi(z)$ satisfies this equation, then only the third term will be preserved in the expression for the difference $I(\Phi) - I(\varphi)$. Introducing for $\varepsilon\omega(z)$ its value $\Phi - \varphi$, we obtain the following relation:

$$I(\Phi) - I(\varphi) = \iint_B (\Phi - \varphi)(\overline{\Phi} - \overline{\varphi})\, dx\, dy.$$

We shall prove the statement:

Any minimizing sequence $\varphi_1(z), \varphi_2(z), \ldots$ converges in the entire region B to the derivative $\varphi(z)$ of the function $f(z)$ that effects the conformal transformation of this region into a circle; the convergence will be uniform in any interior part of the region B.

Indeed, let us take any interior part B' of the region B. We shall choose a positive number η less than the distance from the points of B' to the contour L of the region B. The circumference described by the radius η about any point of the region B' lies wholly within B.

Now let z be any point of B'. Let us describe about it a circumference of radius η and denote it by C_η. By Cauchy's theorem

$$[\varphi_n(z) - \varphi(z)]^2 = \frac{1}{2\pi i} \int_{C_\eta} \frac{[\varphi_n(t) - \varphi(t)]^2}{t - z}\, dt.$$

We shall estimate the integral. On the circumference C_η the variable of integration will be: $t = z + \eta e^{i\vartheta}$.

$$|\varphi_n(z) - \varphi(z)|^2 \leqslant \frac{1}{2\pi} \int_0^{2\pi} |\varphi_n(t) - \varphi(t)|^2\, d\vartheta.$$

On multiplying both sides by $r\, dr$ and integrating with the limits 0 to η, we obtain:

$$|\varphi_n(z) - \varphi(z)|^2 \frac{\eta^2}{2} = \frac{1}{2\pi} \int_0^\eta \int_0^{2\pi} (\varphi_n - \varphi)(\overline{\varphi}_n - \overline{\varphi}) r\, dr\, d\vartheta \leqslant$$

$$\leqslant \frac{1}{2\pi} \iint_B (\varphi_n - \varphi)(\overline{\varphi}_n - \overline{\varphi}) r\, dr\, d\vartheta.$$

The integral taken over the square of the modulus of the difference $\varphi_n - \varphi$ over the entire region B is, as we know, equal to $I(\varphi_n) - I(\varphi)$. We have, therefore,

$$|\varphi_n(z) - \varphi(z)| \leqslant \frac{1}{\sqrt{\pi} \cdot \eta} \sqrt{I(\varphi_n) - I(\varphi)},$$

and since $\varphi_1, \varphi_2, \ldots$ is, on assumption, a minimizing sequence, we have $I(\varphi_n) \to I(\varphi)$ and accordingly $\varphi_n(z)$ will uniformly converge to $\varphi(z)$.

We recall that $\varphi(z)$ is the derivative of the function $f(z)$ that transforms B into a circle, so that

$$f(z) = \int_0^z \varphi(z)\, dz.$$

From the fact of the uniform convergence of $\varphi_n(z)$ to $\varphi(z)$ follows also, of course, the convergence of the approximations $f_n(z) = \int_0^z \varphi_n(z)\, dz$ to the sought transforming function $f(z)$, for

$$f_n(z) - f(z) = \int_0^z [\varphi_n(z) - \varphi(z)]\, dz.$$

Without going into details beyond the framework of the book, we shall make a number of observations on the convergence of the process of successive approximations constructed in accordance with Ritz. Starting with the previously chosen system of coordinate functions $u_0(z)$, $u_1(z)$, ..., we have constructed a sequence of functions $\varphi_1(z)$, $\varphi_2(z)$, ... Whether these approximations do or do not form a minimizing sequence obviously depends on what kind of functions $u_n(z)$ we chose and what kind of region B is. In order that the functions $\varphi_n(z)$ definitely give a minimizing sequence, it is necessary that the system of functions $u_n(z)$ be sufficiently complete for the region B.

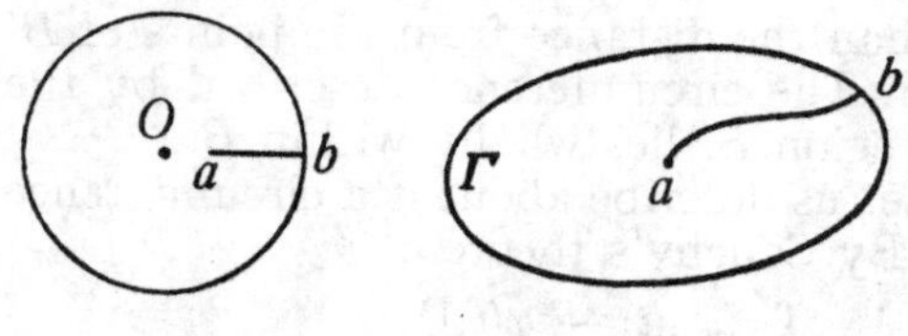

Fig. 29.

We shall adduce without proof the most important and the simplest result pertaining to this. In his memoir previously cited, L. Bieberbach showed that for any region B, the boundary L of which is simultaneously the boundary of the exterior region too, the positive integral powers of z: $1, z, z^2, \ldots, z^n, \ldots$ form a complete system of functions. To this class of regions will belong all regions bounded by Jordan curves. But if the region B has cuts entering it, as for example the circle depicted in Fig. 29 with the cut into it introduced along the radius, the powers of the independent variable cease to give a complete system of functions. The cause of this is very simple. Were we to choose as the coordinate functions in this case not only the powers of z but also other functions that are single-valued in the plane of the complex variable, the cuts entering the region would have no influence on the magnitude of the integrals $\iint_B u_k \bar{u}_j\, dx\, dy$ that figure in system of equations (10) as coefficients; and all computations would therefore be performed as though we were mapping a region without cuts.

We have dwelt in detail on the case of entering cuts because this case is of known practical interest. We shall explain by what the powers of z must be complemented here if a complete system is to be obtained. Let us assume for the sake of simplicity that the boundary L of the region B consists of a simple curve Γ and of one cut passing from the point a within the curve Γ to a point b lying on this curve (Fig. 29). We will introduce for the variable z a new variable $w = \sqrt{a-z}$. On the w-plane the image of region B will be a certain region Δ bounded by a contour without multiple points; for it the powers of w give a complete system. Reverting to the former variable z, we see that a complete system for the region B is formed by the powers of the variable z and the products of them and the root $\sqrt{a-z}$.

5. Exterior regions. The method indicated above can also be applied to the transformation of the exterior of some curve into the interior of a circle. Indeed, let B be a region containing the point at infinity. We shall consider functions which are regular in it and which have at infinity an expansion of the form:

$$t = f(z) = \frac{1}{z} + \frac{a^2}{z^2} + \dots \tag{15}$$

They transform B into a certain region B' which contains the origin of coordinates; moreover the zero of the t-plane will correspond to the infinitely distant point of the z-plane. Among these functions there will be included that which transforms region B into a circle with center at the origin of coordinates. Let us denote it by $\zeta = \varphi(z) = \dfrac{1}{z} + \dfrac{b_2}{z^2} + \dots,$ and the inverse to it by $z = \psi(\zeta)$.

The area of the image B' of the region B will be equal to the integral

$$\iint_B f'(z)\overline{f'(z)}\, dx\, dy. \tag{16}$$

We shall show that this area is greater than the area of the circle G into which B is transformed by the function $\varphi(z)$, i.e., that $\varphi(z)$ possesses the same extremal property as for finite regions.

For this purpose let us consider the function $t = \Phi(\zeta)$, transforming G into B' and formed in the following manner. First by means of $z = \psi(\zeta)$ we transform G into B, and next by $t = f(z)$ the region B into B'

$$t = \Phi(\zeta) = f[\psi(\zeta)].$$

Obviously $\Phi(0) = 0$.

It can easily be established in addition that the derivative of $\Phi(\zeta)$ at the origin of coordinates is equal to unity. Indeed:

$$\Phi'(\zeta) = f'(z)\psi'(\zeta) = f'(z)\frac{1}{\varphi'(z)} = \frac{-\dfrac{1}{z^2} - \dfrac{2a^2}{z^3} - \dots}{-\dfrac{1}{z^2} - \dfrac{2b_2}{z^3} - \dots} = 1 + \frac{c_1}{z} + \dots$$

and for $z \to \infty$, which corresponds to $\zeta \to 0$, $\Phi(\zeta) \to 1$.

But if the function $\Phi(z)$ transforming G into B' has its derivative equal to unity for $\zeta = 0$, then, by what was proved earlier, the area of B' must be greater than the area of the circle.

The problem of finding $\varphi(z)$ thus reduces to the problem of finding among the functions of type (15) that which gives the minimum for integral (16). The methods for the approximate construction of it are analogous to the methods applied in the case of finite regions, and we shall not dwell on them.

§ 3. THE PROPERTY OF MINIMUM CONTOUR LENGTH IN THE TRANSFORMATION OF A REGION INTO A CIRCLE

1. The extremal property of the transforming function. Along with the extremal property characterizing the function conformally transforming the given region into a circle that has been indicated in § 2, there can be given one more extremal characteristic of this function, which in practical applications leads to computations that are analogous to those that are indicated in the preceding section.

Let the circle of radius R in the ζ-plane be transformed by means of a function, holomorphic in the circle,

$$z = f(\zeta), \quad f(0) = 0, \quad f'(0) \neq 0$$

into a certain region B', the nature of which is immaterial to us at the moment. We will call its contour L' and will assume it to be a rectifiable curve.

We shall obtain the relation of dependence, which is fundamental for us, of its length S' upon the value of $f'(0)$.

Let us apply to $f'(\zeta)$ the Cauchy formula

$$f'(\zeta) = \frac{1}{2\pi i} \int\limits_{|t|=R} \frac{f'(t)}{t - \zeta}\, dt \text{ [1]}.$$

Setting $\zeta = 0$ here and having in view that $t = Re^{i\theta}$, we find:

$$|f'(0)| \leqslant \frac{1}{2\pi R} \int\limits_{|t|=R} |f'(t)|\,|dt| = \frac{1}{2\pi R} \int\limits_{L'} |dz|.$$

But the integral standing on the right gives the length S' of the contour L', and therefore:

$$S' \geqslant 2\pi R\, |f'(0)|. \tag{1}$$

[1] By $f'(t)$ we here understand the limit value to which $f'(\zeta)$ tends when ζ approaches the point t on the circumference $|\zeta| = R$. We shall not elucidate the possibility of applying Cauchy's formula to $f'(\zeta)$.

The equality sign can hold here, as is known from the theory of functions of a complex variable, only in case $f'(\zeta) = \text{const.}$, i.e., when the transformation is linear, of the form $z = \alpha\zeta$.

Let us assume that in the transformation the scale at the origin of coordinates is not distorted, i.e., $|f'(0)| = 1$. From inequality (1) it then follows that the length of L' is in any case not less than the length $2\pi R$ of the circumference of the original circle, and can equal it only in case the transformation reduces to the simple rotation of the circle about its center. We have thus arrived at the following result, which is fundamental to the method:

In the transformation of the circle $|\zeta| < R$ by means of a regular function

$$z = f(\zeta), \quad f(0) = 0, \quad f'(0) = 1,$$

the length of the contour of the region being mapped is increased, if the trivial case $f(\zeta) = \zeta$ be excluded.

Let us now consider all possible conformal transformations of the region B accomplished by means of functions regular in B,

$$\zeta = f(z), \quad f(0) = 0, \quad f'(0) = 1. \tag{2}$$

By reasoning exactly like that which we carried through in No. 1, § 2, it can without difficulty be established that *of them, the function giving the conformal transformation of B into a circle possesses the property that it gives the least value for the integral*

$$I = \int_L |f'(z)|\, ds, \tag{3}$$

which is equal to the length of the image of the contour L.

2. The application of Ritz's method. In computations it is more convenient, in view of the linearity of the integrand with respect to $|f'(z)|$, to find the approximations to $\sqrt{f'(z)}$ rather than directly to $f'(z)$. Admissible to the competition will be all functions $\varphi(z)$ which reduce to unity at the origin of coordinates. Of them we shall seek that which gives the least value for the integral:

$$I = \int_L |\varphi(z)|^2\, ds = \int_L \varphi(z)\overline{\varphi(z)}\, ds. \tag{4}$$

If we can succeed in finding the solution to our variational problem, the function effecting the sought transformation will be determined by the formula:

$$f(z) = \int_0^z \varphi^2(z)\, dz.$$

The reasoning connected with the application of Ritz's method to the determination of $\varphi(z)$ will be literally the same as that above; it is enough to repeat word for word what was said by us on this subject in No. 2, § 2, merely replacing all the double integrals over the region

B which figure there by curvilinear integrals along the contour L. We will not reproduce it, but will direct the reader's attention to those parts of the exposition that will present certain divergencies from the previous one.

System of equations (10) No. 2, § 2, from which must be determined the coefficients c_k of the linear combination $\varphi_n(z)$, which is the best approximation, for fixed n, to $\varphi(z)$, is considerably simplified if one starts with an orthogonal system of functions $u_k(z)$; but here the orthogonality must be understood not over the region B, as before, but along the contour L in the sense of the equation:

$$\int_L u_k(z)\overline{u_j(z)}ds = 0, \text{ for } k \neq j. \tag{5}$$

How the construction of an orthogonal system is accomplished, concretely, given the choice as the coordinate functions $u_k(z)$ of the positive integral powers of z, and how the polynomials obtained as the result of the orthogonalization are applied to conformal transformation, we shall see later, when we speak of Szegö's investigations. The apparatus of determinants applied there remains, of course, suitable, given appropriate modifications, for the construction of any orthogonal system of functions.

The proof of the convergence that we have adduced in No. 4, § 2, must, for the method under exposition, be subjected to certain modifications.

Let the functions $\varphi_1(z), \varphi_2(z), \ldots$ form a minimizing sequence, i.e., the values of integral (4) corresponding to them converge to the minimum:

$$\lim_{n\to\infty} I(\varphi_n) = I(\varphi).$$

We shall show that $\varphi_n(z)$ will then uniformly converge in any interior part of the region B to the function $\varphi(z)$ which solves the problem.

Indeed, just as in No. 4, § 2, it can be shown that the difference between the value of the integral $I(\Phi)$ for any admissible function Φ and its minimum value $I(\varphi)$ has the following expression:

$$I(\Phi) - I(\varphi) = \int_L (\Phi - \varphi)(\overline{\Phi} - \overline{\varphi})\, ds.$$

Let us take any interior part B' of the region B. Let z be a point of B' and t a point on the contour L. The difference $t - z$ will be not less than some quantity η in modulus.

By Cauchy's formula we have:

$$\varphi_n(z) - \varphi(z) = \frac{1}{2\pi i} \int_L \frac{\varphi_n(t) - \varphi(t)}{t - z} dt,$$

$$|\varphi_n(z) - \varphi(z)| \leqslant \frac{1}{2\pi\eta} \int_L |\varphi_n(t) - \varphi(t)|\, ds \leqslant$$

$$\leqslant \frac{1}{2\pi\eta} \left\{ \int_L 1^2\, ds \right\}^{\frac{1}{2}} \left\{ \int_L |\varphi_n(t) - \varphi(t)|^2\, ds \right\}^{\frac{1}{2}} = \frac{\sqrt{s}}{2\pi\eta} \sqrt{I(\varphi_n) - I(\varphi)}.$$

Since $I(\varphi_n) \to I(\varphi)$ on assumption, we see that $\varphi_n(z)$ converges to $\varphi(z)$ uniformly in the region B'.

Thus if the adopted system of coordinate functions $u_k(z)$ is sufficiently complete in the sense that the successive approximations $\varphi_n(z)$ constructed by means of it form a minimizing sequence, we can be certain that with n sufficiently large we will approach $\varphi(z)$ within B as closely as we please.

The properties that the coordinate functions must have if the approximations constructed by Ritz's method are to form a minimizing sequence will depend on the properties of the contour L of the region B.

For the overwhelming majority of contours encountered in practice, the simplest functions — the integral powers of z — form a complete system. Thus, the results of the investigations of V. I. Smirnov on the theory of orthogonal polynomials [1]) have shown that if L be a simple curve, and the logarithm of the modulus of the derivative of the function transforming the unit circle into the region B be representable as a Poisson integral, then the polynomials form a complete system on the contour L.

In particular, this holds for all simple piecewise-smooth contours.

Finally, a last observation, touching upon the computations for the case when for the coordinate functions the powers of z are adopted,

$$\varphi_n(z) = 1 + c_1 z + c_2 z^2 + \ldots + c_n z^n.$$

The coefficients c_k are determined from the system:

$$\alpha_{j0} + \alpha_{j1}c_1 + \alpha_{j2}c_2 + \ldots + \alpha_{jn}c_n = 0 \qquad (j = 1, 2, \ldots, n)$$

$$\alpha_{jk} = \int_L \bar{z}^j z^k \, ds = \int_L r^{k+j} e^{i(k-j)\vartheta} \, ds =$$

$$= \int_L r^{k+j} \frac{ds}{d\vartheta} \cos (k - j)\vartheta \, d\vartheta + i \int_L r^{k+j} \frac{ds}{d\vartheta} \sin (k - j)\vartheta \, d\vartheta.$$

The computation of the coefficients of the system leads, accordingly, to the determination of the Fourier coefficients for the $r^{k+j} \dfrac{ds}{d\vartheta}$.

3. The transformation of exterior regions. Let B be a simply connected region containing the point at infinity and bounded by a rectifiable contour L.

Let us consider a function $\zeta = f(z)$, holomorphic in B and having near the point at infinity an expansion of the form:

$$f(z) = \frac{1}{z} + \frac{c_2}{z^2} + \ldots + \frac{c_n}{z^n} + \ldots \tag{6}$$

The following statement is a direct consequence of the result obtained by us in No. 1.

---

[1]) See V. I. Smirnov, [2].

Among the functions of form (6) there will be a function giving the conformal transformation of region B into a circle of the ζ-plane with center at the origin of coordinates. It will differ from all the functions of the same form in that it gives the minimum value for the integral

$$I = \int_L |f'(z)|\, ds.$$

An analogous property also obtains in the case of the transformation of exterior regions into one another. Indeed, let $z = f(\zeta)$ be a function holomorphic outside the circle $|\zeta| \leqslant R$ and having an expansion near the point at infinity of the form:

$$f(\zeta) = c\zeta + c_0 + \frac{c_1}{\zeta} + \ldots + \frac{c_n}{\zeta^n} + \ldots$$

We shall estimate the length S of the curve into which the circumference $|\zeta| = R$ is transformed by means of this function:

$$f'(\zeta) = c - \frac{c_1}{\zeta^2} - \frac{2c_2}{\zeta^3} - \ldots$$

Multiplying this equality by $\dfrac{1}{2\pi i}\,\dfrac{d\zeta}{\zeta}$ and integrating along the circumference $|\zeta| = R$, we have:

$$c = f'(\infty) = \frac{1}{2\pi i} \int_{|\zeta|=R} f'(\zeta)\, \frac{d\zeta}{\zeta}.$$

From this we obtain

$$|c| = |f'(\infty)| \leqslant \frac{1}{2\pi R} \int_{|\zeta|=R} |f'(\zeta)|\, |d\zeta| = \frac{S}{2\pi R}, \quad S \geqslant 2\pi R\, |f'(\infty)|.$$

Here the equality sign is possible only in case $f'(\zeta) = \text{const.}$, i.e., where the transformation is linear.

If $f'(\infty) = 1$, $S \geqslant 2\pi R$. Thus in transformations of this type the length of the image of the circumference will be greater than the length of the circumference itself.

From this, in the same way as previously, we obtain the following theorem.

Among all functions holomorphic in B and having near the point at infinity an expansion of the form

$$\zeta = f(z) = z + c_0 + \frac{c_1}{z} + \frac{c_2}{z^2} + \ldots, \tag{7}$$

the function furnishing the conformal transformation of the region B into the exterior of a circumference with center at the origin of coordinates of the ζ-plane is distinguished by the fact that it gives the minimum value for the integral:

$$I = \int_L |f'(z)|\, ds.$$

We shall limit ourselves to these brief references to the extremal property of the function furnishing the sought conformal transformation, for the methods of its approximate determination are very similar to those that we have employed for the case of finite regions, and the reader will be able to transfer them without difficulty to the case under analysis.

§ 4. ORTHOGONAL POLYNOMIALS AND CONFORMAL TRANSFORMATION

1. Polynomials orthogonal on the contour. Taylor's series, which is the universal tool for the analytic representation of functions holomorphic in a circle, becomes completely inadequate when the region is not a circle. There arises, therefore, the question of the creation of an analytic apparatus that plays the same role for a given region as Taylor's series does in the circle. More exactly speaking, we must give a sequence of functions such as will depend only on the form of the region, and such that any function holomorphic in this region — and in particular the function $\zeta = \varphi(z)$ conformally transforming region B into the circle $|\zeta| < R$, which is of interest to us — can be expanded in a series involving the functions of the sequence. It goes without saying that for the coordinate functions elected, the condition will naturally be stipulated that they be of the greatest simplicity, it being demanded, for instance, that there be adopted for them certain polynomials of special form, or another kind of well-known function. In the Twenties of the current century, two efforts to construct such an apparatus were made, by Carleman, Szegö, Bochner and Bergmann [1]). They have a close contact with the problem of conformal transformation, and we shall expound them in the measure in which this is necessary for the approximate practical effectuation of such transformations.

We will begin our exposition with polynomials orthogonal on the contour.

Let L be an arbitrary rectifiable curve of length l; for the time being we have no need to assume it to be closed.

Let us construct a system of polynomials:

$$P_0(z),\ P_1(z),\ \ldots,\ P_n(z),\ \ldots, \tag{1}$$

possessing the following properties:

1) $P_n(z)$ is a polynomial of degree n in z;
2) the coefficient of z^n in $P_n(z)$ is positive;

---

[1]) T. Carleman, [1]; Szegö, [1]; Bochner, [1]; Bergmann, [1]; V. I. Smirnov, [2].

3) the polynomials $P_n(z)$ are orthogonal and normalized along the curve L:

$$\frac{1}{l}\int_l P_m(z)\overline{P_n(z)}\,ds = \begin{cases} 0 & \text{for } m \neq n \\ 1 & ,, \quad n = m. \end{cases} \tag{2}$$

Such polynomials may be found successively by E. Schmidt's familiar process of orthogonalization. We prefer to give at once a ready expression for them in the form of determinants that can be computed easily, for all the variable quantities in them are concentrated in one row.

Let us introduce the following symbols:

$$h_{pq} = \frac{1}{l}\int_L z^p \bar{z}^q \cdot ds. \tag{3}$$

Obviously

$$h_{pq} = \bar{h}_{qp}.$$

Let us compose the determinants D_n of the numbers h_{pq} in the following manner:

$$D_0 = 1,\ D_n = \begin{vmatrix} h_{00}, & h_{10}, & \dots, & h_{n0} \\ h_{01}, & h_{11}, & \dots, & h_{n1} \\ \cdot & \cdot & \cdot & \cdot \\ h_{0n}, & h_{1n}, & \dots, & h_{nn} \end{vmatrix}.$$

They will be the determinants of the following positive-definite Hermitian quadratic forms:

$$H_n(t) = \sum_{p,q=0}^{n} h_{pq} t_p \bar{t}_q = \frac{1}{l}\int_L |t_0 + t_1\zeta + \dots + t_n\zeta^n|^2\, ds,$$

and therefore all will certainly be positive real numbers:

$$D_n > 0.$$

The Szegö polynomial can be represented in the following form:

$$P_n(z) = \frac{1}{\sqrt{D_{n-1}D_n}} \begin{vmatrix} h_{00}, & h_{10}, & \dots, & h_{n0} \\ h_{01}, & h_{11}, & \dots, & h_{n1} \\ \cdot & \cdot & \cdot & \cdot \\ h_{0,n-1}, & h_{1,n-1}, & \dots, & h_{n,n-1} \\ 1, & z, & \dots, & z^n \end{vmatrix}. \tag{4}$$

One can most simply satisfy oneself of this by verifying the fact that functions (4) satisfy the three requirements, enumerated above, defining Szegö polynomials.

The question of the expansion of an arbitrary function in a series involving the orthogonal Szegö polynomials had been investigated under very circumscribed assumptions about the form of the contour L and about the character of the boundary values of the function being expanded, requiring their analyticity. In the work cited above, V. I. Smirnov somewhat later investigated this question very thoroughly, and gave general criteria for the convergence of expansions of this kind. The following result, presented by us without proof, will be sufficient for us.

If L is a piece-wise smooth closed simple curve, and if the function $f(z)$, holomorphic within L, has almost everywhere boundary values on L and is representable in terms of them by means of the Cauchy integral, then $f(z)$ can be expanded in a series involving orthogonal polynomials, of the form:

$$f(z) = A_0P_0(z) + A_1P_1(z) + \ldots + A_nP_n(z) + \ldots, \tag{5}$$

which is uniformly convergent everywhere within L, and whose coefficients A_n are determined by the formulas:

$$A_n = \frac{1}{l} \int_L f(z)\overline{P_n(z)}\, ds. \tag{6}$$

2. Application to conformal transformation. In the plane of the complex variable z there is given a simply connected region B bounded by a piece-wise smooth contour L. Let us call $\zeta = \gamma(z)$ the function conformally transforming it into a circle with center at the origin of coordinates, so that the point a of B passes to the center of the circle: $\gamma(a) = 0$, and $\gamma'(a) = 1$.

By what was proved in § 3, $\sqrt{\gamma'(z)}$ gives the minimum for the integral

$$I = \frac{1}{i} \int_L |f(z)|^2\, ds \tag{7}$$

by comparison with all the functions $f(z)$ which are regular in B and normalized at the point a:

$$f(a) = 1.$$

Let us expand $f(z)$ in a series of polynomials orthogonal in the Szegö sense:

$$f(z) = t_0P_0(z) + t_1P_1(z) + \ldots + t_nP_n(z) + \ldots$$

Here it is assumed that for $f(z)$ there are fulfilled the sufficient conditions, formulated in No. 1, of its expansibility in a series in the $P_n(z)$.

In particular, for the function $f(z) = \sqrt{\gamma'(z)}$ these conditions, under our assumption of the piece-wise smoothness of L, are certainly fulfilled, and the series for it will converge uniformly in B.

In view of the fact that $f(a) = 1$, the coefficients of the expansion must satisfy the condition

$$t_0P_0(a) + t_1P_1(a) + \ldots + t_nP_n(a) + \ldots = 1. \tag{8}$$

Substituting its expansion for $f(z)$ under the integral sign of (7), we shall obtain:

$$\frac{1}{l}\int_L f(z)\overline{f(z)}\,ds = t_0\bar{t}_0 + t_1\bar{t}_1 + \ldots + t_n\bar{t}_n + \ldots \tag{9}$$

The system of coefficients corresponding to the case $f(z) = \sqrt{\gamma'(z)}$ must give the minimum value for the sum (9) under condition (8). Let us denote these coefficients by δ and set $t_\nu = \delta_\nu + \varepsilon\eta$. Since for δ_ν the condition $\sum\limits_{\nu=0}^{\infty} \delta_\nu P_\nu(a) = 1$ is certainly observed, the numbers η_ν must satisfy the requirement

$$\sum_{\nu=0}^{\infty} \eta_\nu P_\nu(a) = 0. \tag{10}$$

Substitution, in (9), for the t_ν of their values gives

$$\frac{1}{l}\int_L f(z)\overline{f(z)}\,ds = \sum_{\nu=0}^{\infty} \delta_\nu\bar{\delta}_\nu + \varepsilon\sum_{\nu=0}^{\infty} \eta_\nu\bar{\delta}_\nu + \bar{\varepsilon}\sum_{\nu=0}^{\infty} \bar{\eta}_\nu\delta_\nu + \varepsilon\bar{\varepsilon}\sum_{\nu=0}^{\infty} \eta_\nu\bar{\eta}_\nu.$$

In order that the cited expression be, for any ε, not less than $\sum\limits_{\nu=0}^{\infty} \delta_\nu\bar{\delta}_\nu$, equal to the minimum value of the integral, it is necessary and sufficient that the coefficients of ε and $\bar{\varepsilon}$ be equal to zero for any η_ν subject to equation (10):

$$\sum_{\nu=0}^{\infty} \eta_\nu\bar{\delta}_\nu = 0,\quad \sum_{\nu=0}^{\infty} \bar{\eta}_\nu\delta_\nu = 0.$$

Let us introduce into the first of them for η_0 its value from (10)

$$\eta_0 = -\sum_{\nu=1}^{\infty} \eta_\nu P_\nu(a), \text{ since } P_0(z) = 1,$$

$$\sum_{\nu=1}^{\infty} \eta_\nu[\bar{\delta}_\nu - \bar{\delta}_0 P_\nu(a)] = 0.$$

In view of the arbitrariness of $\eta_1, \eta_2, \ldots$, this equality can hold only in the case when:

$$\bar{\delta}_\nu = \bar{\delta}_0 P_\nu(a).$$

Moreover, substituting for δ_ν the values found, in the condition

$$\sum_{\nu=0}^{\infty} \delta_\nu P_\nu(a) = 1,$$

we have:

$$\delta_0 \sum_{\nu=0}^{\infty} P_\nu(a)\overline{P_\nu(a)} = 1.$$

Accordingly, if we put

$$\sum_{\nu=0}^{\infty} \overline{P_\nu(a)} P_\nu(z) = K(a, z),$$

we shall have

$$\bar{\delta}_0 = \delta_0 = \frac{1}{K(a, a)} = \frac{\overline{P_0(a)}}{K(a, a)},$$

$$\delta_\nu = \frac{\overline{P_\nu(a)}}{K(a, a)},$$

$$\sqrt{\gamma'(z)} = \frac{1}{K(a, a)} \cdot \sum_{\nu=0}^{\infty} \overline{P_\nu(a)} P_\nu(z) = \frac{K(a, z)}{K(a, a)},$$

$$\gamma(z) = \frac{1}{K^2(a, a)} \int_a^z K^2(a, z)\, dz. \tag{11}$$

And this is the sought expression for the transforming function $\gamma(z)$.

From this is easily obtained an approximate formula. If there are known only n of the polynomials $P_\nu(z)$ and $K_n(a, z) = \sum_{\nu=0}^{n} \overline{P_\nu(a)} P_\nu(a)$, we obtain:

$$\gamma(z) \cong \frac{1}{K_n^2(a, a)} \int_a^z K_n^2(a, z)\, dz. \tag{12}$$

One can compute simply the radius R of the circle into which the region B is mapped. The length of the circumference of the circle has the following expression.

$$2\pi R = \int_L |\gamma'(z)|\, ds = \frac{1}{K^2(a, a)} \int_L K(a, z) \overline{K(a, z)}\, ds =$$

$$= \frac{l}{K^2(a, a)} \sum_{\nu=0}^{\infty} \overline{P_\nu(a)} P_\nu(a) = \frac{l}{K(a, a)},$$

whence

$$R = \frac{l}{2\pi} \frac{1}{K(a, a)}.$$

In order to obtain the function $\gamma^*(z)$ giving the transformation of B into the unit circle, it is sufficient to divide $\gamma(z)$ by R,

$$\gamma^*(z) = \frac{2\pi}{l} \frac{1}{K(a, a)} \int_a^z K^2(a, z)\, dz. \tag{13}$$

Example. We will compute the Szegö polynomials for the square

$$-1 \leqslant x \leqslant +1, \quad -1 \leqslant y \leqslant +1.$$

The numbers $h_{p,q}$ will here have the following values:

$$h_{p,q} = \frac{1}{l}\int_L z^p \bar{z}^q\, ds = \tfrac{1}{8}\Bigg\{\int_{-1}^{+1}(x+i)^p(x-i)^q dx +$$

$$+\int_{-1}^{+1}(x-i)^p(x+i)^q\, dx + \int_{-1}^{+1}(1+iy)^p(1-iy)^q\, dy +$$

$$+\int_{-1}^{+1}(-1+iy)^p(-1-iy)^q\, dy\Bigg\} = \frac{1+i^{p-q}}{4}\,\mathrm{Re}\int_{-1}^{+1}(x+i)^p(x-i)^q\, dx.$$

From this it is evident that

$$h_{p,q} = \begin{cases} 0, \text{ when } p-q \neq 4k \\ \int_0^1 (x^2+1)^q\, \mathrm{Re}\,[(x+i)^{p-q}]dx, \text{ when } p-q = 4k. \end{cases}$$

Performing the calculation of these quantities, we shall have a table of their values:

$$\begin{matrix} 1, & 0, & 0, & 0, & -\tfrac{4}{5} & \dots \\ 0, & \tfrac{4}{3}, & 0, & 0, & 0 & \dots \\ 0, & 0, & \tfrac{28}{15}, & 0, & 0 & \dots \\ 0, & 0, & 0, & \tfrac{96}{35}, & 0 & \dots \\ -\tfrac{4}{5}, & 0, & 0, & 0, & \tfrac{1328}{315} & \dots \end{matrix}$$

Proceeding from this, we compute by formula (4) the Szegö polynomials for our square:

$$P_0(z) = 1, \quad P_1(z) = \frac{\sqrt{3}}{2} z, \quad P_2(z) = \tfrac{1}{2}\sqrt{\tfrac{15}{7}} z^2,$$

$$P_3(z) = \tfrac{1}{4}\sqrt{\tfrac{35}{6}} z^3, \quad P_4 = \tfrac{15}{16}\sqrt{\tfrac{7}{22}}(z^4 + \tfrac{4}{5}), \dots .$$

We shall find approximately the function $\zeta = \gamma(z)$ transforming a rectangle into a circle, so that $\gamma(0) = 0$, $\gamma'(0) = 1$. Equation (11) gives its exact expression if we put $a = 0$ in it. We shall determine $\gamma(z)$ approximately, preserving only the first five terms in $K(0, z)$:

$$K(0, z) = \sum_{\nu=0} P_\nu(0)P_\nu(z) = 1 + \tfrac{315}{1408}(z^4 + \tfrac{4}{5}), \quad K(0, 0) = \tfrac{415}{352},$$

$$\gamma(z) = z + \tfrac{63}{830}z^5 + \tfrac{441}{110224}z^9.$$

Let us compare the approximate value of $\gamma(z)$ that we have obtained with its known exact series expansion.

For this we shall consider not the function $\zeta = \gamma(z)$ itself, but its inverse $z = \varphi(\zeta)$. It will give the transformation of a circle of a certain radius R: $|\zeta| < R$ into our square, and moreover so that

$$\varphi(0) = 0, \quad \varphi'(0) = 1.$$

Its expression in terms of an elliptic integral of the lemniscate type is known [1]):

$$z = \int_0^{\zeta} \frac{dt}{\sqrt{1 + \alpha^4 t^4}} = \zeta - \frac{\alpha^4}{10}\zeta^5 + \frac{\alpha^8}{24}\zeta^9 - \ldots,$$

where

$$\alpha = \int_0^1 \frac{dt}{\sqrt{1 + t^4}} = 0.926.$$

Invert this function and we find the exact expression of $\gamma(z)$ which we require:

$$\zeta = \gamma(z) = z + \frac{\alpha^4}{10} z^5 + \frac{\alpha^8}{120} z^9 + \ldots$$

The exact value for the coefficient of z^5 is

$$\frac{\alpha^4}{10} = 0.074,$$

the approximate value which we have found, however, is

$$\tfrac{63}{830} = 0.076$$

which differs from the exact by less than 3%.

The polynomial that we have constructed will transform the contour of the square into some curve that does not coincide with the circumference. To characterize the proximity of the curve to the circumference, we will compute the value of the modulus of our polynomial at the middle of a side of the square, at the point $z = 1$, and at a vertex of the square, $z = 1 + i$,

$$|\gamma(1)| = 1.078, \quad |\gamma(1 + i)| = 1.089.$$

The exact value of the radius of the circle into which our square is mapped is equal to

$$R = 1.080.$$

The values of the modulus of $\gamma(z)$ that we have found differ from this exact value by 0.002 and 0.009, which constitute 0.2 and 0.9% of the value of the radius.

3. Polynomials orthogonal in a region. Let us turn to the theory of orthogonal polynomials in a region.

Let B be a simply connected region of the plane of the complex variable z, bounded by a Jordan curve. Let us construct a system of polynomials

$$\Pi_0(z), \Pi_1(z), \ldots, \Pi_n(z), \ldots$$

[1]) See V. I. Smirnov, [1], vol. III, p. 353 (1933).

possessing the following properties:

1) $\Pi_n(z)$ is a polynomial of degree n;

2) the coefficient of z^n in $\Pi_n(z)$ is a positive number;

3) the polynomials $\Pi_n(z)$ are orthogonal and normalized in the region B:

$$\frac{1}{S}\iint\limits_B \Pi_m(z)\cdot\overline{\Pi_n(z)}\,dx\,dy = \begin{cases} 0 \text{ for } m \neq n, \\ 1 \quad ,, \quad m = n, \end{cases} \tag{14}$$

where S is the area of region B.

As were the Szegö polynomials, polynomials orthogonal in the region may be represented in the form of determinants. The distinction will consist solely in the fact that, rather than integrals along the contour, integrals over the region will go to form them.

Let us introduce the constants:

$$\gamma_{pq} = \frac{1}{S}\iint\limits_B z^p\bar{z}^q\,dx\,dy \tag{15}$$

and compose from them the determinants Δ_n, analogous to the D_n of No. 1:

$$\Delta_n = \begin{vmatrix} \gamma_{00}, \gamma_{10}, \ldots, \gamma_{n0} \\ \gamma_{01}, \gamma_{11}, \ldots, \gamma_{n1} \\ \cdot \quad \cdot \quad \cdot \quad \cdot \quad \cdot \\ \gamma_{0n}, \gamma_{1n}, \ldots, \gamma_{nn} \end{vmatrix}. \tag{16}$$

The polynomials $\Pi_n(z)$ have the form

$$\Pi_n(z) = \frac{1}{\sqrt{\Delta_{n-1}\cdot\Delta_n}}\begin{vmatrix} \gamma_{00}, & \gamma_{10}, & \ldots, \gamma_{n0} \\ \gamma_{01}, & \gamma_{11}, & \ldots, \gamma_{n1} \\ \cdot & \cdot & \cdot \quad \cdot \\ \gamma_{0,n-1}, & \gamma_{1,n-1}, & \ldots, \gamma_{n,n-1} \\ 1, & z, & \ldots, z^n \end{vmatrix}. \tag{17}$$

It can be shown that polynomials constructed in such a fashion form, for regions bounded by simple Jordan curves, a complete closed system, and that any function $f(z)$, holomorphic in B, for which the integral

$$\iint\limits_B f(z)\overline{f(z)}\,dx\,dy$$

has a finite value, is uniquely expansible in a series involving the polynomials $\Pi_n(z)$:

$$f(z) = A_0\Pi_0(z) + A_1\Pi_1(z) + \ldots + A_n\Pi_n(z) + \ldots, \tag{18}$$

whose coefficients A_n are defined by the equations:

$$A_n = \frac{1}{S} \iint_B f(z)\, \overline{\Pi_n(z)}\, dx\, dy. \tag{19}$$

4. Application to conformal transformation. From § 2 we know that the function $\zeta = \gamma(z)$ furnishing the conformal transformation of the region B into a circle with center at the origin of coordinates so that the point a passes to the center and $\gamma'(a) = 1$, gives the minimal value for the integral

$$I = \iint_B f'(z)\overline{f'(z)}\, dx\, dy$$

with respect to all functions $f(z)$ holomorphic in B and satisfying the conditions

$$f(a) = 0, \; f'(a) = 1.$$

Completely analogously to the way it was done in No. 2, it can be established that $\gamma(z)$ is expressed in terms of the polynomials $\Pi_n(z)$ in the following fashion:

$$\gamma(z) = \frac{1}{\varkappa(a, a)} \int_a^z \varkappa(a, z)\, dz, \tag{20}$$

where

$$\varkappa(a, z) = \overline{\Pi_0(a)}\Pi_0(z) + \overline{\Pi_1(a)}\Pi_1(z) + \dots$$

The area of the circle that is the image of the region B will be

$$\pi R^2 = \iint_B \gamma'(z)\overline{\gamma'(z)}\, dx\, dy = \frac{S}{\varkappa^2(a, a)} \sum_{\nu=0}^{\infty} \overline{\Pi_\nu(a)}\Pi_\nu(a) = \frac{S}{\varkappa(a, a)};$$

consequently

$$R = \sqrt{\frac{S}{\pi\varkappa(a, a)}}.$$

The function $\gamma^*(z)$ transforming the region B into the unit circle therefore has the form:

$$\gamma^*(z) = \sqrt{\frac{\pi}{S\varkappa(a, a)}} \int_a^z \varkappa(a, z)\, dz.$$

§ 5. EXPANSION IN A SERIES OF THE POWERS OF A SMALL PARAMETER IN THE CASE OF THE TRANSFORMATION OF A REGION INTO A CIRCLE [1])

1. Statement of the problem. Reduction to a system of equations. Let us consider in the z-plane a family of simple closed curves depending on one real parameter λ. Let the parametric equation of this family be

$$z = z(t, \lambda).$$

To each value of the parameter λ there corresponds some curve of the family. We shall consider that all these curves contain the origin of coordinates $z = 0$ within themselves. We shall take any value of the parameter λ and the curve L_λ corresponding to it. Consider the function

$$\zeta = \varphi(z, \lambda), \tag{1}$$

accomplishing the conformal transformation of the region B_λ bounded by this curve, into a circle of the ζ-plane with center at the origin of coordinates.

We shall consider for definiteness that $\varphi(z, \lambda)$ is in addition subject to the requirements

$$\varphi(0, \lambda) = 0 \text{ and } \varphi'_z(0, \lambda) = 1. \tag{2}$$

The geometric meaning of these conditions is obvious. The transforming function $\varphi(z, \lambda)$ will depend, of course, on λ, since the boundary L_λ of the region B_λ depends on λ.

If the function $z(t, \lambda)$ standing on the right side of the equation of the boundary curve L_λ is an analytic function of the parameter λ in the neighbourhood of some value of λ, for example, near $\lambda = 0$, then one might expect that — at least in certain cases — the transforming function (1) would also be an analytic function of λ in the neighbourhood of $\lambda = 0$.

In what measure this expectation is justified we shall see below, in the next section, where we will give sufficient conditions for the analytic character of the dependence of the transforming function on λ.

But if $\varphi(z, \lambda)$ is, for all z of B_λ, an analytic function of λ near $\lambda = 0$, it can be expanded in a power series in λ:

$$\varphi(z, \lambda) = \varphi_0(z) + \lambda\varphi_1(z) + \lambda^2\varphi_2(z) + \ldots, \tag{3}$$

---

[1]) The methods expounded in this and the following sections, for the approximate construction of a function effecting the necessary conformal transformation, make possible, by the nature of the case, the separation from such a function of its principal part, of some or other order of accuracy, by an expansion of the function in a series involving the powers of a "small" parameter figuring in the equation of the boundary of the region.

Acad. M. A. Lavrent'ev has indicated a method of constructing the principal part of the transforming function which is based not on its expansion in a series involving powers of a parameter, but on the results of a study of its dependence on the geometrical properties of the region and the integral representations of analytic functions. See M. A. Lavrent'ev, [1].

convergent for $|\lambda|$ sufficiently small. The construction of the function $\varphi(z, \lambda)$ reduces, therefore, to the computation of the coefficients $\varphi_0(z)$, $\varphi_1(z), \ldots$ of this series.

In view of conditions (2), which must be fulfilled for all values of λ, these coefficients will have to satisfy the following requirements:

$$\varphi_n(0) = 0, \qquad (n = 0, 1, 2, \ldots)$$

$$\varphi_0'(0) = 1, \quad \varphi_n'(0) = 0. \qquad (n = 1, 2, \ldots)$$

The first of the coefficients, $\varphi_0(z)$, has a very simple signification. It is obviously equal to $\varphi(z, \lambda)$ for $\lambda = 0$ and therefore the function $\zeta = \varphi_0(z)$ must give the conformal transformation of the region B_0 bounded by the curve L_0, into a circle of the ζ-plane.

Below, in No. 2, we shall show how, by the known coefficient $\varphi_0(z)$, in certain particular cases of the equation of the boundary L_λ all the remaining coefficients $\varphi_1(z), \varphi_2(z), \ldots$ of series (3) can be computed in succession.

At the moment, however, we shall take a somewhat different point of view, and shall show how the determination of the transforming function can be reduced to the solution of a system of equations.

We will elucidate the idea of this reduction by the following non-rigorous general considerations. Consider an infinite system of functions

$$u_1(z), \quad u_2(z), \ldots,$$

satisfying the conditions

$$u_n(0) = 0, \qquad (n = 1, 2, \ldots)$$

$$u_1'(0) = 1, \quad u_n'(0) = 0. \qquad (n = 2, 3, \ldots)$$

We shall consider that the $u_n(z)$ are, first, regular in a certain region D containing all B_λ for $|\lambda|$ sufficiently small, and, second, that any function $f(z)$ that is regular in D can be expanded in a series involving these functions $u_n(z)$. In particular, one can expand in such a series even $\varphi(z, \lambda)$, furnishing the sought transformation,

$$\varphi(z, \lambda) = u_1(z) + \alpha_2(\lambda) u_2(z) + \alpha_3(\lambda) u_3(z) + \ldots \tag{4}$$

The coefficients $\alpha_n(\lambda)$ of this expansion will depend on the parameter λ. The problem of the construction of $\varphi(z, \lambda)$ reduces to the determination of these coefficients. One can indicate the infinite system of equations which they must satisfy and from which they must be found.

Let us assume that the parameter t defining the position of the point z on the boundary L_λ is varied within the limits $[0, 2\pi]$.

Consider any function $f(z)$ that is subject to the requirements

$$f(0) = 0, \quad f'(0) = 1,$$

and is regular in B_λ.

If we introduce here for z the latter's expression in terms of t for L_λ, we shall obtain the values of $f(z)$ on the contour of the region as a function of the parameter t. The function $f[z(t, \lambda)]$ will be periodic in t, of period 2π. Let us expand $|f[z(t, \lambda)]|^2$ in a trigonometric series involving multiple arcs,

$$|f[z(t, \lambda)]|^2 = C_0 + \sum_{n=1}^{\infty} (C_n \cos nt + C'_n \sin nt).$$

But if $f(z)$ coincides with the function $\varphi(z, \lambda)$ giving the transformation of B_λ into a circle, the square of the modulus of its boundary values on L_λ will be a constant quantity, so that in the preceding equation there must be preserved just the first term C_0 alone, which will be equal to the square of the radius of the circle, and all the remaining coefficients C_n and C'_n must equal zero:

$$\left.\begin{aligned} C_n &= 0 \\ C'_n &= 0 \end{aligned}\right\} \qquad (n = 1, 2, 3, \ldots)$$

If C_n and C'_n are formed for $\varphi(z, \lambda)$, using its expression (4), it will be evident that C_n and C'_n are certain quadratic functions of $\alpha_j(\lambda)$, and the preceding equations will give an infinite system of quadratic equations that all the $\alpha_j(\lambda)$ must satisfy.

In practical computations the function $\varphi(z, \lambda)$ is represented approximately by means of the sum of a finite number of terms of series (4). One then obtains not the infinite system for the determination of the $\alpha_j(\lambda)$ of which we have just spoken, but some finite system.

Let us proceed thus: form a linear combination of the first n functions $u_j(z)$:

$$U_n(z, \lambda) = u_1(z) + \sum_{j=2}^{n} \alpha_j(\lambda) u_j(z),$$

and consider the square of the modulus of $U_n(z, \lambda)$ on the boundary L_λ. The coefficients $C_n(\alpha_j)$ and $C'_n(\alpha_j)$ of its expansion in a trigonometric series in t will be quadratic functions of the $\alpha_j(\lambda)$. We cannot, by a choice of them, reduce to zero all the coefficients $C_n(\alpha_j)$, $C'_n(\alpha_j)$, $n = 1, 2, \ldots$, as would have to be the case for the exact determination of the $\varphi_n(z, \lambda)$. We can, however, so select them that the first $n - 1$ coefficients $C_k(\alpha_j)$ and $C'_k(\alpha_j)$ vanish.

Equating these coefficients to zero, we shall obtain a system of $2n - 2$ equations

$$C_k(\alpha_j) = 0, \; C'_k(\alpha_j) = 0 \qquad (k = 1, 2, \ldots, n - 1)$$

for the determination of the $n - 1$ complex unknowns $\alpha_j(\lambda) = \alpha'_j(\lambda) + i\alpha''_j(\lambda)$.

Its form will depend on what kind of system of functions $u_n(z)$ we choose.

In order to have something definite in view, we shall consider the case most convenient for practical applications, when

$$u_1(z) = z, \quad u_2(z) = z^2, \ldots \quad u_n(z) = z^n, \ldots$$

and $U_n(z, \lambda)$ is the polynomial:

$$U_n(z, \lambda) = P_n(z) = z + \alpha_2(\lambda)z^2 + \ldots + \alpha_n(\lambda)z^n.$$

We shall assume, moreover, that the contour L of the region is a curve close to a circle and defined by the equation

$$z(t) = e^{it}[1 + \lambda F(e^{it}, \lambda)], \tag{5}$$

where $F(\tau, \lambda)$ is an analytic function of its arguments, and is regular for $|\tau|$ close to unity and λ close to zero. We can expand it in a Laurent series in $\tau = e^{it}$, the coefficients of which, $\tilde{\beta}_\nu$, will be analytic functions of λ:

$$z(t) = e^{it}[1 + \lambda \sum_{\nu=-\infty}^{+\infty} \tilde{\beta}_\nu(\lambda) e^{it\nu}]. \tag{6}$$

For simplifying the notation of series (6) we shall write it below in the form

$$z(t) = \sum_{\nu=-\infty}^{+\infty} \beta_\nu^{(1)}(\lambda) e^{it\nu}, \tag{7}$$

where we have put

$$\lambda\tilde{\beta}_{\nu-1}(\lambda) = \beta_\nu^{(1)}(\lambda) \text{ for } \nu \neq 1 \text{ and } \beta_1^{(1)}(\lambda) = 1 + \lambda\tilde{\beta}_0(\lambda). \tag{7'}$$

We still need the powers of $z(t)$. After raising $z(t)$ to the kth power we obtain a Laurent series of the form

$$z^k(t) = \sum_{\nu=-\infty}^{+\infty} \beta_\nu^{(k)}(\lambda) e^{i\nu t},$$

whose coefficients can easily be expressed in terms of $\beta_\nu^{(1)}(\lambda)$. With respect to them one can affirm the following: since for $\lambda = 0$, $z(t) = e^{it}$ and $z^k(t) = e^{itk}$, all $\beta_\nu^{(k)}(\lambda)$ vanish for $\lambda = 0$ if $\nu \neq k$, and $\beta_k^{(k)}(\lambda)$ reduces to unity:

$$\beta_\nu^{(k)}(0) = 0 \text{ for } \nu \neq k, \text{ and } \beta_k^{(k)}(0) = 1.$$

Let us now consider the square of the modulus of the polynomial:

$$P_n(z)\overline{P_n(z)} = \sum_{k,j=1}^{n} \alpha_k\bar{\alpha}_j \cdot z^k\bar{z}^j, \quad \alpha_1 = 1.$$

Substituting here for z its value on the contour of the region B, we obtain:

$$P_n(z)\overline{P_n(z)} = \sum_{\nu=-\infty}^{+\infty} \sum_{k,j=1}^{n} \alpha_k\bar{\alpha}_j \sum_{p,q=-\infty;\ p-q=\nu}^{+\infty} \beta_p^{(k)}(\lambda)\bar{\beta}_q^{(j)}(\lambda) e^{i\nu t}.$$

On the right side of the equation stands a trigonometric series whose coefficients depend on α_k; equating the free term in this expression to ϱ^2, we obtain an approximate equation for the determination of the radius of the circle onto which our region is mapped:

$$\varrho^2 = \sum_{k,j=1}^{n} \alpha_k \alpha_j \sum_{p=-\infty}^{+\infty} \beta_p^{(k)} \bar{\beta}_p^{(j)}. \tag{8}$$

We will choose the numbers $\alpha_2, \ldots, \alpha_n$ so that the coefficients of e^{it}, $e^{2it}, \ldots, e^{(n-1)it}$ vanish:

$$\sum_{k,j=1}^{n} \alpha_k \bar{\alpha}_j \sum_{p=-\infty}^{+\infty} \beta_p^{(k)} \beta_{p-m}^{(j)} = 0, \quad (m = 1, \ldots, n-1). \tag{9}$$

The coefficients of $e^{-it}, e^{-2it}, \ldots, e^{-(n-1)it}$ obviously also vanish, for they are quantities conjugate to the coefficients adopted by us.

System (9) is, then, the one we require.

Let us have succeeded in some way in finding from it $\alpha_2, \alpha_3, \ldots, \alpha_n$. By them we form the linear combination $U_n(z, \lambda)$ and put, approximately,

$$\varphi(z, \lambda) = U_n(z, \lambda) = z + \alpha_2 z^2 + \ldots + \alpha_n z^n.$$

It would be expected that the difference between $\varphi(z, \lambda)$ and $U_n(z, \lambda)$, at least in particular cases, would tend to zero with the increase of n.

Just when this will be the case, i.e., under what conditions the approximations constructed by us, $U_n(z, \lambda)$, converge to $\varphi(z, \lambda)$ as $n \to \infty$, we shall see in the next section.

Right now, however, we will consider in greater detail the computational process for the particular case of regions where system (9) and the computations connected with its solution are considerably simplified.

Let the contour of the region be such that the equation corresponding to it has the form

$$z(t) = e^{it} \sum_{\nu=-\infty}^{+\infty} \lambda^{|\nu|} \delta_\nu^{(1)}(\lambda) e^{i\nu t} = e^{it}\{\delta_0^{(1)}(\lambda) + \sum_{\nu=1}^{\infty} \lambda^\nu [\delta_\nu^{(1)}(\lambda) e^{i\nu t} + \delta_{-\nu}^{(1)}(\lambda) e^{-i\nu t}]\}, \tag{10}$$

where $\delta_\nu^{(1)}(\lambda)$ are holomorphic functions of λ for $|\lambda|$ close to zero and $\delta_0^{(1)}(0) = 1$.

The series

$$\sum_{\nu=-\infty}^{\infty} \lambda^{|\nu|} \delta_\nu^{(1)}(\lambda) \tau^\nu = 1 + \lambda F(\tau, \lambda),$$

in view of the assumption made about the function $F(\tau, \lambda)$ that figures in equation (5) of the contour L_λ, will converge for all τ that are sufficiently close to unity in modulus.

Several lines below we shall have to do with the powers of $z(t)$. We will perform now the calculations necessary to us. Let us consider the square of the series cited above, $1 + \lambda F(\tau, \lambda)$. It will be an analytic function that is regular in λ for $|\lambda|$ sufficiently small, and in τ for $|\tau|$ sufficiently close to 1. The expansion of the function in a power series can be obtained from the series for $1 + \lambda F(\tau, \lambda)$ by means of the ordinary rule for the multiplication of power series. Moreover the coefficients of the powers of τ in the series will be regular functions of λ near the point $\lambda = 0$.

Simple calculations show that the coefficient of τ^ν in the expansion of $[1 + \lambda F(\tau, \lambda)]^2$ will be

$$\sum_{\substack{p,q=-\infty \\ p+q=\nu}}^{+\infty} \lambda^{|p|+|q|}\delta_p^{(1)}(\lambda)\delta_q^{(1)}(\lambda).$$

Since $|p| + |q| \geqslant |p + q| = |\nu|$, all the terms of this sum will have the common factor $\lambda^{|\nu|}$, and the coefficient itself can therefore be represented in the form $\lambda^{|\nu|}\delta_\nu^{(2)}(\lambda)$, where $\delta_\nu^{(2)}(\lambda)$ is a holomorphic function of λ near $\lambda = 0$.

Hence we arrive at the conclusion that $z^2(t)$ is representable in the form

$$z^2(t) = e^{2it} \sum_{\nu=-\infty}^{+\infty} \lambda^{|\nu|}\delta_\nu^{(2)}(\lambda)e^{i\nu t},$$

where the $\delta_\nu^{(2)}(\lambda)$ possess the property of regularity, indicated above, and the series converges for any τ close to 1 in modulus.

If these considerations be applied several times, we find that $z^k(t)$ is representable in the form

$$z^k(t) = e^{ikt} \sum_{\nu=-\infty}^{+\infty} \lambda^{|\nu|}\delta_\nu^{(k)}(\lambda)e^{i\nu t},$$

where $\delta_\nu^{(k)}(\lambda)$ are holomorphic functions of λ near $\lambda = 0$, and $\delta_0^{(k)}(0) = 1$.

Let us see now what form system (9) will take for a contour of type (10). In this case

$$\beta_\nu^{(k)}(\lambda) = \lambda^{|\nu-k|}\delta_{\nu-k}^{(k)}(\lambda),$$

and consequently the equations of system (9) will take the form:

$$\sum_{k,j=1}^{n} \alpha_k\bar{\alpha}_j \sum_{p=-\infty}^{+\infty} \lambda^{|p-k|}\lambda^{|p-m-j|}\delta_{p-k}^{(k)}(\lambda)\delta_{p-m-j}^{(j)}(\lambda) = 0$$
$$(m = 1, 2, \ldots, n-1). \quad (9_1)$$

Let us consider the free term of this equation, not dependent on $\alpha_2, \alpha_3, \ldots, \alpha_n$. It is obtained for $k = 1$, $j = 1$, and is equal to

$$\sum_{p=-\infty}^{+\infty} \beta_p^{(1)}\beta_{p-m}^{(1)} = \sum_{p=-\infty}^{+\infty} \lambda^{|p-1|+|p-m-1|}\delta_{p-1}\bar{\delta}_{p-m-1}^{(1)}.$$

Since $|p - 1| + |p - m - 1|$ is not less than m for any p, the free term will certainly contain the factor λ to a power not lower than m [1]). This fact will be important for us in the subsequent reasoning.

The terms of equations (9_1) containing the sought coefficients linearly are obtained either for $j = 1$ or for $k = 1$. In the first case we obtain α_k with the coefficient

$$\sum_{p=-\infty}^{+\infty} \lambda^{|p-k|+|p-m-1|}\delta_{p-k}^{(k)}(\lambda)\bar{\delta}_{p-m-1}^{(1)}(\lambda).$$

We shall make clear in what case the series expansion in powers of λ of the cited expression will have terms free of λ.

In order that this be so, it is necessary that the exponent $|p - k| + |p - m - 1|$ of the power of λ be equal to zero, whence it is evident that we must have

$$k = p = m + 1.$$

The corresponding term in the sum will equal

$$\delta_0^{(k)}(\lambda)\bar{\delta}_0^{(1)}(\lambda).$$

For $\lambda = 0$ it reduces to unity, since $\delta_0^{(k)}(0) = 1$ for any k.

In the second case, when $k = 1$, we obtain $\bar{\alpha}_j$ with the coefficient

$$\sum_{p=-\infty}^{+\infty} \lambda^{|p-1|+|p-m-j|}\delta_{p-1}^{(1)}(\lambda)\bar{\delta}_{p-m-j}^{(j)}(\lambda).$$

Since

$$|p - 1| + |m + j - p| \geqslant |p - 1 + m + j - p| \geqslant m + j - 1 > 0,$$

because $m \geqslant 1$ and $j \geqslant 1$, all the terms of the cited sum will contain the factor λ to at least the first power, and the expansion in a power series in λ of this coefficient will not contain a term free of λ.

On the left side of (9_1) let us withdraw the term corresponding to $k = m + 1$, $j = 1$. It has the form:

$$\alpha_{m+1}[1 + \lambda Q_{n,m+1}(\lambda)],$$

where $Q_{n,m+1}(\lambda)$ is some power series in λ.

Equation (9_1) can therefore be rewritten in the form:

$$\alpha_{m+1} = -\alpha_{m+1}\lambda Q_{n,m+1}(\lambda) - \\ - \sum_{k,j=1}^{n}{}' \alpha_k\alpha_j \sum_{p=-\infty}^{+\infty} \lambda^{|p-k|+|p-m-j|} \cdot \delta_{p-k}^{(k)}(\lambda)\bar{\delta}_{p-m-j}^{(j)}(\lambda). \qquad (11)$$

The sign $'$ on the sum $\sum\limits_{k,j=1}^{n}$ indicates that the summand for $k = m + 1$, $j = 1$ must be omitted from it.

---

[1]) Indeed,

$$|p - 1| + |m - p + 1| \geqslant |(p - 1) + (m - p + 1)| = |m| = m.$$

The method of successive approximations can be conveniently applied to the solution of system (11). Adopting as the initial approximations $\alpha_2^{(0)} = \alpha_3^{(0)} = \ldots = \alpha_n^{(0)} = 0$, we substitute them in the right sides of system (11); they will reduce to the free terms of which we spoke above. The expressions found in this manner we adopt as the first approximations, $\alpha_2^{(1)}, \alpha_3^{(1)}, \ldots, \alpha_n^{(1)}, \ldots$, substitute them again in the right sides, find the second approximations, and so on.

It goes without saying that in finding the first approximations we need conduct all computations with an accuracy of the first power of λ only. In finding the second approximations we discard all powers of λ higher than the second, etc. In the result we obtain the expansion of the coefficients $\alpha_2, \alpha_3, \ldots, \alpha_n$ in power series in λ. As we have already said, the free term of equation (11) will contain the factor λ^m, and consequently the first approximation for α_{m+1} will contain the powers of λ not lower than the mth.

It can be established without particular difficulties that then any approximation for α_{m+1} will contain powers of λ not lower than the mth. It is sufficient for us to show that if this is true for the sth approximation, then it will be true for the next one too, since it is valid for the first.

For this let us consider the right side of equation (11) and there substitute for α_k, $\bar{\alpha}_j$ the sth approximations to them. The first term on the right side will contain, as is seen from its structure, λ to not lower than the power $m + 1$. Let us reveal the lowest power of λ figuring in the terms of the sum $\sum\limits_{k,j=1}^{n}{}'$ on the right side. The product $\alpha_k\bar{\alpha}_j$ contains, on assumption, λ as a factor to a power not lower than $k + j - 2$.

In the general term of the sum $\sum\limits_{k,j=1}^{n}{}'$ the parameter λ will appear as a factor to at least the power $k + j - 2 + |p - k| + |p - m - j| = N$. Let us now estimate this quantity:

$$
\begin{aligned}
N = k + j - 2 + |p - k| + |m + j - p| &\geqslant \\
\geqslant k + j - 2 + |(m + j - p) + (p - k)| &\geqslant \\
\geqslant k + j - 2 + m + j - k = m + 2(j - 1) &\geqslant m.
\end{aligned}
$$

Thus if one substitutes for the α_k in the right side of (11) the sth approximations, all terms will contain λ in powers not lower than m. And this proves that the $(s + 1)$th approximation to α_{m+1} will also contain in its expansion with respect to λ powers of λ not lower than m.

We will now give several examples of the approximate determination of the transforming function in accordance with the method that has been indicated here.

Example. We will find the function giving the conformal transformation of the interior of the ellipse

$$x = (1 + \lambda^2)\cos t, \quad y = (1 - \lambda^2)\sin t$$

into a circle.

The equation of the ellipse in complex form

$$z = e^{it}(1 + \lambda^2 e^{-2it})$$

will be of exactly the type of which we were just now speaking. Let us set ourselves the task of finding the transforming function accurate to the tenth power of λ. In conformity with what has been said, the last coefficient to be taken into account will be α_{11}, since all α_k with greater indices will contain λ only to higher powers than the tenth. In view of the fact that the ellipse has two axes of symmetry, all the α_k will be real and the α_k with even indices will equal zero:

$$P(z) = z + \alpha_3 z^3 + \alpha_5 z^5 + \alpha_7 z^7 + \alpha_9 z^9 + \alpha_{11} z^{11},$$

$$\begin{aligned} P(z)\overline{P(z)} = z\bar{z} &+ \alpha_3(z^3\bar{z} + \bar{z}^3 z) + [\alpha_5(z^5\bar{z} + \bar{z}_5 z) + \alpha_3^2 z^3\bar{z}^3] + \\ &+ [\alpha_7(z^7\bar{z} + \bar{z}^7 z) + \alpha_5\alpha_3(z^5\bar{z}^3 + \bar{z}^5 z^3)] + \\ &+ [\alpha_9(z^9\bar{z} + \bar{z}^9 z) + \alpha_7\alpha_3(z^7\bar{z}^3 + \bar{z}^7 z^3) + \alpha_5^2 z^5\bar{z}^5] + \\ &+ [\alpha_{11}(z^{11}\bar{z} + \bar{z}^{11} z) + \alpha_9\alpha_3(z^9\bar{z}^3 + \bar{z}^9 z^3) + \alpha_7\alpha_5(z^7\bar{z}^5 + \bar{z}^7 z^5)]. \end{aligned}$$

We shall now form the combinations $z^k\bar{z}^j + \bar{z}^k z^j$ figuring in the square of the modulus of the polynomial. α_3 certainly contains the factor λ^2, and therefore in computing $z^3\bar{z} + \bar{z}^3 z$ we need take into consideration only up to the eighth power of λ. In our case z^3 will contain λ in powers not above the sixth:

$$z^3 = e^{3it} + 3\lambda^2 e^{it} + 3\lambda^4 e^{-it} + \lambda^6 e^{-3it}.$$

α_5 will contain λ in powers not lower than the fourth, and in z^5 we therefore keep the powers of λ up to the sixth only:

$$z^5 = e^{5it} + 5\lambda^2 e^{3it} + 10\lambda^4 e^{it} + 10\lambda^6 e^{-it};$$

in computing the combination $\alpha_3^2 z^3\bar{z}^3$ we keep in the product $z^3\bar{z}^3$ the powers not higher than the sixth, for α_3^2 contains the factor λ^4, etc.

If we take into account all simplifying circumstances of a similar kind, we obtain the following expressions for our combinations:

$$\begin{aligned} z\bar{z} &= 1 + \lambda^4 + 2\lambda^2 \cos 2t, \\ z^3\bar{z} + \bar{z}^3 z &= 2[(\lambda^6 + \lambda^2)\cos 4t + (\lambda^8 + 6\lambda^4 + 1)\cos 2t + (3\lambda^6 + 3\lambda^2)], \\ z^5\bar{z} + \bar{z}^5 z &= 2[\lambda^2 \cos 6t + (5\lambda^4 + 1)\cos 4t + (20\lambda^6 + 5\lambda^2)\cos 2t + 10\lambda^4], \\ z^3\bar{z}^3 &= 2[\lambda^6 \cos 6t + 3\lambda^4 \cos 4t + (9\lambda^6 + 3\lambda^2)\cos 2t + \tfrac{1}{2}(9\lambda^4 + 1)], \\ z^7\bar{z} + \bar{z}^7 z &= 2[\lambda^2 \cos 8t + (7\lambda^4 + 1)\cos 6t + 7\lambda^2 \cos 4t + 21\lambda^4 \cos 2t], \\ z^5\bar{z}^3 + \bar{z}^5 z^3 &= 2[3\lambda^4 \cos 6t + 3\lambda^2 \cos 4t + (25\lambda^4 + 1)\cos 2t + 5\lambda^2], \\ z^9\bar{z} + \bar{z}^9 z &= 2[\lambda^2 \cos 10t + \cos 8t + 9\lambda^2 \cos 6t], \\ z^7\bar{z}^3 + \bar{z}^7 z^3 &= 2[3\lambda^2 \cos 6t + \cos 4t + 7\lambda^2 \cos 2t], \\ z^5\bar{z}^5 &= 2[5\lambda^2 \cos 2t + \tfrac{1}{2}], \\ z^{11}\bar{z} + \bar{z}^{11} z &= 2\cos 10t, \\ z^9\bar{z}^3 + \bar{z}^9 z^3 &= 2\cos 6t, \\ z^7\bar{z}^5 + \bar{z}^7 z^5 &= 2\cos 2t. \end{aligned}$$

Substituting in $P(z)\overline{P(z)}$ the expressions obtained and equating the free term to the square of the radius of the circle, ϱ^2, and the coefficients of the different cosines to zero, we obtain the system of equations:

$$1 + \lambda^4 + 2\alpha_3(3\lambda^6 + 3\lambda^2) + 2\alpha_5\, 10\lambda^4 + \alpha_3^2(9\lambda^4 + 1) + 2\alpha_3\alpha_5 5\lambda^2 + \alpha_5^2 = \varrho^2,$$

$$\lambda^2 + \alpha_3(1 + 6\lambda^4 + \lambda^8) + \alpha_3(5\lambda^2 + 20\lambda^6) + \alpha_3^2(3\lambda^2 + 9\lambda^6) +$$
$$+ \alpha_7 21\lambda^4 + \alpha_3\alpha_5(1 + 25\lambda^4) + \alpha_3\alpha_7 7\lambda^2 + \alpha_5^2 5\lambda^2 + \alpha_5\alpha_7 = 0,$$

$$\alpha_3(\lambda^2 + \lambda^6) + \alpha_5(1 + 5\lambda^4) + \alpha_3^2 3\lambda^4 + \alpha_7 7\lambda^2 + \alpha_3\alpha_5 3\lambda^2 + \alpha_3\alpha_7 = 0,$$

$$\alpha_5\lambda^2 + \alpha_3^2\lambda^6 + \alpha_7(1 + 7\lambda^4) + \alpha_3\alpha_5 3\lambda^4 + \alpha_9 9\lambda^2 + \alpha_3\alpha_7 3\lambda^2 + \alpha_3\alpha_9 = 0,$$

$$\alpha_7\lambda^2 + \alpha_9 = 0,$$

$$\alpha_9\lambda^2 + \alpha_{11} = 0.$$

Successively transposing $\alpha_3, \alpha_5, \ldots$ in the equations of this system beginning with the second, we obtain:

$$\left.\begin{aligned}
\alpha_3 &= -\,[\lambda^2 + \alpha_3(6\lambda^4 + \lambda^8) + \alpha_5(5\lambda^2 + 20\lambda^6) + \alpha_3^2(3\lambda^2 + 9\lambda^6) + \\
&\qquad + \alpha_7 21\lambda^4 + \alpha_3\alpha_5(1 + 25\lambda^4) + \alpha_3\alpha_7 7\lambda^2 + \alpha_5^2 5\lambda^2 + \alpha_5\alpha_7], \\
\alpha_5 &= -\,[\alpha_3(\lambda^2 + \lambda^6) + \alpha_5 5\lambda^4 + \alpha_3^2 3\lambda^4 + \alpha_7 7\lambda^2 + \alpha_3\alpha_5 3\lambda^2 + \alpha_3\alpha_7], \\
\alpha_7 &= -\,[\alpha_5\lambda^2 + \alpha_3^2\lambda^6 + \alpha_7 7\lambda^4 + \alpha_3\alpha_5 3\lambda^4 + \alpha_9 9\lambda^2 + \alpha_3\alpha_7 3\lambda^2 + \alpha_3\alpha_9], \\
\alpha_9 &= -\,\alpha_7\lambda^2, \\
\alpha_{11} &= -\,\alpha_9\lambda^2.
\end{aligned}\right\} \qquad (*)$$

In solving this system we start, as always, with a null approximation; the first approximation, inasmuch as the system contains only even powers of λ, we compute accurate to λ^2; the second approximation, to λ^4, etc.

We summarize the results of the computations in the following table:

Coefficient	Initial Value	First Approx.	Second Approx.	Third Approx.	Fourth Approx.	Fifth Approx.
α_3	0	$-\lambda^2$	$-\lambda^2$	$-\lambda^2-\lambda^6$	$-\lambda^2-\lambda^6$	$-\lambda^2-\lambda^6-4\lambda^{10}$
α_5	0	0	λ^4	λ^4	$\lambda^4+3\lambda^8$	$\lambda^4+3\lambda^8$
α_7	0	0	0	$-\lambda^6$	$-\lambda^6$	$-\lambda^6-5\lambda^{10}$
α_9	0	0	0	0	λ^8	λ^8
α_{11}	0	0	0	0	0	λ^{10}

Thus accurate to λ^{10} we have:

$$P(z) = z - (\lambda^2 + \lambda^6 + 4\lambda^{10})z^3 + (\lambda^4 + 3\lambda^8)z^5 - (\lambda^6 + 5\lambda^{10})z^7 + \lambda^8 z^9 - \lambda^{10}z^{11}.$$

As regards the approximate value for ϱ^2, we obtain it by substituting, for α_3, $\alpha_5, \ldots$, their values into the expression found earlier by us for the free term of $P(z)\overline{P(z)}$. Simple computations show that

$$\varrho^2 = 1 - 4\lambda^4 + 10\lambda^8.$$

We will indicate one method of checking the computations. If in $P(z)$ one substitutes for z its value for the contour of the region and next sets t equal to some value, say $t = 0$, we must obtain a number whose modulus must, accurate to λ^{10}, coincide with ϱ.

In our case for $t = 0$ we have

$$z = 1 + \lambda^2$$

and

$$P(1 + \lambda^2) = 1 - 2\lambda^4 + 3\lambda^8,$$

which fully coincides with

$$\varrho = (1 - 4\lambda^4 + 10\lambda^8)^{\frac{1}{2}} = 1 - 2\lambda^4 + 3\lambda^8.$$

If the equation of the contour is given with the numerical value of the parameter λ, it is often simpler to solve the system of equations for the coefficients α_k directly numerically, without introducing the literal parameter, thus avoiding the difficulties connected with the algebraic character of the calculations with an undefined parameter λ.

We will elucidate this in an example. Let us return to the problem of the mapping of the ellipse considered in the foregoing example. Let us have to do with an ellipse the ratio of whose semi-axes is 3 : 2, i.e.

$$x = \tfrac{6}{5} \cos t, \quad y = \tfrac{4}{5} \sin t.$$

This occurs for $\lambda^2 = \frac{1}{5}$. Let us substitute the value of λ in system (*); it will take the form:

$$\alpha_3 = -[0.2 + 0.241\alpha_3 + 1.16\alpha_5 + 0.672\alpha_3^2 + 0.84\alpha_7 + 2\alpha_3\alpha_5 + 1.4\alpha_3\alpha_7 + \alpha_5^2 + \alpha_5\alpha_7]$$

$$\alpha_5 = -[0.208\alpha_3 + 0.2\alpha_5 + 0.12\alpha_3^2 + 1.4\alpha_7 + 0.6\alpha_3\alpha_5 + \alpha_3\alpha_7],$$

$$\alpha_7 = -[0.2\alpha_5 + 0.008\alpha_3^2 + 0.28\alpha_7 + 0.12\alpha_3\alpha_5 + 1.8\alpha_9 + 0.6\alpha_3\alpha_7 + \alpha_3\alpha_9],$$

$$\alpha_9 = -0.2\alpha_7,$$

$$\alpha_{11} = -0.2\alpha_9.$$

Commencing as usual with $\alpha_3 = \alpha_5 = \ldots = 0$, we employ the method of successive approximations.

Coefficient	Initial Value	First Approx.	Second Approx.	Third Approx.	Fourth Approx.
α_3	0	−0.2	−0.179	−0.210	−0.198
α_5	0	0	0.037	0.030	0.046
α_7	0	0	0	−0.008	−0.010
α_9	0	0	0	0	0.002
α_{11}	0	0	0	0	0

$$P(z) = z - 0.198z^3 + 0.046z^5 - 0.010z^7 + 0.002z^9.$$

2. The method of successive approximations. In the investigations of this section we shall assume that the contour of the region is given by an equation of the form:

$$z = e^{it} \sum_{\nu=-\infty}^{+\infty} \lambda^{|\nu|} \delta_\nu(\lambda) e^{i\nu t}, \tag{12}$$

where the $\delta_\nu(\lambda)$ are holomorphic functions of λ in the neighborhood of $\lambda = 0$ and $\delta_0(0) = 1$. We shall assume that the series $\sum_{\nu=-\infty}^{+\infty} \lambda^{|\nu|}\delta_\nu(\gamma)\zeta^\nu$ is a holomorphic function of λ and ζ near $\lambda = 0$ and $|\zeta| = 1$.

The function effecting the conformal transformation of the region into a circle we shall seek in the form of a series

$$\zeta = P_1(z) + \lambda P_2(z) + \lambda^2 P_3(z) + \ldots + \lambda^n P_{n+1}(z) + \ldots, \tag{13}$$

where the $P_n(z)$ are regular functions of z, subject to determination. We shall consider that the origin of coordinates of the z-plane passes to the origin of coordinates of the ζ-plane. In accordance with this we must have $P_n(0) = 0$. Let us form

$$|\zeta|^2 = P_1(z)\overline{P_1(z)} + \lambda[P_2(z)\overline{P_1(z)} + \overline{P_2(z)}P_1(z)] + \ldots$$
$$\ldots + \lambda^n[P_{n+1}(z)\overline{P_1(z)} + P_n(z)\,\overline{P_2(z)} + \ldots + P_1(z)\overline{P_{n+1}(z)}] + \ldots \tag{14}$$

If there be substituted in this for z its value (12) for the contour of the region, we must obtain, for the function effecting the sought conformal transformation, a constant quantity equal to the square of the radius of the circle onto which the region is mapped.

If the power series (13) has, with respect to λ, a radius of convergence different from zero, the result of the substitution will, generally speaking, be a holomorphic function of λ near $\lambda = 0$, the expansion of which in a Taylor's series in powers of λ we shall obtain by collecting the terms in expression (14) with like powers of λ. The sum of this series, as was said above, must preserve a constant value equal to the square of the radius of the circle. Therefore its term free of λ must be equal to this constant, and the coefficients of all the powers of λ must be equal to zero.

In order to have in view something definite, we shall consider the radius of the circle to be equal to unity. Equating to unity the term in expression (14) — where for z has been introduced its value (12) — that is free of λ, we obtain the following equation:

$$P_1(e^{it})\overline{P_1(e^{it})} = 1.$$

The function $P_1(z)$ corresponds to the conformal transformation of the circle of radius 1, with center at the origin of coordinates, that is obtained from the region bounded by the contour (12), for $\lambda = 0$, into just such a circle of the ζ-plane. Discarding the useless rotation of the circle in the ζ-plane, we put

$$P_1(z) = z. \tag{15}$$

Now equating to zero the coefficient of the first power of λ in (14), we have:

$$P_2(e^{it})e^{-it} = \overline{P_2(e^{it})}e^{it} + \left\{\frac{d}{d\lambda}z\bar{z}\right\}_{\lambda=0} = 0.$$

It is seen from (13) that

$$\left\{\frac{d}{d\lambda} z\bar{z}\right\}_{\lambda=0} = [\delta_1(0)e^{it} + \delta_0'(0) + \delta_{-1}(0)e^{-it}] +$$
$$+ [\overline{\delta_1(0)}e^{-it} + \overline{\delta_0'(0)} + \overline{\delta_{-1}(0)}e^{it}] = -Q_1$$

is a linear polynomial in e^{it} and e^{-it}:

$$2\mathrm{Re}[P_2(e^{it})e^{-it}] = Q_1.$$

To ease the notation henceforth, let us agree upon the following symbols and terms: the two trigonometric series

$$\varphi(t) = \sum_{n=0}^{\infty} (a_n \cos nt - b_n \sin nt)$$

$$\psi(t) = \sum_{n=0}^{\infty} (b_n \cos nt + a_n \sin nt)$$

we will call conjugate, denoting this fact as follows:

$$\psi(t) = \widetilde{\varphi(t)}.$$

The assignment of one of them, for example $\varphi(t)$, defines the other, accurate to a constant term — in our example b_0. As has already been remarked above, all the functions $P_n(z)$ must vanish for $z = 0$. Therefore $\frac{1}{z} P_2(z)$ must be a holomorphic function. On the boundary of the unit circle, however, for $z = e^{it}$, twice its real part is a trigonometric polynomial of the form $Q_1 = a_0 + a_1 \cos t - b_1 \sin t$. Going over to the conjugate polynomial, we find the imaginary part accurate to a constant term

$$2\mathrm{I}\,[P_2(e^{it})e^{-it}] = \widetilde{Q_1}, \quad P_2(e^{it})e^{-it} = \tfrac{1}{2}(Q_1 + i\widetilde{Q_1}). \tag{16}$$

Hence the function $P_2(e^{it})$ is simply found in the form of a polynomial of the second degree in e^{it}. We shall assume that for any λ the direction of the real axis at the origin of coordinates of the z-plane is carried into the direction of the real axis at the origin of coordinates of the ζ-plane. The coefficients of the first power of z in the expansions of all the functions $P_n(z)$ must be real. Thus the constant accurate to which $\widetilde{Q_1}$ is found by the known Q_1 must be equal to zero.

By equating to zero the coefficients of $\lambda, \lambda^2, \ldots, \lambda^{n-1}$ in the expression that we shall have obtained after substituting in (14) for z its value (12), let us have found $P_1(z), P_2(z), \ldots, P_n(z)$ in the form of polynomials in z of degree not higher than $1, 2, \ldots, n$, respectively, without free terms. We shall now show that by equating to zero the coefficient of λ^n, the function $P_{n+1}(z)$ is defined as a polynomial of degree $n + 1$ in z without a free term. Indeed, this coefficient has the form:

$$\frac{1}{n!}\left\{\frac{d^n}{d\lambda^n}\sum_{p=0}^{n}\lambda^p P_{p+1}(z)\sum_{q=0}^{n}\lambda^q \overline{P_{q+1}(z)}\right\}_{\lambda=0},$$

whence it follows that

$$P_{n+1}(e^{it})e^{-it} + \overline{P_{n+1}(e^{it})e^{it}} = 2\text{Re}\,[P_{n+1}(e^{it})e^{-it}] =$$

$$= -\frac{1}{n!}\left\{\frac{d^n}{d\lambda^n}\sum_{p=0}^{n-1}\lambda^p P_{p+1}(z)\cdot\sum_{q=0}^{n-1}\lambda^q \overline{P_{q+1}(z)}\right\}_{\lambda=0} = Q_n. \quad (17)$$

Let us determine the leading powers of e^{it} and e^{-it} appearing on the right side. In computing the latter we shall have to take the nth derivative with respect to λ of expressions of the form $\lambda^{p+q}z^k\bar{z}^l$, where

$$p, q \leqslant n-1,\ 1 \leqslant k \leqslant p+1,\ 1 \leqslant l \leqslant q+1,$$

and, in the result of the differentiation, put $\lambda = 0$.

We shall reconstruct series (12) in powers of λ. As the result we obtain a Taylor's series

$$z(t) = e^{it}\sum_{\nu=0}^{\infty}\Pi_\nu^{(1)}(e^{it}, e^{-it})\lambda^\nu,$$

where $\Pi_\nu(e^{it}, e^{-it})$ will obviously be a polynomial of degree not higher than ν in e^{it} and e^{-it}. Hence it follows that $z^k(t) = e^{ikt}\sum_{\nu=0}^{\infty}\Pi_\nu^{(k)}(e^{it}, e^{-it})\lambda^\nu$, where $\Pi_\nu^{(k)}$ are also polynomials in e^{it} and e^{-it} of degree not higher than ν. On the strength of this,

$$\lambda^{p+q}z^k\bar{z}^l = e^{i(k-l)t}\sum_{\nu=0}^{\infty}\sum_{m=0}^{\nu}\Pi_m^{(k)}\Pi_{\nu-m}^{(l)}\lambda^{\nu+p+q};$$

the coefficient of λ^n will be

$$e^{i(k-l)t}\sum_{m=0}^{n-p-q}\Pi_m^{(k)}\overline{\Pi}_{n-p-q-m}^{(l)}.$$

And since the product $\Pi_m^{(k)}\cdot\overline{\Pi}_{n-p-q-m}^{(l)}$ contains e^{it}, e^{-it} in powers not higher than $n-p-q$, which follows from the structure of these functions, the entire coefficient will contain these quantities in degrees not higher than $k-l+n-p-q \leqslant p+1-1+n-p-q \leqslant n$.

Thus Q_n must be a trigonometric polynomial containing arcs of multiplicity not higher than n.

Going over to the conjugate functions in equation (17) and discarding the arbitrary constant that appears in the composition of $\tilde{Q}_n$, we have:

$$2\text{I}\,[P_{n+1}(e^{it})e^{-it}] = \tilde{Q}_n,$$

whence

$$P_{n+1}(e^{it}) = \frac{e^{it}}{2}(Q_n + i\widetilde{Q_n}). \tag{81}$$

$P_{n+1}(z)$ will be, accordingly, a polynomial of degree not higher than $n + 1$.

Computing $P_1(z), P_2(z), \ldots$ consecutively in the way indicated, we can construct series (13) by them. It remains for us to determine what properties the right side of equation (12) of the contour L_λ must possess if the series (13) constructed by us is to converge, and to estimate the interval of variation of λ for which the convergence can be guaranteed. We will leave aside for the moment the investigation of this question, referring the reader to No. 7 § 6, where this question is considered.

We will give an example of such a calculation. We will take the ellipse considered by us in the preceding No.,

$$z(t) = e^{it}(1 + \lambda^2 e^{-2it}).$$

In view of the fact that only λ^2 figures in the equation of the contour, we set

$$\zeta = z + \lambda^2 P_3(z) + \lambda^4 P_5(z) + \lambda^6 P_7(z) + \lambda^8 P_9(z) + \ldots$$

$$\zeta\bar{\zeta} = z\bar{z} + \lambda^2[P_3\,\bar{z} + \bar{P}_3\,z] + \lambda^4[P_5\bar{z} + P_3\bar{P}_3 + \bar{P}_5 z] + \ldots$$

We shall seek the transforming function approximately, limiting ourselves to the eighth power of λ. In accordance with this, in all computations we shall discard all powers of λ higher than the eighth. Since on the contour

$$z\bar{z} = 1 + \lambda^4 + 2\lambda^2 \cos 2t,$$

equating to zero the coefficient of λ^2 gives

$$P_3(e^{it})e^{-it} + \overline{P_3(e^{it})}e^{it} = -2\cos 2t,$$

whence

$$P_3(e^{it})e^{-it} = -e^{2it} \text{ and } P_3(z) = -z^3.$$

Let us find $P_5(z)$. On the contour

$$z^3 = e^{3it}(1 + 3\lambda^2 e^{-2it} + 3\lambda^4 e^{-4it} + \lambda^6 e^{-6it}),$$

therefore if we discard the powers higher than the sixth:

$$P_3(z)\bar{z} + \overline{P_3(z)}z = -2[3\lambda^2 + 3\lambda^6 + (1 + 6\lambda^4)\cos 2t + (\lambda^2 + \lambda^6)\cos 4t].$$

In calculating $P_3(z)\overline{P_3(z)}$ it is sufficient for us to limit ourselves to the fourth power of λ only, for the result must be multiplied by λ^4. An analogous remark holds for all the subsequent calculations, and we shall not repeat it, simply discarding the superfluous powers of λ.

$$P_3(z)\overline{P_3(z)} = 1 + 9\lambda^4 + 6\lambda^2 \cos 2t + 6\lambda^4 \cos 4t.$$

Collecting like terms and equating to zero the coefficient of λ^4, we obtain the equation for the determination of $P_5(z)$:

$$P_5 e^{-it} + \overline{P}_5 e^{it} = 4 + 2\cos 4t, \quad P_5(z) = 2z + z_5.$$

Let us pass to $P_7(z)$. On the boundary of the region

$$z^5 = e^{5it}[1 + 5\lambda^2 e^{-2it} + 10\lambda^4 e^{-4it}],$$

$$P_5 e^{-it} + \overline{P}_5 e^{it} = 2[2 + 12\lambda^4 + 9\lambda^2 \cos 2t + (1 + 5\lambda^4)\cos 4t + \lambda^2 \cos 6t],$$

$$P_5\overline{P}_3 + \overline{P}_5 P_3 = -2[11\lambda^2 + 3\cos 2t + 5\lambda^2 \cos 4t].$$

Collecting the coefficients of λ^6 and equating the result to zero, we have the equation for P_7:

$$P_7 e^{-it} + \overline{P}_7 e^{it} = -6\cos 2t - 2\cos 6t,$$

$$P_7(z) = -(3z^3 + z^7).$$

Finally $P_9(z)$.

$$P_7 e^{-it} + \overline{P}_7 e^{it} = -2[9\lambda^2 + 3\cos 2t + 10\lambda^2 \cos 4t + \cos 6t + \lambda^2 \cos 8t]$$

$$P_7\overline{P}_3 + \overline{P}_7 P_3 = 6 + 2\cos 4t$$

$$P_5\overline{P}_5 = 5 + 4\cos 4t,$$

whence for P_9 we obtain this equation:

$$P_9 e^{-it} + \overline{P}_9 e^{it} = 2 + 10\cos 2t + 2\cos 8t,$$

and accordingly

$$P_9(z) = z + 5z^5 + z^9,$$

$$\zeta = z - \lambda^2 z^3 + \lambda^4(2z + z^5) - \lambda^6(3z^3 + z^7) + \lambda^8(z + 5z^5 + z^9).$$

3. The conformal transformation of exterior regions. We revert to contours defined by equation (5) or (7), and shall seek the function conformally transforming the exterior of this contour into the exterior or the interior of a certain circle. We shall here assume that the point at infinity of the z-plane passes to the point at infinity or the zero of the ζ-plane. The mapping function will have an expansion near infinity in the case of the mapping onto the exterior of the circumference:

$$\zeta = z + \alpha_2 + \frac{\alpha_3}{z} + \frac{\alpha_4}{z^2} + \cdots \tag{19}$$

We can always contrive that the coefficient of the first power of z be equal to unity, as shown in formula (19), by suitably choosing the radius of the circumference and establishing correspondence between the directions of the real axes of the z- and ζ-planes at the points at infinity.

If we have in view the transformation of the exterior of the contour into the interior of the circumference, the transforming function can be given, owing to considerations analogous to those expounded above, the following form:

$$\zeta = \frac{1}{z} + \frac{\alpha_2}{z^2} + \frac{\alpha_3}{z^3} + \ldots \tag{20}$$

In both cases, for the approximate construction of the function effecting the conformal transformation, the same device can be employed as before.

The reasoning and the calculations for mapping onto the exterior and the interior of the circle are similar to each other and analogous to the case already investigated of the transformation of interior regions; we shall therefore limit ourselves in all the subsequent exposition to the first case, as an example. As previously, we shall seek the function we require approximately, taking the first n terms of the expansion (19):

$$\zeta = z + \alpha_2 + \frac{\alpha_3}{z} + \ldots + \frac{\alpha_n}{z^{n-2}}. \tag{21}$$

To find the square of the modulus of the function ζ on the contour (5), we shall have to compute $\frac{1}{z}$. This quantity can be expanded on the contour in a series of type (7):

$$\frac{1}{z} = \sum_{\nu=-\infty}^{+\infty} \beta_\nu^{(-1)}(\lambda) e^{i\nu t}, \tag{22}$$

the coefficients of which $\beta_\nu^{(-1)}$ are regular functions of the parameter λ. They must obviously satisfy the following equations:

$$\beta_\nu^{(-1)}(0) = 0 \text{ for } \nu \neq -1, \; \beta_{-1}^{(-1)}(0) = 1.$$

Let us now form $|\zeta|^2$. If we use formula (19) and substitute for z and $\frac{1}{z}$ their expressions (7) and (22), we shall obtain some trigonometric series. We equate to the constant ϱ^2 the free term in it and set the coefficients of the first n terms of the series equal to zero; we obtain a system of equations exactly like that of which we have spoken in detail in No. 1. And from it must be found the coefficients $\alpha_2, \alpha_3, \ldots, \alpha_n$ and the approximate value of the radius ϱ of the circle onto whose exterior is mapped the given region. This system is solved by quite the same methods as was (9).

We shall not stop for a study of it, therefore. We shall yet say a few words with respect to the case when the contour of the region is given by equation (10).

We shall show that, as previously, the expansions in power series in λ of the coefficients $\alpha_2, \alpha_3, \ldots$ will in this case begin respectively with the first, second, etc., powers of λ. Indeed, if

$$z = e^{it} \sum_{\nu=-\infty}^{+\infty} \lambda^{|\nu|} \delta_\nu^{(1)} e^{i\nu t},$$

then $\frac{1}{z}$ will be represented by a series of the same form:

$$\frac{1}{z} = e^{-it} \sum_{\nu=-\infty}^{+\infty} \lambda^{|\nu|} \delta_\nu^{(-1)} e^{i\nu t}, \tag{23}$$

where the $\delta_\nu^{(-1)}$, as were the $\delta_\nu^{(1)}$, are holomorphic functions of λ. In fact, after reconstructing the series for z by the powers of λ we obtain:

$$z = e^{it} \sum_{\nu=0}^{\infty} \Pi_\nu^{(1)} \lambda^\nu,$$

where the $\Pi_\nu^{(1)}$ are polynomials of degree not higher than ν in e^{it} and e^{-it}.

We shall now seek successively the coefficients of the expansion of $\frac{1}{z}$ in powers of λ. We set

$$\frac{1}{z} = e^{-it} \sum_{\nu=0}^{\infty} \Pi_\nu^{(-1)} \lambda^\nu,$$

$$e^{-it}\Pi_0^{(-1)} = \left\{\frac{1}{z}\right\}_{\lambda=0} = \frac{1}{e^{it}} \cdot \Pi_0^{(-1)} = 1,$$

$$e^{-it}\Pi_1^{(-1)} = \left\{\frac{\partial}{\partial\lambda} \frac{1}{z}\right\}_{\lambda=0} = -\frac{\Pi_1^{(1)}}{e^{it}},$$

whence

$$\Pi_1^{(-1)} = -\Pi_1^{(1)}.$$

$\Pi_1^{(-1)}$ is certainly a polynomial of the first degree in e^{it} and e^{-it}. Furthermore,

$$e^{-it}\Pi_2^{(-1)} = \tfrac{1}{2}\left\{\frac{\partial^2}{\partial\lambda^2} \frac{1}{z}\right\}_{\lambda=0} = e^{-it}(\Pi_1^{(1)2} - \Pi_2^{(1)}).$$

Hence it follows that $\Pi_2^{(-1)}$ is a polynomial of not higher than the second degree in e^{it} and e^{-it}, etc. Now reconstructing the expansion in powers of λ for $\frac{1}{z}$ in Laurent's expansion in powers of e^{it}, we obtain the required result (23). Thus for any positive and negative integral k we will have

$$z^k(t) = e^{ikt} \sum_{\nu=-\infty}^{+\infty} \lambda^{|\nu|} \delta_\nu^{(k)}(\lambda) e^{i\nu t},$$

with holomorphic $\delta_\nu^{(k)}(\lambda)$. To ease the notation we will put $\alpha_1 = 1$:

$$\zeta\bar{\zeta} = \sum_{k,j=-1}^{n-2} \frac{\alpha_{k+2}\bar{\alpha}_{j+2}}{z^k\bar{z}^j} =$$

$$= \sum_{\nu=-\infty}^{+\infty} \sum_{k,j=-1}^{n-2} \alpha_{k+2}\bar{\alpha}_{j+2} \sum_{p=-\infty}^{+\infty} \lambda^{|p|+|p-\nu|}\delta_p^{(-k)}\bar{\delta}_{p-\nu}^{(-j)}e^{i(\nu-k+j)t}.$$

Equating to the constant ϱ^2 the term of the right side free of e^{it}, we obtain an approximate equation for the determination of the radius of the circle onto whose exterior is mapped the region:

$$\sum_{k,j=-1}^{n-2} \alpha_{k+2}\bar{\alpha}_{j+2} \sum_{p=-\infty}^{+\infty} \lambda^{|p|+|p-k+j|}\delta_p^{(-k)}\bar{\delta}_{p-k+j}^{(-j)} = \varrho^2.$$

If, however, the coefficients of $e^{-it}, e^{-2it}, \ldots, e^{-(n-1)it}$ be equated to zero, we obtain the system of equations for the determination of the numbers $\alpha_2, \alpha_3, \ldots, \alpha_n$:

$$\sum_{k,j=-1}^{n-2} \alpha_{k+2}\bar{\alpha}_{j+2} \sum_{p=-\infty}^{+\infty} \lambda^{|p|+|p+m+j-k|}\delta_p^{(-k)}\bar{\delta}_{p+m+j-k}^{(-j)} = 0,$$
$$(m = 1, \ldots, n-1)$$

which, after simple transformations, may be reduced to the form:

$$\alpha_{m+1} = -\alpha_{m+1}\lambda q_{n,m+1}(\lambda) -$$
$$- \sum_{k,j=-1}^{n-2}{}' \alpha_{k+2}\bar{\alpha}_{j+2} \cdot \sum_{p=-\infty}^{+\infty} \lambda^{|p|+|p+m+j-k|}\delta_p^{(-k)}\bar{\delta}_{p+m+j-k}^{(-j)},$$

where $q_{n,m+1}(\lambda)$ is a certain power series in λ, and the sign $'$ on the sum $\sum_{k,j=-1}^{n-2}{}'$ is put there because the summand corresponding to $k = m - 1$ and $j = -1$ must be omitted from it. This system is solved by successive approximations, beginning with $\alpha_2^{(0)} = \alpha_3^{(0)} = \ldots = \alpha_n^{(0)} = 0$. It is sufficient for us to establish that if the preceding approximation has the required structure, i.e., $\alpha_2, \alpha_3, \ldots, \alpha_n$ contain powers of λ not lower than $\lambda, \lambda^2, \ldots, \lambda^{n-1}$ respectively, then the next approximation will have the same structure. Indeed, if $\alpha_2, \ldots, \alpha_n$ on the right side satisfy our conditions, the first term certainly contains a factor not lower than λ^{m+1}. And in the general term of the sum, λ will appear to a power not lower than $k + 1 + j + |p| + |p + m + j - k|$, with respect to which it can easily be established that for no k, j, p is it lower than m; therefore the next approximation for α_{m+1} will contain powers of λ not lower than the mth. Our statement is thus completely proved.

Example. To illustrate the method that has been expounded, we will consider the transformation of the exterior of the contour

$$z(t) = e^{it}(1 + \lambda^2 e^{2it})$$

into the exterior of a circle.

By virtue of the symmetry of the region with respect to the coordinate axes, the expansion of the transforming function in the neighborhood of the point at infinity will be:

$$\zeta = z + \frac{a_3}{z} + \frac{a_5}{z^3} + \ldots,$$

where the a_k are real numbers.

We shall seek the transforming function with an accuracy to λ^8. In accordance with this we shall adopt as ζ a polynomial of the form:

$$\zeta = z + \frac{a_3}{z} + \frac{a_5}{z^3} + \frac{a_7}{z^5} + \frac{a_9}{z^7}.$$

The square of the modulus of ζ will be

$$\zeta\bar{\zeta} = z\bar{z} + a_3\left(\frac{z}{\bar{z}} + \frac{\bar{z}}{z}\right) + \left[a_5\left(\frac{z}{\bar{z}^3} + \frac{\bar{z}}{z^3}\right) + a_3^2\frac{1}{z\bar{z}}\right] +$$

$$+ \left[a_7\left(\frac{z}{\bar{z}^5} + \frac{\bar{z}}{z^5}\right) + a_3a_5\left(\frac{1}{z^3\bar{z}} + \frac{1}{\bar{z}^3z}\right)\right] +$$

$$+ \left[a_9\left(\frac{z}{\bar{z}^7} + \frac{\bar{z}}{z^7}\right) + a_3a_7\left(\frac{1}{z\bar{z}^5} + \frac{1}{z^5\bar{z}}\right) + a_5^2\frac{1}{z^3\bar{z}^3}\right].$$

If in computing the terms of the cited expression one keeps in mind the fact that $a_3, \ldots$ contain the factors $\lambda^2, \lambda^4, \ldots$ and discards the summands giving λ to powers higher than the eighth, one obtains:

$$z\bar{z} = 2\lambda^2 \cos 2t + 1 + \lambda^4,$$

$$\frac{\bar{z}}{z} + \frac{z}{\bar{z}} = 2[(\lambda^2 - \lambda^6) \cos 4t + \cos 2t - (\lambda^2 - \lambda^6)],$$

$$\frac{z}{\bar{z}^3} + \frac{\bar{z}}{z^3} = 2[\lambda^2 \cos 6t + (1 - 3\lambda^4) \cos 4t - 3\lambda^2 \cos 2t + 6\lambda^4],$$

$$\frac{1}{z\bar{z}} = 2[\lambda^4 \cos 4t - \lambda^2 \cos 2t + \tfrac{1}{2}(1 + \lambda^4)],$$

$$\frac{z}{\bar{z}^5} + \frac{\bar{z}}{z^5} = 2[\lambda^2 \cos 8t + \cos 6t - 5\lambda^2 \cos 4t],$$

$$\frac{1}{z^3\bar{z}} + \frac{1}{\bar{z}^3z} = 2[-\lambda^2 \cos 4t + \cos 2t - 3\lambda^2],$$

$$\frac{z}{\bar{z}^7} + \frac{\bar{z}}{z^7} = 2 \cos 8t,$$

$$\frac{1}{z\bar{z}^5} + \frac{1}{\bar{z}z^5} = 2 \cos 4t, \quad \frac{1}{z^3\bar{z}^3} = 1.$$

Introducing the values obtained into $\zeta\bar{\zeta}$, equating to ϱ^2 the term not dependent on e^{it}, and to zero the coefficients of the cosines, we obtain the following system of equations:

$$1 + \lambda^4 - 2(\lambda^2 - \lambda^6)a_3 + 12\lambda^4 a_5 + (1 + \lambda^4)a_3^2 - 6\lambda^2 a_3 a_5 + a_5^2 = \varrho^2,$$

$$\lambda^2 + a_3 - 3\lambda^2 a_5 - \lambda^2 a_3^2 + a_3 a_5 = 0,$$

$$(\lambda^2 - \lambda^6)a_3 + (1 - 3\lambda^4)a_5 + \lambda^4 a_3^2 - 5\lambda^2 a_7 - \lambda^2 a_3 a_5 + a_3 a_7 = 0,$$

$$\lambda^2 a_5 + a_7 = 0,$$

$$\lambda^2 a_7 + a_9 = 0.$$

Let us rewrite the last four equations in the form:

$$a_3 = -\lambda^2 + 3\lambda^2 a_5 + \lambda^2 a_3^2 - a_3 a_5,$$

$$a_5 = 3\lambda^4 a_5 - (\lambda^2 - \lambda^6)a_3 - \lambda^4 a_3^2 + 5\lambda^2 a_7 + \lambda^2 a_3 a_5 - a_3 a_7,$$

$$a_7 = -\lambda^2 a_5,$$

$$a_9 = -\lambda^2 a_7,$$

and apply the method of successive approximations, preserving, in the computation of the second approximation, powers of λ up to the fourth, in that of the third approximation, up to the sixth, etc.

Coefficient	Initial Value	First Approx.	Second Approx.	Third Approx.	Fourth Approx.
a_3	0	$-\lambda^2$	$-\lambda^2$	$-\lambda^2+5\lambda^6$	$-\lambda^2+5\lambda^6$
a_5	0	0	λ^4	λ^4	$\lambda^4-11\lambda^8$
a_7	0	0	0	$-\lambda^6$	$-\lambda^6$
a_9	0	0	0	0	λ^8

$$\zeta = z - \frac{\lambda^2 - 5\lambda^6}{z} + \frac{\lambda^4 - 11\lambda^8}{z^3} - \frac{\lambda^6}{z^5} + \frac{\lambda^8}{z^7}.$$

Substitution in the equation for ϱ^2 of the values found gives

$$\varrho^2 = 1 + 4\lambda^4 - 2\lambda^8, \quad \varrho = 1 + 2\lambda^4 - 3\lambda^8.$$

For a control of the computations we substitute z in the constructed approximate polynomial for ζ and put $t = 0$ there. After simplifications we obtain for ϱ the value:

$$\varrho = 1 + 2\lambda^4 - 3\lambda^8,$$

which fully agrees with the preceding result.

To the solution of the same problem there may also be applied the method of successive approximations. We shall, as in No. 2, try to find the function effecting the conformal transformation, in the form of a series arrayed in powers of λ. Let the contour L of the region have an equation of type (12). We set

$$\zeta = z + \lambda P_0(z) + \lambda^2 P_1(z) + \ldots + \lambda^{n+1} P_n(z) + \ldots,$$

where $P_n(z)$ is a function holomorphic outside the contour L and on it, and having a pole of the first order at infinity:

$$P_n(z) = \alpha_{n,-1}z + \alpha_{n,0} + \frac{\alpha_{n,1}}{z} + \ldots = \sum_{\nu=-1}^{\infty} \frac{\alpha_{n,\nu}}{z^\nu}.$$

We shall moreover assume that to the direction of the real axis at the infinity of the z-plane corresponds the direction of the real axis at the infinity of the ζ-plane. Therefore the coefficients $\alpha_{n,-1}$ of z in the functions $P_n(z)$ must be real

$$\zeta\bar{\zeta} = z\bar{z} + \lambda[P_0\bar{z} + \bar{P}_0 z] + \ldots + \lambda^n[P_{n-1}\bar{z} + \ldots + \bar{P}_{n-1}z] + \ldots$$

Let us substitute here for z its value (12) on the contour L and equate to zero the coefficients of λ, λ^2, ...

$$P_0 e^{-it} + \bar{P}_0 e^{it} + \left\{\frac{d}{d\lambda} z\bar{z}\right\}_{\lambda=0} = 0.$$

Hence

$$2\text{Re}[P_0\,(e^{it})e^{-it}] = -\,[\delta_0'(0) + \bar{\delta}_0'(0) + \delta_1(0)e^{it} + \bar{\delta}_1(0)e^{-it} + \\ + \delta_{-1}(0)e^{-it} + \bar{\delta}_{-1}(0)e^{it}] = Q_0.$$

But $P_0(e^{it})e^{-it}$ is the value of the function $\dfrac{P_0(z)}{z}$ on the circumference of the unit circle. Its real part is a linear trigonometric polynomial, as is evident from the preceding equation. On the other hand, this function has in its expansion about the point at infinity only negative powers of z:

$$\frac{P_0(z)}{z} = \alpha_{0,-1} + \frac{\alpha_{0,0}}{z} + \ldots$$

Therefore the imaginary part of the function $\dfrac{P_0(z)}{z}$ on the unit circle will be equal to the trigonometric polynomial, taken with opposite sign, that is the conjugate of Q_0 [1])

$$2\text{I}[P_0(e^{it})e^{-it}] = -\,\widetilde{Q}_0.$$

The arbitrary constant, accurate to which $\widetilde{Q}_0$ is defined by the known polynomial Q_0, we discard, remembering that the free term of the function $\dfrac{P_0(z)}{z}$ must be real.

---

[1]) Indeed, let the function $\Phi = \dfrac{a - ib}{z^n}$. On the boundary of the unit circle

$$\Phi = a \cos nt + b \sin nt + i(b \cos nt - a \sin nt),$$

and therefore

$$\text{I}[\Phi] = b \cos nt - \sin nt = -\,\widetilde{\text{Re}\,[\Phi]}.$$

Q_0 and $\widetilde{Q}_0$ are linear trigonometric polynomials in t. Therefore $Q_0 - i\widetilde{Q}_0$ is representable in the form

$$Q_0 - i\widetilde{Q}_0 = \alpha_{0,-1} + \alpha_{0,0}e^{-it}$$

and consequently

$$P_0(e^{it}) = \frac{e^{it}}{2}(Q_0 - i\widetilde{Q}_0) = \alpha_{0,-1}e^{it} + \alpha_{0,0}.$$

Equating to zero the coefficient of λ^2 gives

$$\text{Re}\,[P_1(e^{it})e^{-it}] = Q_1,$$

where Q_1 is a polynomial of the second degree in e^{it} and e^{-it}, whence

$$P_1(e^{it}) = \frac{e^{it}}{2}(Q_1 - i\widetilde{Q}_1).$$

The arbitrary constant figuring in $\widetilde{Q}_1$ we again discard on the same basis as for $\widetilde{Q}_0$. Since Q_1 and $\widetilde{Q}_1$ are trigonometric polynomials of the second degree in t, the combination $Q_1 - i\widetilde{Q}_1$ may be represented in the form

$$Q_1 - i\widetilde{Q}_1 = \alpha_{1,-1} + \alpha_{1,0}e^{-it} + \alpha_{1,1}e^{2it},$$

and consequently

$$P_1(e^{it}) = \alpha_{1,-1}e^{it} + \alpha_{1,0} + \alpha_{1,1}e^{-it},$$

etc.

Proceeding in this way, we determine the function $P_n(e^{it})$ as a polynomial linear in e^{it} and of the nth degree in e^{-it}:

$$P_n(e^{it}) = \alpha_{n,-1}e^{it} + \alpha_{n,0} + \frac{\alpha_{n,1}}{e^{it}} + \ldots + \frac{\alpha_{n,n}}{e^{int}}.$$

Example. We shall seek the function effecting the transformation of the exterior region of the contour $z(t) = e^{it}(1 - \lambda e^{it})$.

Let us say that we have to determine the required function to the fourth power of the parameter λ, inclusive. In accordance with this we set

$$\zeta = z + \lambda P_0(z) + \lambda^2 P_1(z) + \lambda^3 P_2(z) + \lambda^4 P_3(z).$$

In all the calculations that follow, we discard without comment all terms that either contain λ to higher than the fourth power themselves, or such as will give, after the multiplication of them by other expressions required in the course of the reasoning, a power of λ higher than the fourth.

Form $|\zeta|^2$:

$$\zeta\bar{\zeta} = z\bar{z} + \lambda[P_0\bar{z} + z\bar{P}_0] + \lambda^2[P_1\bar{z} + P_0\bar{P}_0 + z\bar{P}_1] + \lambda^3[P_2\bar{z} + P_1\bar{P}_0 + P_0\bar{P}_1 + z\bar{P}_2] + \lambda^4[P_3\bar{z} + P_2\bar{P}_0 + P_1\bar{P}_1 + P_0\bar{P}_2 + z\bar{P}_3].$$

Substitute here for z its value, collect the terms with λ in the first degree, and equate the coefficient to zero. Since $z\bar{z} = 1 - 2\lambda \cos t + \lambda^2$, the coefficient of λ will be

$$2\text{Re}[P_0(e^{it})e^{-it}] - 2\cos t = 0,$$

whence

$$P_0(e^{it})e^{-it} = e^{-it}.$$

Consequently

$$P_0(z) = 1,$$

$$P_0\bar{z} + z\bar{P}_0 = 2[\cos t - \lambda \cos 2t], \quad P_0\bar{P}_0 = 1.$$

We now collect the coefficient of λ^2:

$$P_1(e^{it})e^{-it} + e^{it}\overline{P_1(e^{it})} + 1 - 2\cos 2t + 1 = 0.$$

Therefore

$$P_1(e^{it})e^{-it} = e^{-2it} - 1, \quad P_1(z) = \frac{1}{z} - z.$$

Moreover,

$$P_1\bar{z} + z\bar{P}_1 + P_0\bar{P}_0 = -1 + 6\lambda \cos t + 2(1 - \lambda^2)\cos 2t - 2\lambda \cos 3t,$$

$$P_1\bar{P}_0 + P_0\bar{P}_1 = 2(\lambda + \lambda \cos 2t).$$

We now equate to zero the coefficient of λ^3; this gives

$$P_2(e^{it})e^{-it} + e^{it}\overline{P_2(e^{it})} + 6\cos t - 2\cos 3t = 0,$$

$$P_2(e^{it})e^{-it} = e^{-3it} - 3e^{-it},$$

$$P_2(z) = \frac{1}{z^2} - 3.$$

Consequently

$$P_2\bar{z} + P_1\bar{P}_0 + P_0\bar{P}_1 + z\bar{P}_2 = 2[\lambda - 3\cos t + 6\lambda \cos 2t + \cos 3t - \lambda \cos 4t]$$

$$\bar{P}_2P_0 + P_0\bar{P}_2 = 2(\cos 2t - 3),$$

$$P_1\bar{P}_1 = 2(1 - \cos t).$$

Equating to zero the coefficient of λ^4, we find:

$$P_3(e^{it})e^{-it} + e^{it}\overline{P_3(e^{it})} + 2(\cos 2t - 3) + 2(1 - \cos t) + $$
$$+ 2[1 + 6\cos 2t - \cos 4t] - 2\cos 2t = 0,$$

$$P_3(e^{it})e^{-it} = 1 + e^{-it} - 6e^{-2it} + e^{-4it},$$

$$P_3(z) = z + 1 - \frac{6}{z} + \frac{1}{z^3}.$$

Consequently the mapping function, accurate to the fourth power of λ, is

$$\zeta = z + \lambda + \lambda^2\left(\frac{1}{z} - z\right) + \lambda^3\left(\frac{1}{z^2} - 3\right) + \lambda^4\left(\frac{1}{z^3} - \frac{6}{z} + 1 + z\right) =$$

$$= (1 - \lambda^2 + \lambda^4)z + (\lambda - 3\lambda^3 + \lambda^4) + (\lambda^2 - 6\lambda^4)\frac{1}{z} + \frac{\lambda^3}{z^2} + \frac{\lambda^4}{z^3}.$$

§ 6. EXPANSION IN A SERIES OF THE POWERS OF A SMALL PARAMETER IN THE CASE OF THE TRANSFORMATION OF A CIRCLE INTO A REGION

1. The normal representation of a contour. The methods we have expounded up till now have related to the construction of the function effecting the conformal transformation into a circle of a region given beforehand. However in solving many problems of the theory of elasticity, the theory of electricity, etc. [1]), if is often useful to know, conversely, the function transforming a circle into the given region.

If the function accomplishing the transformation of the region into a circle were known, the new problem would of course be solved by the simple inversion of it. However in a majority of cases, to proceed in this way is useless, for the new problem is, as we shall find below, simpler than the former, generally speaking. In its general formulation, the problem has until lately little attracted the attention of mathematicians, having been considered mainly for particular regions (linear and circular polygons), chiefly because the function effecting the transformation of the circle into a region does not possess such simple extremal properties as does the inverse to it, and is indeed usually solved in theory as a problem of inversion. And only in recent years have practically convenient methods been given for the direct solution of this problem.

We shall first dwell on a method due to L. V. Kantorovich [2]). It has its source in the following simple consideration. Let us map the circle of unit radius onto the region B bounded by a simple closed contour L. The function effecting the required transformation is regular in the unit circle and can be represented there by a Taylor's series

$$z = x + iy = \sum_{n=0}^{\infty} \alpha_n \zeta^n = \alpha_0 + \alpha_1 \zeta + \alpha_2 \zeta^2 + \dots, \tag{1}$$

$$\alpha_n = a_n + ib_n, \quad |\zeta| < 1.$$

For almost all regions B that are encountered in practical problems, the cited power series will converge on the circumference of the circle too, i.e., when $|\zeta| = 1$. The point z corresponding to such a ζ must lie on the contour L. Therefore putting $\zeta = e^{i\vartheta}$ in equation (1), we must obtain the equation of the contour L in complex form, in the form of the series

$$z = \sum_{n=0}^{\infty} \alpha_n e^{in\vartheta}. \tag{2}$$

Or, if the real part here be separated from the imaginary, we shall obtain x and y in the form of two conjugate trigonometric series:

$$\left.\begin{aligned} x = \varphi(\vartheta) &= \sum_{n=0}^{\infty} (a_n \cos n\vartheta - b_n \sin n\vartheta), \\ y = \psi(\vartheta) &= \sum_{n=0}^{\infty} (b_n \cos n\vartheta + a_n \sin n\vartheta). \end{aligned}\right\} \tag{3}$$

[1]) See, for example, N. Muskhelishvili, [2] or [3].

[2]) See these works of L. V. Kantorovich: [16], [17], [18], [19], [20], [21].

The fact of the conjugacy of the series $\varphi(\vartheta)$ and $\psi(\vartheta)$ we have agreed earlier to denote by the following symbol:

$$\psi(\vartheta) = \widetilde{\varphi(\vartheta)}.$$

The parametric representation of the contour L in form (2) or (3), under the condition that when ϑ varies from 0 to 2π, the point $z = x + iy$ runs once around the contour L, we shall below call the normal one.

The normal representation (2) and (3) is not, of course, completely defined by the indication of only the curve L. In order to fully define the coefficients of series (2) and (3), it is sufficient to indicate, first, into what point α_0 the origin of coordinates is carried in transforming the circle $|\zeta| < 1$ into the interior of L. This is equivalent to the assignment of the free terms α_0, a_0, b_0 of our series, since $z(0) = \alpha_0 = a_0 + ib_0$. Second, there must be given the direction at the point α_0 into which the direction of the real axis of the ζ-plane is carried under the transformation. This direction is defined by indicating the value of the argument of the derivative $z'(0)$ of the transforming function at the origin of coordinates. But, as is evident from expansion (1), $z'(0) = \alpha_1$, and we shall still have to assign $\arg \alpha_1$. If α_0 and $\arg \alpha_1$ are known, then the normal representations (2) and (3) are fully defined.

Conversely, if the normal representation of the contour L is known to us, we find at once the analytic function effecting the sought conformal transformation of the unit circle into the region, either in the form of a series

$$z = \sum_{n=0}^{\infty} (a_n + ib_n)\zeta^n, \tag{4}$$

or in the form of a Cauchy integral

$$z = \frac{1}{2\pi} \int_0^{2\pi} \frac{\varphi(\vartheta) + i\varphi(\vartheta)}{e^{i\vartheta} - \zeta} e^{i\vartheta}\, d\vartheta. \tag{5}$$

The problem of the conformal transformation of the circle into a region is consequently equivalent, under suitable conditions of convergence of series (3) and (4), to finding a special parametric representation of the contour L of the region — its normal representation.

It is for the solution of the latter problem that we shall develop here two methods that differ from one another only in the form in which the computations are conducted, leading to one and the same result: the method of infinite systems of equations and the method of successive approximations.

2. The method of infinite systems. One can arrive at this, for example, when the curve L is given by an implicit equation,

$$F(x, y) = 0, \tag{6}$$

where $F(x, y)$ is an analytic function of its arguments for the values of x and y lying on the contour L, if the representation (3) of the curve L be sought by the method of undetermined coefficients. Namely, by substituting in (6) for x and y their values (3), and expanding the result in a Fourier series, we find:

$$F(x, y) = F[\varphi(\vartheta), \psi(\vartheta)] =$$
$$= F_0(a_j, b_j) + \sum_{n=1}^{\infty} (F_n(a_j, b_j) \cos n\vartheta + F_n^*(a_j, b_j) \sin n\vartheta)\ ^{1)}. \quad (7)$$

Hence, equating to zero the coefficients of $\cos n\vartheta$ and $\sin n\vartheta$, we obtain an infinite system of equations relating to the coefficients a_j and b_j:

$$F_0(a_j, b_j) = 0, \quad F_n(a_j, b_j) = 0, \quad F_n^*(a_j, b_j) = 0, \quad (n = 1, 2, 3, \ldots) \quad (8)$$

Matters are analogous for the case when the curve is given in implicit but complex form:

$$\Phi(z, \bar{z}) = 0. \quad (6')$$

Substituting in (6′) for z series (2) and expanding everything in the positive and negative powers of $e^{i\vartheta}$, we obtain

$$\Phi\Big(\sum_{n=0}^{\infty} \alpha_n e^{in\vartheta}, \sum_{n=0}^{\infty} \bar{\alpha}_n e^{-in\vartheta}\Big) = \sum_{n=-\infty}^{+\infty} \Phi_n(a_j, b_j) e^{in\vartheta}, \quad (7')$$

whence, rather than system (8), we shall have:

$$\Phi_n(a_j, b_j) = 0 \qquad (n = \ldots -1, 0, 1, \ldots). \quad (8')$$

For the solution of the posed problem of the normal representation of the contour L it is sufficient that there be found such a system of numbers $a_0, b_0, a_1, b_1, \ldots$ as will satisfy equations (8) or (8′) and for which all the series we have been obliged to use shall converge, and moreover that the derivative of the series $\sum_{n=0}^{\infty} \alpha_n \zeta^n$ not vanish in the circle.

Systems (8) and (8′) will be especially simple for the particular case of regions B close to a circle, i.e., those for which the contour has the equation

$$x^2 + y^2 + \lambda P(x, y) = 1 \quad (9)$$

or

$$z\bar{z} + \lambda \Pi(z, \bar{z}) = 1, \quad (9')$$

where λ is some "small" parameter, and $P(x, y)$ and $\Pi(z, \bar{z})$ satisfy the same conditions that $F(x, y)$ and $\Phi(z, \bar{z})$ do. Both of these equations

---

[1]) By the symbol $F_n(a_j, b_j)$ we denote the fact that the function F_n depends on the numbers $a_0, b_0, a_1, b_1, \ldots$.

are equivalent, one of them passing into the other if z be replaced by $x + iy$ or, conversely, x and y be replaced by $\frac{z+\bar{z}}{2}$ and $\frac{z-\bar{z}}{2i}$.

It is therefore immaterial to us for which of these equations the computations are conducted. A system of equations of form (8), which we shall find by performing the calculations for (9), is obtained if in equations of form (8′) found for (9′), the real and imaginary parts be separated.

Let us consider, for example, equation (9′) in complex form. Obviously

$$z\bar{z} = \ldots + e^{-i\vartheta}(\alpha_0\bar{\alpha}_1 + \alpha_1\bar{\alpha}_2 + \ldots) + (\alpha_0\bar{\alpha}_0 + \alpha_1\bar{\alpha}_1 + \ldots) + \\ + e^{i\vartheta}(\alpha_1\bar{\alpha}_0 + \alpha_2\bar{\alpha}_1 + \ldots) + \ldots$$

In the expansion of the function $\Pi(z, \bar{z})$ in a series of the powers of $e^{i\vartheta}$, the coefficients of the conjugate quantities $e^{i\vartheta}$ and $e^{-i\vartheta}$ must also be conjugates, since the parameter λ is real and from equation (9′) it is evident that $\Pi(z, \bar{z})$ must be real. Thus $\Pi(z, \bar{z})$ will have, after their values (2) are substituted for z and $\bar{z}$, the form:

$$\Pi(z, \bar{z}) = \sum_{n=0}^{\infty} \tau_n(a_j, b_j)e^{in\vartheta} + \sum_{n=1}^{\infty} \overline{\tau_n(a_j, b_j)}e^{-in\vartheta},$$

$$\tau_n(a_j, b_j) = t_n(a_j, b_j) + it_n^*(a_j, b_j), \quad t_0^*(a_j, b_j) = 0.$$

If in (9′) there be substituted for $z\bar{z}$ and $\Pi(z, \bar{z})$ the cited series, and the coefficients of the positive powers of $e^{i\vartheta}$ be compared, we arrive at the system:

$$\left.\begin{aligned} \alpha_0\bar{\alpha}_0 + \alpha_1\bar{\alpha}_1 + \alpha_2\bar{\alpha}_2 + \ldots + \lambda t_0(a_j, b_j) &= 1 \\ \alpha_1\bar{\alpha}_0 + \alpha_2\bar{\alpha}_1 + \alpha_3\bar{\alpha}_2 + \ldots + \lambda\tau_1(a_j, b_j) &= 0 \\ \alpha_2\bar{\alpha}_0 + \alpha_3\bar{\alpha}_1 + \alpha_4\bar{\alpha}_2 + \ldots + \lambda\tau_2(a_j, b_j) &= 0 \\ \ldots\ldots\ldots\ldots\ldots\ldots\ldots\ldots\ldots& \end{aligned}\right\}. \tag{10}$$

The equations that we shall obtain by equating the coefficients of the negative powers of $e^{i\vartheta}$ will be the conjugates of equations (10) and will give nothing new by comparison with them.

System (10) may be written in the form:

$$\left.\begin{aligned} \alpha_1\bar{\alpha}_1 &= 1 - \alpha_0\bar{\alpha}_0 - \alpha_2\bar{\alpha}_2 - \ldots - \lambda t_0(a_j, b_j) \\ \alpha_2 &= -\frac{\alpha_1\bar{\alpha}_0}{\bar{\alpha}_1} - \frac{\alpha_3\bar{\alpha}_2}{\bar{\alpha}_1} - \ldots - \frac{\lambda}{\bar{\alpha}_1}\tau_1(a_j, b_j) \\ \alpha_3 &= -\frac{\alpha_2\bar{\alpha}_0}{\bar{\alpha}_1} - \frac{\alpha_4\bar{\alpha}_2}{\bar{\alpha}_1} - \ldots - \frac{\lambda}{\bar{\alpha}_1}\tau_2(a_j, b_j) \\ \ldots\ldots\ldots&\ldots\ldots\ldots\ldots\ldots\ldots \end{aligned}\right\}. \tag{11}$$

The method of successive approximations may be employed for its solution.

The assumptions that we will now make with respect to the curve L_λ with equation (9′) and about the transforming function, are not necessary to our problem, being made only to lighten the computations. We shall consider that the origin of coordinates $\zeta = 0$ lies within all curves L_λ when λ varies in some interval about $\lambda = 0$. In addition, we will seek a transforming function satisfying the conditions

$$z(0) = 0, \quad z'(0) > 0. \tag{12}$$

In accordance with this, we must put $\alpha_0 = 0$ in system (11) and consider α_1 to be a positive real number.

As the initial system of values for the a_j in our problem it is natural to take that which corresponds to the case $\lambda = 0$. Then the curve L will be a circle of unit radius, and the matter will accordingly be one of the identical transformation of this circle:

$$z = \zeta.$$

It is therefore possible to start from the following initial values of the α_i in the computations:

$$\alpha_1^{(0)} = 1, \ \alpha_2^{(0)} = \alpha_3^{(0)} = \ldots = 0.$$

Substituting them in the right side of system (11), we shall obtain first approximations for the α_j, which will differ from the initial ones by a first correction of order λ.

The first equation presents a certain peculiarity: from it we can find the square of α_1:

$$\alpha_1^2 = 1 - \lambda t_0(a_j^{(0)}, b_j^{(0)}).$$

Hence

$$\alpha_1 = (1 - \lambda t_0)^{\frac{1}{2}}.$$

In our computations it will probably be expedient to expand the value of α_1 found into a power series in λ, and to preserve only those terms in the expansion with not higher than the first power of λ:

$$\alpha_1 = 1 - \tfrac{1}{2}\lambda t_0(a_j^{(0)}, b_j^{(0)}).$$

We substitute the first approximation anew in the right side and compute the second approximation, in which we preserve a correction of the order λ^2 only. Continuing this process, we find the kth approximation in the form:

$$\alpha_1 = 1 + \lambda\alpha_1^{(1)} + \ldots + \lambda^k\alpha_1^{(k)},$$

$$\alpha_j = \lambda\alpha_j^{(1)} + \ldots + \lambda^k\alpha_j^{(k)}. \qquad (j = 2, 3, \ldots).$$

As regards the convergence of this process it can be shown, as we shall see somewhat later, that it will obtain at least for small values of the parameter λ.

3. Examples. 1. We will illustrate our reasoning in some examples. We will seek the function mapping the unit circle onto the interior of the ellipse:

$$x^2 + y^2 - \lambda(x^2 - y^2) = 1,$$

whose semi-axes are $(1 - \lambda)^{-\frac{1}{2}}$ and $(1 + \lambda)^{-\frac{1}{2}}$, the function being such that the center shall pass to the center and the real axis to the real axis:

$$\alpha_0 = 0, \quad \alpha_1 \text{ is a real number.}$$

By virtue of the fact that the x-axis is an axis of symmetry of the ellipse, all α_j will be real numbers:

$$b_1 = b_2 = \ldots = 0.$$

Finally, since the axis y is also an axis of symmetry, all a_j with even indices must equal zero:

$$a_2 = a_4 = \ldots = 0.$$

The equation of the ellipse in complex form is:

$$z\bar{z} - \lambda \frac{z^2 + \bar{z}^2}{2} = 1,$$

$$\Pi(z, \bar{z}) = -\frac{z^2 + \bar{z}^2}{2},$$

and since on the contour

$$z^2 = (a_1 e^{i\vartheta} + a_3 e^{3i\vartheta} + \ldots)^2 = a_1^2 e^{2i\vartheta} + 2a_1 a_3 e^{4i\vartheta} + (2a_1 a_5 + a_3^2) e^{6i\vartheta} + \\ + (2a_1 a_7 + 2a_3 a_5) e^{8i\vartheta} + (2a_1 a_9 + 2a_3 a_7 + a_5^2) e^{10i\vartheta} + \ldots,$$

the functions τ_j are defined by the equations:

$$\tau_0 = \tau_1 = \tau_3 = \tau_5 = \ldots = 0,$$

$$\tau_2 = -\tfrac{1}{2}a_1^2, \ \tau_4 = -a_1 a_3, \ \tau_6 = -(a_1 a_5 + \tfrac{1}{2}a_3^2), \ \tau_8 = -(a_1 a_7 + a_3 a_5), \ \ldots.$$

System (11) will therefore be:

$$a_1^2 = 1 - a_3^2 - a_5^2 - a_7^2 - a_9^2 - a_{11}^2 - \ldots,$$

$$a_3 = -\frac{1}{a_1}\left(a_5 a_3 + a_7 a_5 + a_9 a_7 + a_{11} a_9 + \ldots - \frac{\lambda}{2} a_1^2\right),$$

$$a_5 = -\frac{1}{a_1}(a_7 a_3 + a_9 a_5 + a_{11} a_7 + \ldots - \lambda a_1 a_3),$$

$$a_7 = -\frac{1}{a_1}[a_9 a_3 + a_{11} a_5 + \ldots - \lambda(a_1 a_5 + \tfrac{1}{2}a_3^2)],$$

$$a_9 = -\frac{1}{a_1}[a_{11} a_3 + \ldots - \lambda(a_1 a_7 + a_3 a_5)],$$

$$a_{11} = -\frac{1}{a_1}[a_{13} a_3 + \ldots - \lambda(a_1 a_9 + a_3 a_7 + \tfrac{1}{2}a_5^2)],$$

. . . .

Let us introduce the new unknowns

$$\varrho_0 = a_1, \quad \varrho_1 = \frac{a_3}{a_1}, \quad \varrho_2 = \frac{a_5}{a_1}, \quad \varrho_3 = \frac{a_7}{a_1} \dots$$

For them the system acquires the form:

$$\begin{aligned}
\varrho_0 &= (1 + \varrho_1^2 + \varrho_2^2 + \varrho_3^2 + \varrho_4^2 + \varrho_5^2 + \dots)^{-\frac{1}{2}},\\
\varrho_1 &= \tfrac{1}{2}\lambda - \varrho_1\varrho_2 - \varrho_2\varrho_3 - \varrho_3\varrho_4 - \varrho_4\varrho_5 - \dots,\\
\varrho_2 &= \lambda\varrho_1 - \varrho_1\varrho_3 - \varrho_2\varrho_4 - \varrho_3\varrho_5 - \dots,\\
\varrho_3 &= \lambda(\varrho_2 + \tfrac{1}{2}\varrho_1^2) - \varrho_1\varrho_4 - \varrho_2\varrho_5 - \dots,\\
\varrho_4 &= \lambda(\varrho_3 + \varrho_1\varrho_2) - \varrho_1\varrho_5 - \dots,\\
\varrho_5 &= \lambda(\varrho_4 + \varrho_1\varrho_3 + \tfrac{1}{2}\varrho_2^2) - \dots,\\
&\dots\dots\dots\dots\dots\dots
\end{aligned}$$

Leaving ϱ_0 undetermined for the time being and discarding the first equation of the system, we can find from the rest, by successive approximations, $\varrho_1, \varrho_2, \varrho_3, \dots$, adopting as the initial approximations

$$\varrho_1 = \varrho_2 = \varrho_3 = \dots = 0.$$

We summarize the results of the computations in the table:

Quantity	Initial Value	First Approx.	Second Approx.	Third Approx.	Fourth Approx.	Fifth Approx.
ϱ_1	0	$\frac{1}{2}\lambda$	$\frac{1}{2}\lambda$	$\frac{1}{2}\lambda - \frac{1}{4}\lambda^3$	$\frac{1}{2}\lambda - \frac{1}{4}\lambda^3$	$\frac{1}{2}\lambda - \frac{1}{4}\lambda^3 + \frac{3}{32}\lambda^5$
ϱ_2	0	0	$\frac{1}{2}\lambda^2$	$\frac{1}{2}\lambda^2$	$\frac{1}{2}\lambda^2 - \frac{9}{16}\lambda^4$	$\frac{1}{2}\lambda^2 - \frac{9}{16}\lambda^4$
ϱ_3	0	0	0	$\frac{5}{8}\lambda^3$	$\frac{5}{8}\lambda^3$	$\frac{5}{8}\lambda^3 - \frac{9}{8}\lambda^5$
ϱ_4	0	0	0	0	$\frac{7}{8}\lambda^4$	$\frac{7}{8}\lambda^4$
ϱ_5	0	0	0	0	0	$\frac{21}{16}\lambda^5$

Using these values of $\varrho_1, \varrho_2, \dots$. we find ϱ from the first equation, accurate to the fifth power of λ:

$$\varrho_0 = 1 - \tfrac{1}{2}(\varrho_1^2 + \varrho_2^2 + \dots) + \frac{\frac{1}{2}\cdot\frac{3}{2}}{1\cdot 2}(\varrho_1^2 + \varrho_2^2 + \dots)^2 + \dots = 1 - \tfrac{1}{8}\lambda^2 + \tfrac{3}{128}\lambda^4.$$

The mapping function will be, accurate to λ^5,

$$\begin{aligned}
z = \varrho_0\zeta(1 + \varrho_1\zeta^2 + \varrho_2\zeta^4 + \dots) &= \\
&= (1 - \tfrac{1}{8}\lambda^2 + \tfrac{3}{128}\lambda^4)\Big[\zeta + (\tfrac{1}{2}\lambda - \tfrac{1}{4}\lambda^3 + \tfrac{3}{32}\lambda^5)\zeta^3 + \\
&\quad + (\tfrac{1}{2}\lambda^2 - \tfrac{9}{16}\lambda^4)\zeta^5 + (\tfrac{5}{8}\lambda^3 - \tfrac{9}{8}\lambda^5)\zeta^7 + \tfrac{7}{8}\lambda^4\zeta^9 + \tfrac{21}{16}\lambda^5\zeta^{11}\Big].
\end{aligned}$$

In the particular case of the ellipse

$$\frac{x^2}{5} + \frac{y^2}{3} = \frac{4}{15},$$

we find $\lambda = \frac{1}{4}$, thanks to which the mapping function will be, if we limit ourselves to two decimal places in the coefficients,

$$z = 0.99[\zeta + 0.12\zeta^3 + 0.03\zeta^5 + 0.01\zeta^7].$$

More exact calculations show that the curve that is the image of the circle deviates from the ellipse by a quantity not greater than 0.02.

2. The method developed by us for determining the mapping function can be applied in many cases when the contour bounding the region is composed of pieces of several algebraic curves. As an example of such applications we will solve the problem of the mapping of the circle onto a square. We shall have in view a square with sides of length 2 parallel to the axes of coordinates, with center at the origin of coordinates.

Let us again assume a passage from center to center:

$$\alpha_0 = 0,$$

and the preservation of the real axes:

$$\arg \alpha_1 = 0.$$

The equation of the sides of the square will be:

$$(x^2 - 1)(y^2 - 1) = 0$$

or

$$x^2 + y^2 - x^2y^2 = 1.$$

In complex form it will be

$$z\bar{z} + \left(\frac{z^2 - \bar{z}^2}{4}\right)^2 = 1.$$

By introducing the parameter λ, we shall be considering the family of curves for which the square is obtained for $\lambda = 1$:

$$z\bar{z} + \lambda\left(\frac{z^2 - \bar{z}^2}{4}\right)^2 = 1.$$

The axes of symmetry here will be not only the coordinate axes, but also the bisectors of the coordinate angles; besides the equations $b_1 = b_2 = b_3 = \ldots = 0$ and $a_2 = a_4 = a_6 = \ldots = 0$ there will also be fulfilled here $a_3 = a_7 = a_{11} = \ldots = 0$. The normal representation (2) for z we shall therefore seek in the form:

$$z = a_1e^{i\vartheta} + a_5e^{5i\vartheta} + a_9e^{9i\vartheta} + a_{13}e^{13i\vartheta} + \ldots,$$

$$\Pi(z, \bar{z}) = \left(\frac{z^2 - \bar{z}^2}{4}\right)^2 = \frac{1}{16}[a_1^2e^{2i\vartheta} + 2a_1a_5e^{6i\vartheta} + (2a_1a_9 + a_5^2)e^{10i\vartheta} + \ldots - a_1^2e^{-2i\vartheta} -$$

$$- 2a_1a_5e^{-6i\vartheta} - (2a_1a_9 + a_5^2)e^{-10i\vartheta} + \ldots]^2,$$

whence it is evident that

$$\tau_0\,(a_j, b_j) = -\tfrac{1}{2}\left[\left(\frac{a_1^2}{2}\right)^2 + (a_1a_5)^2 + (a_1a_9 + \tfrac{1}{2}a_5^2)^2 + (a_1a_{13} + a_5a_9)^2 + \dots\right],$$

$$\tau_4\,(a_j, b_j) = -\tfrac{1}{2}\left[-\tfrac{1}{2}\left(\frac{a_1^2}{2}\right)^2 + (a_1a_5)\,\frac{a_1^2}{2} + (a_1a_9 + \tfrac{1}{2}a_5^2)(a_1a_5) + \dots\right],$$

$$\tau_8\,(a_j, b_j) = -\tfrac{1}{2}\left[-\tfrac{1}{2}a_1^2(a_1a_5) + (a_1a_9 + \tfrac{1}{2}a_5^2)\,\frac{a_1^2}{2} + (a_1a_{13} + a_5a_9)(a_1a_5) + \dots\right],$$

$$\tau_{12}(a_j, b_j) = -\tfrac{1}{2}\left[-\frac{a_1^2}{2}\,(a_1a_9 + \tfrac{1}{2}a_5^2) - \tfrac{1}{2}(a_1a_5)^2 + (a_1a_{13} + a_5a_9)\,\frac{a_1^2}{2} + \dots\right].$$

System of equations (11) for the determination of the a_j takes the form:

$$a_1^2 = 1 - a_5^2 - a_9^2 - a_{13}^2 - \dots +$$
$$+ \frac{\lambda}{2}\left[\left(\frac{a_1^2}{2}\right)^2 + (a_1a_5)^2 + (a_1a_9 + \tfrac{1}{2}a_5^2)^2 + (a_1a_{13} + a_5a_9)^2 + \dots\right],$$

$$a_5 = -\frac{a_5a_9}{a_1} - \frac{a_9a_{13}}{a_1} - \dots + \frac{\lambda}{2a_1}\left[-\tfrac{1}{2}\left(\frac{a_1^2}{2}\right)^2 + a_1a_5\,\frac{a_1^2}{2} + \right.$$
$$\left. + (a_1a_9 + \tfrac{1}{2}a_5^2)a_1a_5 + (a_1a_{13} + a_5a_9)(a_1a_9 + \tfrac{1}{2}a_1a_5^2) + \dots\right],$$

$$a_9 = -\frac{a_5a_{13}}{a_1} - \dots +$$
$$+ \frac{\lambda}{2a_1}\left[-\frac{a_1^2}{2}\,(a_1a_5) + (a_1a_9 + \tfrac{1}{2}a_5^2)\tfrac{1}{2}a_1^2 + (a_1a_{13} + a_5a_9)a_1a_5 + \dots\right],$$

$$a_{13} = -\frac{a_5a_{17}}{a_1} - \dots +$$
$$+ \frac{\lambda}{2a_1}\left[-\frac{a_1^2}{2}\,(a_1a_9 + \tfrac{1}{2}a_5^2) - \tfrac{1}{2}(a_1a_5)^2 + (a_1a_{13} + a_5a_9)\frac{a_1^2}{2} + \dots\right].$$

The result of its solution by the method of successive approximations, if

$$a_1 = 1,\ a_5 = a_9 = \dots = 0$$

be taken as the initial values and one limits oneself to three approximations, is summarized in the following table:

Quantity	Initial Value	First Approx.	Second Approx.	Third Approx.
a_1	1	$1 + \frac{1}{16}\lambda$	$1 + \frac{1}{16}\lambda + \frac{3}{256}\lambda^2$	$1 + \frac{1}{16}\lambda + \frac{1}{256}\lambda^2 + \frac{3}{1024}\lambda^3$
a_5	0	$-\frac{1}{16}\lambda$	$-\frac{1}{16}\lambda - \frac{7}{256}\lambda^2$	$-\frac{1}{16}\lambda - \frac{7}{256}\lambda^2 - \frac{11}{1024}\lambda^3$
a_9	0	0	$\frac{1}{64}\lambda^2$	$\frac{1}{64}\lambda^2 + \frac{27}{2048}\lambda^3$
a_{13}	0	0	0	$-\frac{11}{2048}\lambda^3$

from which it is seen that the mapping function will be, approximately,

$$z = \left(1 + \frac{\lambda}{16} + \frac{3\lambda^2}{256} + \frac{3\lambda^3}{1024}\right)\zeta - \left(\frac{\lambda}{16} + \frac{7\lambda^2}{256} + \frac{11\lambda^3}{1024}\right)\zeta^5 + \\ + \left(\frac{\lambda^2}{64} + \frac{27\lambda^3}{2048}\right)\zeta^9 - \frac{11\lambda^3}{2048}\zeta^{13}.$$

Putting $\lambda = 1$ in particular, we find approximately the function mapping the unit circle onto the square under consideration:

$$z = 1.077\zeta - 0.1006\zeta^5 + 0.0288\zeta^9 - 0.0054\zeta^{13}.$$

It can be shown that the exact solution of the problem is given by the following elliptic integral:

$$z = \frac{\int_0^\zeta (1 + t^4)^{-\frac{1}{2}} dt}{\int_0^1 (1 + t^4)^{-\frac{1}{2}} dt} = 1.080\left(\zeta - \tfrac{1}{10}\zeta^5 + \tfrac{1}{24}\zeta^9 - \tfrac{5}{208}\zeta^{13} + \ldots\right).$$ [1]

Comparison of it with the preceding shows that the approximate value of the first coefficient differs from the exact by not more than 0.003, and the second by not more than 0.01.

In the majority of practical problems the parameter λ is given numerically. The introduction of it as an indeterminate, followed by the determination of the expansions of the coefficients in powers of λ is not entirely convenient, since the determination of these expansions requires substantial algebraic calculations. To avoid this inconvenience, one can elect a somewhat different path, namely, that of substituting at once for λ its numerical value required in the given problem, and afterwards, beginning with a definite system of first approximations for the sought coefficients, substitute them in the right side of the system, find the second approximations, again substitute them in the right side, find the third approximations, and so forth, until the required accuracy in the computation of the coefficients has been attained.

We will employ this method for the solution of the system of equations of Example 2, substituting there the value $\lambda = 1$ and performing the computations with an accuracy of 0.0001. We have summarized the results of the computations in the following table:

Quantity	Initial Value	First Approx.	Second Approx.	Third Approx.	Fourth Approx.	Fifth Approx.
a_1	1	1.0607	1.0748	1.0785	1.0795	1.0807
a_5	0	−0.0625	−0.0922	−0.1014	−0.1069	−0.1081
a_9	0	0	0.0181	0.0319	0.0402	0.0450
a_{13}	0	0	−0.0022	−0.0074	−0.0167	−0.0242
a_{17}	0	0	0	0.0021	0.0050	0.0174
a_{21}	0	0	0	−0.0003	−0.0013	−0.0126

$$z = 1.0807\zeta - 0.1081\zeta^5 + 0.0450\zeta^9 - 0.0242\zeta^{13} + 0.0174\zeta^{17} - 0.0126\zeta^{21}.$$

[1] See, for example, V. I. Smirnov, [1], vol. III, 1933, p. 353.

4. The method of successive approximations for regions close to a circle, the contour of the regions being given by an implicit equation. We shall first expound the method of successive approximations in its simplest form, in application to regions close to a circle. We shall give a generalization of this method for regions of another form in the following Nos. Let the contour L_λ bounding the region B_λ depend on the parameter λ and have for its equation

$$x^2 + y^2 + \lambda P(x, y) = 1 \tag{13}$$

or, in complex form,

$$z\bar{z} + \lambda \Pi(z, \bar{z}) = 1, \quad \Pi(x + iy, x - iy) = P(x, y). \tag{13'}$$

$P(x, y)$ and $\Pi(z, \bar{z})$, as before, are analytic functions of x and y near the points of the circumference $x^2 + y^2 = 1$.

We shall seek the function accomplishing the conformal transformation of the unit circle into the region B_λ. For this it is sufficient, as we know, to find the normal representation of the curve L_λ, i.e., such a parametric representation of it,

$$x = x(\vartheta, \lambda), \quad y = y(\vartheta, \lambda), \tag{14}$$

that $x(\vartheta, \lambda)$ and $y(\vartheta, \lambda)$ are conjugate functions of ϑ.

The method of successive approximations is applicable to finding the required representation, but not to computing the separate coefficients of the expansions of these functions in trigonometric series in ϑ, nor to finding the expansions into series of powers of the parameter λ, either of the functions $x(\vartheta, \lambda)$, $y(\vartheta, \lambda)$ themselves, or of $z(\vartheta, \lambda) = x(\vartheta, \lambda) + iy(\vartheta, \lambda)$.

For this purpose let us put

$$\begin{aligned} x(\vartheta, \lambda) &= x_0(\vartheta) + \lambda x_1(\vartheta) + \ldots + \lambda^n x_n(\vartheta) + \ldots, \\ y(\vartheta, \lambda) &= y_0(\vartheta) + \lambda y_1(\vartheta) + \ldots + \lambda^n y_n(\vartheta) + \ldots, \\ z(\vartheta, \lambda) &= z_0(\vartheta) + \lambda z_1(\vartheta) + \ldots + \lambda^n z_n(\vartheta) + \ldots, \end{aligned} \tag{15}$$

where $x_n(\vartheta)$ and $y_n(\vartheta)$ are conjugate functions:

$$y_n(\vartheta) = \widetilde{x_n(\vartheta)} \quad (n = 0, 1, 2, \ldots) \tag{16}$$

and $x_0(\vartheta) + iy_0(\vartheta) = z_0(\vartheta)$ is the function giving the normal representation of the curve L_λ for $\lambda = 0$, i.e., the circumference of the unit circle. The form of this function will depend on which of the transformations of the unit circle into itself we choose. To simplify all the calculations, we shall stop at once with the simplest of such transformations — the identical transformation $z = \zeta$; we shall encounter transformations of general form in the next No.

Then

$$x_0(\vartheta) = \cos\vartheta, \quad y_0(\vartheta) = \sin\vartheta \text{ and } z_0(\vartheta) = e^{i\vartheta}.$$

Moreover, we shall assume for the sake of simplicity that for any λ the origin of coordinates $\zeta = 0$ passes to the origin of coordinates of the z-plane. All the functions $z_n(\vartheta)$ will then be boundary values of functions regular in the unit circle and vanishing for $\zeta = 0$.

In addition, we shall consider that $z'(0) > 0$ for all λ. Therefore in the expansion of $z_n(\vartheta)$ in a series of the powers of $e^{i\vartheta}$ the coefficient of $e^{i\vartheta}$ must be real.

We shall seek the functions $x_n(\vartheta)$, $y_n(\vartheta)$ and $z_n(\vartheta)$ by successive approximations. Since the formulas for the determination of $x_n(\vartheta)$ and $y_n(\vartheta)$ will be obtained from the corresponding formulas for $z_n(\vartheta)$ by separating the real and imaginary parts in them, it will be sufficient for us to perform all calculations for $z_n(\vartheta)$.

Let us substitute in equation (13′) the expansion for $z(\vartheta, \lambda)$ and compare the coefficients of λ:

$$(z_0 + \lambda z_1 + \ldots)(\bar{z}_0 + \lambda \bar{z}_1 + \ldots) + \\ + \lambda \Pi(z_0 + \lambda z_1 + \ldots, \bar{z}_0 + \lambda \bar{z}_1 + \ldots) = 1$$

$$\bar{z}_0 z_1 + \bar{z}_1 z_0 = -\Pi(z_0, \bar{z}_0) = -\tau_1.$$

In the left side of the equation stands the sum of two conjugate numbers and it can accordingly be rewritten in the form

$$2\mathrm{Re}\,[\bar{z}_0 z_1] = -\tau_1.$$

For the construction of the number $\bar{z}_0 z_1$ it remains only for us to indicate its imaginary part. But to do this is very easy. Since each of the z_k must have the form

$$z_k = \sum_{n=1}^{\infty} \alpha_n^{(k)} e^{in\vartheta},$$

the product of any z_k by $\bar{z}_0 = e^{-i\vartheta}$ will not contain negative powers of $e^{i\vartheta}$ and therefore for the construction of $\mathrm{I}\,(\bar{z}_0 z_1)$ is it sufficient for us to take just the function $-\tilde{\tau}_1$ [1]), conjugate, in our sense of the word, to $-\tau_1$:

$$2\mathrm{I}\,(\bar{z}_0 z_1) = -\tilde{\tau}_1.$$

Consequently

$$2\bar{z}_0 z_1 = -(\tau_1 + i\tilde{\tau}_1) \text{ and } z_1 = -\tfrac{1}{2} e^{i\vartheta}(\tau_1 + i\tilde{\tau}_1). \tag{17}$$

In the same way as we have found z_1, the value of z_n can be found by the known $z_1, z_2, \ldots, z_{n-1}$. To this end we compare the coefficients of λ^n in equation (13′) after having introduced into it series (15) for z.

---

[1]) The function $\tilde{\tau}_1$ will contain one arbitrary quantity, namely the free term in its Fourier series. The same thing can be said with respect to all the other quantities τ_n that we will encounter below. However, all these constants are equal to zero by virtue of the fact that the coefficients of $e^{i\vartheta}$ in the expansion of all $z_n(\vartheta)$ in powers of $e^{i\vartheta}$ must be real.

We obtain an equation of the form:

$$\bar{z}_0 z_n + \bar{z}_n z_0 = -\tau_n,$$

where

$$\tau_n = (\bar{z}_1 z_{n-1} + \ldots + \bar{z}_{n-1} z_1) + \\ + \frac{1}{(n-1)!}\left\{\frac{d^{n-1}}{d\lambda^{n-1}} \Pi(z_0 + \lambda z_1 + \ldots + \lambda^{n-1} z_{n-1}, \\ \bar{z}_0 + \lambda \bar{z}_1 + \ldots + \lambda^{n-1} \bar{z}_{n-1})\right\}_{\lambda=0}.$$

In accordance with the previous considerations,

$$2\mathrm{Re}\,[z_0 z_n] = -\tau_n,$$
$$2\mathrm{I}\,[\bar{z}_0 z_n] = -\widetilde{\tau}_n, \tag{18}$$
$$z_n = -\tfrac{1}{2} e^{i\vartheta}(\tau_n + i\widetilde{\tau}_n).$$

We shall now indicate the formulas from which $x_n(\vartheta)$ and $y_n(\vartheta)$ can be calculated. For this we separate the real and imaginary parts in (17) and (18).

Since $P(x, y) = \Pi(z, \bar{z})$, we have

$$\tau_1 = \Pi(z_0, \bar{z}_0) = P(x_0, y_0),$$
$$\cdots\cdots\cdots\cdots$$
$$\tau_n = \frac{1}{n!}\left\{\frac{d^n}{d\lambda^n}\left[(\lambda x_1 + \ldots + \lambda^{n-1} x_{n-1})^2 + (\lambda y_1 + \ldots + \lambda^{n-1} y_{n-1})^2 + \\ + \lambda P(x_0 + \ldots + \lambda^{n-1} x_{n-1}, y_0 + \ldots + \lambda^{n-1} y_{n-1})\right]\right\}_{\lambda=0}.$$

Therefore formulas (17) and (18) will give for $x_1, y_1, \ldots, x_n, y_n$:

$$x_1 = -\frac{\tau_1 \cos\vartheta - \widetilde{\tau}_1 \sin\vartheta}{2}, \quad y_1 = -\frac{\tau_1 \sin\vartheta + \widetilde{\tau}_1 \cos\vartheta}{2}, \tag{17'}$$
$$\cdots\cdots\cdots\cdots\cdots\cdots\cdots\cdots$$
$$x_n = -\frac{\tau_n \cos\vartheta - \widetilde{\tau}_n \sin\vartheta}{2}, \quad y_n = -\frac{\tau_n \sin\vartheta + \widetilde{\tau}_n \cos\vartheta}{2}. \tag{18'}$$

Example. We will solve by the method of successive approximations the problem of the mapping of a circle onto an ellipse, which has already been considered by us earlier (see No. 3):

$$z\bar{z} - \frac{\lambda}{2}(z^2 + \bar{z}^2) = 1, \quad z_0 = e^{i\vartheta},$$

$$\tau_1 = -\tfrac{1}{2}(e^{2i\vartheta} + e^{-2i\vartheta}) = -\cos 2\vartheta,$$

$$\widetilde{\tau}_1 = -\sin 2\vartheta = -\frac{1}{2i}(e^{2i\vartheta} - e^{-2i\vartheta}),$$

$$\tau_1 + i\widetilde{\tau}_1 = -e^{2i\vartheta}, \quad z_1 = \tfrac{1}{2} e^{3i\vartheta};$$

$$\tau_2 = \frac{1}{2!}\left\{\frac{d^2}{d\lambda^2}\left[\left(e^{i\vartheta} + \frac{\lambda}{2}e^{3i\vartheta}\right)\left(e^{-i\vartheta} + \frac{\lambda}{2}e^{-3i\vartheta}\right) - \right.\right.$$

$$\left.\left. - \frac{\lambda}{2}\left(e^{i\vartheta} + \frac{\lambda}{2}e^{3i\vartheta}\right)^2 - \frac{\lambda}{2}\left(e^{-i\vartheta} + \frac{\lambda}{2}e^{-3i\vartheta}\right)^2\right]\right\}_{\lambda=0} =$$

$$= \tfrac{1}{4} - \tfrac{1}{2}e^{4i\vartheta} - \tfrac{1}{2}e^{-4i\vartheta} = \tfrac{1}{4} - \cos 4\vartheta,$$

$$\widetilde{\tau_2} = -\sin 4\vartheta, \quad \tau_2 + i\widetilde{\tau_2} = \tfrac{1}{4} - e^{4i\vartheta},$$

$$z_2 = -\tfrac{1}{8}e^{i\vartheta} + \tfrac{1}{2}e^{5i\vartheta}.$$

Continuing by this method, we find:

$$z = \zeta + \frac{\lambda}{2}\zeta^3 + \lambda^2(-\tfrac{1}{8}\zeta + \tfrac{1}{2}\zeta^5) + \lambda^3(-\tfrac{5}{16}\zeta^3 + \tfrac{5}{8}\zeta^7) + \lambda^4(\tfrac{3}{128}\zeta - \tfrac{5}{8}\zeta^5 + \tfrac{7}{8}\zeta^9) + \ldots,$$

which coincides perfectly with the result obtained earlier.

5. The method of successive approximations for regions close to those into which the conformal transformation of a circle is known. Let

$$F(x, y, \lambda) = 0, \tag{19}$$

or, in complex form,

$$\Phi(z, \bar{z}, \lambda) = 0 \tag{19'}$$

be the equation of the contour L_λ bounding a finite region B_λ that depends on the parameter λ. For $\lambda = 0$ curve (19) reduces to the curve L_0:

$$F(x, y, 0) = 0,$$

which bounds the region B_0. We shall assume that L_0 is a simple closed curve without singular points, i.e., such that

$$[F_x(x, y, 0)]^2 + [F_y(x, y, 0)]^2 \neq 0,$$

if the point (x, y) lies on L_0.

Let us in addition assume that there is known to us the function $\chi(\zeta)$ accomplishing the conformal transformation of the unit circle into the region B_0. Then we can give the normal parametric representation of the contour L_0:

$$x = x_0(\vartheta), \; y = y_0(\vartheta), \text{ where } y_0(\vartheta) = \widetilde{x_0}(\vartheta), \tag{21}$$

or

$$z = z_0(\vartheta).$$

We shall also make the assumption that $F(x, y, \lambda)$ is an analytic function of its arguments near the values $\lambda = 0$ and x and y lying on the curve L_0. In the theory of functions of a complex variable, for contours of the form of L_0 it is proved that the derivative of the function effecting the conformal transformation will have a modulus different from zero not only within the circumference, but also on it, i.e., that

$$|z_0'|^2 = x_0'^2(\vartheta) + y_0'^2(\vartheta) > 0.\ ^{1)} \tag{22}$$

We shall now seek the function effecting the conformal transformation of the circle into the region B_λ. For the determination of the normal parametric representation of its contour L_λ we shall put, following the method of the preceding No.,

$$\begin{aligned} x(\vartheta, \lambda) &= x_0(\vartheta) + \lambda x_1(\vartheta) + \ldots + \lambda^n x_n(\vartheta) + \ldots, \\ y(\vartheta, \lambda) &= y_0(\vartheta) + \lambda y_1(\vartheta) + \ldots + \lambda^n y_n(\vartheta) + \ldots, \end{aligned} \tag{22'}$$

or

$$z(\vartheta, \lambda) = z_0(\vartheta) + \lambda z_1(\vartheta) + \ldots + \lambda^n z_n(\vartheta) + \ldots. \tag{22''}$$

The functions $x_0(\vartheta), y_0(\vartheta), z_0(\vartheta)$ correspond to the case $\lambda = 0$ and coincide with those indicated in equation (21).

We shall now find $x_n(\vartheta), y_n(\vartheta), z_n(\vartheta)$ in succession. It is again sufficient to construct just the formulas for the determination of $z_n(\vartheta)$, since the corresponding equations for $x_n(\vartheta)$ and $y_n(\vartheta)$ are obtained by separating the real and the imaginary parts in them.

To simplify the computations we shall consider that the point $z = 0$ lies within B_λ, that the origin of coordinates of the ζ-plane passes to the point $z = 0$, and that the direction of the real axis at the point $\zeta = 0$ passes to the direction of the real axis at the point $z = 0$. In accordance with this, in the expansions of the $z_n(\vartheta)$ in powers of $e^{i\vartheta}$ the free terms will be absent, and the coefficients of the first powers of $e^{i\vartheta}$ will be real.

Let us first substitute for z in equation (19') expansion (22'') and equate to zero the coefficient of the first power of λ:

$$\left\{\frac{d}{d\lambda}\,\Phi(z_0 + \lambda z_1,\ \bar{z}_0 + \lambda \bar{z}_1,\ \lambda)\right\}_{\lambda=0} = 0,$$

$$\Phi_z(z_0, \bar{z}_0, 0) z_1 + \Phi_{\bar{z}}(z_0, \bar{z}_0, 0)\bar{z}_1 + \Phi_\lambda(z_0, \bar{z}_0, 0) = 0.$$

Denoting, for the sake of brevity,

$$\Phi_z(z_0, \bar{z}_0, 0) = \Psi,\ \Phi_{\bar{z}}(z_0, \bar{z}_0, 0) = \theta,\ \Phi_\lambda(z_0, \bar{z}_0, 0) = \tau_1, \tag{23}$$

we shall have

$$\Psi z_1 + \theta \bar{z}_1 = -\tau_1. \tag{24}$$

---

[1]) Here and everywhere below in this section we understand by z_0' the limit values on the unit circumference of the derivative $\frac{dz_0}{d\zeta}$ of the function $z_0 = \chi(\zeta)$ that gives the transformation of the unit circle into L_0:

$$z_0' = \frac{dz_0}{d\zeta} = \frac{dz_0}{d\vartheta}\cdot\frac{d\vartheta}{d\zeta} = -ie^{-i\vartheta}\frac{dz_0}{d\vartheta} = -ie^{-i\vartheta}(x_0'(\vartheta) + iy_0'(\vartheta)).$$

On the other hand, substituting z_0 and $\bar{z}_0$ for z and $\bar{z}$ in the equation of the curve L_0: $\Phi(z_0, \bar{z}_0, 0) = 0$, differentiating the result with respect to ϑ and multiplying by $(-i)$, we obtain

$$\Psi z_0' e^{i\vartheta} - \theta \bar{z}_0' e^{-i\vartheta} = 0,$$

where z_0' is the boundary value of the derivative $\dfrac{dz}{d\zeta}$.

Using this equation, we can rewrite equation (24) in the form:

$$\frac{z_1}{z_0' e^{i\vartheta}} + \frac{\bar{z}_1}{\bar{z}_0' e^{-i\vartheta}} = -\frac{\tau_1}{\Psi z_0' e^{i\vartheta}} = -\frac{\tau_1}{\theta \bar{z}_0' e^{-i\vartheta}} = \Lambda_1,$$

or

$$2\mathrm{Re}\left(\frac{z_1}{z_0' e^{i\vartheta}}\right) = \Lambda_1.$$

The function z_0', on the strength of the conformality of the transformation, will be the boundary value of a function different from zero everywhere in the unit circle. Therefore the quotient $\dfrac{z_1}{z_0' e^{i\vartheta}}$ will be the boundary value of a regular function:

$$\mathrm{I}\left[\frac{z_1}{z_0' e^{i\vartheta}}\right] = \tfrac{1}{2}\widetilde{\Lambda}_1,$$

and

$$z_1 = \frac{z_0' e^{i\vartheta}}{2}(\Lambda_1 + i\widetilde{\Lambda}_1). \tag{25}$$

$\widetilde{\Lambda}_1$ is found accurate to a constant summand. It must be determined from the condition that in the expansion of z_1 in powers of $e^{i\vartheta}$ the coefficient of the first power of $e^{i\vartheta}$ shall be real. If one takes into account the fact that the expansion of z_0' in powers of $e^{i\vartheta}$ does not contain negative powers of that quantity, it will be clear from the preceding formula that our condition for the determination of the constant is equivalent to the following one: if $\widetilde{\Lambda}_1$ be expanded in a trigonometric series in ϑ, the free term of the expansion must be absent.

We proceed in quite the same way when we wish to determine $z_n(\vartheta)$. Substituting for z in (19′) series (22″) and equating to zero the coefficient of λ^n in the expression obtained, we find:

$$\frac{1}{n!}\left\{\frac{d^n}{d\lambda^n}\Phi(z, \bar{z}, \lambda)\right\}_{\lambda=0} =$$

$$= \frac{1}{n!}\left\{\frac{d^n}{d\lambda^n}\Phi(z_0 + \lambda z_1 + \ldots + \lambda^n z_n,\ \bar{z}_0 + \lambda\bar{z}_1 + \ldots + \lambda^n \bar{z}_n, \lambda)\right\}_{\lambda=0} =$$

$$= \frac{1}{n!}\left\{\frac{d^n}{d\lambda^n}\Phi(z_0 + \ldots + \lambda^{n-1} z_{n-1},\ \bar{z}_0 + \ldots + \lambda^{n-1}\bar{z}_{n-1}, \lambda) + \right.$$

$$+ \Phi_z(z_0 + \ldots + \lambda^{n-1} z_{n-1}, \bar{z}_0 + \ldots + \lambda^{n-1}\bar{z}_{n-1}, \lambda)\lambda^n z_n +$$
$$+ \Phi_{\bar{z}}(z_0 + \ldots + \lambda^{n-1} z_{n-1}, \bar{z}_0 + \ldots + \lambda^{n-1}\bar{z}_{n-1}, \lambda)\lambda^n \bar{z}_n +$$
$$+ \Phi_{zz}(z_0 + \ldots + \lambda^{n-1} z_{n-1}, \bar{z}_0 + \ldots + \lambda^{n-1}\bar{z}_{n-1}, \lambda)(\lambda^n z_n)^2 +$$
$$+ \ldots\Big\}_{\lambda=0} =$$
$$= \Phi_z(z_0, \bar{z}_0, 0) z_n + \Phi_{\bar{z}}(z_0, \bar{z}_0, 0)\bar{z}_n +$$
$$+ \frac{1}{n!}\left\{\frac{d^n}{d\lambda^n}\Phi(z_0 + \ldots + \lambda^{n-1} z_{n-1}, \bar{z}_0 + \ldots + \lambda^{n-1}\bar{z}_{n-1}, \lambda)\right\}_{\lambda=0},$$

or, if we call

$$\tau_n = \frac{1}{n!}\left\{\frac{d^n}{d\lambda^n}\Phi(z_0 + \ldots + \lambda^{n-1} z_{n-1}, \bar{z}_0 + \ldots + \lambda^{n-1}\bar{z}_{n-1}, \lambda)\right\}_{\lambda=0},$$

$$\Psi z_n + \theta\bar{z}_n = -\tau_n.$$

Again using the equation $\Psi z_0' e^{i\vartheta} - \theta\bar{z}_0' e^{-i\vartheta} = 0$, we shall have the following set of equations, which requires no explanations,

$$\frac{z_n}{z_0' e^{i\vartheta}} + \frac{\bar{z}_n}{\bar{z}_0' e^{-i\vartheta}} = -\frac{\tau_n}{\Psi z_0' e^{i\vartheta}} = -\frac{\tau_n}{\theta\bar{z}_0' e^{-i\vartheta}} = \Lambda_n$$

$$2Re\left[\frac{z_n}{z_0' e^{i\vartheta}}\right] = \Lambda_n, \quad 2I\left[\frac{z_n}{z_0' e^{i\vartheta}}\right] = \tilde{\Lambda}_n$$

and we obtain at last

$$z_n = \frac{z_0' e^{i\vartheta}}{2}(\Lambda_n + i\tilde{\Lambda}_n). \tag{26}$$

The constant summand accurate to which $\tilde{\Lambda}_n$ is defined is again determined from the condition that in the expansion of z_n in powers of $e^{i\vartheta}$ the coefficient of the first power shall be real, or, what is the same thing, that the free term in the expansion of $\tilde{\Lambda}_n$ in a trigonometric series in ϑ shall be equal to zero.

Such will be the formulas permitting one to compute $z_1, z_2, \ldots$ in succession. For the determination of the corresponding formulas for $x_1, y_1, \ldots, x_n, y_n, \ldots$, we separate the real and imaginary parts in them.

Since $\Phi(z, \bar{z}, \lambda) = F(x, y, \lambda)$,

$$\tau_n = \frac{1}{n!}\left\{\frac{d^n}{d\lambda^n}\Phi(z_0 + \ldots + \lambda^{n-1} z_{n-1}, \bar{z}_0 + \ldots + \lambda^{n-1}\bar{z}_{n-1}, \lambda)\right\}_{\lambda=0} =$$
$$= \frac{1}{n!}\left\{\frac{d^n}{d\lambda^n}F(x_0 + \ldots + \lambda^{n-1} x_{n-1}, y_0 + \ldots + \lambda^{n-1} y_{n-1}, \lambda)\right\}_{\lambda=0} = T_n$$

and, by virtue of the obvious equalities

$$F_x = \Phi_z \frac{\partial z}{\partial x} + \Phi_{\bar z} \frac{\partial \bar z}{\partial x} = \Phi_z + \Phi_{\bar z},$$

$$F_y = \Phi_z \frac{\partial z}{\partial y} + \Phi_{\bar z} \frac{\partial \bar z}{\partial y} = i\Phi_z - i\Phi_{\bar z},$$

if we introduce the symbols $F_x(x_0, y_0, 0) = u$, $F_y(x_0, y_0, 0) = v$, we shall have:

$$\Psi = \Phi_z(z_0, \bar z_0, 0) = \tfrac{1}{2}(u - iv), \quad \theta = \tfrac{1}{2}(u + iv),$$

$$\Lambda_n = \frac{T_n}{\frac{i}{2}(u - iv)(x_0' + iy_0')} = -\frac{T_n}{\frac{i}{2}(u + iv)(x_0' - iy_0')};$$

but

$$\frac{y_0'}{x_0'} = -\frac{F_x(x_0, y_0, 0)}{F_y(x_0, y_0, 0)} = -\frac{u}{v},$$

and therefore

$$\Lambda_n = 2\frac{T_n x_0'}{v(x_0'^2 + y_0'^2)} = 2S_n.$$

Equations (25) and (26) then give

$$x_n = \operatorname{Re}\left[\frac{z_0' e^{i\theta}}{2}(2S_n + 2i\tilde{S}_n)\right] = x_0'\tilde{S}_n + y_0' S_n, \quad y_n = -x_0' S_n + y_0' \tilde{S}_n$$

$$(n = 1, 2, \ldots).$$

6. The method of successive approximations for curves given in parametric form. We shall now develop the method of successive approximations for the case when the curve bounding the region is given in parametric form. The trend of the reasoning is, in its basic idea, close to that already expounded in the preceding No., but it has its specific features, which compel us to carry through the reasoning over again for this case, although more briefly. We omit the simplest case of the region close to a circle, passing immediately to the more general case.

Thus let us assume that a simple closed curve L_λ, bounding a simply connected region B_λ, is given by the equations

$$x = \varphi(t, \lambda), \; y = \psi(t, \lambda). \tag{27}$$

We can assume the functions φ and ψ to have been expanded in trigonometric series

$$x = p_0 + \sum_{n=1}^{\infty} (p_n \cos nt + r_n \sin nt),$$

$$y = q_0 + \sum_{n=1}^{\infty} (q_n \cos nt - s_n \sin nt),$$

which, generally speaking, will not be conjugates. If, therefore, we form the quantity

$$z = x + iy = \pi_0 + \sum_{n=1}^{\infty} (\pi_n \cos nt + \varrho_n \sin nt)$$

$$\pi_n = p_n + iq_n, \quad \varrho_n = r_n - is_n$$

and in this series replace cos nt and sin nt by the powers of the number e^{it}, by Euler's formulas, we will obtain z as some function of e^{it}, the expansion of which into a series will contain not only the positive but also the negative powers of e^{it}:

$$z = \omega(e^{it}, \lambda) = \pi_0 + \sum_{n=1}^{\infty} \left(\frac{\pi_n}{2} - \frac{\varrho_n}{2} i\right) e^{int} + \sum_{n=1}^{\infty} \left(\frac{\pi_n}{2} + \frac{\varrho_n}{2} i\right) e^{-int}. \quad (28)$$

Below we shall assume the curve to have been given in precisely the latter form. The coefficients π_n and ϱ_n will, of course, depend on λ. Let us lay down the condition that for $\lambda = 0$ series (28) shall not contain negative powers of λ and that the expansion of $\omega(e^{it}, 0)$ in a trigonometric series shall give the normal representation of the curve L_0. We shall consider in addition that the curve L_0 does not have singular points.

As regards the function $\omega(w, \lambda)$, we shall assume that it is analytic with respect to λ and w near the values $\lambda = 0$ and $|w| = 1$.

If it should happen that the parameter t is the polar angle in the ζ-plane, i.e., $\zeta = |\zeta| e^{it}$, then there would be no negative powers of e^{it} in expansion (28), and we would have a normal series for z by which the mapping function could be constructed immediately. But the parameter t that we have adopted generally does not coincide with the polar angle ϑ, and our problem will be to pass, by some substitution, from t to ϑ, i.e., so to change the parameter in (28) that series (28) becomes normal after the substitution. We will take the dependence of t upon ϑ in the form

$$t = \vartheta + \lambda\varphi_1(\vartheta) + \lambda^2\varphi_2(\vartheta) + \ldots, \quad (29)$$

where $\varphi_1(\vartheta), \varphi_2(\vartheta), \ldots$ are certain real, periodic functions, with the determination of which we shall occupy ourselves below. The term free of λ we have taken in the form ϑ because, under the condition, series (28) shall be normal for $\lambda = 0$, and then t must certainly coincide with ϑ.

Let all $\varphi_j(\vartheta)$ for $j \leqslant n - 1$ be so defined that the coefficients of λ^j in the expansion of $\omega(e^{it}, \lambda)$ in powers of λ shall not contain negative powers of $e^{i\vartheta}$ in their Fourier series.

We now choose $\varphi_n(\vartheta)$ so that this holds also for $j = n$. Substitute in $\omega(e^{it}, \lambda)$ series (29) for t, and compute the coefficient of λ^n. It will equal:

$$\frac{1}{n!}\left\{\frac{d^n}{d\lambda^n}\,\omega(e^{it},\lambda)\right\}_{\lambda=0} = \frac{1}{n!}\left\{\frac{d^n}{d\lambda^n}\,\omega(e^{i(\vartheta+\lambda\varphi_1+\ldots+\lambda^n\varphi_n)},\lambda)\right\}_{\lambda=0} =$$

$$= \frac{1}{n!}\left\{\frac{d^n}{d\lambda^n}\,\omega(e^{i(\vartheta+\lambda\varphi_1+\ldots+\lambda^{n-1}\varphi_{n-1})}\,(1+i\lambda^n\varphi_n+\ldots),\lambda)\right\}_{\lambda=0}$$

Hence, calling

$$\frac{1}{n!}\left\{\frac{d^n}{d\lambda^n}\,\omega(e^{i(\vartheta+\ldots+\lambda^{n-1}\varphi_{n-1})},\lambda)\right\}_{\lambda=0} = A_n,$$

we find that the sought coefficient will be:

$$\omega_\zeta(e^{i\vartheta},0)ie^{i\vartheta}\cdot\varphi_n + A_n = i\omega_\zeta(e^{i\vartheta},0)e^{i\vartheta}\left(\varphi_n + \frac{A_n e^{-i\vartheta}}{i\omega_\zeta(e^{i\vartheta},0)}\right).\text{[1]}$$

By the choice of φ_n, we must contrive that the expression obtained for the coefficient shall not contain negative powers of $e^{i\vartheta}$ in its Fourier series. But this will certainly be the case if the function standing in parentheses does not contain them, since the multiplier standing before it does not contain the negative powers of $e^{i\vartheta}$, by the condition. Let

$$\frac{A_n e^{-i\vartheta}}{i\omega_\zeta(e^{i\vartheta},0)} = \sum_{n=0}^{\infty} c_n e^{in\vartheta} + \sum_{n=1}^{\infty} c'_n\cdot e^{-in\vartheta}.$$

From this it is evident that if we want to contrive the absence of negative powers of $e^{i\vartheta}$ in the parentheses, there must figure in the composition of the function φ_n the series $-\sum_{n=1}^{\infty} c'_n e^{-in\vartheta}$, containing all the negative powers of $e^{i\vartheta}$ of the second summand of the parentheses. On the other hand, since φ_n is to be real, its part containing positive powers of $e^{i\vartheta}$ must reduce without fail to $-\sum_{n=1}^{\infty}\bar{c}'_n e^{in\vartheta}$. There thus remains arbitrary in the expression for φ_n only the real free term:

$$\varphi_n = \gamma_n - \sum_{n=1}^{\infty}\bar{c}'_n e^{in\vartheta} - \sum_{n=1}^{\infty} c'_n e^{-in\vartheta}. \tag{30}$$

By the choice of it we must manage to attain the given conditions of the correspondence of the points. In the example solved below we consider $\gamma_n = 0$.

Example. Let us take the curve given by the polar equation $r = 1 + \lambda\cos nt$. Its equation in complex form will be:

$$z = e^{it} + \lambda(\tfrac{1}{2}e^{i(n+1)t} + \tfrac{1}{2}e^{-i(n-1)t}).$$

---

[1]) By the symbol ω_ζ we understand the derivative $\dfrac{d\omega}{d\zeta}$.

Let us substitute here series (29) for t and collect the coefficients of like powers of λ:

$$e^{i(\vartheta+\lambda\vartheta_1+\dots)} + \frac{\lambda}{2}\{e^{i(n+1)(\vartheta+\lambda\varphi_1+\dots)} + e^{-i(n-1)(\vartheta+\lambda\varphi_1+\dots)}\} =$$

$$= e^{i\vartheta} + \lambda[e^{i\vartheta}i\varphi_1 + \tfrac{1}{2}(e^{i(n+1)\vartheta} + e^{-i(n-1)\vartheta})] + \lambda^2[e^{i\vartheta}(-\tfrac{1}{2}\varphi_1^2 + i\varphi_2) +$$
$$+ \tfrac{1}{2}e^{i(n+1)\vartheta}\cdot i(n+1)\varphi_1 - \tfrac{1}{2}e^{-i(n-1)\vartheta}i(n-1)\varphi_1] + \dots,$$

whence it is evident that in order that negative powers of $e^{i\vartheta}$ be absent from the first brackets, φ_1 must equal (if we discard the arbitrary real summand)

$$\frac{i}{2}\,e^{-in\vartheta} - \frac{i}{2}\,e^{in\vartheta} = \sin n\vartheta.$$

As regards φ_2, for it there is obtained the equation:

$$\varphi_2 = i\,\frac{2n-1}{8}\,(e^{-2in\vartheta} - e^{2in\vartheta}) = \frac{2n-1}{4}\sin 2n\vartheta,$$

etc.

After substitution in the preceding series of the values of $\varphi_1, \varphi_2, \dots$, we shall have, if we limit ourselves to the second power of λ, for the boundary value of z the following normal representation:

$$z = e^{i\vartheta} + \lambda e^{i(n+1)\vartheta} + \frac{2n+1}{4}\lambda^2(e^{i(2n+1)\vartheta} - e^{i\vartheta}).$$

Thus, accurate to λ^2, the mapping function will be approximately equal to

$$z = \left(1 - \frac{2n+1}{4}\lambda^2\right)\zeta + \lambda\zeta^{n+1} + \frac{2n+1}{4}\lambda^2\zeta^{2n+1}.$$

Observation. In the method we have expounded, to set up the function giving the conformal transformation of the circle into the region, it is initially necessary to find a family of curves, dependent on the parameter λ, such that for the curve obtained from this family for $\lambda = 0$ the mapping function shall be known, and such that for some value of the parameter $\lambda = \lambda_1$ there will be obtained the given curve. Such a family can usually be chosen in a variety of ways. We shall note one of them, which proves to be useful in the solution of particular examples. We shall consider the case when the contour of the region can be given by a polar equation:

$$\varrho = f(t) \qquad (0 \leqslant t \leqslant 2\pi).$$

Let a be the radius of a certain circle with center at the pole, and which is wholly situated within L. Then as the sought family one can adopt the following:

$$\varrho = a + \lambda[f(t) - a] \qquad (0 \leqslant t \leqslant 2\pi).$$

Indeed, for $\lambda = 0$ this family gives the circumference, and for $\lambda = 1$, the given curve.

7. Proof of the convergence of the process of successive approximations [1]. We shall prove that the series (15), (22′), (29) constructed by us converge, in any case for values of the parameter λ that are of sufficiently small modulus, absolutely and uniformly with respect to the variable ϑ. We shall first concern ourselves with the case when the contour of the region is given in implicit form.

In view of the fact that series (15) are a particular case of series (22′), and, finally, since series (22′) are obtained from series (22″) by separating real and imaginary parts, it will be sufficient for us to establish the convergence of series (22″) only.

We will introduce at the outset the terminology that we shall use in the proofs. We will call the norm of the function $\varphi(t)$, expanded in a trigonometric series:

$$\varphi(t) = \sum_{n=0}^{\infty} (a_n \cos nt + b_n \sin nt),$$

the sum of the moduli of its Fourier coefficients:

$$\|\varphi(t)\| = \sum_{n=0}^{\infty} (|a_n| + |b_n|).$$

We note some of the properties of the norm, proof of which we leave to the reader:

1) $\|\varphi_1(t) + \varphi_2(t) + \ldots\| \leqslant \|\varphi_1(t)\| + \|\varphi_2(t)\| + \ldots,$

2) $\|\varphi_1(t) \cdot \varphi_2(t)\| \leqslant \|\varphi_1(t)\| \cdot \|\varphi_2(t)\|,$

3) $\|\widetilde{\varphi(t)}\| \leqslant \|\varphi(t)\|$ [2].

We present in addition the following lemma, which also will prove to be useful to us.

LEMMA I. *If the coefficients of the two power series*

$$\sum_{n=1}^{\infty} a_n \xi^n, \quad \sum_{n=1}^{\infty} b_n \xi^n$$

satisfy the inequalities:

$$|a_n| \leqslant \frac{A}{(n+1)^\alpha}, \quad |b_n| \leqslant \frac{B}{(n+1)^\alpha}, \quad (n = 1, 2, 3, \ldots, \alpha > 1),$$

then the coefficients c_n of the power series that is their product satisfy the

---

[1]) In recent years the problem of the convergence of successive approximations on the basis of the same ideas as are applied in this No. has been investigated by G. M. Goluzin, [1].

[2]) By $(\widetilde{\varphi t})$ we understand here the trigonometric series conjugate to the trigonometric series for $\varphi(t)$, but without a free term.

inequalities:

$$|c_n| \leqslant \frac{AB}{(n+1)^\alpha} \cdot M_\alpha,$$

where M_α is a constant depending only on α.

Indeed:

$$|c_n| = \left| \sum_{\nu=1}^{n-1} a_\nu b_{n-\nu} \right| \leqslant AB \sum_{\nu=1}^{n-1} \frac{1}{(\nu+1)^\alpha (n+1-\nu)^\alpha} =$$

$$= AB \left\{ 2 \sum_{\nu=1}^{\left[\frac{n-1}{2}\right]} \frac{1}{(\nu+1)^\alpha (n+1-\nu)^\alpha} + (\text{when } n \text{ is even}) \frac{1}{\left(\frac{n}{2}+1\right)^{2\alpha}} \right\} =$$

$$= \frac{AB}{(n+1)^\alpha} \left\{ 2 \sum_{\nu=1}^{\left[\frac{n-1}{2}\right]} \frac{1}{(\nu+1)^\alpha \left(1 - \frac{\nu}{n+1}\right)^\alpha} + \right.$$

$$\left. + (\text{when } n \text{ is even}) \frac{1}{\left(\frac{n}{2}+1\right)^\alpha \left(1 - \frac{n}{2(n+1)}\right)^\alpha} \right\} <$$

$$< \frac{AB}{(n+1)^\alpha} \cdot 2^{\alpha+1} \sum_{\nu=1}^{\infty} \frac{1}{(\nu+1)^\alpha} < \frac{2^{\alpha+1} AB}{(n+1)^\alpha} \int_1^\infty \frac{dx}{x^\alpha} =$$

$$= \frac{2^{\alpha+1} AB}{(n+1)^\alpha (\alpha - 1)} = \frac{AB}{(n+1)^\alpha} M_\alpha,$$

where $M_\alpha = \dfrac{2^{\alpha+1}}{\alpha - 1}$, Q.E.D.

In particular, for $\alpha = 2$ we find $M_2 = 8$. It is not difficult to verify by a more exact estimate that this constant can be replaced by a lesser one: $M_2 = 1.520$.

We shall now state in greater detail the assumptions about the function $\Phi(z, \bar{z}, \lambda)$, figuring in equation (19′) of the contour L_λ, under which we will prove the convergence of the process of successive approximations.

For $\lambda = 0$ equation (19′) gives the curve L_0, as regards which we considered that it contained the origin of coordinates, was devoid of singular points, and that there was known the function $\chi(z)$ conformally transforming the circle $|\zeta| < 1$ into the region B_0 bounded by L_0, such that

$$\chi(0) = 0, \quad \chi'(0) < 0.$$

We will denote by D_0 a certain neighborhood of the curve L_0 and by $\bar{D}_0$ the region obtained from D_0 by reflecting it in the real axis x, i.e., formed of the points conjugate to the points of region D_0.

We shall assume that $\Phi(z, w, \lambda)$ is a regular analytic function of its arguments when z varies in D_0, w in $\bar{D}_0$, and λ is less in modulus than a certain λ_0:

$$z \in D_0, \quad w \in \bar{D}_0, \quad |\lambda| < \lambda_0.$$

In view of the analyticity of the contour L_0 and the absence from it of singular points, the function $\chi(z)$ will be regular not only in the unit circle but also in a circle of radius greater than unity.

The boundary values of $\chi(z)$ on the unit circumference we have previously designated by $z_0(\vartheta)$:

$$z_0(\vartheta) = \chi(e^{i\vartheta}).$$

Let us consider the function $\Phi(z_0(\vartheta) + \xi, \overline{z_0(\vartheta)} + \eta, \lambda)$. It will be a regular function of ξ and η when the points $z_0(\vartheta) + \xi$ and $\overline{z_0(\vartheta)} + \eta$ belong to D_0 and $\bar{D}_0$ respectively. The point $z_0(\vartheta)$ lies on L_0 and $\overline{z_0(\vartheta)}$ on $\bar{L}_0$. Let us denote by r_1 the distance from L_0 to the boundary of D_0. The points $z_0(\vartheta) + \xi$ and $\overline{z_0(\vartheta)} + \eta$ will certainly belong to D_0 and $\bar{D}_0$ respectively if it turns out that ξ and η are less in modulus than r_1.

We can therefore say that for any values of ϑ the function we have under consideration will be a regular analytic function of ξ, η and λ in the region

$$|\xi| < r_1, \quad |\eta| < r_1, \quad |\lambda| < \lambda_0. \tag{31}$$

Let us expand it in a power series in ξ, η, λ,

$$\Phi(z_0(\vartheta) + \xi, \overline{z_0(\vartheta)} + \eta, \lambda) = \sum_{i,j,k=0}^{\infty} A_{ik}^{(j)}(\vartheta)\xi^i\eta^k\lambda^j. \tag{31'}$$

Its coefficients will be some functions of ϑ,

$$A_{ik}^{(j)}(\vartheta) = \frac{1}{i!k!j!}\left\{\frac{\partial^{i+k+j}\Phi}{\partial z^i \partial w^k \partial \lambda^j}\right\}_{z=z_0,\ w=\bar{z}_0,\ \lambda=0}.$$

It is seen directly that these coefficients are analytic periodic functions of ϑ. They have a finite norm.

Lemma II. *If the function $\Phi(z, w, \lambda)$ satisfies the requirements enumerated above, then the series*

$$\sum_{i,j,k=0}^{\infty} \|A_{ik}^{(j)}(\vartheta)\| \xi^i\eta^k\lambda^j \tag{32}$$

converges in region (31).

We shall take an arbitrary positive number r_2, less than r_1, and consider the function [1] $\Phi\left(\chi(\zeta)+\xi, \bar{\chi}\left(\frac{1}{\zeta}\right)+\eta, \lambda\right)$. When ζ varies in the neighborhood of the unit circumference, the points $\chi(\zeta)$ and $\bar{\chi}\left(\frac{1}{\zeta}\right)$ will belong to the neighborhoods of L_0 and $\bar{L}_0$. We can choose such a narrow annular neighborhood of the unit circumference,

$$\frac{1}{\varrho}<|\zeta|<\varrho, \ \varrho>1,$$

that the points $\chi(\zeta)$ and $\bar{\chi}\left(\frac{1}{\zeta}\right)$ are removed from L_0 and $\bar{L}_0$ by a distance less than $r_1 - r_2$, when ζ is varied in this annulus. Given such a choice of ϱ the points $\chi(\zeta)+\xi$ and $\bar{\chi}\left(\frac{1}{\zeta}\right)+\eta$ will certainly belong to D_0 and $\bar{D}_0$ as soon as the moduli of ξ and η are less than r_2.

We can therefore affirm that the function under consideration will be a regular analytic function of ζ, ξ, η, λ in the region

$$\frac{1}{\varrho}<|\zeta|<\varrho, \ |\xi|<r_2, \ |\eta|<r_2, \ |\lambda|<\lambda_0.$$

Let us expand it in a power series in all its variables. Here the series with respect to ξ, η, λ will be Taylor's series, but the series with respect to ζ will be of the Laurent type,

$$\Phi\left(\chi(\zeta)+\xi, \bar{\chi}\left(\frac{1}{\zeta}\right)+\eta, \lambda\right)=\sum_{i,j,k=0}^{\infty} \sum_{l=-\infty}^{\infty} A_{i,k,l}^{(j)} \xi^i \eta^k \lambda^j \zeta^l. \tag{33}$$

If in this series each of the coefficients $A_{i,k,l}^{(j)}$ be replaced by its modulus, the newly obtained power series

$$\sum_{i,j,k=0}^{\infty} \sum_{l=-\infty}^{+\infty} |A_{i,k,l}^{(j)}| \xi^i \eta^k \lambda^j \zeta^l$$

will converge in the same region as does the previous series. In particular, even the following series, obtained for $\zeta = 1$, will converge:

$$\sum_{i,k,j=0}^{\infty} \left(\sum_{l=-\infty}^{+\infty} |A_{i,k,l}^{(j)}|\right) \xi^i \eta^k \lambda^j. \tag{34}$$

We shall show that this latter series, multiplied by 2, is dominant for (32). Indeed, for $\zeta = e^{i\vartheta}$ we shall have

$$\chi(e^{i\vartheta}) = z_0(\vartheta), \ \bar{\chi}\left(\frac{1}{e^{i\vartheta}}\right) = \bar{\chi}(e^{-i\vartheta}) = \overline{z_0(\vartheta)}.$$

---

[1] If $\chi(\zeta)=\sum_{\nu=0}^{\infty} \alpha_\nu \zeta^\nu$, then by $\bar{\chi}(\zeta)$ we understand the series

$$\bar{\chi}(\zeta)=\sum_{\nu=0}^{\infty} \bar{\alpha}_\nu \zeta^\nu.$$

The function $\Phi\left(\chi(\zeta) + \xi,\ \bar{\chi}\left(\frac{1}{\zeta}\right) + \eta,\ \lambda\right)$ reduces to $\Phi(z_0 + \xi, \bar{z}_0 + \eta, \lambda)$, and therefore the coefficients $A^{(j)}_{i,k}(\vartheta)$ and $A^{(j)}_{i,k,l}$ are connected by the equation

$$A^{(j)}_{i,k}(\vartheta) = \sum_{l=-\infty}^{\infty} A^{(j)}_{i,k,l}\, e^{il\vartheta}.$$

Hence it is evident that

$$\|A^{(j)}_{i,k}(\vartheta)\| \leqslant 2 \sum_{l=-\infty}^{\infty} |A^{(j)}_{i,k,l}|,$$

and, accordingly, the series (34), doubled, is indeed dominant for (32).

As we said above, series (34) converges in the region $|\xi| < r_2$, $|\eta| < r_2$, $|\lambda| < \lambda_0$, where r_2 is any number less than r_1. Its coefficients will not depend on r_2, and therefore it will also be convergent in region (31). But if the majorant series (34) converges in this region, then our series (32) will also converge at least in the same region. With this Lemma II is proved.

Simultaneously with series (32), there will also converge in the same region (31) the series

$$\Omega(\xi, \eta, \lambda) = \sum_{\substack{i,j,k=0 \\ i+j+k>2}}^{\infty} B^{(j)}_{i,k}\, \xi^i \eta^k \lambda^j, \tag{32$_1$}$$

where

$$B^{(j)}_{i,k} = \|A^{(j)}_{i,k}\|\, (j+1)^\alpha.$$

We note that the functions $\dfrac{e^{-i\vartheta}}{z'_0 \Phi_z(z_0, \bar{z}_0, 0)}$ and $z'_0 e^{i\vartheta}$ are analytic periodic functions of ϑ. Therefore the norms of these functions are finite;[1]) we will denote them by

$$\mu = \left\| \frac{e^{-i\vartheta}}{z'_0 \Phi_z(z_0, \bar{z}_0, 0)} \right\|, \quad \mu_1 = \|z'_0 e^{i\vartheta}\|. \tag{35}$$

Let us now proceed to the proof of the convergence of series (22″), for which we shall estimate the function $z_n(\vartheta)$. We shall assume that for $l < n$ the estimate

$$\|z_l(\vartheta)\| \leqslant \frac{CR^{l-1}}{(l+1)^\alpha} \qquad (\alpha > 1) \tag{36}$$

---

[1]) As regards the function $\dfrac{1}{z'_0}$ this is obvious because $|z'_0| \neq 0$. And from the equality $F(x, y, \lambda) = \Phi(z, \bar{z}, \lambda)$ there follow: $F_x = \Phi_z + \Phi_{\bar{z}}$, $F_y = i\Phi_z - i\Phi_{\bar{z}}$, whence $\Phi_z = \frac{1}{2}(F_x - iF_y)$ and $|\Phi_z|^2 = \frac{1}{4}(F_x^2 + F_y^2)$, and therefore from the inequality assumed earlier, $F_x^2(x_0, y_0, 0) + F_y^2(x_0, y_0, 0) > 0$, follows the validity of the statement.

is true, and prove that this estimate is true also for $l = n \geqslant 2$. We shall speak below about how the constants C and R are to be chosen.

Let us estimate $\|\tau_n\|$ at the outset. Using (31′) and (36), we have:

$$\|\tau_n\| = \left\| \frac{1}{n!} \left\{ \frac{d^n}{d\lambda^n} \Phi(z_0 + \ldots + \lambda^{n-1} z_{n-1}, \bar{z}_0 + \ldots + \lambda^{n-1} \bar{z}_{n-1}, \lambda) \right\}_{\lambda=0} \right\| \leqslant$$

$$\leqslant \sum_{i,j,k=0}^{\infty} \|A_{ik}^{(j)}\| \left\| \left\{ \frac{d^n}{d\lambda^n} \frac{1}{n!} (\lambda z_1 + \ldots + \lambda^{n-1} z_{n-1})^i (\lambda \bar{z}_1 + \ldots + \lambda^{n-1} \bar{z}_{n-1})^k \lambda^j \right\}_{\lambda=0} \right\| \leqslant$$

$$\leqslant \sum_{\substack{i,j,k=0 \\ i+j+k \geqslant 2}}^{j \leqslant n} \|A_{ik}^{(j)}\| \left\{ \frac{1}{(n-j)!} \frac{d^{n-j}}{d\lambda^{n-j}} \left(\frac{C}{2^\alpha} \lambda + \frac{CR}{3^\alpha} \lambda^2 + \ldots + \frac{CR^{n-2}}{n^\alpha} \lambda^{n-1} \right)^{i+k} \right\}_{\lambda=0} \text{[1]}.$$

The last brace standing under the sum sign is the coefficient of λ^{n-j} in the expression:

$$\left(\frac{C}{2^\alpha} \lambda + \frac{CR}{3^\alpha} \lambda^2 + \ldots + \frac{CR^{n-2}}{n^\alpha} \lambda^{n-1} \right)^{i+k} =$$

$$= \left(\frac{C}{R} \right)^{i+k} \left(\frac{R\lambda}{2^\alpha} + \frac{(R\lambda)^2}{3^\alpha} + \ldots + \frac{(R\lambda)^{n-1}}{n^\alpha} \right)^{i+k}.$$

On the basis of Lemma I, applying it to the case of $i + k$ factors, it is clear that this coefficient will be not greater than

$$\left(\frac{C}{R} \right)^{i+k} \frac{R^{n-j}}{(n - j + 1)^\alpha} M_\alpha^{i+k}.$$

Using this, and also (32), we can continue the estimate of the expression of interest to us, $\|\tau_n\|$ [2]:

$$\|\tau_n\| \leqslant \sum_{\substack{i,j,k=0 \\ i+j+k \geqslant 2}}^{j \leqslant n} \frac{1}{(j+1)^\alpha} B_{ik}^{(j)} R^{n-i-j-k} (CM_\alpha)^{i+k} \frac{1}{(n - j + 1)^\alpha} \leqslant$$

$$\leqslant \frac{R^n}{(n+1)^\alpha} \sum_{\substack{i,j,k=0 \\ i+j+k \geqslant 2}}^{\infty} B_{ik}^{(j)} \left(\frac{1}{R} \right)^j \left(\frac{CM_\alpha}{R} \right)^i \left(\frac{CM_\alpha}{R} \right)^k =$$

$$= \frac{R^n}{(n+1)^\alpha} \Omega\left(\frac{CM_\alpha}{R}, \frac{CM_\alpha}{R}, \frac{1}{R} \right).$$

---

[1]) In the transition to the last part of the inequality the condition $i + j + k \geqslant 0$ accompanying the sum sign has been replaced by $i + j + k \geqslant 2$ on this basis, that the terms with powers of λ up to $n - 1$ which are obtained for $i + j + k < n$ upon differentiating n times with respect to λ will certainly fall out, and n is, by the condition, not less than 2.

[2]) To obtain the last part of the inequality, we have replaced the quantity $\frac{1}{(j+1)(n-j+1)}$ by the not less $\frac{1}{n+1}$.

Moreover, from the equality

$$\Delta_n = -\frac{\tau_n e^{-i\vartheta}}{z_0'\Psi} = -\frac{\tau_n e^{-i\vartheta}}{z_0'\Phi_z(z_0, \bar{z}_0, 0)}$$

it follows that

$$\|\Delta_n\| \leqslant \mu \|\tau_n\| \leqslant \mu \frac{R^n}{(n+1)^\alpha} \Omega\left(\frac{CM_\alpha}{R}, \frac{CM_\alpha}{R}, \frac{1}{R}\right),$$

and, accordingly [see formula (26)]:

$$\|z_n\| \leqslant \mu_1 \|\Delta_n\| \leqslant \mu\mu_1 \frac{R^n}{(n+1)^\alpha} \Omega\left(\frac{CM_\alpha}{R}, \frac{CM_\alpha}{R}, \frac{1}{R}\right). \tag{37}$$

Let us now deal with the choice of the constants C and R. We shall take the quantity C so large that it will fulfill the inequality:

$$\|z_1\| \leqslant \frac{C}{2^\alpha}, \tag{38}$$

i.e., that inequality (36) will be fulfilled for $l = 1$.

In order that this inequality be fulfilled for any l, it is sufficient to choose, as is clear from (37), R so large that we will have:

$$\frac{CM_\alpha}{R} < r_1, \quad \frac{1}{R} < \lambda_0,$$

$$\mu\mu_1 \frac{R^n}{(n+1)^\alpha} \Omega\left(\frac{CM_\alpha}{R}, \frac{CM_\alpha}{R}, \frac{1}{R}\right) \leqslant \frac{CR^{n-1}}{(n+1)^\alpha}. \tag{39}$$

For the fulfillment of the last of these inequalities, it is required that there be satisfied the relation:

$$\Omega\left(\frac{CM_\alpha}{R}, \frac{CM_\alpha}{R}, \frac{1}{R}\right) \leqslant \frac{C}{\mu\mu_1} \frac{1}{R}, \tag{40}$$

which can always be satisfied by choosing R sufficiently large, since for $\frac{1}{R} = 0$ the function Ω, series (32_1) for which contains terms of not lower than the second degree, has a double root, whereas on the right stands $\frac{1}{R}$ to the first power.

Thus inequality (36) has been established for all l for the C and R that we have chosen, and by that fact the convergence of series (22″) has been established for

$$|\lambda| \leqslant \frac{1}{R}. \tag{41}$$

Example. In No. 4 we found the function conformally transforming the circle into the ellipse $z\bar{z} - \frac{\lambda}{2}(z^2 + \bar{z}^2) = 1$. Let us estimate in this problem the region of convergence of our algorithm:

$$\Phi(z, \bar{z}, \lambda) = z\bar{z} - \frac{\lambda}{2}(z^2 + \bar{z}^2) - 1 = (z - e^{it})(\bar{z} - e^{-it}) +$$

$$+ e^{it}(\bar{z} - e^{-it}) + e^{-it}(z - e^{it}) - \frac{\lambda}{2}(z - e^{it})^2 - \frac{\lambda}{2}(\bar{z} - e^{-it})^2 -$$

$$- \lambda e^{it}(z - e^{it}) - \lambda e^{-it}(\bar{z} - e^{-it}) - \frac{\lambda}{2}(e^{2it} + e^{-2it}).$$

From this we find, if we take $\alpha = 2$,

$$\Omega(\xi, \eta, \lambda) = \xi\eta + 2\lambda(\xi^2 + \eta^2 + 2\xi + 2\eta);$$

since in the given case, as we saw earlier, $z_1 = \frac{1}{2}e^{3i\vartheta}$, we can adopt $C = 4$.

Moreover, $z_0 = e^{i\vartheta}$, $z_0' = 1$; therefore $\mu_1 = 2$. On the other hand $\Phi_z(z_0, \bar{z}_0, 0) = \bar{z}_0$ and consequently

$$\mu = \left\| \frac{e^{i\vartheta}}{z_0' \Phi_z(z_0, \bar{z}_0, 0)} \right\| = 2.$$

Earlier we mentioned the fact that $M_2 = 1.52$; inequality (40) will consequently give:

$$\frac{(1.52 \cdot 4)^2}{R^2} + \frac{2}{R}\left[\frac{(1.52 \cdot 4)^2}{R^2} + \frac{(1.52 \cdot 4)^2}{R^2} + 2\frac{1.52 \cdot 4}{R} + 2\frac{1.52 \cdot 4}{R}\right] \leqslant \frac{1}{R},$$

which is equivalent to the following:

$$\frac{147.9}{R^2} + \frac{85.61}{R} \leqslant 1,$$

for which it is sufficient to put $R \geqslant 87.3$. Thus the successive approximations will certainly converge for

$$|\lambda| \leqslant 0.011 < \frac{1}{R}.$$

In fact the boundary of convergence will of course be broader.

Observation. We shall yet note the following. Let us have found, in any other way, the parametric representation of the contour L_λ in the form of a series arrayed in powers of λ:

$$z(t, \lambda) = \sum_{n=1}^{\infty} \lambda^n z_n^*(t),$$

where $z_n^*(t)$ are functions the Fourier series expansion of which contain powers of e^{it} greater than the zero-th power. We saw above that in such a case the $z_n(t)$ are uniquely determined, since the arbitrariness arises only from the constants that appear in Λ_n, which are here all taken as equalling zero, and therefore the $z_n(t)$ must coincide with the $z_n^*(t)$. In particular, the method of infinite systems that we have developed in Nos 1—4 must give the same results; therefore, by having established the convergence of series (22′), we have simultaneously also proved the convergence of the method of infinite systems — that method in which we found power series in λ for the coefficients of the expansion in a power series of the transforming function.

Let us pass on to the proof of the convergence for sufficiently small λ of expansion (29) for the problem in parametric form.

Expanding $\omega(e^{it}, \lambda)$ in a power series in λ and $t - \vartheta$, we find:

$$\omega(e^{it}, \lambda) = \sum_{j=0}^{\infty} \sum_{i=0}^{\infty} A_{i,j}(\vartheta)\, \lambda^j (t - \vartheta)^i. \tag{42}$$

Moreover, putting

$$B_{ij} = \|A_{ij}(\vartheta)\| (j + i)^\alpha,$$

let us form the function:

$$\Omega^*(\xi, \lambda) = \sum_{\substack{i,j=0 \\ i+j \geqslant 2}}^{\infty} B_{ij} \xi^i \lambda^j.$$

As in the preceding case, it can be shown that the series on the right side will converge for ξ, λ sufficiently small in modulus.

Let us now assume that for $l < n$ there has been established the inequality:

$$\|\varphi_l(\vartheta)\| \leqslant \frac{CR^{l-1}}{(l+1)^\alpha}. \tag{43}$$

Then on performing transformations analogous to those that we made in estimating τ_n, we find that

$$\|A_n(\vartheta)\| \leqslant \frac{R^n}{(n+1)^\alpha}\, \Omega^*\left(\frac{CM_\alpha}{R}, \frac{1}{R}\right).$$

Introducing yet the symbol

$$\left\| \frac{e^{-i\vartheta}}{\omega_\zeta(e^{i\vartheta}, 0)} \right\| = \gamma,$$

we obtain, in accordance with equation (30), if we again discard the constant γ_n:

$$\|\varphi_n\| \leqslant 2\gamma \frac{R^n}{(n+1)^\alpha}\, \Omega^*\left(\frac{CM_\alpha}{R}, \frac{1}{R}\right). \tag{44}$$

Choosing C so as to fulfill the inequality:

$$C \geqslant 2^\alpha \|\varphi_1(\vartheta)\|,$$

and next R so large that

$$\Omega^*\left(\frac{CM_\alpha}{R}, \frac{1}{R}\right) \leqslant \frac{C}{2\gamma R}, \tag{45}$$

such a choice obviously being always possible, since Ω^* has for $\frac{1}{R} = 0$

a zero of the second order, we shall have established that (43) holds also for $l = n$.

Inequality (43) shows that series (29) certainly converges for those values of λ for which

$$|\lambda| < \frac{1}{R}.$$

In all these cases we have succeeded in establishing the convergence only for small values of λ, and the boundary given by inequalities (41) and (45) which we have found generally departs widely from the real region of convergence of the expansions in powers of λ. The actual region of convergence will be considerably wider than that indicated by us. Efforts to find the exact region of convergence of the expansions here obtained have until now yielded no results, the question remaining open.

It may happen that the value of the parameter λ that appears in the computations lies outside of the boundaries we have obtained. The method that has been developed here frequently leads, however, to the result even in this case. Then after having found the mapping function it will be necessary to verify the accuracy of the result. For this purpose it will be necessary to discover first of all how much the curve that is obtained as the result of the mapping of the unit circle by means of the function found, departs from the curve bounding the given region. This can either be done analytically, using the equations of these curves, or graphically, by constructing both curves and comparing them. One judges the accuracy of the approximate expression by the maximum distance between the two curves [1]).

In § 5, where we considered the method of expansion in a series involving a small parameter of the function giving the conformal transformation of the region into a circle, we did not carry out an investigation of the convergence of the power series in the parameter λ that was constructed there. We now find it possible to prove this proposition without difficulty.

Indeed, the type of contour having an equation of form (12) § 5, which we were considering in the investigations of that section, is a particular case of the contours for which we are conducting the investigation of the convergence of the process of successive approximations in the method just expounded for the case of the parametrically given contour, when the latter is close to the circumference.

If λ is small in modulus, the contour will be a simple analytic curve without angular points. The function effecting the transformation of the circle into the region within the contour will be holomorphic with respect to ζ both within the circle and on its boundary. We can invert the function single-valuedly, by virtue of the fact that the region is

[1]) See the note on p. 387.

single-sheeted. By the familiar theorems about implicit functions, the inverse function will be regular with respect to z in the entire region B, including the contour, and regular with respect to λ for small $|\lambda|$. One can consequently expand it in a power series in λ whose coefficients will be holomorphic functions of z in the region B.

It thus remains for us to prove that expansion (13), constructed by us in No. 2 § 5, coincides with the actual expansion. To establish this presents no difficulty of any kind. Indeed, let us assume that with the transformation the origin of coordinates passes to origin, and the direction of the real axis again passes to the direction of the real axis. Let us take the actual expansion of the transforming function in a power series in λ. It will have the form (13), where $P_k(z)$ will be certain functions of z which are regular in B_λ, including even the boundary L_λ.

Let us form the square of the modulus of ζ on the boundary of the region B. It must be identically equal to unity there. Equating to zero the coefficients of $\lambda, \lambda^2, \ldots,$ we shall obtain the system of equations that the coefficients of the expansion must satisfy. We formed this system before, in No. 2 § 5, and since the system constructed has, as we saw, a unique solution, the expansion we have constructed will coincide with the actual expansion. And by this is proved its convergence for small values of the parameter λ.

8. Observations on the mapping of a circle onto the exterior of a curve. Examples. The same idea can, of course, be used for the construction of the function conformally transforming the circle of unit radius into the exterior of some curve.

Let the boundary curve of the region possess the same properties as before, and be given by an equation of form (9), which we have considered in the preceding Nos.

The function accomplishing the necessary transformation of the circle into the exterior of the curve, if it be required that the center of the circle pass to the point at infinity, has here the following form:

$$z = x + iy = \alpha_1\zeta^{-1} + \alpha_2 + \alpha_3\zeta + \alpha_4\zeta^2 + \ldots, \quad \alpha_n = a_n + ib_n. \tag{46}$$

We can, as above, consider $b_1 = 0$. For $|\zeta| = 1$ the affix of the corresponding value of z must be situated on the boundary curve, and z will therefore satisfy the contour equation (9′). If in the expression

$$z = x + iy = \alpha_1 e^{-i\vartheta} + \alpha_2 + \alpha_3 e^{i\vartheta} + \ldots$$

the real and imaginary parts be separated, the trigonometric series for x and y will have to satisfy the contour equation (9). Carrying out the substitution in (9′) of the series for z or of the series for x and y in equation (9) of the contour and comparing the coefficients of like powers of $e^{i\vartheta}$ or of the cosines and sines of the angles of like multiplicity of ϑ, we shall obtain the system of equations for the determination of $\alpha_1, \alpha_2, \alpha_3, \ldots,$ which can be solved by the previous method. The calculations and the investigation of the convergence here differ little from those already presented in the preceding Nos. We shall therefore limit ourselves here to some examples only.

Example. We will map a circle onto the exterior of a square. Let this again be a square with sides parallel to the axes of coordinates, of length 2. The equation of its sides is

$$(x^2 - 1)(1 - y^2) = 0, \quad x^2 + y^2 - x^2y^2 = 1,$$

or in complex form

$$z\bar{z} + \left(\frac{z^2 - \bar{z}^2}{4}\right)^2 = 1.$$

We introduce the auxiliary family of curves dependent on the parameter λ,

$$z\bar{z} + \lambda\left(\frac{z^2 - \bar{z}^2}{4}\right)^2 = 1,$$

in which our square figures for $\lambda = 1$.

In view of the presence in the curves of four axes of symmetry, we seek the function z in the form:

$$z = a_1 e^{-i\vartheta} + a_5 e^{3i\vartheta} + a_9 e^{7i\vartheta} + \ldots,$$

where the a_k are real numbers.

Since

$$\frac{z^2 - \bar{z}^2}{4} = \tfrac{1}{2}\{-(\tfrac{1}{2}a_9^2 + \ldots)e^{-14i\vartheta} - (a_5a_9 + \ldots)e^{-10i\vartheta} -$$
$$- (a_1a_9 + \tfrac{1}{2}a_5^2)e^{-6i\vartheta} - (a_1a_5 - \tfrac{1}{2}a_1^2)e^{-2i\vartheta} + (a_1a_5 - \tfrac{1}{2}a_1^2)e^{2i\vartheta} +$$
$$+ (a_1a_9 + \tfrac{1}{2}a_5^2)e^{6i\vartheta} + (a_5a_9 + \ldots)e^{10i\vartheta} + (\tfrac{1}{2}a_9^2 + \ldots)e^{14i\vartheta} + \ldots\},$$

after substituting in the contour equation and comparing the coefficients we shall have the system of equations:

$$a_1^2 + a_5^2 + a_9^2 + \ldots = 1 + \frac{\lambda}{2}[(a_1a_5 - \tfrac{1}{2}a_1^2)^2 +$$
$$+ (a_1a_9 + \tfrac{1}{2}a_5^2)^2 + a_5^2a_9^2 + \tfrac{1}{4}a_9^4 + \ldots],$$

$$a_1a_5 + a_5a_9 + \ldots = -\frac{\lambda}{2}[\tfrac{1}{2}(a_1a_5 - \tfrac{1}{2}a_1^2)^2 - (a_1a_9 + \tfrac{1}{2}a_5^2)(a_1a_5 - \tfrac{1}{2}a_1^2) -$$
$$- a_5a_9(a_1a_9 + \tfrac{1}{2}a_5^2) - \tfrac{1}{2}a_9^2a_5a_9 - \ldots],$$

$$a_1a_9 + \ldots = -\frac{\lambda}{2}[(a_1a_5 - \tfrac{1}{2}a_1^2)(a_1a_9 + \tfrac{1}{2}a_5^2) - a_5a_9(a_1a_5 - \tfrac{1}{2}a_1) -$$
$$- \tfrac{1}{2}a_9^2(a_1a_9 + \tfrac{1}{2}a_5^2) - \ldots].$$

Taking as the initial values $a_1 = 1$, $a_5 = a_9 = \ldots = 0$, we solve this by the usual method of successive approximations. The respective steps of the calculations are exhibited in the table:

Quantity	Initial Value	First Approx.	Second Approx.	Third Approx.
a_1	1	$1 + \frac{1}{16}\lambda$	$1 + \frac{1}{16}\lambda + \frac{7}{256}\lambda^2$	$1 + \frac{1}{16}\lambda + \frac{7}{256}\lambda^2 + \frac{9}{1024}\lambda^3$
a_5	0	$-\frac{1}{16}\lambda$	$-\frac{1}{16}\lambda - \frac{7}{256}\lambda^2$	$-\frac{1}{16}\lambda - \frac{7}{256}\lambda^2 - \frac{19}{2048}\lambda^3$
a_9	0	0	0	$\frac{\lambda^3}{2048}$

Accurate to λ^3, the mapping function will thus be:

$$z = (1 + \tfrac{1}{16}\lambda + \tfrac{7}{256}\lambda^2 + \tfrac{9}{1024}\lambda^3)\frac{1}{\zeta} - (\tfrac{1}{16}\lambda + \tfrac{7}{256}\lambda^2 + \tfrac{19}{2048}\lambda^3)\zeta^3 + \frac{\lambda^3}{2048}\zeta^7.$$

Setting $\lambda = 1$ in this expression, we obtain:

$$z = \tfrac{1125}{1024}\frac{1}{\zeta} - \tfrac{203}{2048}\zeta^3 + \tfrac{1}{2048}\zeta^7.$$

As in Nos. 4—5, we can seek the expansion in powers of λ of the complex coordinate $z = x + iy$ of the points of the contour directly, rather than that of the separate coefficients of the Fourier series for $x(\vartheta)$ and $y(\vartheta)$.

Let us assume that the contour L_λ of the region is given by equation (19), or, what is the same thing, by (19′), and that the function effecting the transformation of the unit circle into the exterior of L_0 is known to us. We set:

$$z(\vartheta, \lambda) = z_0(\vartheta) + \lambda z_1(\vartheta) + \ldots + \lambda^n z_n(\vartheta) + \ldots \tag{47}$$

The expansion in a power series in $e^{i\vartheta}$ of each of the coefficients must contain the positive powers of this quantity and only the first negative power; $z_0(\vartheta)$, corresponding to the case $\lambda = 0$, we shall assume to be known.

Let us substitute series (47) in the equation of the contour (19′) and equate to zero the coefficients of the powers of λ.

For the coefficient of the first power of λ we obtain:

$$\left\{\frac{d}{d\lambda}\Phi(z_0 + \lambda z_1, \bar{z}_0 + \lambda\bar{z}_1, \lambda)\right\}_{\lambda=0} =$$

$$= \Phi_z(z_0, \bar{z}_0, 0)\, z_1 + \Phi_{\bar{z}}(z_0, \bar{z}_0, 0)\, \bar{z}_1 + \Phi_\lambda(z_0, \bar{z}_0, 0) = 0,$$

or, in the notation of (23):

$$\Psi z_1 + \theta\bar{z}_1 = -\tau_1.$$

If, though, we differentiate with respect to ϑ the equation of the curve L_0

$$\Phi(z_0, \bar{z}_0, 0) = 0,$$

we find:

$$\Psi z_0' e^{i\vartheta} - \theta \bar{z}_0' e^{-i\vartheta} = 0.$$

From that and the preceding equation it follows that

$$\frac{z_1}{z_0' e^{i\vartheta}} + \frac{\bar{z}_1}{\bar{z}_0' e^{-i\vartheta}} = - \frac{\tau_1}{\Psi z_0' e^{i\vartheta}} = - \frac{\tau_1}{\theta \bar{z}_0' e^{-i\vartheta}} = \Lambda_1.$$

We recall that z_0, and also z_1, are the boundary values of functions holomorphic everywhere in the unit circle with the exception of the point $\zeta = 0$, where the first of them certainly has a pole of the first order and the second may perhaps have one.

Therefore the ratio

$$\frac{z_1}{z_0' e^{i\vartheta}}$$

will be the boundary value of a function holomorphic everywhere for $|\zeta| \leqslant 1$. The boundary value of its real part is found from the preceding equation, which can be rewritten in the form:

$$2\mathrm{Re}\left[\frac{z_1}{z_0' e^{i\vartheta}}\right] = \Lambda_1.$$

The boundary value of the imaginary part will be:

$$\mathrm{I}\left[\frac{z_1}{z_0' e^{i\vartheta}}\right] = \tfrac{1}{2}\widetilde{\Lambda}_1.$$

Hence

$$z_1 = \frac{z_0' e^{i\vartheta}}{2} (\Lambda_1 - i\widetilde{\Lambda}_1).$$

Equating to zero the coefficient of λ^2, we find $z_2(\vartheta)$, and so forth.

In quite the same way, the question of the conformal transformation of a circle into the exterior of a curve is solved in the case when the latter is given parametrically. We shall briefly repeat the reasoning. Let the contour be given by the equations:

$$x = \varphi(t, \lambda), \; y = \psi(t, \lambda)$$

or

$$z = x + iy = F(e^{it}, \lambda).$$

We shall assume in addition that for $\lambda = 0$ we obtain a curve $z = F(e^{it}, 0)$, such that the expansion of the function $F(e^{it}, 0)$ in a series of the powers of e^{it} contains the positive powers of e^{it} and only the first of the negative powers, i.e., has the form:

$$z = \alpha_1 e^{-it} + \alpha_2 + \alpha_3 e^{it} + \alpha_4 e^{2it} + \ldots; \ \alpha_1 \neq 0.$$

If the function $F(e^{it}, \lambda)$ should also turn out to be expansible in this kind of series in the powers of $e^{it} = \zeta$, then this series, after e^{it} is replaced in it by ζ, would give us the mapping function. Here the parameter t would be the polar angle of the ζ-plane. Since this is generally speaking not the case, the expansion of $F(e^{it}, \lambda)$ may contain negative powers of e^{it} higher than the first, and our problem will be so to replace the parameter

$$t = \vartheta + \lambda\varphi_1(\vartheta) + \lambda^2\varphi_2(\vartheta) + \ldots,$$

where $\varphi_1(\vartheta)$, $\varphi_2(\vartheta)$, ... are real functions of ϑ, that these negative powers of e^{it} in the expansion fall out.

We substitute for t in $F(e^{it}, \lambda)$ its expression in terms of ϑ. In the expansion in powers of λ of the expression obtained, the term free of λ is $F(e^{it}, 0)$, and, by the condition, it contains only the first negative power of $e^{i\vartheta}$.

Let us find the coefficient of the first power of λ:

$$\left\{\frac{d}{d\lambda}[F(e^{i\vartheta+\lambda\varphi_1+\cdots}, \lambda)]\right\}_{\lambda=0} = F_\zeta(e^{i\vartheta}, 0)\, ie^{i\vartheta}\varphi_1 + F_\lambda(e^{i\vartheta}, 0) =$$

$$= ie^{i\vartheta}F_\zeta(e^{i\vartheta}, 0)\left\{\varphi_1 + \frac{F_\lambda(e^{i\vartheta}, 0)}{iF_\zeta(e^{i\vartheta}, 0)}e^{-i\vartheta}\right\}.$$

Since $F(e^{i\vartheta}, 0)$ contains $e^{-i\vartheta}$, the derivative $F_\zeta(e^{i\vartheta}, 0)$ will contain $e^{-2i\vartheta}$, and therefore the factor in front of the brace will contain $e^{-i\vartheta}$ in its expansion. Thus in order that the entire coefficient of λ contain no negative powers of $e^{i\vartheta}$ higher than the first, φ_1 must be so chosen that the expression in braces contains none of the negative powers in its expansion. Let the series expansion of the second term in the braces have the form:

$$\frac{F_\lambda(e^{i\vartheta}, 0)}{iF_\zeta(e^{i\vartheta}, 0)}e^{-i\vartheta} = \sum_{n=0}^{+\infty} c_n e^{in\vartheta} + \sum_{n=1}^{\infty} c'_n e^{-in\vartheta}.$$

In order that the whole brace not contain $e^{-i\vartheta}$, $e^{-2i\vartheta}$, ..., it is necessary that the sum $\sum_{n=1}^{\infty} c'_n e^{-in\vartheta}$ appear with reversed sign in the composition of the function φ_1. On the other hand, since φ_1 is a real function, it is necessary that the second component part of φ_1 be conjugate to the first, and, accordingly, $\varphi_1(\vartheta)$ must have this form [1]):

$$\varphi_1(\vartheta) = -\sum_{n=1}^{\infty} c'_n e^{-in\vartheta} - \sum_{n=1}^{\infty} \bar{c}'_n e^{in\vartheta}.$$

---

[1]) In the determination of $\varphi_1(\vartheta)$ there remains undetermined the free term of the expansion of $\varphi_1(\vartheta)$ in powers of $e^{i\vartheta}$. In our reasoning we consider it to be equal to zero.

Analogously, $\varphi_2(\vartheta)$ is chosen on the stipulation that the coefficient of λ^2 in the expansion of $F(e^{i\vartheta}, \lambda)$ in powers of λ shall contain no powers of the quantity $e^{-i\vartheta}$ higher than the first, and so forth. Proceeding in such a fashion, we successively determine all the coefficients $\varphi_n(\vartheta)$. The proof of the convergence of the expansion obtained in this way is carried out in the same fashion as for the case of the mapping of a circle onto the interior of a curve.

Example. We will find the function mapping the circle onto the exterior of the curve

$$z = e^{-it} + \lambda e^{-ipt}.$$

For $\lambda = -\dfrac{1}{p}$ there are obtained from the equation epicycloids of different order.

In the expression for z we substitute the series for t, limiting ourselves in the computations to the third power of the parameter λ:

$$z = e^{-i(\vartheta+\lambda\varphi_1+\lambda^2\varphi_2+\lambda^3\varphi_3+\dots)} + \lambda e^{-ip(\vartheta+\lambda\varphi_1+\lambda^2\varphi_2+\dots)} =$$

$$= e^{-i\vartheta}\Big[1 - i(\lambda\varphi_1 + \lambda^2\varphi_2 + \lambda^3\varphi_3 + \dots) + \frac{i^2}{2!}(\lambda\varphi_1 + \lambda^2\varphi_2 + \dots)^2 -$$

$$- \frac{i^3}{3!}\lambda\varphi_1 + \dots)^3 + \dots\Big] + \lambda e^{-ip\vartheta}\Big[1 - ip(\lambda\varphi_1 + \lambda^2\varphi_2 + \dots) +$$

$$+ \frac{i^2p^2}{2!}(\lambda\varphi_1 + \dots)^2 + \dots\Big] = e^{-i\vartheta} + (-ie^{-i\vartheta}\varphi_1 + e^{-ip\vartheta})\lambda +$$

$$+ \left(-ie^{-i\vartheta}\varphi_2 - \frac{1}{2!}e^{-i\vartheta}\varphi_1^2 - ipe^{-ip\vartheta}\varphi_1\right)\lambda^2 +$$

$$+ \left[e^{-i\vartheta}\left(-i\varphi_3 - \varphi_1\varphi_2 + \frac{i}{6}\varphi_1^3\right) + e^{-ip\vartheta}\left(-ip\varphi_2 - \frac{p^2}{2}\varphi_1^2\right)\right]\lambda^3 + \dots$$

Let us consider the coefficient of λ:

$$-ie^{-i\vartheta}(\varphi_1 + ie^{-i(p-1)\vartheta}).$$

From this it is seen that the parentheses will not contain a negative power of $e^{i\vartheta}$ when

$$\varphi_1 = -ie^{-i(p-1)\vartheta} + ie^{i(p-1)\vartheta} = -2\sin(p-1)\vartheta,$$

and will then have the form:

$$-ie^{-i\vartheta}ie^{i(p-1)\vartheta} = e^{i(p-2)\vartheta}.$$

Let us pass to the coefficient of λ^2:

$$-ie^{-i\vartheta}\left(\varphi_2 - \frac{i}{2}\varphi_1^2 + pe^{-i(p-1)\vartheta}\varphi_1\right) =$$

$$= -ie^{-i\vartheta}\left[\varphi_2 + \frac{i}{2}e^{2i(p-1)\vartheta} + i(p-1) - i(p-\tfrac{1}{2})e^{-2i(p-1)\vartheta}\right].$$

Therefore

$$\varphi_2 = i(p - \tfrac{1}{2})(e^{-2i(p-1)\vartheta} - e^{2i(p-1)\vartheta}) = (2p - 1) \sin 2(p - 1)\vartheta.$$

The coefficient itself will take the form:

$$(p - 1)(e^{-i\vartheta} - e^{i(2p-3)\vartheta}).$$

Finally, the coefficient of λ^3 will be:

$$- ie^{-i\vartheta}\left(\varphi_3 - i\varphi_1\varphi_2 - \tfrac{1}{6}\varphi_1^3 + pe^{-i(p-1)\vartheta}\varphi_2 - \frac{i}{2} p^2 e^{-i(p-1)\vartheta}\varphi_1^2\right) =$$

$$= - ie^{-i\vartheta}\left[\varphi_3 - i(p - \tfrac{2}{3})e^{-3i(p-1)\vartheta} - i\left(\frac{p^2}{2} + 1 - \tfrac{3}{2}p\right)e^{i(p-1)\vartheta} - \right.$$

$$\left. - ip(p - 1)e^{-i(p-1)\vartheta} + i(\tfrac{3}{2}p^2 - \tfrac{3}{2}p + \tfrac{1}{3})e^{-3i(p-1)\vartheta}\right];$$

consequently

$$\varphi_3 = ip(p - 1)(e^{-i(p-1)\vartheta} - e^{i(p-1)\vartheta}) - i(\tfrac{3}{2}p^2 - \tfrac{3}{2}p + \tfrac{1}{3})(e^{-3i(p-1)\vartheta} -$$

$$- e^{3i(p-1)\vartheta}) = 2p(p - 1) \sin (p - 1)\vartheta - (3p^2 - 3p + \tfrac{2}{3}) \sin 3 (p - 1)\vartheta.$$

Substituting for φ_3 its value in the preceding expression, we shall have for the coefficient of λ^3 the following value:

$$- (\tfrac{3}{2}p^2 - \tfrac{5}{2}p + 1) (e^{i(p-2)\vartheta} - e^{i(3p-4)\vartheta}).$$

So approximately — accurate to the third power of λ — the equation of the contour as a function of the parameter ϑ has the form:

$$z = e^{-i\vartheta} + \lambda e^{i(p-2)\vartheta} + \lambda^2(p - 1)(e^{-i\vartheta} - e^{i(2p-3)\vartheta}) -$$

$$- \lambda^3(\tfrac{3}{2}p^2 - \tfrac{5}{2}p + 1)(e^{i(p-2)\vartheta} - e^{i(3p-4)\vartheta}),$$

and the mapping function will thus be approximately equal to

$$z = [1 + \lambda^2(p - 1)]\frac{1}{\xi} + [\lambda - \lambda^3(\tfrac{3}{2}p^2 - \tfrac{5}{2}p + 1)]\zeta^{p-2} -$$

$$- \lambda^2(p - 1)\zeta^{2p-3} - \lambda^3(\tfrac{3}{2}p^2 - \tfrac{5}{2}p + 1)\zeta^{3p-4}.$$

§ 7. MELENTIEV'S METHOD OF APPROXIMATE CONFORMAL TRANSFORMATION [1)]

1. The algorithm of successive approximations. In engineering practice the contour of the region onto which one is obliged to map the circle is very frequently given graphically, and the application of the methods of conformal transformation requiring an analytical statement of the contour therefore becomes impossible in practice, or requires the expenditure of additional labor in setting up its equation. Moreover, if the equation of the contour is complicated, the application

---

1) P. V. Melent'ev, [1].

of these methods may indeed be possible in principle, but the computations connected with them frequently become so difficult that it is usually preferable to resort to graphical methods of solution of the problem, if this is admitted by the required accuracy in the result — for here that will be limited by the accuracy of the instruments we employ in the work, whereas the analytic methods permit, generally speaking, the result to be obtained with any degree of accuracy.

We shall now give one of such methods, which is due to Engineer P. V. Melentiev, who has successfully applied it to the solution of a number of practical problems.

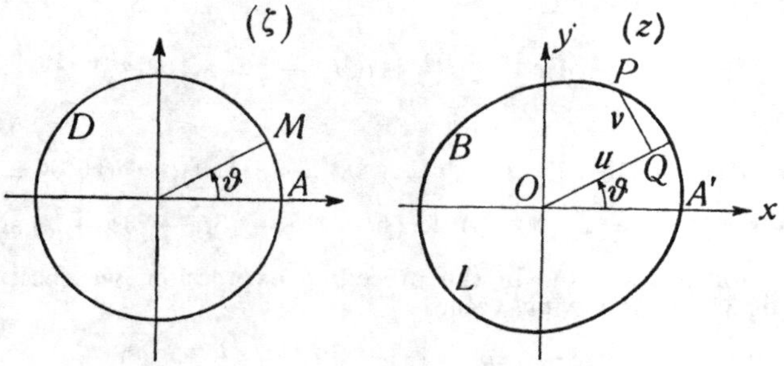

Fig. 30.

Let the circle D of unit radius be conformally mapped onto a simply connected single-sheeted region B by an analytic function

$$z = \sum_{n=0}^{\infty} \alpha_n \zeta^{n+1} = \alpha_0 \zeta + \alpha_1 \zeta^2 + \dots, \tag{1}$$

where

$$\alpha_n = a_n + ib_n.$$

With this it is assumed that the origin of coordinates $\zeta = 0$ passes to the point $z = 0$ within the region B and that the point A, the intersection of the circumference of the circle with the real axis, passes to the point A', the intersection of the contour L of the region B with the axis x (Fig. 30).

The right side of equation (1) gives for $\zeta = 1$ the expansion in a Fourier series of the real and imaginary components of the number z:

$$x = \sum_{n=0}^{\infty} [a_n \cos (n+1)\vartheta - b_n \sin (n+1)\vartheta],$$

$$y = \sum_{n=0}^{\infty} [b_n \cos (n+1)\vartheta + a_n \sin (n+1)\vartheta].$$

For the sake of computational convenience, P. V. Melentiev proposes to consider the relation, not between x, y and ϑ, but between the polar angle ϑ and other quantities that are more uniform in respect of their behavior than are x and y for simple regions. He considers the ratio:

$$u + iv = w = \frac{z}{\zeta} = \sum_{n=0}^{\infty} \alpha_n \zeta^n. \tag{2}$$

On the boundary of the circle D the distinction between w and z will consist solely in that the argument of z will be greater by ϑ than the argument of w. In order to find u and v by a known $z = P$ lying on the contour L, it is necessary to draw the ray OQ at an angle ϑ from the origin of coordinates of the z-plane and drop a perpendicular to it from P. The segment OQ of the ray will equal u and the segment QP of the perpendicular will equal v (see Fig. 30). The following is at once evident: when the contour L of the region B is close to a circle, the quantity u will be close to the radius of this circle, while v will always be close to zero. By contrast, the behavior of the quantities x and y will be considerably more complicated — it will be close to the behavior of the trigonometric functions.

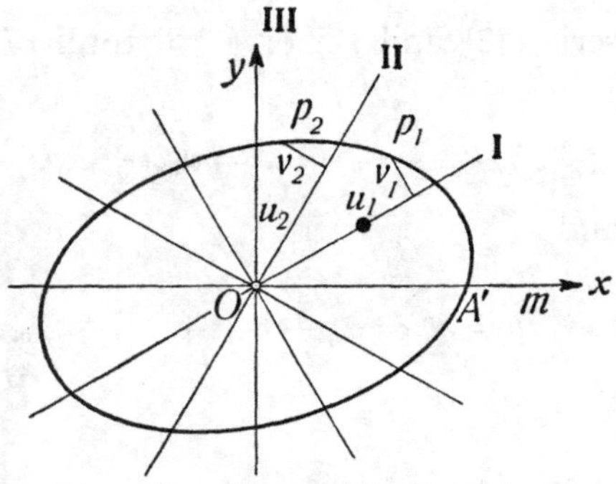

Fig. 31.

For $|\zeta| = 1$ it follows from equation (2) that:

$$\left.\begin{aligned} u &= \sum_{n=0}^{\infty} a_n \cos n\vartheta - b_n \sin n\vartheta, \\ v &= \sum_{n=0}^{\infty} b_n \cos n\vartheta + a_n \sin n\vartheta. \end{aligned}\right\} \tag{3}$$

And all the computations will be constructed upon the use of these last two relations. For our purposes it is sufficient to know one of the functions u or v; for, in view of their conjugacy, knowledge of u, say, permits v to be found accurate to the coefficient b_0; but the latter is equal, as one can see at once, to the sum of all the rest of the b_n, taken with sign reversed, because for $\zeta = 1$, u must be equal to the real number corresponding to the point A', and the imaginary part of the number w must be zero; but it equals the sum of all b_n, and we have, therefore,

$$\sum_{n=0}^{\infty} b_n = 0,$$

whence it follows that

$$b_0 = -\sum_{n=1}^{\infty} b_n. \tag{4}$$

We shall utilize this fact below in constructing the algorithm of successive approximations for the determination of the numbers a_n and b_n.

If the series standing on the right side of equations (3) are uniformly convergent, one can choose segments of them such that the error shall be beyond the adopted accuracy of the computations for any ϑ, i.e., within the limits of the adopted accuracy, the circle D shall be mapped onto the region B by some polynomial.

Let us assume that in order to attain this it is necessary to take in series (3) and (2) or (1) a total of $\frac{m}{2}+1$ terms, putting

$$z = \alpha_0\zeta + \alpha_1\zeta^2 + \ldots + \alpha_{\frac{m}{2}}\zeta^{\frac{m}{2}+1} \tag{5}$$

and

$$u + iv = w = \alpha_0 + \alpha_1\zeta + \ldots + \alpha_{\frac{m}{2}}\zeta^{\frac{m}{2}} = \sum_{k=0}^{\frac{m}{2}} [(a_k \cos k\vartheta - b_k \sin k\vartheta) + i(b_k \cos k\vartheta + a_k \sin k\vartheta)]. \tag{6}$$

It is known from the theory of Fourier series that if the function is reducible to a trigonometric polynomial, then to know its Fourier coefficients it is sufficient to know the values of this function at certain points only. To avoid burdening the reader with references to the literature, we shall carry out the derivation of the appropriate equations. For this purpose we shall divide the angle 2π about the origin of coordinates in the ζ-plane into m equal parts, constructing the rays $I, II, \ldots, m$ (see Fig. 31). By taking, for the respective values of $\vartheta = 2\frac{\pi}{m}, 4\frac{\pi}{m}, \ldots, 2\pi$, one of the quantities u or v, for instance u, we shall have:

$$\left.\begin{aligned}
u_1 &= \sum_{k=0}^{\frac{m}{2}} \left(a_k \cos\frac{2\pi}{m}k - b_k \sin\frac{2\pi}{m}k\right),\\
&\ldots\ldots\ldots\ldots\ldots\ldots\\
u_n &= \sum_{k=0}^{\frac{m}{2}} \left(a_k \cos\frac{2\pi}{m}nk - b_k \sin\frac{2\pi}{m}nk\right),\\
&\ldots\ldots\ldots\ldots\ldots\ldots\\
u_m &= \sum_{k=0}^{\frac{m}{2}} (a_k \cos 2\pi k - b_k \sin 2\pi k).
\end{aligned}\right\} \tag{7}$$

These equations can be solved for the coefficients $a_k \left(k = 0, 1, \ldots, \frac{m}{2}\right)$

and $b_k\left(\text{for } k = 1, 2, \ldots, \frac{m}{2} - 1\right)$. The dependence of the coefficients upon u_n is given by the following formulas:

$$\left.\begin{aligned} a_k &= \frac{2}{m}\sum_{n=1}^{m} u_n \cos\frac{2\pi}{m} nk, \\ b_k &= -\frac{2}{m}\sum_{n=1}^{m} u_n \sin\frac{2\pi}{m} nk, \quad \left(k = 1, 2, \ldots, \frac{m}{2} - 1\right), \\ a_0 &= \frac{1}{m}\sum_{n=1}^{m} u_n, \quad a_{\frac{m}{2}} = \frac{1}{m}\sum_{n=1}^{m} (-1)^n u_n. \end{aligned}\right\} \tag{8}$$

As regards the quantities b_0 and $b_{\frac{m}{2}}$, to determine them from system of equations (7) is obviously impossible, for their coefficients will be zeros. One of them, b_0, we can find from the condition of correspondence of the points A and A', expressed by equation (4), which for our case reduces to:

$$b_0 = -\sum_{k=1}^{\frac{m}{2}} b_k; \tag{9}$$

the coefficient $b_{\frac{m}{2}}$ generally cannot be found, however. By taking m sufficiently large, we can always consider that $b_{\frac{m}{2}} = 0$.

Had we adopted not the u_n but the v_n for the initial system of values, we would have obtained, since $v_m = 0$, as is seen from the Figure, the following equations for the determination of the coefficients:

$$\left.\begin{aligned} a_k &= \frac{2}{m}\sum_{n=1}^{m-1} v_n \sin\frac{2\pi}{m} nk, \\ b_k &= \frac{2}{m}\sum_{n=1}^{m-1} v_n \cos\frac{2\pi}{m} nk, \quad \left(k = 1, 2, \ldots, \frac{m}{2} - 1\right), \\ b_0 &= \frac{1}{m}\sum_{n=1}^{m-1} v_n, \quad b_{\frac{m}{2}} = \frac{1}{m}\sum_{n=1}^{m-1} (-1)^n v_n. \end{aligned}\right\} \tag{10}$$

Equation (9) is here automatically satisfied, for, as is seen from the expression for v_m:

$$\sum_{k=0}^{\frac{m}{2}} b_k = v_m = 0.$$

The two quantities a_0 and $a_{\frac{m}{2}}$ cannot be expressed in terms of v_n. By the choice of m we can again make $a_{\frac{m}{2}}$ equal to zero. The coefficient

a_0 we find here, though, from the requirement that u_m must equal a given quantity — the length of the segment $OA' = l$. But

$$u_m = \sum_{k=0}^{\frac{m}{2}} a_k,$$

and accordingly:

$$a_0 = l - \sum_{k=1}^{\frac{m}{2}} a_k. \tag{11}$$

There is no need, since the proofs are of the same type, to establish the validity of all the groups of formulas (8) and (10); we shall dwell only on the first group of formulas, (8). Multiplying equations (7) successively by $\cos\frac{2\pi}{m}s$, $\cos\frac{2\pi}{m}2s$, ..., $\cos 2\pi s$ and adding, we obtain:

$$\sum_{n=1}^{m} u_n \cos\frac{2\pi}{m}ns =$$

$$= \sum_{k=0}^{\frac{m}{2}} \left[a_k \sum_{n=1}^{m} \cos\frac{2\pi}{m}nk \cos\frac{2\pi}{m}ns - b_k \sum_{n=1}^{m} \sin\frac{2\pi}{m}nk \cos\frac{2\pi}{m}ns\right]$$

or, by replacing the trigonometric functions in accordance with the Euler formulas, we find that the right side is equal to

$$\tfrac{1}{4}\sum_{k=0}^{\frac{m}{2}} [a_k \sum_{n=1}^{m} (e^{i\frac{2\pi}{m}n(k+s)} + e^{i\frac{2\pi}{m}n(k-s)} + e^{i\frac{2\pi}{m}n(s-k)} + e^{i\frac{2\pi}{m}n(-s-k)}) +$$

$$+ ib_k \sum_{n=1}^{m} (e^{i\frac{2\pi}{m}n(k+s)} + e^{i\frac{2\pi}{m}n(k-s)} - e^{i\frac{2\pi}{m}n(s-k)} - e^{i\frac{2\pi}{m}n(-s-k)})].$$

All the sums in brackets are computed without difficulty; obviously if p is an integer and $-m < p < m$, $p \neq 0$:

$$\sum_{n=1}^{m} e^{i\frac{2\pi}{m}np} = e^{i\frac{2\pi}{m}p} \frac{e^{i2\pi p} - 1}{e^{i\frac{2\pi}{m}p} - 1} = 0;$$

therefore, since the index k satisfies the inequality $0 \leqslant k \leqslant \frac{m}{2}$ and we are considering that $0 < s < \frac{m}{2}$ and consequently that $0 < s + k < m$, we shall always have:

$$\sum_{n=1}^{m} e^{i\frac{2\pi}{m}(k+s)n} = \sum_{n=1}^{m} e^{-i\frac{2\pi}{m}(k+s)n} = 0.$$

There remain only the two sums:

$$\sum_{n=1}^{m} e^{i\frac{2\pi}{m}(k-s)n}$$

and

$$\sum_{n=1}^{m} e^{i\frac{2\pi}{m}(s-k)n}.$$

For $s \neq k$ they both equal zero; when $s = k$, though, they both equal m. Consequently the right side of the equation we are considering reduces to

$$\tfrac{1}{4}a_s(m+m) + \frac{i}{4}b_s(m-m) = \frac{m}{2}a_s,$$

which indeed proves the required result.

Let us now assume that we have succeeded, in one way or another, in finding some approximate values for $u_1, \ldots, u_m$, which we shall denote by $u_1^{(0)}, u_2^{(0)}, \ldots, u_m^{(0)}$. Using them, by formulas (8) and (9) and the requirement $b_{\frac{m}{2}} = 0$ we quickly find the approximate values $a_k^{(0)}$, $b_k^{(0)}$ for the coefficients a_k and b_k. And introducing their values into the equations

$$v_n = \sum_{k=0}^{\frac{m}{2}} \left(b_k \cos\frac{2\pi}{m}kn + a_k \sin\frac{2\pi}{m}kn\right), \tag{12}$$

we compute the approximate values $v_n^{(0)}$ for the v_n; then by the $u_n^{(0)}$ and $v_n^{(0)}$ we construct the points $M_1^{(0)}, \ldots, M_m^{(0)}$, approximations to the points $M_1, M_2, \ldots, M_m$. If all of them lay on the curve, we should have obtained the exact positions of the points $M_1, M_2, \ldots, M_m$ and by that fact should have constructed the mapping polynomial exactly. Generally, however, this will not be the case. Had the initial values for the u_n differed little from the real values, the points $M_n^{(0)}$ found by us would depart little from their real positions on the curve L. The points $M_n^{(0)}$ must somehow be transferred to the curve L so that if they do not coincide with their real positions, they are as close as possible to them. Three methods of making the transfer can be proposed.

1. The first, and apparently the most expedient, is the method of transferring via the shortest distance: from the $M_n^{(0)}$ perpendiculars are dropped to the curve L, and the ends of these perpendiculars taken as the new positions of the points M_n (Fig. 32).

2. More simply executed, but probably less expedient is the method of transfer via the radius. Through $M_n^{(0)}$ and the origin of coordinates one draws a straight line, and adopts the point of intersection of it with the curve L that is closest to $M_n^{(0)}$ as the new position of M_n (Fig. 33).

3. The simplest (Fig. 34), if one resorts to such a transparent draughts-

man's tool as is depicted in Fig. 35, is the method of transferring $M_n^{(0)}$ to the curve L along a straight line parallel to the ray $\vartheta = \frac{2\pi}{m} n$, i.e., with a constant v_n. The tool is so superimposed upon the diagram that its

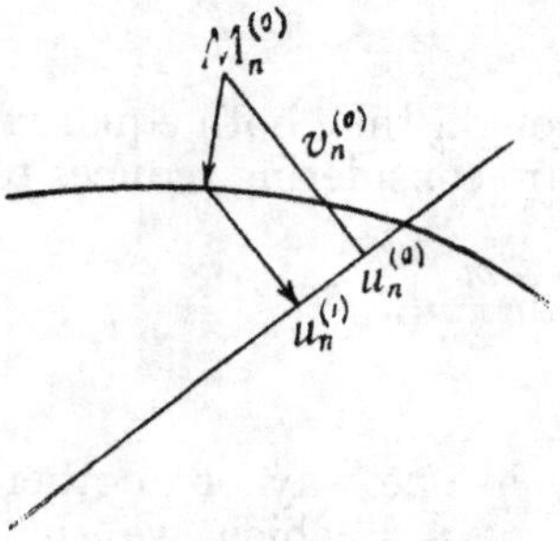

Fig. 32.

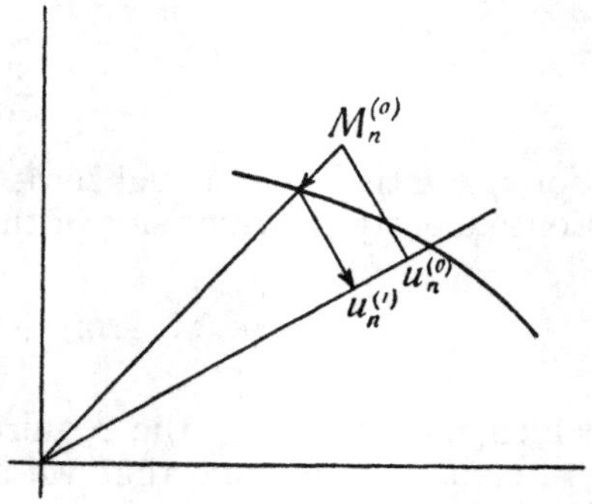

Fig. 33.

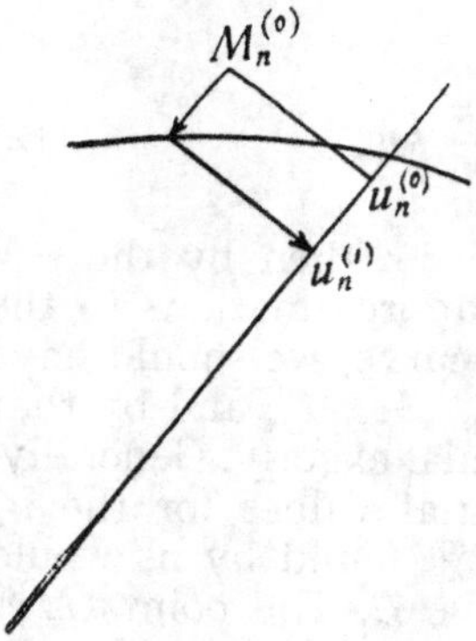

Fig. 34.

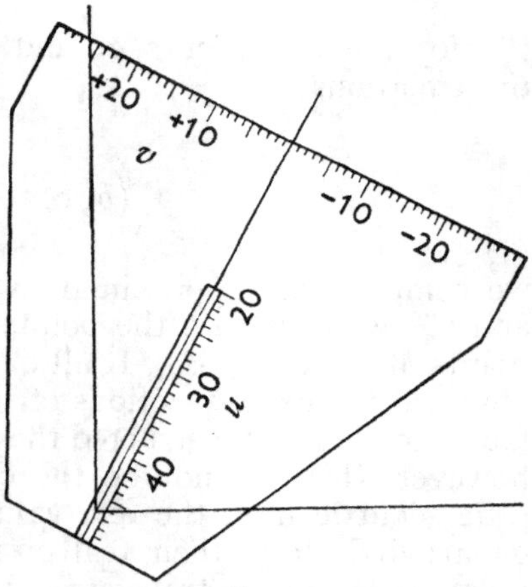

Fig. 35.

base line u coincides with the ray $\vartheta = \frac{2\pi}{m} n$ and the scale v passes through $M_n^{(0)}$. One marks the position of $M_n^{(0)}$ and moves along the ray until the mark strikes the curve L at some point, which will be the new position sought for M_n.

We remark further that in the third method of transferring the points $M_n^{(0)}$ $\left(\text{parallel to the ray } \vartheta = \frac{2\pi}{m} n\right)$, in using the draughtsman's tool for the construction of the u_n there is no need to actually drop the perpendiculars from the points $M_n^{(0)}$ to the rays, it being sufficient to simply take a reading of $u_n^{(1)}$ on the scale.

Because of its simplicity, this method has been adopted in all the examples exhibited below.

In all three cases one would think that the new positions of the points would be closer to the actual ones than the prior positions were. From them we drop perpendiculars onto the rays $\vartheta = \frac{2\pi}{m} n$, adopt as the new values $u_n^{(1)}$ for u_n the segments of the rays from the origin to the bases of the perpendiculars, and then repeat the computations.

A few words about the convergence. By iterating these computations, we shall obtain for each point M_n a sequence of approximations $M_n^{(0)}$, $M_n^{(1)}, M_n^{(2)}, \ldots$. There then arises the question of defining the conditions that the contour L must satisfy, sufficient to secure the convergence to the limit point M_n of each of our sequences of approximations $M_n^{(0)}$, $M_n^{(1)}, \ldots$. This question unfortunately still remains open, and the convergence must be investigated for the solution of each separate problem.

But we shall assume that these sequences have all turned out to be convergent to some limiting points M_n lying on L. At each step of the calculations we shall have computed approximations to the coefficients a_k and b_k. These sequences will also be convergent. Let us call the limiting values of the coefficients $\tilde{a}_k$ and $\tilde{b}_k$.

Substituting for the α_k in equation (5) the values $\tilde{\alpha}_k = \tilde{a}_k + i\tilde{b}_k$ which we have found, we shall have constructed a polynomial in ζ of degree $\frac{m}{2} + 1$,

$$z = \tilde{\alpha}_0 \zeta + \tilde{\alpha}_1 \zeta^2 + \ldots + \tilde{\alpha}_{\frac{m}{2}+1} \zeta^{\frac{m}{2}+1}.$$

It will give the transformation of the unit circumference into some curve L_m, which has m common points with the given curve L. For large m this curve L_m will, generally speaking, differ little from L, and it can be expected that as $m \to \infty$ the curves L_m will tend to coincide with L. If this happens, then the polynomials constructed by us will converge to the function transforming the circle $|\zeta| < 1$ into the interior of the curve L, as is shown in the theory of conformal transformations. But when this will be the case, i.e., for what kind of L it can be expected that L_m will tend to coincide with L, has not been discovered up to the present time.

2. Choice of the first approximation. Computational scheme. In order to begin the computation of the successive approximations of which we spoke in No. 1, it is still necessary to indicate a rule for the graphical construction of the initial values of $u_1, u_2, \ldots, u_m$. As stands to reason, the rapidity of the convergence, and together with it the quantity of necessary computations, will be closely connected with how felicitous the choice of these approximations has been. Proceeding in a crude way, one could — particularly if the contour L is close to a circle — take all u_n as constants equal to the radius of some circle best fitting the region B. Such a circle can be constructed by the most varied methods. Let R_1 be the greatest distance of the points of the contour L from the origin of coordinates, and R_2 the least distance; as the radius of the aforementioned circle one could, for example, adopt the arithmetic mean, $\frac{R_1 + R_2}{2}$, of the numbers R_1 and R_2. The same device can be used also for a somewhat more exact determination of the initial values for the u_n. Divide the entire curve into parts between the maxima and minima of its distances from the origin of coordinates. For each of these parts separately construct its initial values for the u_n in the manner indicated.

In some cases the following construction leads to more accurate results. All rays required in the computation of the successive approximations are drawn, and the lengths of their segments from the origin of coordinates to the intersection of them with the contour L are found. As the initial values of the u_n the arithmetic mean of the lengths of all segments may be adopted.

In both of these cases we replace the values of the u_n for the contour L by their values for a certain circle into which the unit circle is carried under the linear transformation

$$z = \alpha\zeta.$$

The distinction will consist solely in the choice of the number α. More exact methods of constructing the values $u_n^{(0)}$ can be indicated which use more complicated functions of the complex variable.

The idea of the constructions will be this. We shall divide the curve L into parts and for each of them choose the function that will transform the segment of circle corresponding to this part into a curve as close as possible to the adopted part of the curve. One could then have confidence that the auxiliary function would differ little, generally speaking, from the sought transforming function on the corresponding arc of the circle and near it, although it might differ very greatly from it for other values.

Let us consider initially the function

$$z = R\zeta + r\zeta^{p+1},$$

where R, r and p are positive real numbers. We do not assume the number p to be an integer. We will now study the transformation of the unit circle effected by this function.

Let the point ζ be located on the circumference of the unit circle: $\zeta = e^{i\vartheta}$. From the structure of the function it is evident that

$$\frac{z}{\zeta} = R + re^{ip\vartheta}$$

will be a periodic function of ϑ of period $\frac{2\pi}{p}$. It will therefore be sufficient for us to study a piece of the curve only, corresponding to the variation of ϑ in the interval $-\frac{\pi}{p} \leqslant \vartheta \leqslant +\frac{\pi}{p}$. Obviously, since R and r are real, the real axis will be the axis of symmetry of the segment, and we can accordingly limit our study to just the positive half-interval.

From the formula

$$|z|^2 = |w|^2 = R^2 + 2Rr \cos p\vartheta + r^2$$

it is seen that $|z|$ has its greatest value in our interval of variation of ϑ for $\vartheta = 0$, it being equal to

$$R + r = R_1.$$

For ϑ increasing from 0 to $\frac{\pi}{p}$, the modulus of the number z will monotonically diminish and attain its least value $R - r = R_2$ for $\vartheta = \frac{\pi}{p}$.

For the determination of the point z corresponding to a given value of $\zeta = e^{i\vartheta}$ one must lay out a segment of length R from the origin of coordinates along the ray forming the angle ϑ with the real axis, draw a straight line from the point obtained at an angle $\omega = p\vartheta$ to the ray and lay off on it a segment of length r; the end of this segment will be the affix z.

The construction of the point z will be particularly simple if the curve into which the unit circle is carried by means of the function under consideration has been found. Then after having drawn the ray and laid out the distance R from the origin on it, it is enough to describe a circle of radius r from the point obtained. It will intersect the curve in two points. We have to choose the one of them that corresponds to the angle ω at the constructed circumference, the origin of the reading being the ray of which we were just speaking. As is easily understood, it will always lie on the side of the ray closest to the axes of symmetry $\vartheta = \pm \frac{\pi}{p}$, i.e., will always be the closer to the origin of coordinates.

We shall now show how the function we are studying can be applied to the determination of the initial approximate position of the points M_n. We first divide the curve into segments between the maximum and minimum distances of its points from the origin of coordinates. We take any of them and measure the angle about the origin of coordinates which it subtends. Let it equal χ. Putting $p = \frac{\pi}{\chi}$ and adopting the radius vector corresponding to the more distant end of the segment of the curve as the

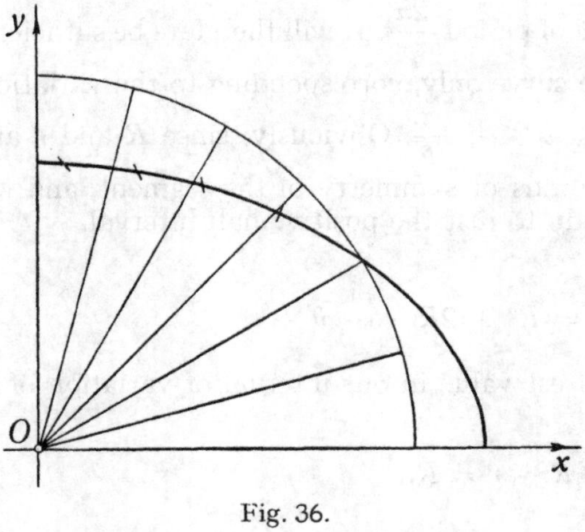

Fig. 36.

larger axis of symmetry of the auxiliary curve, we perform, for the rays lying within the part under consideration, the construction of the points M.

The initial approximations for the points constructed in this fashion not only do not coincide with their actual positions, but generally speaking even lie outside the curve L. But we can try to find positions of the points M_n that are closer to the actual ones than those we have constructed. With this purpose we shall transfer the points we have constructed to the curve L. Such a transfer can be performed by one of the methods that we applied in No. 1, but here it is simpler to employ the method of transference along the arc of the circumference of radius r with center on the respective ray, taken at a distance R from the origin. The constructions are much simplified, since the necessity of finding the angle $\omega = p\vartheta$ is then eliminated.

To construct the approximate position of the point M_n on L it will be necessary only to describe, with the point $Re^{i\vartheta}$ on the ray as center, a circle of radius r, taking the closest to the origin of coordinates of the points of its intersection with the curve L.

We have given an example of such a transfer in Fig. 36, for the ellipse with semi-axes 3.75 and 2.3 cm. As a second example we have taken the curve depicted in Fig. 37. The parts between the maximum and minimum distances of its points from the origin of co-

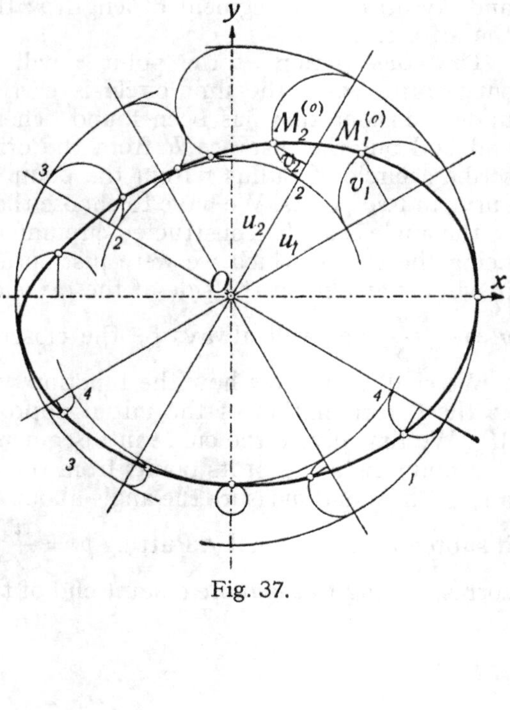

Fig. 37.

ordinates here number four. We have no need of measuring the angles subtended by these parts, for they were needed in the construction for finding the angle $p\vartheta$ to be laid out from the ray for the determination of the direction in which we had to lay out the segment of length r, whereas in making the transfer along the circumference the construction of this ray is superfluous, it being sufficient for us to know merely r and R for each part, as was explained several lines above.

After having found the initial positions for the points M_n, we drop a perpendicular from them to the respective rays and determine the initial values for u_n.

We shall use below, in all examples, $M_n^{(0)}$ constructed in this manner. The simplicity of the geometrical constructions that we have had to perform in finding the $M_n^{(0)}$ has its explanation in the fact that the additional summand $r\zeta^p$ in the function

$$z = \zeta(R + r\zeta^p),$$

with which we have tried to take into account the effect of the contour's deviation from a circle, has a constant modulus when ζ varies on the circumference.

To the same trend of thought pertains another method of choosing the auxiliary function that takes into account the departure of L from a circle. One can utilize any of the functions transforming the unit circle into itself. The presence of an additional parameter — the coordinates of the point to which the center passes under such a transformation — permits the more exact determination of the $M_n^{(0)}$.

This complication loses its sense, however, if after the construction the M_n be transferred to the curve L along the circle already mentioned, for then it is completely immaterial which of the functions preserving constant modulus on the circle we take, and we therefore naturally use the simplest of them: $r\zeta^p$.

We shall here indicate the possibility of using still another function, $\frac{\zeta - c}{1 - \zeta c}$, giving the transformation of the unit circle into itself with the passing of the point c to the center, and requiring comparatively few computations for its application. Replacing ζ by ζ^p in it, we shall consider the function:

$$z = \zeta\left(R + z\,\frac{\zeta^p - c}{1 - c\zeta^p}\right) = a\zeta\,\frac{1 + b\zeta^p}{1 - c\zeta^p},$$

where

$$a = R - cr$$

and

$$b = \frac{r - cR}{R - cr}.$$

As previously, we shall consider R, r and c to be real and positive, and $c < 1$.

$$\frac{z}{\zeta} = w = R + r\frac{|\zeta|^p e^{ip\vartheta} - c}{1 - c\,|\zeta|^p e^{ip\vartheta}},$$

from which it is evident that w is a periodic function of ϑ of period $\frac{2\pi}{p}$. It will therefore be sufficient for us to consider, as above, only values of w for the angle $-\frac{\pi}{p} \leqslant \vartheta \leqslant +\frac{\pi}{p}$. Since the real axis is the axis of symmetry of the function, we can limit ourselves to positive ϑ only.

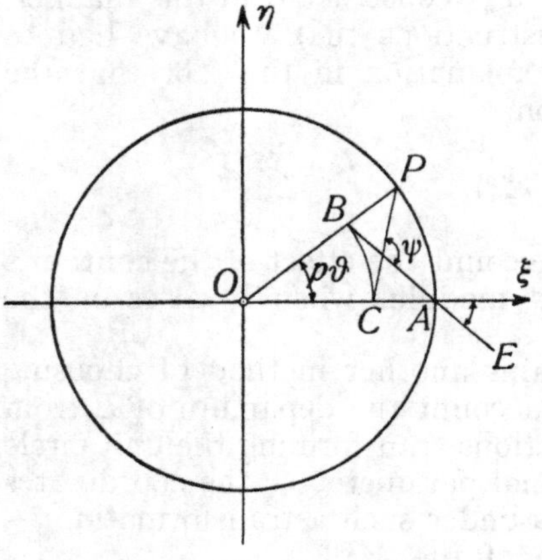

Fig. 38.

Let us trace the variation of w on the circumference of the circle. We shall put

$$w = R + r\frac{e^{ip\vartheta} - c}{1 - ce^{ip\vartheta}}.$$

The function $\frac{t - c}{1 - ct}$ gives a transformation of the unit circle into itself such that the real axis passes to real axis, and the point $+1$ passes to the point $+1$. Hence it follows that the argument of the fraction $\frac{e^{ip\vartheta} - c}{1 - ce^{ip\vartheta}}$ will increase from zero to π when ϑ increases from zero to $\frac{\pi}{p}$, and, accordingly, $|w|$ will all the time decrease and will attain its least value $R - r$ for $\vartheta = +\frac{\pi}{p}$. Its greatest value is $R + r$, corresponding to $\vartheta = 0$.

To find the point z corresponding to $\zeta = e^{i\vartheta}$, we can proceed in the following way, as follows from the analytical expression for z. As is seen in Fig. 38, the denominator $1 - ce^{ip\vartheta}$ of the fraction $\alpha = \frac{e^{ip\vartheta} - c}{1 - ce^{ip\vartheta}}$ is the vector BA. Its argument is the angle $\widehat{\xi AE}$, taken with the minus sign. The numerator, however, $e^{ip\vartheta} - c$, is the vector CP, and its argument will be the angle $\widehat{ACP}$. The argument of the entire fraction, however, being equal to the difference between the arguments of numerator and denominator, is obtained if we add the absolute values of the angles $\widehat{\xi AE}$ and $\widehat{ACP}$. Let us denote the result of the addition by ψ. Thus for the determination of z it is sufficient to lay out, along a ray from the origin at an angle ϑ to the x-axis, a segment of length R, and then to lay out a segment of length r on the line making an angle ψ with this ray; its end will be the affix z.

The construction of the point z corresponding to $\zeta = e^{i\vartheta}$ will be particularly simple in case the curve corresponding to the arc $-\frac{\pi}{p} \leqslant \vartheta \leqslant \frac{\pi}{p}$ of the circle has been found geometrically; there is then no need to construct the angle ψ, it being sufficient to describe, from the point $Re^{i\vartheta}$, a circle with radius r, and, of its two points of intersection with the curve, to take the closer to the origin of coordinates.

We shall now give formulas permitting the finding of the parameters defining the curve under consideration by its geometrical elements. Two of them, R and r, have a simple geometric meaning: R is half the sum of the maximum and minimum distances of the points of the curve from the origin, and r, half their difference.

To determine c, we shall resort to the following method. We have previously designated $\arg \frac{e^{ip\vartheta} - c}{1 - ce^{ip\vartheta}}$ by ψ. Therefore

$$w = R + re^{i\psi}.$$

Here let us put $\vartheta = \frac{\pi}{2p}$ and denote the corresponding value of ψ by $\psi_{\frac{1}{2}}$:

$$e^{i\psi_{\frac{1}{2}}} = \frac{e^{i\frac{\pi}{2}} - c}{1 - ce^{i\frac{\pi}{2}}} = \frac{i - c}{1 - ic} = \frac{-2c}{1 + c^2} + i\frac{1 - c^2}{1 + c^2},$$

whence

$$\sin \psi_{\frac{1}{2}} = \frac{1 - c^2}{1 + c^2}, \quad \cos \psi_{\frac{1}{2}} = -\frac{2c}{1 + c^2}$$

$$c = -\frac{1 - \sin \psi_{\frac{1}{2}}}{\cos \psi_{\frac{1}{2}}}.$$

Having measured the angle $\psi_{\frac{1}{2}}$, we find c by this formula.

After the initial values for the u_n have been constructed, the calculations go on in the way we are now familiar with.

Formulas (8), (9) and (12), by which they must be conducted, are, properly speaking, formulas for the approximate computation of the coefficients of a Fourier series, or combinations of them, and therefore for them there may be employed any of the great number of computational methods used in approximate harmonic analysis. For acquaintance with them we refer the reader to books specially treating of this question [1]).

---

[1]) See, for example, Acad. A. N. Krylov, [2], Whittaker and Robinson, [1], or Smirnov, V. I., [3], vol. II, where a sufficiently complete exposition of this question is to be found.

Here, however, for an example of this kind of calculation, we shall dwell on a method that has been applied by the author of the method to the solution of problems of diverse kinds.

Let us put

$$m = 12, \quad \frac{2\pi}{m} = 30^\circ.$$

Formulas (8) and (9) for the determination of the a_n and b_n take in this case the form:

$$a_0 = \tfrac{1}{12}(u_1 + u_2 + \dots + u_{12}),$$

$$a_1 = \tfrac{2}{12}(u_1 \cos 30^\circ + u_2 \cos 60^\circ - u_4 \cos 60^\circ - u_5 \cos 30^\circ - u_6 - u_7 \cos 30^\circ - u_8 \cos 60^\circ + u_{10} \cos 60^\circ + u_{11} \cos 30^\circ + u_{12}),$$

$$a_2 = \tfrac{2}{12}(u_1 \cos 60^\circ - u_2 \cos 60^\circ - u_3 - u_4 \cos 60^\circ + u_5 \cos 60^\circ + u_6 + u_7 \cos 60^\circ - u_8 \cos 60^\circ - u_9 - u_{10} \cos 60^\circ + u_{11} \cos 60^\circ + u_{12}),$$

$$a_3 = \tfrac{2}{12}(- u_2 + u_4 - u_6 + u_8 - u_{10} + u_{12}),$$

$$a_4 = \tfrac{2}{12}(- u_1 \cos 60^\circ - u_2 \cos 60^\circ + u_3 - u_4 \cos 60^\circ - u_5 \cos 60^\circ + u_3 - u_7 \cos 60^\circ - u_8 \cos 60^\circ + u_9 - u_{10} \cos 60^\circ - u_{11} \cos 60^\circ + u_{12}),$$

$$a_5 = \tfrac{2}{12}(- u_1 \cos 30^\circ + u_2 \cos 60^\circ - u_4 \cos 60^\circ + u_5 \cos 30^\circ - u_6 + u_7 \cos 30^\circ - u_8 \cos 60^\circ + u_{10} \cos 60^\circ - u_{11} \cos 30^\circ + u_{12}),$$

$$a_6 = \tfrac{1}{12}(- u_1 + u_2 - u_3 + \dots + u_{12}).$$

The formulas for $b_1, b_2, \dots, b_5$ are constructed just as those are for $a_1, a_2, \dots, a_6$, with the sole distinction that instead of the function cos stands sin and the factor in front of the parentheses is not $\frac{2}{12}$ but $-\frac{2}{12}$; we shall therefore not copy them. We adopt b_6 as equal to zero; equality (9) for b_0 gives

$$b_0 = -(b_1 + b_2 + \dots + b_5).$$

U_k	1.732	1.00	2.00
$k=0$			
1			
2			
3			
4			
5			
6			
7			
8			
9			
10			
11			

Fig. 39.

From this it is evident that all coefficients a_n and b_n are obtained from $u_1, \dots, u_{12}$ by means of multiplications (we neglect for the time being the factor common to all of them, $\frac{1}{12}$): by 1, by $\sqrt{3} = 1.732 = 2 \cos 30^\circ = 2 \sin 60^\circ$, and by 2, followed by the addition or subtraction of them.

Accordingly we prepare beforehand a table of the products of the u_n by the numbers indicated (Fig. 39), and next a series of masks conforming in their dimensions with the table, and shown in Fig. 40a. These last are so designed that by superimposing them upon the table there will be visible only those numbers of the table that figure in the formation of the coefficient corresponding to the mask.

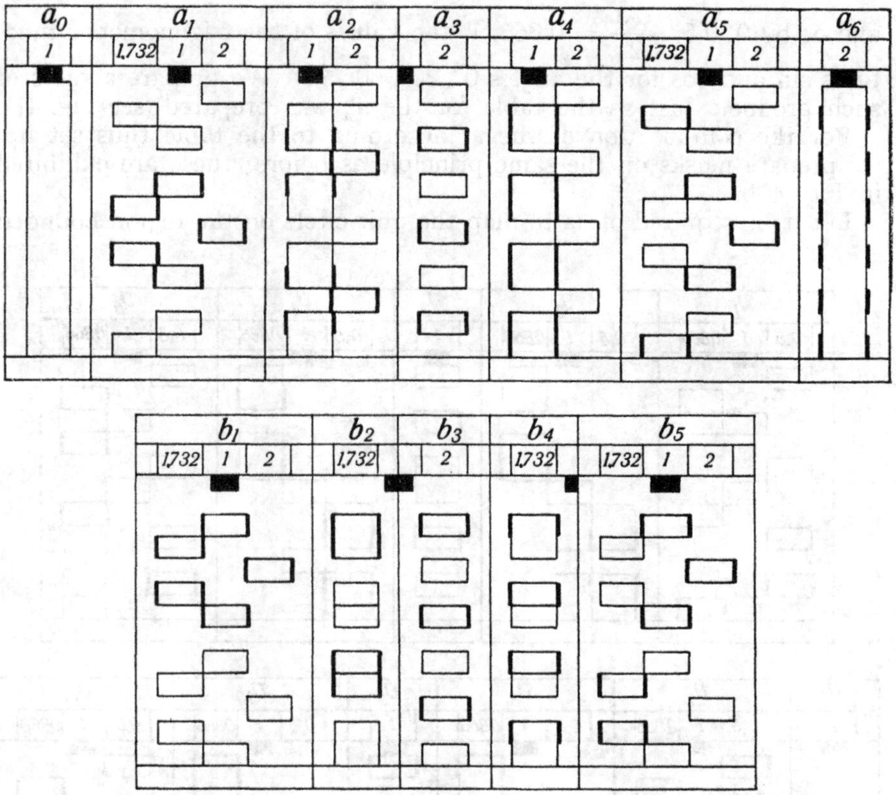

Fig. 40a.

Thus, for instance, in the mask for a_0 windows are made for all the numbers of the second column, containing the numbers u_n; in the mask for a_1 in the first column, $u_n\sqrt{3}$, are made windows for u_1, u_5, u_7 and u_{11}; in the second column, $u_n \cdot 1$, are made windows for u_2, u_4, u_8, u_{10}; and in the third column, $u_n \cdot 2$, windows for u_6 and $u_0 = u_{12}$, and so forth. Windows that give numbers to be taken with a minus sign we distinguish in some way from the other windows of the masks, for instance, as is done in our Figures, we mark them with heavied lines placed along the sides.

We proceed in the same way exactly during the computation of v_1, v_2, ...

$$v_1 = b_0 + (b_1 \cos 30° + a_1 \sin 30°) + (b_2 \cos 60° + a_2 \sin 60°) + \dots$$

. . . .

All v_n are formed of the sums and differences of the products of a_n

and b_n by 0, 0.5, $\frac{\sqrt{3}}{2} = 0.866$, 1, the values of the trigonometric functions sin and cos for the angles 0°, 30°, 60°, 90°. We prepare a table of such products just as the table for the u_n was prepared (see Fig. 41).

For the computation of the v_n according to the table thus set up, we prepare masks on the same principle as before; these are exhibited in Fig. 40b.

Let it be required of us to map the unit circle on the region bounded

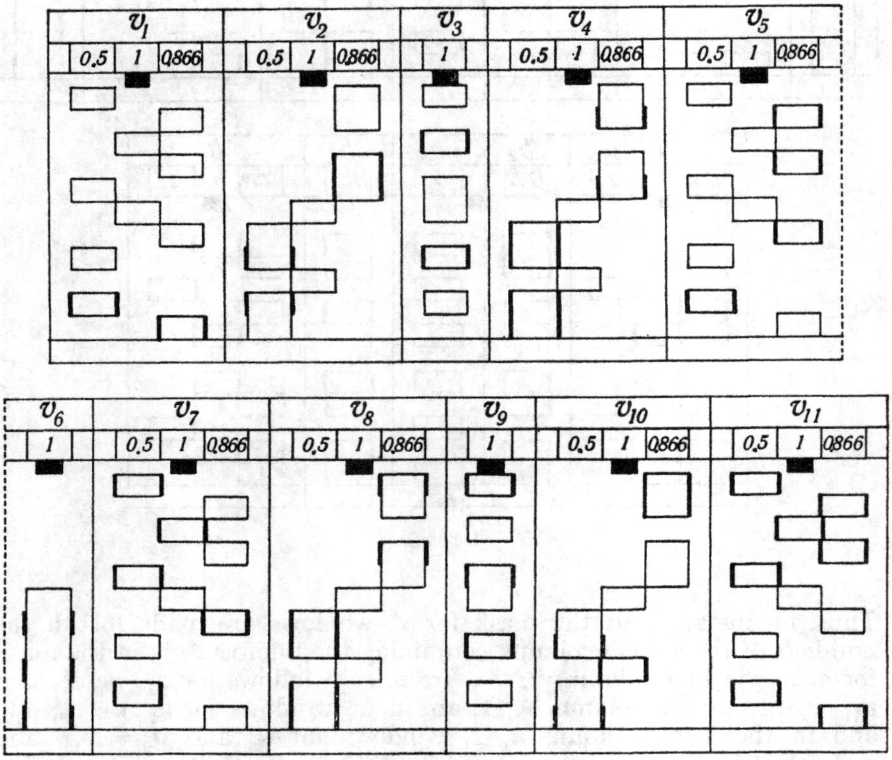

Fig. 40b.

by the curve depicted in Fig. 42 in reduced form. The construction of the initial values for the u_n and the multiplication of them by 1, 1.732, 2 give us the first three columns located in the upper left corner of Table 1, in which we record the computations. We superimpose on these columns the mask for a_0, sum all the numbers showing in the windows, and enter the number obtained (982.5) in the row a_0 in the first column marked by the sign $\Sigma +$, and, since there are in this case no negative summands, transfer the result of its division to the fourth column opposite a_0, marked by the sign :12. After this we superimpose the mask for a_1, sum all numbers in windows not marked by heavied lines and enter the result (758) in the column $\Sigma +$ opposite a_1, sum all numbers in marked windows and enter the result (491) in the column $\Sigma -$, subtract the second sum from the first, enter the difference (267) in column D, divide by 12, and enter the quotient (22.2) in the fourth column, etc.

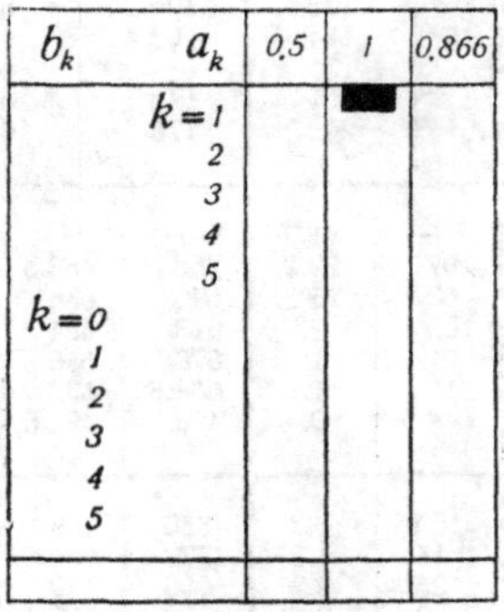

Fig. 41.

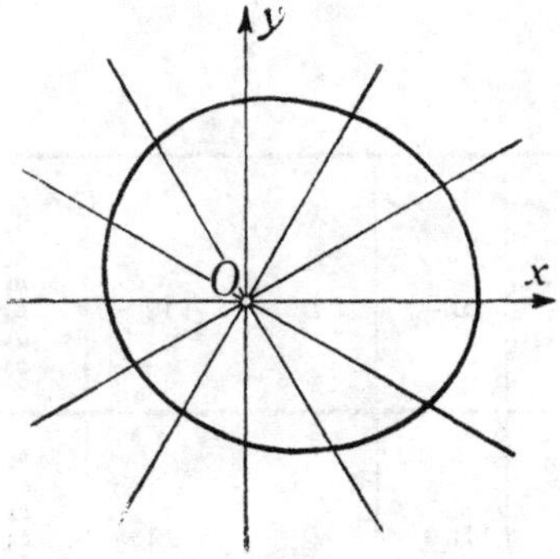

Fig. 42.

After b_1, b_2, b_3, b_4, b_5 are found, we sum them and, by taking the result with reversed sign, obtain b_0, which is indicated by us in the last line of the second section of the Table. On the basis of the values obtained for a_n and b_n we set up the table for the products of a_n, b_n and 0.5, $\frac{1}{2}\sqrt{3}$, 1. It occupies the beginning of the third section. Using the masks for the v_n, we find the first approximations for them in the usual way. The entries of the form need no explanations. On obtaining the v_n, we graphically determine new values for the u_n and continue the computations by the same method. The computations have been continued to the fifth approximation. One can judge the accuracy obtained by the fact that the fifth approximations for u_n differ from the fourth in only two cases (u_1 and u_{11}), the two coinciding in the rest; the relative error is less than 0.5%.

For the real parts a_n of the coefficients of the Taylor's series, the fourth and fifth approximations do not differ from one another. Matters are worse with the b_n; this fact can apparently be explained by the fact that it is those of them, with the exception of b_0, that are quantities of the order of the last figure of the adopted computational accuracy that show the strongest oscillation, and the oscillations are probably the result of rounding.

The approximate value of the mapping function will be

$$z = (82.2 + 4.5i)\zeta + (23.6 - 7.9i)\zeta^2 + (5.0 + 0.5i)\zeta^3 + \\ + (2.2 + 2.2i)\zeta^4 + (1.2 + 0.5i)\zeta^5 + (0.9 + 0.2i)\zeta^6 + 0.4\zeta^7.$$

u_k	1.732	1	2	1.732	1	2
$k = 0(12)$	—	■ 115.5	231	—	■ 115.5	231
1	176	101.5	203	181	104.5	209
2	164	94.5	189	169	97.5	195
3	—	88	176	—	89.5	179
4	135	78	156	136	78.5	157
5	148.5	68.5	137	117	67.5	135
6	—	63	126	—	62	124
7	106.5	61.5	123	106.5	61.5	123
8	107.5	62	124	108	62.5	125
9	—	68	136	—	68	136
10	139.5	80.5	161	143	82.5	165
11	176	101.5	203	176	101.5	203

Σ: $\Sigma+$, $\Sigma-$, D, $:12 =$	$\Sigma+$	$\Sigma-$	D	$:12$	$\Sigma+$	$\Sigma-$	D	$:12$
$12 = a_0$	982.5	—	—	81.9	991	—	—	82.6
a_1	758	491	267	22.2	768	488.5	279.5	23.3
a_2	690	627	63	5.2	690	636	54	4.5
a_3	511	476	35	2.9	513	484	29	2.4
a_4	669	648	21	1.7	670	656	14	1.2
a_5	631	618	13	1.1	634.5	622	12.5	1.0
a_6	493.5	489	4.5	0.4	498.5	492.5	6	0.5
b_1	545	645	100	−8.3	550	656	−106	−8.8
b_2	569	554	15	1.2	572	564.5	7.5	0.6
b_3	502	476	26	2.2	505	480	25	2.1
b_4	566	557	9	0.8	570	566.5	3.5	0.3
b_5	597	593	4	0.3	604	602	2	0.2
$-\Sigma b = b_0$				3.8				5.6

b_k / a_k	0.5	1	0.866	0.5	1	0.866
a_k: $k = 1$	11.1	■ 22.2	19.2	11.6	■ 23.3	20.2
2	—	5.2	4.5	—	4.5	3.9
3	—	2.9	—	—	2.4	—
4	—	1.7	1.5	—	1.2	1.1
5	0.5	1.1	1.0	0.5	1.0	0.9
b_k: $k = 0$	—	3.8	—	—	5.6	—
1	−4.1	−8.3	−7.2	−4.4	−8.8	−7.6
2	0.6	1.2	—	0.3	0.6	—
3	—	2.2	—	—	2.1	—
4	0.4	0.8	—	0.1	0.3	—
5	0.1	0.3	0.3	0.1	0.2	0.2

$\Sigma+$, $\Sigma-$, $D =$						
v_1	17.7	0.7	17.0	17.8	0.3	17.5
v_2	23.8	5.4	18.4	25.4	4.5	20.9
v_3	27.9	4.1	23.8	30.2	3.0	27.2
v_4	26.7	2.5	24.2	29.0	0.9	28.1
v_5	19.2	−0.8	20.0	20.6	−2.5	23.1
v_6	5.8	−5.8	11.6	6.5	−6.5	13.0
v_7	10.7	7.7	3.0	11.1	7.0	4.1
v_8	11.5	17.7	− 6.2	12.5	17.4	− 4.9
v_9	7.5	24.5	−17.0	8.3	24.9	−16.6
v_{10}	2.6	26.6	−24.0	3.3	26.6	−23.3
v_{11}	−2.8	21.2	−24.0	−1.7	19.8	−21.5

Table 1

1.732	1	2	1.732	1	2	1.732	1	2	1.732	1	2
	■			■			■			■	
—	115.5	231	—	115.5	231	—	115.5	231	—	115.5	231
181	104.5	209	181	104.5	209	181	104.5	209	180	104	208
168	97	194	168	97	194	167	96.5	193	167	96.5	193
—	88	176	—	88	176	—	88	176	—	88	176
131.1	76	152	132.5	76.5	153	132.5	76.5	153	132.5	76.5	153
113.5	65.5	131	116	67	134	115	66.5	133	115	66.5	133
—	62	124	—	62.5	125	—	62	124	—	62	124
106.5	61.5	123	106.5	61.5	123	106.5	61.5	123	106.5	61.5	123
108	62.5	125	108	62.5	125	108	62.5	125	108	62.5	125
—	68	136	—	67.5	135	—	68	136	—	68	136
143.5	83	166	141	81.5	163	143	82.5	165	143	82.5	165
178	103	206	176.5	102	204	176.5	102	204	177.5	102.5	205

986.5	—	—	82.2	986	—	—	82.2	986	—	—	83.2	986	—	—	82.2
770	482.5	287.5	24.0	767	486.5	280.5	23.4	767.5	484.5	283	23.6	767.5	484.5	283	23.6
689.5	630.5	59	4.9	691	628.5	62.5	5.2	689.5	630	59.5	5.0	689.5	630	59.5	5.0
508	484	24	2.0	509	482	22	2.2	509	482	27	2.2	509	482	27	2.2
667	653	14	1.2	667	652.5	14.5	1.2	667	652.5	14.5	1.2	667	652.5	14.5	1.2
631	621.5	9.5	0.8	632	621.5	10.5	0.9	631.5	620.5	11	0.9	631.5	620.5	11	0.9
496	490.5	5.5	0.5	495.5	490.5	5	0.4	495.5	490.5	5	0.4	495.5	490.5	5	0.4

552	645.5	−93.5	−7.8	547.5	648	−100.5	−8.4	550.5	646.5	−96	−8.0	551	646	−95	−7.9
566.5	563.5	3	0.2	566	563.5	2.2	0.2	557	562.5	4.5	0.4	568	561.5	6.5	0.5
505	476	29	2.4	503	478	2.5	2.1	503	478	2.5	2.1	504	477	2.5	2.2
567.5	562.5	5	0.4	568.5	561	7.5	0.6	566.5	563	3.5	0.3	567.5	562	5.5	0.5
600	597.5	2.5	0.2	599	596.5	2.5	0.2	599	598	1	0.1	599.5	597.5	2	0.2
			4.6				5.3				5.1				4.5

0.5	1	0.866	0.5	1	0.866	0.5	1	0.866	0.5	1	0.866
	■			■			■				
12.0	24	20.8	11.7	23.4	20.3	11.8	23.6	20.5			
—	4.9	4.2	—	5.2	4.5	—	5.0	4.3			
—	2.0	—	—	2.2	—	—	2.2	—			
—	1.2	1.0	—	1.2	1.0	—	1.2	1.0			
0.4	0.8	0.7	0.4	0.9	0.8	0.4	0.9	0.8			
—	4.6	—	—	5.3	—	—	5.1	—			
−3.9	−7.8	−6.8	−4.2	−8.4	−7.3	−4.0	−8.0	−7.0			
0.1	0.2	—	0.1	0.2	—	0.2	0.4	—			
—	2.4	—	—	2.1	—	—	2.1	—			
0.2	0.4	—	0.3	0.6	—	0.1	0.3	—			
0.1	0.2	0.2	0.1	0.2	0.2	0	0.1	0.1			

17.5	0.4	17.1	17.9	0.5	17.4	18.0	0.2	17.8			
25.9	4.3	21.6	26.2	4.1	22.1	26.0	4.1	21 9			
29.8	2.2	27.6	29.2	2.4	26.8	29.9	2.6	27.3			
28.8	1.4	27.4	28.7	1.6	27.1	28.7	1.4	27.3			
19.3	−1.4	20.7	19.9	−1.5	21.4	19.8	−1.6	21.4			
5.2	−5.2	10.4	6.1	−6.1	12.2	5.8	−5.8	11.6			
10.1	7.8	3.3	11.1	7.3	3.8	10.7	7.5	3.2			
11.9	18.3	− 6.4	12.7	17.6	− 4.9	12.3	17.8	− 5.5			
7.0	25.0	−18.0	8.1	24.5	−16.4	7.6	24.9	−17.3			
2.6	27.6	−25.0	3.2	27.1	−23.9	3.0	27.1	−24.1			
−2.1	20.5	−22.1	−1.9	20.3	−22.2	19.9	−1.7	−21.6			

3. The mapping of exterior regions. We shall seek the function conformally transforming the exterior of the unit circle into the exterior of a curve.

In case of the preservation of the point at infinity under the transformation, it is known to have the form:

$$z = \alpha_0 \zeta + \alpha_1 + \alpha_2 \zeta^{-1} + \ldots + \alpha_n \zeta^{1-n} + \ldots \tag{13}$$

We shall demand, for definiteness, that the point A of the real axis of the ζ-plane shall again pass to the point A' of the real axis of the z-plane.

We shall again assume that, with the adopted computational accuracy, the mapping is accomplished by the following rational function, which is a segment of series (13):

$$z = \alpha_0 \zeta + \alpha_1 + \alpha_2 \zeta^{-1} + \ldots + \alpha_{\frac{m}{2}} \zeta^{1-\frac{m}{2}}. \tag{14}$$

Let w have its previous meaning.

The relation of dependence between a_n, b_n and the real and imaginary parts u, v for $|\zeta| = 1$ will be given here by the equations:

$$\left.\begin{aligned} u &= \sum_{k=0}^{m/2} (a_k \cos k\vartheta + b_k \sin k\vartheta), \\ v &= \sum_{k=0}^{m/2} (b_k \cos k\vartheta - a_k \sin k\vartheta). \end{aligned}\right\} \tag{15}$$

From these equations it is evident that compared with (8) and (3):

1) the dependence of a_k on u_n does not change;

2) in obtaining b_k we must take a different sign in the sum than the former one;

3) in computing v_n all a_k are taken with opposite sign.

All the rest is preserved. The computations are performed by the same scheme and with the use of the same masks.

To find the first approximation a device can be indicated quite like that applied by us to the mapping of interior regions.

As the auxiliary function one can here take:

$$z = a\zeta \frac{1 + b\zeta^{-p}}{1 - c\zeta^{-p}}.$$

All the previous formulas are preserved, now with only one distinction: $(-\theta)$ must be taken for $\theta = p\vartheta$ everywhere in them. This, as is at once evident, has as its consequence the circumstance that of the two points of intersection of the auxiliary circle of radius r with the curve L we will have to take that which is the farther from the origin of coordinates.

Table 2

u_k	1.732	1	2	1.732	1	2
		■			■	
$k = 0(12)$	—	115.5	231	—	115.5	231
1	190.5	117	220	190.5	110	220
2	174	100.5	201	174	100.5	201
3	—	95	190	—	96	192
4	154	89	178	155	89.5	179
5	138.5	80	160	138.5	80	160
6	—	68.5	137	—	68.5	137
7	106.5	61.5	123	106.5	61.5	123
8	116	67	134	116	67	134
9	—	84	168	—	84	168
10	173	100	200	173	100	200
11	191	110.5	221	191	110.5	221

Σ			: 12 =	a_0	1081.0			90.1	1083			90.2
				a_1	813	538	275	22.9	813	538.5	274.5	22.9
				a_2	730	714.5	15.5	1.3	730	717	13	1.1
Σ +	Σ −	D	: 12	a_3	543	538	5	0.4	544	538	5	0.5
				a_4	726	718.5	7.5	0.6	728	719	9	0.7
				a_5	676.5	674.5	2	0.2	676.5	675	1.5	0.1
				a_6	540.5	541	−0.5	0	541	542	−1	−0.1

				b_1	629	708	79	6.6	629	711	82	6.8
				b_2	656.5	587	−69.5	−5.8	657.5	587	−70.5	−5.9
Σ +	Σ −	D	: 12	b_3	534	548	14	1.2	536	548	12	1.0
				b_4	619.5	624	4.5	0.4	619.5	625	5.5	0.5
				b_5	668	669	1	0.1	669	671	2	0.2
			$- \Sigma b = b_0$					−2.5				−2.6

b_k	$-a_k$	0.5	1	0.866	0.5	1	0.866
			■			■	
	$k = 1$	−11.4	−22.9	−19.8			
	2	—	− 1.3	− 1.1			
	3	—	− 0.4	—			
	4	—	− 0.6	− 0.5			
	5	− 0.1	− 0.2	− 0.2			
$k = 0$		—	− 2.5	—			
1		3.3	6.6	5.7			
2		− 2.9	− 5.8	—			
3		—	1.2	—			
4		0.2	0.4	—			
5		0	0.1	0.1			

			v_1	−13.2	0.3	−13.5			
			v_2	−19.9	− 2.4	−17.5			
			v_3	−25.2	− 6.2	−19.0			
			v_4	−21.6	− 0.7	−20.9			
			v_5	−17.2	4.3	−21.5			
Σ +	Σ +	$D =$	v_6	− 7.9	7.9	−15.8			
			v_7	− 6.9	− 6.0	− 0.9			
			v_8	− 2.6	−19.7	17.1			
			v_9	− 2.5	−28.9	26.4			
			v_{10}	0.3	−22.6	22.9			
			v_{11}	0.3	−13.2	13.5			

For the case of mapping the exterior of the circle onto the exterior of the curve of Fig. 42 we exhibit the calculations.

As has already been stated, to obtain b_n we take in it not $\frac{D}{12}$, but $-\frac{D}{12}$, and to determine v_n in Table 2, we copy the products, not of a_n but of $-a_n$, by 0.5, $\frac{1}{2}\sqrt{3}$, 1.

As is seen in Table 2, the convergence is very rapid: the initial graphical approximation has already given a sufficiently accurate result.

The mapping function has approximately the following expression:

$$z = (90.2 - 2.6i)\zeta + (22.9 + 6.8i) + (1.1 - 5.9i)\zeta^{-1} + \\ + (0.5 + 0.1i)\zeta^{-2} + (0.7 + 0.5i)\zeta^{-3} + (0.1 + 0.2i)\zeta^{-4} - 0.1\zeta^{-5}.$$

4. The case of symmetric contours. Examples. If the region B bounded by the curve L has p axes of symmetry, the mapping function will have the form:

$$z = \sum_{k=0}^{\infty} a_k \zeta^{1+kp}, \tag{16}$$

for interior regions and

$$z = \sum_{k=0}^{\infty} a_k \zeta^{1-kp}, \tag{17}$$

for exterior regions.

Relations (6) and (15) will take the form:

$$u = \sum_{k=0}^{m} a_k \cos kp\vartheta,$$

$$v = \pm \sum_{k=0}^{m} a_k \sin kp\vartheta,$$

where the sign + is taken for interior regions and − for exterior. Let us consider the angle between the real axis and the nearest axis of symmetry lying in the direction of positive rotation from it. This angle is equal to $\frac{\pi}{p}$.

We divide it into m equal parts $\frac{\pi}{pm}$ and put

$$\vartheta = 0, \quad \frac{\pi}{pm}, \ldots, \frac{\pi}{pm}(m-1).$$

Denote the values of the functions u and v that correspond to these rays by $u_0, v_0, u_1, v_1, \ldots$

$$\left.\begin{aligned} u_n &= \sum_{k=0}^{m} a_k \cos kp\frac{\pi}{pm} n = \sum_{k=0}^{m} a_k \cos\frac{\pi}{m} kn, \\ v_n &= \pm \sum_{k=0}^{m} a_k \sin kp\frac{\pi}{pm} n = \pm \sum_{k=0}^{m} a_k \sin\frac{\pi}{m} kn. \end{aligned}\right\} \tag{18}$$

From them are simply obtained the equations for the determination of the coefficients a_k:

$$\left.\begin{aligned} a_0 &= \frac{1}{2m}(u_0 + u_m) + \frac{1}{m}\sum_{n=1}^{m-1} u_n, \\ a_m &= \frac{1}{2m}[u_0 + (-1)^m u_m] + \frac{1}{m}\sum_{n=1}^{m-1}(-1)^n u_n, \\ a_k &= \frac{1}{m}[u_0 + (-1)^k u_m] + \frac{2}{m}\sum_{n=1}^{m-1} u_n \cos\frac{\pi}{m} nk, \\ &\qquad (1 \leqslant k \leqslant m-1). \end{aligned}\right\} \quad (19)$$

Since p does not appear in formulas (18) and (19), the calculations remain the same for all cases of axes of symmetry. The number p influences only the disposition of the rays.

To illustrate the arrangement of the computations in using formulas (18) and (19) for the solution of problems, we shall dwell on the case $m = 6$. Equations (19) will then take the form:

$$\left.\begin{aligned} a_0 &= \tfrac{1}{12}[u_0 + u_6 + 2(u_1 + u_2 + \ldots + u_5)], \\ a_6 &= \tfrac{1}{12}[u_0 + u_6 + 2(-u_1 + u_2 - u_3 + u_4 - u_5)], \\ a_1 &= \tfrac{1}{6}[u_0 - u_6 + \sqrt{3}(u_1 - u_5) + (u_2 - u_4)], \\ a_2 &= \tfrac{1}{6}[u_0 + u_6 + u_1 - u_2 - 2u_3 - u_4 + u_5], \\ a_3 &= \tfrac{1}{6}[u_0 - u_6 - 2u_2 + 2u_4], \\ a_4 &= \tfrac{1}{6}[u_0 + u_6 - u_1 - u_2 + 2u_3 - u_4 - u_5], \\ a_5 &= \tfrac{1}{6}[u_0 - u_6 - \sqrt{3}u_1 + u_2 - u_4 + \sqrt{3}u_5]. \end{aligned}\right\} \quad (20)$$

The equations from which the v_n must be determined will have the form:

$$\left.\begin{aligned} v_1 &= \pm\left[0.5(a_1 + a_5) + \frac{\sqrt{3}}{2}(a_2 + a_4) + a_3\right], \\ v_2 &= \pm\frac{\sqrt{3}}{2}[a_1 + a_2 - a_4 - a_5], \\ v_3 &= \pm[a_1 - a_3 + a_5], \\ v_4 &= \pm\frac{\sqrt{3}}{2}[a_1 - a_2 + a_4 - a_5], \\ v_5 &= \pm\left[0.5(a_1 + a_5) - \frac{\sqrt{3}}{2}(a_2 + a_4) + a_3\right]. \end{aligned}\right\} \quad (21)$$

For the computation of the successive approximations to the coefficients a_k we prepare a table of the values of the u_n and their products with 2, $\sqrt{3}=1.732$ (Fig. 43), and masks for each of the a_k (Fig. 44).

To determine the v_n, we form for the same purpose a table (Fig. 45)

Table 3

u_k	1.732	1	2	1.732	1	2	1.732	1	2
$k = 0$	—	■ 150	—	—	■ 150	—	—	■ 150	
1	232	134	268	233	134.5	269	—	134.5	
2	—	118.5	237	—	120	240	—	120	
3	—	109.5	219	—	110	220	—	110	
4	—	104	208	—	104.5	209	—	104.5	
5	175	101	202	175	101	202	—	101	
6	—	100	—	—	100	—	—	100	

Σ / Σ+	Σ−	D		Σ, Σ+	Σ−	D		Σ, Σ+	Σ−	D	
Σ			$: 12 = a_0$	1384			115.3	1390			115.8
Σ+	Σ−		a_1	500.5	379	121.5	20.2	503	379.5	123.5	20.6
			a_2	485	441.5	43.5	7.2	485.5	444.5	41.0	6.8
		$: 6$	a_3	358	337	21	3.5	359	340	19	3.2
		D	a_4	469	457.5	11.5	1.9	470	460	10	1.7
			a_5	443.5	436	7.5	1.2	445	437.5	7.5	1.2
			$: 12 = a_6$	695	689	6	0.5	699	691	8	0.7

a_k	0.5	1	0.866	0.5	1	0.866			
$k = 1$	10.1	■ 20.2	17.5	10.3	■ 20.6	17.9			
2	—	7.2	6.2	—	6.8	5.8			
3	—	3.5	—	—	3.2	—			
4	—	1.9	1.6	—	1.7	1.5			
5	0.6	1.2	1	0.6	1.2	1.0			

Σ+	Σ−	D =										
			v_1			22.0			21.4			
			v_2	23.7	2.6	21.1	23.7	2.5	21.2			
			v_3	21.4	3.5	17.9	21.8	3.2	18.6			
			v_4	19.1	7.2	11.9	19.4	6.8	12.6			
			v_5	14.2	7.8	6.4	14.1	7.3	6.8			

Table 4

u_k	1.732		1	2	1.732		1	2
$k = 0$	—		■ 100	—	—		■ 100	—
1	176.5		102	204	176.5		102	204
2	—		108	216	—		107.5	215
3	—		116.5	233	—		116.5	233
4	—		126.5	253	—		126	252
5	235.5		136	172	235.5		136	272
6	—		141.5	—	—		141.5	—
$\Sigma \quad : 12 = a_0$	1419.5			118.3	1417.5			118.1
a_1	384.5	503.5	−119	−19.8	384	503	−119	−19.8
a_2	479.5	467.5	12	2.0	479.5	466.5	13	2.2
$\Sigma + \quad \Sigma - \quad D \quad : 6 = a_3$	353	357.5	−4.5	−0.7	352	356.5	−4.5	−0.7
a_4	474.5	472.5	2	0.3	474.5	471.5	3	0.5
a_5	443.5	444.5	−1	−0.2	443	444	−1	−0.2
$: 12 = a_6$	710.5	709	1.5	0.1	708.5	709	0.5	0
$-a_k$	0.5		1	0.866	0.5		1	0.866
$k = 1$	9.9		■ 19.8	17.1			■	
2	—		−2.0	−1.7				
3	—		0.7	—				
4	—		−0.3	−0.3				
5	0.1		0.2	−0.2				
v_1				8.7				
v_2	15.0		−0.1	15.5				
$\Sigma + \quad \Sigma - \quad D = \quad v_3$	20.0		0.7	19.3				
v_4	16.8		−0.5	18.3				
v_5	10.7		2.0	12.7				

u_k	1.732	1	2
$k=0$			
1			
2			
3			
4			
5			
6			

Fig. 43.

and the masks depicted in Fig. 46. The principle of their formation remains the previous one, requiring no explanations.

Table 3 gives the computations of the transformation of a circle of radius 1 mm. into the interior of an ellipse of semi-axes 100 and 150 mm. The transforming function has the form

$$z = 115.8\zeta + 20.6\zeta^3 + 6.8\zeta^5 + 3.2\zeta^7 + \\ + 1.7\zeta^9 + 1.2\zeta^{11} + 0.7\zeta^{13}.$$

In Table 4 are exhibited the computations for determining the function giving the conformal transformation of the exterior of the circle into the exterior of a square with side 200 mm. As the result of the computations we obtain:

$$z = 118.1\zeta - 19.8\zeta^{-3} + 2.2\zeta^{-7} - 0.7\zeta^{-11} + 0.5\zeta^{-15} - 0.2\zeta^{-19}.$$

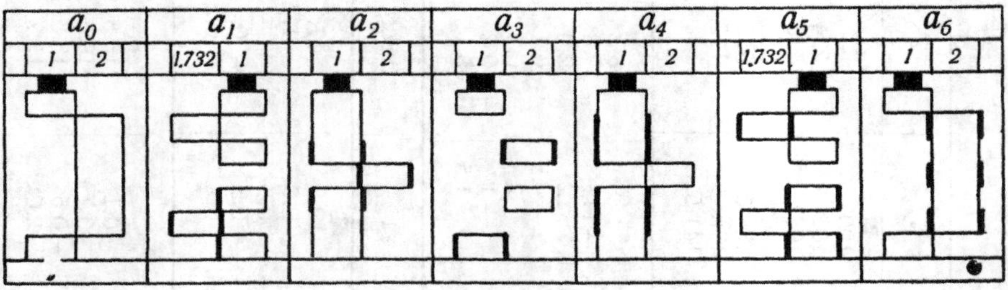

Fig. 44.

a_k	1.732	1	2
$k=1$			
2			
3			
4			
5			

Fig. 45.

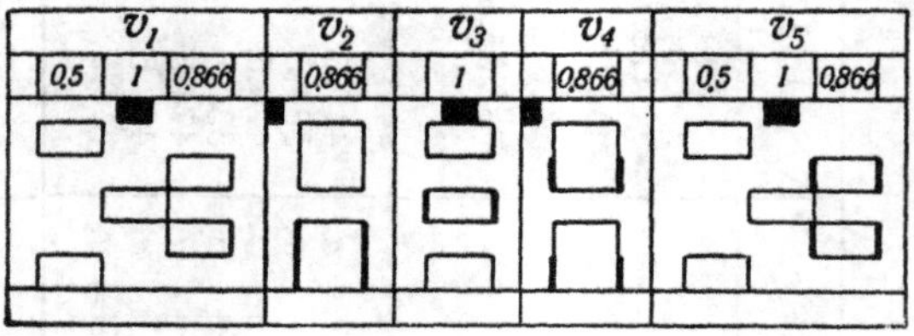

Fig. 46.

§ 8. GREEN'S FUNCTIONS AND THE CONFORMAL TRANSFORMATION OF REGIONS

1. Introduction. Green's function for the Dirichlet problem. In the solution of problems of mathematical physics and engineering connected with harmonic functions, in many cases special harmonic functions, which possess certain singularities of which we shall speak below, are frequently found to be useful. These functions are called Green's functions, after the English physicist who first employed them.

Their definitions and properties depend essentially on what kind of problem is being investigated, first, and second, on what is the number of dimensions of the space in which this problem is being solved. We shall presently interest ourselves in the plane case, where the harmonic function depends on two arguments; it is this case for which we shall consider Green's functions for the most frequently encountered problems.

We shall elucidate the connection between Green's function for one or another problem that we solve in the given region B, and the problem of the conformal transformation of that region into some other region of canonical type. The form of these canonical regions for each problem we will make clear below.

In the exposition we shall limit ourselves to simply connected regions only, since the case of multiply connected regions is more complicated and we shall not find it possible to dwell on it in this book.

To facilitate the reading, we shall here repeat the familiar definitions of Green's functions, and shall derive several necessary formulas. We remark that the definitions will of course not depend on the connectivity of the region, being identical for regions of any connectivity. The simple connectivity of the region will have significance only in establishing the connection between Green's function and the problem of the conformal transformation of the region.

We shall begin with the Dirichlet problem. In the plane referred to the coordinate system xy let there be given a region B bounded by the contour L. We shall choose some point in it, $M_0(x_0, y_0)$, and fix it. By Green's function for the Dirichlet problem in the region B with pole at M_0 there is understood a function $G(x, y; x_0, y_0)$ satisfying the following requirements:

1. $G(x, y; x_0, y_0)$, as a function of x and y, is harmonic everywhere in B with the exception of the point M_0.

2. Near the point M_0 the function $G(x, y; x_0, y_0)$ has the structure

$$G = \frac{1}{2\pi} \ln \frac{1}{r} + g(x, y, x_0, y_0),$$

where $g(x, y; x_0, y_0)$ is a function harmonic in B everywhere, including even the point M_0, and r is the distance M_0M from the point M_0 chosen by us to the variable point M:

$$r = \sqrt{(x - x_0)^2 + (y - y_0)^2}.$$

3. $G(x, y; x_0, y_0)$ vanishes when the point M lies on the contour L.

Relying on the theorem of the uniqueness of the solution of the Dirichlet problem, it can easily be shown that by these three requirements the function $G(x, y; x_0, y_0)$ is completely defined.

By means of $G(x, y; x_0, y_0)$, as we shall show forthwith, the Dirichlet problem can be solved in explicit form for the region B for any continuous boundary values. To obtain the necessary expression for the harmonic function at any point of the region B in terms of its values on the boundary and $G(x, y; x_0, y_0)$, we shall consider Green's formula. If $u(x, y)$ and $v(x, y)$ are continuous in B, together with their derivatives of up to the second order inclusive, then the following equation holds:

$$\iint\limits_B (u\Delta v - v\Delta u)d\sigma = \int\limits_L \left(u\frac{\partial v}{\partial n} - v\frac{\partial u}{\partial n}\right) ds. \tag{1}$$

Here $\Delta = \dfrac{\partial^2}{\partial x^2} + \dfrac{\partial^2}{\partial y^2}$ is the Laplacian operator, $d\sigma$ is the element of area of the region B, ds is the element of length of the arc of the contour L and n is the exterior normal to the contour L at the point $M(x, y)$.

In the formula let u be the harmonic function we require, and let v be Green's function $G(x, y; x_0, y_0)$ with pole at M_0.

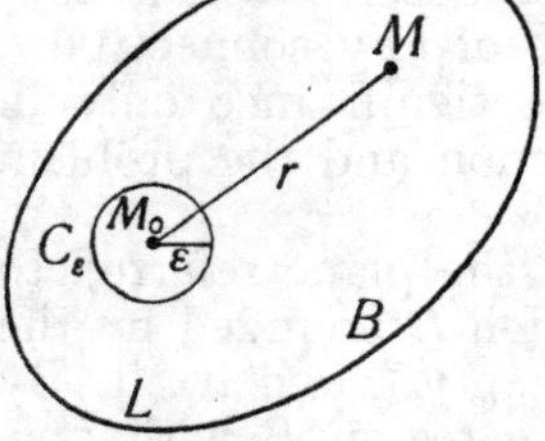

Fig. 47.

Green's formula (1) cannot be directly applied to u and G, since G has a discontinuity at the point M_0; it follows, namely, from condition (2) of the definition of Green's function, that at M_0 it becomes infinite. Let us describe about M_0 as a center the circle C_ε of small radius ε and exclude from the region the circle with boundary C_ε (Fig. 47). We shall have obtained some new region B_ε whose boundary consists of the curves L and C_ε. In B_ε both u and $G(x, y; x_0, y_0)$ will satisfy all the conditions of continuity that Green's formula (1) demands, and the formula can be applied to u and $G(x, y; x_0, y_0)$ in B_ε.

If one takes it into account that in B_ε $\Delta u = 0$ and $\Delta G = 0$, then from (1) it follows that

$$\int\limits_{L+C_\varepsilon} \left(u\frac{\partial G}{\partial n} - G\frac{\partial u}{\partial n}\right) ds = 0. \tag{2}$$

We recall that by n must here be understood the normal to the contour L that is exterior to B_ε. On L this will be the normal to L exterior to B, and on C_ε this is the interior normal to C_ε.

The integral over $L + C_\varepsilon$ we shall separate into two integrals, around L and around C_ε. The first of them does not depend on ε; we shall leave it aside for the time being and investigate the integral around C_ε. If for $G(x, y; x_0, y_0)$ in it, its value $G = \dfrac{1}{2\pi} \ln \dfrac{1}{r} + g$ be substituted, this integral will be separated into the following four integrals:

$$\int_C \left(u\frac{\partial G}{\partial n} - G\frac{\partial u}{\partial n}\right) ds = \frac{1}{2\pi}\int_{C_\varepsilon} u \frac{\partial \ln \frac{1}{r}}{\partial n} ds +$$

$$+ \int_{C_\varepsilon} u \frac{\partial g}{\partial n} ds - \frac{1}{2\pi}\int_{C_\varepsilon} \ln\frac{1}{r}\frac{\partial u}{\partial n} ds - \int_{C_\varepsilon} g\frac{\partial u}{\partial n} ds.$$

Here we shall diminish ε without limit and determine the limit to which each of the cited integrals tends.

The interior normal to C_ε is directed along the radius $r = M_0M$. But since the direction of this normal is opposite to the direction of increase of r, we have

$$\frac{\partial}{\partial n} = -\frac{\partial}{\partial r},$$

and consequently

$$\frac{\partial \ln \frac{1}{r}}{\partial n} = -\frac{\partial \ln \frac{1}{r}}{\partial r} = \frac{1}{r}.$$

Moreover on the circle, r has the constant value ε, and therefore the first of the integrals we are interested in will be equal to

$$\frac{1}{2\pi}\int_{C_\varepsilon} u \frac{\partial \ln \frac{1}{r}}{\partial n} ds = \frac{1}{2\pi\varepsilon}\int_{C_\varepsilon} u\, ds.$$

On the right side of the equation stands the arithmetic mean of the values of the harmonic function u on the circle C_ε. By Gauss's theorem, it is equal to the value $u(x_0, y_0)$ u acquires at the center M_0 of the circumference C_ε. This means that for any ε the integral under consideration equals $u(x_0, y_0)$.

Let us now take the third of the integrals under consideration. We take into account the fact that here $r = \varepsilon$, so that

$$\frac{1}{2\pi}\int_{C_\varepsilon} \ln\frac{1}{r}\frac{\partial u}{\partial n} ds = \frac{\ln\frac{1}{\varepsilon}}{2\pi}\int_{C_\varepsilon}\frac{\partial u}{\partial n} ds.$$

But, by a familiar property of an harmonic function, $\int_{C_\varepsilon}\frac{\partial u}{\partial n} ds = 0$, and accordingly this integral is equal to zero for any ε.

As regards the two remaining integrals, it is easily proved that they both tend to zero as $\varepsilon \to 0$. Indeed:

$$\left| \int_{C_\varepsilon} u \frac{\partial g}{\partial n} ds \right| \leqslant \left| u \frac{\partial g}{\partial n} \right|_{\max} \cdot 2\pi\varepsilon,$$

$$\left| \int_{C_\varepsilon} g \frac{\partial u}{\partial n} ds \right| \leqslant \left| g \frac{\partial u}{\partial n} \right|_{\max} \cdot 2\pi\varepsilon.$$

Since u, g, $\frac{\partial u}{\partial n}$ and $\frac{\partial g}{\partial n}$ in the neighborhood of the point M_0 are bounded functions, it is clear that the right members of these inequalities, and consequently the integrals themselves, will tend to zero as $\varepsilon \to 0$.

At the limit, when $\varepsilon = 0$, equation (2) gives the following result:

$$\int_L \left(u \frac{\partial G}{\partial n} - G \frac{\partial u}{\partial n} \right) ds + u(x_0, y_0) = 0.$$

Hence, in view of the fact that, by condition 3, G vanishes on L, we obtain the value we need, $u(x_0, y_0)$, of the harmonic function $u(x, y)$ at any interior point M_0 of the region B, in terms of the values that $u(x, y)$ takes on the boundary L of the region:

$$u(x_0, y_0) = -\int_L u \frac{\partial G}{\partial n} ds. \tag{3}$$

If Green's function $G(x, y; x_0, y_0)$ is known for B, this formula permits one to solve the Dirichlet problem for any continuous, or even piecewise continuous, boundary values of the harmonic function $u(x, y)$.

Let us now go on to establish the connection between Green's function and conformal transformation.

We shall now consider the region B to be simply connected. We shall introduce the complex variable $z = x + iy$, calling $z_0 = x_0 + iy_0$ the complex coordinate of the pole M_0 of the Green function $G(x, y; x_0, y_0)$.

Let us consider the function $\zeta = \varrho e^{i\vartheta} = f(z)$ giving the conformal transformation of the region B into a circle of unit radius $|\zeta| < 1$ of the ζ-plane, such that the point z_0 transforms into the center of the circle: $f(z_0) = 0$.

We have in view showing that:

$$G(x, y; x_0, y_0) = \frac{1}{2\pi} \ln \frac{1}{\varrho} = \frac{1}{2\pi} \ln \frac{1}{|f(z)|}. \tag{4}$$

The point z_0 is the sole zero of the function $f(z)$. Its multiplicity being unity, the ratio

$$\frac{f(z)}{z - z_0} = \chi(z)$$

will be a function regular everywhere in B and different from zero. The logarithm of $\chi(z)$ is accordingly also a regular function in B. Denote it by $u + iv$; then

$$f(z) = (z - z_0)e^{u+iv}.$$

Hence

$$\frac{1}{2\pi} \ln \frac{1}{|f(z)|} = \frac{1}{2\pi} \ln \frac{1}{r} - \frac{1}{2\pi} u, \; z - z_0 = re^{i\varphi}.$$

It is easily verified that $\frac{1}{2\pi} \ln \frac{1}{|f(z)|}$ satisfies all three requirements that we formulated above and which define uniquely Green's function $G(x, y; x_0, y_0)$. Indeed, as the logarithm of the modulus of the regular function $\frac{1}{f(z)}$ with a single simple pole at z_0, this function will be harmonic everywhere in B, with the exception of the point z_0 itself. In the neighborhood of this point, however, it has the required singularity of logarithmic type, $\frac{1}{2\pi} \ln \frac{1}{r} + g(x, y)$, where $g(x, y) = -\frac{1}{2\pi} u$.

On the boundary of the region, L, $|f(z)|$ reduces to unity, and therefore $\frac{1}{2\pi} \ln \frac{1}{|f(z)|}$ vanishes there.

By this it has been proved that Green's function $G(x, y; x_0, y_0)$ and the function $f(z)$ furnishing the conformal transformation of B into the unit circle $|\zeta| < 1$ are connected with each other by relation (4).

For the questions of conformal transformation with which we are engaged in this chapter, it is important to note that if Green's function $G(x, y; x_0, y_0)$ for the simply connected region is known, then by it one can construct the function $f(z)$ conformally transforming B into a circle of unit radius.

Indeed, let the function $G = \frac{1}{2\pi} \ln \frac{1}{r} + g(x, y)$ be known. For each of the two terms of G, we construct the conjugate harmonic function. For

$$\frac{1}{2\pi} \ln \frac{1}{r}$$

this will be

$$-\frac{1}{2\pi} \varphi,$$

where φ is the argument of $z - z_0$. As regards the function conjugate to $g(x, y)$, it is recovered by means of the well-known curvilinear integral, accurate to an arbitrary constant summand,

$$h(x, y) = \int_{(x_0, y_0)}^{(x, y)} \left(\frac{\partial u}{\partial x} dy - \frac{\partial u}{\partial y} dx \right) + C.$$

The function $f(z)$ effecting the required transformation of the region B will be given by the following equation:

$$\zeta = f(z) = re^{i\varphi} \cdot e^{-2\pi(g+ih)} = (z - z_0)\, e^{-2\pi(g+ih)}.$$

The arbitrary constant appearing in $h(x, y)$ corresponds to the rotation of the circle $|\zeta| < 1$ about the origin of coordinates.

The construction of the Green function $G(x, y; x_0, y_0)$ itself is equivalent to the determination of the harmonic function $g(x, y)$. Its boundary values are determined from the consideration that on L $G(x, y; x_0, y_0)$ must vanish and consequently $g(x, y)$ must take on L values equal to $-\dfrac{1}{2\pi} \ln \dfrac{1}{r}$. Thus the problem of the conformal transformation of the region B into a circle is reduced to the solution of the Dirichlet problem for the boundary condition

$$g(x, y) \Big|_L = \frac{1}{2\pi} \ln r.$$

We will make one more simple observation, which it is useful to keep in view in practical computations. Knowledge of the Green function with pole at the point $z_0 = x_0 + iy_0$ will permit the finding, by formula (3), of the value of the solution of the Dirichlet problem at the point z_0. To compute by this formula the solution at another point, for example at $z_1 = x_1 + iy_1$, Green's function must be recomputed, now placing the pole at the point z_1; let us denote Green's function for this new position of the pole by $G(x, y; x_1, y_1)$. The two functions with poles at z_0 and z_1 prove to be connected with one another by a very simple relation, which we now intend to establish. Let $G(x, y; x_0, y_0)$ be known; we shall show how $G(x, y; x_1, y_1)$ can be established by it.

By the known $G(x, y; x_0, y_0)$ we find the function $\zeta = f(z)$ corresponding to it and effecting the conformal transformation of B into the unit circle $|\zeta| < 1$ so that z_0 transforms into the center of the circle. These two functions are connected with each other by relation (4).

In the transformation $\zeta = f(z)$ the point z_1 transforms into some point $\zeta_1 = f(z_1)$ that can be found by the known function $f(z)$. In order to determine $G(x, y; x_1, y_1)$ by formula (4), we have to find the function $\zeta = f_1(z)$ that would transform the region B into the same unit circle $|\zeta| < 1$, but such that the point z_1 is transformed into the center of the circle. If $f(z)$ is known, then for the determination of $f_1(z)$ it is only necessary to transform the unit circle $|\zeta| < 1$ into itself, in such a way that the point ζ_1 transforms into the origin of coordinates $\zeta = 0$. This transformation can be accomplished by means of the linear fractional function $\dfrac{\zeta - \zeta_1}{1 - \zeta\bar{\zeta}_1}$. Accordingly one can consider that the functions $f(z)$ and $f_1(z)$ are connected with one another by the equation

$$f_1(z) = \frac{f(z) - \zeta_1}{1 - \bar{\zeta}_1 f(z)}.$$

From this is at once obtained the expression we require for Green's function with pole at z_1:

$$G(x, y; x_1, y_1) = \frac{1}{2\pi} \ln \frac{1}{|f_1(z)|} = \frac{1}{2\pi} \ln \frac{|1 - \bar{\zeta}_1 f(z)|}{|f(z) - \zeta_1|}. \tag{5}$$

This equation permits the computation of Green's function for the Dirichlet problem for any position z_1 of the pole, if Green's function is known for one completely determined position, z_0, of the pole.

2. The approximate construction of Green's function. Let the region B with contour L be simply connected, finite, and let $z_0 = x_0 + iy_0$ be an interior point of this region.

The construction of Green's function with pole at the point z_0 reduces, as has been explained in the preceding No., to the determination of a function $g(x, y)$ such as to be harmonic in B and to satisfy on the contour L the condition: $g(x, y) = \frac{1}{2\pi} \ln r$, when z lies on L. Here $r = |z - z_0|$ is the distance from the pole z_0 to the variable point $z = x + iy$. If $g(x, y)$ has been found, Green's function is found by it in accordance with the equation

$$G(x, y; x_0, y_0) = \frac{1}{2\pi} \ln \frac{1}{r} + g(x, y).$$

For the approximate construction of the function $g(x, y)$ there can be given the following interpolation device, whose idea is very simple [1]).

Let us take an harmonic polynomial of degree n with arbitrary coefficients. In polar coordinates, if z_0 be taken as the pole and the polar axis be directed parallel to the x-axis, so that $z - z_0 = re^{i\varphi}$, it will have the form:

$$P_n(r, \varphi) = a_0 + \sum_{k=1}^{n} r^k (a_k \cos k\varphi - b_k \sin k\varphi). \tag{6}$$

It involves $2n + 1$ arbitrary coefficients, $a_0, a_1, b_1, \ldots, a_n, b_n$. On L let

1) A. M. Chufistova, [1].

us take, arbitrarily, $2n + 1$ points $z_1, z_2, \ldots, z_{2n+1}$ and try to choose the coefficients a_k and b_k of the polynomial $P_n(r, \varphi)$ so that at these $2n + 1$ points $P_n(r, \varphi)$ takes the same values as does $g(x, y)$. If we take it into account that the boundary values for $g(x, y)$ are equal to $\frac{1}{2\pi} \ln r$, we obtain for the determination of the a_k and b_k the following system of equations:

$$\left.\begin{aligned} a_0 + \sum_{k=1}^{n} r_1^k(a_k \cos k\varphi_1 - b_k \sin k\varphi_1) &= \frac{1}{2\pi} \ln r_1, \\ a_0 + \sum_{k=1}^{n} r_2^k(a_k \cos k\varphi_2 - b_k \sin k\varphi_2) &= \frac{1}{2\pi} \ln r_2, \\ \cdots\cdots\cdots\cdots\cdots\cdots\cdots\cdots\cdots\cdots \\ a_0 + \sum_{k=1}^{n} r_{2n+1}^k (a_k \cos k\varphi_{2n+1} - b_k \sin k\varphi_{2n+1}) &= \frac{1}{2\pi} \ln r_{2n+1}, \\ z_j - z_0 = r_j e^{i\varphi_j}. \end{aligned}\right\} \tag{7}$$

The determinant of this system will depend on the choice of the points $z_1, z_2, \ldots$. If it is different from zero, the system will have a unique solution. We will explain the geometric meaning of the determinant's equalling zero. Let us consider the points $z_1, z_2, \ldots$ that we have chosen and see what condition the coordinates of these points must satisfy if they are to lie on an equipotential line of an harmonic polynomial of degree not higher than n, i.e., on a line with equation

$$Q_n(r, \varphi) = a_0 + \sum_{k=1}^{n} r^k(a_k \cos k\varphi - b_k \sin k\varphi) = 0, \tag{8}$$

where the a_k and b_k are not simultaneously equal to zero.

For the existence of such a line it is necessary and sufficient that the homogeneous system

$$a_0 + \sum_{k=1}^{n} r_j^k(a_k \cos k\varphi_j - b_k \sin k\varphi_j) = 0 \quad (j = 1, 2, \ldots, 2n + 1)$$

have, with respect to the numbers $a_0, a_1, b_1, \ldots$ a solution different from zero. If the determinant of this system is different from zero, the system has only a null solution, and consequently there is not even one curve with equation (8) passing through all the points z_j. If the determinant of this system is equal to zero, however, the system will have a solution different from zero, and consequently there is a curve with an equation of form (8) passing through all the adopted points $z_1, z_2, \ldots, z_{2n+1}$.

But the determinant of the last system is exactly the determinant of system (7). Thus the fact that the determinant of system (7) is equal to zero signifies that the points chosen by us, $z_1, z_2, \ldots$ all lie on an equipotential line of a harmonic polynomial of degree not higher than n.

We observe that the points $z_1, z_2, \ldots$ in our problem can always be so taken on the curve L that the determinant of system (7) for them is different from zero.

Let us, indeed, assume the contrary. For any position of the points z_j on L let the determinant of system (7) equal zero. Geometrically this signifies that any $2n + 1$ points taken on L always turn out to lie on an equipotential line of a harmonic polynomial (8) of degree not higher than n.

It can be shown that in such a case the curve L must lie wholly on an equipotential line of such a polynomial, i.e., the coordinates of all its points must satisfy an equation of form (8), and of the coefficients $a_0, a_1, b_1, \ldots$ at least one is different from zero.

The problem of which we are here speaking is a particular case of the following more general problem. Let there be given n functions $u_1(x, y), u_2(x, y), \ldots, u_n(x, y)$, which we shall consider to be linearly independent, i.e., such that the equation

$$c_1u_1(x, y) + c_2u_2(x, y) + \ldots + c_nu_n(x, y) = 0$$

holds identically with respect to x and y only if all the coefficients $c_1, \ldots, c_n$ are equal to zero.

Let us consider the curve with the equation

$$P_n(x, y) = a_1u_1(x, y) + a_2u_2(x, y) + \ldots + a_nu_n(x, y) = 0. \qquad (*)$$

We shall now choose arbitrarily n points of the plane, $M_k(x_k, y_k)$ $(k = 1, 2, \ldots, n)$, and try to determine what condition the coordinates of these points must satisfy in order that they all lie on a curve of form (*), i.e., we shall try to determine when the coefficients $a_1, \ldots, a_n$ of equation (*) can be so chosen that the coordinates of all the points M_k satisfy equation (*).

It can be verified that in order that there exist a line of form (*), passing through all the points M_k, it is necessary and sufficient that the coordinates of these points fulfill the condition:

$$\begin{vmatrix} u_1(x_1, y_1), & u_2(x_1, y_1), & \ldots, & u_n(x_1, y_1) \\ u_1(x_2, y_2), & u_2(x_2, y_2), & \ldots, & u_n(x_2, y_2) \\ \cdot & \cdot & \cdot & \cdot \\ u_1(x_n, y_n), & u_2(x_n, y_n), & \ldots, & u_n(x_n, y_n) \end{vmatrix} = 0. \qquad (**)$$

Let us take some arc l in the plane, and let it turn out that for any choice of n points M_k on this arc equation (**) is fulfilled. Geometrically this signifies that whatever n points on the arc l we may take, it is always possible to pass a line with the equation (*) through them. We shall show that this can be true only in case the arc l lies wholly on some line of form (*).

Indeed, let us take, on l, n arbitrary points $M_k(x_k, y_k)$ and consider the table

$$\begin{matrix} u_1(x_1, y_1), & u_2(x_1, y_1), & \ldots, & u_n(x_1, y_1) \\ u_1(x_2, y_2), & u_2(x_2, y_2), & \ldots, & u_n(x_2, y_2) \\ \cdot & \cdot & \cdot & \cdot \\ u_1(x_n, y_n), & u_2(x_n, y_n), & \ldots, & u_n(x_n, y_n) \end{matrix} \qquad (***)$$

The determinant of this table is, on assumption, equal to zero, and the rank of the table for any choice of the n points on l is less than n. The rank will, generally speaking, depend on the choice of the points M_k. Let the greatest value that it will attain when M_k are chosen in all possible manners on l be r. We shall take that system of points M_k for which the rank of the table will be greatest. Performing a renumbering of the points if this is required, we can without limiting the generality consider that the principal determinant of the table will be the following:

$$\Delta = \begin{vmatrix} u_1(x_1, y_1), & \ldots, & u_r(x_1, y_1) \\ \cdot & \cdot\ \cdot\ \cdot & \cdot \\ u_1(x_r, y_r), & \ldots, & u_r(x_r, y_r) \end{vmatrix} \neq 0.$$

Let us take any function $u_k(x, y)$ $(k > r)$ and form the equation:

$$\begin{vmatrix} u_1(x_1, y_1), & \ldots, & u_r(x_1, y_1), & u_k(x_1, y_1) \\ \cdot & \cdot\ \cdot\ \cdot & \cdot & \cdot \\ u_1(x_r, y_r), & \ldots, & u_r(x_r, y_r), & u_k(x_r, y_r) \\ u_1(x, y), & \ldots, & u_r(x, y), & u_k(x, y) \end{vmatrix} = 0. \qquad (****)$$

All the elements of the determinant on the left side of the equality, with the exception of the last row, will be numerical; the elements of the last row contain, however, the variable coordinates x and y. It is easily seen that to this equation there corresponds a curve of form (*). Indeed, our equation at once reduces to form (*) if the determinant be expanded by the elements of the last row. And not all the coefficients of $u_1(x, y), \ldots, u_r(x, y), u_k(x, y)$ will turn out to equal zero, because, for instance, the coefficient of $u_k(x, y)$ will be the principal determinant Δ, which is different from zero.

The equation we have constructed is determined by the choice of the points $M_1, M_2, \ldots, M_r$ only; the rest of the points, $M_{r+1}, \ldots, M_n$, figuring in the table do not appear in the equation in any way, and can be taken arbitrarily.

We assert that our arc l belongs wholly to the curve (****). Indeed, we shall take any point on l and show that its coordinates satisfy equation (****). Let us take as this point $M_k(x_k, y_k)$. Substitute x_k, y_k for x and y in equation (****). On the left side, after the substitution, we obtain some determinant of order $r + 1$ belonging to table (***). The rank of this table is at most equal to r, and consequently the determinant we have obtained must equal zero. Accordingly the coordinates of the point M_k satisfy equation (****), Q.E.D.

We shall return to our fundamental question. The harmonic polynomial $Q_n(r, \varphi)$ on the left side of the equation must be equal to zero everywhere along the closed curve L. But then it must be equal to zero everywhere in the region B bounded by the curve L. The last, however, can be true only if all its coefficients $a_0, a_1, \ldots, b_n$ equal zero, which contradicts the preceding. Thus the assumption that for any choice of the points z_1, z_2 on L the determinant of system (7) equals zero leads to a contradiction, and on L there are certainly points $z_1, z_2, \ldots, z_{2n+1}$ for which the determinant of the system is different from zero.

By solving system (7) with respect to the coefficients $a_0, a_1, \ldots, b_n$ and introducing in $P_n(r, \varphi)$ the values obtained, we construct an harmonic polynomial that at the $2n + 1$ points, $z_1, z_2, \ldots, z_{2n+1}$ acquires the same values as does $g(x, y)$, and can be adopted as approximately the latter. Green's function will then have the following approximate representation:

$$G(x, y; x_0, y_0) = \frac{1}{2\pi} \ln \frac{1}{r} + a_0 + \sum_{k=1}^{n} r^k(a_k \cos k\varphi - b_k \sin k\varphi). \qquad (9)$$

One would expect that the difference between $P_n(r, \varphi)$ and $g(x, y)$ will, generally speaking, be the less, the greater n is, and that when n increases without limit, the error of the approximate equation (9) tends to zero.

What properties the contour L must possess, and how the points z_j must be disposed on it, in order that the interpolation process described above converge, i.e., in order that there hold the equality:

$$\lim_{n\to\infty} P_n(r, \varphi) = g(x, y),$$

remains unclarified up to the present, as far as we know.

Example. We will consider in the z-plane the square

$$-1 < x < +1, \quad -1 < y < +1$$

and find Green's function for the case when the pole z_0 lies at the origin of coordinates: $z = 0$.

The presence of axes of symmetry in the region we have under consideration, and the choice of the pole at the center of symmetry of the region permit one to simplify the determination of the regular harmonic component $g(x, y)$ of Green's function. We expand $g(x, y)$ in a series of harmonic polynomials about the origin of coordinates:

$$g(x, y) = a_0 + \sum_{k=1}^{\infty} r^k(a_k \cos k\varphi - b_k \sin k\varphi).$$

Since the values of the function $g(x, y)$ must be arranged symmetrically with respect to the x-axis, in this expansion only the terms with cosines must be preserved, all the coefficients b_k $(k = 1, 2, \ldots,)$ of the sines equalling zero.

Moreover, the values of $g(x, y)$ must be disposed symmetrically with respect to the y-axis and both bisectors of the coordinate angles. Hence it follows that in the expansions there must be preserved only terms containing trigonometric functions of the angles $0, 4\varphi, 8\varphi, \ldots,$ the expansion therefore having the form

$$g(x, y) = a_0 + a_4 r^4 \cos 4\varphi + a_8 r^8 \cos 8\varphi + \ldots$$

In this expansion we shall preserve only the first three terms, discarding all the rest. In accordance with this, on the contour of the square we must take three points of interpolation. As these we will take the points $z_1 = 1$, $z_2 = \frac{2}{\sqrt{3}} e^{i\frac{\pi}{6}}$ and $z_3 = \sqrt{2}e^{i\frac{\pi}{4}}$.

For the determination of a_0, a_4, a_8 we shall have the following system of three equations:

$$a_0 + a_4 + a_8 = 0,$$

$$a_0 - \tfrac{8}{9}a_4 - \tfrac{128}{81}a_8 = \frac{1}{2\pi} \ln \frac{2}{\sqrt{3}},$$

$$a_0 - 4a_4 + 16a_8 = \frac{1}{2\pi} \ln \sqrt{2},$$

from which the coefficients a_0, a_4, a_8 are found to be:

$$a_0 = \frac{1}{2\pi}\,\frac{1}{668}\left[324 \ln \frac{2}{\sqrt{3}} + \frac{56}{5} \ln \sqrt{2}\right] = \frac{0.075577}{2\pi},$$

$$a_4 = -\frac{1}{2\pi}\,\frac{1}{668}\left[243 \ln \frac{2}{\sqrt{3}} + \frac{209}{5} \ln \sqrt{2}\right] = -\frac{0.074012}{2\pi},$$

$$a_8 = \frac{1}{2\pi}\,\frac{1}{668}\left[\frac{153}{5} \ln \sqrt{2} - 81 \ln \frac{2}{\sqrt{3}}\right] = -\frac{0.001566}{2\pi}.$$

By them we construct approximately Green's function:

$$G(x, y;\ 0, 0) = \frac{1}{2\pi} \ln \frac{1}{r} + g(x, y) =$$

$$= \frac{1}{2\pi}\left[\ln \frac{1}{r} + 0.075577 - 0.074012 r^4 \cos 4\varphi - 0.001566 r^8 \cos 8\varphi\right] =$$

$$= \frac{1}{2\pi} Re\left[\ln \frac{1}{z} + 0.075577 - 0.074012 z^4 - 0.001566 z^8\right].$$

Using the Green function found we shall yet construct the function $\zeta = f(z)$ transforming the square into the unit circle. They are connected with each other, as we explained above, by equality (4). For our case there follows from it:

$$\ln |f(z)| = -2\pi G(x, y;\ 0, 0) = \mathrm{Re}\,[\ln z - 0.075577 + 0.074012 z^4 + 0.001566 z^8].$$

From this it is evident that $\ln f(z)$ can differ from the expression standing under the sign of the real part, only by a pure imaginary constant term. We discard it. This means that we shall consider the derivative $f'(0)$ of the transforming function at the origin of coordinates to be real and positive:

$$f(z) = z e^{-0.075577 + 0.074012 z^4 + 0.001566 z^8}.$$

For a characterization of the accuracy of the result obtained, we compute $f'(0)$. From the approximate expression for $f(z)$ obtained by us we find

$$f'(0) = e^{-0.075577} = 0.9272.$$

The exact value, however, is

$$f'(0) = \int_0^1 \frac{dt}{\sqrt{1 + t^4}} = 0.9257.$$

The error of the approximate value equals 0.16%.

3. Green's function for the Neumann problem. It is known that the Neumann problem in a plane region reduces to the Dirichlet problem for the conjugate harmonic function. We shall elucidate the idea of this reduction in the simplest case of a simply connected region.

In the simply connected region B with boundary L let it be required to find a harmonic function $u(x, y)$ whose normal derivative acquires assigned values on the boundary,

$$\frac{\partial u}{\partial n} = f(M) \text{ on } L, \tag{10}$$

n being the exterior normal. $f(M)$ must satisfy the necessary condition of the solubility of the Neumann problem,

$$\int_L f(M)\,ds = 0. \tag{11}$$

Together with the function $u(x, y)$ let us consider the harmonic function conjugate to it, $v(x, y)$. We shall adopt as the positive direction of motion along L the counter-clockwise direction. From the Cauchy-Riemann equations it is seen that

$$\frac{\partial u}{\partial n} = \frac{\partial v}{\partial s}.$$

Boundary condition (10) for the function $v(x, y)$ conjugate to $u(x, y)$ will have the form:

$$\frac{\partial v}{\partial s} = f(M).$$

Hence, assigning $v(x, y)$ arbitrarily at any point $M_0(x_0, y_0)$ of the boundary L, we recover by integration the value of $v(x, y)$ everywhere on L:

$$v(x, y) = v_0 + (L)\int_{(x_0, y_0)}^{(x, y)} f(M)\,ds = F(M). \tag{12}$$

$F(M)$ will be on L a single-valued continuous function, since in view of (11), on circuiting L we will return to the initial value. Thus the Neumann problem for the function $u(x, y)$ under boundary condition (10) reduces to the Dirichlet problem for the function $v(x, y)$ under condition (12).

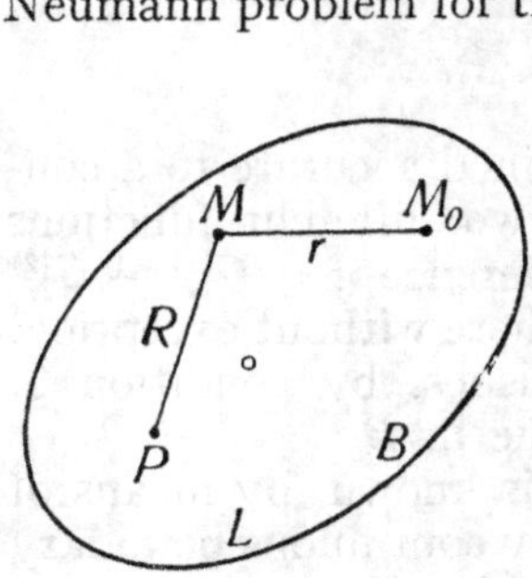

Fig. 48.

Such a reduction, theoretically always admissible, has two defects, which can in practical computations require excessive work, to wit: first, it is necessary to compute the boundary values of $v(x, y)$ from the known boundary values of $u(x, y)$; and, second, using the function found on solving the Dirichlet problem, $v(x, y)$, it is necessary to recover the function conjugate to it, $u(x, y)$.

It is therefore useful, at least from the point of view of practical computations, to indicate such methods for the solution of the Neumann problem as would not require its reduction to the Dirichlet problem. We shall now draw the attention of the reader to one of such methods.

We now discard the assumption of the simple connectivity of the region, reverting for a while to the general case.

Let us choose, within the region B, two points $M_0(x_0, y_0)$ and $P(a, b)$. The variable point of this region we shall designate by M, and its coordinates, by x, y.

The distance from M_0 to M we shall designate by r, the distance from P to M, by R (Fig. 48):

$$r = \sqrt{(x - x_0)^2 + (y - y_0)^2} = |z - z_0|,$$
$$R = \sqrt{(x - a)^2 + (y - b)^2} = |z - \alpha|,$$
$$\alpha = a + ib.$$

A function $G_N\left(x, y; \begin{matrix} x_0, y_0 \\ a, b \end{matrix}\right) = G_N(x, y)$ is called Green's function for the Neumann problem in the region B with poles at M_0 and P, if it satisfies the conditions:

1. $G_N(x, y)$, as a function of x and y, is harmonic everywhere in B, with the exception of the points M_0 and P.
2. $G_N(x, y)$ can be represented in the form:

$$G_N(x, y) = \frac{1}{2\pi} \ln \frac{1}{r} - \frac{1}{2\pi} \ln \frac{1}{R} + g_N(x, y),$$

where $g_N(x, y)$ is a harmonic function everywhere in B, including even the points M_0 and P.

3. The normal derivative of $G_N(x, y)$ on the contour of the region B vanishes:

$$\left.\frac{\partial G_N}{\partial n}\right|_L = 0.$$

By these three conditions Green's function is defined accurate to a constant summand. Indeed, let $G_N^{(1)}$ and $G_N^{(2)}$ be two different functions satisfying the requirements enumerated. The difference $u = G_N^{(1)} - G_N^{(2)}$ between them is a function harmonic in B everywhere without exception. Its normal derivative on the boundary L vanishes, by condition 3. Therefore u itself must preserve a constant value in B.

It can be shown that if the function $G_N(x, y)$ is known, by means of it can be solved the Neumann problem in B for any continuous boundary values of the normal derivative of the harmonic function.

For proof we shall again employ Green's formula (1). In it we shall take as u the sought harmonic function, and as v Green's function G_N. We do not have the right to apply this formula directly in B, for $v = G_N$ has in B the two points of discontinuity M_0 and P. About them as centers we describe circles C_ε and C_η of small radii ε and η. The regions lying within these circles we exclude from B. We obtain a new region $B_{\varepsilon,\eta}$

whose boundary consists of L, C_ε and C_η. If formula (1) be applied to $B_{\varepsilon,\eta}$ and ε and η be later diminished to zero, then by reasoning quite analogous to that carried through in No. 1, one can show that in the limit there is obtained the following equation:

$$\int_L \left(u \frac{\partial G_N}{\partial n} - G_N \frac{\partial u}{\partial n}\right) ds + u(x_0, y_0) - u(a, b) = 0.$$

We shall yet take into consideration that on L, by the third condition, $\dfrac{\partial G_N}{\partial n} = 0$.

$$u(x_0, y_0) = u(a, b) + \int_L G_N \frac{\partial u}{\partial n} ds. \tag{13}$$

This equation makes it possible to find the value of the harmonic function at any interior point $M_0(x_0, y_0)$ of the region B, if there are known the values of its normal derivative on L and the value $u(a, b)$ of the function $u(x, y)$ itself at some fixed point $P(a, b)$.

We shall now establish the connection between Green's function for the Neumann problem and the conformal transformation of B. For simplicity of exposition we shall again limit ourselves to the case of a simply connected region B.

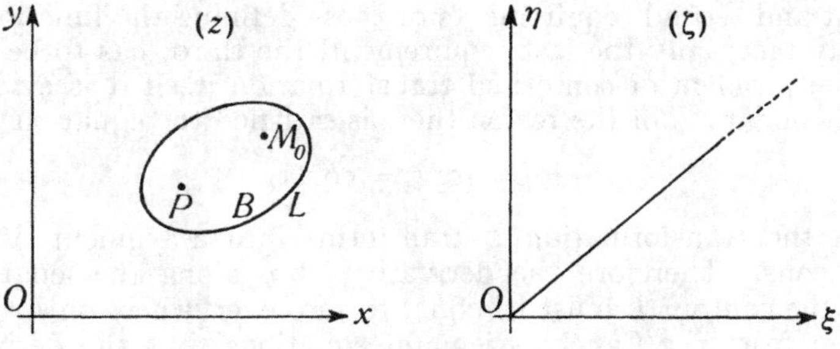

Fig. 49.

We shall effect the conformal transformation of the region B into the ζ-plane so that the point M_0 passes to the point at infinity of the ζ-plane, the point P passes to the zero point of the ζ-plane, and the contour L of the region B transforms into some straight-line cut made along the ray $\arg \zeta = \text{const.}$ (Fig. 49). We shall call the transforming function $\zeta = g(z)$. That the function $g(z)$ effecting such a transformation exists, we shall see below, when we give its expression in terms of the function $\zeta = f(z)$ effecting the conformal transformation of the region B into the unit circle. Now, however, we shall just note that $g(z)$ in our problem is defined only accurate to an arbitrary complex factor, which in our reasoning is immaterial.

Put

$$g(z) = \varrho e^{i\psi}.$$

We shall now show that as Green's function $G_N(x, y)$ one can adopt

$$G_N(x, y) = \frac{1}{2\pi} \ln \varrho = \frac{1}{2\pi} \ln |g(z)|. \tag{14}$$

Indeed, $g(z)$ has a simple pole and a simple zero, at the points $z_0 = x_0 + iy_0$ and $\alpha = a + ib$ respectively. Therefore the function $\chi(z) = \frac{(z - z_0)g(z)}{z - \alpha}$ will be a regular function different from zero everywhere in B. Any branch of its logarithm will also be a regular function in B. Let us call it

$$u + iv = \ln \chi(z).$$

Here u and v are functions harmonic everywhere in B,

$$g(z) = \frac{z - \alpha}{z - z_0} e^{u+iv}$$

$$\frac{1}{2\pi} \ln \varrho = \frac{1}{2\pi} \ln \frac{1}{r} - \frac{1}{2\pi} \ln \frac{1}{R} + \frac{1}{2\pi} u.$$

The first and second requirements of those defining the function G_N are obviously met; only the last requirement, the third, has to be checked. From the problem of conformal transformation itself it is evident that on the boundary L of the region there is satisfied the equation

$$\psi = \arg \zeta = \text{const},$$

since in the transformation L transforms into a segment of the ray $\arg \zeta = \text{const}$. Therefore the derivative of ψ along the length of the arc s of the contour L must be equal to zero everywhere on L. But then it is seen from the Cauchy-Riemann equations that the derivative of the harmonic function conjugate to ψ, $\ln \varrho$, in the direction orthogonal to the element of arc s, i.e., in the direction of the normal n to L, must also be equal to zero:

$$\frac{\partial}{\partial n} \frac{1}{2\pi} \ln \varrho = 0 \text{ everywhere on } L.$$

This last signifies the fulfillment of the third condition for $\frac{1}{2\pi} \ln \varrho$, and consequently we can adopt:

$$G_N = \frac{1}{2\pi} \ln \varrho = \frac{1}{2\pi} \ln |g(z)|.$$

Conversely, if Green's function $G_N(x, y)$ is known, we can construct by it the function $g(z)$, analytic in the region B, and effecting the conformal transformation of this region into a plane with a cut along the ray made from the origin of coordinates, which transformation is such that the points α and z_0 are carried to 0 and ∞. Indeed, let us find, by the known function G_N, the harmonic function conjugate to it, H_N. The methods of constructing it are known. Then, as may be readily verified,

$$g(z) = e^{G_N + iH_N}.$$

Finally we shall indicate how $g(z)$ can be constructed if a conformal transformation of B into the unit circle is known.

Of all the possible functions transforming B into the unit circle $|w| < 1$, we shall choose that which carries the point M_0 to the origin of coordinates and the point P to a certain point p lying on the negative part of the real axis. Let this be the function $w = f(z)$ (Fig. 50). Let us now perform the transformation

$$w' = \tfrac{1}{2}\left(w + \frac{1}{w}\right).$$

The interior of the unit circle $|w| < 1$ will transform into the entire plane w', with the exception of a straight-line segment along the real axis between the points $-1, +1$, the transformation being such that $w = 0$ is carried to the infinitely distant point $w' = \infty$, the point $-p$ to a certain point $-p'$ on the negative real semi-axis, where, moreover,

$$p' = \tfrac{1}{2}\left(p + \frac{1}{p}\right) > 1,$$

and the circle $|w| = 1$ transforms into the segment $(-1, +1)$, passed over twice.

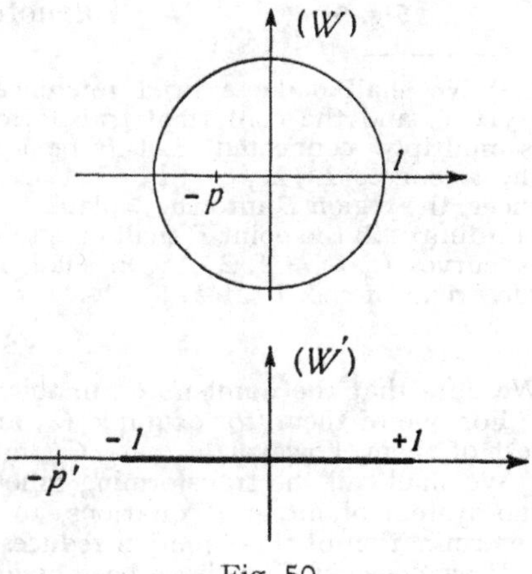

Fig. 50.

It remains for us only to transfer the point $-p'$ to the origin of coordinates, and this is accomplished by the transformation:

$$\zeta = w' + p'.$$

Thus we can consider that

$$g(z) = \tfrac{1}{2}\left(f(z) + \frac{1}{f(z)}\right) + p'.$$

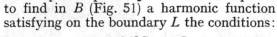

This equation, by the way, also shows that the solution of the Neumann problem in B reduces to the solution of the Dirichlet problem, for by using this equation it is always possible, from one of the two functions $G_N\left(x, y; \begin{matrix} x_0, y_0 \\ a, b \end{matrix}\right)$ and $G(x, y; x_0, y_0)$, to find the other [1]).

4. Green's function for the mixed problem. Let the boundary L of the region B be separated into two parts L_1 and L_2, and let it be required to find in B (Fig. 51) a harmonic function satisfying on the boundary L the conditions:

$$u = f_1(M) \text{ on } L_1,$$

$$\frac{\partial u}{\partial n} = f_2(M) \text{ on } L_2.$$

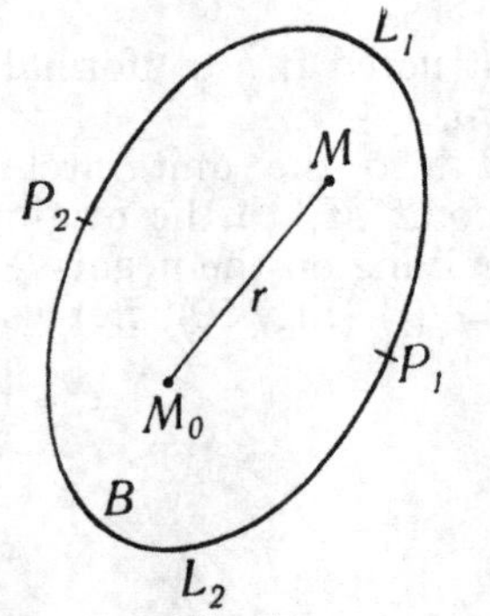

Fig. 51.

As in the case of the Dirichlet and Neumann problems, the solution of this problem for any $f_1(M)$ and $f_2(M)$ can be reduced to finding some singular particular solution of the problem, which we shall call Green's function for the mixed problem in the region B with pole at $M_0(x_0, y_0)$, and denote by $G_g(x, y; x_0, y_0) = G_g(x, y)$.

---

[1]) We shall make a brief reference to the connection between the function $G_N(x, y)$ and the conformal transformation of the region B in the case when B is multiply connected. Let B be n-ply connected, its boundary consisting of the n curves $L_1, L_2, \ldots, L_n$. Let us transform, conformally and onto a single sheet, the region B into the ζ-plane so that 1) the point M_0 shall pass to the point at infinity; 2) the point P shall pass to the origin of coordinates; and 3) the boundary curves L_k $(k = 1, 2, \ldots, n)$ shall pass to cuts along certain rays issuing from the origin of coordinates:

$$\arg \zeta = C_k.$$

We note that the numbers C_k in this problem cannot all be assigned arbitrarily. For one of them, for example C_1, any value can be prescribed in advance; the rest of them, however, $C_2, \ldots, C_n$, must be determined from the problem itself. We shall call the transforming function $\zeta = g(z)$. In § 9 No. 5 we shall indicate the system of integral equations, to the solution of which the problem of the determination of this function reduces.

If we consider C_1 to have been assigned, then $g(z)$ is determined accurate to an arbitrary real factor, to the variation of which corresponds a similarity transformation of all the cuts from the origin of coordinates. It can easily be seen that

$$G_N(x, y) = \frac{1}{2\pi} \ln |g(z)|.$$

To satisfy oneself of this, it is sufficient to verify that all three conditions defining G_N are met for $\frac{1}{2\pi} \ln |g(z)|$.

This is very simply done by means of the reasoning that we have carried through in the text proper of the book for a simply connected region, and we leave this to the reader.

We define this Green's function by the following requirements:

1. $G_g(x, y)$, as a function of x and y, must be harmonic in B, with the exception of the point M_0.

2. In the neighborhood of the point M_0, $G_g(x, y)$ must have the representation:

$$G_g(x, y) = \frac{1}{2\pi} \ln \frac{1}{r} + h(x, y),$$

where $r = \sqrt{(x - x_0)^2 + (y - y_0)^2}$ is the distance between the variable point M and the pole M_0, and $h(x, y)$ is a function harmonic everywhere in B including even the point M_0.

3. On the contour L of region B, $G_g(x, y)$ satisfies the boundary conditions

$$G_g(x, y) = 0 \text{ on } L_1$$

and

$$\frac{\partial}{\partial n} G_g(x, y) = 0 \text{ on } L_2.$$

It can be shown that $G_g(x, y)$ is fully determined by these three requirements.

By means of this Green's function can be solved the mixed problem posed by us, for any continuous $f_1(M)$ and $f_2(M)$. Indeed, let us take Green's formula (1) No. 1 and adopt u in it as the harmonic function we are seeking, setting v equal to $G_g(x, y)$. We do not have the right to apply Green's formula to the entire region B for the pair of functions we have chosen, since for one of them the point M_0 will be singular. We describe about M_0 a circumference C_ε of small radius ε and exclude from B the circle bounded by this circumference. If we now apply Green's formula to the remaining region and afterwards carry out a passage to the limit as $\varepsilon \to 0$, we obtain — repeating literally the reasoning of No. 1:

$$\int_L \left(u \frac{\partial G_g}{\partial n} - G_g \frac{\partial u}{\partial n}\right) ds + u(x_0, y_0) = 0.$$

From this, if we take into account the boundary conditions to which we have subjected $G_g(x, y)$, it will follow that

$$u(x_0, y_0) = -\int_{L_1} u \frac{\partial G_g}{\partial n} ds + \int_{L_2} G_g \frac{\partial u}{\partial n} ds, \tag{15}$$

and this last equation permits the computation of $u(x, y)$ at any interior point $M_0(x_0, y_0)$ of the region if the Green function $G_g(x, y)$ and the values of u and $\frac{\partial u}{\partial n}$ on L_1 and L_2 respectively are known.

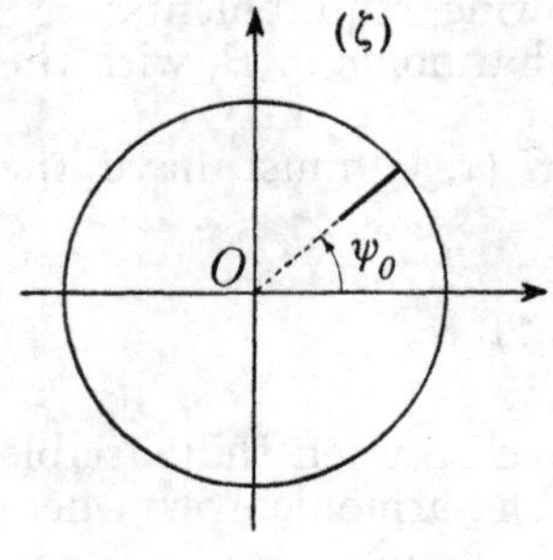

Fig. 52.

We will now explain the connection between the function $G_g(x, y)$ and the conformal transformation of the region B.

We shall consider the region B to be simply connected, and its boundary L subdivided, so that L_1 and L_2 each consists of one segment of the curve. We shall effect the conformal transformation into the unit circle of the ζ-plane with a cut, so that

1) the point $M_0(x_0, y_0)$ shall pass to the center of the circle $|\zeta| < 1$;

2) the arc L_1 of the contour shall transform into the circumference of the circle $|\zeta| = 1$;

3) the arc L_2 of the contour shall transform into a cut from the circumference of the unit circle inward along the radius $\arg \zeta = \psi_0 = \text{const.}$ (Fig. 52).

We note here that the length of this cut cannot be assigned and must be determined from the problem itself.

We shall call the transforming function

$$\zeta = \zeta(z).$$

Its existence can easily be proved if one proceeds from the Riemann theorem on the transformation of simply connected regions into a circle. A little below, we shall give an explicit expression for $\zeta(z)$ in terms of the function transforming the region B into the unit circle.

If we fix the ray along which the cut must be disposed, i.e., consider ψ_0 to have been given, then the transforming function $\zeta(z)$ is uniquely defined by the stipulated requirements.

We shall now show that the Green function $G_g(x, y)$ and the transforming function $\zeta(z)$ are connected with each other by this equation:

$$G_g(x, y) = \frac{1}{2\pi} \ln \frac{1}{|\zeta(z)|}. \tag{16}$$

Since $\zeta(z)$ is a bounded function and has a simple zero at the point z_0, that the first two conditions of the three defining $G_g(x, y)$ are met for $\frac{1}{2\pi} \ln \frac{1}{|\zeta(z)|}$ is obvious; only the last condition, the third, requires checking. The arc L_1 of the contour transforms into the unit circumference, therefore everywhere on L_1 $|\zeta(z)| = 1$ and consequently:

$$\frac{1}{2\pi} \ln \frac{1}{|\zeta(z)|} = 0 \text{ everywhere on } L_1.$$

The arc L_2 of the contour transforms into the cut lying on the ray $\arg \zeta = \psi_0$. Therefore everywhere on L_2 $\arg \zeta(z)$ has the constant value ψ_0. But then we must have $\frac{\partial}{\partial s} \arg \zeta(z) = 0$ everywhere on L_2, and, as is seen from the Cauchy-Riemann equations, for $\frac{1}{2\pi} \ln \frac{1}{|\zeta(z)|}$ there will be fulfilled the equation

$$\frac{\partial}{\partial n} \frac{1}{2\pi} \ln \frac{1}{|\zeta(z)|} = 0 \text{ everywhere on } L_2.$$

And by this the relation needed by us, (16), is proved.

We will now establish the connection between $\zeta(z)$ and the function furnishing the conformal transformation of B into a circle.

Of all the functions realizing the conformal transformation of B into the unit circle $|w| < 1$, we shall choose that which carries the point z_0 to the origin of coordinates $w = 0$ and transforms the part L_1 of the contour into an arc of the unit circle having its origin at the point $w = 1$ and going counter-clockwise from this point. The central angle corresponding to this arc we shall designate by α, so that for this arc

$$0 \leqslant \arg w \leqslant \alpha, \qquad \alpha < 2\pi.$$

We shall call the transforming function

$$w = f(z).$$

It now remains for us to transform the circle $|w| < 1$ into the unit circle of the ζ-plane with a cut along some ray, for example, along the real positive semi-axis from some point to $\zeta = 1$, and so that the arc $0 \leqslant \arg w \leqslant \alpha$, corresponding to L_1, shall transform into the circumference $|\zeta| = 1$, and the arc $\alpha \leqslant \arg w \leqslant 2\pi$, corresponding to L_2, shall transform into both edges of the cut. The origins of both planes, $w = 0$ and $\zeta = 0$, must transform the one into the other.

It can be shown [1]) that the necessary transformation is furnished by

---

[1]) This can be accomplished via the following elementary transformations:

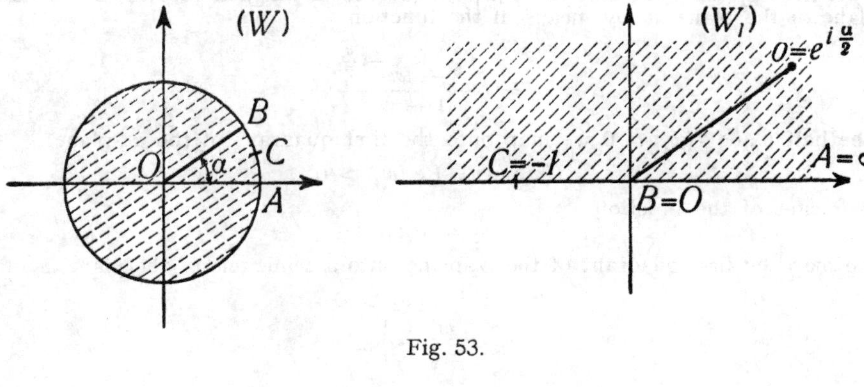

Fig. 53.

the function

$$\zeta = \frac{\left(\sqrt{\dfrac{e^{i\frac{\alpha}{2}} - e^{-i\frac{\alpha}{2}} w}{1-w}} - 1\right)^2 + \operatorname{tg}^2\dfrac{\alpha}{8}\left(\sqrt{\dfrac{e^{i\frac{\alpha}{2}} - e^{-i\frac{\alpha}{2}} w}{1-w}} + 1\right)^2}{\left(\sqrt{\dfrac{e^{i\frac{\alpha}{2}} - e^{-i\frac{\alpha}{2}} w}{1-w}} + 1\right)^2 + \operatorname{tg}^2\dfrac{\alpha}{8}\left(\sqrt{\dfrac{e^{i\frac{\alpha}{2}} - e^{-i\frac{\alpha}{2}} w}{1-w}} - 1\right)^2}.$$

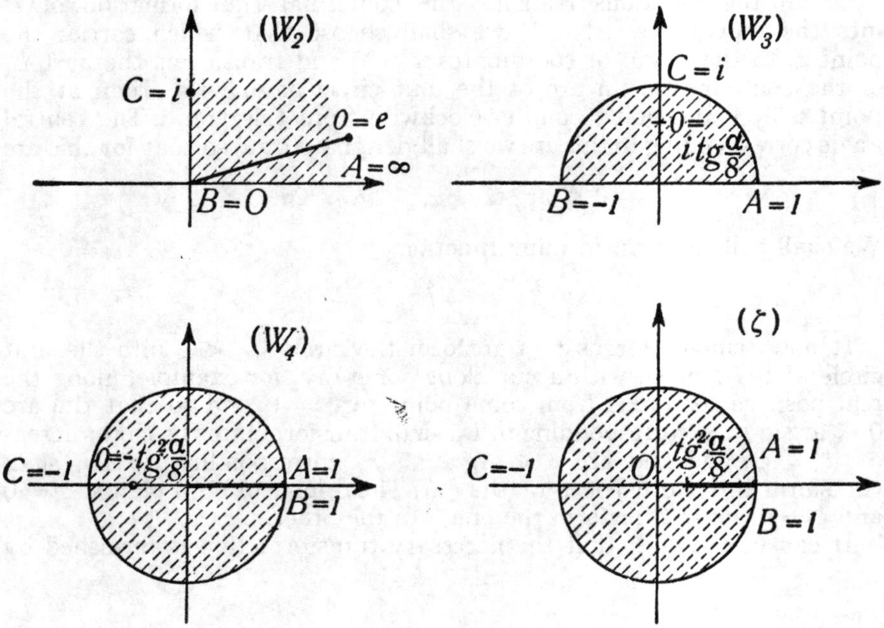

Fig. 53.

At the beginning the unit circle $|w| < 1$ (Fig. 53) is mapped onto the upper half plane of the plane w_1 by means of the function

$$w_1 = \frac{e^{i\frac{\alpha}{2}} - we^{-i\frac{\alpha}{2}}}{1-w}.$$

The half plane $I(w_1) > 0$ we map onto the first quadrant of the w_2-plane:

$$I(w_2) > 0, \ \mathrm{Re}\,(w_2) > 0,$$

by means of the function

$$w_2 = \sqrt{w_1}.$$

We map the first quadrant of the w_2-plane onto a semicircle of the plane w_3:

$$|w_3| < 1, \ I(w_3) > 0,$$

$$w_3 = \frac{w_2 - 1}{w_2 + 1}.$$

§ 9. THE APPLICATION OF INTEGRAL EQUATIONS TO CONFORMAL TRANSFORMATION [1])

1. The integral equation for the transformation of interior regions. We shall have in view the transformation of a simply connected region B of the z-plane, bounded by a simple contour L, into the unit circle with center at the origin of coordinates of the plane $\zeta = \varrho e^{i\vartheta}$.

If we know the relation between the points of the contour L and the circumference of the unit circle, i.e., either know ζ as a function of z on L, or know z as a function of ζ on the circle, the mapping function can be promptly found, for example by means of the Cauchy integral or in another way. Therefore for the solution of the problem of the con-

---

The semicircle of the w_3-plane we transform into the circle $|w_4| < 1$ of the w_4-plane with a cut along the radius (0, 1):

$$w_4 = w_3^2.$$

And, finally, we map the circle with cut in the w_4-plane onto the circle $|\zeta| < 1$ of the ζ-plane with a cut along the real axis from the point $\operatorname{tg}^2 \frac{\alpha}{8}$ to the point 1:

$$\zeta = \frac{w_4 + \operatorname{tg}^2 \frac{\alpha}{8}}{1 + \operatorname{tg}^2 \frac{\alpha}{8} \cdot w_4}.$$

The relation between the points of interest to us through the successive transformations is given by the following table:

w	w_1	w_2	w_3	w_4	ζ
$A = 1$	∞	∞	1	1	1
$C = e^{i\frac{\alpha}{2}}$	-1	i	i	-1	-1
$B = e^{i\alpha}$	0	0	-1	1	1
0	$e^{i\frac{\alpha}{2}}$	$e^{i\frac{\alpha}{4}}$	$i \operatorname{tg} \frac{\alpha}{8}$	$-\operatorname{tg}^2 \frac{\alpha}{8}$	0

In the accompanying Figure 53, the corresponding points are identified by the same letters.

If the auxiliary variables $w_1, \ldots, w_4$ are eliminated from all the preceding formulas and the expression for ζ is found directly in terms of w, the function indicated in the text proper is obtained.

Here, of the two possible values of the square root, that one is chosen which for $w = 0$ reduces to $e^{i\frac{\alpha}{4}}$.

This equation gives the relation of dependence between the functions $w = f(z)$ and $\zeta(z)$, and permits the rapid computation, using one of the Green functions $G(x,y)$ and $G_g(x, y)$, of the other.

[1]) In writing Nos. 1 and 3 of this section, we have utilized a work of S. A. Gershgorin, [3]. An example of the application of integral equations to conformal transformation can be found in a work of A. M. Banin, [1].

formal transformation of the region B and the circle into one another, knowledge of just the relation mentioned is sufficient.

To establish it one can resort to the following method. The mapping function will be fully determined if the relation between one interior and one boundary point of the region B and one interior and one boundary point of the circle is known.

For definiteness we shall assume that a certain point z_0 of B is carried to the center $\zeta = 0$ and that a point z_1 on L is carried to $\zeta = 1$.

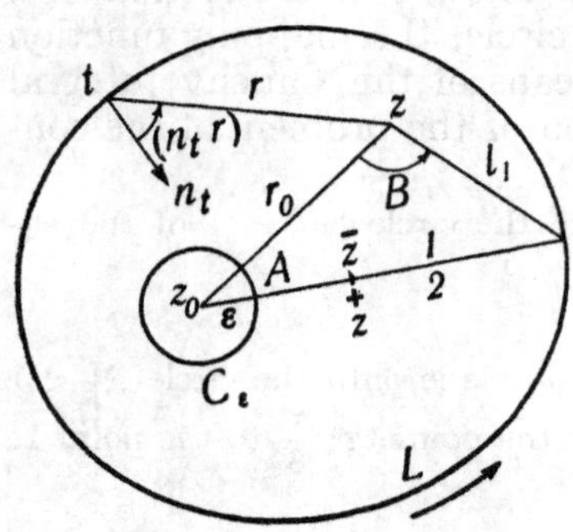

Fig. 54.

We shall denote the function transforming B into the circle by $\zeta = \varphi(z)$, and consider

$$F(z) = \ln \zeta = \ln \varrho + i\vartheta, \quad \zeta = \varrho e^{i\vartheta}. \tag{1}$$

In B the function $F(z)$ is many-valued and has at the point z_0 a logarithmic singularity; on the contour L, however, since there $\varrho = 1$, its real part $\ln \varrho$ vanishes.

In order to make $F(z)$ single-valued, we make a cut from z_0 to z_1. In Fig. 54 it is depicted as being a straight line. In the cut region, $F(z)$ will be a single-valued function; its values at respective points $\overset{-}{z}$ and $\overset{+}{z}$ on the sides of the cut will differ from one another by a quantity equal to the increment of $i\vartheta$ in circuiting z_0, which corresponds to a circuit about the origin of coordinates in the ζ-plane, i.e., they will differ by $2\pi i$:

$$F(\overset{+}{z}) = F(\overset{-}{z}) + 2\pi i. \tag{2}$$

Let us now construct a contour L', consisting of L, the cut from z_1 to A, and a small circle C_ε of radius ε about z_0. By Cauchy's theorem:

$$F(z) = \frac{1}{2\pi i} \int\limits_{L'} \frac{F(t)}{t - z}\, dt =$$

$$= \frac{1}{2\pi i} \int\limits_{L} \frac{F(t)}{t - z}\, dt - \frac{1}{2\pi i} \int\limits_{C_\varepsilon} \frac{F(t)}{t - z}\, dt + \frac{1}{2\pi i} \int\limits_{2-1} \frac{F(t)}{t - z}\, dt.$$

We shall now diminish ε to zero. We shall show that the integral around the circle C_ε tends to zero. Indeed, as has already been said above, $F(z)$ has at z_0 a logarithmic singularity and will differ from $\ln (z - z_0)$ by a holomorphic summand.

It is therefore always possible to indicate a number M such that on

C_ε there holds the estimate

$$|F(z)| \leqslant M\,|\ln \varepsilon|.$$

Hence it follows, since $|t - z|$ is not less than some $\delta > 0$:

$$\left| \int\limits_{C_\varepsilon} \frac{F(t)}{t - z}\,dt \right| \leqslant \frac{2\pi M}{\delta}\,\varepsilon\,|\ln \varepsilon| \to 0.$$

Moreover, along the positive side of the cut we integrate from z_1 to A, along the negative, in the opposite direction. On the strength of (2), the integrals containing $F(z)$ will cancel, there remaining only the integral of $2\pi i$ from z_1 to A, and after passing to the limit we shall have:

$$\frac{1}{2\pi i} \int\limits_{2-1} \frac{F(t)}{t - z}\,dt = \frac{1}{2\pi i} \int\limits_{z_1}^{z_0} \frac{2\pi i}{t - z}\,dt = -\ln \frac{z_1 - z}{z_0 - z}.$$

Finally, if we take it into account that on the contour L $F(z) = i\vartheta$, we shall obtain:

$$F(z) = \frac{1}{2\pi} \int\limits_L \frac{\vartheta\,dt}{t - z} - \ln \frac{z_1 - z}{z_0 - z}.$$

We shall rewrite this result in another form. Let us put $dt = e^{i\alpha}d\sigma$, $d\sigma$ here being the element of arc length of the curve L, reckoned from some point, and α the angle formed by the tangent to L at the point t and the x-axis:

$$t - z = re^{i\varphi},\ z_0 - z = r_0 e^{i\varphi_0},\ z_1 - z = r_1 e^{i\varphi_1}, \tag{3}$$

$$\beta = \varphi_1 - \varphi_0.$$

β is the angle that the segment $z_0 z_1$ subtends at the point z, the angle being read from z_0 to z_1. Moreover let us denote by (n_t, r) the angle at the point t between the interior normal n_t to the contour L and r, the angle being read from n_t to r. Obviously

$$\alpha - \varphi = \frac{\pi}{2} - (n_t, r).$$

$$F(z) = \frac{1}{2\pi} \int\limits_L \frac{\vartheta e^{i\pi/2} e^{-i(n_t, r)}}{r}\,d\sigma - \ln \frac{r_1}{r_0} - \beta i =$$

$$= \frac{i}{2\pi} \int\limits_L \frac{\vartheta e^{-i(n_t, r)}}{r}\,d\sigma - \ln \frac{r_1}{r_0} - \beta i. \tag{4}$$

If we separate the real part from the imaginary here, we find:

$$\ln \varrho = \frac{1}{2\pi} \int_L \frac{\vartheta \sin (n_t, r)}{r} d\sigma - \ln \frac{r_1}{r_0},$$

$$\vartheta = \frac{1}{2\pi} \int_L \frac{\vartheta \cos (n_t, r)}{r} d\sigma - \beta. \tag{5}$$

From the formulas obtained it is clear that if the boundary values of the function ϑ are known, $\ln \varrho$ is found immediately.

By using the second equation, one can easily construct the integral equation for the boundary values of ϑ.

We shall bring up the point z to the point of the contour corresponding to a length s of the arc. The integral on the right side of the equation represents the potential of a double layer with density ϑ. The nature of this function has been well studied, and we shall therefore not enter into a discussion of its limit values on the boundary of the region, referring the reader to special textbooks on mathematical physics or the theory of differential equations [1]). We shall here present without proof the result, graphic as regards its geometrical content: with the approach of z to the contour L from within, the aforementioned integral tends to

$$\int_L \frac{\vartheta(\sigma) \cos (n_t, r)}{r} d\sigma + \pi\vartheta(s),$$

if s is not an angular point; if s is an angular point, though, the second term must be replaced by $(2\pi - \delta)\vartheta(s)$, where δ is the interior angle of the region at the point s. We shall below require to know that for an approach to the contour L from without, the potential of the double layer will tend to

$$\int_L \frac{\vartheta(\sigma) \cos (n_t, r)}{r} d\sigma - \pi\vartheta(s);$$

if s is an angular point, then the second term must be replaced by $\delta\vartheta(s)$. In both cases the integral term is equal to the value of the potential at the point of the contour with the coordinate s.

Substituting this expression in the second equation of (5), we arrive at the conclusion that $\vartheta(s)$ must be a solution of the following integral equation of the Fredholm type, of the second kind:

$$\vartheta(s) = \int_L K(s, \sigma)\vartheta(\sigma)\, d\sigma - 2\beta(s), \tag{6}$$

[1]) E. Goursat, [1], vol. III, ed. 1 (1936), chap. XXVII, par. 505. See also § 3 chap. II of the present book.

where

$$K(s, \sigma) = \frac{1}{\pi} \frac{\cos (n_t, r)}{r}.$$

For an angular point this equation must be replaced by the following:

$$\frac{\delta}{\pi} \vartheta(s) = \int_L K(s, \sigma)\vartheta(\sigma)\, d\sigma - 2\beta(s). \qquad (6')$$

Equation (6) is the particular case of the equation:

$$\vartheta(s) = \lambda \int_L K(s, \sigma)\vartheta(\sigma)\, d\sigma - 2\beta(s) \qquad (7)$$

for $\lambda = 1$, which will indeed be a proper number for this equation. Indeed, up to the present time we have utilized in our reasoning only the fact that the point z_0 is carried to the center of the circle; we have nowhere yet established a correspondence between two points on the boundaries of the regions being transformed, assuming by this fact a rotation of the circle about the origin of coordinates, which corresponds to the addition to $\vartheta(s)$ of an arbitrary constant. The solution of equation (6) must therefore certainly be determined accurate to a constant summand, which can only be the case if 1 is a proper number and the homogeneous equation

$$\vartheta(s) = \int_L K(s, \sigma)\vartheta(\sigma)\, d\sigma$$

has an arbitrary constant for its solution, to satisfy oneself of which presents no difficulty, since

$$\int_L K(s, \sigma)\, d\sigma = 1.$$

This is easily verified, for the integral cited above differs from the Gauss integral for the plane by the factor $\frac{1}{\pi}$ only. It can be shown that the homogeneous equation does not have a solution different from a constant.

Integral equation (6) can accordingly have a solution if and only if its free term, $-2\beta(s)$, is orthogonal to any solution of the adjoint homogeneous equation

$$\mu(s) = \int_L K(\sigma, s)\mu(\sigma)\, d\sigma.$$

We shall leave it to the reader to convince himself of this by direct calculation, and shall take the existence of the solution as an obvious fact, a simple consequence of Riemann's theorem on the existence of the function effecting the conformal transformation we require.

The problem of finding the function $\vartheta(s)$ satisfying relation (6) becomes completely determined if an initial value for $\vartheta(s)$ be assigned, for example:

$$\vartheta(0) = 0.$$

2. Observations on the solution of the integral equation and the approximate construction of the mapping function. The approximate solution of our integral equation can be effected by one of the methods developed above, in Chapter II, whither we refer the reader for an acquaintance with them.

We shall now make a few remarks which may prove useful in practical computations.

The expression

$$\frac{\cos(n_t, r)}{r}\,d\sigma$$

has a direct geometric signification. It is equal to the angle that the element of length of the arc, $d\sigma$, subtends at the point s. We shall denote by $\omega(s, \sigma)$ the angle between the tangent to the curve L at the point s (the tangent being directed to the side of increasing arc), and r; then

$$\int_L K(s, \sigma)\vartheta(\sigma)\,d\sigma$$

will be equal to the integral

$$\frac{1}{\pi}\int_L \vartheta(\sigma)\,d\omega,$$

taken along the entire curve L, and the integral equation will take the form:

$$\vartheta(s) = \frac{1}{\pi}\int_L \vartheta(\sigma)\,d\omega - 2\beta(s). \tag{8}$$

The last form of the equation is particularly important in that it is far simpler in structure than (6). When the required accuracy permits graphical evaluation, the quantity ω can be very simply taken from the figure, as can the angle β, equalling the angle which the segment z_0z_1 (in the direction from z_0 to z_1 [1])) subtends at the point s.

In one way or another let us have reduced the task of solving the integral equation to that of solving a system of linear algebraic equations.

By virtue of the fact that the solution of the original problem is not unique, the algebraic system obtained will either be indeterminate or will have a determinant close to zero. We must yet add to it the condition fixing the initial value for ϑ: $\vartheta(0) = 0$.

For example, if we replace the integral by a finite sum, then in the system of equations obtained

$$\vartheta(s_i) = \sum_{j=1}^{n} K(s_i, s_j)\vartheta(s_j)\Delta s_j - 2\beta(s_i), \qquad (i = 1, 2, \ldots, n),$$

[1]) All this will hold for all the integral equations that we shall encounter below.

we can omit one of the equations, adding to the remaining ones the supplementary equation:

$$\sum_{j=1}^{n} K(0, s_j)\vartheta(s_j)\Delta s_j - 2\beta(0) = 0.$$

Finally, if the region has m axes of symmetry, then, on quite understandable considerations, the number of equations must be reduced to $\dfrac{1}{m+1}$ times its value owing to the symmetry of the values of $\vartheta(s)$.

On solving the integral equation, we find the relation between the points of the contour L and the circumference $|\zeta| = 1$ in the form of a dependence of the central angle ϑ on the length of arc s of the contour L. By using it we can construct either the function transforming the region B into the circle, or, conversely, the function transforming the circle into the region.

In the first case, for the construction of $F(z)$ we can utilize, for instance, equation (3). The integral on its right side can be approximately computed by one of the familiar methods, for example, by replacing it by the sum:

$$\frac{1}{2\pi i}\sum_{k=1}^{n}\vartheta_k\int_{t_k}^{t_{k+1}}\frac{dt}{t-z} = \frac{1}{2\pi i}\sum_{k=1}^{n}\vartheta_k \ln\frac{t_{k+1}-z}{t_k - z}.$$

Conversely, let it be required to find the inverse function, transforming the circle into the region. Then we can utilize the conjugate trigonometric series.

By the known relation between the points of the contour L and those of the unit circumference $|\zeta| = 1$, we find, by whatever method, the dependence of the x- and y-coordinates of the points of the contour on the polar angle ϑ of the ζ-plane. We choose one of these coordinates, for example x, and represent it approximately by a trigonometric polynomial

$$x = \sum_{\nu=0}^{n} (a_\nu \cos \nu\vartheta + b_\nu \sin \nu\vartheta).$$

The second coordinate, y, in view of the conjugacy of x and y, must then be represented by the conjugate trigonometric polynomial

$$y = \sum_{\nu=0}^{n} (-\, b_\nu \cos \nu\vartheta + a_\nu \sin \nu\vartheta),$$

in which all the coefficients, with the

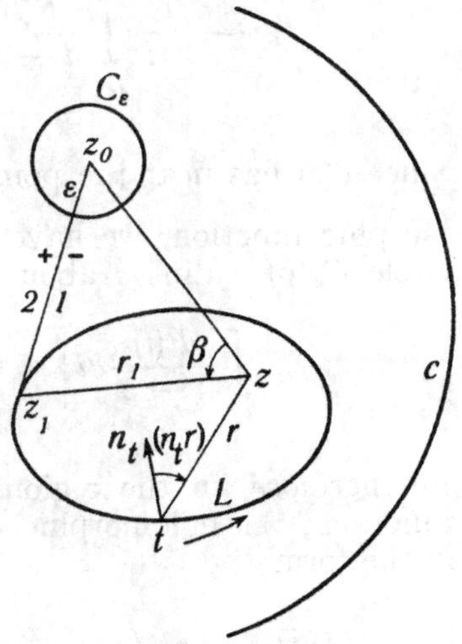

Fig. 55.

exception of b_0, are known. This free term, b_0, must be found from the known relation between the points of the circumference and of the contour L.

By the x and y found, one then constructs the polynomial of degree n approximately representing the function mapping the circle $|\zeta| < 1$ onto B,

$$z = \sum_{\nu=0}^{n} (a_\nu - ib_\nu)\zeta^\nu.$$

3. The integral equation for the transformation of exterior regions. Let us now turn to the case of the transformation of a region B exterior to a contour L of the z-plane into the exterior of the unit circle of the ζ-plane. In the transformation (Fig. 55) let the point $z_0 \neq \infty$ within the region B be carried to the infinitely distant point of the ζ-plane, and the point z_1 of the contour L, to $\zeta = 1$. As in the preceding case, it is convenient for us to consider not the function ζ itself, effecting the required transformation, but its logarithm, $\ln \zeta = \ln \varrho + i\vartheta$. To establish the integral equation for ϑ in the case we are analysing, we shall construct a contour L' consisting of the given curve L, of a cut passing from z_1 to z_0, of a small circle C_ε about z_0, and of a circle C of large radius R.

Within L' each branch of $\ln \zeta = F(z)$ will be a regular and single-valued function:

$$F(z) = \frac{1}{2\pi i}\int_{L'} \frac{F(t)}{t-z}\,dt = \frac{1}{2\pi i}\int_{C} \frac{F(t)}{t-z}\,dt -$$

$$- \frac{1}{2\pi i}\int_{L} \frac{F(t)}{t-z}\,dt - \frac{1}{2\pi i}\int_{C_\varepsilon} \frac{F(t)}{t-z}\,dt + \frac{1}{2\pi i}\int_{2-1} \frac{F(t)}{t-z}\,dt.$$

Since $\zeta(z)$ has near the point z_0 the structure $\zeta(z) = \dfrac{a}{z-z_0} +$ a holomorphic function, we have the valid estimate $|F(t)| \leqslant M\,|\ln \varepsilon|$ on the circle C_ε of radius ε about z_0, and consequently:

$$\left|\int_{C_\varepsilon} \frac{F(t)}{t-z}\,dt\right| \leqslant \frac{2\pi M}{\min|t-z|}\,\varepsilon\,|\ln \varepsilon| \to 0 \text{ as } \varepsilon \to 0.$$

Furthermore, in the region of the point at infinity of the plane, the function $\zeta(z)$ is holomorphic, and $\zeta(z)$ and $F(z)$ can therefore be represented in the form:

$$\zeta(z) = \zeta_\infty\left(1 + \frac{\alpha}{z} + \dots\right),\quad F(z) = \ln \zeta_\infty + \frac{\beta}{z} + \dots.$$

The factor $\frac{1}{t-z}$ has there the representation:

$$\frac{1}{t-z} = \frac{1}{t} + \frac{z}{t^2} + \dots$$

Therefore

$$\frac{F(t)}{t-z} = \frac{\ln \zeta_\infty}{t} + \frac{\gamma}{t^2} + \dots \text{ and } \frac{1}{2\pi i}\int_C \frac{F(t)}{t-z}\,dt = \ln \zeta_\infty.$$

A circuit about the point z_0 from side 2 of the cut to side 1 corresponds to a circuit of the point at infinity of the ζ-plane in the positive direction, and consequently:

$$\overset{-}{F}(z) = \overset{+}{F}(z) + 2\pi i.$$

Hence

$$\frac{1}{2\pi i}\int_{2-1} \frac{F(t)}{t-z}\,dt = -\frac{1}{2\pi i}\int_{z_1}^{z_0} \frac{2\pi i}{t-z}\,dt = \ln \frac{z_1 - z}{z_0 - z}.$$

Finally, in view of the fact that on the contour L, $F(t) = i\vartheta$, we obtain:

$$F(z) = -\frac{1}{2\pi}\int_L \frac{\vartheta\,dt}{t-z} + \ln \frac{z_1 - z}{z_0 - z} + \ln \zeta_\infty.$$

If we now put $\zeta_\infty = \varrho_\infty e^{i\vartheta_\infty}$ and in the rest keep the old symbols, we shall have

$$F(z) = -\frac{i}{2\pi}\int_L \frac{\vartheta e^{-i(n_t, r)}}{r}\,d\sigma + \ln \frac{r_1 \varrho_\infty}{r_0} e^{i(\beta + \vartheta_\infty)}. \tag{9}$$

Separating the real part from the imaginary, we find:

$$\left.\begin{aligned} \ln \varrho &= -\frac{1}{2\pi}\int_L \frac{\vartheta \sin (n_t, r)}{r}\,d\sigma + \ln \frac{r_1 \varrho_\infty}{r_0}, \\ \vartheta &= -\frac{1}{2\pi}\int_L \frac{\vartheta \cos (n_t, r)}{r}\,d\sigma + \beta + \vartheta_\infty. \end{aligned}\right\} \tag{9'}$$

β has here the same value as it did in No. 1 — being the angle which the segment $z_0 z_1$, directed from z_0 to z_1, subtends at the point z.

Let the point z be brought towards the point s of the contour L. The limit which the integral $\int_L \frac{\vartheta \cos (n_t, r)}{r}\,d\sigma$ approaches — as we are approaching the contour L from without — will equal:

$$-\pi\vartheta(s) + \int_L \frac{\vartheta \cos(n_t, r)}{r} d\sigma.$$

In the limit, from the second equation of (9′) we shall obtain the following integral equation for $\vartheta(s)$:

$$\vartheta(s) = -\frac{1}{\pi}\int_L \frac{\vartheta \cos(n_t, r)}{r} d\sigma + 2[\beta(s) + \vartheta_\infty]. \qquad (10)$$

It will be a particular case of the equation

$$\vartheta(s) = \lambda \int_L K(s, \sigma)\vartheta(\sigma)d\sigma + 2[\beta(s) + \vartheta_\infty], \qquad (11)$$

where $K(s, \sigma)$ has its former value, with the value of the parameter $\lambda = -1$ not being its proper number; it therefore has a unique solution.

In its right side there appears the quantity ϑ_∞, unknown to us and equalling the argument of the point that corresponds to the point at infinity of the ζ-plane. To any change of ϑ_∞ there corresponds a rotation of the entire ζ-plane about the origin of coordinates [1]). Previously we made the conformal transformation definite by establishing the relation between a point of the contour L and a point of the circle $|\zeta| = 1$. The same thing can be attained by assigning the value of ϑ_∞. In the solution of equation (11) we can therefore choose ϑ_∞ arbitrarily.

After $\vartheta(s)$ has been found, the function effecting the transformation of the exterior of the curve into the exterior of the circle can be found approximately, either by using formula (9), or by resorting to interpolation by means of rational functions.

Let us now pass on to the case when the infinitely distant point of the z-plane passes to the point of the ζ-plane that is also infinitely distant.

We shall designate as $\zeta = f(z)$ the function effecting the conformal transformation of B into the exterior of a certain circle with center at the point $\zeta = 0$ of the ζ-plane. By the conditions stipulated, the transforming function is defined accurate to a constant complex factor. To simplify the computations we shall so choose this factor that the expansion of $f(z)$ in the neighborhood of the point at infinity has the form

$$\zeta = f(z) = z + a_0 + \frac{a_1}{z} + \frac{a_2}{z^2} + \dots$$

---

[1]) This is seen from equation (10), since with a change of ϑ_∞ by a certain number, by virtue of the fact that $\int_L \frac{\cos(n_t, r)}{r} d\sigma = \pi$, the quantity $\vartheta(s)$ will change by this same amount.

On L we choose an arbitrary point z_1 and consider the function $\varphi(z) = \dfrac{f(z)}{z - z_1}$. It is regular everywhere within B and differs from zero there. Any branch of its logarithm is a regular function in B. Since $\varphi(\infty) = \lim\limits_{z\to\infty} \dfrac{f(z)}{z - z_1} = 1$, there will exist a branch of $\ln \varphi(z)$ that vanishes at the point at infinity. We shall choose this branch and apply to it Cauchy's formula

$$\ln \varphi(z) = -\frac{1}{2\pi i} \int\limits_L \frac{\ln \varphi(t)}{t - z}\, dt.$$

The integral along L is taken in the counter-clockwise direction.

If for $\ln \varphi(z)$ there be substituted here its value $\ln f(z) - \ln (z - z_1)$, after simple calculations we obtain the following equation:

$$\ln \zeta = \ln f(z) = \ln (z - z_1) - \frac{1}{2\pi i} \int\limits_L \frac{\ln f(t)}{t - z}\, dt.$$

Here put $\zeta = f(z) = \varrho e^{i\vartheta}$, $z - z_1 = r_1 e^{i\beta}$ and separate the real part from the imaginary:

$$\ln \varrho = \ln r_1 - \frac{1}{2\pi} \int\limits_L \frac{\vartheta(\sigma) \sin (n_t, r)}{r}\, d\sigma,$$

$$\vartheta = \beta - \frac{1}{2\pi} \int\limits_L \frac{\vartheta(\sigma) \cos (n_t, r)}{r}\, d\sigma.$$

If in the second of these equations z tends to the point of the contour L corresponding to the length s of arc, we obtain on the contour, after passing to the limit, the following integral equation for the boundary values of $\vartheta(s)$ on L:

$$\vartheta(s) = 2\beta(s) - \frac{1}{\pi} \int\limits_L \frac{\vartheta(\sigma) \cos (n_t, r)}{r}\, d\sigma, \tag{12}$$

where $\beta(s)$ is the angle at which the vector $z - z_1$ is inclined to the positive direction of the real axis.

After $\vartheta(s)$ has been determined, the rest presents no difficulty. Let us be required, for example, to transform the region $|\zeta| \geqslant 1$ into the exterior of the contour L. The coordinates x and y of the contour L we find as functions $x(\vartheta)$, $y(\vartheta)$ of the angle ϑ. We represent the first of them approximately by a trigonometric polynomial

$$x(\vartheta) = \sum_{k=0}^{n} (a_k \cos k\vartheta + b_k \sin k\vartheta). \tag{13}$$

In the expansion for $y(\vartheta) = \sum_{k=0}^{n} (c_k \cos k\vartheta + d_k \sin k\vartheta)$, however, we find only the first two terms, i.e., we compute the coefficients c_0, c_1, d_1.

The transforming function will then be:

$$z = \alpha_{-1}\zeta + \sum_{k=0}^{n} \frac{\alpha_k}{\zeta^k}, \tag{14}$$

where

$$\alpha_{-1} = \frac{a_1 + d_1}{2} + i\frac{c_1 - b_1}{2},$$

$$\alpha_0 = a_0 + ic_0,$$

$$\alpha_1 = \frac{a_1 - d_1}{2} + i\frac{b_1 + c_1}{2}, \tag{15}$$

$$\alpha_k = a_k + ib_k, \; k \geqslant 2. \tag{16}$$

4. The transformation of a region into a plane with parallel cuts. In certain questions — for instance, in the study of two-dimensional potential flow of a fluid past various obstacles, it is usually necessary to find the function effecting the conformal transformation of the given multiply connected region into a plane with cuts parallel to some direction, which can always be considered as coinciding with the direction of the real axis.

We shall now obtain the integral equation for the real part of the transforming function [1]).

Let us concern ourselves with the determination of the function $\zeta = \xi + i\eta$ transforming the exterior B of a system of contours L_1, $L_2, \ldots, L_n$ into the ζ-plane with cuts parallel to the real axis (Fig. 56):

$$\eta = \eta_k \quad (k = 1, 2, \ldots, n), \tag{17}$$

this being such a transformation that the points at infinity correspond to one another:

$$\zeta(\infty) = \infty.$$

Let us describe a circle C of such a large radius R that all the contours L_k fall within it, and apply the Cauchy theorem to the parts of the region B lying within C:

$$\zeta = f(z) = \frac{1}{2\pi i}\left\{\int_C \frac{f(t)}{t - z}\,dt - \sum_{k=1}^{n} \int_{L_k} \frac{f(t)}{t - z}\,dt\right\}.$$

---

[1]) The task of constructing the transforming function has been reduced by G. M. Goluzin to the solution of a system of functional equations of very simple form, and the possibility has been established of solving it by successive approximations, on the assumption that the boundary curves are sufficiently distant from one another. In certain cases this method makes possible the effective construction of the function being sought. The computations become particularly simple when the region is the exterior of several circles. See G. M. Goluzin, [2].

Since in the neighborhood of the point at infinity $f(z)$ obviously has an expansion of the form:

$$f(z) = cz + c_0 + \frac{c_1}{z} + \ldots, \quad c_0 = \gamma_0 + i\delta_0,$$

the integral around the circumference C will give:

$$\frac{1}{2\pi i} \int_C \frac{f(t)}{t - z}\, dt = cz + c_0.$$

If, however, we also take into consideration the fact that on the contour

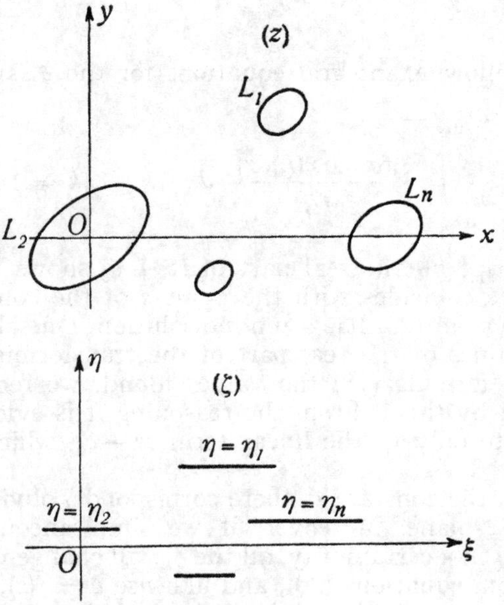

Fig. 56.

L_k, $\zeta = \xi_k + i\eta_k$, and that the point z lies outside all contours L_k, so that

$$\int_{L_k} \frac{\eta_k}{t - z}\, dt = 0,$$

we obtain the following expression for the transforming function in terms of the values of its real part on the contours:

$$f(z) = cz + c_0 - \sum_{k=1}^{n} \frac{1}{2\pi i} \int_{L_k} \frac{\xi_k}{t - z}\, dt. \tag{18}$$

Separation of the real and imaginary parts leads to the equations:

$$\left.\begin{aligned} \xi &= \mathrm{Re}\,[cz] + \gamma_0 - \sum_{k=1}^{n} \frac{1}{2\pi} \int_{L_k} \frac{\xi_k \cos(n_t, r)}{r}\, d\sigma, \\ \eta &= \mathrm{I}\,[cz] + \delta_0 + \sum_{k=1}^{n} \frac{1}{2\pi} \int_{L_k} \frac{\xi_k \sin(n_t, r)}{r}\, d\sigma. \end{aligned}\right\} \quad (19)$$

Passing, in the first of them, to the limit as z tends to the point s of the contour L_k, and having in view the fact that

$$\int_{L_k} \frac{\xi_k \cos(n_t, r)}{r}\, d\sigma \to \int_{L_k} \frac{\xi_k \cos(n_t, r)}{r}\, d\sigma - \pi\xi_k(s),$$

we obtain the following integral equations for the $\xi_k(s)$:

$$\xi_k(s) = 2\mathrm{Re}\,[cz + c_0]_s -$$

$$- \sum_{j=1}^{n} \frac{1}{\pi} \int_{L_j} \frac{\xi_j(\sigma) \cos(n_t, r)}{r}\, d\sigma\,^{1)}, \qquad (k = 1, 2, \ldots, n), \quad (20)$$

where the subscript on the real part of $cz + c_0$ shows that it must be computed when z coincides with the point s of the contour L_k. It can be shown that system (20) has a unique solution. On solving it, we find the boundary values of the real part of the transforming function. On introducing into formula (18) the values found, we recover the transforming function by them. From the reasoning it is evident that it will be found accurate only to the linear term $cz + c_0$, which remains arbitrary.

To a change of the constant c_0 there corresponds, obviously, a parallel transfer of the ζ-plane. Moreover, if we simultaneously change the moduli of c and c_0 in a certain way, all the ξ_k will change in the same way, which follows from equations (20), and likewise $\zeta = f(z)$, as is seen from (18). Finally, the argument of c defines the direction at the point $\zeta = \infty$ to which is carried the direction of the real axis at the infinity of the z-plane after the transformation. c and c_0 can be assigned arbitrarily. It is simplest to consider that $c_0 = 0$, $c = 1$. After the determination of the $\xi_k(s)$ from system (20), the imaginary part η of the function $f(z)$ is found from the second equation of (19); the function itself can be found by formula (18).

For the approximate construction of ζ one can utilize interpolation by means of rational functions. Within each of the contours L_k we take

---

1) When s is an angular point of the contour, on the left side there must stand $\frac{\delta}{\pi}\xi_k(s)$, where δ is the exterior angle of the contour at the point s. A like observation relates to all the other integral equations of this No. as well.

a point a_k. For a sufficiently broad class of regions B one can choose a rational function of the form

$$\zeta = cz + c_0 + \sum_{k=1}^{n} \sum_{s=1}^{m_k} \frac{a_s^{(k)}}{(z - a_k)^s}, \tag{21}$$

such as to differ by an arbitrarily small number from the function $f(z)$ effecting the required conformal transformation. We shall assume that this difference lies beyond the bound of admissible error, so that on the contours the real part of the cited rational function coincides with ξ_k and the imaginary part with η_k. For the determination of the coefficients corresponding to such a choice of the rational function one can proceed as follows. Multiply the preceding expression by $(z - a_k)^{s-1}$ and integrate around the contour L_k; utilizing the familiar residue theorems, we can write:

$$\int_{L_k} \zeta (z - a_k)^{s-1} dz = 2\pi i \alpha_s^{(k)}.$$

But on L_k, $\zeta = \xi_k + i\eta_k$, and since $\int_{L_k} \eta_k (z - a_k)^{s-1} dz = 0$ and η_k is constant, we have

$$\alpha_s^{(k)} = \frac{1}{2\pi i} \int_L \xi_k (z - a_k)^{s-1} dz.$$

By a like method one can construct the integral equation for the case when a point z_0 lying at a finite distance is carried to the point at infinity of the ζ-plane by the transformation. In the neighborhood of z_0, $f(z)$ must have an expansion of the form

$$f(z) = \frac{c}{z - z_0} + c_0 + c_1(z - z_0) + \dots.$$

Moreover let ζ_∞ be that point of the ζ-plane to which passes $z = \infty$, so that $f(\infty) = \zeta_\infty$.

Application of the Cauchy formula leads us here to the equation

$$f(z) = \frac{c}{z - z_0} + \zeta_\infty - \sum_{k=1}^{n} \frac{1}{2\pi i} \int_{L_k} \frac{\xi_k}{t - z} dt,$$

equivalent to the two following ones:

$$\xi = \mathrm{Re}\left[\frac{c}{z - z_0} + \zeta_\infty\right] - \sum_{k=1}^{n} \frac{1}{2\pi} \int_{L_k} \frac{\xi_k(\sigma) \cos (n_t, r)}{r} d\sigma,$$

$$\eta = \mathrm{I}\left[\frac{c}{z - z_0} + \zeta_\infty\right] + \sum_{k=1}^{n} \frac{1}{2\pi} \int_{L_k} \frac{\xi_k(\sigma) \sin (n_t, r)}{r} d\sigma.$$

From this by passing to the limit on the contour L_k is obtained the system of equations for the $\xi_k(s)$:

$$\xi_k(s) = 2\mathrm{Re}\left[\frac{c}{z - z_0} + \zeta_\infty\right]_s - \sum_{j=1}^{n} \frac{1}{\pi} \int_{L_j} \frac{\xi_j(\sigma) \cos (n_t, r)}{r}\, d\sigma. \quad (22)$$

$$(k = 1, 2, \ldots, n)$$

This system contains in its right side the expression $\dfrac{c}{z - z_0} + \zeta_\infty$, which remains arbitrary in the computations. The constants c and ζ_∞ have meanings analogous to the meanings of c and c_0 in the preceding problem. The system has, for a fixed free term, a unique solution.

We shall yet dwell on that case of the transformation of a multiply connected region into a plane with cuts where the point at infinity does not belong to the region.

Let the region B be bounded by the exterior contour L_0 and by several interior contours $L_1, \ldots, L_n$. We shall seek a function transforming it into a cut plane such that an assigned point z_0 passes to the point at infinity of the ζ-plane. Here the boundaries of the region transform to cuts:

$$\eta = \eta_k \qquad (k = 0, 1, 2, \ldots, n).$$

We describe about the point z_0 the circle C_ε of small radius ε and to the region obtained after this we apply the Cauchy integral:

$$\zeta = \frac{1}{2\pi i}\left\{\int_{L_0} \frac{f(t)}{t - z}\, dt - \sum_{k=1}^{n} \int_{L_k} \frac{f(t)}{t - z}\, dt - \int_{C_\varepsilon} \frac{f(t)}{t - z}\, dt\right\}.$$

In the neighborhood of the point z_0, $f(z)$ must have the form:

$$f(z) = \frac{c}{z - z_0} + c_0 + c_1(z - z_0) + \ldots,$$

and therefore

$$\int_{C_\varepsilon} \frac{f(t)}{t - z}\, dt = -\, 2\pi i \frac{c}{z - z_0};$$

the remaining integrals, however, have here the form:

$$\frac{1}{2\pi i} \int_{L_0} \frac{f(t)}{t - z}\, dt = i\eta_0 + \frac{1}{2\pi i} \int_{L_0} \frac{\xi_0}{t - z}\, dt,$$

$$\frac{1}{2\pi i} \int_{L_k} \frac{f(t)}{t - z}\, dt = \frac{1}{2\pi i} \int_{L_k} \frac{\xi_k}{t - z}\, dt,$$

and we obtain the following expression for the transforming function in terms of the boundary values of its real part:

$$f(z) = \frac{c}{z - z_0} + i\eta_0 + \frac{1}{2\pi i} \int_{L_0} \frac{\xi_0}{t - z} dt - \sum_{k=1}^{n} \frac{1}{2\pi i} \int_{L_k} \frac{\xi_k}{t - z} dt. \quad (23)$$

Separation of the real and imaginary parts gives:

$$\left.\begin{aligned} \xi &= \text{Re}\left[\frac{c}{z - z_0}\right] + \\ &\quad + \frac{1}{2\pi} \int_{L_0} \frac{\xi_0 \cos (n_t, r)}{r} d\sigma - \sum_{k=1}^{n} \frac{1}{2\pi} \int_{L_k} \frac{\xi_k \cos (n_t, r)}{r} d\sigma, \\ \eta &= \text{I}\left[\frac{c}{z - z_0}\right] + \eta_0 - \frac{1}{2\pi} \int_{L_0} \frac{\xi_0 \sin (n_t, r)}{r} d\sigma + \\ &\qquad + \sum_{k=1}^{n} \frac{1}{2\pi} \int_{L_k} \frac{\xi_k \sin (n_t, r)}{r} d\sigma. \end{aligned}\right\} \quad (24)$$

In the first of these relations passage to the limit on the contour L_k permits the construction of the integral equations for the $\xi_k(s)$:

$$\xi_k(s) = 2\text{Re}\left[\frac{c}{z - z_0}\right]_s + \\ + \frac{1}{\pi} \int_{L_0} \frac{\xi_0(\sigma) \cos (n_t, r)}{r} d\sigma - \sum_{j=1}^{n} \frac{1}{\pi} \int_{L_j} \frac{\xi_j(\sigma) \cos (n_t, r)}{r} d\sigma, \quad (25)$$

$$(k = 0, 1, 2, \ldots, n).$$

In the right sides of the system is contained the coefficient c, which remains undetermined, and the meaning of which we mentioned above. After the choice of it, the system will still not have a unique solution, for the corresponding homogeneous system will, as is easily seen, have the solution $\xi_0 = \xi_1 = \ldots = \xi_n =$ const. The attentive reader will probably already have observed the cause of this. In all of our preceding formulas there has figured only one constant summand η_0, corresponding to a translation of the cuts along the imaginary axis; the possibility of transferring them along the ξ-axis has obviously served as the cause of the indeterminateness of system (25). To make the problem definite, we must still add to (25) the initial value for one of the ξ_k, putting $\xi_0(0) = 0$, for instance.

5. The transformation of a multiply connected region into a plane with cuts lying on rays originating at one point. In § 8, while elucidating the connection between Green's function for the Neumann problem and the conformal transformation of a given region, we showed that this Green function is closely connected with the problem of transforming a region into a plane with cuts situated on rays issuing from a single point. We shall now construct the system of integral equations, to whose solution reduces the construction of the analytic function furnishing such a transformation.

Let the region B be $(n+1)$-ply connected, bounded by the exterior curve L_0 and by n interior curves $L_1, L_2, \ldots, L_n$ (Fig. 57).

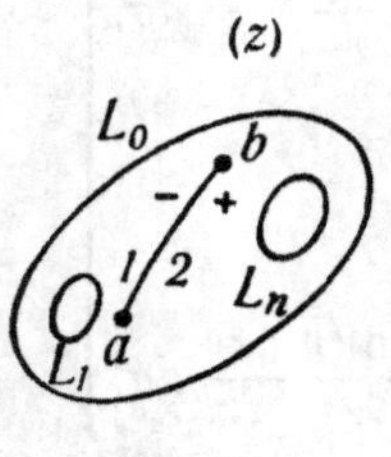

Fig. 57.

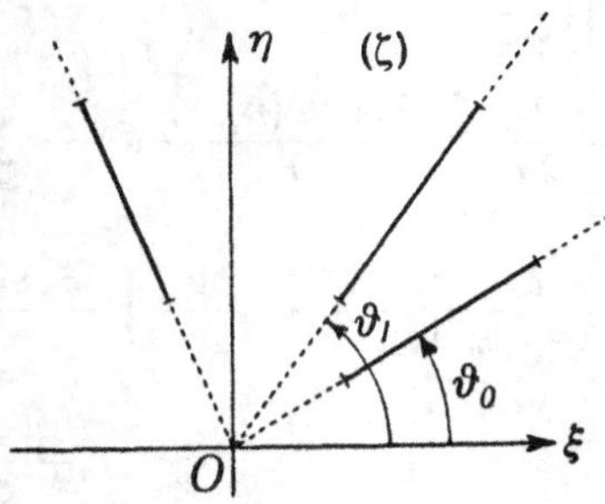

Fig. 58.

We shall choose two arbitrary points in B, a and b, and shall seek the function

$$\zeta = \varrho e^{i\vartheta} = f(z),$$

effecting a single-sheeted transformation of the region B into the ζ-plane such that:

1) the point a is carried to the origin of coordinates of the ζ-plane:

$$f(a) = 0;$$

2) the point b is carried to the point at infinity of the ζ-plane:

$$f(b) = \infty;$$

3) the boundary curves $L_0, \ldots, L_n$ transform into cuts along some rays issuing from the origin of coordinates:

$$\arg \zeta = \vartheta = \vartheta_k \text{ on } L_k \qquad (k = 0, 1, 2, \ldots, n).$$

We note here that one of the angles ϑ_k, for example ϑ_0, can be assigned arbitrarily; the rest, however, must be found.

The transforming function $\zeta = f(z)$ is defined by the stipulated conditions, as is explained in the theory of conformal transformation, accurate to an arbitrary constant complex factor. To a change of the latter there will correspond a similarity transformation of the ζ-plane with center at the point $\zeta = 0$ and the rotation of it about this point.

In B let us make a cut joining the points a and b; we mark the positive and negative sides of this cut so that in circuiting the point a counter-clockwise we shall pass from the negative side of the cut to the positive.

Let us consider the function $\ln f(z)$. Any branch of it in the region B with the cut will be a single-valued function.

We shall choose any one branch and apply Cauchy's formula to it:

$$\ln f(z) = \frac{1}{2\pi i}\int_{L_0} \frac{\ln f(t)}{t-z}\,dt - \sum_{k=1}^{n} \frac{1}{2\pi i}\int_{L_k} \frac{\ln f(t)}{t-z}\,dt + \frac{1}{2\pi i}\int_{1-2} \frac{\ln f(t)}{t-z}\,dt.$$

The last integral is taken first along the negative side of the cut from a to b and then along the positive side of it from b to a. Since the values of $\ln f(z)$ along the different sides of the cut are related to each other by the equation

$$\overset{+}{\ln f(z)} = \overset{-}{\ln f(z)} + 2\pi i,$$

the integral along the cut is simply calculated, equaling

$$\frac{1}{2\pi i}\int_{1-2} \frac{\ln f(t)}{t-z}\,dt = \frac{1}{2\pi i}\int_a^b \frac{\overset{-}{\ln f(t)} - \overset{+}{\ln f(t)}}{t-z}\,dt = -\ln\frac{b-z}{a-z}.$$

We shall yet take into consideration the fact that on each of the boundary curves

$$\ln f(z) = \ln \varrho_k(s) + i\vartheta_k,$$

where ϑ_k is some constant and $\varrho_k(s)$ is a function, unknown to us, of the position of the point on L_k. Moreover, since the point z lies within L_0 and outside of all the L_k $(k = 1, 2, \ldots, n)$:

$$\int_{L_0} \frac{\vartheta_0}{t-z}\,dt = 2\pi i\vartheta_0, \quad \int_{L_k} \frac{\vartheta_k}{t-z}\,dt = 0.$$

Therefore

$$\ln f(z) = i\vartheta_0 + \frac{1}{2\pi i}\int_{L_0} \frac{\ln \varrho_0(\sigma)\,dt}{t-z} - \sum_{k=1}^{n} \frac{1}{2\pi i}\int_{L_k} \frac{\ln \varrho_k(\sigma)\,dt}{t-z} - \ln\frac{b-z}{a-z}. \quad (26)$$

If the real and the imaginary parts be separated here, we obtain the following two equations:

$$\ln \varrho = \ln\left|\frac{a-z}{b-z}\right| + \frac{1}{2\pi}\int_{L_0} \frac{\ln \varrho_0(\sigma)\cos(n_t, r)}{r}\,d\sigma - \sum_{k=1}^{n} \frac{1}{2\pi}\int_{L_k} \frac{\ln \varrho_k(\sigma)\cos(n_t, r)}{r}\,d\sigma, \quad (27)$$

$$\vartheta = \vartheta_0 + \arg \frac{a - z}{b - z} - \frac{1}{2\pi} \int\limits_{L_0} \frac{\ln \varrho_0(\sigma) \sin (n_t, r)}{r} d\sigma + \\ + \sum_{k=1}^{n} \frac{1}{2\pi} \int\limits_{L_k} \frac{\ln \varrho_k(\sigma) \sin (n_t, r)}{r} d\sigma.$$

Here, as before, n_t is the interior normal to the boundary curve at the point t [1]), $r = |t - z|$, (n_t, r) is the angle between the normal n_t and r, reckoned from n_t to r, and σ is the length of the arc, reckoned from some initial point to the variable point t.

If in the first of these equations the point z approach some point of the curve L_0 or L_j with length of arc s, and it be taken into account that with the approach of z to L_0

$$\int\limits_{L_0} \frac{\ln \varrho_0(\sigma) \cos (n_t, r)}{r} d\sigma \to \pi \ln \varrho_0(s) + \int\limits_{L_0} \frac{\ln \varrho_0(\sigma) \cos (n_t, r)}{r} d\sigma,$$

and with the approach of z to L_j $(j = 1, 2, \ldots, n)$

$$\int\limits_{L_j} \frac{\ln \varrho_j(\sigma) \cos (n_t, r)}{r} d\sigma \to - \pi \ln \varrho_j(s) + \int\limits_{L_j} \frac{\ln \varrho_j(\sigma) \cos (n_t, r)}{r} d\sigma,$$

we will obtain the following system of integral equations for the determination of the values of the logarithm of the modulus of the transforming function on the boundary curves L_j:

$$\ln \varrho_j(s) = 2 \ln \left| \frac{a - z}{b - z} \right| + \frac{1}{\pi} \int\limits_{L_0} \frac{\ln \varrho_0(\sigma) \cos (n_t, r)}{r} d\sigma - \\ - \sum_{k=1}^{n} \frac{1}{\pi} \int\limits_{L_k} \frac{\ln \varrho_k(\sigma) \cos (n_t, r)}{r} d\sigma. \qquad (28)$$

This system differs from (25) of the preceding No. only by its free term.

The corresponding homogeneous system has the non-zero solution $\ln \varrho_0(s) = \ln \varrho_1(s) = \ldots = C$, where C is an arbitrary constant. To make the problem of its solution determinate, the initial value of one of the functions $\varrho_j(s)$ must be assigned. After the limiting values of the modulus have been found, the logarithm of the transforming function is determined by formula (26), accurate to the pure imaginary summand $i\vartheta_0$, corresponding to a rotation of the ζ-plane about the origin of coordinates.

We shall limit ourselves to the derivation of the equations for this fundamental case. In the rest of the cases the equations are derived analogously.

---

[1]) For L_0 it will be the interior normal to the region B, and for the L_k $(k = 1, 2, \ldots, n)$, the exterior normal to the region B.

§ 10. MAPPING A POLYGON ON A HALF PLANE

1. Derivation of the Christoffel-Schwarz formula. The conformal transformation of the upper half plane into a polygon with given interior angles is effected by means of a definite function whose explicit expression can be obtained. We shall obtain this expression. Let it be the function $z = f(t)$. It effects the conformal transformation of the upper half of the t-plane into the polygon A of the z-plane. The magnitudes of the interior angles of this polygon we shall denote by $\alpha_1\pi, \alpha_2\pi, \ldots, \alpha_n\pi$, and the vertices by $A_1, A_2, \ldots, A_n$. They correspond to certain points $a_1, a_2, \ldots, a_n$ of the real axis of the t-plane. We shall consider all these points to lie in a finite distance and to be so chosen that

$$a_1 < a_2 < \ldots < a_n.$$

We shall investigate the behavior of the function $f(t)$ in the neighborhood of these points [1]). Let us consider a certain point a_k. The function $z = f(t)$ maps the neighborhood of the point a_k on the neighborhood of the point b_k (b_k is the complex coordinate of the vertex A_k). The angle π in the t-plane corresponds to the angle $\alpha_k\pi$ in the z-plane.

We shall yet consider the function

$$w = (z - b_k)^{\frac{1}{\alpha_k}}.$$

This function will map the neighborhood of the point b_k of the z-plane on the neighborhood of the origin of the w-plane. To the angle $\alpha_k\pi$ in the z-plane will correspond the angle π in the w-plane. The part of the polygon A lying near the apex A_k will be mapped onto the part of the w-plane lying about the origin along one side of the straight line l passing through the origin. This straight line is the image of the sides $A_{k-1}A_k$ and A_kA_{k+1}.

Two successive mappings by means of the functions

$$z = f(t) \text{ and } w = (z - b_k)^{\frac{1}{\alpha_k}}$$

transform conformally and in a one-to-one manner the neighborhood of the point a_k lying along one side of the real axis into the neighborhood of the point $w = 0$ lying along one side of the straight line l. Using the principle of symmetry, we can affirm that the function

$$w = [f(t) - b_k]^{\frac{1}{\alpha_k}}$$

will map conformally and in a one-to-one manner the complete neighborhood of the point $t = a_k$ on the complete neighborhood of the point

[1]) This question is expounded in greater detail in the books of V. I. Smirnov, [1], vol. III, 1933, p. 343, and I. I. Privalov, [3], 1932, p. 295.

$w = 0$. w must therefore be a regular function of t in the neighborhood of $t = a_k$. Thus there holds the expansion:

$$w = c_1(t - a_k) + c_2(t - a_k)^2 + \dots \qquad (c_1 \neq 0)$$

or

$$(z - b_k)^{\frac{1}{\alpha_k}} = c_1(t - a_k) + c_2(t - a_k)^2 + \dots.$$

Let us raise both sides of the last equation to the power α_k. We then obtain

$$z = b_k + d_1(t - a_k)^{\alpha_k}\left\{1 + \frac{c_2}{c_1}(t - a_k) + \dots\right\}^{\alpha_k},$$

or

$$f(t) = b_k + d_1(t - a_k)^{\alpha_k} f_1(t),$$

where the function $f_1(t)$ is regular about the point $t = a_k$:

$$f_1(t) = 1 + d_2(t - a_k) + \dots$$

Let us form

$$\varphi(t) = \frac{f''(t)}{f'(t)},$$

$$f'(t) = \alpha_k d_1(t - a_k)^{\alpha_k - 1} f_1(t) + d_1(t - a_k)^{\alpha_k} f_1'(t),$$

$$f''(t) = \alpha_k(\alpha_k - 1) d_1(t - a_k)^{\alpha_k - 2} f_1(t) + 2\alpha_k d_1(t - a_k)^{\alpha_k - 1} f_1'(t) + \\ + d_1(t - a_k)^{\alpha_k} f_1''(t),$$

$$\varphi(t) = \frac{1}{t - a_k} \cdot \frac{\alpha_k(\alpha_k - 1) f_1(t) + 2\alpha_k(t - a_k) f_1'(t) + (t - a_k)^2 f_1''(t)}{\alpha_k f_1(t) + (t - a_k) f_1'(t)}.$$

The second factor is a regular function in the neighborhood of $t = a_k$. Its value at this point is equal to $\alpha_k - 1$. Consequently the point $t = a_k$ is a simple pole of the function $\varphi(t)$ with residue equal to $\alpha_k - 1$. The function $\varphi(t)$ in the neighborhood of the point a_k has the expansion:

$$\varphi(t) = \frac{\alpha_k - 1}{t - a_k} + \text{a regular function.}$$

$f'(t)$ reduces to zero or infinity only at the points a_k; therefore $\varphi(t)$ has no other singular points in a finite distance. Thus over the entire t-plane there holds the expansion

$$\varphi(t) = \sum_{k=1}^{n} \frac{\alpha_k - 1}{t - a_k} + \varphi_1(t),$$

where $\varphi_1(t)$ is a function regular over the entire t-plane. Moreover, the point at infinity of the t-plane is carried over to a definite finite point on a side of the polygon and is a point of regularity of the function $f(t)$. Hence one is readily convinced that $\varphi_1(t) = 0$.

For $\frac{f''(t)}{f'(t)}$ we have the expression:

$$\frac{f''(t)}{f'(t)} = \sum_{k=1}^{n} \frac{\alpha_k - 1}{t - a_k}.$$

Integrating twice, we obtain in succession:

$$\ln f'(t) = \sum_{k=1}^{n} (\alpha_k - 1) \ln (t - a_k) + \ln C_1 = \\ = \ln C_1(t - a_1)^{\alpha_1 - 1} \dots (t - a_n)^{\alpha_n - 1},$$

$$f(t) = C_1 \int_0^t (t - a_1)^{\alpha_1 - 1}(t - a_2)^{\alpha_2 - 1} \dots (t - a_n)^{\alpha_n - 1} dt + C_2. \qquad (1)$$

And this gives the required analytic expression for $f(t)$.

The limitation that the points $a_1, a_2, \dots, a_n$ shall lie in a finite distance is easily removed. If the point corresponding to the vertex A_n of the polygon is located at infinity, the expression for the mapping function can be obtained from (1) by the change of variable:

$$t = -\frac{1}{\tau} + a_n.$$

Using the equality

$$\alpha_1 + \alpha_2 + \dots + \alpha_n = n - 2,$$

we obtain, after elementary transformations,

$$f(\tau) = C_1' \int_0^\tau (\tau - a_1')^{\alpha_1 - 1}(\tau - a_2')^{\alpha_2 - 1} \dots (\tau - a_{n-1}')^{\alpha_n - 1} d\tau + C_2', \qquad (2)$$

where $a_1', a_2', \dots, a_{n-1}'$ are the points of the real axis of the τ-plane corresponding to the points $a_1, a_2, \dots, a_{n-1}$ of the t-plane. The formulas (1) and (2) derived by us are usually called the Christoffel-Schwarz integral [1]).

2. The value of the parameters in the Christoffel-Schwarz integral. The function

$$z = C_1 \int_0^t (t - a_1)^{\alpha_1 - 1}(t - a_2)^{\alpha_2 - 1} \dots (t - a_n)^{\alpha_n - 1} dt + C_2, \qquad (3)$$

mapping the upper half of the t-plane onto a polygon in the z-plane, contains $(2n + 2)$ parameters: $\alpha_1, \alpha_2, \dots, \alpha_n, a_1, \dots, a_n, C_1, C_2$. If the values of all these parameters are known, the elements of the polygon are simply determined. The interior angles of the polygon are equal to: $\alpha_1\pi, \alpha_2\pi, \dots, \alpha_n\pi$. The coordinates b_k of the vertices A_k are defined

1) H. A. Schwarz, [1]; E. B. Christoffel, [1].

by the equations:

$$b_k = C_1 \int_0^{a_k} (t - a_1)^{\alpha_1 - 1}(t - a_2)^{\alpha_2 - 1} \dots (t - a_n)^{\alpha_n - 1} dt + C_2. \tag{4}$$

Here k runs over the values 1, 2, ..., n. The length of the side $A_k A_{k+1}$ can be determined as follows:

length of $A_k A_{k+1} =$

$$= |b_{k+1} - b_k| = |C_1 \int_0^{a_{k+1}} (t - a_1)^{\alpha_1 - 1}(t - a_2)^{\alpha_2 - 1} \dots (t - a_n)^{\alpha_n - 1} dt +$$

$$+ C_2 - [C_1 \int_0^{a_k} (t - a_1)^{\alpha_1 - 1}(t - a_2)^{\alpha_2 - 1} \dots (t - a_n)^{\alpha_n - 1} dt + C_2]| =$$

$$= |C_1| \int_{a_k}^{a_{k+1}} (t - a_1)^{\alpha_1 - 1}(t - a_2)^{\alpha_2 - 1} \dots (t - a_k)^{\alpha_k - 1}(a_{k+1} - t)^{\alpha_{k+1} - 1} \dots$$

$$\dots (a_n - t)^{\alpha_n - 1} dt. \tag{5}$$

On the basis of formulas (4) and (5) it is very easy to explain the role of the parameters $\alpha_1, \alpha_2, \dots, \alpha_n, C_1, C_2$. The role of the parameters $\alpha_1, \alpha_2, \dots, \alpha_n$, which are the ratios of the angles of the polygon to the angle π, is immediately evident. The parameter C_2 figures as a term in the expression for the coordinate of each vertex, therefore with a change of C_2 the coordinates of all vertices will be changed by the same quantity, and the entire polygon, as a unit, will be displaced. The parameter C_1 figures as a factor in the expression for the length of each side of the polygon, therefore with a change in C_1 the lengths of all sides are changed proportionately and in addition the polygon is rotated, and we obtain a similar polygon. To a change in the parameters $a_1, \dots, a_n$ there corresponds a relative change of the lengths of the sides of the polygon.

Until now we have been occupied only with a polygon with given angles, not having been at all concerned with the sides of the polygon. In practical questions it is usually required to find the function effecting the conformal transformation of the half plane (or circle) into a given definite form of the polygon A. To form this function it is necessary to define appropriately the parameters

$$\alpha_1, \alpha_2, \dots, \alpha_n, a_1, a_2, \dots, a_n, C_1, C_2.$$

The angles of the polygon, as before, we shall designate as $\alpha_1\pi, \alpha_2\pi, \dots, \alpha_n\pi$, and the ratio of the 2nd, 3rd, ..., $(n - 2)$th side to the first side we shall designate as $\lambda_2, \lambda_3, \dots, \lambda_{n-2}$, i.e.,

$$\frac{\text{length } A_2A_3}{\text{length } A_1A_2} = \lambda_2, \frac{\text{length } A_3A_4}{\text{length } A_1A_2} = \lambda_3, \dots, \frac{\text{length } A_{n-2}A_{n-1}}{\text{length } A_1A_2} = \lambda_{n-2}, \tag{6}$$

where $A_1, A_2, \dots, A_n$ are the vertices of the polygon A. We note that the ratios of the last two sides to the first side are defined on the basis of these data. If in the function

$$z^* = \int_0^t (t-a_1)^{\alpha_1-1}(t-a_2)^{\alpha_2-1} \ldots (t-a_n)^{\alpha_n-1}\,dt$$

we so choose the numbers $a_1, a_2, \ldots, a_n$ that relations (6) are satisfied, and take as $\alpha_1, \alpha_2, \ldots, \alpha_n$ the ratios of the angles of the given polygon to the angle π, then this function will effect the conformal transformation of the upper half plane into the rectangle A^*, similar to A. We can pass from A^* to A in the following manner. We transfer any vertex, for instance A_k^*, of the polygon A^* to the corresponding vertex A_k of the polygon A, next turn the polygon A^* about this vertex so that its sides become parallel to the sides of the polygon A, and, lastly, without changing the position of the vertex A_k^*, we change the length of all the sides of A^* in such a way that it will coincide with A. All of this can be accomplished by the linear transformation

$$z = C_1 z^* + C_2.$$

The constants C_1 and C_2 are determined by comparing the positions and linear dimensions of the polygons A^* and A.

For the determination of the parameters $a_1, a_2, \ldots, a_n$ we have only $(n-3)$ equations. Three parameters can therefore be taken completely arbitrarily. This corresponds to the fact that by means of a linear fractional substitution in an integral of form (3) we can carry the upper half plane into itself, and moreover so that three points of the real axis are carried into any three preassigned points of the same axis. As these arbitrary points we shall take $t = a_1$, $t = a_2$, and $t = a_n$ and designate them by p_1, p_2 and p_3, in contradistinction to the other points, which are subject to determination. We shall write relations (6) *in extenso*:

$$\left.\begin{aligned} I_2(a_3, a_4, \ldots, a_{n-1}) &= \lambda_2 I_1(a_3, a_4, \ldots, a_{n-1}),\\ I_3(a_3, a_4, \ldots, a_{n-1}) &= \lambda_3 I_1(a_3, a_4, \ldots, a_{n-1}),\\ &\ldots\ldots\ldots\ldots\ldots\\ I_{n-2}(a_3, a_4, \ldots, a_{n-1}) &= \lambda_{n-2} I_1(a_3, a_4, \ldots, a_{n-1}), \end{aligned}\right\} \quad (7)$$

where

$$\left.\begin{aligned} I_1 &= \int_{p_1}^{p_2} (t-p_1)^{\alpha_1-1}(p_2-t)^{\alpha_2-1}(a_3-t)^{\alpha_3-1}\ldots\\ &\qquad \ldots (a_{n-1}-t)^{\alpha_{n-1}-1}(p_3-t)^{\alpha_n-1}\,dt,\\ I_2 &= \int_{p_2}^{a_3} (t-p_1)^{\alpha_1-1}(t-p_2)^{\alpha_2-1}(a_3-t)^{\alpha_3-1}\ldots\\ &\qquad \ldots (a_{n-1}-t)^{\alpha_{n-1}-1}(p_3-t)^{\alpha_n-1}\,dt,\\ &\ldots\ldots\ldots\ldots\ldots\\ I_{n-2} &= \int_{a_{n-2}}^{a_{n-1}} (t-p_1)^{\alpha_1-1}(t-p_2)^{\alpha_2-1}(t-a_3)^{\alpha_3-1}\ldots\\ &\ldots (t-a_{n-2})^{\alpha_{n-2}-1}(a_{n-1}-t)^{\alpha_{n-1}-1}(p_3-t)^{\alpha_n-1}\,dt. \end{aligned}\right\} \quad (8)$$

3. On the Newton-Fourier method for system of equations (7) and on the computation of the improper integrals. In the preceding paragraph it was shown that to set up the function effecting the conformal transformation of the upper half plane into a polygon with given angles and given ratios of sides, it is necessary to know the constants $a_3, a_4, \ldots, a_{n-1}$ that are the solution of the system of equations:

$$\begin{aligned} I_2(a_3, a_4, \ldots, a_{n-1}) &= \lambda_2 I_1(a_3, a_4, \ldots, a_{n-1}), \\ I_3(a_3, a_4, \ldots, a_{n-1}) &= \lambda_3 I_1(a_3, a_4, \ldots, a_{n-1}), \\ &\ldots\ldots\ldots\ldots \\ I_{n-2}(a_3, a_4, \ldots, a_{n-1}) &= \lambda_{n-2} I_1(a_3, a_4, \ldots, a_{n-1}). \end{aligned}$$

On the basis of the theorem on the uniqueness of the function effecting the conformal transformation of one given region into another, it can be affirmed that for the chosen p_1, p_2 and p_3 this system has a solution, and that the solution is moreover unique. We shall solve this system by the method usually called the Newton-Fourier method [1]).

Let $a_3^*, a_4^*, \ldots, a_{n-1}^*$ be the solution of the system. As the initial values we shall take numbers $a_3^{(0)}, a_4^{(0)}, \ldots, a_{n-1}^{(0)}$ differing little from $a_3^*, a_4^*, \ldots, a_{n-1}^*$. We expand each equation of system (7) in a Taylor's series in powers of the differences

$$a_3^* - a_3^{(0)},\ a_4^* - a_4^{(0)},\ \ldots,\ a_{n-1}^* - a_{n-1}^{(0)}, \tag{9}$$

and terminate these series on the first powers of the differences. We then obtain the system of equations:

$$\left.\begin{aligned} & I_2^{(0)} + \alpha_3^{(1)} \frac{\partial I_2^{(0)}}{\partial a_3} + \alpha_4^{(1)} \frac{\partial I_2^{(0)}}{\partial a_4} + \ldots + \alpha_{n-1}^{(1)} \frac{\partial I_2^{(0)}}{\partial a_{n-1}} = \\ & \quad = \lambda_2 \left[I_1^{(0)} + \alpha_3^{(1)} \frac{\partial I_1^{(0)}}{\partial a_3} + \alpha_4^{(1)} \frac{\partial I_1^{(0)}}{\partial a_4} + \ldots + \alpha_{n-1}^{(1)} \frac{\partial I_1^{(0)}}{\partial a_{n-1}} \right], \\ & I_3^{(0)} + \alpha_3^{(1)} \frac{\partial I_3^{(0)}}{\partial a_2} + \alpha_4^{(1)} \frac{\partial I_3^{(0)}}{\partial a_4} + \ldots + \alpha_{n-1}^{(1)} \frac{\partial I_3^{(0)}}{\partial a_{n-1}} = \\ & \quad = \lambda_3 \left[I_1^{(0)} + \alpha_3^{(1)} \frac{\partial I_1^{(0)}}{\partial a_3} + \alpha_4^{(1)} \frac{\partial I_1^{(0)}}{\partial a_4} + \ldots + \alpha_{n-1}^{(1)} \frac{\partial I_1^{(0)}}{\partial a_{n-1}} \right], \\ & \ldots\ldots\ldots\ldots\ldots\ldots\ldots\ldots \\ & I_{n-2}^{(0)} + \alpha_3^{(1)} \frac{\partial I_{n-2}^{(0)}}{\partial a_3} + \alpha_4^{(1)} \frac{\partial I_{n-2}^{(0)}}{\partial a_4} + \ldots + \alpha_{n-1}^{(1)} \frac{\partial I_{n-2}^{(0)}}{\partial a_{n-1}} = \\ & \quad = \lambda_{n-2} \left[I_1^{(0)} + \alpha_3^{(1)} \frac{\partial I_1^{(0)}}{\partial a_3} + \alpha_4^{(1)} \frac{\partial I_1^{(0)}}{\partial a_4} + \ldots + \alpha_{n-1}^{(1)} \frac{\partial I_1^{(0)}}{\partial a_{n-1}} \right]. \end{aligned}\right\} \tag{10}$$

In this system of equations the unknowns are the perturbed values of differences (9): $\alpha_3^{(1)}, \alpha_4^{(1)}, \ldots, \alpha_{n-1}^{(1)}$ and the free terms and coefficients

---

[1]) P. P. Kufarev, [1], has proposed a method of determining the constants that is essentially different from that expounded here, being based on the utilization of the differential equation of Loewner.

of the unknowns are the values of $I_1, I_2, \ldots, I_{n-2}$ and the values of of their first derivatives with respect to $a_3, a_4, \ldots, a_{n-1}$ for $a_3 = a_3^{(0)}$, $a_4 = a_4^{(0)}, \ldots, a_{n-1} = a_{n-1}^{(0)}$ [1]).

We reduce this system to the usual form and solve. Using the values $\alpha_3^{(1)}, \alpha_4^{(1)}, \ldots, \alpha_{n-1}^{(1)}$ found, and the initial values $a_3^{(0)}, a_4^{(0)}, \ldots, a_{n-1}^{(0)}$, we construct the first approximations:

$$a_3^{(1)} = a_3^{(0)} + \alpha_3^{(1)},\ a_4^{(1)} = a_4^{(0)} + \alpha_4^{(1)},\ \ldots,\ a_{n-1}^{(1)} = a_{n-1}^{(0)} + \alpha_{n-1}^{(1)}.$$

Next, expanding the equations of system (7) in Taylor's series in powers of the differences

$$a_3^* - a_3^{(1)},\ a_4^* - a_4^{(1)},\ \ldots,\ a_{n-1}^* - a_{n-1}^{(1)} \tag{11}$$

and terminating these series on the first powers of the differences, we construct system of equations (10′) analogous to system (10). In system (10′) the unknowns will be $\alpha_3^{(2)}, \alpha_4^{(2)}, \ldots, \alpha_{n-1}^{(2)}$ — the perturbed values of differences (11), and the free terms and coefficients of the unknowns will be the same expressions as in system (10), computed, however, for $a_3 = a_3^{(1)}, a_4 = a_4^{(1)}, \ldots, a_{n-1} = a_{n-1}^{(1)}$.

From system (10′) we find $\alpha_3^{(2)}, \alpha_4^{(2)}, \ldots, \alpha_{n-1}^{(2)}$ and next construct the second approximations:

$$a_3^{(2)} = a_3^{(1)} + \alpha_3^{(2)},\ a_4^{(2)} = a_4^{(1)} + \alpha_4^{(2)},\ \ldots,\ a_{n-1}^{(2)} = a_{n-1}^{(1)} + \alpha_{n-1}^{(2)}.$$

It can be shown that the determinant of system (10) is different from zero, i.e., that the equations for the determination of the corrections α are always soluble, and that if the initial values are taken sufficiently close to the ones sought, we can approach arbitrarily closely [2]) to the solution $a_3^*, a_4^*, \ldots, a_{n-1}^*$ by using the Newton-Fourier method.

For the determination of the coordinates of the vertices of the polygon, the lengths of its sides, and for the execution of the method just expounded for solving system (7), one must be able to calculate integrals of the form

$$E = \int_{a_k}^{a_{k+1}} (t-p_1)^{\alpha_1-1}(t-p_2)^{\alpha_2-1}(t-a_3)^{\alpha_3-1} \ldots \\ \ldots (t-a_k)^{\alpha_k-1}(a_{k+1}-t)^{\alpha_{k+1}-1} \ldots (p_3-t)^{\alpha_n-1}\,dt.$$

These integrals are improper. The integrand of each of them can become infinite at two points: for t equal to the lower and the upper limits

---

[1]) In forming $\frac{\partial I_2}{\partial a_3}, \frac{\partial I_3}{\partial a_3}, \frac{\partial I_3}{\partial a_4}, \ldots, \frac{\partial I_{n-2}}{\partial a_{n-2}}, \frac{\partial I_{n-2}}{\partial a_{n-1}}$ it is convenient, in the integrals $I_2, I_3, \ldots, I_{n-2}$, to pass to limits not depending on those quantities with respect to which the differentiation is being performed.

[2]) N. P. Stenin, [1]. In this work are also given criteria by means of which one can state, with respect to the given initial values, that these initial values guarantee a convergent process if the application of the Newton-Fourier method be begun with them. More exact conditions of the convergence of this method are indicated in Chap. IV of L. V. Kantorovich's work [3].

of the integration. However these improper integrals exist. This follows directly from the Cauchy test, since the numbers $\alpha_1, \alpha_2, \ldots, \alpha_n$ are always greater than zero. Any integral of such a form can be computed by a method proposed by L. V. Kantorovich [1]. The idea of this method consists in the following. Let us denote the integrand by $F(t)$. It can be represented as follows:

$$F(t) = (t - p_1)^{\alpha_1-1}(t - p_2)^{\alpha_2-1}(t - a_3)^{\alpha_3-1} \ldots$$
$$\ldots (t - a_k)^{\alpha_k-1}(a_{k+1} - t)^{\alpha_{k+1}-1} \ldots (p_3 - t)^{\alpha_n-1} =$$
$$= (t - a_k)^{\alpha_k-1}[\varphi(a_k) + \varphi'(a_k)(t - a_k)] +$$
$$+ (a_{k+1} - t)^{\alpha_{k+1}-1}[\psi(a_{k+1}) - \psi'(a_{k+1})(a_{k+1} - t)] +$$
$$+ \{F(t) - (t - a_k)^{\alpha_k-1}[\varphi(a_k) + \varphi'(a_k)(t - a_k)] -$$
$$- (a_{k+1} - t)^{\alpha_{k+1}-1}[\psi(a_{k+1}) - \psi'(a_{k+1})(a_{k+1} - t)]\},$$

where

$$\varphi(t) = (t - p_1)^{\alpha_1-1}(t - p_2)^{\alpha_2-1}(t - a_3)^{\alpha_3-1} \ldots$$
$$\ldots (t - a_{k-1})^{\alpha_{k-1}-1}(a_{k+1} - t)^{\alpha_{k+1}-1} \ldots (p_3 - t)^{\alpha_n-1},$$
$$\psi(t) = (t - p_1)^{\alpha_1-1}(t - p_2)^{\alpha_2-1}(t - a_3)^{\alpha_3-1} \ldots$$
$$\ldots (t - a_k)^{\alpha_k-1}(a_{k+2} - t)^{\alpha_{k+2}-1} \ldots (p_3 - t)^{\alpha_n-1}.$$

The integral E can be separated into two integrals E_1 and E_2:

$$E_1 = \int_{a_k}^{a_k+1} \{(t - a_k)^{\alpha_k-1}[\varphi(a_k) + \varphi'(a_k)(t - a_k)] +$$
$$+ (a_{k+1} - t)^{\alpha_{k+1}-1}[\psi(a_{k+1}) - \psi'(a_{k+1})(a_{k+1} - t)]\}\,dt,$$
$$E_2 = \int_{a_k}^{a_k+1} \{F(t) - (t - a_k)^{\alpha_k-1}[\varphi(a_k) + \varphi'(a_k)(t - a_k)] -$$
$$- (a_{k+1} - t)^{\alpha_{k+1}-1}[\psi(a_{k+1}) - \psi'(a_{k+1})(a_{k+1} - t)]\}\,dt.$$

The integral E_1 is easily calculated in finite form, and the integral E_2 is entirely free of singularities, since the integrand is continuous, together with its derivative. Therefore E_1 is computed exactly, and E_2 can be found approximately by any of the familiar formulas for the approximate computation of definite integrals.

4. Examples. We shall apply the reasoning of the foregoing paragraphs to particular cases.

1. The polygon A is a triangle with angles $\alpha_1\pi$, $\alpha_2\pi$, $\alpha_3\pi$. Since two triangles having equal angles are always similar, there is in this case no need to concern ourselves with the choice of the constants a. As the points $t = a_1$, $t = a_2$, $t = a_3$, there can be taken three arbitrary points of the real axis.

Let us put $a_1 = 0$, $a_2 = 1$, $a_3 = \infty$. In accordance with formula (2), the mapping

[1]) L. V. Kantorovich, [7].

function will have the form

$$z = C_1 \int_0^t t^{\alpha_1-1}(1-t)^{\alpha_2-1}dt + C_2.$$

For the complete determination of the mapping function the constants C_1 and C_2 must be found.

We shall determine C_1 and C_2 for the case of an isosceles right triangle whose vertices are the points $A_1(0)$, $A_2(1)$, $A_3(i)$.

To the point $t = 0$ let there correspond the vertex A_1, and to the point $t = 1$, the vertex A_2. Then $\alpha_1 = \frac{1}{2}$, $\alpha_2 = \frac{1}{4}$, and the mapping function will be.

$$z = C_1 \int_0^t t^{-\frac{1}{2}}(1-t)^{-\frac{3}{4}}dt + C_2.$$

Since, by the condition, $t = 0$ must correspond to $z = 0$, we have $C_2 = 0$. For $t = 1$ we must obtain the coordinate of the vertex A_2, consequently C_1 must be such that

$$C_1 \int_0^1 t^{-\frac{1}{2}}(1-t)^{-\frac{3}{4}}dt = 1,$$

whence

$$C_1 = \frac{1}{\int_0^1 t^{-\frac{1}{2}}(1-t)^{-\frac{3}{4}}dt}.$$

Let us undertake the evaluation of the integral in the denominator. We shall designate it by I:

$$I = \int_0^1 t^{-\frac{1}{2}}(1-t)^{-\frac{3}{4}}dt = \int_0^{\frac{1}{2}} t^{-\frac{1}{2}}(1-t)^{-\frac{3}{4}}dt + \int_{\frac{1}{2}}^1 t^{-\frac{1}{2}}(1-t)^{-\frac{3}{4}}dt = I_1 + I_2.$$

We shall first compute I_1. We represent the integrand as follows:

$$t^{-\frac{1}{2}}(1-t)^{-\frac{3}{4}} = t^{-\frac{1}{2}}(1+\tfrac{3}{4}t) + t^{-\frac{1}{2}}[(1-t)^{-\frac{3}{4}} - 1 - \tfrac{3}{4}t].$$

Then

$$I_1 = \int_0^{\frac{1}{2}} t^{-\frac{1}{2}}(1+\tfrac{3}{4}t)dt + \int_0^{\frac{1}{2}} t^{-\frac{1}{2}}[(1-t)^{-\frac{3}{4}} - 1 - \tfrac{3}{4}t]dt.$$

The first integral is calculated directly; it equals 1.5910. The second integral we compute by Simpson's formula. We shall call

$$y(t) = t^{-\frac{1}{2}}[(1-t)^{-\frac{3}{4}} - 1 - \tfrac{3}{4}t];$$

then

$$y(0) = 0, \qquad y(\tfrac{1}{16}) = 0.0104, \qquad y(\tfrac{1}{8}) = 0.0328,$$

$$y(\tfrac{3}{16}) = 0.0644, \qquad y(\tfrac{1}{4}) = 0.1066, \qquad y(\tfrac{5}{16}) = 0.1612,$$

$$y(\tfrac{3}{8}) = 0.2309, \qquad y(\tfrac{7}{16}) = 0.3198, \qquad y(\tfrac{1}{2}) = 0.4339,$$

$$\int_0^{\frac{1}{2}} y(t)dt = 0.0707,$$

$$I_1 = 1.5910 + 0.0707 = 1.6617.$$

Let us go on to the computation of I_2. We represent the integrand as follows:

$$t^{-\frac{1}{2}}(1-t)^{-\frac{3}{4}} = (1-t)^{-\frac{3}{4}}[1+\tfrac{1}{2}(1-t)] + (1-t)^{-\frac{3}{4}}[t^{-\frac{1}{2}} - 1 - \tfrac{1}{2}(1-t)];$$

then

$$I_2 = \int_{\frac{1}{2}}^{1} (1-t)^{-\frac{3}{4}}[1+\tfrac{1}{2}(1-t)]dt + \int_{\frac{1}{2}}^{1} (1-t)^{-\frac{3}{4}}[t^{-\frac{1}{2}} - 1 - \tfrac{1}{2}(1-t)]dt.$$

The first integral is equal to 3.5318.

The second integral is computed the same way as was the second part of the integral I_1; it is equal to 0.0506.

$$I_2 = 3.5824, \quad I = I_1 + I_2 = 5.2441.$$

Thus

$$C_1 = \frac{1}{5.2441} = 0.1907.$$

The problem is solved in full.

For a check we yet compute the hypotenuse of the triangle

$$A_2A_3 = 0.1907 \int_1^{\infty} t^{-\frac{1}{2}}(t-1)^{-\frac{3}{4}}dt,$$

$$\int_1^{\infty} t^{-\frac{1}{2}}(t-1)^{-\frac{3}{4}}dt = 2\int_0^{\frac{1}{2}} \tau^{-\frac{3}{4}}(1-\tau)^{-\frac{3}{4}}d\tau \ \left(\text{substitution } \tau = \frac{1}{t}\right);$$

$$\int_0^{\frac{1}{2}} \tau^{-\frac{3}{4}}(1-\tau)^{-\frac{3}{4}}d\tau = \int_0^{\frac{1}{2}} \tau^{-\frac{3}{4}}(1+\tfrac{3}{4}\tau)d\tau + \int_0^{\frac{1}{2}} \tau^{-\frac{3}{4}}[(1-\tau)^{-\frac{3}{4}} - 1 - \tfrac{3}{4}\tau]d\tau =$$
$$= 3.6158 + 0.0922 = 3.7080,$$

$$A_2A_3 = 0.1907 \cdot 2 \cdot 3.7080 = 1.4142.$$

The exact value of A_2A_3 is equal to $\sqrt{2} = 1.4142\ldots$.

2. The polygon A is a rectangle, whose vertices are the points:

$$A_1(-\omega),\ A_2(\omega),\ A_3(\omega+\omega i) \text{ and } A_4(-\omega+\omega i),$$

where ω is a real constant. Let us first assume that the right half of the upper semiplane ($\mathrm{Re}(t) > 0$, $I(t) > 0$) is mapped onto the right side of the rectangle so that the point $t = 0$ passes to $z = 0$, $t = 1$ to $z = \omega$, and $t = \infty$ to $z = \omega i$. We shall make a reflection in the imaginary axes in the t-plane and in the z-plane. Then on the principle of symmetry we will have the mapping of the entire upper half plane onto the entire rectangle, and $t = 0$ will correspond to $z = 0$, $t = 1$ to $z = \omega$, $t = -1$ to $z = -\omega$, $t = \infty$ to $z = \omega i$.

The vertices A_3 and A_4 will correspond to points symmetric with respect to the imaginary axis; let these be the points $\frac{1}{K}$ and $-\frac{1}{K}$, where $K < 1$. The mapping function will have the form

$$z = C_1 \int_0^t (t+1)^{-\frac{1}{2}}(t-1)^{-\frac{1}{2}}\left(t-\frac{1}{K}\right)^{-\frac{1}{2}}\left(t+\frac{1}{K}\right)^{-\frac{1}{2}} dt$$

or

$$z = C_1' \int_0^t (1 - t^2)^{-\frac{1}{2}}(1 - K^2t^2)^{-\frac{1}{2}}dt.$$

We shall define K in such a way that the ratio

$$\frac{A_1A_2}{A_2A_3} = \frac{2\omega}{\omega} = 2.$$

In this case system (7) reduces to one equation:

$$\int_{-1}^{1} (1 - t^2)^{-\frac{1}{2}}(1 - K^2t^2)^{-\frac{1}{2}}dt = 2\int_{1}^{\frac{1}{K}} (t^2 - 1)^{-\frac{1}{2}}(1 - K^2t^2)^{-\frac{1}{2}}dt.$$

$$\int_{-1}^{1} (1 - t^2)^{-\frac{1}{2}}(1 - K^2t^2)^{-\frac{1}{2}}dt$$

can be replaced by

$$2\int_{0}^{1} (1 - t^2)^{-\frac{1}{2}}(1 - K^2t^2)^{-\frac{1}{2}}dt.$$

On substituting in the equation for the determination of K, we obtain

$$\int_{0}^{1} (1 - t^2)^{-\frac{1}{2}}(1 - K^2t^2)^{-\frac{1}{2}}dt = \int_{1}^{\frac{1}{K}} (t^2 - 1)^{-\frac{1}{2}}(1 - K^2t^2)^{-\frac{1}{2}}dt.$$

We shall introduce the notation

$$\int_{0}^{1} (1 - t^2)^{-\frac{1}{2}}(1 - K^2t^2)^{-\frac{1}{2}}dt = I_1(K),$$

$$\int_{0}^{\frac{1}{K}} (t^2 - 1)^{-\frac{1}{2}}(1 - K^2t^2)^{-\frac{1}{2}}dt = I_2(K).$$

We shall solve the equation $I_1(K) = I_2(K)$ by the Newton-Fourier method. We shall have to deal with only one correction, and we shall therefore call it h, with a subscript indicating the number of the approximation. As the initial approximation we adopt $K = \frac{1}{2}$. For the determination of h_1 we shall have the equation

$$I_1(\tfrac{1}{2}) + h_1\frac{dI_1(\frac{1}{2})}{dK} = I_2(\tfrac{1}{2}) + h_1\frac{dI_2(\frac{1}{2})}{dK}.$$

Let us compute the free term and the coefficients of this equation. We have

$$I_1(\tfrac{1}{2}) = \int_{0}^{1} (1 - t)^{-\frac{1}{2}}\left[(1 + t)^{-\frac{1}{2}}\left(1 - \frac{t^2}{4}\right)^{-\frac{1}{2}}\right]dt.$$

Let

$$(1+t)^{-\frac{1}{2}}\left(1-\frac{t^2}{4}\right)^{-\frac{1}{2}} = \psi(t).$$

Then $\psi_1(1) = \sqrt{\frac{2}{3}}$, $\psi'(1) = \frac{1}{12}\sqrt{\frac{2}{3}}$. The integrand of $I_1(\frac{1}{2})$ can be represented in the following manner:

$$(1-t)^{-\frac{1}{2}}\left[(1+t)^{-\frac{1}{2}}\left(1-\frac{t^2}{4}\right)^{-\frac{1}{2}}\right] = \sqrt{\tfrac{2}{3}}(1-t)^{-\frac{1}{2}}\left[1-\tfrac{1}{12}(1-t)\right] +$$

$$+ (1-t)^{-\frac{1}{2}}\left[(1+t)^{-\frac{1}{2}}\left(1-\frac{t^2}{4}\right)^{-\frac{1}{2}} - \sqrt{\tfrac{2}{3}} + \tfrac{1}{12}\sqrt{\tfrac{2}{3}}\,(1-t)\right].$$

Then

$$I_1(\tfrac{1}{2}) = \sqrt{\tfrac{2}{3}}\int_0^1 (1-t)^{-\frac{1}{2}}\left[1-\tfrac{1}{12}(1-t)\right]dt +$$

$$+ \int_0^1 (1-t)^{-\frac{1}{2}}\left[(1+t)^{-\frac{1}{2}}\left(1-\frac{t^2}{4}\right)^{-\frac{1}{2}} - \sqrt{\tfrac{2}{3}} + \tfrac{1}{12}\sqrt{\tfrac{2}{3}}(1-t)\right]dt.$$

The first integral = 1.5877.

We compute the second integral by Simpson's rule with three intermediate points. We shall designate by $y(t)$ the integrand. Then

$$y(0) = 0.2515,\ y(\tfrac{1}{4}) = 0.1571,\ y(\tfrac{1}{2}) = 0.0860,\ y(\tfrac{3}{4}) = 0.0318,\ y(1) = 0.$$

$$\int_0^1 y(t)dt = \tfrac{1}{12}[0.2515 + 4\cdot 0.1571 + 2\cdot 0.0860 + 4\cdot 0.0318] = 0.0983,$$

$$I_1(\tfrac{1}{2}) = 1.6860.$$

Differentiating $I_1(K)$ with respect to K, we obtain:

$$\frac{dI_1(K)}{dK} = \int_0^1 \frac{Kt^2dt}{(1-t^2)^{1/2}(1-K^2t^2)^{3/2}},$$

$$\frac{dI_1(\frac{1}{2})}{dK} = \frac{1}{2}\int_0^1 \frac{t^2dt}{(1-t^2)^{1/2}\left(1-\frac{t^2}{4}\right)^{3/2}}.$$

Computing it by the same method, we find $\dfrac{dI_1(\frac{1}{2})}{dK} = 0.54$.

Let us pass on to the computation of $I_2(\frac{1}{2})$ and $\dfrac{dI_2(\frac{1}{2})}{dK}$.

Having in view the differentiation of $I_2(K)$ with respect to K, we transform this integral from the interval $\left(1, \dfrac{1}{K}\right)$ to the interval (0, 1): making a change of variable, we put

$$t = \frac{(1-K)t_1 + K}{K}.$$

Then

$$I_2(K) = \int_0^1 [(1-K)t_1^2 + 2Kt_1]^{-\frac{1}{2}}[(1+K) - 2Kt_1 - (1-K)t_1^2]^{-\frac{1}{2}}dt_1,$$

$$\frac{dI_2(K)}{dK} = \int_0^1 [(1-K)t_1^2 - (1-3K)t_1 - (1+2K)]t_1^{-\frac{1}{2}}(1-t_1)^{-\frac{1}{2}} \times$$
$$\times [(1-K)t_1 + 2K]^{-3/2} \cdot [(1+K) + (1-K)t_1]^{-3/2}dt_1.$$

$$I_2(\tfrac{1}{2}) = \int_0^1 (\tfrac{1}{2}t_1^2 + t_1)^{-\frac{1}{2}}(\tfrac{3}{2} - t_1 - \tfrac{1}{2}t_1^2)^{-\frac{1}{2}}dt_1 =$$
$$= 2\int_0^1 t_1^{-\frac{1}{2}}(2+t_1)^{-\frac{1}{2}}(1-t_1)^{-\frac{1}{2}}(3+t_1)^{-\frac{1}{2}}dt_1 =$$
$$= 2\int_0^{\frac{1}{2}} t_1^{-\frac{1}{2}}(2+t_1)^{-\frac{1}{2}}(1-t_1)^{-\frac{1}{2}}(3+t_1)^{-\frac{1}{2}}dt_1 +$$
$$+ 2\int_{\frac{1}{2}}^1 t^{-\frac{1}{2}}(2+t_1)^{-\frac{1}{2}}(1-t_1)^{-\frac{1}{2}}(3+t_1)^{-\frac{1}{2}}dt_1.$$

Each of the last integrals we separate into two, the first of which are computed directly, the second by Simpson's rule. On performing all the computations, we obtain:

$$I_2(\tfrac{1}{2}) = 2.1570.$$

The computation of $\frac{dI_2(\frac{1}{2})}{dK}$ is performed in exactly the same way as the computation of all the foregoing integrals, and we therefore omit it:

$$\frac{dI_2(\tfrac{1}{2})}{dK} = -1.76.$$

On substituting in the equation for h_1 the values found for $I_1(\frac{1}{2})$, $\frac{dI_1(\frac{1}{2})}{dK}$, $I_2(\frac{1}{2})$ and $\frac{dI_2(\frac{1}{2})}{dK}$, we obtain:

$$1.6860 + 0.54h_1 = 2.1570 - 1.76h_1,$$

whence $h_1 = 0.2$. The first approximation will be equal to $0.5 + 0.2 = 0.7$.

For the determination of h_2 we shall have the equation:

$$I_1(0.7) + h_2\frac{dI(0.7)}{dK} = I_2(0.7) + h_2\frac{dI_2(0.7)}{dK}.$$

The computation of the free term and coefficient of this equation is performed as for the case $K = \frac{1}{2}$. On completing these computations we obtain [1])

$$I_1(0.7) = 1.8458, \quad I_2(0.7) = 1.8631,$$

$$\frac{dI_1(0.7)}{dK} = 1.18, \quad \frac{dI_2(0.7)}{dK} = -1.21.$$

---

[1]) We will remark that in the given particular example the required integrals can be found without using the method indicated above, by means of tables of elliptic integrals.

The equation for h_2 can therefore be written in the following form:

$$1.8458 + 1.18h_2 = 1.8631 - 1.21h_2,$$

whence $h_2 = 0.0072$. The second approximate value of K is equal to 0.7072.

It is easy to find the exact value of K. In the second integral of our equation for the determination of K, i.e., in

$$I_2(K) = \int_1^{\frac{1}{K}} (t^2 - 1)^{-\frac{1}{2}}(1 - K^2t^2)^{-\frac{1}{2}}dt,$$

we introduce a new variable of integration x by the formula:

$$t = \frac{1}{\sqrt{1 - K_1^2x^2}},$$

where $K_1^2 = 1 - K^2$.

Then

$$I_2(K) = \int_0^1 (1 - x^2)^{-\frac{1}{2}}(1 - K_1^2x^2)^{-\frac{1}{2}}dx.$$

Our equation will have the form:

$$\int_0^1 (1 - t^2)^{-\frac{1}{2}}(1 - K^2t^2)^{-\frac{1}{2}}dt = \int_0^1 (1 - x^2)^{-\frac{1}{2}}(1 - K_1^2x^2)^{-\frac{1}{2}}dx.$$

Consequently

$$K = K_1 \text{ or } K^2 = 1 - K^2,$$

whence

$$K = \sqrt{\tfrac{1}{2}} = 0.7071.$$

Comparing the approximate value of K with the exact, we see that in the second approximation we admit an error less than 0.0002.

5. The mapping of a half plane on an arbitrary quadrilateral. We shall show that in the case of a quadrilateral there is no need to bother with the choice of the initial values.

Let the upper half of the t-plane be conformally mapped onto a quadrilateral with the angles $\alpha_1\pi$, $\alpha_2\pi$, $\alpha_3\pi$, $\alpha_4\pi$ and ratio of the fourth side to the first equal to λ.

The numbering of the angles we shall assume to have been so done that the angle $\alpha_1\pi$ will not be less than any of the rest. To the vertices of the quadrilateral let there correspond the points -1, 1, a and 3. This does not limit the generality. By a linear fractional transformation the upper half plane can be mapped onto itself in such a way that -1, 1, 3 will pass to any points of the real axis. Therefore in all the further reasoning of this paragraph we can assume a to be between 1 and 3.

The mapping function will have the form:

$$z(t) = C_1\int_1^t (t + 1)^{\alpha_1-1}(t - 1)^{\alpha_2-1}(t - a)^{\alpha_3-1}(t - 3)^{\alpha_4-1}dt + C_2.$$

For the determination of a we have the equation

$$I_4(a) = \lambda I_1(a),$$

where

$$I_4(a) = \int_3^\infty (t+1)^{\alpha_1-1}(t-1)^{\alpha_2-1}(t-a)^{\alpha_3-1}(t-3)^{\alpha_4-1}\,dt + \\ + \int_{-\infty}^{-1} (-1-t)^{\alpha_1-1}(1-t)^{\alpha_2-1}(a-t)^{\alpha_3-1}(3-t)^{\alpha_4-1}\,dt$$

and

$$I_1(a) = \int_{-1}^{1} (t+1)^{\alpha_1-1}(1-t)^{\alpha_2-1}(a-t)^{\alpha_3-1}(3-t)^{\alpha_4-1}\,dt.$$

In the integrals expressing $I_4(a)$ we shall pass to a finite interval of integration. For this we make a change of variables, setting $t = 2 + \frac{1}{x}$ in the first integral, and $t = -\frac{1}{x}$ in the second.

Then

$$I_4(a) = \int_0^1 (1+3x)^{\alpha_1-1}(1+x)^{\alpha_2-1}[1+(2-a)x]^{\alpha_3-1}(1-x)^{\alpha_4-1}\,dx + \\ + \int_0^1 (1-x)^{\alpha_1-1}(1+x)^{\alpha_2-1}(1+ax)^{\alpha_3-1}(1+3x)^{\alpha_4-1}\,dx.$$

We shall put $f(a) = I_4(a) - \lambda I_1(a)$ and shall show that $f'(a)$ is positive.

$$f'(a) = (1-\alpha_3)\{\int_0^1 x(1+3x)^{\alpha_1-1}(1+x)^{\alpha_2-1}\times [1+(2-a)x]^{\alpha_3-2}(1-x)^{\alpha_4-1}\,dx - \\ - \int_0^1 x(1-x)^{\alpha_1-1}(1+x)^{\alpha_2-1}(1+ax)^{\alpha_3-2}(1+3x)^{\alpha_4-1}\,dx + \\ + \lambda \int_{-1}^1 (t+1)^{\alpha_1-1}(1-t)^{\alpha_2-1}(a-t)^{\alpha_3-2}(3-t)^{\alpha_4-1}\,dt\},$$

or

$$f'(a) = (1-\alpha_3)\ \{\int_0^1 x(1+3x)^{\alpha_4-1}(1+x)^{\alpha_2-1}(1-x)^{\alpha_4-1}\times \\ \times\left[\frac{(1+3x)^{\alpha_1-\alpha_4}}{(1+(2-a)x)^{2-\alpha_3}} - \frac{(1-x)^{\alpha_1-\alpha_4}}{(1+ax)^{2-\alpha_3}}\right]dx + \\ + \lambda \int_{-1}^1 (t+1)^{\alpha_1-1}(1-t)^{\alpha_2-1}(a-t)^{\alpha_3-2}(3-t)^{\alpha_4-1}\,dt\}.$$

It is immediately clear that the expression

$$\frac{(1+3x)^{\alpha_1-\alpha_4}}{[1+(2-a)x]^{2-\alpha_3}} - \frac{(1-x)^{\alpha_1-\alpha_4}}{(1+ax)^{2-\alpha_3}}$$

is positive for all $0 \leqslant x \leqslant 1$ and $1 \leqslant a \leqslant 3$, since $\alpha_1 \geqslant \alpha_4$ on assumption. Consequently $f'(a)$ is also positive [1]).

For $a = 1$ and $a = 3$ respectively the third and the first integral which figure in the expression for $f'(a)$ will cease to exist.

Therefore $f'(1)$ and $f'(3)$ equal $+\infty$. We shall yet find $f''(a)$ and $f'''(a)$.

$$f''(a) = (1-\alpha_3)(2-\alpha_3)\{\int_0^1 x^2(1+3x)^{\alpha_1-1}(1+x)^{\alpha_2-1} \times$$
$$\times [1+(2-a)x]^{\alpha_3-3}(1-x)^{\alpha_4-1}dx +$$
$$+\int_0^1 x^2(1-x)^{\alpha_1-1}(1+x)^{\alpha_2-1}(1+ax)^{\alpha_3-3}(1+3x)^{\alpha_4-1}dx -$$
$$-\lambda\int_1^{-1}(t+1)^{\alpha_1-1}(1-t)^{\alpha_2-1}(a-t)^{\alpha_3-3}(3-t)^{\alpha_4-1}dt\},$$

$$f'''(a) = (1-\alpha_3)(2-\alpha_3)(3-\alpha_3) \times$$
$$\times \{\int_0^1 x^3(1+3x)^{\alpha_1-1}(1+x)^{\alpha_2-1}[1+(2-a)x]^{\alpha_3-4}(1-x)^{\alpha_4-1}dx -$$
$$-\int_0^1 x^3(1-x)^{\alpha_1-1}(1+x)^{\alpha_2-1}(1+ax)^{\alpha_3-4}(1+3x)^{\alpha_4-1}dx +$$
$$+\lambda\int_{-1}^{1}(t+1)^{\alpha_1-1}(1-t)^{\alpha_2-1}(a-t)^{\alpha_3-4}(3-t)^{\alpha_4-1}dt\}.$$

$f'''(a)$, as was $f'(a)$, is positive in the interval $(1, 3)$ [2]). For $a = 1$, $f(a)$ and $f''(a)$ equal $-\infty$, and for $a = 3$ equal $+\infty$. Therefore with the change of a from 1 to 3, $f(a)$ and $f''(a)$ will continuously increase from $-\infty$ to $+\infty$.

It is easy to give for $f'(a)$ an estimate from below [3])

$$\min\left\{\frac{(1+3x)^{\alpha_1-\alpha_4}}{[1+(2-a)x]^{2-\alpha_3}} - \frac{(1-x)^{\alpha_1-\alpha_4}}{(1+ax)^{2-\alpha_3}}\right\} =$$
$$= \frac{(1+3x)^{\alpha_1-\alpha_4}}{(1+x)^{2-\alpha_3}} - \frac{(1-x)^{\alpha_1-\alpha_4}}{(1+x)^{2-\alpha_3}},$$

$$\min\,(a-t)^{\alpha_3-2} = \frac{1}{(3-t)^{2-\alpha_3}}.$$

---

[1]) This same conclusion can be drawn from consideration of the function

$$\Phi(a) = \frac{I_4(a)}{I_1(a)}.$$

From the uniqueness of the mapping function it follows that with the change of a from 1 to 3, $\Phi(a)$ will monotonically increase or monotonically decrease.

[2]) The same thing can be said with respect to all other odd derivatives of the function $f(a)$.

[3]) One can also give an estimate from below for $f'(a)$ in the root of the equation $f(a) = 0$.

Therefore

$$f'(a) > (1 - a_3)\{\int_0^1 x(1+3x)^{\alpha_4-1}(1+x)^{\alpha_2+\alpha_3-3}(1-x)^{\alpha_4-1} \times$$
$$\times [(1+3x)^{\alpha_1-\alpha_4} - (1-x)^{\alpha_1-\alpha_4}]\,dx +$$
$$+ \lambda \int_{-1}^{1} (t+1)^{\alpha_1-1}(1-t)^{\alpha_2-1}(3-t)^{\alpha_3+\alpha_4-3}\,dt\} = K.$$

From the consideration of the derivatives of the function $f(a)$ it follows that the equations $f(a) = 0$ and $f''(a) = 0$ have in the interval (1, 3) one root apiece. We shall designate as a^* the root of the first of these equations, and as a^{**} the root of the second.

We shall solve the equation $f(a) = 0$ by the Newton-Fourier method.

When the equation is solved by this method, a very essential role, generally speaking, is played by the choice of the initial value. We shall show that in our case the initial value will not depend on the relative position of the numbers a^{**} and a^*, and that it can be taken quite arbitrarily in the interval (1, 3).

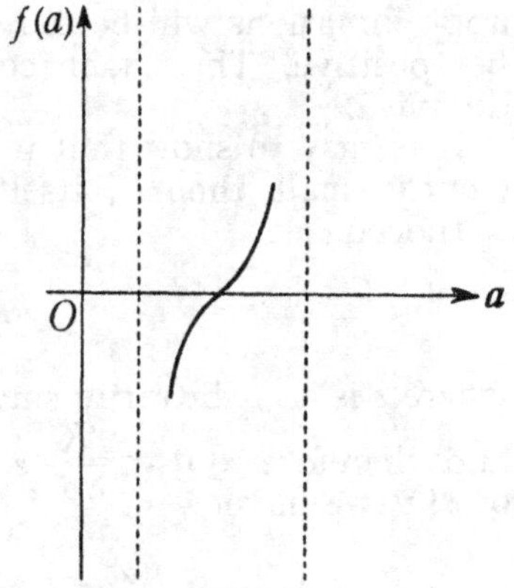

Fig. 59.

The number a^{**} may be 1) equal to the number a^*; 2) greater than a^*; 3) less than a^*. We shall analyse all these cases.

1. Geometrically it is clear (Fig. 59) that as the initial value in this case any number a_0 of the interval (1, 3) can be taken. We shall prove this analytically. For definiteness we shall say that a_0 is an approximation from below. In this case $f(a)$ and $f''(a)$ will be negative in the interval (a_0, a^*). There obviously exists a number M such that for all $a_0 \leqslant a \leqslant a^*$ we shall have $f'(a) < M$. The first correction

$$\alpha^{(1)} = -\frac{f(a_0)}{f'(a_0)}$$

will be positive. We shall in addition introduce into the discussion the exact corrections, denoting them by the letter h.

The correction h_1 will be positive (once a_0 is an approximation from below) and will satisfy the equation

$$f(a_0) + h_1 f'(a_0) + \frac{h_1^2}{2} f''(a^c) = 0,$$

where a^c is some value of a between a_0 and a^*.

From the last equation let us subtract the equation for the determination of $\alpha^{(1)}$. We then obtain

$$(h_1 - \alpha^{(1)})f'(a_0) = -\frac{h_1^2}{2} f''(a^c).$$

But $h_1 - \alpha^{(1)} = h_2$, $f''(a^c) < 0$, and therefore

$$h_2 = -\frac{h_1^2}{2} \frac{f''(a^c)}{f'(a_0)}$$

is positive.

Accordingly, the first approximation a_1, like a_0, is an approximation from below. From this it follows also that all the subsequent approximations will be too small. The corresponding corrections will be positive. The exact corrections h, while remaining positive, will diminish.

It is easy to show that if the difference between h_N and h_{N+1} is sufficiently small, then h_N itself is small too.

Indeed, let

$$h_N - h_{N+1} < \frac{K}{M} \varepsilon,$$

where ε is an arbitrarily small positive quantity. But $h_N - h_{N+1} = \alpha^{(N)}$, and therefore $\alpha^{(N)} < \frac{K}{M} \varepsilon$. From the equation for the determination of $\alpha^{(N)}$ we have:

$$\alpha^{(N)} = -\frac{f(a_{N-1})}{f'(a_{N-1})} = \frac{|f(a_{N-1})|}{f'(a_{N-1})}, \quad |f(a_{N-1})| = \alpha^{(N)} f'(a_{N-1});$$

h_N can be determined from the equation

$$f(a_{N-1}) + h_N f'(a^c) = 0, \text{ where } a_{N-1} < a^c < a^*.$$

Therefore

$$h_N = \frac{|f(a_{N-1})|}{f'(a^c)} = \frac{\alpha^{(N)} f'(a_{N-1})}{f'(a^c)}, \quad h_N < \frac{K}{M} \varepsilon \frac{M}{K} = \varepsilon.$$

Accordingly the limit of the sequence of the diminishing positive numbers $h_1, h_2, \ldots, h_N$ will be zero.

The discussion can be carried through by the same method in the case where a_0 is an approximation from above. All the successive approximations in this case will be approximations with an excess, too.

2. $a^{**} > a^*$.

For initial values less than a^*, the convergence is already established (case 1). In the interval (a_0, a^*), $f(a)$ and $f''(a)$ will be negative, as in case 1. The condition $f''(a^*) = 0$ remained unused in case 1.

Henceforth let a_0 be such that $a^* < a_0 \leqslant a^{**}$.

Then $f(a_0) > 0$, and $f''(a_0) < 0$ and for all $a < a_0$ will be negative;

$\alpha^{(1)}$ and h_1 will be negative too. Let us see what h_2 will be:

$$h_2 = -\tfrac{1}{2}h_1^2 \frac{f''(a^c)}{f'(a_0)} \cdot f''(a^c) < 0,$$

and therefore h_2 will be positive. This signifies that the first approximation will be an approximation from below [1]). All further approximations will also be such, and we have convergence to a^* from below (again case 1).

Consider now the last possibility

$$a_0 > a^{**}.$$

$f(a)$ will be positive in the entire interval (a^*, a_0); $f''(a)$ will be negative in the interval (a^*, a^{**}), and positive in the interval (a^{**}, a_0); $\alpha^{(1)}$ and h_1 will be negative; a_1 can turn up in one of the following intervals: (a^{**}, a_0); (a^*, a^{**}); $(1, a^*)$.

If a_1 turns up in the second or third intervals, then, by what has been proved, we have convergence to a^* from below.

Thus an additional investigation is still required for the case when a_1 and all the approximations subsequent to it are greater than a^{**}. However, all the successive approximations cannot be greater than a^{**}. Continuously diminishing, meanwhile remaining greater than a^{**}, they would approach some limit greater than or equal to a^{**}. For sufficiently large N the difference between h_N and h_{N+1} would be arbitrarily small, less, for example, than

$$\frac{a^{**} - a^*}{2} \cdot \frac{K}{M},$$

where M is the greatest value of $f'(a)$ in the interval (a^*, a_0). Just as in case 1, it can be shown that $|h_N|$ will here be less than $\frac{a^{**} - a^*}{2}$. By assumption, however, $|h_N| \geqslant a^{**} - a^*$.

The contradiction obtained convinces us that if the execution of the Newton-Fourier method is begun with $a_0 > a^{**}$, we shall cross a^{**} and fall in the interval in which the convergence has already been established.

3. $a^{**} < a^*$.

In this case, by the same method as in the two preceding cases, it can be shown that as a_0 there can be taken any number from the interval under consideration, (1, 3).

Example. Let us set up the function mapping the upper half of the t-plane onto a trapezoid $A_1A_2A_3A_4$ of the z-plane (Fig. 60). The angles of the trapezoid equal $\frac{5}{6}\pi$, $\frac{2}{3}\pi$, $\frac{1}{3}\pi$, $\frac{1}{6}\pi$ respectively. The lengths of the sides of the trapezoid we shall assume to be equal to 1, $\sqrt{3}$, $1 + 2\sqrt{3}$, 3.

---

[1]) If a_1 proves to be outside the interval (1, 3), i.e., is less than 1, then for the first approximation it is necessary to take another number, greater than a_1 and 1.

The mapping function will be

$$z = C_1 \int_0^t (t+1)^{-\frac{1}{6}}(t-1)^{-\frac{1}{3}}(t+a)^{-\frac{2}{3}}(t-3)^{-\frac{5}{6}}dt + C_2.$$

We shall first attend to the determination of the parameter a, determining it from the equation

$$I_4(a) = 3I_1(a), \tag{*}$$

where

$$I_4(a) = \int_0^1 (1+3x)^{-\frac{1}{6}}(1+x)^{-\frac{1}{3}}[1+(2-a)x]^{-\frac{2}{3}}(1-x)^{-\frac{5}{6}}dx +$$
$$+ \int_0^1 (1-x)^{-\frac{1}{6}}(1+x)^{-\frac{1}{3}}(1+ax)^{-\frac{2}{3}}(1+3x)^{-\frac{5}{6}}dx,$$

$$I_1(a) = \int_0^1 (t+1)^{-\frac{1}{6}}(1-t)^{-\frac{1}{3}}(a-t)^{-\frac{2}{3}}(3-t)^{-\frac{5}{6}}dt.$$

The first term of $I_4(a)$ we shall denote by $I_{41}(a)$, and the second by $I_{42}(a)$, whereby $I_4(a) = I_{41}(a) + I_{42}(a)$. The integral $I_1(a)$ we also represent in the form of the sum of the two integrals

$$\int_{-1}^0 (t+1)^{-\frac{1}{6}}(1-t)^{-\frac{1}{3}}(a-t)^{-\frac{2}{3}}(3-t)^{-\frac{5}{6}}dt$$

and

$$\int_0^1 (t+1)^{-\frac{1}{6}}(1-t)^{-\frac{1}{3}}(a-t)^{-\frac{2}{3}}(3-t)^{-\frac{5}{6}}dt,$$

the first of which we denote by $I_{11}(a)$ and second by $I_{12}(a)$.

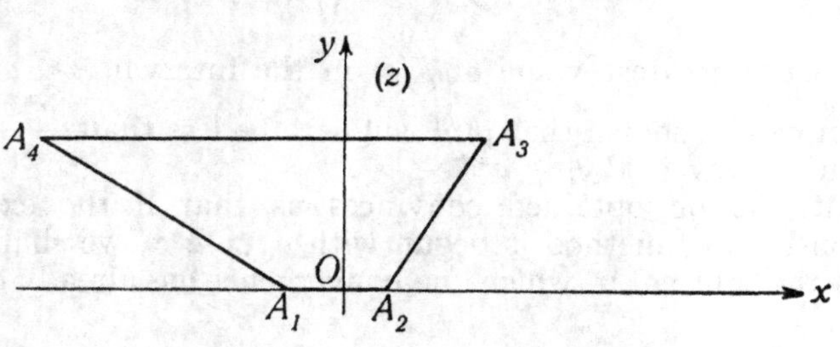

Fig. 60.

We solve equation (*) by the Newton-Fourier method. We perform the computation of the particular values of I_{11}, I_{12}, I_{41}, I_{42} and their derivatives with respect to the parameter a by the method of L. V. Kantorovich. The integrals free of singularities (we henceforth call them proper integrals) we compute by Simpson's rule, with three intermediate ordinates. We perform the computations in the following order: we first set up general expressions for the singular integrals, the Simpson ordinates and the proper integrals, afterwards substituting the numerical values of a in the formulas just obtained.

If we adopt 2 as the initial value of a, the first correction $\alpha^{(1)}$ will be equal to -0.65 and the first approximation $a_1 = 1.35$. For the second correction $\alpha^{(2)}$ and for the second approximation a_2 we obtain the values 0.015 and 1.365 respectively.

The third correction $\alpha^{(3)}$ we compute using the values of $\frac{dI_1}{da}$ and $\frac{dI_4}{da}$ for $a = 1.35$. For $\alpha^{(3)}$ and a_3 we obtain the values 0.0005 and 1.3655.

The results of the computations are to be seen in the table:

Integral	Element of the integral	For $a = 2$	For $a = 1.35$	For $a = 1.365$	For $a = 1.3655$
I_{11}	Singular integral	0.1834	0.2205	0.2195	0.21944
	Proper integral	0.0180	0.0272	0.0269	0.02687
I_{12}	Singular integral	0.4500	0.1582	0.1998	0.20110
	Proper integral	0.1101	0.7618	0.7038	0.70201
I_1	I_1	0.7615	1.1677	1.1500	1.14942
I_{41}	Singular integral	3.9372	2.9213	2.9377	2.93826
	Proper integral	0.0631	0.1019	0.1012	0.10111
I_{42}	Singular integral	0.2253	0.2603	0.2593	0.25935
	Proper integral	0.1517	0.1494	0.1495	0.14954
I_4	I_4	4.3773	3.4329	3.4478	3.44826
$3I_1$	$3I_1$	2.2845	3.5031	3.4500	3.44826
$\frac{3}{2}I'_{11}$	Singular integral	−0.0684	−0.1078	—	—
	Proper integral	−0.0140	−0.0322	—	—
$\frac{3}{2}I'_{12}$	Singular integral	−0.1500	−4.4790	—	—
	Proper integral	−0.2871	−6.1551	—	—
I'_1	I'_1	−0.3463	−1.2107	—	—
$\frac{3}{2}I'_{41}$	Singular integral	3.3972	1.6285	—	—
	Proper integral	−0.0667	−0.0791	—	—
$\frac{3}{2}I'_{42}$	Singular integral	−0.0678	−0.0968	—	—
	Proper integral	0.0022	0.0092	—	—
I'_4	I'_4	2.1766	0.9745	—	—

Comparing $I_4(a)$ and $3I_1(a)$, we see that for $a = 1.3655$, $I_4(a)$ equals $3I_1(a)$, accurate to the fifth decimal place. We therefore adopt for a the value 1.3655.

Let us go on to the determination of the parameters C_1 and C_2. The function

$$z^* = \int_0^t (t+1)^{-\frac{1}{3}}(t-1)^{-\frac{1}{3}}(t-1.3655)^{-\frac{2}{3}}(t-3)^{-\frac{2}{3}}\,dt$$

will map the upper half of the t-plane onto the trapezoid $A_1^*A_2^*A_3^*A_4^*$, similar to the given $A_1A_2A_3A_4$. Let us determine the complex coordinate z_1^* of the vertex A_1^*, corresponding to the point $t = -1$.

$$z_1^* = \int_0^{-1} (t+1)^{-\frac{1}{6}}(t-1)^{-\frac{1}{3}}(t-1.3655)^{-\frac{2}{3}}(t-3)^{-\frac{5}{6}}dt =$$

$$= -(-1)^{-\frac{11}{6}}\int_{-1}^{0} (t+1)^{-\frac{1}{6}}(1-t)^{-\frac{1}{3}}(1.3655-t)^{-\frac{2}{3}}(3-t)^{-\frac{5}{6}}dt =$$

$$= -(-1)^{-\frac{1}{6}}\int_{-1}^{0} (t+1)^{-\frac{1}{6}}(1-t)^{-\frac{1}{3}}(1.3655-t)^{-\frac{2}{3}}(3-t)^{-\frac{5}{6}}dt =$$

$$= -(-1)^{\frac{1}{6}}\ 0.24631.$$

If $(-1)^{\frac{1}{6}}$ be taken as equalling i, we have

$$z^* = -0.24631i.$$

For the complex coordinate z_2^* of the vertex A_2^*, corresponding to the point $t = 1$, we obtain the value

$$z_2^* = 0.90311i.$$

We pass from the trapezoid $A_1^*A_2^*A_3^*A_4^*$ to the trapezoid $A_1A_2A_3A_4$ in the following manner. We transform the z^*-plane into the z-plane in such a way that the vertex A_1^* passes to A_1 and the vertex A_2^* to A_2. The function effecting this transformation will have the form

$$z = C_1z^* + C_2.$$

For the determination of C_1 and C_2 we shall have the system of equations:

$$-0.5 = -0.24631iC_1 + C_2$$

$$0.5 = 0.90311iC_1 + C_2,$$

whence

$$C_1 = \frac{1}{1.14942i} = -0.87000i \text{ and } C_2 = -0.28571.$$

Thus for the mapping function we obtain the expression.:

$$z = -0.87000i\int_0^{t} (t+1)^{-\frac{1}{6}}(t-1)^{-\frac{1}{3}}(t-1.3655)^{-\frac{2}{3}}(t-3)^{-\frac{5}{6}}dt - 0.28571.$$

Chapter VI

PRINCIPLES OF THE APPLICATION OF CONFORMAL TRANSFORMATION TO THE SOLUTION OF THE FUNDAMENTAL PROBLEMS FOR CANONICAL REGIONS

§ 1. INTRODUCTION

1. The transformation of the Laplace operator. In the preceding chapter we have investigated the conformal transformation of regions into one another. Now we have in view giving the principles of its application to the solution of a number of problems of mathematical physics.

As we have already had occasion to remark, in the type of problems of interest to us the solution is determined by three circumstances: by the equation defining the class of functions corresponding to the process under study; by the region for which the solution is sought, which is dependent on the form of the object under study; and, lastly, by the boundary conditions.

Instead of the old independent variables x and y, let us have introduced the new variables ξ and η by the equations:

$$\xi = \xi(x, y),$$
$$\eta = \eta(x, y).$$

The sought function $u(x, y)$ becomes a new function u of ξ and η. Generally speaking, not only the differential equation that the function u must satisfy is thereby changed, but also the region in which it is sought and the conditions that must be observed on the boundary of the region. Given a suitable choice of the variables ξ and η, the new, transformed problem may prove to be simpler than the original one — the posed problem, difficult to solve in the old variables, becoming easily soluble in the new.

Leaving aside the general theory of this kind of transformation, we shall now concern ourselves with conformal transformations, provided by pairs of conjugate harmonic functions, or, what is the same thing, by analytic functions of a complex variable.

Let $\xi(x, y)$ and $\eta(x, y)$ satisfy the Cauchy-Riemann conditions:

$$\frac{\partial \xi}{\partial x} = \frac{\partial \eta}{\partial y}, \quad \frac{\partial \xi}{\partial y} = -\frac{\partial \eta}{\partial x};$$

$f(z) = \xi + i\eta$ will be an analytic function of the complex variable $z = x + iy$. Let us see how the Laplace operation is transformed. It is easily seen that

$$\frac{\partial^2 u}{\partial x^2} + \frac{\partial^2 u}{\partial y^2} = \left[\left(\frac{\partial \xi}{\partial x}\right)^2 + \left(\frac{\partial \xi}{\partial y}\right)^2\right]\frac{\partial^2 u}{\partial \xi^2} + 2\left[\frac{\partial \xi}{\partial x}\cdot\frac{\partial \eta}{\partial x} + \frac{\partial \xi}{\partial y}\cdot\frac{\partial \eta}{\partial y}\right]\frac{\partial^2 u}{\partial \xi \partial \eta} +$$

$$+\left[\left(\frac{\partial \eta}{\partial x}\right)^2 + \left(\frac{\partial \eta}{\partial y}\right)^2\right]\frac{\partial^2 u}{\partial \eta^2} + \frac{\partial u}{\partial \xi}\left(\frac{\partial^2 \xi}{\partial x^2} + \frac{\partial^2 \xi}{\partial y^2}\right) + \frac{\partial u}{\partial \eta}\left(\frac{\partial^2 \eta}{\partial x^2} + \frac{\partial^2 \eta}{\partial y^2}\right)$$

or, if the Cauchy-Riemann equations and the harmonicity of ξ and η are used:

$$\frac{\partial^2 u}{\partial x^2} + \frac{\partial^2 u}{\partial y^2} = \left[\left(\frac{\partial \xi}{\partial x}\right)^2 + \left(\frac{\partial \xi}{\partial y}\right)^2\right]\left(\frac{\partial^2 u}{\partial \xi^2} + \frac{\partial^2 u}{\partial \eta^2}\right).$$

Thus the Laplace operation $\Delta_{xy}u$ on the old variables takes, after the transformation, the form

$$\Delta_{xy}u = |f'(z)|^2\, \Delta_{\xi\eta}u. \tag{1}$$

Hence follows the familiar result that a function harmonic before the transformation will be harmonic after it too, for $|f'(z)|$ is generally different from zero.

We will note one more fact, important from the point of view of practical applications. Let us have been given in the variables x, y the Poisson equation

$$\Delta_{xy}u = \varphi,$$

where φ is a known function of x and y.

After transformation this equation will take the form:

$$\Delta_{\xi\eta}u = \frac{1}{|f'(z)|^2}\varphi = \left|\frac{dz}{d\zeta}\right|^2 \varphi, \quad \zeta = \xi + i\eta.$$

We have obtained the Poisson equation anew, although with a changed right side, it is true. This circumstance makes possible the wide application of conformal transformation to the investigation of problems connected with the Poisson equation, for example, problems of the torsion of homogeneous isotropic beams and rods.

2. The transformation of the biharmonic operator. Goursat's formula. The connection of biharmonic functions with the plane problem of the theory of elasticity. By the appellation *biharmonic operator* is understood the operator of form

$$\Delta_{xy}\Delta_{xy} = \frac{\partial^4}{\partial x^4} + 2\frac{\partial^4}{\partial x^2 \partial y^2} + \frac{\partial^4}{\partial y^4}.$$

Functions satisfying the equation

$$\Delta_{xy}\Delta_{xy}u = 0, \tag{2}$$

are accordingly called biharmonic.

From the result established by us with respect to the Laplace operator in the last No. it follows that after transformation of the region B of the plane $z = x + iy$ by means of the function $\zeta = f(z) = \xi + i\eta$, the biharmonic operator will take the form:

$$\Delta_{xy}\Delta_{xy}u = |f'(z)|^2 \Delta_{\xi\eta}(|f'(z)|^2 \Delta_{\xi\eta}u). \tag{3}$$

From this it is clear that a function $u(x, y)$ biharmonic before the transformation will generally speaking not be biharmonic in the variables ξ, η. Its biharmonic character is preserved only in case $|f'(z)| =$ const., i.e., under a linear transformation, of the form $\zeta = \alpha z + \beta$.

E. Goursat first established a formula, important in the theory of biharmonic functions, permitting the expression of these functions very simply in terms of two analytic functions of a complex variable.

According to (2), $\Delta_{xy}u = r(x, y)$ must be a harmonic function. We will denote by $s(x, y)$ the function conjugate to it.

$$f(z) = r + is$$

will be an analytic function of z. Let us introduce the function:

$$\varphi(z) = \tfrac{1}{4}\int f(z)\,dz = p + iq,$$

where

$$\Delta p = \Delta q = 0, \quad \frac{\partial p}{\partial x} = \frac{\partial q}{\partial y} = \tfrac{1}{4}r, \quad \frac{\partial p}{\partial y} = -\frac{\partial q}{\partial x} = -\tfrac{1}{4}s.$$

On the basis of these equations it can be verified without difficulty that $u - xp - yq$ will be an harmonic function. We will call it p_1 and denote by $\chi(z)$ the function of a complex variable for which p_1 is the real part:

$$u = xp + yq + p_1 = \operatorname{Re}\,[\bar{z}\varphi(z) + \chi(z)]. \tag{4}$$

It can be shown that conversely, for any analytic functions $\varphi(z)$ and $\chi(z)$ formula (4) gives a biharmonic function [1]).

We will note in addition that for a given function u, the $\varphi(z)$ and $\chi(z)$ that figure in formula (4) are not fully defined, retaining a certain arbitrariness.

Indeed, by the known function $r(x, y)$ the function $f(z)$ is defined accurate to a pure imaginary summand ai, and therefore $\varphi(z)$ is found accurate only to a summand of the form

$$C + aiz,$$

where C is an arbitrary complex constant. We can fix it by arbitrarily

---

[1]) E. Goursat, [3].

assigning at some point z_0 the value of the function $\varphi(z)$ itself and the imaginary part of its derivative, for instance:

$$\varphi(z_0) = 0, \ \mathrm{I}\,[\varphi'(z_0)] = 0.$$

Moreover the function $\chi(z)$ is found accurate to a pure imaginary constant, which can be defined by assigning the value of the imaginary part of $\chi(z)$ at z_0, for instance, subjecting it to the requirement:

$$\mathrm{I}\,[\chi(z_0)] = 0.$$

Equation (4) permits one to determine simply the form of the functions into which biharmonic functions in the plane $z = x + iy$ pass after conformal transformation.

We will denote by $z = \omega(\zeta)$ the function inverse to $\zeta = f(z)$, and set

$$\varphi[\omega(\zeta)] = \Phi(\zeta), \quad \chi[\omega(\zeta)] = \mathrm{X}(\zeta);$$

$u(x, y)$, after the transformation, will take the form:

$$u = \mathrm{Re}\,[\overline{\omega(\zeta)}\,\Phi(\zeta) + \mathrm{X}(\zeta)], \tag{5}$$

where $\Phi(\zeta)$ and $X(\zeta)$ are arbitrary analytic functions of the variable ζ, and the first of them is determined for the given function u accurate to a summand of the form $C + ai\omega(\zeta)$, while the second is determined accurate to an imaginary constant.

We shall refer briefly to the connection between the plane problem of the theory of elasticity and functions of a complex variable. For a more thorough acquaintance with this question we refer the reader to a book of N. I. Muskhelishvili, [4], Chap. II.

In case external forces are absent, the equations of the plane problem have the following form, if we preserve the usual symbols [1]):

$$\frac{\partial X_x}{\partial x} + \frac{\partial X_y}{\partial y} = 0, \quad \frac{\partial Y_x}{\partial x} + \frac{\partial Y_y}{\partial y} = 0,$$

$$X_x = \lambda\theta + 2\mu\frac{\partial u}{\partial x}, \quad Y_y = \lambda\theta + 2\mu\frac{\partial v}{\partial y}, \tag{6}$$

$$X_y = Y_x = \mu\left(\frac{\partial v}{\partial x} + \frac{\partial u}{\partial y}\right), \quad \theta = \frac{\partial u}{\partial x} + \frac{\partial v}{\partial y}.$$

In addition, there must be fulfilled the Beltrami-Michell compatibility conditions, which in our case reduce to one equation

$$\Delta(X_x + Y_y) = 0.$$

The first of the cited equations gives a necessary and sufficient condition of the existence of a function B satisfying the condition:

$$\frac{\partial B}{\partial x} = -X_y, \quad \frac{\partial B}{\partial y} = X_x. \tag{7}$$

---

[1]) See N. I. Muskhelishvili, [4], p. 84.

In quite the same way there follows from the second the existence of a function A for which

$$\frac{\partial A}{\partial x} = Y_y, \quad \frac{\partial A}{\partial y} = -Y_x = -X_y.$$

Comparison of the two expressions for X_y shows that

$$\frac{\partial A}{\partial y} = \frac{\partial B}{\partial x}.$$

Consequently there must exist a function of two variables, $U(x, y)$, such that

$$A = \frac{\partial U}{\partial x}, \quad B = \frac{\partial U}{\partial y}.$$

The stress components are expressible in terms of it by the formulas

$$X_x = \frac{\partial^2 U}{\partial y^2}, \quad X_y = -\frac{\partial^2 U}{\partial x \partial y}, \quad Y_y = \frac{\partial^2 U}{\partial x^2}; \tag{8}$$

$U(x, y)$ is called the Airy stress function, after the name of the eminent English astronomer G. Biddell Airy, who first called attention to its existence.

From the Beltrami-Michell condition it follows, since $X_x + Y_y = \Delta U$, that $U(x, y)$ must be biharmonic:

$$\Delta\Delta U = 0. \tag{9}$$

By virtue of the results obtained in this No, there can be found two analytic functions of a complex variable, $\varphi(z)$ and $\chi(z)$, such that

$$U = \mathrm{Re}\,[\bar{z}\varphi(z) + \chi(z)]. \tag{10}$$

Below we shall also require expressions for the partial derivatives of $U(x, y)$.

Differentiating (10) with respect to x and y and setting $\chi'(z) = \psi(z)$ everywhere, we obtain:

$$\frac{\partial U}{\partial x} = \mathrm{Re}\,[\varphi(z) + \bar{z}\varphi'(z) + \psi(z)],$$

$$\frac{\partial U}{\partial y} = \mathrm{Re}\,[-i\varphi(z) + \bar{z}i\varphi'(z) + i\psi(z)].$$

It is often more convenient, however, to use not these formulas but the following combinations of them:

$$\left.\begin{aligned} \frac{\partial U}{\partial x} + i\frac{\partial U}{\partial y} &= \varphi(z) + z\overline{\varphi'(z)} + \overline{\psi(z)}, \\ \frac{\partial U}{\partial x} - i\frac{\partial U}{\partial y} &= \overline{\varphi(z)} + \bar{z}\varphi'(z) + \psi(z). \end{aligned}\right\} \tag{11}$$

We shall give the complex representation of the displacements and stresses:

$$\left.\begin{aligned} X_x + iX_y &= \frac{\partial^2 U}{\partial y^2} - i\frac{\partial^2 U}{\partial x\,\partial y} = -i\frac{\partial}{\partial y}\left(\frac{\partial U}{\partial x} + i\frac{\partial U}{\partial y}\right), \\ X_x + iX_y &= \varphi'(z) + \overline{\varphi'(z)} - \overline{z\varphi''(z)} - \overline{\psi'(z)}, \end{aligned}\right\} \quad (12)$$

$$\left.\begin{aligned} Y_y - iX_y &= \frac{\partial^2 U}{\partial x^2} + i\frac{\partial^2 U}{\partial x\,\partial y} = \frac{\partial}{\partial x}\left(\frac{\partial U}{\partial x} + i\frac{\partial U}{\partial y}\right), \\ Y_y - iX_y &= \varphi'(z) + \overline{\varphi'(z)} + \overline{z\varphi''(z)} + \overline{\psi'(z)}. \end{aligned}\right\} \quad (12')$$

From formulas (12) and (12′) are easily obtained the following two simpler expressions:

$$X_x + Y_y = 2[\varphi'(z) + \overline{\varphi'(z)}] = 4\mathrm{Re}\,[\varphi'(z)],$$

$$Y_y - X_x + 2iX_y = 2[\bar{z}\varphi''(z) + \psi'(z)].$$

The question of the expression of the components of the displacement u and v in terms of the biharmonic function U reduces to the problem of the determination of u and v from the equations:

$$\lambda\theta + 2\mu\frac{\partial u}{\partial x} = \frac{\partial^2 U}{\partial y^2}, \quad \lambda\theta + 2\mu\frac{\partial v}{\partial y} = \frac{\partial^2 U}{\partial x^2},$$

$$\mu\left(\frac{\partial v}{\partial x} + \frac{\partial u}{\partial y}\right) = -\frac{\partial^2 U}{\partial x\,\partial y}.$$

The first two of them, solved for $\dfrac{\partial u}{\partial x}$ and $\dfrac{\partial v}{\partial y}$, give

$$2\mu\frac{\partial u}{\partial x} = -\frac{\partial^2 U}{\partial x^2} + \frac{\lambda + 2\mu}{2(\lambda + \mu)}\Delta U,$$

$$2\mu\frac{\partial v}{\partial y} = -\frac{\partial^2 U}{\partial^2 y} + \frac{\lambda + 2\mu}{2(\lambda + \mu)}\Delta U,$$

or, since

$$\Delta U = r = 4\frac{\partial p}{\partial x} = 4\frac{\partial q}{\partial y},$$

$$2\mu\frac{\partial u}{\partial x} = -\frac{\partial^2 U}{\partial x^2} + \frac{2(\lambda + 2\mu)}{\lambda + \mu}\frac{\partial p}{\partial x},$$

$$2\mu\frac{\partial v}{\partial y} = -\frac{\partial^2 U}{\partial y^2} + \frac{2(\lambda + 2\mu)}{\lambda + \mu}\frac{\partial q}{\partial y}.$$

Integrating these equations with respect to x and y respectively, we obtain:

$$2\mu u = -\frac{\partial U}{\partial x} + \frac{2(\lambda + 2\mu)}{\lambda + \mu} p + F_1(y),$$

$$2\mu v = -\frac{\partial U}{\partial y} + \frac{2(\lambda + 2\mu)}{\lambda + \mu} q + F_2(x).$$

Substituting in the third of the equations the expressions found, we have, after cancellations,

$$F_1'(y) + F_2'(x) = 0,$$

whence it follows that

$$F_1(y) = -\gamma y + \alpha, \quad F_2(x) = \gamma x + \beta,$$

where α, β and γ are arbitrary constants. These expressions give the rigid displacement of the body, having no significance for the investigation of its elastic properties. Discarding them, we obtain finally

$$\left.\begin{aligned} 2\mu u &= -\frac{\partial U}{\partial x} + \frac{2(\lambda + 2\mu)}{\lambda + \mu} p, \\ 2\mu v &= -\frac{\partial U}{\partial y} + \frac{2(\lambda + 2\mu)}{\lambda + \mu} q. \end{aligned}\right\} \tag{14}$$

We multiply the second formula by i and add to the first:

$$2\mu(u + iv) = -\left(\frac{\partial U}{\partial x} + i\frac{\partial U}{\partial y}\right) + \frac{2(\lambda + 2\mu)}{\lambda + \mu}\varphi(z), \tag{15}$$

whence, in view of (11), if we put

$$\varkappa = \frac{\lambda + 3\mu}{\lambda + \mu},$$

there emerges the following very convenient formula:

$$2\mu(u + iv) = \varkappa\varphi(z) - z\overline{\varphi'(z)} - \overline{\psi(z)}.$$

The fundamental problems of the theory of elasticity are the two following ones:

1) the determination of the elastic equilibrium, given external stresses applied to the boundary L of the body B;

2) the determination of the elastic equilibrium, given displacements of the points of the boundary L of the body B.

We shall obtain the boundary conditions for the first of them.

We take as the positive direction of the normal that exterior to the region B, as is ordinarily done in the theory of elasticity, so that the directions of normal and tangent are related to each other as the axes x and y (see Fig. 61).

By the stress $X_n ds$ and $Y_n ds$ acting on an element of the boundary, we understand, as usual, the stress acting from the side of the positive normal:

$$X_n = X_x \cos(n, x) + X_y \cos(n, y) = \frac{\partial^2 U}{\partial y^2} \cos(n, x) - \frac{\partial^2 U}{\partial x \partial y} \cos(n, y),$$

$$Y_n = Y_x \cos(n, x) + Y_y \cos(n, y) = -\frac{\partial^2 U}{\partial x \partial y} \cos(n, x) + \frac{\partial^2 U}{\partial x^2} \cos(n, y),$$

or, since for the exterior normal

$$\cos(n, x) = \cos(t, y) = \frac{dy}{ds}, \quad \cos(n, y) = -\cos(t, x) = -\frac{dx}{ds},$$

we have

$$X_n = \frac{\partial^2 U}{\partial y^2} \frac{dy}{ds} + \frac{\partial^2 U}{\partial x \partial y} \frac{dx}{ds} = \frac{d}{ds}\left(\frac{\partial U}{\partial y}\right),$$

$$Y_n = -\frac{d}{ds}\left(\frac{\partial U}{\partial x}\right).$$

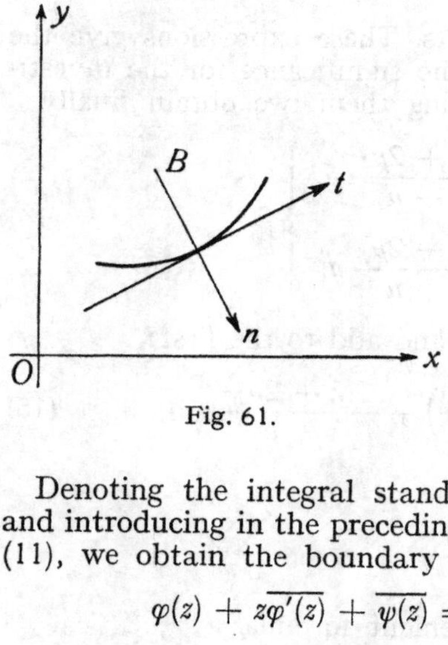

Fig. 61.

In complex form,

$$X_n + iY_n = -i\frac{d}{ds}\left(\frac{\partial U}{\partial x} + i\frac{\partial U}{\partial y}\right),$$

whence

$$\frac{\partial U}{\partial x} + i\frac{\partial U}{\partial y} = i\int_{s_0}^{s} (X_n + iY_n) ds + \text{const.}$$

Denoting the integral standing on the right side by $f_1(s) + if_2(s)$, and introducing in the preceding equation for the left side its expression (11), we obtain the boundary condition of the problem in the form:

$$\varphi(z) + z\overline{\varphi'(z)} + \overline{\psi(z)} = f_1(s) + if_2(s) + \text{const. on } L. \qquad (16)$$

The combination $\frac{\partial U}{\partial x} + i\frac{\partial U}{\partial y}$ is defined only accurate to a constant summand, and therefore the constant on the right side can be fixed arbitrarily.

The boundary conditions in the second fundamental problem are a direct corollary to formula (15). If by $g_1(s)$ and $g_2(s)$ we denote the components of the displacement of the points of the contour, we obtain:

$$\varkappa\varphi(z) - z\overline{\varphi'(z)} - \overline{\psi(z)} = 2\mu(g_1 + ig_2). \qquad (17)$$

Besides the cited boundary conditions (16) or (17), within the region the functions $\varphi(z)$ and $\psi(z)$ must still satisfy some requirements following from the physical meaning of the problem. Without going into detail, we shall enumerate them.

If the region is bounded and simply connected, $\varphi(z)$ and $\psi(z)$ must be single-valued and regular in the entire region.

Let the region be multiply connected and bounded by the interior contours $L_1, L_2, \ldots, L_m$ and the exterior contour L_{m+1}. The functions $\varphi(z)$, $\psi(z)$ will in this case be multi-valued, generally speaking.

Let us denote by X_k and Y_k the resultant vector of the external forces applied to the contour L_k. It can be shown that, on condition of the single-valuedness of the displacements and stresses, here:

$$\left.\begin{aligned}\varphi(z) &= -\frac{1}{2\pi(1+\varkappa)}\sum_{k=1}^{m}(X_k+iY_k)\ln(z-z_k)+\varphi^*(z),\\ \psi(z) &= \frac{\varkappa}{2\pi(1+\varkappa)}\sum_{k=1}^{m}(X_k-iY_k)\ln(z-z_k)+\psi^*(z),\end{aligned}\right\}\quad(18)$$

where $\varphi^*(z)$, $\psi^*(z)$ are single-valued analytic functions in B, and z_k is a point lying within the contour L_k.

If, however, the single-valuedness of stresses only be required, admitting multi-valuedness in the displacements, $\varphi(z)$ and $\psi(z)$ will have the form:

$$\left.\begin{aligned}\varphi(z) &= z\sum_{k=1}^{m}A_k\ln(z-z_k)+\sum_{k=1}^{m}(a_k+ib_k)\ln(z-z_k)+\varphi^*(z),\\ \psi(z) &= z\sum_{k=1}^{m}(a_k'+ib_k')\ln(z-z_k)+\sum_{k=1}^{m}(a_k''+ib_k'')\ln(z-z_k)+\psi^*(z);\end{aligned}\right\}\quad(18')$$

A_k, a_k, b_k, a_k', b_k', a_k'', b_k'' are certain constants, whose significance we shall not go into; they must be determined from the conditions of the problem.

The case of an infinite region is of interest. We shall consider that the region consists of the exterior of several closed contours $L_1, L_2, \ldots, L_m$.

Let X, Y be the components of the resultant vector of the external forces applied to the boundary of the region, so that

$$X=\sum_{k=1}^{m}X_k,\quad Y=\sum_{k=1}^{m}Y_k.$$

If one assumes the boundedness of the components of the stress near the point at infinity, the functions $\varphi(z)$ and $\psi(z)$ must have the form:

$$\left.\begin{aligned}\varphi(z) &= -\frac{X+iY}{2\pi(1+\varkappa)}\ln z+(B+iC')z+\varphi_0(z),\\ \psi(z) &= \frac{\varkappa(X-iY)}{2\pi(1+\varkappa)}\ln z+(B'+iC')z+\psi_0(z).\end{aligned}\right\}\quad(19)$$

Here $\varphi_0(z)$ and $\psi_0(z)$ are functions holomorphic in the region of the point at infinity:

$$\varphi_0(z) = a_0 + \frac{a_1}{z} + \frac{a_2}{z^2} + \dots$$

$$\psi_0(z) = a_0' + \frac{a_1'}{z} + \frac{a_2'}{z^2} + \dots$$

and the origin of coordinates is considered to lie outside the region.

The constants B, C, B', C' figuring in formula (19) have a simple physical meaning. If we denote by $X_x(\infty)$, $Y_y(\infty)$, $X_y(\infty)$ the limits to which X_x, Y_y, X_y tend on removing the point to infinity, then

$$X_x(\infty) = 2B - B', \quad Y_y(\infty) = 2B + B', \quad X_y(\infty) = C'.$$

Hence it follows, if N_1 and N_2 be the principal stresses at infinity and α be the angle that the principal axis corresponding to N_1 makes with the x-axis, that

$$B = \frac{N_1 + N_2}{4}, \quad B' = -\frac{N_1 - N_2}{2}\cos 2\alpha, \quad C' = \frac{N_1 - N_2}{2}\sin 2\alpha.$$

The constant C does not influence the stresses, and is connected with a rotation at the infinitely distant point of the plane:

$$\frac{1+\varkappa}{2\mu} C = \lim_{z\to\infty}\left(\frac{\partial v}{\partial x} - \frac{\partial u}{\partial y}\right).$$

We had occasion above to remark that the functions φ and ψ contain arbitrary elements. To make the problem definite, this arbitrariness must somehow be fixed. As stands to reason, the number of arbitrary elements will depend on the kind of problem.

Thus in the first fundamental problem for bounded regions we can arbitrarily fix the values of φ, ψ and the imaginary part of φ' at some point of the region, putting, for example,

$$\varphi(0) = 0, \quad \psi(0) = 0, \quad \mathrm{I}\,[\varphi'(0)] = 0.$$

If, however, we have fixed the constant in equation (16), only one of the constants $\varphi(0)$ and $\psi(0)$ can be chosen.

We shall also remark here that when the region is multiply connected, we have the right to assign arbitrarily the constant of equation (16) on only one of the contours L_k. Its values on the rest of the contours must be determined when solving the problem.

When the region is infinite, $\varphi(z)$ and $\psi(z)$ have the form (19). In these formulas we can consider:

$$\varphi_0(\infty) = 0, \quad \psi_0(\infty) = 0, \quad C = 0,$$

with an analogous proviso for the constant of condition (16).

In the second fundamental problem for bounded regions one can fix only one of the values $\varphi(0)$ or $\psi(0)$, considering, for instance,

$$\varphi(0) = 0.$$

Analogously for an infinite region:

$$\varphi_0(\infty) = 0.$$

In the book of N. I. Muskhelishvili that we have mentioned, the reader can find a more detailed exposition of all the results cited.

3. Transformation of the boundary conditions. The simplest case of the changing of the boundary conditions under conformal transformation will be that where these conditions relate only to the sought functions, as is the case, for example, in the Dirichlet and Hilbert problems.

To set up the boundary conditions that the sought function must satisfy after the transformation, it is sufficient here to know the relation between the points of the boundaries of the regions being transformed.

Somewhat more complicated is the case where derivatives of the sought function along some direction figure in the boundary conditions.

Let us consider the derivative $\frac{\partial u}{\partial l}$ of the function u in the direction l. Let us effect the conformal transformation of the z-plane by means of the function $\zeta = f(z)$. With the transformation, the direction l passes to some direction λ of the ζ-plane, and the derivatives in these directions will be connected with each other by the following obvious equality:

$$\frac{\partial u}{\partial l} = \frac{\partial u}{\partial \lambda} \cdot \frac{d\lambda}{dl},$$

where $\frac{d\lambda}{dl}$ is the stretching coefficient in the respective directions. By virtue of the conformality it will not depend on the direction, but will depend only on the position in the plane of the point in question. Numerically it is equal, as we know, to the modulus of the derivative $f'(z)$.

Thus we have:

$$\frac{\partial u}{\partial l} = \frac{\partial u}{\partial \lambda} |f'(z)| = \frac{\partial u}{\partial \lambda} \left| \frac{d\zeta}{dz} \right|.$$

Hence it follows, for example, that if the sought function must before the transformation satisfy on the contour L of the region B the relation

$$k \frac{\partial u}{\partial n} + l \frac{\partial u}{\partial s} + mu = d,$$

where k, l, m, d are given functions of the position of the point of the contour, then on the boundary of the transformed region there must be observed the equation:

$$k\frac{\partial u}{\partial \nu} + l\frac{\partial u}{\partial \sigma} + m\left|\frac{dz}{d\zeta}\right| u = d\left|\frac{dz}{d\zeta}\right|,$$

where $\frac{\partial}{\partial \nu}$ and $\frac{\partial}{\partial \sigma}$ are the derivatives along the normal and the arc of the transformed contour.

Of the vast class of problems reducing to biharmonic functions, and the diverse boundary conditions corresponding to them, we shall dwell only on the problems most frequently encountered.

Let us consider the first fundamental problem:

1. To find the function u, biharmonic within the region, given on the contour the values of the function itself and of its normal derivative:

$$u = \omega_1(s), \quad \frac{\partial u}{\partial n} = \omega_2(s). \tag{20}$$

We have formulated the problem in the usual form. For us, though, a somewhat different formulation will be more convenient. It is easily seen that conditions (20) give us the values on the contour of the region of the partial derivatives of the sought function along the coordinate axes. Indeed:

$$\left.\begin{aligned} \frac{\partial u}{\partial x} &= \frac{\partial u}{\partial s}\cos(s, x) + \frac{\partial u}{\partial n}\cos(n, x) = \\ &= \omega_1'(s)\cos(s, x) + \omega_2(s)\cos(n, x) = \omega_3(s), \\ \frac{\partial u}{\partial y} &= \frac{\partial u}{\partial s}\cos(s, y) + \frac{\partial u}{\partial n}\cos(n, y) = \\ &= \omega_1'(s)\cos(s, y) + \omega_2(s)\cos(n, y) = \omega_4(s). \end{aligned}\right\} \tag{21}$$

Conversely, let us be given on the boundary of the region the values of the partial derivatives along the coordinate axes:

$$\frac{\partial u}{\partial x} = \omega_3(s), \quad \frac{\partial u}{\partial y} = \omega_4(s).$$

By them the value of the normal derivative can be found:

$$\frac{\partial u}{\partial n} = \frac{\partial u}{\partial x}\cos(n, x) + \frac{\partial u}{\partial y}\cos(n, y).$$

As regards the values on the contour of the region of the function u itself, they are determined accurate to an arbitrary constant term from the equation

$$u = \int_{(x_0, y_0)}^{(x, y)} \left(\frac{\partial u}{\partial x}dx + \frac{\partial u}{\partial y}dy\right) + C.$$

The presence of the arbitrary additive constant in the boundary values of u in our problem is unimportant, since with a change of all the values of u on the boundary of the region by a constant, u will change by the same constant everywhere within the region.

We can therefore replace boundary conditions (20) by conditions (21).

Using Goursat's formula and equations (11), we can write the new boundary conditions in the form:

$$\frac{\partial u}{\partial x} + i\frac{\partial u}{\partial y} = \varphi(z) + z\overline{\varphi'(z)} + \overline{\psi(z)} = \omega_3 + i\omega_4 = F, \tag{22}$$

which fully coincides with (16).

Let us turn to the second fundamental problem.

2. It is required to find the biharmonic function if it is known that the functions $\varphi(z)$ and $\psi(z)$ that enter into its composition satisfy on the boundary of the region the condition:

$$\varkappa\varphi(z) - z\overline{\varphi'(z)} - \overline{\psi(z)} = \Phi, \tag{23}$$

where Φ is a given function of a complex variable.

Let us see how the boundary conditions (22) and (23) change under conformal transformation.

We shall introduce for the variable z a new variable ζ by the relation:

$$z = \omega(\zeta),$$

transforming the region G into B.

Let us denote by $\varphi_1(\zeta)$ and $\psi_1(\zeta)$ the result of the substitution in $\varphi(z)$ and $\psi(z)$ of the indicated expression for z:

$$\varphi[\omega(\zeta)] = \varphi_1(\zeta), \quad \psi[\omega(\zeta)] = \psi_1(\zeta).$$

Since

$$\varphi'(z) = \frac{d\varphi}{dz} = \varphi_1'(\zeta)\,\frac{1}{\omega'(\zeta)},$$

equations (22) and (23) will take the form:

$$\varphi_1(\zeta) + \frac{\omega(\zeta)}{\overline{\omega'(\zeta)}}\overline{\varphi_1'(\zeta)} + \overline{\psi_1(\zeta)} = F, \tag{24}$$

$$\varkappa\varphi_1(\zeta) - \frac{\omega(\zeta)}{\overline{\omega'(\zeta)}}\overline{\varphi_1'(\zeta)} - \overline{\psi(\zeta)} = \Phi. \tag{25}$$

Observation. For the existence of a solution of the first fundamental problem for bounded regions it is necessary that the resultant vector of all external forces applied to the contour of the region, and the resultant moment of this vector, be equal to zero.

The resultant vector $X + iY$ obviously has the following expression:

$$X + iY = \int_L (X_n + iY_n)ds = \frac{1}{i}\Big|_L (f_1 + if_2) = \frac{1}{i}\Big|_L F,$$

therefore its equality to zero reduces to the equality to zero of the integral:

$$\int_L (X_n + iY_n)ds, \tag{26}$$

taken along the entire contour of the region, or, what is the same thing, to the equality to zero of the increment of $f_1(s) + if_2(s) = F$ for a circuit of the contour:

$$|_L F = |_L (f_1(s) + if_2(s)) = 0. \tag{26'}$$

After conformal transformation this reduces to the vanishing of the increment of the function F standing on the right side of equation (24).

As regards the resultant moment, its equality to zero after integration by parts takes the following form:

$$0 = \int_L (xY_n - yX_n)ds = -\int_L (xdf_1 + ydf_2) = -|_L (xf_1 + yf_2) + \int_L (f_1dx + f_2dy),$$

or, if the resultant vector is equal to zero and the increments of the functions f_1 and f_2 along the contour are equal to zero:

$$\int_L (f_1dx + f_2dy) = 0. \tag{27}$$

On the other hand, as is at once evident, for the function F standing on the right side of equation (24), this reduces to the following:

$$\operatorname{Re}\,[\int_L F\overline{dz}] = 0. \tag{28}$$

After conformal transformation of the region B by the equation $z = \omega(\zeta)$, the last condition reduces to the following:

$$\operatorname{Re}\,[\int_L F\overline{\omega'(\zeta)}\overline{d\zeta}] = 0. \tag{29}$$

4. Integrals of the Cauchy type; their computation. Later, in investigating the question of the reduction of the problems of mathematical physics to functional, in particular, integral equations, we shall have to deal with integrals of a special form, which, in view of the analogy between their structure and the structure of the generally known Cauchy integral, have obtained the name "integrals of the Cauchy type".

Let L be an arbitrary rectifiable curve, and let $F(z)$ be an arbitrary function of the complex variable z, defined at the points of L and absolutely integrable along it, i.e., such that the following integral converges:

$$\int_L |F(z)|\,ds.$$

By "integral of the Cauchy type" is understood an integral of the following form:

$$f(z) = \frac{1}{2\pi i} \int_L \frac{F(t)}{t - z} \, dt. \tag{30}$$

It is obviously absolutely convergent for all z lying outside L, and will be an analytic function of z. However the character of its dependence on z may be very complicated.

Thus if L is a simple arc, the integral is only one analytic function, defined at all points of the plane with the exception of the points of L, which are singular points for this function, generally speaking. If L is a simple closed curve, the integral will be two functions differing from each other — the one inside the contour L, the other outside it. Neither of them is the continuation of the other, excluding the trivial case $F(t) = 0$. However, between their boundary values on L and the function $F(t)$ there exists a definite relation, valid for particular assumptions about the function $F(t)$.

We shall give it without proof, referring the reader who wishes to acquaint himself with its proof to special literature [1]).

Now and everywhere below where we have to deal with integrals of the Cauchy type and pass to the limit on the contour, we shall assume the function $F(t)$ standing under the sign of this integral to be continuous and to satisfy a Lipschitz condition of order α, which consists in the following: there exist two positive numbers M and $\alpha \leqslant 1$, such that for every pair of points z' and z'' on L there is fulfilled the inequality

$$|F(z') - F(z'')| \leqslant M \, |z' - z''|^{\alpha}.$$

For the sake of simplicity we shall consider the contour L to be piecewise smooth. When the point z lies on the contour itself, integral (30) is devoid of sense, for the function $\dfrac{F(t)}{t - z}$ is not integrable along L. By the value of the integral one is then to understand its principal value. On either side of the point z we remove an arc of L of length ε, calling the remainder L_ε. Along L_ε the integral has a meaning, because L_ε does not contain z. Now let ε tend to zero.

The limit to which the integral along L_ε tends will then be the principal value of (30) at the point z on L:

$$\lim_{\varepsilon \to 0} \frac{1}{2\pi i} \int_{L_\varepsilon} \frac{F(t)}{t - z} \, dt = \frac{1}{2\pi i} \int_L \frac{F(t)}{t - z} \, dt.$$

We shall call $f_i(z_0)$ and $f_e(z_0)$ the limits to which (30) tends with the approach of z to z_0 from within and from without L respectively [2]).

---

[1]) N. I. Muskhelishvili, [5], Chap. I, §§ 9—22.

[2]) It can be shown that these limits exist for an approach of z to z_0 along any non-tangential path.

The proposition that we are going to formulate is given by the following two equations:

$$\left.\begin{aligned} f_i(z_0) &= \tfrac{1}{2}F(z_0) + \frac{1}{2\pi i}\int\limits_L \frac{F(t)}{t-z_0}\,dt, \\ f_e(z_0) &= -\tfrac{1}{2}F(z_0) + \frac{1}{2\pi i}\int\limits_L \frac{F(t)}{t-z_0}\,dt. \end{aligned}\right\} \tag{31}$$

Formulas (31) hold if z_0 is not an angular point. In the contrary case they must be replaced by the following two:

$$\left.\begin{aligned} f_i(z_0) &= \frac{\delta_e}{2\pi}F(z_0) + \frac{1}{2\pi i}\int\limits_L \frac{F(t)}{t-z_0}\,dt, \\ f_i(z_0) &= -\frac{2\pi-\delta_e}{2\pi}F(z_0) + \frac{1}{2\pi}\int\limits_L \frac{F(t)}{t-z_0}\,dt, \end{aligned}\right\} \tag{31'}$$

where δ_e is the exterior angle of the region at the point z_0.

We will investigate the behavior of the derivative

$$f'(z) = \frac{1}{2\pi i}\int\limits_L \frac{F(t)}{(t-z)^2}\,dt$$

near L. Let us assume that $F(t)$ has along the contour L a continuous derivative with respect to t that satisfies the Lipschitz condition. We shall calculate the integral by the following device:

$$f'(z) = -\frac{1}{2\pi i}\int\limits_L F(t)\frac{d}{dt}\left(\frac{1}{t-z}\right)dt =$$

$$= -\frac{1}{2\pi i}\left|\frac{F(t)}{t-z}\right|_L + \frac{1}{2\pi i}\int\limits_L \frac{F'_t(t)}{t-z}\,dt.$$

The first term on the right side gives zero, in view of the single-valuedness of $F(t)$ on L; the integral has the form already considered, with the sole distinction that under its sign now stands not $F(t)$ but $F'(t)$. In an approach to the contour L, $f'(z)$ will tend to a limit, to compute which, using formulas (31), presents no difficulty.

We can obviously apply the same device to the derivative of the nth order, too. If $F(t)$ has a derivative of the nth order along the contour L, satisfying the Lipschitz condition, then

$$f^{(n)}(z) = \frac{n!}{2\pi i}\int\limits_L \frac{F(t)}{(t-z)^{n+1}}\,dt = \frac{1}{2\pi i}\int\limits_L \frac{F_t^{(n)}(t)}{t-z}\,dt$$

will tend to a definite limit when z approaches the point z_0 of the contour L along any non-tangential path.

A few words on the calculation of integrals of the Cauchy type. If one puts $dt = e^{i\alpha} ds$, $t - z = re^{i\varphi}$ and denotes by (n_t, r) the angle formed by the interior normal n_t and r, the segment connecting the points z and t, the angle being read in the direction from n_t to r, then, as we have explained in detail in Chap. V, § 9, No. 1:

$$\frac{dt}{t - z} = \frac{i}{r} e^{-i(n_t, r)} ds = \frac{i}{r} [\cos (n_t, r) - i \sin (n_t, r)].$$

Therefore

$$\frac{1}{2\pi i} \int\limits_L \frac{F(t)}{t - z} dt = \frac{1}{2\pi} \int\limits_L \frac{F \cos (n_t, r)}{r} ds - \frac{i}{2\pi} \int\limits_L \frac{F \sin (n_t, r)}{r} ds.$$

The first integral, representing the logarithmic potential of a double layer, much simplifies as regards computation if it be represented in the form of a Stieltjes integral. We have, namely,

$$\frac{\cos (n_t, r)}{r} ds = d\omega,$$

where $d\omega$ is the elementary angle that the element of arc ds subtends at the point z, and therefore the first integral can be written in the following more simple form, and one more convenient for computations:

$$\frac{1}{2\pi} \int\limits_L F(t) \, d\omega.$$

If ω is a monotone function of the length of arc s, the last integral reduces to just a Riemann integral, taken along the arc ω, within limits corresponding to a circuit of the entire curve L. If L is a closed curve, ω will vary from zero to 2π, and then

$$\frac{1}{2\pi} \int\limits_L \frac{F \cos (n_t, r)}{r} ds = \frac{1}{2\pi} \int\limits_0^{2\pi} F(t) \, d\omega.$$

Expressing it differently, the magnitude of the integral is the arithmetic mean value of the function $F(t)$, taken with respect to the angle ω.

The second integral presents somewhat greater difficulties in the matter of computations. If it be transformed, as was the first, to the angle ω, it will take the form:

$$\frac{1}{2\pi} \int\limits_L \frac{F \sin (n_t, r)}{r} ds = \frac{1}{2\pi} \int\limits_L F \operatorname{tg} (n_t, r) \, d\omega.$$

For the particular case when ω as a function of s is monotonic, the integral will also be equal to the arithmetic mean, but no longer of the function $F(t)$ but of the product $F(t)\,\mathrm{tg}\,(n_t, r)$.

A similar transformation is particularly advantageous in case r is not tangent to the curve L, i.e., $(n_t, r) \neq \frac{\pi}{2}$; in the contrary case $\mathrm{tg}\,(n_t, r)$ becomes infinite and the integral becomes improper.

In applications of the Cauchy integral to the problems of mathematical physics and mechanics, a very important role is played by the theorem first enunciated by Harnack in 1855, which states the following[1]):

Let L be a circle of radius R with center at the origin of coordinates, and let $F(\vartheta)$ be a real, piece-wise continuous function of the polar angle ϑ.

If the equation

$$f(z) = \frac{1}{2\pi i} \int\limits_L \frac{F(\vartheta)}{t - z}\, dt = 0, \quad t = Re^{i\vartheta},$$

holds for all z lying within the circle $|z| < R$, then $F(\vartheta) = 0$ everywhere, with the possible exception of points of discontinuity.

If, however, the cited equation holds for all z lying outside the circle, then $F(\vartheta)$ is equal to a constant, again with the possible exception of points of discontinuity.

In view of the analogy between the proofs of both parts of the theorem, we shall establish only the first part of it as an example.

Let us expand $f(z)$ in the powers of z. Since

$$\frac{1}{t - z} = \frac{1}{t} + \frac{z}{t^2} + \frac{z^2}{t^3} + \ldots,$$

we have:

$$\frac{1}{2\pi i} \int\limits_L \frac{F(\vartheta)}{t - z}\, dt = \sum_{\nu=0}^{\infty} \alpha_\nu z^\nu,$$

where

$$\alpha_\nu = \frac{1}{2\pi i} \int\limits_L F(\vartheta)\, t^{-\nu-1}\, dt = \frac{1}{2\pi R^\nu} \int\limits_0^{2\pi} F(\vartheta)\, e^{-i\nu\vartheta}\, d\vartheta.$$

Since the sum of the power series is to be equal to zero, all of its coefficients α_ν must be equal to zero, i.e., all the coefficients of the Fourier series for the function $F(\vartheta)$ must be equal to zero, which could be true only in case $F(\vartheta)$ either is zero everywhere, or is different from it only at points of discontinuity. And by this everything is proved.

From Harnack's theorem there follow two corollaries, simple, but essential to us:

---

1) We give a formulation of Harnack's theorem in simplified form, but one sufficiently general for applications.

1) If $F_1(\vartheta)$ and $F_2(\vartheta)$ are two real, piece-wise continuous functions and if the equation

$$\frac{1}{2\pi i}\int_L \frac{F_1(\vartheta)}{t-z}\,dt = \frac{1}{2\pi i}\int_L \frac{F_2(\vartheta)}{t-z}\,dt$$

holds for all $|z| < R$, then

$$F_1(\vartheta) = F_2(\vartheta),$$

with the possible exception of points of discontinuity of the functions $F_1(\vartheta)$ and $F_2(\vartheta)$.

If this equation holds for all $|z| > R$, then

$$F_1(\vartheta) = F_2(\vartheta) + \text{const},$$

with the possible exception of points of discontinuity.

2) If $F_1(\vartheta)$ and $F_2(\vartheta)$ are two arbitrary piece-wise continuous complex functions of ϑ and the equations:

$$\frac{1}{2\pi i}\int_L \frac{F_1(\vartheta)}{t-z}\,dt = \frac{1}{2\pi i}\int_L \frac{F_2(\vartheta)}{t-z}\,dt,$$

$$\frac{1}{2\pi i}\int_L \frac{\overline{F_1(\vartheta)}}{t-z}\,dt = \frac{1}{2\pi i}\int_L \frac{\overline{F_2(\vartheta)}}{t-z}\,dt$$

hold for all $|z| < R$, then

$$F_1(\vartheta) = F_2(\vartheta).$$

If, though, they hold for $|z| > R$, then

$$F_1(\vartheta) = F_2(\vartheta) + \text{const}.$$

Points of discontinuity may constitute an exception in both cases.

The first corollary is perfectly obvious. The second can be reduced to the first by the addition and subtraction of the integral equations that figure in its formulation.

The importance of Harnack's theorem in applications is due to the fact that it makes possible the reduction of boundary conditions expressed by some functional equation to integral conditions, now valid not on the contour but in the entire region under consideration.

§ 2. THE DIRICHLET PROBLEM

1. The Poisson integral. The Dirichlet problem of the determination of the values of a harmonic function within a circle, if its values on the boundary are known, is solved, as we know, by the Poisson integral:

$$u(r, \varphi) = \int_{-\pi}^{+\pi} u(\psi) \frac{R^2 - r^2}{R^2 - 2Rr\cos(\psi - \varphi) + r^2} d\psi, \tag{1}$$

where r, φ are the polar coordinates of the point where the value of the solution is sought, R is the radius of the circle, and $u(\vartheta)$ are the boundary values of $u(r, \varphi)$ as a function of the polar angle ϑ.

When it is necessary to find a function harmonic everywhere in the circle and differing from $u(r, \varphi)$ by not more than an assigned quantity, it will probably be simplest to utilize trigonometric series. It can be verified without difficulty that the kernel of the integral, i.e., the fraction standing as the multiplier of $u(\psi)$ under the integral sign, is the real part of the following analytic function:

$$\frac{R^2 - r^2}{R^2 - 2Rr\cos(\psi - \varphi) + r^2} = \mathrm{Re}\left(\frac{t + z}{t - z}\right), \quad (z = re^{i\varphi},\ t = Re^{i\psi}).$$

On expanding the right side in a Taylor's series about the origin of coordinates and separating the real part in the result, we obtain:

$$\frac{R^2 - r^2}{R^2 - 2Rr\cos(\psi - \varphi) + r^2} = 1 + 2\sum_{n=1}^{\infty}\left(\frac{r}{R}\right)^n \cos n(\psi - \varphi).$$

Therefore $u(r, \varphi)$ is representable by the following series:

$$u(r, \varphi) = \frac{1}{2\pi}\int_{-\pi}^{+\pi} u(\psi)\, d\psi + \sum_{n=1}^{\infty}\left(\frac{r}{R}\right)^n \frac{1}{\pi}\int_{-\pi}^{+\pi} u(\psi)\cos n(\psi - \varphi)\, d\psi =$$

$$= a_0 + \sum_{n=1}^{\infty}\left(\frac{r}{R}\right)^n (a_n \cos n\varphi + b_n \sin n\varphi); \tag{2}$$

a_n and b_n are the Fourier coefficients of $u(\vartheta)$:

$$a_0 = \frac{1}{2\pi}\int_{-\pi}^{+\pi} u(\psi)\, d\psi, \quad a_n = \frac{1}{\pi}\int_{-\pi}^{+\pi} u(\psi)\cos n\psi\, d\psi,$$

$$b_n = \frac{1}{\pi}\int_{-\pi}^{+\pi} u(\psi)\sin n\psi\, d\psi.$$

By taking a sufficiently large segment of the series, we can, generally speaking, represent $u(r, \varphi)$ as exactly as we please over the entire circle. Indeed, the difference between $u(r, \varphi)$ and the sum of the first N terms of the series will be a function harmonic within and on the boundary, acquiring the values:

$$\sum_{n=N+1}^{\infty} (a_n \cos n\varphi + b_n \sin n\varphi).$$

Since the harmonic function attains its maximum on the boundary of the region, we have

$$\left| u(r, \varphi) - \sum_{n=0}^{N} \left(\frac{r}{R}\right)^n (a_n \cos n\varphi + b_n \sin n\varphi) \right| \leqslant$$

$$\leqslant \max \left| \sum_{n=N+1}^{\infty} (a_n \cos n\varphi + b_n \sin n\varphi) \right|,$$

and if the Fourier series for $u(\vartheta)$ converges uniformly, the right side can, by the choice of N, be made as small as one pleases for all φ.

When not a general approximate representation of $u(r, \varphi)$ is required, but its values at separate points within, it would be pointless to resort to the trigonometric series. Here it is simpler to utilize the integral formula, (1). We shall not dwell on the methods for the direct computation of the Poisson integral in the form it has in formula (1), since it has a perfectly lucid structure, and its application requires no explanations.

We shall focus our attention on another method, connected with a preliminary transformation, or, if you please, with the representation of the integral in Stieltjes form. This path seems to us more effective than the direct use of formula (1).

The fundamental idea of the method is very simple and is based on the reduction of (1) to the simplest case.

At the center of the circle $r = 0$ we obtain:

$$u(0, \varphi) = \frac{1}{2\pi} \int_{-\pi}^{+\pi} u(\psi)\, d\psi. \tag{3}$$

This equation represents the familiar Gauss theorem stating that the value of a harmonic function at the center of a circle is the arithmetic mean of its values on the circle itself.

The integral on the right side has a structure simpler than (1), and we shall try to reduce (1) to the computation of integrals of such a kind in the general case.

Let us assume that by means of a linear fractional transformation we have mapped the circle $|t| \leqslant R$ onto the unit circle of the plane $\zeta = \varrho e^{i\vartheta}$ so that the point $z = re^{i\varphi}$ has passed to the center $\zeta = 0$. The function u, harmonic in the circle $|t| \leqslant R$, will transform to a function harmonic in the unit circle of the ζ-plane. Its boundary values on the circumference $|t| = R$ will be carried to the boundary values on the unit circumference $|\zeta| = 1$. We shall know their dependence on the polar angle ϑ if we determine the dependence of ψ on ϑ. This last can be represented in analytic form as follows.

The function furnishing the required conformal transformation, if we lay down the condition that the ray $\psi = \varphi$ shall lie on the real axis, has the form:

$$\zeta = Re^{-i\varphi} \frac{t - z}{R^2 - t\bar{z}}.$$

On the boundary of the circle, when $t = Re^{i\psi}$:

$$e^{i\vartheta} = e^{-i\varphi} \frac{\frac{R}{r} e^{i\psi} - e^{i\varphi}}{\frac{R}{r} - e^{i(\psi-\varphi)}},$$

or, if we put $\frac{R}{r} = e^{\alpha}$, $\alpha = \ln \frac{R}{r}$,

$$e^{i\vartheta} = \frac{e^{\alpha} e^{i(\psi-\varphi)} - 1}{e^{\alpha} - e^{i(\psi-\varphi)}}.$$

Hence is obtained the following formula:

$$\frac{e^{i\vartheta} - 1}{e^{i\vartheta} + 1} = \frac{[e^{i(\psi-\varphi)} - 1][e^{\alpha} + 1]}{[e^{i(\psi-\varphi)} + 1][e^{\alpha} - 1]} \quad \text{or} \quad \operatorname{tg} \frac{\psi - \varphi}{2} = \operatorname{tg} \frac{\vartheta}{2} \operatorname{th} \frac{\alpha}{2}. \tag{4}$$

It gives the relation between ϑ, ψ and α in a very simple form.

In the ζ-plane that point for which we must find the value of the harmonic function u is located at the center of the circle, and therefore by the mean value theorem we have:

$$u = \frac{1}{2\pi} \int_{-\pi}^{+\pi} u[\psi(\vartheta)] \, d\vartheta. \tag{5}$$

The last formula permits one to give a method of computing the integral simpler than those preceding. We divide the interval $(-\pi, \pi)$ of variation of ϑ into equal parts, by the points $\vartheta_0 = -\pi$, $\vartheta_1 = -\frac{(n-2)\pi}{n}, \ldots, \vartheta_{n-1} = \frac{(n-2)\pi}{n}$. We shall denote the values of ψ and $u(\psi)$ corresponding to them by

$$\psi_0, \psi_1, \ldots, \psi_{n-1} \text{ and } u_0 = u(\psi_0),\ u_1 = u(\psi_1),\ \ldots,\ u_{n-1} = u(\psi_{n-1}).$$

Let us now apply to the computation of the integral one of the approximate methods, for instance, the rectangular rule, putting:

$$u = \frac{1}{2\pi} \cdot \frac{2\pi}{n} (u_0 + u_1 + \ldots + u_{n-1}) = \frac{u_0 + u_1 + \ldots + u_{n-1}}{n}. \tag{6}$$

$u(r, \varphi)$ will be equal to the arithmetic mean of the values of the function $u(\psi)$ on the contour; these values, however, will be taken not at equidistant points, as we would have obtained in computing u at the center of the circle of the z-plane, using formula (3), but at points whose location on the circle depends on the position of the point where we are seeking the solution, and which correspond to equidistant values of ϑ. We can easily find them, using relation (4). In practice, a simple graphical device, based on application of non-uniform scales, will doubtless lead most rapidly to the objective [1]).

To an increase of the argument of z, i.e., to a shift of the point z along the circle with center at the origin of coordinates, there corresponds the subtraction from ψ of a constant number equal to the angle of rotation. The latter in turn is equivalent simply to a shift of the scale $\psi(\vartheta)$ along itself. Therefore we always have the right to consider $\varphi = 0$, and to take formula (4) in the form:

$$\operatorname{tg} \frac{\psi}{2} = \operatorname{th} \frac{\alpha}{2} \operatorname{tg} \frac{\vartheta}{2}.$$

We prepare a template representing in the coordinate system ψ, r lines of constant ϑ. For this we divide the interval $(-\pi, +\pi)$ of variation of ϑ into an even number of parts, n, and in the rectangle $0 \leqslant r \leqslant 1$, $-\pi \leqslant \psi \leqslant +\pi$ we trace the lines for which ϑ preserves the constant values

$$\vartheta = 0, \quad \vartheta = \pm \frac{2\pi}{n}, \quad \vartheta = \pm 2 \frac{2\pi}{n}, \ldots, \vartheta = \pm \pi.$$

Juxtaposing to such a template at the proper height a uniform scale for ψ in the interval $(-\pi, +\pi)$, we can read on this scale those values of ψ which correspond to the values of ϑ adopted above, for any r.

After this, $u_0, u_1, \ldots, u_{n-1}$ are found, and u at the point adopted is determined as their arithmetic mean by formula (6).

Similar simplifications can also be effected for the other methods of the approximate computation of integral (5).

2. The Poisson integral for the exterior of a circle. Let it be required to find a function harmonic and bounded outside the circle $|z| = R$, and acquiring given values on the circumference:

$$u(R, \varphi) = u(\varphi).$$

We shall now show that the sought function can be represented by an integral of the Poisson type, which can be obtained from (1) in a very simple way. Set $z = re^{i\varphi}$, as before, and introduce the new inde-

---

[1]) On this see the article of C. Runge, [2].

pendent variable

$$\zeta = \frac{1}{z} = \varrho e^{i\vartheta}.$$

The function $u(r, \varphi)$, harmonic outside the circle $|z| = R$, will pass to the function $u^*(\varrho, \vartheta) = u\left(\frac{1}{\varrho}, -\vartheta\right)$, harmonic within the circle of radius $\frac{1}{R}$, acquiring on its boundary the values:

$$u^*\left(\frac{1}{R}, \vartheta\right) = u(R, -\vartheta) = u(-\vartheta).$$

By formula (1), for $\varrho < \frac{1}{R}$ it is representable by the Poisson integral:

$$u^*(\varrho, \vartheta) = \frac{1}{2\pi} \int_{-\pi}^{+\pi} u(-\psi^*) \frac{\frac{1}{R^2} - \varrho^2}{\frac{1}{R^2} - 2\frac{1}{R}\varrho \cos(\psi^* - \vartheta) + \varrho^2} d\psi^*.$$

If in this equation one substitutes for ϱ and ϑ their expressions in terms of r and φ and replaces the variable of integration, setting $\psi^* = -\psi$, one obtains the formula:

$$u(r, \varphi) = \frac{1}{2\pi} \int_{-\pi}^{+\pi} u(\psi) \frac{r^2 - R^2}{r^2 - 2Rr\cos(\psi - \varphi) + R^2} d\psi, \tag{7}$$

solving the problem posed. It differs from (1) only by the fact that in it R and r have changed places, so that the kernel of integral (7) differs from the kernel of Poisson integral (1) only in sign.

The expansion of the sought function in a trigonometric series like series (2), representing it outside the circle, can therefore be obtained from (2) by replacing r by R, and conversely:

$$u(r, \varphi) = a_0 + \sum_{n=1}^{\infty} \left(\frac{R}{r}\right)^n (a_n \cos n\varphi + b_n \sin n\varphi), \tag{7'}$$

a_n and b_n having their previous values.

In equation (7') let us increase r to infinity. In the limit we obtain the equation representing the Gauss theorem for the exterior of the circle:

$$u_\infty = \frac{1}{2\pi} \int_{-\pi}^{+\pi} u(\psi)\, d\psi, \tag{8}$$

i.e., the value of the harmonic function at infinity is the arithmetic mean of its values on the bounding circle.

On this formula can be based, as in the preceding No, a simple method of determining $u(r, \varphi)$ at any point of the exterior of the circle.

We transform the exterior of the circle $|t| \leqslant R$ of the plane in which we are computing the values of $u(r, \varphi)$ into the exterior of the unit circle of the ζ-plane so that the point z passes to the point at infinity. We shall require in addition that the ray $\arg t = \varphi$ shall pass to the real axis. The transforming function has the form:

$$\zeta = \frac{e^{i\varphi}}{R} \frac{t\bar{z} - R^2}{t - z};$$

on the circumference $t = Re^{i\psi}$

$$d\vartheta = |d\zeta| = \frac{1}{R} \frac{r^2 - R^2}{|t - z|^2} |dt| = \frac{r^2 - R^2}{r^2 - 2Rr\cos(\psi - \varphi) + R^2} d\psi.$$

The relation between ψ and ϑ, if we introduce the designation $\alpha = \ln \frac{r}{R}$, can be reduced to the equation:

$$\operatorname{tg} \frac{\psi - \varphi}{2} = - \operatorname{th} \frac{\alpha}{2} \operatorname{ctg} \frac{\vartheta}{2}. \tag{9}$$

Application of formula (8) in the ζ-plane gives us the following expression for $u(r, \varphi)$:

$$u(r, \varphi) = \frac{1}{2\pi} \int_{-\pi}^{+\pi} u[\psi(\vartheta)]\, d\vartheta. \tag{10}$$

To it can be applied all that we said about integral (5) in the preceding No.

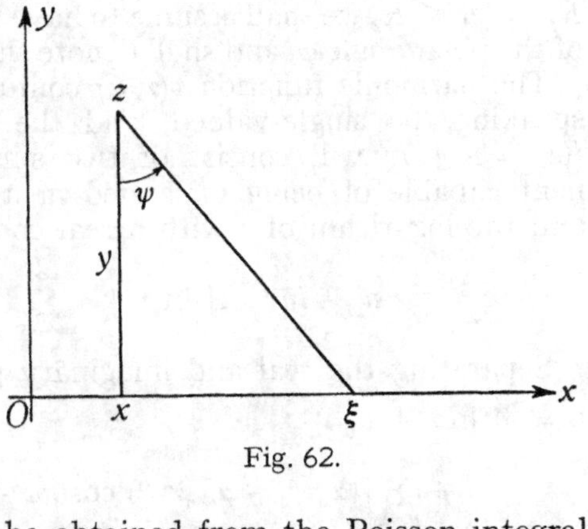

Fig. 62.

3. The Dirichlet problem for a half plane. In some problems it is often simpler to transform the given region not into a circle, but into a half plane, or, conversely, the half plane into the region. The analytical apparatus permitting one to represent the harmonic function within the upper half plane by its known boundary values on the real axis is therefore of a certain interest. It can be obtained from the Poisson integral by transforming the circle of radius R of the z-plane into the upper half of the ζ-plane by means of the function $\zeta = \frac{1}{i} \frac{z - R}{z + R}$. The boundary values on the circumference will pass to boundary values on the real axis. After calculations devoid of complication, we obtain the sought formula in the form:

$$u(x, y) = \frac{1}{\pi} \int_{-\infty}^{+\infty} u(\xi) \frac{y}{(x-\xi)^2 + y^2} d\xi = \frac{y}{\pi} \int_{-\infty}^{+\infty} u(x+\xi') \frac{d\xi'}{\xi'^2 + y^2}. \tag{11}$$

The first of the integrals is transformed into the second by the change of variable $\xi = x + \xi'$.

In inexact, graphical calculations, it is most conveniently employed in another form, taking as the variable of integration not ξ, but the angle ψ, which is formed by the straight line $z\xi$ and the perpendicular zx dropped from the point z to the ξ-axis (Fig. 62):

$$\operatorname{arctg} \frac{\xi - x}{y} = \operatorname{arctg} \frac{\xi'}{y} = \psi, \quad d\psi = \frac{y\,d\xi}{\xi'^2 + y^2}.$$

Equation (11) therefore takes the following simple form:

$$u(x, y) = \frac{1}{\pi} \int_{-\pi}^{+\pi} u[x + \xi'(\psi)]\, d\psi. \tag{12}$$

The sought quantity $u(x, y)$ is again determined as the arithmetic mean of the contour values, the mean here being taken in terms of the angle ψ. All the elements needed for the computation of the integral can be taken from the drawing.

4. Dirichlet's problem for the annulus. The boundary values of the harmonic function $u(r, \varphi)$ on the circumferences of the annulus $R_1 \leqslant |z| \leqslant R_2$ we shall assume to have been given in the form of functions of the polar angle φ, and shall denote them as $u_1(\varphi)$ and $u_2(\varphi)$ respectively.

The harmonic function $v(r, \varphi)$ conjugate to $u(r, \varphi)$ will not, generally speaking, be single-valued, and the function of a complex variable $f(z) = u + iv$ will consist of two summands: a single-valued component capable of being expanded in a Laurent's series in the annulus, and the logarithm of z with a real coefficient:

$$u + iv = A \ln z + \sum_{n=-\infty}^{+\infty} \gamma_\nu z^\nu, \quad \gamma_\nu = \alpha_\nu + i\beta_\nu. \tag{13}$$

Separating the real and imaginary parts, we have:

$$\left.\begin{aligned} u &= A \ln r + \alpha_0 + \\ &\quad + \sum_{n=1}^{\infty} [(\alpha_n r^n + \alpha_{-n} r^{-n}) \cos n\varphi - (\beta_n r^n - \beta_{-n} r^{-n}) \sin n\varphi], \\ v &= A\varphi + \beta_0 + \\ &\quad + \sum_{n=1}^{\infty} [(\beta_n r^n + \beta_{-n} r^{-n}) \cos n\varphi + (\alpha_n r^n - \alpha_{-n} r^{-n}) \sin n\varphi]. \end{aligned}\right\} \tag{14}$$

If in the first of the equations obtained we pass to the boundaries of the annulus, putting $r = R_1$ and $r = R_2$ there, we must obtain the Fourier series expansion of the functions $u_1(\varphi)$ and $u_2(\varphi)$. This gives:

$$\left.\begin{aligned}
A \ln R_1 + \alpha_0 &= a_0^{(1)} = \frac{1}{2\pi}\int_{-\pi}^{+\pi} u_1(\psi)\,d\psi,\\
A \ln R_2 + \alpha_0 &= a_0^{(2)} = \frac{1}{2\pi}\int_{-\pi}^{+\pi} u_2(\psi)\,d\psi,\\
\alpha_n R_1^n + \alpha_{-n} R_1^{-n} &= a_n^{(1)} = \frac{1}{\pi}\int_{-\pi}^{+\pi} u_1(\psi)\cos n\psi\,d\psi,\\
\alpha_n R_2^n + \alpha_{-n} R_2^{-n} &= a_n^{(2)} = \frac{1}{\pi}\int_{-\pi}^{+\pi} u_2(\psi)\cos n\psi\,d\psi,\\
\beta_{-n} R_1^{-n} - \beta_n R_1^n &= b_n^{(1)} = \frac{1}{\pi}\int_{-\pi}^{+\pi} u_1(\psi)\sin n\psi\,d\psi,\\
\beta_{-n} R_2^{-n} - \beta_n R_2^n &= b_n^{(2)} = \frac{1}{\pi}\int_{-\pi}^{+\pi} u_2(\psi)\sin n\psi\,d\psi.
\end{aligned}\right\} \tag{15}$$

Hence we obtain (cf. Chap. I, § 1, No. 1):

$$\left.\begin{aligned}
A &= \frac{a_0^{(2)} - a_0^{(1)}}{\ln R_2 - \ln R_1}, & \alpha_0 &= \frac{a_0^{(1)} \ln R_2 - a_0^{(2)} \ln R_1}{\ln R_2 - \ln R_1},\\
\alpha_n &= \frac{a_n^{(1)} R_2^{-n} - a_n^{(2)} R_1^{-n}}{R_1^n R_2^{-n} - R_1^{-n} R_2^n}, & \alpha_{-n} &= \frac{a_n^{(1)} R_2^n - a_n^{(2)} R_1^n}{R_2^n R_1^{-n} - R_2^{-n} R_1^n},\\
\beta_n &= \frac{b_n^{(2)} R_1^{-n} - b_n^{(1)} R_2^{-n}}{R_1^n R_2^{-n} - R_1^{-n} R_2^n}, & \beta_{-n} &= -\frac{b_n^{(2)} R_1^n - b_n^{(1)} R_2^n}{R_2^n R_1^{-n} - R_2^{-n} R_1^n}.
\end{aligned}\right\} \tag{16}$$

Series (13) is summed in more complicated functions than the analogous series for the circle [1]).

5. Schwarz's formula. Determination of the conjugate harmonic function. By the given boundary values of the real part, the function of a complex variable is found, as we know, accurate to a pure imaginary summand. The analytical apparatus giving the expression for the func-

---

[1]) Schwarz's formula for the annulus was obtained, as far as we know, by H. Villat. To familiarize himself with it, the reader is referred to a memoir of this scientist, [1], or to his book [2], p. 12.

tion $f(z) = u + iv$ regular in the region, in terms of the values of u on the contour, in the case when the region is a circle of radius R, is known; it is spoken of as the Schwarz integral [1]):

$$u + iv = \frac{1}{2\pi} \int_{-\pi}^{+\pi} u(\psi) \frac{Re^{i\psi} + z}{Re^{i\psi} - z} d\psi, \quad z = re^{i\varphi}. \tag{17}$$

Putting $z = 0$ here, we shall find for $u + iv$ a pure real value, and the formula cited above gives a value of $f(z)$ for which the imaginary part vanishes at the origin of coordinates. To obtain the general solution, we must add to the right side the arbitrary imaginary term Ci:

$$u + iv = \frac{1}{2\pi} \int_{-\pi}^{+\pi} u(\psi) \frac{Re^{i\psi} + z}{Re^{i\psi} - z} d\psi + Ci. \tag{18}$$

In (17) separate real and imaginary parts; since

$$\frac{Re^{i\psi} + z}{Re^{i\psi} - z} = \frac{R^2 - r^2}{R^2 - 2Rr\cos(\psi - \varphi) + r^2} - i\frac{2Rr\sin(\psi - \varphi)}{R^2 - 2Rr\cos(\psi - \varphi) + r^2},$$

the real part will give us the Poisson integral for u, whereas the imaginary part furnishes the expression of $v(r, \varphi)$ in terms of $u(\psi)$:

$$v(r, \varphi) = -\frac{1}{2\pi} \int_{-\pi}^{+\pi} u(\psi) \frac{2Rr\sin(\psi - \varphi)}{R^2 - 2Rr\cos(\psi - \varphi) + r^2} d\psi. \tag{19}$$

The trigonometric series for $v(r, \varphi)$ will be conjugate to the series for $u(r, \varphi)$, without the free term:

$$v(r, \varphi) = \sum_{n=1}^{\infty} \left(\frac{r}{R}\right)^n (-b_n \cos n\varphi + a_n \sin n\varphi).$$

Setting it up requires no new operations of any kind.

If v need be known approximately not everywhere in the circle but only at separate points within the circle, integral formula (19) will naturally be utilized. Computations involving it will nevertheless be performed more simply if it be transformed beforehand, by introducing rather than the variable of integration ψ the new variable ϑ, as is indicated in No. 2.

Since

$$d\vartheta = \frac{R^2 - r^2}{R^2 - 2Rr\cos(\psi - \varphi) + r^2} d\psi,$$

---

[1]) See, for example, V. I. Smirnov, [1], Vol. III, p. 378 (1933), or R. Courant, [1], p. 95 (1934).

which is evident from a comparison of formulas (1) and (5) § 2, we obtain the following expression for $v(r, \varphi)$:

$$v(r, \varphi) = -\frac{Rr}{\pi(R^2 - r^2)} \int_{-\pi}^{+\pi} u[\psi(\vartheta)] \sin(\psi - \varphi)\, d\vartheta. \tag{20}$$

Like formulas can be given for the exterior of the circle, too. From (7') No. 2 it is seen that the function $u(r, \varphi)$ harmonic outside the circle $|z| = R$, acquiring the given values for $r = R$, is the real part of the following function of a complex variable:

$$u + iv = \sum_{n=0}^{\infty} (a_n + ib_n) \left(\frac{R}{z}\right)^n.$$

The conjugate function $v(r, \varphi)$ is accordingly represented by the series

$$v(r, \varphi) = b_0 + \sum_{n=0}^{\infty} \left(\frac{R}{r}\right)^n (b_n \cos n\varphi - a_n \sin n\varphi).$$

We have already had occasion, in No. 2, to remark that the kernel of the Poisson integral for the exterior of the circle will differ from the kernel for the interior only in sign; therefore Schwarz's formula for the exterior is obtained from (18) by changing the sign in the right side to its opposite:

$$u + iv = \frac{1}{2\pi} \int_{-\pi}^{+\pi} u(\psi) \frac{z + Re^{i\psi}}{z - Re^{i\psi}}\, d\psi + Ci$$

and

$$v(r, \varphi) = \frac{1}{\pi} \int_{-\pi}^{+\pi} u(\psi) \frac{Rr \sin(\psi - \varphi)}{r^2 - 2Rr \cos(\psi - \varphi) + R^2}\, d\psi + C.$$

Finally, if instead of the variable ψ the new variable of integration ϑ be introduced, as we did at the end of No. 2, we obtain for the numerical determination of $v(r, \varphi)$ the following more convenient formula:

$$v(r, \varphi) = \frac{Rr}{\pi(r^2 - R^2)} \int_{-\pi}^{+\pi} u[\psi(\vartheta)] \sin(\psi - \vartheta)\, d\vartheta.$$

Let us turn to the case of the half plane. For constructing the Schwarz integral we utilize formula (11) No. 3. Since

$$\frac{y}{(x - \xi)^2 + y^2} = Re\left[i\left(\frac{1}{z - \xi} + c(\xi)\right)\right],$$

where $c(\xi)$ is an arbitrary real function of the real variable ξ, the function $f(z) = u + iv$, holomorphic in the upper half plane, is represented in terms of the values of its real part on the axis $y = 0$ by the following integral:

$$f(z) = \frac{i}{\pi} \int_{-\infty}^{+\infty} u(\xi) \left\{ \frac{1}{z - \xi} + c(\xi) \right\} d\xi.$$

As $c(\xi)$ one can choose a function that is regular on the real axis and which strengthens the convergence of the integral as much as possible. Near the point at infinity, the first term of the brace in the integrand has the expansion:

$$\frac{1}{z - \xi} = -\frac{1}{\xi} - \frac{z}{\xi^2} - \ldots$$

We take as $c(\xi)$ a function such that the entire brace shall have for $\xi = \infty$ a zero of the second order, i.e., such that $c(\xi)$ shall have at infinity the principal term $\frac{1}{\xi}$. The simplest function possessing the properties enumerated will be $c(\xi) = \frac{\xi}{1 + \xi^2}$. We shall therefore put:

$$f(z) = \frac{i}{\pi} \int_{-\infty}^{+\infty} u(\xi) \frac{1 + \xi z}{z - \xi} \frac{d\xi}{1 + \xi^2}. \tag{21}$$

If, however, $u(\xi)$ decreases sufficiently rapidly as $\xi \to \infty$, satisfying, for example, the inequality $|u(\xi)| \leqslant \frac{M}{|\xi|^\alpha}$, $\alpha > 0$, then it is simpler to put $c(\xi) = 0$ and take $f(z)$ in the form:

$$f(z) = \frac{i}{\pi} \int_{-\infty}^{+\infty} u(\xi) \frac{d\xi}{z - \xi}.$$

6. Solution of the Poisson equation in the circle. Let us consider the Poisson equation

$$\Delta u = \frac{\partial^2 u}{\partial x^2} + \frac{\partial^2 u}{\partial y^2} = -f(x, y), \tag{22}$$

and let it be required to find in the circle $|z| < R$ the solution of this equation taking given values on the boundary:

$$u|_{r=R} = F(\varphi).$$

We will simplify our problem at the outset, conceiving of the sought function u in the form of two terms

$$u = u_1 + u_2,$$

where u_1 satisfies the Laplace equation $\Delta u_1 = 0$ and takes on the circle the given values, $u_1|_{r=R} = F(\varphi)$. The second function, u_2, however, satisfies the Poisson equation (22) and vanishes on the boundary:

$$u_2|_{r=R} = 0.$$

The function u_1 can be found by means of the Poisson integral (1). Only the determination of u_2 is now of interest to us. In conformity with this we shall alter our original problem, seeking in the circle $|z| < R$ the solution of equation (22) satisfying on the circumference of the circle the boundary condition

$$u|_{r=R} = 0. \tag{23}$$

The function u furnishing the solution of the problem can be simply constructed on the basis of the properties of Green's function. We, however, will not proceed from Green's function in constructing u but will take as the basis of our reasoning a more elementary fact from the theory of logarithmic potential.

Let there be given a function $\mu(\xi, \eta)$ in the region B of the z-plane. We will designate as $t = \xi + i\eta$ the variable point of the region B. The logarithmic potential of masses continuously distributed in B with density $\mu(\xi, \eta)$ at the point $z = x + iy$ is what the following function of x and y is called:

$$v(x, y) = \iint\limits_B \mu(\xi, \eta) \ln \frac{1}{|z - t|} \, d\sigma.$$

It is known [1]) that outside the attracting masses, i.e., outside the region B, the potential $v(x, y)$ is an harmonic function, whereas within the region it will satisfy not the Laplace but the Poisson equation

$$\Delta v = -2\pi\mu(x, y).$$

If this equation be compared with (22), it will be clear that if we take the density $\mu(x, y)$ as equalling $\dfrac{1}{2\pi} f(x, y)$ and take our circle $|z| < R$ as the region B, then the potential with such a density

$$v(x, y) = \frac{1}{2\pi} \iint\limits_{|t|<R} f(\xi, \eta) \ln \frac{1}{|z - t|} \, d\sigma \tag{24}$$

will be a solution of equation (22). But $v(x, y)$ will not satisfy boundary condition (23). Let us construct in the circle a harmonic function $w(x, y)$

---

[1]) See, for example, E. Goursat, [1], Vol. III, No. 536 (1936).

with the boundary values

$$w|_{r=R} = -v|_{r=R}.$$

It is clear that the sum $u = v + w$ will satisfy both equation (22) and boundary requirement (23).

We shall construct w by means of a series, in accordance with the rule given by equation (2) No. 1 of this Section. In accordance with it, we must expand the boundary values of w in a Fourier series involving the polar angle and then prefix to harmonic number n the factor $\left(\frac{r}{R}\right)^n$.

Set $z = re^{i\varphi}$ and $t = \varrho e^{i\psi}$. The boundary values of w on the circle will be equal to

$$w|_{r=R} = \frac{1}{2\pi} \iint\limits_{|t|<R} f \ln |Re^{i\varphi} - \varrho e^{i\psi}|\, d\sigma.$$

The trigonometric series for them is obtained immediately if the logarithm standing under the integral sign be expanded in a series. Since $\varrho < R$, we shall have:

$$\ln |Re^{i\varphi} - \varrho e^{i\psi}| = \operatorname{Re} \{\ln (Re^{i\varphi} - \varrho e^{i\psi})\} =$$

$$= \operatorname{Re} \left\{\ln Re^{i\varphi} + \ln \left(1 - \frac{\varrho}{R} e^{(i\psi - \varphi)}\right)\right\} =$$

$$= \operatorname{Re} \left\{\ln Re^{i\varphi} - \sum_{n=1}^{\infty} \frac{1}{n} \frac{\varrho^n}{R^n} e^{in(\psi-\varphi)} \right\} = \ln R - \sum_{n=1}^{\infty} \frac{1}{n} \frac{\varrho^n}{R^n} \cos n\,(\varphi - \psi).$$

If this series be substituted in the expression for $w|_{r=R}$ and the integration be performed term by term, we shall obtain the following Fourier series for the boundary values we need:

$$w|_{r=R} = \frac{\ln R}{2\pi} \iint\limits_{|t|<R} f\, d\sigma - \sum_{n=1}^{\infty} \frac{1}{2\pi n} \iint\limits_{|t|<R} f \frac{\varrho^n}{R^n} \cos n\,(\varphi - \psi)\, d\sigma.$$

To obtain the value of w at any interior point $z = re^{i\varphi}$ of the circle, we must, by the rule given above, prefix the factor $\frac{r^n}{R^n}$ to the nth term of the series, so that

$$w(r, \varphi) = \frac{\ln R}{2\pi} \iint\limits_{|t|<R} f\, d\sigma - \sum_{n=1}^{\infty} \frac{1}{2\pi n} \frac{r^n}{R^n} \iint\limits_{|t|<R} f \frac{\varrho^n}{R^n} \cos n\,(\varphi - \psi)\, d\sigma. \quad (25)$$

To represent $w(r, \varphi)$ in more compact form we transform the series obtained into an integral. We first change the order of the summation and integration:

$$w(r, \varphi) = \frac{1}{2\pi} \iint\limits_{|t|<R} f \left\{\ln R - \sum_{n=1}^{\infty} \frac{1}{n} \frac{r^n \varrho^n}{R^{2n}} \cos n(\varphi - \psi)\right\} d\sigma.$$

If it be observed, moreover, that

$$-\sum_{n=1}^{\infty}\frac{1}{n}\,\frac{r^n\varrho^n}{R^{2n}}\cos n\,(\varphi-\psi)=\operatorname{Re}\left\{-\sum_{n=1}^{\infty}\frac{1}{n}\,\frac{r^n\varrho^n}{R^{2n}}\,e^{i(\varphi-\psi)}\right\}=$$

$$=\operatorname{Re}\left\{-\sum_{n=1}^{\infty}\frac{1}{n}\,\frac{z^n\bar{t}^n}{R^{2n}}\right\}=\operatorname{Re}\left\{\ln\left(1-\frac{z\bar{t}}{R^2}\right)\right\}=\ln\left|1-\frac{z\bar{t}}{R^2}\right|,$$

we shall obtain for w

$$w(r,\varphi)=\frac{1}{2\pi}\iint\limits_{|t|<R} f\ln R\left|1-\frac{z\bar{t}}{R^2}\right|d\sigma. \tag{26}$$

Consequently the function u sought by us, equal to the sum of v and w, is represented by the following integral:

$$u(r,\varphi)=\frac{1}{2\pi}\iint\limits_{|t|<R} f\ln\frac{R\left|1-\dfrac{z\bar{t}}{R^2}\right|}{|z-t|}\,d\sigma. \tag{27}$$

§ 3. THE NEUMANN PROBLEM

1. Dini's formula. Along with Dirichlet's problem, it is often necessary to solve Neumann's problem, consisting in the determination of the values of the harmonic function within the region by the known values of its normal derivative on the boundary.

It is known that the solution does not always exist. A condition necessary and sufficient for its existence is the fulfillment — obvious from a physical point of view — of the requirement that the integral of the boundary values of the normal derivative be equal to zero. If it is satisfied, then the solution is defined accurate to an arbitrary constant summand.

For the circle the Neumann problem has a simple solution, and the sought function is represented in closed form by means of the so-called Dini integral, which can be obtained by elementary means from the theory of trigonometric series. We shall assume below that on the boundary we are given the derivative along the interior normal.

Any function harmonic in the circle $|z| < R$ can be represented there by the series:

$$u=a_0+\sum_{k=1}^{\infty}\left(\frac{r}{R}\right)^k(a_k\cos k\varphi+b_k\sin k\varphi). \tag{1}$$

In the case of the circle, the derivative along the interior normal will differ from the derivative along r only in sign, since the direction of the normal is opposite to the direction of increase of r. Differentiating the preceding equation with respect to r, next putting $r = R$ in it and equating the result to $-\dfrac{\partial u}{\partial n}$, we obtain:

$$\sum_{k=1}^{\infty} \frac{k}{R} (a_k \cos k\varphi + b_k \sin k\varphi) = -\frac{\partial u}{\partial n},$$

whence

$$\left. \begin{aligned} a_k = -\frac{R}{k\pi} \int_{-\pi}^{+\pi} \frac{\partial u}{\partial n} \cos k\psi \, d\psi, \quad b_k = -\frac{R}{k\pi} \int_{-\pi}^{+\pi} \frac{\partial u}{\partial n} \sin k\psi \, d\psi, \\ (k = 1, 2, \ldots), \end{aligned} \right\} \tag{2}$$

and consequently if a_0, which remains indeterminate, be discarded:

$$u = -\frac{R}{\pi} \sum_{k=1}^{\infty} \frac{1}{k} \left(\frac{r}{R}\right)^k \int_{-\pi}^{+\pi} \frac{\partial u}{\partial n} \cos k(\psi - \varphi) \, d\psi =$$

$$= -\frac{R}{\pi} \int_{-\pi}^{+\pi} \frac{\partial u}{\partial n} \sum_{k=1}^{\infty} \frac{1}{k} \left(\frac{r}{R}\right)^k \cos k(\psi - \varphi) \, d\psi.$$

Noting that the sum under the integral sign is the real part of the series

$$\sum_{k=1}^{\infty} \frac{1}{k} \left(\frac{z}{t}\right)^k = -\ln\left(1 - \frac{z}{t}\right), \; z = re^{i\varphi}, \; t = Re^{i\psi},$$

we have:

$$u = \frac{R}{\pi} \int_{-\pi}^{+\pi} \frac{\partial u}{\partial n} \operatorname{Re}\left[\ln\left(1 - \frac{z}{t}\right)\right] d\psi \tag{3}$$

or, since

$$\operatorname{Re}\left[\ln\left(1 - \frac{z}{t}\right)\right] = \ln |t - z| - \ln R$$

and

$$\int_{-\pi}^{+\pi} \frac{\partial u}{\partial n} \ln R \, d\psi = \ln R \int_{-\pi}^{+\pi} \frac{\partial u}{\partial n} d\psi = 0,$$

we obtain the following integral, that of Dini,

$$u = \frac{R}{\pi} \int_{-\pi}^{+\pi} \frac{\partial u}{\partial n} \ln |t - z| \, d\psi. \tag{3'}$$

Series (1) and formula (3′) have a very simple form, and the use of them presents no difficulty of any kind.

By the known $\frac{\partial u}{\partial n}$ the function conjugate to u, v, is determined accurate to a constant summand and can be found either as the trigonometric series conjugate to (1), or from formula (3):

$$v = \frac{R}{\pi} \int_{-\pi}^{+\pi} \frac{\partial u}{\partial n} \mathrm{I}\left[\ln\left(\frac{t-z}{t}\right)\right] d\psi = \frac{R}{\pi} \int_{-\pi}^{+\pi} \frac{\partial u}{\partial n} \{\arg (t-z) - \arg t\} d\psi. \quad (4)$$

However if we discard the constant summand

$$\frac{R}{\pi} \int_{-\pi}^{+\pi} \frac{\partial u}{\partial n} \psi d\psi,$$

we shall then have

$$v = \frac{R}{\pi} \int_{-\pi}^{+\pi} \frac{\partial u}{\partial n} \arg (t-z)\, d\psi.$$

Finally, the analytic function of a complex variable for which are given the values of the normal derivative of its real part, $\frac{\partial u}{\partial n}$, on the boundary, is defined accurate to an arbitrary complex summand and in closed form is represented either by the integral:

$$f(z) = u + iv = \frac{R}{\pi} \int_{-\pi}^{+\pi} \frac{\partial u}{\partial n} \ln\left(1 - \frac{z}{t}\right) d\psi, \quad (5)$$

when we have in view the normalizing condition $f(0) = 0$, or, if we discard the constant $\frac{R}{\pi} \int_{-\pi}^{+\pi} \frac{\partial u}{\partial n} \ln t\, d\psi$, by the integral:

$$f(z) = \frac{R}{\pi} \int_{-\pi}^{+\pi} \frac{\partial u}{\partial n} \ln (t-z)\, d\psi. \quad (5')$$

Observation. Integral (3′), if $\frac{\partial u}{\partial n}$ be replaced in it by an arbitrary real function $F(\psi)$, will give a harmonic function

$$u = \frac{R}{\pi} \int_{-\pi}^{+\pi} F(\psi) \ln |t-z|\, d\psi. \quad (*)$$

However the boundary values of its normal derivative will not, generally speaking, coincide with $F(\varphi)$, but will equal:

$$F(\varphi) - \frac{1}{2\pi}\int_{-\pi}^{+\pi} F(\psi)d\psi,$$

which is easily verified by expanding the preceding integral in a power series in z.

2. The exterior of a circle. Dini's formula for the exterior of a circle can be obtained from (3) or (3′) in a way similar to that which we employed for the Poisson integral in No. 2, § 2.

Outside the circle $|z| = R$ let there be sought the harmonic function, regular at the point at infinity and satisfying on the boundary for $r = R$ the condition

$$\left(\frac{\partial u}{\partial n}\right)_{r=R} = F(\varphi),$$

where n is the interior normal of the region. In our case the derivative along the normal will simultaneously be the derivative along the radius, and the condition on the contour can therefore be rewritten in the form:

$$\left(\frac{\partial u}{\partial r}\right)_{r=R} = F(\varphi).$$

By means of the function $\zeta = \varrho e^{i\vartheta} = \frac{1}{z}$ we transform the given region into the interior of the circle $|\zeta| = \frac{1}{R}$.

The sought harmonic function $u(r, \varphi)$ will become the function $u\left(\frac{1}{\varrho}, -\vartheta\right)$, the boundary conditions for which will be:

$$\left(\frac{\partial u}{\partial \nu}\right)_{\varrho=\frac{1}{R}} = -\left(\frac{\partial u}{\partial \varrho}\right)_{\varrho=\frac{1}{R}} = \left(r^2 \frac{\partial u(r, -\vartheta)}{\partial r}\right)_{r=R} = R^2F(-\vartheta).$$

Applying formula (3) to it, we obtain:

$$u\left(\frac{1}{\varrho}, -\vartheta\right) = \frac{1}{\pi R}\int_{-\pi}^{+\pi} R^2F(-\chi) \ln\left|1 - \frac{\zeta}{\tau}\right| d\chi, \quad \tau = \frac{1}{R}e^{i\chi},$$

or, if we replace the variable of integration, setting $\tau = \frac{1}{t} = \frac{1}{R}e^{-i\psi}$ and return to the former variable z:

$$u(r, \varphi) = \frac{R}{\pi}\int_{-\pi}^{+\pi} F(\psi) \ln\left|1 - \frac{t}{z}\right| d\psi. \tag{6}$$

Finally, by virtue of the fact that $\int_{-\pi}^{+\pi} F(\psi)\,d\psi = 0$,

$$u(r, \varphi) = \frac{R}{\pi} \int_{-\pi}^{+\pi} F(\psi) \ln |t - z|\, d\psi. \tag{6'}$$

As regards the function of a complex variable having $u(r, \varphi)$ as its real part, it is represented, as is seen from these formulas, by integrals of form (5) and (5').

3. Neumann's problem for a half plane. Let the region for which we solve the Neumann problem be the half plane:

$$\mathrm{I}(z) = y > 0.$$

The derivative along the normal to the boundary is here also the derivative along y; the boundary condition can therefore be taken in the form

$$u_y(x, 0) = F(x). \tag{7}$$

For the sake of simplicity we will assume the function $F(x)$ to be continuous and to satisfy, for sufficiently large $|x|$, the condition:

$$|F(x)| \leqslant \frac{M}{|x|^p}, \text{ where } p > 1.$$

A much simpler formula, analogous to Dini's formula for the circle, can evidently be obtained for the case in hand by the application of equation (11) No. 3, § 2, solving the Dirichlet problem for the half plane.

Let us consider the harmonic function $\dfrac{\partial u}{\partial y}$. By its known values (7) on the real axis it can be recovered by applying the equation indicated by us:

$$\frac{\partial u}{\partial y} = \frac{1}{\pi} \int_{-\infty}^{+\infty} F(\xi) \frac{y}{(\xi - x)^2 + y^2}\, d\xi.$$

The integral on the right side is the derivative with respect to y of $\dfrac{1}{\pi} \displaystyle\int_{-\infty}^{+\infty} F(\xi) \ln \sqrt{(\xi - x)^2 + y^2}\, d\xi$, of which one can readily satisfy oneself on the strength of the limitations imposed upon $F(x)$. Therefore the sought function $u(x, y)$ differs from the last integral by the summand $\omega(x)$, depending only on x; on the other hand, $\omega(x)$, as the difference between two harmonic functions — $u(x, y)$ and the integral cited above — must also be a harmonic function, and since among functions of one variable only the linear function will be harmonic, we must have

$$u(x, y) = \frac{1}{\pi} \int_{-\infty}^{+\infty} F(\xi) \ln \sqrt{(\xi - x)^2 + y^2}\, d\xi + Ax + B =$$

$$= \operatorname{Re}\left[\frac{1}{\pi} \int_{-\infty}^{+\infty} F(\xi) \ln (\xi - z)\, d\xi + Az + B\right]. \qquad (8)$$

From this follows the analog of the Dini formula,

$$f(z) = u + iv = \frac{1}{\pi} \int_{-\infty}^{+\infty} F(\xi) \ln (\xi - z)\, d\xi + Az + B. \qquad (8')$$

We have carried through our reasoning with the simplest assumptions with respect to the boundary values of $\frac{\partial u}{\partial n}$. If one utilizes the circumstance that in recovering $u(x, y)$ by the known $\frac{\partial u}{\partial y}$ one can, without changing the harmonic character of u, add under the integral sign of (8) an arbitrary function of ξ, linearly dependent on x, one can construct a formula permitting the solution of Neumann's problem for the half plane under more general assumptions.

Near the point at infinity, $\ln \sqrt{(x - \xi)^2 + y^2}$ has an expansion of the form:

$$\ln \sqrt{(x - \xi)^2 + y^2} = \ln \sqrt{\xi^2} - \frac{x}{\xi} + \frac{y^2 - x^2}{2\xi^2} + \cdots$$

Let us subtract from this the function $\ln \sqrt{1 + \xi^2} - \frac{x\xi}{1 + \xi^2}$, having in its expansion about infinity the same two first terms as does $\ln \sqrt{(x - \xi)^2 + y^2}$. As the result we obtain a function having for all x and y at the point $\xi = \infty$ a zero of the second order.

The harmonic function

$$u(x, y) = \frac{1}{\pi} \int_{-\infty}^{+\infty} F(\xi) \left[\ln \sqrt{(x - \xi)^2 + y^2} - \ln \sqrt{1 + \xi^2} + \frac{x\xi}{1 + \xi^2}\right] d\xi \qquad (9)$$

obviously has the same derivative with respect to y as does (8), and, accordingly, satisfies the stipulated requirements. However formula (9) preserves its meaning under more general assumptions than does (8); the integral on its right side remains absolutely convergent in case

$$|F(\xi)| \leqslant M\, |\xi|^{\alpha}, \text{ for } \alpha < 1.$$

4. Neumann's problem for the annulus. Denote by $F_1(\varphi)$ and $F_2(\varphi)$ the values of the normal derivative of the harmonic function $u(r, \varphi)$ on the boundary circumferences $|z| = R_1$ and $|z| = R_2$ of the annulus: $R_1 < |z| < R_2$.

The function $v(r, \varphi)$ conjugate to $u(r, \varphi)$ will not, generally speaking, be single-valued in the annulus, and the analytic function $f(z) = u + iv$ will have within the annulus an expansion of the form:

$$f(z) = A \ln z + \sum_{k=-\infty}^{\infty} \gamma_k z^k, \quad (\gamma_k = \alpha_k + i\beta_k). \tag{9'}$$

Separation of the real and imaginary parts leads to equations (14) No. 4, § 2. We recall that in our investigations we have in view now the interior normal; the derivative along it on the lesser circumference of the annulus will coincide with the derivative along the radius, and on the larger will differ from it in sign.

Thus the function u defined by the first series of (14) § 2 must on the outer and inner contours of the annulus satisfy the conditions:

$$F_1(\varphi) = \frac{A}{R_1} + \sum_{k=1}^{\infty} k[(\alpha_k R_1^{k-1} - \alpha_{-k} R_1^{-k-1}) \cos k\varphi - \\ - (\beta_k R_1^{k-1} + \beta_{-k} R_1^{-k-1}) \sin k\varphi],$$

$$F_2(\varphi) = -\frac{A}{R_2} - \sum_{k=1}^{\infty} k[(\alpha_k R_2^{k-1} - \alpha_{-k} R_2^{-k-1}) \cos k\varphi - \\ - (\beta_k R_2^{k-1} + \beta_{-k} R_2^{-k-1}) \sin k\varphi].$$

The coefficient A is found in two ways, and the equality of both the expressions obtained is the familiar condition for the existence of a solution of the Neumann problem:

$$A = \frac{R_1}{2\pi} \int_{-\pi}^{+\pi} F_1(\psi)\, d\psi = -\frac{R_2}{2\pi} \int_{-\pi}^{+\pi} F_2(\psi)\, d\psi. \tag{10}$$

As regards the rest of the coefficients, they must satisfy the following system of equations:

$$\left.\begin{aligned} k(\alpha_k R_1^{k-1} - \alpha_{-k} R_1^{-k-1}) &= a_k^{(1)} = \frac{1}{\pi} \int_{-\pi}^{+\pi} F_1(\psi) \cos k\psi\, d\psi, \\ -k(\alpha_k R_2^{k-1} - \alpha_{-k} R_2^{-k-1}) &= a_k^{(2)} = \frac{1}{\pi} \int_{-\pi}^{+\pi} F_2(\psi) \cos k\psi\, d\psi, \\ -k(\beta_k R_1^{k-1} + \beta_{-k} R_1^{-k-1}) &= b_k^{(1)} = \frac{1}{\pi} \int_{-\pi}^{+\pi} F_1(\psi) \sin k\psi\, d\psi, \\ k(\beta_k R_2^{k-1} + \beta_{-k} R_2^{-k-1}) &= b_k^{(2)} = \frac{1}{\pi} \int_{-\pi}^{+\pi} F_2(\psi) \sin k\psi\, d\psi. \end{aligned}\right\} \tag{11}$$

On solving them for α_k, α_{-k}, β_k, β_{-k} we obtain:

$$\alpha_k = \frac{1}{k}\frac{a_k^{(1)}R_1R_2^{-k} + a_k^{(2)}R_2R_1^{-k}}{R_1^kR_2^{-k} - R_2^kR_1^{-k}}, \quad \beta_k = -\frac{1}{k}\frac{b_k^{(1)}R_1R_2^{-k} + b_k^{(2)}R_2R_1^{-k}}{R_1^kR_2^{-k} - R_2^kR_1^{-k}},$$

$$\alpha_{-k} = \frac{1}{k}\frac{a_k^{(1)}R_1R_2^k + a_k^{(2)}R_2R_1^k}{R_1^kR_2^{-k} - R_2^kR_1^{-k}}, \quad \beta_{-k} = \frac{1}{k}\frac{b_k^{(2)}R_2R_1^k + b_k^{(1)}R_1R_2^k}{R_1^kR_2^{-k} - R_2^kR_1^{-k}}.$$

§ 4. THE GENERAL BOUNDARY-VALUE PROBLEM FOR HARMONIC FUNCTIONS

1. The problem stated. The case of constant coefficients in the boundary condition. The Neumann and Dirichlet problems are the simplest boundary-value problems of the theory of harmonic functions. We shall now investigate methods of solving the general boundary-value problem. Let it be required to find the harmonic function within the region B if it is known that on the contour L it satisfies a boundary condition of the form:

$$a\frac{\partial u}{\partial x} + b\frac{\partial u}{\partial y} + cu = g \tag{1}$$

or of the form:

$$k\frac{\partial u}{\partial n} + l\frac{\partial u}{\partial s} + mu = g, \tag{1'}$$

where a, b, c, g, k, l, m are given real functions of the point of the contour and n is the interior normal.

Equations (1) and (1′) differ from each other only in form; in the first of them we take the derivatives along the coordinate axes, and in the second along directions connected with the boundary of the region. But, as we well know, the one derivative can be simply expressed in terms of the other.

In applied problems one most frequently uses boundary conditions of form (1′).

We shall everywhere assume the region B to be a circle with center at the origin of coordinates, whose radius we shall, for the sake of simplicity, set equal to unity. We shall begin our analysis with the simplest case, of constant coefficients.

In condition (1′), let k, l, m be constants, and g a known function of the polar angle φ. The case $k = 0$ is obviously of no interest; the boundary condition then degenerates into the differential equation for the boundary values of u, for there will appear in it only the function u and the derivative of it along the length of arc. Solving it for u, we find the values of the harmonic function on L and obtain simply the Dirichlet problem, which we have just analysed.

By dividing (1′) by k, we can take the condition on the boundary in the form:

$$\frac{\partial u}{\partial n} + l\frac{\partial u}{\partial s} + mu = g(\varphi). \tag{2}$$

Let us put:

$$f(z) = u + iv = \sum_{n=0}^{\infty} \alpha_n z^n,$$

$$\alpha_n = a_n + ib_n,$$

moreover

$$u = \sum_{n=0}^{\infty} r^n(a_n \cos n\varphi - b_n \sin n\varphi). \tag{3}$$

Let us expand the function $g(\varphi)$ on the right side of (2) in a trigonometric series:

$$g(\psi) = A_0 + \sum_{n=1}^{\infty} (A_n \cos n\varphi + B_n \sin n\varphi). \tag{4}$$

We said above that

$$\frac{\partial}{\partial n} = -\frac{\partial}{\partial r}, \quad \frac{\partial}{\partial s} = \frac{\partial}{\partial \varphi};$$

having observed this, we substitute series (3) in condition (2) and compare the coefficients of the sines and cosines of like angles. As the result of the comparison we obtain the following system of equations, from which a_k and b_k must be determined:

$$\left.\begin{aligned} ma_0 &= A_0, \\ -na_n - lnb_n + ma_n &= A_n, \\ nb_n - lna_n - mb_n &= B_n. \end{aligned}\right\} \tag{5}$$

If $m = 0$, the first equation gives the necessary condition for the solubility of the problem:

$$A_0 = 0$$

or

$$\int_{-\pi}^{+\pi} g(\varphi)\, d\varphi = 0; \tag{6}$$

the remaining equations will take the form:

$$\left.\begin{aligned} a_n + lb_n &= -\frac{A_n}{n}, \\ -la_n + b_n &= \frac{B_n}{n}. \end{aligned}\right\} \tag{7}$$

The determinant of each system is equal to $1 + l^2 \neq 0$, and the systems can certainly be solved for a_n, b_n:

$$a_n = -\frac{A_n + lB_n}{n(1 + l^2)}, \quad b_n = \frac{B_n - lA_n}{n(1 + l^2)}. \tag{8}$$

The coefficient a_0 remains arbitrary, and the sought harmonic function can thus be found only accurate to an arbitrary constant term.

We shall now investigate the second case of the degeneration of the problem, when $l = 0$ and m is a positive integer; we will denote the latter by k.

System (5) for $n = k$ gives the equations:

$$A_k = 0, \quad B_k = 0,$$

or

$$\int_{-\pi}^{+\pi} g(\varphi) \cos k\varphi \, d\varphi = 0, \quad \int_{-\pi}^{+\pi} g(\varphi) \sin k\varphi \, d\varphi = 0, \tag{9}$$

which are the necessary conditions for the solubility of the problem. The remaining systems have a determinant different from zero, and can be solved for a_n and b_n:

$$a_0 = \frac{A_0}{k}, \quad a_n = -\frac{A_n}{n - k}, \quad b_n = \frac{B_n}{n - k}, \quad n \neq k, \tag{10}$$

a_k and b_k remaining arbitrary. The solution of the problem will be found, accordingly, only accurate to a summand of the form:

$$r^k(a_k \cos k\varphi - b_k \sin k\varphi).$$

In all the remaining cases the problem has a unique solution, and the coefficients a_n and b_n are determined by the formulas:

$$a_0 = \frac{A_0}{m}, \quad a_n = -\frac{(n - m)A_n + lnB_n}{(n - m)^2 + l^2n^2},$$

$$b_n = \frac{(n - m)B_n - lnA_n}{(n - m)^2 + l^2n^2}, \qquad (n = 1, 2, \ldots).$$

Combining all the results, we can enunciate the following proposition.

A harmonic function $u(x, y)$ satisfying condition (2) is represented by the series:

$$u(x, y) = \frac{A_0}{m} +$$

$$+ \sum_{n=1}^{\infty} r^n \left\{ -\frac{(n - m)A_n + lnB_n}{(n - m)^2 + l^2n^2} \cos n\varphi - \frac{(n - m)B_n - lnA_n}{(n - m)^2 + l^2n^2} \sin n\varphi \right\}. \tag{11}$$

If one of the terms of the series loses meaning in view of the denominator's vanishing, i.e., when $m = 0$, or $l = 0$ and $m = k$, then the corresponding term in the series must be omitted, and the function $u(x, y)$ can be found only providing conditions (6) or (9) respectively are observed, and the solution ceases to be unique. Here for constructing the function of general form satisfying the imposed condition, we must add to series (11) either an arbitrary constant for $m = 0$, or the term $r^k(a_k \cos k\varphi - b_k \sin k\varphi)$ with arbitrary a_k and b_k.

Series (11) is the real part of the following function of the complex variable z:

$$f(z) = \frac{A_0}{m} + \sum_{n=1}^{\infty} \frac{A_n - iB_n}{m - n(1 - li)} z^n. \tag{12}$$

If for A_n and B_n here one substitutes their values

$$\frac{1}{\pi}\int_{-\pi}^{+\pi} g(\psi) \cos n\psi \, d\psi, \qquad \frac{1}{\pi}\int_{-\pi}^{+\pi} g(\psi) \sin n\psi \, d\psi$$

and interchanges the order of integration and summation, we obtain

$$f(z) = \frac{A_0}{m} + \frac{1}{\pi}\int_{-\pi}^{+\pi} g(\psi) \sum_{n=1}^{\infty} \frac{e^{-in\psi} z^n}{m - n(1 - li)} \, d\psi.$$

Moreover, if we agree to discard the arbitrary constant of integration, we can write:

$$\frac{z^n}{m - n(1 - li)} = -\frac{1}{1 - li} z^p \int^{z} \zeta^{n-p-1} d\zeta, \quad p = \frac{m}{1 - li},$$

and accordingly:

$$f(z) = \frac{A_0}{m} - \frac{1}{1 - li}\frac{z^p}{\pi}\int_{-\pi}^{+\pi} g(\psi) \sum_{n=1}^{\infty} e^{-in\psi} \int^{z} \zeta^{n-p-1} d\zeta \, d\psi =$$

$$= \frac{A_0}{m} - \frac{1}{1 - li}\frac{z^p}{\pi}\int_{-\pi}^{+\pi} g(\psi) \int^{z} \frac{\zeta^{-p}}{e^{i\psi} - \zeta} d\zeta \, d\psi,$$

or, in view of the fact that

$$A_0 = \frac{1}{2\pi}\int_{-\pi}^{+\pi} g(\psi) \, d\psi \quad \text{and} \quad \frac{1}{m} = -\frac{pz^p}{m}\int^{z} \zeta^{-p-1} d\zeta,$$

$$f(z) = \frac{z^p}{\pi(1 - li)}\int_{-\pi}^{+\pi} g(\psi) \int^{z} \left\{-\frac{1}{2\zeta} - \frac{1}{e^{i\psi} - \zeta}\right\} \zeta^{-p} d\zeta \, d\psi =$$

$$= \frac{z^p}{2\pi(li - 1)}\int_{-\pi}^{+\pi} g(\psi) \int^{z} \zeta^{-p-1} \frac{e^{i\psi} + \zeta}{e^{i\psi} - \zeta} d\zeta \, d\psi. \tag{12'}$$

Finally, if it be noted that

$$\frac{1}{2\pi}\int_{-\pi}^{+\pi} g(\psi)\,\frac{e^{i\psi}+z}{e^{i\psi}-z}\,d\psi = F(z)$$

is the Schwarz integral for a function analytic in the unit circle, having on the circumference the real part $g(\varphi)$, we obtain the formula:

$$f(z) = \frac{z^p}{li-1}\int^{z} F(\zeta)\zeta^{-p-1}\,d\zeta. \tag{13}$$

In the particular case when m is a negative number, the real part of p will be negative, for $p = \frac{m}{1+l^2} + \frac{mli}{1-l^2}$, and the integration with respect to the variable ζ can be performed from 0 to z:

$$f(z) = \frac{z^p}{2\pi(li-1)}\int_{-\pi}^{+\pi} g(\psi)\int_0^z \zeta^{-p-1}\,\frac{e^{i\psi}+\zeta}{e^{i\psi}-\zeta}\,d\zeta\,d\psi. \tag{14}$$

For the exterior of the circle the problem being analysed can be formulated as follows:

To find, a function $f(z) = \sum_{n=0}^{\infty} \frac{a_n + ib_n}{z^n}$, holomorphic everywhere outside the unit circle, including the point at infinity, the real part of which $u = a_0 + \sum_{n=1}^{\infty} r^{-n}(a_n \cos n\varphi + b_n \sin n\varphi)$ satisfies condition (2) on the boundary.

Differentiation along the interior normal will coincide with differentiation along r, and therefore condition (2) can be rewritten in the form:

$$\frac{\partial u}{\partial r} + l\frac{\partial u}{\partial \varphi} + mu = g(\varphi).$$

Let us transform variables, setting $z = \frac{1}{\zeta} = \frac{1}{\varrho}e^{-i\vartheta}$. In the ζ-plane we shall have an interior problem. Let us discover how the boundary conditions change.

Since $r = \frac{1}{\varrho}$ and $\varphi = -\vartheta$, we have on the boundary

$$\frac{\partial u}{\partial r} = -\frac{\partial u}{\partial \varrho}, \quad \frac{\partial u}{\partial \varphi} = -\frac{\partial u}{\partial \vartheta},$$

and (2) will therefore take the form:

$$-\frac{\partial u}{\partial \varrho} - l\frac{\partial u}{\partial \vartheta} + mu = A_0 + \sum_{n=1}^{\infty}(A_n \cos n\vartheta - B_n \sin n\vartheta) = g(-\vartheta).$$

The new boundary condition differs from that considered earlier by the change of the sign on l and B_n only.

The sought function $f(z)$ is thus represented by the following series:

$$f(z) = \frac{A_0}{m} + \sum_{n=1}^{\infty} \frac{A_n + iB_n}{m - n(1 + li)} \cdot \frac{1}{z^n}. \tag{15}$$

If one of the denominators of the terms of the series vanishes, i.e., when either $m = 0$, or $l = 0$, $m = k$, then the solution only exists either under the conditions $A_0 = 0$ or $A_k = B_k = 0$, respectively; it then ceases to be unique. We obtain the general form of the sought functions by omitting from series (15) the term that loses meaning and adding to it an arbitrary expression of the form $\dfrac{a_k + ib_k}{z^k}$.

The series on the right side can easily be summed. As the result we obtain:

$$f(z) = \frac{1}{2\pi(1 + li)} \cdot \frac{1}{z^p} \int_{-\pi}^{+\pi} g(\psi) \int^{z} \zeta^{p-1} \frac{\zeta + e^{i\psi}}{\zeta - e^{i\psi}} \, d\zeta \, d\psi, \tag{16}$$

where

$$p = \frac{m}{1 + li}.$$

2. Hilbert's problem. Let us pass on to the investigation of problems with variable coefficients. It is now more convenient for us to take the boundary condition in form (1). We shall begin with the simplest case, when $c = 0$:

$$a \frac{\partial u}{\partial x} + b \frac{\partial u}{\partial y} = g.$$

If u is the real part of a regular function $f(z)$, then, in view of the fact that its derivative will be equal to $\dfrac{\partial u}{\partial x} - i \dfrac{\partial u}{\partial y}$, $\dfrac{\partial u}{\partial x}$ and $-\dfrac{\partial u}{\partial y}$ will be the real and the imaginary parts of the regular function $f'(z)$, and our problem is consequently equivalent to the following problem of Hilbert:

To find, a function $f(z) = u + iv$, regular in the circle $|z| < 1$ and such that its real and imaginary parts on the circumference of the circle satisfy the linear condition:

$$\alpha(\varphi)u + \beta(\varphi)v = g(\varphi). \tag{17}$$

We will consider the coefficients $\alpha(\varphi)$ and $\beta(\varphi)$ to be continuous functions of the polar angle φ, and not to vanish simultaneously. Dividing equation (17) by $\sqrt{\alpha^2(\varphi) + \beta^2(\varphi)}$, one can always contrive to satisfy the requirement.:

$$\alpha^2(\varphi) + \beta^2(\varphi) = 1.$$

Let us introduce the angle $\omega(\varphi)$ defined by the equations:

$$\alpha = \cos \omega, \quad \beta = -\sin \omega, \quad \omega = -\operatorname{arctg} \frac{\beta}{\alpha}.$$

Boundary condition (17) will take the form:

$$u \cos \omega - v \sin \omega = \operatorname{Re}[e^{i\omega(\varphi)} f(z)] = g(\varphi). \tag{17'}$$

1. Let us dwell on the simplest case, when $\omega(\varphi)$ is a single-valued function of φ.

We will consider $\omega(\varphi)$ as the boundary values of the real part of some holomorphic function $\Pi(z)$. Using Schwarz's integral, we can write:

$$\Pi(z) = \frac{1}{2\pi} \int_{-\pi}^{+\pi} \omega(\psi) \frac{e^{i\psi} + z}{e^{i\psi} - z} d\psi.$$

Let $\omega_1(\varphi)$ be the boundary values of its imaginary part on the circle. Consider the holomorphic function $e^{i\Pi(z)} f(z)$. Since on the circle $e^{i\Pi(z)} = e^{-\omega_1(\varphi) + i\omega(\varphi)}$, the real part of it for $|z| = 1$ will be equal to

$$e^{-\omega_1(\varphi)} \operatorname{Re}[e^{i\omega(\varphi)} f(z)].$$

Our problem has thus reduced to finding the function $f(z)$ satisfying the boundary requirement:

$$\operatorname{Re}[e^{i\Pi(z)} f(z)] = g(\varphi) e^{-\omega_1(\varphi)}. \tag{17''}$$

Applying Schwarz's formula, we obtain:

$$e^{i\Pi(z)} f(z) = \frac{1}{2\pi} \int_{-\pi}^{+\pi} g(\psi) e^{-\omega_1(\psi)} \frac{e^{i\psi} + z}{e^{i\psi} - z} d\psi + Ci,$$

and finally for $f(z)$ we have the following formula:

$$f(z) = e^{-i\Pi(z)} \left\{ \frac{1}{2\pi} \int_{-\pi}^{+\pi} g(\psi) e^{-\omega_1(\varphi)} \frac{e^{i\psi} + z}{e^{i\psi} - z} d\psi + Ci \right\}. \tag{18}$$

The solution of the problem is determined accurate to a summand of the form $Cie^{-i\Pi(z)}$.

Let us now pass to more complicated cases, when $\omega(\varphi)$ is a many-valued function of φ.

2. Let $\omega(\varphi)$ increase by $-2n\pi$ $(n = 1, 2, 3, \ldots)$ in circuiting the

circumference:

$$\omega(\pi) - \omega(-\pi) = -2n\pi.$$

Let us construct the function, single-valued on the bounding circumference:

$$\chi(\varphi) = \omega(\varphi) + n\varphi.$$

Considering $\chi(\varphi)$ as the boundary values of the real part of the regular function $\sigma(z)$, we shall recover this function $\sigma(z)$ by Schwarz's formula. Moreover, the real part of the function

$$\sigma_1(z) = \sigma(z) + in \ln z$$

on the unit circle will be equal to $\omega(\varphi)$. The boundary values of the imaginary part will, however, be the same as those of $\sigma(z)$. Denote them by $\omega_1(z)$ and form the function:

$$e^{i\sigma_1(z)} f(z) = z^{-n} e^{i\sigma(z)} f(z).$$

For $|z| = 1$ it satisfies the requirement:

$$\operatorname{Re}[z^{-n} e^{i\sigma(z)} f(z)] = e^{-\omega_1(\varphi)} \operatorname{Re}[e^{i\omega(\varphi)} f(z)] = g(\varphi) e^{-\omega_1(\varphi)}. \tag{19}$$

In view of the presence of the factor z^{-n}, the function standing under the sign Re can have at the origin of coordinates a pole of order n. Let us first construct a regular function having on the boundary the real part (19). This will be

$$\frac{1}{2\pi} \int_{-\pi}^{+\pi} g(\psi) e^{-\omega_1(\psi)} \frac{e^{i\psi} + z}{e^{i\psi} - z} d\psi + Ci.$$

To this we must add a function having on the circumference a real part equal to zero, and capable of having at the origin of coordinates a pole of order n.

It has the form:

$$\sum_{k=1}^{n} \{A_k(z^{-k} - z^k) + iB_k(z^{-k} + z^k)\},$$

where A_k and B_k are arbitrary real constants.

We finally obtain for $f(z)$ the following expression:

$$f(z) = z^n e^{-i\sigma(z)} \left\{ \frac{1}{2\pi} \int_{-\pi}^{+\pi} g(\psi) e^{-\omega_1(\psi)} \frac{e^{i\psi} + z}{e^{i\psi} - z} d\psi + Ci + \right.$$

$$\left. + \sum_{k=1}^{n} [A_k(z^{-k} - z^k) + iB_k(z^{-k} + z^k)] \right\}. \tag{20}$$

The general solution will thus contain $2n + 1$ arbitrary constants.

3. Let us now assume that in circuiting the circumference, $\omega(\varphi)$ acquires the summand $2n\pi$ ($n = 1, 2, 3, \ldots$).

The single-valued part $\chi(\varphi)$ of the function $\omega(\varphi)$ will here be:

$$\chi(\varphi) = \omega(\varphi) - n\varphi.$$

Let $\sigma(z)$ be a regular function having for $|z| = 1$ the real part $\chi(\varphi)$.

The boundary values of the real part of the function

$$\sigma_1(z) = \sigma(z) - in \ln z$$

will coincide with $\omega(\varphi)$. As regards the imaginary part for $|z| = 1$, it will be the same as that of $\sigma(z)$. We shall preserve for it the designation $\omega_1(z)$.

In exactly the same way it can be shown that

$$e^{i\sigma_1(z)} f(z) = z^n e^{i\sigma(z)} f(z)$$

satisfies on the boundary the requirement:

$$\mathrm{Re}\,[z^n e^{i\sigma(z)} f(z)] = g(\varphi)\, e^{-\omega_1(\varphi)}. \tag{21}$$

The expression in brackets obviously has at the origin of coordinates a root of multiplicity n; therefore in the Fourier series expansion of the right side the free term and the coefficients of $\cos k\varphi$ and $\sin k\varphi$ must vanish for all k from 1 to $n - 1$.

The problem will accordingly have a solution only in case the following requirements are observed:

$$\int_{-\pi}^{+\pi} g(\psi)\, e^{-\omega_1(\psi)}\, d\psi = 0, \quad \int_{-\pi}^{+\pi} g(\psi)\, e^{-\omega_1(\psi)} \cos k\psi\, d\psi = 0, \tag{22}$$

$$\int_{-\pi}^{+\pi} g(\psi)\, e^{-\omega_1(\psi)} \sin k\psi\, d\psi = 0, \qquad [k = 1, \ldots, n - 1],$$

and the solution is given by

$$f(z) = z^{-n} e^{-i\sigma(z)} \left\{ \frac{1}{2\pi} \int_{-\pi}^{+\pi} g(\psi)\, e^{-\omega_1(\psi)} \frac{e^{i\psi} + z}{e^{i\psi} - z}\, d\psi + Ci \right\}. \tag{23}$$

4. In circuiting the circumference let $\omega(\varphi)$ obtain an increment of the form $(2n + 1)\pi$, $(n = 0, \pm 1, \ldots)$; here it is obvious that $e^{i\omega(\varphi)}$ will change sign.

From (17′) it is evident that this problem will have a holomorphic solution only in case $g(\varphi)$ is not a single-valued function of φ and upon the circuit changes its sign.

If $g(\varphi)$ should be a single-valued function of φ, $f(z)$, which solves the problem, must obviously have the form $f(z) = \sqrt{z}P(z)$, where $P(z)$ is single-valued in the circle.

In both cases we reduce the problem to questions already analysed, if for the variable z we introduce the new variable $\zeta = \sqrt{z}$.

3. The general boundary-value problem. The general boundary-value problem with boundary condition (1) and (1′) can be reduced to an integral equation if one utilizes the formulas solving the Hilbert problem in the general case, obtained in the preceding No.

However the investigations connected with this course are cumbersome and difficult in the general case. We shall limit ourselves to a particular case, which is of the most interest from the point of view of applications, when the conditions on the boundary of the circle $|z| = 1$ are expressed by the equation:

$$\frac{\partial u}{\partial n} + m(\varphi)u = f(\varphi), \tag{24}$$

where $m(\varphi)$, $f(\varphi)$ are given continuous functions of the polar angle [1]).

Transpose the second term of the left side to the right, apply formula (3′) § 3, and we recover the function u within the circle:

$$u = -\frac{1}{\pi}\int_{-\pi}^{+\pi} mu \ln|t - z|\, d\psi + \frac{1}{\pi}\int_{-\pi}^{+\pi} f \ln|t - z|\, d\psi + C,$$

where C is some quite definite constant, equal, as we know, to the value of the function u at the center of the circle.

In this equation let us pass to the circle $z = e^{i\varphi}$.

$$u = -\frac{1}{\pi}\int_{-\pi}^{+\pi} m(\psi)\, u \ln|t - z|\, d\psi + F(\varphi) + C, \tag{25}$$

$$F(\varphi) = \frac{1}{\pi}\int_{-\pi}^{+\pi} f(\psi) \ln|t - z|\, d\psi$$

We have obtained an integral equation for the boundary values of the sought function. Its solution is obviously not equivalent, however, to problem (24), because of the integral of the Dini type that we have utilized in its derivation. The latter is indeed a harmonic function, but its normal derivative on the boundary coincides with the function standing under the sign of the Dini integral only if its integral around the circumference equals zero, which is the known necessary condition of the solubility of the Neumann problem; in the contrary case, the values of the normal derivative on the contour of the Dini-type integral [(*) § 3] will give, not the function standing under the sign of the Dini integral, but this function minus the arithmetic mean of its values. In

---

[1]) For an acquaintance with the integral equations that arise in solving the general problem, we refer the reader to a work of Noether, [1].

the notation of No. 1 § 3 this will be

$$F(\varphi) - \frac{1}{2\pi} \int_{-\pi}^{+\pi} F(\psi)\, d\psi \; ^{1)}.$$

Consequently, in addition to (24), the sought function must also satisfy the requirement:

$$\int_{-\pi}^{+\pi} [m(\psi)u - f(\psi)]\, d\psi = 0. \tag{26}$$

The last could have been obtained by the direct integration of condition (24) around the circle.

The solution of equation (25) under condition (26) is fully equivalent to the solution of the posed boundary-value problem.

Equation (25) itself contains the unknown constant C, which should be determined from the condition of existence of the solution.

The kernel of the integral equation is not bounded; it is logarithmically infinite for $z = t$. But the second integrated kernel will be bounded, as one can see, and the entire Fredholm theory is applicable to the equation.

Equation (25) is the particular case of the more general equation:

$$u = \lambda \int_{-\pi}^{+\pi} m(\psi) \ln |t - z|\, u\, d\psi + F(\varphi) + C \tag{27}$$

for $\lambda = -\dfrac{1}{\pi}$. First assume that $\lambda = -\dfrac{1}{\pi}$ is not a proper number. Then the equation has a unique solution. Denote by u_F the solution of the equation:

$$u = -\frac{1}{\pi} \int_{-\pi}^{+\pi} m(\psi) \ln |t - z|\, u\, d\psi + F(\varphi), \tag{25'}$$

and by u_1 the solution of the equation:

$$u = -\frac{1}{\pi} \int_{-\pi}^{+\pi} m(\psi) \ln |t - z|\, u\, d\psi + 1. \tag{25''}$$

The solution of the initial equation will obviously equal

$$u = u_F + C u_1. \tag{28}$$

It still remains to choose C so as to satisfy requirement (26). It will give:

$$C \int_{-\pi}^{+\pi} m u_1\, d\varphi + \int_{-\pi}^{+\pi} (m u_F - f)\, d\varphi = 0. \tag{29}$$

---

1) See the Observation in No. 1 § 3.

If

$$\int_{-\pi}^{+\pi} mu_1 \, d\varphi \neq 0,$$

then we find C from this uniquely, and by the contour values (28) of the harmonic function u we recover this function too. The problem has here a unique solution.

In the contrary case, when

$$\int_{-\pi}^{+\pi} mu_1 \, d\varphi = 0,$$

equation (29) is possible only in case the functions m and f satisfy the condition:

$$\int_{-\pi}^{+\pi} (mu_F - f) \, d\varphi = 0. \tag{30}$$

The number C remains arbitrary, and the problem admits of infinitely many solutions.

The case when $\lambda = -\dfrac{1}{\pi}$ is a proper number of equation (27) is more complicated. (25) will then have a solution only for a special form of the right side; for the existence of the solution it is necessary and sufficient, namely, that any solution of the homogeneous adjoint equation:

$$v = -\frac{m(\varphi)}{\pi} \int_{-\pi}^{+\pi} \ln |t - z| \, v \, d\psi \tag{31}$$

be orthogonal to the free term $F(\varphi) + C$.

Let $\bar{v}_1, \bar{v}_2, \ldots, \bar{v}_n$ be the complete system of linearly independent solutions of equation (31). One can readily satisfy oneself that each solution (31) differs only by the factor $m(\varphi)$ from the solution of the homogeneous equation:

$$u = -\frac{1}{\pi} \int_{-\pi}^{+\pi} m(\psi) \ln |t - z| \, u \, d\psi. \tag{32}$$

Indeed, let v satisfy (31). If one divides both sides of it by $m(\varphi)$ and writes $\dfrac{v}{m(\varphi)} = u$, one obtains equation (32). Conversely, if u be considered to satisfy equation (32) and one sets $v = m(\varphi)u$, then v will be a solution of (31).

Therefore the complete system of independent solutions of (32) will be:

$$\bar{u}_1 = \frac{\bar{v}_1}{m}, \quad \bar{u}_2 = \frac{\bar{v}_2}{m}, \quad \ldots, \quad \bar{u}_n = \frac{\bar{v}_n}{m}.$$

The conditions of the existence of the solution of equation (25) will be·

$$C\int_{-\pi}^{+\pi}\bar{v}_k d\varphi + \int_{-\pi}^{+\pi}F\bar{v}_k d\varphi = 0,$$

or

$$\frac{\int_{-\pi}^{+\pi}F\bar{v}_1 d\varphi}{\int_{-\pi}^{+\pi}\bar{v}_1 d\varphi} = \dots = \frac{\int_{-\pi}^{+\pi}F\bar{v}_n d\varphi}{\int_{-\pi}^{+\pi}\bar{v}_n d\varphi}. \tag{33}$$

We shall assume that they are fulfilled and that at least one of the denominators, for instance $\int_{-\pi}^{+\pi}\bar{v}_1 d\varphi$, is different from zero. The number C is found from the preceding equation.

Let us call $\tilde{u}$ any particular solution of the non-homogeneous equation (25). Its general solution has the form:

$$u = \tilde{u} + \sum_{k=1}^{n} A_k\bar{u}_k,$$

where the A_k are arbitrary constants.

Condition (26) here gives:

$$\sum_{k=1}^{n} A_k\int_{-\pi}^{+\pi}m\bar{u}_k d\varphi + \int_{-\pi}^{+\pi}(m\tilde{u} - f)\, d\varphi = 0.$$

From this we can find the constant A_1 as a function of $A_2, \dots, A_n$, for its coefficient will be different from zero, since

$$\int_{-\pi}^{+\pi}m\bar{u}_1 d\varphi = \int_{-\pi}^{+\pi}\bar{v}_1 d\varphi \neq 0.$$

Thus we obtain the solution of the posed problem, containing $n - 1$ arbitrary constants.

Let us go on to the case when all the denominators of equations (33) vanish:

$$\int_{-\pi}^{+\pi}m\bar{u}_k d\varphi = \int_{-\pi}^{+\pi}\bar{v}_k d\varphi = 0.$$

For the fulfillment of (32) there must then be necessary

$$\int_{-\pi}^{+\pi}F\bar{v}_k d\varphi = 0. \qquad (k = 1, 2, \dots, n)$$

This means that equations (25′) and (25″) simultaneously have solutions. Let us call the particular solutions of them $\tilde{u}_F$ and $\tilde{u}_1$, respectively.

The general solution of (25) will be:

$$u = \tilde{u}_F + C\tilde{u}_1 + \sum_{k=1}^{n} A_k\bar{u}_k,$$

where C and A_k are arbitrary constants.

In view of the fact that $\int\limits_{+\pi}^{-\pi} m\tilde{u}_k\,d\varphi = 0$, requirement (26) will give:

$$C\int\limits_{-\pi}^{+\pi} m\tilde{u}_1\,d\varphi + \int\limits_{-\pi}^{+\pi}(m\tilde{u}_F - f)\,d\varphi = 0.$$

From this the number C must be found. If the coefficient of C is different from zero, we obtain a solution containing n arbitrary constants; if it is equal to zero, however, then the problem either fails to have a solution in case

$$\int\limits_{-\pi}^{+\pi}(m\tilde{u}_F - f)\,df \neq 0,$$

or has a solution containing $n + 1$ arbitrary constants in the contrary case.

§ 5. THE FUNDAMENTAL PROBLEMS FOR BIHARMONIC FUNCTIONS

1. First fundamental problem. Reduction to a system of equations. It is necessary to find two functions $\varphi(z)$, $\psi(z)$ holomorphic in the unit circle $|z| < 1$ and satisfying on the boundary the condition:

$$\varphi(z) + \frac{\omega(z)}{\overline{\omega'(z)}}\overline{\varphi'(z)} + \overline{\psi(z)} = F(z), \tag{1}$$

$\omega(z)$ being the function giving the conformal transformation of the circle $|z| < 1$ into the simply connected bounded region B.

Below, in No. 3, we shall find it necessary to consider its contour L to be such that there exist continuous derivatives of up to the second order with respect to the arc.

In No. 2 § 1 we had occasion to remark that the problem can have a unique solution if we fix $\varphi(0)$ and I $\left[\frac{\varphi'(0)}{\omega'(0)}\right]$[1]. We shall consider that $\varphi(0) = 0$; the imaginary part of $\{\varphi'(0) : \omega'(0)\}$ we shall for the time being leave arbitrary.

---

[1] The second of these quantities has a form different from that which was indicated in No. 2 § 1, namely, if in the z-plane there be given a certain region B containing the origin of coordinates, and for it there be solved the first fundamental problem with a boundary condition of form (16) § 1, then the functions $\varphi(z)$ and $\psi(z)$ are defined uniquely when the values of $\varphi(0)$ and $I[\varphi'(0)]$ are assigned. A boundary condition of form (1), which we now have under consideration, however, is obtained from a condition of form (16) § 1 by transforming the circle into B. If it be required that under the transformation the origins pass the one to the other, then the assignment of the quantities $\varphi(0)$ and $I[\varphi'(0)]$ in the region B is equivalent to the assignment, in the circle, of quantities of the form $\varphi(0)$ and $I\left[\frac{\varphi'(0)}{\omega'(0)}\right]$ given by us in the text of the book.

The functions $\varphi(z)$ and $\psi(z)$ can be represented by the series:

$$\varphi(z) = \sum_{\nu=1}^{\infty} \alpha_\nu z^\nu, \quad \psi(z) = \sum_{\nu=0}^{\infty} \beta_\nu z^\nu. \tag{2}$$

We shall seek α_ν, β_ν by the method of undetermined coefficients.

To this end let us expand the free term $F(z)$ and the coefficient $\dfrac{\omega(z)}{\overline{\omega'(z)}}$ in Fourier series in powers of $e^{i\vartheta} = \sigma$:

$$\frac{\omega(\sigma)}{\overline{\omega'(\sigma)}} = \sum_{\nu=-\infty}^{+\infty} \delta_\nu \sigma^\nu,\ ^{1)} \quad F(\sigma) = \sum_{\nu=-\infty}^{+\infty} A_\nu \sigma^\nu.$$

Substituting the series (2) in condition (1) and putting $z = \sigma$ there, we arrive at the following equation:

$$\sum_{\nu=1}^{\infty} \alpha_\nu \sigma^\nu + \sum_{\nu=-\infty}^{+\infty} \delta_\nu \sigma^\nu \cdot \sum_{\nu=1}^{+\infty} \nu \bar{\alpha}_\nu \sigma^{-\nu+1} + \sum_{\nu=0}^{\infty} \bar{\beta}_\nu \sigma^{-\nu} = \sum_{\nu=-\infty}^{+\infty} A_\nu \sigma^\nu. \tag{3}$$

Hence, on comparing the coefficients of the positive powers of σ, we obtain the system of equations:

$$\alpha_\nu + \sum_{k=1}^{\infty} k \bar{\alpha}_k \delta_{\nu+k-1} = A_\nu \qquad (\nu = 1, 2, 3, \ldots). \tag{4}$$

On the other hand, comparison of the coefficients of σ^ν for $\nu = 0$, -1, -2, ... gives the following system:

$$\bar{\beta}_\nu + \sum_{k=1}^{\infty} k \bar{\alpha}_k \delta_{-\nu+k-1} = A_{-\nu} \quad (\nu = 0, +1, +2, \ldots). \tag{5}$$

Obviously everything reduces to the determination of the numbers α_ν from system (4), because after they have been found from this, the numbers β_ν are fully defined by formulas (5).

From the very statement of the problem it is clear that system (4) cannot have a unique solution satisfying the conditions of convergence of the series for $\varphi(z)$, $\varphi'(z)$ and $\psi(z)$ until we fix the imaginary part of $\dfrac{\alpha_1}{\omega'(0)}$.

Moreover, if $\mathrm{I}\left[\dfrac{\alpha_1}{\omega'(0)}\right]$ be fixed, the solution can exist only under definite conditions to be imposed on the right sides of the system, equivalent to conditions (26′) and (29) indicated in the introduction to the present chapter and expressing the equality to zero of the resultant vector of the forces applied to the contour of the region, as well as of its resultant moment.

In the case under analysis, condition (26′) reduces to the equality to zero of the increment of the function $F(z)$ along the circumference of the circle or simply to its single-valuedness.

---

1) Given the assumptions made with respect to L, it can be shown that the series converges absolutely and uniformly.

Equation (29), however, since $\zeta = e^{i\vartheta}$, $d\zeta = ie^{i\vartheta}d\vartheta$, gives:

$$\mathrm{I}\left[\int_{-\pi}^{+\pi} F(e^{i\vartheta})\, \overline{\omega'(e^{i\vartheta})}\, e^{-i\vartheta}\, d\vartheta\right] = 0. \tag{6}$$

If we can succeed in finding a solution of (4) such that the series for $\varphi(z)$, $\varphi'(z)$, $\psi(z)$ converge absolutely for $|z| \leqslant 1$, then we can be assured that the constructed functions will satisfy all the requirements stipulated.

In approximate computations the case is frequently encountered where the unit circle is mapped onto the region B by means of a polynomial of some degree n, if not exactly, then at least approximately, with the error lying beyond the limits of admissible computational error.

Thus let us assume that

$$\omega(z) = \gamma_1 z + \gamma_2 z^2 + \ldots + \gamma_n z^n \qquad (\gamma_1 \neq 0,\ \gamma_n \neq 0). \tag{7}$$

Denote by $\overline{\omega}(z)$ the polynomial whose coefficients are the conjugates of the coefficients of $\omega(z)$:

$$\overline{\omega}(z) = \overline{\gamma}_1 z + \overline{\gamma}_2 z^2 + \ldots + \overline{\gamma}_n z^n. \text{ [1]}$$

Since $\omega'(z) \neq 0$ for $|z| \leqslant 1$, the function $\overline{\omega}'\left(\frac{1}{z}\right)$ will be regular and different from zero everywhere for $|z| \geqslant 1$. Therefore the ratio

$$\frac{\omega(z)}{\overline{\omega}'\left(\frac{1}{z}\right)}$$

represents a function regular for $|z| \geqslant 1$, having at infinity a pole of the nth order.

Its expansion in a Laurent's series about the point at infinity has the form:

$$\frac{\omega(z)}{\overline{\omega}'\left(\frac{1}{z}\right)} = \delta_n z^n + \delta_{n-1} z^{n-1} + \ldots + \delta_1 z + \sum_{\nu=0}^{\infty} \delta_{-\nu} z^{-\nu}. \tag{8}$$

Putting $z = \sigma$ in this equation, we have:

$$\frac{\omega(\sigma)}{\overline{\omega}'\left(\frac{1}{\sigma}\right)} = \frac{\omega(\sigma)}{\overline{\omega'(\sigma)}} = \delta_n \sigma^n + \delta_{n-1}\sigma^{n-1} + \ldots + \delta_1 \sigma + \sum_{\nu=0}^{\infty} \delta_{-\nu}\sigma^{-\nu}. \tag{9}$$

All coefficients δ_ν with index greater than n prove to be equal to zero.

---

[1] We shall repeatedly use a notation of this kind below. Let us agree upon the following: if the function $\varphi(z)$ be represented by the series

$$\varphi(z) = a_0 + a_1 z + a_2 z^2 + \ldots,$$

then by $\overline{\varphi}(z)$ we shall understand the function represented by the series

$$\overline{\varphi}(z) = \bar{a}_0 + \bar{a}_1 z + \bar{a}_2 z^2 + \ldots.$$

System (4) accordingly takes the form:

$$\left.\begin{aligned} \alpha_1 + \bar{\alpha}_1\delta_1 + 2\bar{\alpha}_2\delta_2 + \ldots + n\bar{\alpha}_n\delta_n &= A_1, \\ \alpha_2 + \bar{\alpha}_1\delta_2 + 2\bar{\alpha}_2\delta_3 + \ldots + (n-1)\bar{\alpha}_{n-1}\delta_n &= A_2, \\ \cdots\cdots\cdots\cdots\cdots\cdots\cdots\cdots \\ \alpha_n + \bar{\alpha}_1\delta_n &= A_n, \end{aligned}\right\} \tag{10}$$

$$\alpha_\nu = A_\nu, \quad \nu \geqslant n + 1. \tag{11}$$

And system (5) for the determination of the β_ν will be:

$$\bar{\beta}_\nu + \sum_{k=1}^{n+\nu+1} k\bar{\alpha}_k\delta_{-\nu+k-1} = A_{-\nu} \quad (\nu = 0, 1, 2, 3, \ldots). \tag{12}$$

Equations (10) give a system of $2n$ real equations in the $2n$ unknowns a_k and b_k ($\alpha_k = a_k + ib_k$). Its determinant must equal zero, for $\mathrm{I}\left(\frac{\alpha_1}{\gamma_1}\right)$ remains indeterminate. The rank of the system is equal to $n - 1$, for, if we assign this quantity, the system will have a unique solution (this follows from the uniqueness of the solution of the fundamental biharmonic problem).

Observation I. The integral representation of the function $\psi(z)$ in terms of $\varphi(z)$ in the general case can easily be obtained from boundary condition (1).

Indeed, putting $z = \sigma$ there and going over to the conjugate quantities, we obtain the boundary values of $\psi(z)$ in the form:

$$\psi(\sigma) = \overline{F(\sigma)} - \overline{\varphi(\sigma)} - \frac{\overline{\omega(\sigma)}}{\omega'(\sigma)}\varphi'(\sigma).$$

Let us apply Cauchy's formula to both sides of the equation. From the theory of analytic functions we know that if $f(\sigma)$ represents the boundary values of a function holomorphic in the unit circle, then

$$\frac{1}{2\pi i}\int_{|\sigma|=1} \frac{\overline{f(\sigma)}}{\sigma - z}\, d\sigma = \overline{f(0)},$$ [1]

[1] Indeed, let $f(\sigma) = \sum_{k=0}^{\infty} \bar{a}_k\sigma^k$. On the other hand,

$$\frac{1}{\sigma - z} = \sum_{\nu=0}^{\infty} \frac{z^\nu}{\sigma^{\nu+1}}, \quad d\sigma = ie^{i\vartheta}d\vartheta = i\sigma d\vartheta.$$

If everything be substituted under the integral sign, we reduce it to the computation of the sum:

$$\frac{1}{2\pi}\sum_{k,\nu=0}^{\infty} \bar{a}_k z^\nu \int_{-\pi}^{+\pi} \sigma^{-k-\nu}d\vartheta;$$

but all the integrals of the form $\int_{-\pi}^{+\pi} \sigma^{-k-\nu}d\vartheta$ are equal to zero, with the exception of the one where $k = \nu = 0$, and the latter is equal to 2π; so of the whole sum there is preserved only the term $\bar{a}_0 = \overline{f(0)}$.

and consequently

$$\frac{1}{2\pi i}\int\limits_{|\sigma|=1}\frac{\overline{\varphi(\sigma)}}{\sigma - z}\,d\sigma = \overline{\varphi(0)} = 0.$$

Therefore

$$\psi(z) = \frac{1}{2\pi i}\int\limits_{|\sigma|=1}\frac{\overline{F(\sigma)}}{\sigma - z}\,d\sigma - \frac{1}{2\pi i}\int\limits_{|\sigma|=1}\frac{\overline{\omega(\sigma)}}{\omega'(\sigma)}\,\frac{\varphi'(\sigma)}{\sigma - z}\,d\sigma. \tag{13}$$

The case is of interest where $\omega(z)$ is an integral function

$$\omega(z) = \gamma_1 z + \gamma_2 z^2 + \ldots$$

Then $\overline{\omega(\sigma)}$ will be the boundary values on the unit circle of the function:

$$\bar{\omega}\left(\frac{1}{z}\right) = \frac{\bar{\gamma}_1}{z} + \frac{\bar{\gamma}_2}{z^2} + \ldots,$$

holomorphic everywhere with the exception of the point zero, which will be for it either a pole — when $\omega(z)$ is a polynomial, or an essential singularity — when $\omega(z)$ is a transcendental function.

The second integral of the right side of (13) will be equal to:

$$\frac{1}{2\pi i}\int\limits_{|\sigma|=1}\frac{\bar{\omega}\left(\frac{1}{\sigma}\right)}{\omega'(\sigma)}\,\frac{\varphi'(\sigma)}{\sigma - z}\,d\sigma,$$

and will give the sum of the residues of the function

$$\frac{\bar{\omega}\left(\frac{1}{\zeta}\right)}{\omega'(\zeta)}\,\frac{\varphi'(\zeta)}{\zeta - z} \tag{*}$$

arising from singular points within the unit circle. There are two of such points, one $\zeta = z$ with residue

$$\frac{\bar{\omega}\left(\frac{1}{z}\right)}{\omega'(z)}\,\varphi'(z), \tag{**}$$

and the second the point $\zeta = 0$. We have no need of performing the calculations to find the second residue. It is sufficient to note the following very simple fact. As we know, it will be equal to the coefficient of $\frac{1}{\zeta}$ in the expansion of the function (*) about the origin of coordinates in a Laurent's series. The latter, however, can be obtained by multiplying the power series for each factor. But when we expand in powers of ζ the fraction $\frac{1}{\zeta - z}$, the variable z will appear in negative powers everywhere, since

$$\frac{1}{\zeta - z} = -\frac{1}{z} - \frac{\zeta}{z^2} - \ldots.$$

The residue is therefore represented in the form of a power series in z containing only negative powers of z. This function of z, together with residue (**), must give a function holomorphic in the unit circle. The residue at zero must, consequently, be equal to the principal part, taken with opposite sign, of the Laurent expansion in powers of z of the function:

$$\frac{\bar{\omega}\left(\frac{1}{z}\right)}{\omega'(z)}\varphi'(z);$$

the integral we are studying is therefore equal to the regular part of this series. We will denote it by the symbol Reg:

$$\psi(z) = \frac{1}{2\pi i}\int\limits_{|\sigma|=1} \frac{\bar{\omega}\left(\frac{1}{\sigma}\right)}{\omega'(\sigma)}\frac{\varphi'(\sigma)}{\sigma - z}\,d\sigma = \mathrm{Reg}\left[\frac{\bar{\omega}\left(\frac{1}{z}\right)}{\omega'(z)}\varphi'(z)\right],$$

$$\psi(z) = \frac{1}{2\pi i}\int\limits_{|\sigma|=1} \frac{\overline{F(\sigma)}}{\sigma - z}\,d\sigma - \mathrm{Reg}\left[\frac{\bar{\omega}\left(\frac{1}{z}\right)}{\omega'(z)}\varphi'(z)\right]. \tag{14}$$

We will present an example.

Example 1. We will first solve the simplest problem. Let the region B be a circle of radius R. Here $\omega(z) = Rz$, $\dfrac{\omega(\sigma)}{\overline{\omega'(\sigma)}} = \sigma$, $\delta_1 = 1$, $\delta_\nu = 0$, for $\nu \neq 1$. The equations for the numbers α take the form:

$$\alpha_1 + \bar{\alpha}_1 = A_1, \quad \alpha_\nu = A_\nu, \quad \nu \geqslant 2.$$

A solution is possible only if $I(A_1) = 0$.
Put $b_1 = I(\alpha_1) = 0$; then

$$\alpha_1 = a_1 = \frac{A_1}{2}.$$

The numbers β_ν will be found from the equations:

$$\bar{\beta}_\nu + (\nu + 2)\bar{\alpha}_{\nu+2} = A_{-\nu},$$

$$\beta_\nu = \bar{A}_{-\nu} - (\nu + 2)\alpha_{\nu+2},$$

$$\varphi(z) = \frac{A_1}{2}z + A_2 z^2 + A_3 z^3 + \dots$$

or, since

$$A_\nu = \frac{1}{2\pi}\int\limits_{-\pi}^{+\pi} F(\sigma)\sigma^{-\nu}\,d\vartheta,$$

$$\varphi(z) = \frac{1}{2\pi}\int\limits_{-\pi}^{+\pi} F(\sigma)\left\{\frac{z}{2\sigma} + \frac{z^2}{\sigma^2} + \frac{z^3}{\sigma^3} + \dots\right\} d\vartheta =$$

$$= \frac{1}{2\pi i}\int\limits_{|\sigma|=1} \frac{F(\sigma)}{\sigma - z}\, d\sigma - A_0 - \tfrac{1}{2}A_1 z,$$

$$\psi(z) = \sum_{\nu=0}^{\infty} \bar{A}_{-\nu} z^{\nu} - \sum_{\nu=2}^{\infty} \nu A_{\nu} z^{\nu-2} = \frac{2}{2\pi i}\int\limits_{|\sigma|=1} \frac{\overline{F(\sigma)}}{\sigma - z}\, d\sigma - \frac{\varphi'(z)}{z} + \frac{A_1}{2z}.$$

To obtain the solution in the region B, we must now replace z by $\frac{z}{R}$ everywhere in the formulas. Having put $\varphi\left(\frac{z}{R}\right) = \varphi_1(z)$ and $\psi\left(\frac{z}{R}\right) = \psi_1(z)$, we have:

$$\varphi_1(z) = \frac{A_1}{2R} z + \frac{A_2}{R^2} z^2 + \dots = \frac{1}{2\pi i}\int\limits_{|\sigma|=R} \frac{F\left(\frac{\sigma}{R}\right)}{\sigma - z}\, d\sigma - A_0 - \frac{1}{2}\frac{A_1}{R} z,$$

$$\psi_1(z) = \sum_{\nu=0}^{\infty} \frac{\bar{A}_{-\nu}}{R^{\nu}} z^{\nu} - \sum_{\nu=2}^{\infty} \nu \frac{A_{\nu}}{R^{\nu-2}} z^{\nu-2} =$$

$$= \frac{1}{2\pi i}\int\limits_{|\sigma|=R} \frac{\overline{F\left(\frac{\sigma}{R}\right)}}{\sigma - z}\, d\sigma - \frac{R^2\varphi_1'(z)}{z} + \frac{RA_1}{2z}.$$

Example 2. The boundary of region B is Pascal's limaçon:

$$\omega(z) = bz(1 + az), \quad b > 0, \quad 0 < a < \tfrac{1}{2},$$

$$\frac{\omega(\sigma)}{\overline{\omega'(\sigma)}} = \frac{\sigma(1 + a\sigma)}{1 + 2a\bar{\sigma}} = \frac{\sigma^2(1 + a\sigma)}{2a + \sigma} =$$

$$= a\sigma^2 + (1 - 2a^2)\sigma - 2a(1 - 2a^2)\sum_{\nu=0}^{\infty} (-1)^{\nu}\left(\frac{2a}{\sigma}\right)^{\nu},$$

$$\delta_{\nu} = 0, \quad \nu > 2, \quad \delta_2 = a, \quad \delta_1 = 1 - 2a^2,$$

$$\delta_{-\nu} = (-1)^{\nu+1}(2a)^{\nu+1}(1 - 2a^2), \qquad (\nu = 0, 1, 2, \dots).$$

The system for the numbers α_{ν} reduces to the following:

$$\alpha_1 + \bar{\alpha}_1(1 - 2a^2) + 2\bar{\alpha}_2 a = A_1, \quad \alpha_2 + \bar{\alpha}_1 a = A_2,$$

$$\alpha_{\nu} = A_{\nu}, \quad \nu \geqslant 3.$$

From the second equation we find $\alpha_2 = A_2 - \bar{\alpha}_1 a$. Introducing this into the first equation, we obtain:

$$\alpha_1 + \bar{\alpha}_1 = \frac{A_1 - 2a\bar{A}_2}{1 - 2a^2}.$$

From this it is evident that a solution is possible only under the condition $I[A_1 - 2a\bar{A}_2] = 0$. We will again consider that $b_1 = 0$:

$$\alpha_1 = \frac{A_1 - 2a\bar{A}_2}{2(1 - 2a^2)},$$

$\varphi(z)$ is found very simply. To find $\psi(z)$ we employ formula (14), presented in the Observation:

$$\operatorname{Reg}\left[\frac{\bar{\omega}\left(\frac{1}{z}\right)}{\omega'(z)}\varphi'(z)\right] = \frac{z+a}{z^2(1+2az)}\varphi'(z) - \frac{a\alpha_1}{z^2} - \frac{2a\alpha_2 + \alpha_1(1-2a^2)}{z}.$$

$$\psi(z) = \frac{1}{2\pi i}\int\limits_{|\sigma|=1}\frac{\overline{F(\sigma)}}{\sigma - z}\,d\sigma - \frac{z+a}{z^2(1+2az)}\varphi'(z) + \frac{a\alpha_1}{z^2} + \frac{2a\alpha_2 + \alpha_1(1-2a^2)}{z}.$$

For the exterior of a closed curve the problem can by a simple transformation be reduced to the case already analysed.

Let B be the region exterior to the curve L and not containing the origin of coordinates. We map the unit circle onto it, by means of the function $z = \omega(\zeta)$, so that the zero of the ζ-plane passes to the point at infinity of the z-plane.

The mapping function has here the form:

$$\omega(\zeta) = \frac{\gamma}{\zeta} + \gamma_0 + \gamma_1\zeta + \dots \tag{15}$$

In region B, as we have indicated in No. 2 § 1, the functions $\varphi(z)$ and $\psi(z)$ have in the general case the form:

$$\left.\begin{aligned} \varphi_1(z) &= -\frac{X+iY}{2\pi(1+\varkappa)}\ln z + (B+Ci)z + \varphi^0(z),\\ \psi_1(z) &= \frac{\varkappa(X-iY)}{2\pi(1+\varkappa)}\ln z + (B'+C'i)z + \psi^0(z), \end{aligned}\right\} \tag{16}$$

where X, Y, B, C, B', C' are some constants possessing a known physical meaning, and φ^0, ψ^0 are single-valued analytic functions regular everywhere outside the contour L, including the infinity. Substituting for z here its value, we see that, by virtue of equation (15), the functions φ_1 and ψ_1 will take the following form:

$$\left.\begin{aligned} \varphi(\zeta) &= \frac{X+iY}{2\pi(1+\varkappa)}\ln\zeta + (B+Ci)\frac{\gamma}{\zeta} + \varphi_0(\zeta),\\ \psi(\zeta) &= -\frac{\varkappa(X-iY)}{2\pi(1+\varkappa)}\ln\zeta + (B'+C'i)\frac{\gamma}{\zeta} + \psi_0(\zeta), \end{aligned}\right\} \tag{17}$$

where $\varphi_0(\zeta)$ and $\psi_0(\zeta)$ are holomorphic in the unit circle. The constant C, on the basis of what was stated at the end of No. 2 § 1, we shall consider to be equal to zero. Introduce $\varphi(\zeta)$ and $\psi(\zeta)$ in boundary condition (1). Simple computations show that $\varphi_0(\zeta)$ and $\psi_0(\zeta)$ must satisfy the same condition, but with altered right side:

$$\varphi_0(\sigma) + \frac{\omega(\sigma)}{\overline{\omega'(\sigma)}}\overline{\varphi_0'(\sigma)} + \overline{\psi_0(\sigma)} = F_0(\sigma),$$

where

$$F_0(\sigma) = F(\sigma) - \frac{X + iY}{2\pi} \ln \sigma - \frac{B\gamma}{\sigma} - \frac{\omega(\sigma)}{\overline{\omega'(\sigma)}} \left\{ \frac{X - iY}{2\pi(1 + \varkappa)} \sigma - B\bar{\gamma}\sigma^2 \right\} - \\ - (B' - iC')\gamma\bar{\sigma}. \quad (18)$$

In the statement of the problem itself, we did not assume the resultant vector of the external stresses acting on the contour L to equal zero, and the function $F(\sigma)$ could be many-valued. In circuiting the circumference in the positive direction, $F(\sigma)$ acquires the increment $i(X + iY)$. $\frac{X + iY}{2\pi} \ln \sigma$ is increased by the same quantity, and $F_0(\sigma)$ therefore returns to its former value.

For φ_0 and ψ_0 we have obtained, therefore, a problem we already are familiar with.

As in the preceding problem, the particular case of the mapping function which is of interest is that in which the system of equations for the α_ν is particularly simple.

Let the unit circle be mapped onto the region B with sufficient accuracy by means of the function $\omega(\zeta)$, where

$$\left.\begin{aligned} \omega(\zeta) &= \frac{\gamma}{\zeta} + \gamma_0 + \gamma_1\zeta + \ldots + \gamma_n\zeta^n, \\ \bar{\omega}'\left(\frac{1}{\zeta}\right) &= -\bar{\gamma}\zeta^2 + \bar{\gamma}_1 + \ldots + n\frac{\bar{\gamma}_n}{\zeta^{n-1}}. \end{aligned}\right\} \quad (19)$$

Since $\omega'(\zeta) \neq 0$ in the unit circle, $\dfrac{1}{\bar{\omega}'\left(\frac{1}{\zeta}\right)}$ will be regular everywhere for $|\zeta| \geqslant 1$ and have at the point at infinity a pole of the second order; the ratio

$$\frac{\omega(\zeta)}{\bar{\omega}'\left(\frac{1}{\zeta}\right)}$$

though, will be a function regular outside the unit circle and having at infinity a pole of order $n - 2$. Its expansion about infinity will be:

$$\frac{\omega(\zeta)}{\bar{\omega}'\left(\frac{1}{\zeta}\right)} = \delta_{n-2}\zeta^{n-2} + \ldots + \delta_1\zeta + \delta_0 + \delta_{-1}\zeta^{-1} + \ldots,$$

whence:

$$\frac{\omega(\sigma)}{\overline{\omega'(\sigma)}} = \frac{\omega(\sigma)}{\bar{\omega}'\left(\frac{1}{\sigma}\right)} = \delta_{n-2}\sigma^{n-2} + \ldots + \delta_1\sigma + \delta_0 + \delta_{-1}\sigma^{-1} + \ldots$$

The system of equations for the coefficients α_ν will take the form:

$$\left.\begin{aligned} &\alpha_1 + \bar{\alpha}_1\delta_1 + 2\bar{\alpha}_2\delta_2 + \ldots + (n-2)\bar{\alpha}_{n-2}\delta_{n-2} = A_1^{(0)}, \\ &\alpha_2 + \bar{\alpha}_1\delta_2 + 2\bar{\alpha}_2\delta_3 + \ldots + (n-3)\bar{\alpha}_{n-3}\delta_{n-2} = A_2^{(0)} \\ &\cdots\cdots\cdots\cdots\cdots\cdots\cdots\cdots \\ &\alpha_{n-2} + \bar{\alpha}_1\delta_{n-2} = A_{n-2}^{(0)}, \quad \alpha_\nu = A_\nu^{(0)}, \quad \nu \geqslant n-1. \end{aligned}\right\} \quad (20)$$

Observation II. We will yet focus the attention of the reader on the case when the series for $\omega(\zeta)$ converges over the entire plane:

$$\omega(\zeta) = \frac{\gamma}{\zeta} + \text{an integral function.}$$

The literal repetition of what we said in Observation I regarding $\psi(z)$ is applicable to the function $\psi_0(\zeta)$, and we can therefore write:

$$\psi_0(\zeta) = \frac{1}{2\pi i} \int\limits_{|\sigma|=1} \frac{F_0(\sigma)}{\sigma - \zeta}\, d\sigma - \operatorname{Reg} \left\{ \frac{\overline{\omega}\left(\frac{1}{\zeta}\right)}{\omega'(\zeta)} \varphi_0'(\zeta) \right\}. \tag{21}$$

Example 3. Let B be the exterior of the ellipse with center at the origin of coordinates and semi-axes $a = R(1 + m)$, $b = R(1 - m)$, $R > 0$, $0 \leqslant m < 1$; then

$$\omega(\zeta) = R\left(m\zeta + \frac{1}{\zeta}\right),$$

$$\frac{\omega(\sigma)}{\overline{\omega(\sigma)}} = \frac{1 + m\sigma^2}{\sigma(m - \sigma^2)} = -\frac{1 + m\sigma^2}{\sigma^3}\left(1 + \frac{m}{\sigma^2} + \frac{m^2}{\sigma^4} + \ldots\right).$$

The expansion of $\dfrac{\omega(\sigma)}{\overline{\omega'(\sigma)}}$ in this case is completely free of the positive powers of σ and the equations for the determination of the α_ν here take a very simple form:

$$\alpha_\nu = A_\nu,$$

$$\varphi_0(\zeta) = \sum_{\nu=1}^{\infty} A_\nu \zeta^\nu = \frac{\zeta}{2\pi} \int\limits_{-\pi}^{+\pi} F_0(\sigma) \frac{d\vartheta}{\sigma - \zeta} = \frac{1}{2\pi i} \int\limits_{|\sigma|=1} \frac{\zeta}{\sigma} F_0(\sigma) \frac{d\sigma}{\sigma - \zeta}.$$

We find the function $\psi_0(\zeta)$ by using equation (21). The ratio

$$\frac{\overline{\omega}\left(\frac{1}{\zeta}\right)}{\omega'(\zeta)} = -\frac{\zeta(m + \zeta^2)}{1 - m\zeta^2}$$

is a holomorphic function of ζ for $|\zeta| \leqslant 1$, since $m < 1$;

$$\psi_0(\zeta) = \frac{1}{2\pi i} \int\limits_{|\sigma|=1} \frac{\overline{F_0(\sigma)}}{\sigma - \zeta}\, d\sigma + \frac{\zeta(m + \zeta^2)}{1 - m\zeta^2} \varphi_0'(\zeta).$$

φ_0 and ψ_0 are the solutions of the problem for the case $X = Y = B = B' = C' = 0$. To obtain the general solution, it is sufficient to add to them respectively:

$$\frac{X + iY}{2\pi(1 + \varkappa)} \ln \zeta + B\frac{\gamma}{\zeta}, \quad -\frac{\varkappa(X - iY)}{2\pi(1 + \varkappa)} \ln \zeta + (B' + iC')\frac{\gamma}{\zeta}.$$

Here:

$$F_0(\sigma) = F(\sigma) - \frac{X + iY}{2\pi} \ln \sigma - \frac{BR}{\sigma} +$$

$$+ \frac{1 + m\sigma^2}{\sigma(\sigma^2 - m)} \left\{ \frac{X - iY}{2\pi(1 + \varkappa)} \sigma - BR\sigma^2 \right\} - (B' - iC')R\sigma.$$

2. The second fundamental problem. Reduction to a system of equations. After the mapping of the region B, the boundary conditions of the second fundamental problem are written in the form (see No. 3, § 1):

$$\varkappa\varphi(z) - \frac{\omega(z)}{\overline{\omega'(z)}} \overline{\varphi'(z)} - \overline{\psi(z)} = \Phi(z), \ |z| = 1. \tag{22}$$

The reasoning and calculations connected with the application of the method of undetermined coefficients here are very similar to those advanced for the first problem. Here there remains arbitrary only one of the values of φ or ψ at some point of the region. The value of the imaginary part of the ratio $\frac{\varphi'(z)}{\omega'(z)}$ at that point must be found from the problem.

Let the region B be finite and simply connected; $\varphi(z)$ and $\psi(z)$ will be holomorphic for $|z| < 1$. We shall consider $\varphi(0) = 0$.

$$\varphi(z) = \sum_{\nu=1}^{\infty} \alpha_\nu z^\nu, \quad \psi(z) = \sum_{\nu=0}^{\infty} \beta_\nu z^\nu.$$

After substitution in boundary condition (22), if we put $z = \sigma$ there, we have:

$$\varkappa \sum_{\nu=1}^{\infty} \alpha_\nu \sigma^\nu - \sum_{\nu=-\infty}^{+\infty} \delta_\nu \sigma^\nu \sum_{\nu=1}^{\infty} \nu \bar{\alpha}_\nu \sigma^{-\nu+1} - \sum_{\nu=0}^{\infty} \bar{\beta}_\nu \bar{\sigma}^{-\nu} = \Phi(\sigma) = \sum_{\nu=-\infty}^{+\infty} B_\nu \sigma^\nu.$$

Hence, on comparing the coefficients of the powers of σ, we obtain the system of equations:

$$\varkappa\alpha_\nu - \sum_{k=1}^{\infty} k\bar{\alpha}_k \delta_{\nu+k-1} = B_\nu, \quad (\nu = 1, 2, 3, \ldots) \tag{23}$$

$$- \sum_{k=1}^{\infty} k\bar{\alpha}_k \delta_{k-\nu-1} - \bar{\beta}_\nu = B_{-\nu}, \quad (\nu = 0, 1, 2, \ldots) \tag{24}$$

As in the preceding problem, if we can succeed in finding the solution of system (23), the coefficients β_ν will be determined from system (24) without difficulty.

In case the mapping function is polynomial (7) and the expansion of $\dfrac{\omega(\sigma)}{\overline{\omega'(\sigma)}}$ contains only a finite number of the positive powers of σ, system (23) degenerates, taking the form:

$$\begin{aligned}\varkappa\alpha_1 - \bar{\alpha}_1\delta_1 - 2\bar{\alpha}_2\delta_2 - \ldots - n\bar{\alpha}_n\delta_n &= B_1,\\ \varkappa\alpha_2 - \bar{\alpha}_1\delta_2 - 2\bar{\alpha}_2\delta_3 - \ldots - (n-1)\bar{\alpha}_{n-1}\delta_n &= B_2,\\ \ldots\ldots\ldots\ldots\ldots\ldots\ldots\ldots\ldots\ldots\ldots &\\ \varkappa\alpha_n - \bar{\alpha}_1\delta_n &= B_n,\\ \varkappa\alpha_\nu &= B_\nu,\ (\nu \geqslant n+1).\end{aligned}$$

The integral representation of the function ψ is found by the same method as before (see Observation I, No. 1):

$$\psi(z) = -\frac{1}{2\pi i}\int\limits_{|\sigma|=1}\frac{\overline{\Phi(\sigma)}}{\sigma - z}\,d\sigma - \frac{1}{2\pi i}\int\limits_{|\sigma|=1}\frac{\overline{\omega(\sigma)}}{\omega'(\sigma)}\cdot\frac{\varphi'(\sigma)}{\sigma - z}\,d\sigma. \qquad (25)$$

Finally, when $\omega(z)$ is an integral function, the second term in the preceding equation is computed simply, and it takes the form

$$\psi(z) = -\frac{1}{2\pi i}\int\limits_{|\sigma|=1}\frac{\overline{\Phi(\sigma)}}{\sigma - z}\,d\sigma - \operatorname{Reg}\left[\frac{\bar{\omega}\left(\frac{1}{z}\right)}{\omega'(z)}\varphi'(z)\right]. \qquad (26)$$

Everything will develop analogously for an infinite region bounded by a simple contour.

We will not exhibit the corresponding calculations.

3. First fundamental problem. Reduction to functional equations. We shall assume the region B to be bounded and simply connected. We shall show below how the functional equations change for the case of an unbounded region.

Together with the boundary condition obtained after the conformal transformation of the region into a circle, we shall consider the condition conjugate to it:

$$\left.\begin{aligned}\varphi(\sigma) + \frac{\omega(\sigma)}{\overline{\omega'(\sigma)}}\overline{\varphi'(\sigma)} + \overline{\psi(\sigma)} &= F(\sigma),\\ \overline{\varphi(\sigma)} + \frac{\overline{\omega(\sigma)}}{\omega'(\sigma)}\varphi'(\sigma) + \psi(\sigma) &= \overline{F(\sigma)}.\end{aligned}\right\} \qquad (27)$$

Let us multiply both equations by $\dfrac{1}{2\pi i}\dfrac{d\sigma}{\sigma - z}$ and integrate along the circumference $|\sigma| = 1$. By Harnack's theorem we obtain conditions equivalent to the original ones:

$$\left.\begin{aligned}
&\frac{1}{2\pi i}\int\limits_{|\sigma|=1}\frac{\varphi(\sigma)}{\sigma-z}\,d\sigma+\frac{1}{2\pi i}\int\limits_{|\sigma|=1}\frac{\omega(\sigma)}{\overline{\omega'(\sigma)}}\,\frac{\overline{\varphi'(\sigma)}}{\sigma-z}\,d\sigma+\\
&\qquad\qquad+\frac{1}{2\pi i}\int\limits_{|\sigma|=1}\frac{\overline{\psi(\sigma)}}{\sigma-z}\,d\sigma=A(z),\\
&\frac{1}{2\pi i}\int\limits_{|\sigma|=1}\frac{\overline{\varphi(\sigma)}}{\sigma-z}\,d\sigma+\frac{1}{2\pi i}\int\limits_{|\sigma|=1}\frac{\overline{\omega(\sigma)}}{\omega'(\sigma)}\,\frac{\varphi'(\sigma)}{\sigma-z}\,d\sigma+\\
&\qquad\qquad+\frac{1}{2\pi i}\int\limits_{|\sigma|=1}\frac{\psi(\sigma)}{\sigma-z}\,d\sigma=B(z),
\end{aligned}\right\}\tag{28}$$

where for brevity we have put:

$$A(z)=\frac{1}{2\pi i}\int\limits_{|\sigma|=1}\frac{F(\sigma)}{\sigma-z}\,d\sigma,\quad B(z)=\frac{1}{2\pi i}\int\limits_{|\sigma|=1}\frac{\overline{F(\sigma)}}{\sigma-z}\,d\sigma.\tag{29}$$

After applying Cauchy's formula and utilizing the fact that

$$\frac{1}{2\pi i}\int\limits_{|\sigma|=1}\frac{\overline{\varphi(\sigma)}}{\sigma-z}\,d\sigma=\overline{\varphi(0)}=\bar{\alpha}_0,$$

$$\frac{1}{2\pi i}\int\limits_{|\sigma|=1}\frac{\overline{\psi(\sigma)}}{\sigma-z}\,d\sigma=\overline{\psi(0)}=\bar{\beta}_0,$$ [1]

conditions (28) reduce to the following ones:

$$\left.\begin{aligned}
&\varphi(z)+\frac{1}{2\pi i}\int\limits_{|\sigma|=1}\frac{\omega(\sigma)}{\overline{\omega'(\sigma)}}\,\frac{\overline{\varphi'(\sigma)}}{\sigma-z}\,d\sigma+\bar{\beta}_0=A(z),\\
&\psi(z)+\frac{1}{2\pi i}\int\limits_{|\sigma|=1}\frac{\overline{\omega(\sigma)}}{\omega'(z)}\,\frac{\varphi'(\sigma)}{\sigma-z}\,d\sigma+\bar{\alpha}_0=B(z).
\end{aligned}\right\}\tag{30}$$

The first equation is an integro-differential equation for $\varphi(z)$. Jointly with the condition $\varphi(0)=\alpha_0$, it fully determines the function $\varphi(z)$ if the imaginary part of the ratio $\dfrac{\varphi'(0)}{\omega'(0)}$ is assigned. In its composition there figures the unknown constant $\bar{\beta}_0=\overline{\psi(0)}$. We recall, however, that we can assign one of the quantities α_0 and β_0 arbitrarily; therefore in solving

[1]) See the footnote in No. 1.

it we could arbitrarily fix $\bar{\beta}_0$ rather than considering $\alpha_0 = 0$ as we have ordinarily done.

From (30) is easily obtained an integral equation relating to the boundary values of $\varphi'(z)$.

It is more convenient for us to first transform (30) to remove the singularity of the kernel.

On the strength of the equation

$$\frac{\omega(z)}{2\pi i}\int\limits_{|\sigma|=1}\frac{\overline{\varphi'(\sigma)}}{\overline{\omega'(\sigma)}}\,\frac{d\sigma}{\sigma - z} = \frac{\overline{\varphi'(0)}}{\overline{\omega'(0)}}\,\omega(z),$$

we can rewrite the first equation of (30) in the form:

$$\left.\begin{aligned}\varphi(z) + \frac{1}{2\pi i}\int\limits_{|\sigma|=1}\frac{\omega(\sigma) - \omega(z)}{\overline{\omega'(\sigma)}\,(\sigma - z)}\,\overline{\varphi'(\sigma)}\,d\sigma + k\omega(z) + \bar{\beta}_0 = A(z),\\ k = \overline{\varphi'(0)} : \overline{\omega'(0)}.\end{aligned}\right\} \quad (31)$$

The equation obtained contains the term $k\omega(z)$, essentially connected, through k, with the value of $\varphi'(z)$ at the center of the circle, and from it there cannot be obtained, by passing to the limit on the circle, an equation relating exclusively to the contour values of the function $\varphi'(z)$. To obviate this difficulty, we will introduce in place of $\varphi(z)$ a new unknown function $\varphi_0(z)$, connected with it by the equation:

$$\varphi(z) = -\,k\omega(z) + \varphi_0(z). \quad (32)$$

On the right side both terms are unknown, for we know neither the function $\varphi_0(z)$ nor the number k. After φ_0 has been found, we shall have to choose the number k in conformity with the requirement $k = \dfrac{\overline{\varphi'(0)}}{\overline{\omega'(0)}}$.

Substituting (32) in (31), in view of the fact that

$$\frac{1}{2\pi i}\int\limits_{|\sigma|=1}\frac{\omega(\sigma) - \omega(z)}{\sigma - z}\,d\sigma = 0,$$

we arrive at the equation:

$$\varphi_0(z) + \frac{1}{2\pi i}\int\limits_{|\sigma|=1}\frac{\omega(\sigma) - \omega(z)}{\overline{\omega'(\sigma)}(\sigma - z)}\,\overline{\varphi_0'(\sigma)}\,d\sigma + \bar{\beta}_0 = A(z). \quad (33)$$

Differentiating with respect to z, we obtain:

$$\varphi_0'(z) + \frac{1}{2\pi i}\int\limits_{|\sigma|=1}\frac{\partial}{\partial z}\left\{\frac{\omega(\sigma) - \omega(z)}{\sigma - z}\right\}\frac{\overline{\varphi_0'(\sigma)}}{\overline{\omega'(\sigma)}}\,d\sigma = A'(z),$$

whence, moving z to the arbitrary point z_0 of the circumference, we obtain on passing to the limit an integral equation for $\varphi_0'(z_0)$, of the

Fredholm type, second kind:

$$\varphi_0'(z_0) + \frac{1}{2\pi i}\int\limits_{|\sigma|=1} K(z_0, \sigma)\overline{\varphi_0'(\sigma)}\, d\sigma = A'(z_0), \tag{34}$$

the kernel of which

$$K(z_0, \sigma) = \frac{\omega(\sigma) - \omega(z_0) - (\sigma - z_0)\omega'(z_0)}{\overline{\omega'(\sigma)}(\sigma - z_0)^2}, \tag{35}$$

given our assumptions about the contour L, will be bounded, for as $\sigma \to z_0$ it will tend to

$$K(z_0, z_0) = \frac{1}{2}\frac{\omega''(z_0)}{\overline{\omega'(z_0)}}.$$

In view of the uniqueness of the solution of the problem of the theory of elasticity, equation (34) must also have a unique solution.

Let us assume that we have found $\varphi_0'(\sigma)$ in some way. We construct $\varphi_0(z)$ by it, suitably defining the constant $\varphi_0(0)$; thus, if we consider that $\varphi(0) = 0$ and that the origin of coordinates is preserved in the transformation, i.e., $\omega(0) = 0$, we must consider that $\varphi_0(0) = 0$.

Next form the function

$$\varphi(z) = -k\omega(z) + \varphi_0(z),$$

considering the constant k to be indeterminate. We must choose it in accordance with the condition $k = \frac{\overline{\varphi'(0)}}{\overline{\omega'(0)}}$. This will give

$$k + \bar{k} = \frac{\overline{\varphi_0'(0)}}{\overline{\omega_0'(0)}} = 2\frac{\overline{\varphi'(0)}}{\overline{\omega'(0)}}, \tag{36}$$

which is possible only in case the ratio $\frac{\varphi'(0)}{\omega'(0)}$ is equal to a real number. Thus the equation:

$$\mathrm{I}\left(\frac{\varphi'(0)}{\omega'(0)}\right) = 0$$

will be the condition of the possibility of solving the problem. It must obviously reduce to the requirement that the moment of the external forces vanish. We will not verify this.

From equation (36) one can find only the real part of k; the imaginary part remains arbitrary.

The problem is solved especially simply when the mapping function is rational. We intend to dwell on the simplest case, when $\omega(z)$ is a polynomial, leaving it to the reader wishing to acquaint himself with the problem in its general formulation to repair to the book of Acad. N. I. Muskhelishvili that we have already mentioned several times, [4].

Therefore let

$$\omega(z) = \gamma_1 z + \gamma_2 z^2 + \ldots + \gamma_n z^n.$$

It goes without saying that what we will now expound will differ only in form from what we expounded in No. 1 when developing the method of infinite systems of equations.

The ratio

$$\frac{\omega(z)}{\bar{\omega}'\left(\frac{1}{z}\right)},$$

which reduces to $\frac{\omega(\sigma)}{\omega'(\sigma)}$ for $z = \sigma$, has for $|z| \geqslant 1$ an expansion of the form

$$\frac{\omega(z)}{\bar{\omega}'\left(\frac{1}{z}\right)} = \delta_n z^n + \ldots + \delta_1 z + \sum_{\nu=0}^{\infty} \delta_{-\nu} z^{-\nu}.$$

To solve the problem there is no need of computing all the coefficients δ_ν; it is sufficient to find only $\delta_1, \delta_2, \ldots, \delta_n$.

Functional equation (30) for $\varphi(z)$, if σ be replaced in it by $\frac{1}{\sigma}$, will take the following form:

$$\varphi(z) + \frac{1}{2\pi i} \int\limits_{|\sigma|=1} \frac{\omega(\sigma)}{\bar{\omega}'\left(\frac{1}{\sigma}\right)} \frac{\bar{\varphi}'\left(\frac{1}{\sigma}\right)}{\sigma - z} d\sigma + \beta_0 = A\ (z).$$

The function

$$\frac{\omega(\zeta)}{\bar{\omega}'\left(\frac{1}{\zeta}\right)} \frac{\bar{\varphi}'\left(\frac{1}{\zeta}\right)}{\zeta - z},$$

since $|z| < 1$, is holomorphic with respect to ζ outside the unit circle, and the integral is therefore equal to the residue of this function at the point at infinity, taken with opposite sign, i.e., is equal to the coefficient of $\frac{1}{\zeta}$ in the expansion of this function about $\zeta = \infty$:

$$\left.\begin{aligned} \varphi(z) &= \alpha_1 z + \alpha_2 z^2 + \ldots, \\ \bar{\varphi}'\left(\frac{1}{\zeta}\right) &= \bar{\alpha}_1 + \frac{2\bar{\alpha}_2}{\zeta} + \frac{3\bar{\alpha}_3}{\zeta^2} + \ldots + \frac{n\bar{\alpha}_n}{\zeta^{n-1}} + \ldots, \\ \frac{\omega(\zeta)}{\bar{\omega}'\left(\frac{1}{\zeta}\right)} \bar{\varphi}'\left(\frac{1}{\zeta}\right) &= k_n\zeta^n + k_{n-1}\zeta^{n-1} + \ldots + k_1\zeta + k_0 + \sum_{\nu=1}^{\infty} k_{-\nu}\zeta^{-\nu}, \end{aligned}\right\} \quad (37)$$

where

$$\begin{aligned} k_n &= \bar{\alpha}_1\delta_n, \\ k_{n-1} &= \bar{\alpha}_1\delta_{n-1} + 2\bar{\alpha}_2\delta_n, \\ &\cdots\cdots\cdots \\ k_2 &= \bar{\alpha}_1\delta_2 + 2\bar{\alpha}_2\delta_3 + \ldots + (n-1)\bar{\alpha}_{n-1}\delta_n, \\ k_1 &= \bar{\alpha}_1\delta_1 + 2\bar{\alpha}_2\delta_2 + \ldots + n\bar{\alpha}_n\delta_n. \end{aligned}$$

Moreover

$$\frac{1}{\zeta - z} = \frac{1}{\zeta} + \frac{z}{\zeta^2} + \frac{z^2}{\zeta^3} + \ldots$$

Therefore the sought coefficient will be:

$$k_0 + k_1 z + k_2 z^2 + \ldots + k_n z^n.$$

Substituting this in the last equation for $\varphi(z)$, we at once obtain a ready expression for it:

$$\varphi(z) + \bar{\beta}_0 + k_0 + k_1 z + k_2 z^2 + \ldots + k_n z^n = \frac{1}{2\pi i} \int\limits_{|\sigma|=1} \frac{F(\sigma)}{\sigma - z}\, d\sigma, \quad (38)$$

which contains, it is true, the quantities $\bar{\beta}_0, k_0, \ldots, k_n$, which are for the time being unknown; but the determination of them now presents no difficulties. Expanding the integral on the right side of the equation in powers of z, we have:

$$\frac{1}{2\pi i} \int\limits_{|\sigma|=1} \frac{F(\sigma)}{\sigma - z}\, d\sigma = A_0 + A_1 z + A_2 z^2 + \ldots,$$

$$A_\nu = \frac{1}{2\pi i} \int\limits_{|\sigma|=1} F(\sigma)\sigma^{-\nu-1}\, d\sigma = \frac{1}{2\pi} \int\limits_{-\pi}^{+\pi} F(\sigma)\, e^{-i\nu\vartheta}\, d\vartheta.$$

We now compare the coefficients of $z^0, z^1, z^2, \ldots, z^n$ in (38):

$$\begin{aligned} &\bar{\beta}_0 + k_0 = A_0, \\ &\alpha_\nu + k_\nu = A_\nu, \quad (\nu = 1, 2, \ldots, n) \\ &\qquad \alpha_\nu = A_\nu \quad (\nu = n+1, n+2, \ldots). \end{aligned}$$

Developing the n middle equations, we obtain a system of linear equations for the determination of the first n coefficients α_ν:

$$\begin{aligned} \alpha_1 + \bar{\alpha}_1\delta_1 + \ldots + n\bar{\alpha}_n\delta_n &= A_1, \\ \alpha_2 + \bar{\alpha}_1\delta_2 + \ldots + (n-1)\bar{\alpha}_{n-1}\delta_n &= A_2, \\ \cdots\cdots\cdots\cdots\cdots& \\ \alpha_n + \bar{\alpha}_1\delta_n &= A_n, \end{aligned}$$

which was also constructed by us in No. 1.

After $\varphi(z)$ has been found, the function $\psi(z)$ can be found, using equation (14), which, by virtue of (37), takes the form:

$$\psi(z) = \frac{1}{2\pi i} \int\limits_{|\sigma|=1} \frac{\overline{F(\sigma)}}{\sigma - z}\, d\sigma - \frac{\bar{\omega}\left(\frac{1}{z}\right)}{\omega'(z)} \varphi'(z) + \frac{\bar{k}_1}{z} + \frac{\bar{k}_2}{z^2} + \ldots + \frac{\bar{k}_n}{z^n}.$$

Let us turn to the case of an infinite region B. Under the condition of the correspondence of the zero of the unit circle $|z| \leqslant 1$ and the point at infinity of the region B, the mapping function will be:

$$\omega(z) = \frac{\gamma}{z} + \gamma_0 + \gamma_1 z + \ldots = \frac{\gamma}{z} + \omega_0(z).$$

As we have revealed above, $\varphi(z)$ and $\psi(z)$ have the form:

$$\varphi(z) = \frac{X + iY}{2\pi(1 + \varkappa)} \ln z + \frac{B\gamma}{z} + \varphi_0(z),$$

$$\psi(z) = -\frac{\varkappa(X - iY)}{2\pi(1 + \varkappa)} \ln z + (B' + iC')\frac{\gamma}{z} + \psi_0(z),$$

where $\varphi_0(z)$ and $\psi_0(z)$ are holomorphic and must satisfy on the boundary $z = \sigma$ the condition:

$$\varphi_0(\sigma) + \frac{\omega(\sigma)}{\overline{\omega'(\sigma)}} \overline{\varphi_0'(\sigma)} + \overline{\psi_0(\sigma)} = F_0(\sigma).$$

$F_0(\sigma)$ has the known value (18).

Therefore the Cauchy-type integrals $A(z)$ and $B(z)$ on the right side of equations (28) must at the outset be replaced respectively by

$$A_0(z) = \frac{1}{2\pi i} \int\limits_{|\sigma|=1} \frac{F_0(\sigma)}{\sigma - z}\, d\sigma, \quad B_0(z) = \frac{1}{2\pi i} \int\limits_{|\sigma|=1} \frac{\overline{F_0(\sigma)}}{\sigma - z}\, d\sigma. \tag{39}$$

Moreover, in view of the fact that $\omega'(0) = \infty$, $k = \dfrac{\overline{\varphi'(0)}}{\omega'(0)} = 0$, the term $k\omega(z)$ in equation (31) drops out, and (31) acquires the form:

$$\varphi_0(z) + \frac{1}{2\pi i} \int\limits_{|\sigma|=1} \frac{\omega(\sigma) - \omega(z)}{\overline{\omega'(\sigma)}(\sigma - z)} \overline{\varphi_0'(\sigma)}\, d\sigma + \bar{\beta}_0 = A_0(z). \tag{40}$$

Finally,

$$\frac{\omega(\sigma) - \omega(z)}{\sigma - z} = \frac{\omega_0(\sigma) - \omega_0(z)}{\sigma - z} - \frac{\gamma}{\sigma z},$$

and since

$$\frac{1}{2\pi i}\int\limits_{|\sigma|=1} \frac{\overline{\varphi_0'(\sigma)}}{\overline{\omega'(\sigma)}}\,\frac{d\sigma}{\sigma} = 0,$$

which is evident from the fact that the expansion of the integrand in powers of σ does not contain a free term, equation (40) can be written as follows:

$$\varphi_0(z) + \frac{1}{2\pi i}\int\limits_{|\sigma|=1} \frac{\omega_0(\sigma) - \omega_0(z)}{\overline{\omega'(\sigma)}(\sigma - z)}\,\overline{\varphi_0'(\sigma)}\,d\sigma + \bar{\beta}_0 = A_0(z). \tag{41}$$

From this, by differentiation and passing to the limit on the circle we obtain the integral equation for $\varphi_0'(z)$:

$$\varphi_1'(z_0) + \frac{1}{2\pi i}\int\limits_{|\sigma|=1} K_0(z_0, \sigma)\overline{\varphi_0'(\sigma)}\,d\sigma = A_0'(z_0), \tag{42}$$

where

$$K_0(z_0, \sigma) = \frac{\omega_0(\sigma) - \omega_0(z_0) - (\sigma - z_0)\omega_0'(z_0)}{\overline{\omega'(\sigma)}(\sigma - z_0)^2}.$$

Let us consider the particular case when the mapping function is

$$\omega(z) = \frac{\gamma}{z} + \gamma_0 + \gamma_1 z + \ldots + \gamma_n z^n.$$

Here

$$\frac{\omega(\zeta)}{\overline{\omega}'\left(\dfrac{1}{\zeta}\right)} = \delta_{n-2}\zeta^{n-2} + \ldots + \delta_1\zeta + \sum_{\nu=0}^{\infty} \delta_{-\nu}\zeta^{-\nu}.$$

Therefore for $|\zeta| = 1$,

$$\frac{\omega(\zeta)}{\overline{\omega}'\left(\dfrac{1}{\zeta}\right)}\overline{\varphi_0'(\zeta)} = k_{n-2}^{(0)}\zeta^{n-2} + \ldots + k_1^{(0)}\zeta + k_0^{(0)} + \sum_{\nu=1}^{\infty} k_{-\nu}^{(0)}\zeta^{-\nu},$$

where

$$\begin{aligned} k_{n-2}^{(0)} &= \bar{\alpha}_1\delta_{n-2}, \\ k_{n-3}^{(0)} &= \bar{\alpha}_1\delta_{n-3} + 2\bar{\alpha}_2\delta_{n-2} \\ &\cdots\cdots\cdots\cdots \\ k_1^{(0)} &= \bar{\alpha}_1\delta_1 + 2\bar{\alpha}_2\delta_2 + \ldots + (n-2)\bar{\alpha}_{n-2}\delta_{n-2}. \end{aligned}$$

The integral on the left side of equation (31) will accordingly be equal to:

$$\frac{1}{2\pi i}\int\limits_{|\sigma|=1} \frac{\omega(\sigma) - \omega(z)}{\overline{\omega'(\sigma)}(\sigma - z)}\,\overline{\varphi_0'(\sigma)}\,d\sigma = k_0^{(0)} + k_1^{(0)}z + \ldots + k_{n-2}^{(0)}z^{n-2},$$

and the equation itself takes the form

$$\varphi_0(z) + \bar{\beta}_0 + k_0^{(0)} + k_1^{(0)}z + \ldots + k_{n-2}^{(0)}z^{n-2} = \frac{1}{2\pi i}\int\limits_{|\sigma|=1} \frac{F_0(\sigma)}{\sigma - z}\,d\sigma.$$

As for the coefficients $\alpha_1, \alpha_2, \ldots, \alpha_{n-2}$, they are determined from the system:

$$\begin{aligned} \alpha_1 + \bar{\alpha}_1\delta_1 + 2\bar{\alpha}_2\delta_2 + \ldots + (n-2)\bar{\alpha}_{n-2}\delta_{n-2} &= A_1^{(0)}, \\ \alpha_2 + \bar{\alpha}_1\delta_2 + 2\bar{\alpha}_2\delta_3 + \ldots + (n-3)\bar{\alpha}_{n-3}\delta_{n-2} &= A_2^{(0)}, \\ \ldots\ldots\ldots\ldots\ldots\ldots\ldots\ldots\ldots \\ \alpha_{n-2} + \bar{\alpha}_1\delta_{n-2} &= A_{n-2}^{(0)}, \end{aligned}$$

$$A_\nu^{(0)} = \frac{1}{2\pi}\int\limits_0^{2\pi} F_0(\sigma)\,e^{-i\nu\vartheta}\,d\vartheta.$$

4. Second fundamental problem. Reduction to functional equations. In view of the high degree of analogy between the functional equations for the first and second fundamental problems, we shall touch briefly on bounded regions only.

On multiplying the boundary conditions:

$$\left.\begin{aligned} \varkappa\varphi(\sigma) - \frac{\omega(\sigma)}{\overline{\omega'(\sigma)}}\overline{\varphi'(\sigma)} - \psi(\sigma) &= \Phi(\sigma), \text{ [1]} \\ \overline{\varkappa\varphi(\sigma)} - \frac{\overline{\omega(\sigma)}}{\omega'(\sigma)}\varphi'(\sigma) - \psi(\sigma) &= \overline{\Phi(\sigma)} \end{aligned}\right\} \qquad (43)$$

by $\dfrac{1}{2\pi i}\,\dfrac{d\sigma}{\sigma - z}$ and integrating along the unit circle, we obtain:

$$\varkappa\varphi(z) - \frac{1}{2\pi i}\int\limits_{|\sigma|=1} \frac{\omega(\sigma)}{\overline{\omega'(\sigma)}}\,\frac{\overline{\varphi'(\sigma)}}{\sigma - z}\,d\sigma - \bar{\beta}_0 = C(z), \qquad (44)$$

$$-\psi(z) - \frac{1}{2\pi i}\int\limits_{|\sigma|=1} \frac{\overline{\omega(\sigma)}}{\omega'(\sigma)}\,\frac{\varphi'(\sigma)}{\sigma - z}\,d\sigma + \bar{\alpha}_0\varkappa = D(z),$$

$$C(z) = \frac{1}{2\pi i}\int\limits_{|\sigma|=1} \frac{\Phi(\sigma)}{\sigma - z}\,d\sigma, \quad D(z) = \frac{1}{2\pi i}\int\limits_{|\sigma|=1} \frac{\overline{\Phi(\sigma)}}{\sigma - z}\,dz.$$

---

[1]) We recall that by $\varkappa$ we understand the quantity:

$$\varkappa = \frac{\lambda + 3\mu}{\lambda + \mu},$$

where λ and μ are positive constants depending on the elastic properties of the body under study. Hence it is clear that $\varkappa$ is always a positive quantity.

By using the relation:

$$\frac{1}{2\pi i}\int_{|\sigma|=1}\frac{\overline{\varphi'(\sigma)}}{\overline{\omega'(\sigma)}}\,\frac{d\sigma}{\sigma-z}=\frac{\overline{\varphi'(\sigma)}}{\overline{\omega'(0)}}=k,$$

we transform the first of the equations into the following:

$$\varkappa\varphi(z)-\frac{1}{2\pi i}\int_{|\sigma|=1}\frac{\omega(\sigma)-\omega(z)}{\sigma-z}\cdot\frac{\overline{\varphi'(\omega)}}{\overline{\omega'(\sigma)}}\,d\sigma-k\omega(z)-\bar{\beta}_0=C(z). \quad (45)$$

We introduce the new unknown function $\varphi_0(z)$, defined by the equation:

$$\varkappa\varphi_0(z)=\varkappa\varphi(z)-k\omega(z).$$

Substituting in the preceding equation this value for $\varphi(z)$, we obtain the following equation for $\varphi_0(z)$:

$$\varkappa\varphi_0(z)-\frac{1}{2\pi i}\int_{|\sigma|=1}\frac{\omega(\sigma)-\omega(z)}{\sigma-z}\,\frac{\overline{\varphi_0'(\sigma)}}{\overline{\omega'(\sigma)}}\,d\sigma-\bar{\beta}_0=C(z).$$

From this by differentiation and passing to the limit on the circumference of the circle we find the integral equation for the contour values of $\varphi_0'(z)$:

$$\varkappa\varphi_0'(z_0)-\frac{1}{2\pi i}\int_{|\sigma|=1}K(z_0,\sigma)\overline{\varphi_0'(\sigma)}\,d\sigma=C'(z_0). \quad (46)$$

After having determined $\varphi_0'(z_0)$ from it and recovered by it the function $\varphi_0(z)$, we must still find the number k. Since $\varphi(z)=\varphi_0(z)+\frac{k}{\varkappa}\,\omega(z)$, for k we obtain the equation:

$$k-\frac{\bar{k}}{\varkappa}=\frac{\overline{\varphi_0'(0)}}{\overline{\omega'(0)}},$$

which, since $\varkappa\neq\pm 1$, is always soluble for k.

Chapter VII

SCHWARZ'S METHOD

§ 1. SCHWARZ'S METHOD FOR THE SOLUTION OF THE DIRICHLET PROBLEM FOR THE SUM OF TWO REGIONS

1. Schwarz's method in the general case. Investigation of the convergence. In the sixties of the last century there was developed by Schwarz a method — called by him "alternating" — of solving the Dirichlet problem for harmonic functions.

It consists in the following: in the plane $z = x + iy$ let there be given two regions B_1 and B_2. We shall consider them to overlap one another, and call B' the common part of these regions (the intersection of the regions B_1 and B_2 in the meaning of the theory of sets). How the Dirichlet problem for harmonic functions is solved in each of the regions B_1 and B_2, for any continuous or piece-wise continuous boundary data, we shall consider as known. Schwarz showed how, by the successive solution of the problems for each of the regions, we can solve the Dirichlet problem for the region B obtained by combining both of the given regions (the sum of the regions B_1 and B_2).

Just how the approximations to the solution of the Dirichlet problem in $B = B_1 + B_2$ are constructed by solving this problem in each of the component regions, we shall show below.

For us it is important to note that Schwarz's method permits one to obtain — by known solutions of the Dirichlet problem for a comparatively narrow class of particular regions, by combining these regions — solutions for regions of a more and more complex character. At least in principle, it gives us the unlimited possibility of extending the class of regions for which the explicit solution of the first boundary-value problem can be constructed.

In its idea, and in its computations too, in some cases, Schwarz's method is comparatively simple; despite this, however, it has not obtained a wide enough dissemination, the occasions of its application to the solution of practical problems having until now been solitary ones [1]).

Schwarz himself and many subsequent authors who continued his works considered the application of the alternating method to the

---

[1]) S. L. Sobolev, [2]; S. G. Mikhlin, [4].

Laplace equation

$$\Delta u = \frac{\partial^2 u}{\partial x^2} + \frac{\partial^2 u}{\partial y^2} = 0,$$

and solved by means of it the first boundary-value problem, of Dirichlet. It was perfectly clear, however, that the method was applicable to the determination of functions satisfying not only the Laplace equation, but also more general equations, and not only to the first fundamental problem, when on the boundary of the region are given the values of the function sought, but also to problems with more complicated boundary conditions.

In this No. we will consider the basis of Schwarz's method for the first boundary-value problem in the theory of differential equations of the second order of general form [1]).

We will consider a partial differential equation of the form:

$$F\left(x, y, u, \frac{\partial u}{\partial x}, \frac{\partial u}{\partial y}, \frac{\partial^2 u}{\partial x^2}, \frac{\partial^2 u}{\partial x \partial y}, \frac{\partial^2 u}{\partial y^2}\right) = 0. \tag{1}$$

Let B be some region of the plane $z = x + iy$ bounded by the contour L, and let $f(M)$ be a piece-wise continuous function given at the points M of the contour L [2]).

By the first boundary-value problem, or the Dirichlet problem, for equation (1), we understand the following problem:

To find, the function $u(x, y)$, subject to the requirements: 1) $u(x, y)$ is bounded in B; 2) within B, u satisfies equation (1); 3) at every point M of continuity of $f(M)$, u has boundary values equal to the value of $f(M)$ at this point.

With respect to the equation (1) itself, we make the following

ASSUMPTION I. *Let D be any region of the xy-plane. Two functions $u(x, y)$ and $u^*(x, y)$ satisfying equation* (1) *in D, bounded and having identical values on the boundary of the region D, except, perhaps, at a finite set of points, are equal to each other everywhere in D.*

We shall consider that for equation (1) in region B_1 and B_2 there exists a solution of the Dirichlet problem for any piece-wise continuous values of $u(x, y)$ on the boundaries L_1 and L_2.

We will call $\bar{\alpha}$ that part of the boundary L_1 of the region B_1 which lies within B_2, and α the remaining part of L_1. Analogously we will call

---

[1]) We have spoken here of differential equations only because the book in hand is devoted to the solution of them. For Schwarz's method it is quite immaterial whether the functions under study satisfy a differential or some other equation. The fact only is essential that the class of functions being considered satisfies the five requirements enumerated below. Therefore the result we have obtained is in force not only for differential equations, but also for problems of other kinds, everywhere where our requirements are met.

[2]) Here and below we call a function piece-wise continuous if it is bounded and has a finite number of points of discontinuity.

$\bar{\beta}$ that part of the boundary of the region L_2 which lies within B_1, and β the remaining part.

To simplify the ensuing reasoning we will consider everywhere below that α, $\bar{\alpha}$, β, $\bar{\beta}$ consist of a finite number of arcs of Jordan curves.

The boundary of the sum $B = B_1 + B_2$ of the given regions will obviously be $\alpha + \beta$.

We will assume that at the points of the boundary L of the region B there is given the piece-wise continuous function $f(M)$, and set ourselves the task of finding in this region the solution of equation (1) under the boundary condition

$$u(x, y) = f(M) \text{ on } L. \tag{2}$$

We shall now give a description of the alternating process of Schwarz, which permits the construction of successive approximations to the solution of the problem in B_1 and B_2.

The proof of the convergence of the successive approximations and that the limit function satisfies equation (1) and boundary condition (2) will, as stands to reason, require of us further assumptions about equation (1). We will give them below, when we are investigating this question.

Let us begin with the region B_1. The boundary values here are given only on the part α of its contour L_1. Let us assign arbitrarily on $\bar{\alpha}$ a function $\varphi(M)$, subjecting it to the sole condition that together with the values of $f(M)$ on α it give a piece-wise continuous function on the entire contour of the region B_1.

Let us construct the function $u_1(x, y)$, solving the Dirichlet problem for equation (1) in B_1 under the boundary condition

$$u_1(x, y) = \begin{cases} f(M) & \text{on } \alpha, \\ \varphi(M) & \text{,, } \bar{\alpha}. \end{cases} \tag{3}$$

We will adopt it as the first approximation to $u(x, y)$ in B_1. By the $u_1(x, y)$ found, we construct the function $v_1(x, y)$ solving in B_2 the Dirichlet problem under the boundary conditions

$$v_1(x, y) = \begin{cases} f(M) & \text{on } \beta, \\ u_1(x, y) & \text{,, } \bar{\beta}, \end{cases} \tag{4}$$

and $v_1(x, y)$ is the first approximation to $v(x, y)$ in B_2. By $v_1(x, y)$ we construct the second approximation $u_2(x, y)$ to $u(x, y)$ in B_1, as the solution of the Dirichlet problem for (1) under the boundary condition:

$$u_2(x, y) = \begin{cases} f(M) & \text{on } \alpha, \\ v_1(x, y) & \text{,, } \bar{\alpha}, \end{cases}$$

and so forth.

The kth approximations, $u_k(x, y)$ and $v_k(x, y)$, to $u(x, y)$ in regions B_1 and B_2 we determine in terms of the preceding approximations as solutions of the Dirichlet problems under the boundary requirements

$$u_k(x, y) = \begin{cases} f(M) & \text{on } \alpha, \\ v_{k-1}(x, y) & ,, \ \bar{\alpha}, \end{cases}$$

$$v_k(x, y) = \begin{cases} f(M) & \text{on } \beta, \\ u_k(x, y) & ,, \ \bar{\beta}. \end{cases}$$

In each of the regions B_1 and B_2 we have constructed a sequence of approximations to the sought function $u(x, y)$:

$$\left.\begin{array}{l} u_1(x, y),\ u_2(x, y),\ \ldots,\ u_n(x, y) \ldots \text{ in } B_1 \\ v_1(x, y),\ v_2(x, y),\ \ldots,\ v_n(x, y) \ldots \ ,, \ B_2 \end{array}\right\}. \tag{5}$$

We will undertake an investigation of the convergence of sequences (5) and determine whether the limit functions satisfy equation (1) and the boundary requirements that we impose. We can succeed in doing this under certain limiting assumptions about equation (1).

With respect to it we will make the following additional assumptions.

ASSUMPTION II. *Let D be some region of the xy-plane with the contour Γ; let $u(x, y)$ and $u^*(x, y)$ be two bounded functions satisfying equation (1) in region D and having on its boundary certain boundary values, perhaps excepting a finite number of points. We shall consider that if on the contour Γ there is fulfilled the condition:*

$$u^*(x, y) \geqslant u(x, y),$$

then everywhere in the region D we will also have

$$u^*(x, y) \geqslant u(x, y).$$

ASSUMPTION III. *In D is given a sequence of solutions of equation (1):*

$$u_1(x, y),\ u_2(x, y),\ \ldots,\ u_n(x, y),\ \ldots$$

We will consider this sequence to be monotonically increasing or decreasing, i.e., such that for all x and y of D and for any n there is fulfilled either the inequality

$$u_{n+1}(x, y) \geqslant u_n(x, y),$$

or

$$u_{n+1}(x, y) \leqslant u_n(x, y).$$

In addition we will assume that *the functions $u_n(x, y)$ are uniformly bounded*:

$$|u_n(x, y)| \leqslant M.$$

Our sequence, in view of its monotonicity and boundedness, will converge everywhere in D:

$$\lim_{n\to\infty} u_n(x, y) = u(x, y).$$

We will consider — and the Third Assumption consists in this — that *the limit function also satisfies equation* (1) *everywhere within the region.*

Putting it briefly, we consider that *the limit of any monotone and bounded sequence of solutions of equation* (1) *is also a solution of equation* (1).

ASSUMPTION IV. We state it in circumscribed form, referring to a narrow class of differential equations.

Let $u(x, y)$ be a solution of equation (1) *in region D, bounded, and having boundary values everywhere on Γ, perhaps excepting a finite number of points. We will assume that $u(x, y)$ cannot have within D either a positive maximum or negative minimum.* Putting it differently, *if for all boundary values of $u(x, y)$ on the border Γ of the region D there are fulfilled the inequalities*

$$-h \leqslant u(x, y) \leqslant +g, \qquad (h, g \geqslant 0),$$

then everywhere within D there will be fulfilled the inequalities

$$-h \leqslant u(x, y) \leqslant +g.$$

ASSUMPTION V. Let us consider a certain arc γ, part of the boundary Γ of the region D. On γ let there be given a continuous function $f(M)$ and let P be an interior point of the arc γ. Consider a bounded solution $u(x, y)$ of equation (1). We will assume that *if $u(x, y)$ has at all points of the arc γ, excluding, perhaps, the point P only, boundary values equal to $f(M)$, then with the approach of the point (x, y) to P, $u(x, y)$ will also have a limit value, and this limit value will be precisely $f(P)$.*

The idea of the reasoning to follow consists in the following. We will construct, for sequences (5) of a particular kind, majorant and minorant sequences, and will show that they converge to the same limits. From this it will be clear that sequences (5) will converge to the same limits.

We will begin with the construction of a dominant sequence. Let us denote by m the exact upper bound of the values taken by $|\varphi(M)|$ on $\bar{\alpha}$ and by $f(M)$ on L:

$$m = \max\,[\sup|\varphi(M)|,\ \sup f(M)],$$

and construct, solving the Dirichlet problem in B_1 and B_2, the sequences of functions $u_n^+(x, y)$ and $v_n^+(x, y)$, by the rule:

$$u_1^+(x, y) = \begin{cases} f(M) & \text{on } \alpha, \\ m & \text{,, } \bar{\alpha}, \end{cases}$$

$$v_1^+(x, y) = \begin{cases} f(M) & \text{on } \beta, \\ u_1^+(x, y) & \text{,, } \bar{\beta}, \end{cases}$$

. . . .

$$u_n^+(x, y) = \begin{cases} f(M) & \text{on } \alpha, \\ v_{n-1}^+(x, y) & ,, \ \bar{\alpha}, \end{cases}$$

$$v_n^+(x, y) = \begin{cases} f(M) & \text{on } \beta, \\ u_n^+(x, y) & ,, \ \bar{\beta}. \end{cases}$$

We shall now show that $u_n^+(x, y) \geqslant u_n(x, y)$ and $v_n^+(x, y) \geqslant v_n(x, y)$. Indeed, by the choice of the boundary values for $u_1^+(x, y)$, everywhere on L_1 $u_1^+(x, y) \geqslant u_1(x, y)$. But then, by the second assumption, we will have everywhere in B_1:

$$u_1^+(x, y) \geqslant u_1(x, y),$$

and in particular, on $\bar{\beta}$ $v_1^+(x, y) = u_1^+(x, y) \geqslant u_1(x, y) = v_1(x, y)$, and since on β $v_1^+ = v_1 = f(M)$, it follows that everywhere on the boundary L_2 of the region B_2 $v_1^+ \geqslant v_1$. By virtue of the same assumption we will have everywhere in the region B_2

$$v_1^+(x, y) \geqslant v_1(x, y).$$

Continuing this reasoning, we shall have established that in B_1, or in B_2 respectively, there are fulfilled the inequalities:

$$u_n^+(x, y) \geqslant u_n(x, y) \quad \text{or} \quad v_n^+(x, y) \geqslant v_n(x, y).$$

The minorant sequence is constructed analogously. We define $u_n^-(x, y)$, $v_n^-(x, y)$ by the conditions:

$$u_1^-(x, y) = \begin{cases} f(M) & \text{on } \alpha, \\ -m & ,, \ \bar{\alpha}, \end{cases}$$

$$v_1^-(x, y) = \begin{cases} f(M) & \text{on } \beta, \\ u_1^-(x, y) & ,, \ \bar{\beta}, \end{cases}$$

. . . .

$$u_n^-(x, y) = \begin{cases} f(M) & \text{on } \alpha, \\ v_{n-1}^-(x, y) & ,, \ \bar{\alpha}, \end{cases}$$

$$v_n^-(x, y) = \begin{cases} f(M) & \text{on } \beta, \\ u_n^-(x, y) & ,, \ \bar{\beta}. \end{cases}$$

By reasoning like that in the case of the dominant functions u_n^+ and v_n^+, it can be shown that for u_n^- and v_n^- there are fulfilled, in the regions B_1 and B_2 respectively, the inequalities:

$$u_n^-(x, y) \leqslant u_n(x, y), \quad v_n^-(x, y) \leqslant v_n(x, y).$$

Now we will investigate the convergence of both the auxiliary sequences. First we will consider the majorant sequence. By the choice of the number m, the boundary values of $u_1^+(x, y)$ on L_1 do not exceed m.

Therefore by virtue of the fourth assumption, $u_1^+(x, y) \leqslant m$ everywhere within B_1. In particular, $u_1^+(x, y) \leqslant m$ on $\bar{\beta}$. The boundary values of $v_1^+(x, y)$, equal to $f(M)$ on β and $u_1^+(x, y)$ on $\bar{\beta}$, will also be not greater than m. Consequently $v_1^+(x, y) \leqslant m$ everywhere in B_2. In particular this inequality will be fulfilled on $\bar{\alpha}$. Let us compare the boundary values of $u_1^+(x, y)$ and $u_2^+(x, y)$ with each other. On α they coincide and are equal to $f(M)$; on $\bar{\alpha}$ $u_2^+(x, y) = v_1^+(x, y) \leqslant m$, and $u_1^+(x, y) = m$. Accordingly everywhere on the boundary L_1 $u_1^+(x, y) \geqslant u_2^+(x, y)$. By the third assumption $u_1^+(x, y)$ will be greater than $u_2^+(x, y)$ everywhere in B_1:

$$u_1^+(x, y) \geqslant u_2^+(x, y).$$

From this and from the boundary conditions:

$$v_2^+(x, y) - v_1^+(x, y) = \begin{cases} 0 & \text{on } \beta, \\ u_2^+(x, y) - u_1^+(x, y) & \text{on } \bar{\beta}, \end{cases}$$

. . . .

$$u_{n+1}^+(x, y) - u_n^+(x, y) = \begin{cases} 0 & \text{on } \alpha, \\ v_n^+(x, y) - v_{n-1}^+(x, y) & \text{on } \bar{\alpha}, \end{cases}$$

$$v_{n+1}^+(x, y) - v_n^+(x, y) = \begin{cases} 0 & \text{on } \beta, \\ u_{n+1}^+(x, y) - u_n^+(x, y) & \text{on } \bar{\beta}. \end{cases}$$

using the second assumption, the following inequalities can be derived in succession:

$$u_1^+(x, y) \geqslant u_2^+(x, y) \geqslant \ldots$$

$$v_1^+(x, y) \geqslant v_2^+(x, y) \geqslant \ldots,$$

establishing that the sequences u_n^+ and v_n^+ are monotonically decreasing. Analogously, from the fact that a solution of equation (1) cannot have negative minimum within the region, one can arrive at the conclusion that the sequences u_n^- and v_n^- are monotonically increasing:

$$u_1^-(x, y) \leqslant u_2^-(x, y) \leqslant \ldots$$

$$v_1^-(x, y) \leqslant v_2^-(x, y) \leqslant \ldots$$

Earlier we saw that $u_n^+ \geqslant u_n \geqslant u_n^-$, and accordingly

$$u_n^+(x, y) \geqslant u_1^-(x, y),$$

and also

$$v_n^+(x, y) \geqslant v_1^-(x, y).$$

Thus it turns out that the sequences u_n^+ and v_n^+ are monotonically decreasing and bounded below. They will converge in the regions B_1 and B_2

respectively to some solutions of equation (1):

$$\lim_{n\to\infty} u_n^+(x, y) = u^+(x, y),$$

$$\lim_{n\to\infty} v_n^+(x, y) = v^+(x, y).$$

We will now show that in the common part $B' = B_1B_2$ of the regions B_1 and B_2 both limit functions coincide:

$$u^+(x, y) = v^+(x, y),$$

and consequently that the function $v^+(x, y)$ will be the continuation of $u^+(x, y)$ from B_1 to the region B_2. In addition we will show that the function $u^+(x, y)$ thus continued will be a solution of our boundary-value problem in the region $B = B_1 + B_2$.

On $\bar{\alpha}$ and $\bar{\beta}$ the functions $u_n^+(x, y)$ and $v_n^+(x, y)$ are connected by the relations:

$$u_n^+(x, y) = v_{n-1}^+(x, y) \text{ on } \bar{\alpha},$$

$$u_n^+(x, y) = v_n^+(x, y) \quad ,, \quad \bar{\beta}.$$

If we pass to the limit here as $n \to \infty$, it will be evident that $u^+(x, y)$ and $v^+(x, y)$ have identical values on $\bar{\alpha}$ and $\bar{\beta}$ [1]).

---

[1]) The reasoning showing the possibility of passing to the limit in the equations exhibited in the text has certain peculiarities, and we will therefore indicate it in brief outline.

Let the point M lie on $\bar{\alpha}$. Let us consider the inequality $u^+(x, y) \leqslant u_n^+(x, y)$ and in it move the point (x, y) to M:

$$\overline{\lim_{(x,y)\to M}}\; u^+(x, y) \leqslant \lim_{(x,y)\to M} u_n^+(x, y) = v_{n-1}^+(M).$$

This inequality is true for all n. Let us pass to the limit as $n \to \infty$:

$$\overline{\lim_{(x,y)\to M}}\; u^+(x, y) \leqslant v^+(M). \tag{a}$$

Moreover let us construct the function $u^*(x, y)$, solving the Dirichlet problem in B_1, under the boundary condition

$$u^*(x, y) = \begin{cases} f(M) & \text{on } \alpha, \\ v^+(x, y) & ,, \ \bar{\alpha}. \end{cases}$$

On $\bar{\alpha}$ we shall have $v_{n-1}^+(x, y) \geqslant v^+(x, y)$. Therefore on the entire boundary L_1 there is fulfilled the inequality $u_n^+(x, y) \geqslant u^*(x, y)$.

According to the second assumption, everywhere in B_1 $u_n^+(x, y) \geqslant u^*(x, y)$. Hence it is evident that $u^+(x, y) \geqslant u^*(x, y)$. When the point (x, y) approaches M, it is seen from this inequality that

$$\underline{\lim_{(x,y)\to M}}\; u^+(x, y) \geqslant \lim_{(x,y)\to M} u^*(x, y) = v^+(M).$$

The last, however, is compatible with (a) only if $u^+(x, y)$ has at the point M a limit equal to $v^+(x, y)$.

Let us now consider their behavior on α and β. We take any interior point M of the part α on which the function $f(M)$ is continuous. We know that

$$u_1^+(x, y) \geqslant u^+(x, y) \geqslant u_1^-(x, y);$$

when the point (x, y) tends to the point M, both $u_1^+(x, y)$ and $u_1^-(x, y)$ will tend to $f(M)$ by their construction. Therefore so will $u^+(x, y)$ tend to $f(M)$ when the point $(x, y) \to M$.

The same thing can be said of $v^+(x, y)$. The boundary of B' consists of $\bar{\alpha}$ and $\bar{\beta}$ and perhaps also of a common part of α and β which we shall call γ. In Fig. 63 is depicted the case when B_1 and B_2 are the two overlapping rectangles $ACDH$ and $BCEF$. Here $\bar{\alpha}$ is the segment DG and $\bar{\beta}$ the segment BG. B' is the rectangle $BCDG$. Its boundary consists of $\bar{\alpha}$, $\bar{\beta}$, and the point G and the broken line BCD, which are the common part of α and β. In particular cases it may turn out that γ consists of several points that are the ends of the arcs of the curves figuring in $\bar{\alpha}$ and $\bar{\beta}$. When the point (x, y) approaches an interior point of the part $\bar{\alpha}$ or $\bar{\beta}$ of the boundary of B', $u^+(x, y)$ and $v^+(x, y)$ will tend to one and the same limit. In exactly the same way so will $u^+(x, y)$ and $v^+(x, y)$ tend to the same limit when (x, y) approaches an interior point of γ that is not a place where $f(M)$ is discontinuous.

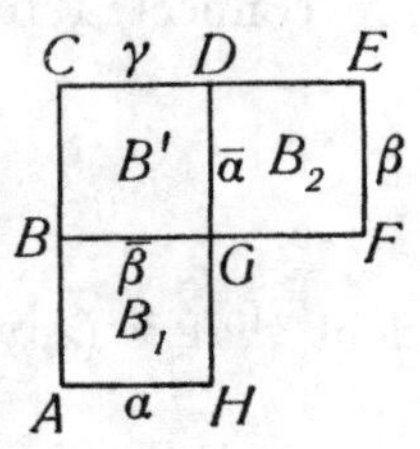

Fig. 63.

We can say nothing about the boundary values of $u^+(x, y)$ and $v^+(x, y)$ at points of discontinuity of $f(M)$ on γ (there is a finite number of such points) and at points that are the ends of the arcs of the curves forming $\bar{\alpha}$ and $\bar{\beta}$ (of them there is also a finite number). Thus $u^+(x, y)$ and $v^+(x, y)$ have identical boundary values everywhere on the boundary of B', perhaps excepting a finite number of points. By Assumption I, it follows from this that everywhere in the region B'

$$u^+(x, y) = v^+(x, y).$$

The function $v^+(x, y)$ is the continuation of $u^+(x, y)$ from the region B_1 to the region B_2. Moreover, from the reasoning it is clear that the solution $u^+(x, y)$ continued thus will have the assigned boundary values $f(M)$ on the boundary of the region $B = B_1 + B_2$, perhaps excepting a finite number of points, which are the ends of the arcs of the curves of which $\bar{\alpha}$ and $\bar{\beta}$ are composed, and the places where $f(M)$ is discontinuous.

It can be shown analogously that the minorant sequences $u_n^-(x, y)$ and $v_n^-(x, y)$ converge in the regions B_1 and B_2 to certain limit functions which will be solutions of equations (1):

$$\lim_{n\to\infty} u_n^-(x, y) = u^-(x, y),$$

$$\lim_{n\to\infty} v_n^-(x, y) = v^-(x, y).$$

Moreover the limit functions will coincide in the common part B' of the regions B_1 and B_2, and the solution of the equation, $v^-(x, y)$, will accordingly be the continuation of the solution $u^-(x, y)$ from B_1 to B_2. Finally, the continued function $u^-(x, y)$ will acquire the assigned values $f(M)$ on the boundary of the region B, perhaps excepting the finite number of points of discontinuity of $f(M)$ and the ends of the arcs forming $\bar{\alpha}$ and $\bar{\beta}$.

The continued solutions $u^+(x, y)$ and $u^-(x, y)$ have identical values on the boundary L, with the possible exception of a finite number of points. But then they must coincide everywhere in B:

$$u^+(x, y) = u^-(x, y) = u(x, y).$$

Thus it turns out that the majorant and minorant sequences u_n^+, v_n^+, u_n^-, v_n^- converge in B_1 and B_2 respectively to the same solution $u(x, y)$. From this it is clear that the original successive approximations (5) constructed by us will converge to the same solution $u(x, y)$:

$$\lim_{n\to\infty} u_n(x, y) = u(x, y) \text{ in } B_1,$$

$$\lim_{n\to\infty} v_n(x, y) = u(x, y) \text{ in } B_2.$$

It remains for us to verify that the limit function $u(x, y)$ gives the solution of the posed Dirichlet problem in the region B, i.e., that at any point of continuity of $f(M)$, $u(x, y)$ has a boundary value on L equal to $f(M)$. At all points of continuity of $f(M)$ that are not the ends of the arcs of $\bar{\alpha}$ and $\bar{\beta}$, $u(x, y)$ has boundary values equal to $f(M)$, as we have seen earlier. It remains only to verify that if the point P is the end of some arc of $\bar{\alpha}$ or $\bar{\beta}$ and if at this point $f(M)$ is continuous, then in approaching this point $u(x, y)$ will tend to $f(P)$. Of such points there is a finite set. Since there is also a finite number of points of discontinuity of $f(M)$, about every point P is located a part of the boundary L that contains the point P and does not contain any point of discontinuity of $f(M)$ and not one end of the arcs of $\bar{\alpha}$ and $\bar{\beta}$ except P.

For all the points M of this arc, excepting perhaps the point P only, $u(x, y)$ will tend to $f(M)$ when the point $x(, y)$ approaches M.

By Assumption V, this will also be the case at the point P. With this it has been proved that $u(x, y)$ will have the necessary boundary values everywhere on L and that it is, consequently, the sought solution of the Dirichlet problem in B.

We will make one more observation on the last, the fifth, assumption. We have utilized it at the end of the reasoning in proving that the limit function $u(x, y)$ acquires the assigned values at the ends of the arcs of $\bar{\alpha}$ and $\bar{\beta}$. It will be superfluous if one assumes the existence of the solution of the Dirichlet problem in the region B and investigates only the question of the convergence of Schwarz's process and the proof that the limit function will coincide with the solution whose existence is assumed.

Indeed, from the first four assumptions it follows that the successive approximations u_n and v_n constructed according to Schwarz will converge to some limit function $u(x, y)$ that will satisfy equation (1) and acquire values equal to $f(M)$ at all the points of the boundary L of the region B, perhaps excepting only the points of discontinuity of $f(M)$ and the ends of the arcs figuring in $\bar{\alpha}$ and $\bar{\beta}$. In any case the boundary values of $u(x, y)$, except at a finite number of points, coincide with the boundary values of the sought solution of the Dirichlet problem. By the first of the assumptions we can say that the limit function $u(x, y)$ will be identically equal to the sought solution everywhere within B.

2. The case of a linear equation of elliptic type. An estimate of the rapidity of convergence of Schwarz's process for the Laplace equation. We shall now consider the possibility of employing the alternating method of Schwarz for the solution of the Dirichlet problem for linear equations of elliptic type. To simplify the reasoning we will limit ourselves to an investigation of an equation of the form:

$$\frac{\partial^2 u}{\partial x^2} + \frac{\partial^2 u}{\partial y^2} - q(x, y)u = 0, \tag{6}$$

although the results remain in force for equations of a more general form than (6).

With respect to the coefficient $q(x, y)$, we will assume that it is non-negative and continuous:

$$q(x, y) \geqslant 0. \tag{7}$$

We shall now obtain a relation for the functions satisfying equation (6) that will be useful to us in what is to follow.

We choose in the region B an arbitrary point $M_0(x_0, y_0)$ and describe about it a circle of radius R lying in the region D.

We can regard equation (6) as the Poisson equation $\Delta u = q(x, y)u$ with free term $q(x, y)u$.

Any solution of it is representable in the form of the sum of two functions $u = v + w$, the first of which, v, satisfies the Laplace equation $\Delta v = 0$ and acquires on the boundary of the region the same values as does u, and the second of which, w, is a solution of the Poisson equation $\Delta w = q(x, y)u$ and vanishes on the boundary of the region.

In the circle constructed by us, v can be represented as a Poisson integral [Chap. VI, § 2, No. 1, (1)]. If we introduce polar coordinates r, φ with pole at M_0 and with polar axis parallel to x-axis, so that

$$x = x_0 + r \cos \varphi, \quad y = y_0 + r \sin \varphi,$$

the Poisson integral will have the form:

$$v(x, y) = \frac{1}{2\pi} \int_{-\pi}^{+\pi} u(x_0 + R \cos \psi, y_0 + R \sin \psi) \frac{R^2 - r^2}{R^2 - 2Rr \cos (\varphi - \psi) + r^2} d\psi.$$

The function w can be represented [see Chap. VI § 2 No. 6 (27)] in terms of Green's function for the circle by means of the following integral:

$$w(x, y) = -\frac{1}{2\pi}\int_{-\pi}^{+\pi}\int_{0}^{R} qu \ln\frac{|R^2 - \bar{t}z|}{R\,|t - z|}\,\varrho\, d\varrho\, d\psi,\ t = \varrho e^{i\psi},\ z = re^{i\varphi}.$$

Therefore the function $u(x, y)$ can be represented in the circle by the following sum of two integrals:

$$u(x, y) = \frac{1}{2\pi}\int_{-\pi}^{+\pi} u \frac{R^2 - r^2}{R^2 - 2Rr\cos(\varphi - \psi) + r^2}\, d\psi - \\ - \frac{1}{2\pi}\int_{-\pi}^{+\pi}\int_{0}^{R} qu \ln\frac{|R^2 - \bar{t}z|}{R\,|t - z|}\,\varrho\, d\varrho\, d\psi. \quad (8)$$

Integral equation (8) for the function u in the circle is equivalent to the fact that in this circle it satisfies differential equation (6).

From this equation it is also seen that on the circumference of the circle u acquires values equal to those that stand under the Poisson integral sign. But $u(x_0 + R\cos\varphi,\ y_0 + R\sin\varphi)$ can be assigned arbitrarily and consequently, by the nature of the case, equation (8) imposes no limitations of any kind on the boundary values of u.

To establish whether Schwarz's method is applicable to equation (6), we must verify that for it are fulfilled all the assumptions that we formulated in No. 1.

We will begin with the fourth of them and show that, given the observance of condition (7), the solution of equation (6) cannot have within the region either positive maxima or negative minima [1]). The case of a negative minimum reduces to the case of a positive maximum with the multiplication of $u(x, y)$ by -1, and it is therefore sufficient to establish the absence of positive maxima for $u(x, y)$ within D.

Let us examine that part of the region D where $u(x, y) > 0$, and consider that the circle of radius R constructed by us above belongs to this part. In formula (8), since for any z and t less than R in modulus, $\frac{|R^2 - \bar{t}z|}{R\,|t - z|} > 1$, the function under the sign of the double integral will be non-negative. Therefore $u(x, y)$ will not exceed the function represented by the Poisson integral. Let us consider $u(x, y)$ at the

[1]) For the reader familiar with the theory of subharmonic functions, this fact is obvious. It is known that if the function $u(x, y)$ is twice continuously differentiable, the criterion of its subharmonicity is the non-negativeness of the Laplace operation on it: $\Delta u \geqslant 0$. Therefore in those parts of the plane where $u \geqslant 0$, a function satisfying the equation $\Delta u = qu$ will be subharmonic, and where $u \leqslant 0$ will be superharmonic. In addition, it is known that a subharmonic function cannot have maxima within the region, nor a superharmonic function minima. Therefore the function satisfying the equation $\Delta u = qu$ cannot have within the region either positive maxima or negative minima.

center of the circle and set $r = 0$. We then obtain:

$$u(x_0, y_0) \leqslant \frac{1}{2\pi} \int_{-\pi}^{+\pi} u(x_0 + R\cos\psi,\ y_0 + R\sin\psi)\, d\psi.$$

This last inequality speaks of the fact that the value of $u(x, y)$ at the center of the circle will not be greater than the arithmetic mean of its values on the circumference of the circle. From this we easily obtain the result we need. Indeed, let us assume that $u(x, y)$ has a positive maximum at the point $M_0(x_0, y_0)$. We can always consider it proper in the sense that in any neighborhood of the point M_0 there will be found values of $u(x, y)$ less than $u(x_0, y_0)$. Let us replace in the last inequality the integrand $u(x_0 + R\cos\psi,\ y_0 + R\sin\psi)$ by $u(x_0, y_0)$. Since it is nowhere greater than $u(x_0, y_0)$, and we can always so choose the circumference that points are to be found on it at which the integrand will be less than $u(x_0, y_0)$, it follows that with such a substitution the right side of the inequality is increased, and after the substitution we shall have:

$$u(x_0, y_0) < \frac{1}{2\pi} \int_{-\pi}^{+\pi} u(x_0, y_0)\, d\psi = u(x_0, y_0).$$

The last inequality is obviously impossible. Thus the assumption that $u(x, y)$ has a positive maximum leads to an impossible consequence and must be discarded.

In the particular case, it follows from this that if $u(x, y)$ vanishes on the contour of the region, then $u(x, y) = 0$ everywhere within the region. Indeed, if we assume that $u(x, y)$ is not identically zero within D, then $u(x, y)$ must have within D either a positive maximum or a negative minimum, which cannot be.

Let us now verify Assumption I. Let $u(x, y)$ and $u^*(x, y)$ be two solutions bounded in D and acquiring identical values on the boundary Γ, with the exception, perhaps, of a finite set of points. We must show that $u(x, y) = u^*(x, y)$ everywhere in D.

We will resolve each of these functions into two summands by the following rule.

Put $u = v + w$, where w satisfies the Poisson equation $\Delta w = qu$ and vanishes at all points of the contour Γ, and v is a harmonic function: $\Delta v = 0$. With this it is seen that at all points of Γ where u has boundary values, v also has boundary values equal to the boundary values of u [1]).

[1]) In our reasoning here and several pages below, we assume the possibility of such a resolution. Putting it differently, we consider that the Poisson equation $\Delta w = q(x, y)u$ has a solution vanishing on the contour Γ. We will not occupy ourselves with an investigation of this question, noting only that for the existence of the desired solution it is sufficient to require that $q(x, y)$ satisfy the Lipschitz condition with some exponent, and that the region D be such that Green's function exists for it. For the overwhelming majority of the problems encountered practically, both of these requirements will certainly be met.

In exactly the same way we put

$$u^* = v^* + w^*,$$

where v^* and w^* possess analogous properties.

Let us now consider the difference between u and u^*:

$$u - u^* = (v - v^*) + (w - w^*).$$

$v - v^*$ is a function harmonic in D, which vanishes on L, with the possible exception of a finite number of points. It is easily seen that it is bounded, moreover. Indeed, since u and u^* are bounded on assumption, $u - u^*$ will be bounded. The functions w and w^* are continuous in D, including the boundary Γ; they will therefore both be bounded in D. Hence it is seen that $v - v^* = (u - u^*) - (w - w^*)$ will be a bounded function. But we know that if a harmonic function is bounded and vanishes on the contour of the region, with the possible exception of a finite set of points, it is identically equal to zero everywhere in D [1])

$$v - v^* = 0,$$

which means that

$$u(x, y) - u^*(x, y) = w(x, y) - w^*(x, y).$$

The function of interest to us, $u - u^*$, as the difference of two functions having null boundary values everywhere on the contour Γ, is equal to zero on the entire contour. On the other hand, it will obviously

---

[1]) It is simplest to convince oneself of this by the following reasoning. We shall consider for simplicity that D is simply connected, and perform the conformal transformation of it into the unit circle. Let us consider in D the harmonic function U. Let it be known that it is bounded and that given an approach to any point of Γ, with the exception of a finite set of points, it tends to zero. We must show that it is equal to zero everywhere in D.

After the transformation it will have passed to some function U_1, defined in the circle, bounded, and having null limit values everywhere on the circumference, except at a finite number of points.

Let us take a circle of radius $R < 1$ and represent the function U_1 in it by a Poisson integral

$$U_1(r, \varphi) = \frac{1}{2\pi} \int_{-\pi}^{+\pi} U_1(R, \psi) \frac{R^2 - r^2}{R^2 - 2Rr\cos(\varphi - \psi) + r^2} d\psi.$$

We fix r and φ in some way and increase R to unity. The kernel of the Poisson integral will tend to $\dfrac{1 - r^2}{1 - 2r\cos(\varphi - \psi) + r^2}$; the function $U_1(R, \psi)$, save at a finite set of values of ψ, will tend to zero. In view of the boundedness of $U_1(R, \psi)$, it will be possible to pass to the limit under the integral sign. Therefore it is clear that in the limit the integral will give zero, and consequently for any r and φ:

$$U_1(r, \varphi) = 0.$$

be a solution of equation (6). Several lines above we spoke of the fact that a solution of equation (6) vanishing on the boundary of a region must be equal to zero within the region too. Consequently

$$u(x, y) = u^*(x, y)$$

and for equation (6) Assumption I is fulfilled.

We turn to Assumption II. Let $u(x, y)$ and $u^*(x, y)$ be solutions of equation (6) satisfying on the boundary Γ of the region D the condition

$$u(x, y) \geqslant u^*(x, y).$$

We have to show that then $u(x, y) \geqslant u^*(x, y)$ everywhere in D. Let us take the difference between them, $u - u^*$. It satisfies equation (6) and on Γ $u - u^* \geqslant 0$. If it acquired negative values within the region D, it would have to have a negative minimum within the region, which cannot be, and accordingly we must have $u - u^* \geqslant 0$ everywhere in D.

Let us verify Assumption III, that the limit of any monotone and bounded sequence of solutions of equation (6) is also a solution of equation (6). In D let there be given a monotone and uniformly bounded sequence of solutions of (6):

$$u_1(x, y),\ u_2(x, y),\ \ldots,\ u_n(x, y),\ \ldots$$

We will call the limit function of this sequence $u(x, y)$:

$$\lim_{n\to\infty} u_n(x, y) = u(x, y).$$

We have to show that $u(x, y)$ satisfies equation (6).

In D let us take any point $M_0(x_0, y_0)$ and describe about it as center a circle of radius R lying wholly within D. Apply equation (8) to u_n:

$$u_n(x, y) = \frac{1}{2\pi}\int_{-\pi}^{+\pi} u_n \frac{R^2 - r^2}{R^2 - 2Rr\cos(\varphi - \psi) + r^2}\, d\psi - \\ - \frac{1}{2\pi}\int_{-\pi}^{+\pi}\int_0^R q u_n \ln \frac{|R^2 - \bar{t}z|}{R\,|t - z|}\, \varrho\, d\varrho\, d\psi.$$

Let us pass to the limit here as $n \to \infty$. Since u_n is uniformly bounded, passage to the limit under the integral sign is admissible. In the limit we find that for the limit function $u(x, y)$ equation (8) is fulfilled. In the derivation of equation (8), however, we remarked that its observance is equivalent to the fulfillment of equation (6). Accordingly the limit function actually satisfies equation (6).

Finally, the last assumption, V. We take an arc γ lying on the boundary Γ of the region D. Let there be given on it the continuous function $f(M)$, and P, some interior point of γ. We shall consider the bounded solution $u(x, y)$ of equation (6) in D and assume that for an approach to any point M of the arc, with the exception of P, $u(x, y)$ tends to $f(M)$. We must show that then, for an approach to the point P, we also have

$$\lim_{(x,y)\to M} u(x, y) = f(P).$$

Here there again comes to our aid the fact that the corresponding proposition for harmonic functions is known. We resolve the solution u taken by us into two terms

$$u = v + w,$$

where w satisfies the Poisson equation $\Delta w = qu$ and with an approach to the contour Γ everywhere tends to zero, and v is a harmonic function. With respect to v it can be said that at all points M where u has boundary values, v will also have boundary values, and they will be equal to the boundary values of u. In particular, on the arc γ it will have at all points, with the exception perhaps of P, boundary values equal to $f(M)$. In addition, v is bounded, since u is bounded on assumption, and w is bounded in view of its continuity in D up to and including the contour Γ. If v has the properties indicated, then one can affirm that with the approach of the point (x, y) to P, v will tend to $f(P)$ [1]).

---

[1]) This is obtained on the basis of the following theorem: let $u(x, y)$ be harmonic and bounded in D. On the boundary Γ of the region is given a continuous function $f(M)$, and it is known that, except at the point P perhaps, $\lim_{(x,y)\to M} u(x, y) = f(M)$. Then $u(x, y) \to f(P)$ with an approach to the point P.

The theorem can be proved by the following reasoning. For simplification we will consider D to be simply connected and effect the conformal transformation of it into the unit circle. The function $f(M)$ will become some function $F(M)$ defined at the points of the circumference. The image of the point P we will call P'. The function $u(x, y)$ becomes the function $U(x, y)$, bounded and harmonic in the circle. With an approach to any point of the circumference, excepting perhaps the point P', $U(x, y)$ will tend to $F(M)$.

If we carry through a passage to the limit like that we made in the preceding note, it can be shown that $U(x, y)$ is representable by a Poisson integral:

$$U(x, y) = \frac{1}{2\pi} \int_0^{2\pi} U(\psi) \frac{1 - r^2}{1 - 2r\cos(\varphi - \psi) + r^2} d\psi.$$

The values $U(\psi)$ acquired by the function $U(x, y)$ on the circumference of the circle, which stand under the integral sign, are equal to $F(M)$ everywhere, with the exception, perhaps, of the one point P'. Without altering the magnitude of the integral, we can substitute $F(M)$ for $U(\psi)$ in it:

$$U(x, y) = \frac{1}{2\pi} \int_0^{2\pi} F(M) \frac{1 - r^2}{1 - 2r\cos(\varphi - \psi) + r^2} d\psi.$$

It is known from the theory of the Poisson integral that if $F(M)$ is continuous at any point of the circumference, then for an approach to this point the integral tends to the value $F(M)$ at this point. In particular, since $F(M)$ is continuous at the point P', for the approach of (x, y) to P', $U(x, y)$ will tend to $F(P')$.

But since for an approach to any point of the contour Γ, w will tend to zero, we can affirm that

$$\lim_{(x,y)\to P} u(x, y) = f(P).$$

Thus it turns out that for equation (6) there are fulfilled all five assumptions sufficient for the convergence of Schwarz's alternating method.

Up to the present we have established only the mere fact of the convergence of the successive approximations constructed according to Schwarz, to the solution of the problem in the sum of two regions, not having concerned ourselves with the question of the estimation of the rate of the convergence of these approximations. It stands to reason that the rate of the convergence will depend first upon what equation is under consideration, and second, upon the geometric properties of the regions in which the successive approximations are constructed.

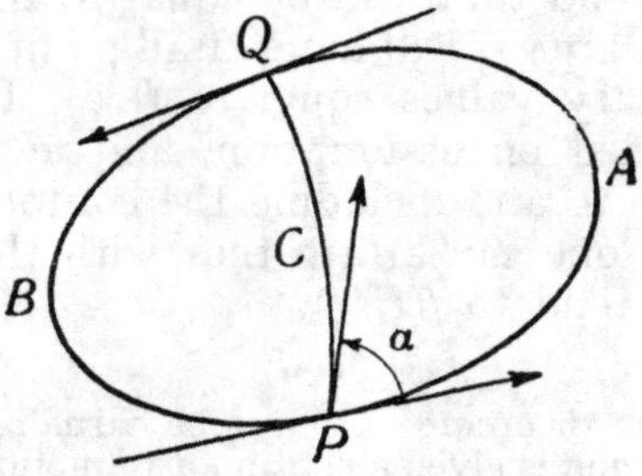

Fig. 64.

We will now make an estimate of the rate of the convergence for the Laplace equation. At the basis of it we place the following lemma, belonging to Schwarz.

Let D be a plane region bounded by the contour Γ. Take on Γ two simple[1]) points P and Q and consider that there exist tangents to Γ at these points. Draw, in the region D, any curve C joining the points P and Q. With respect to it we will assume that it has tangents at its ends P and Q which do not coincide with the tangents to the contour Γ at these points.

In addition we will consider that the section PQ of the region which we have made divides it into two new regions. We will call A and B the parts of Γ into which it is separated by the section drawn (Fig. 64).

Let u be a harmonic function having on the part A boundary values equal to zero, and on the part B boundary values not exceeding in magnitude a certain number m. It is then asserted that there exists a positive number $\vartheta < 1$, depending only on the geometrical properties of the region D and the curve C and not depending on the choice of the function u and the number m, such that on the curve C

$$|u| \leqslant \vartheta m.$$

For proof of the lemma we can limit ourselves to the case $m = 1$, since we could otherwise consider the function $\frac{u}{m}$ rather than u.

---

[1]) In No. 2 § 1 of Chap. V we spoke about what kind of points of the contour are called multiple. Any non-multiple point of the contour is called simple.

We introduce the function U, harmonic and bounded in D, having on A boundary values equal to zero and on B boundary values equal to unity.

On the contour Γ of the region D there is fulfilled the inequality

$$-U \leqslant u \leqslant +U.$$

On the principle of the maximum we can affirm that the same inequality is fulfilled within D too. Putting it differently, we can affirm that everywhere in D

$$|u| \leqslant U.$$

Everywhere within D the function U is less than unity. This will be the case, in particular, at all the interior points of the curve C. We shall have proved Schwarz's lemma if we can satisfy ourselves that for an approach to the points P and Q along the curve C, U will tend to limits less than 1.

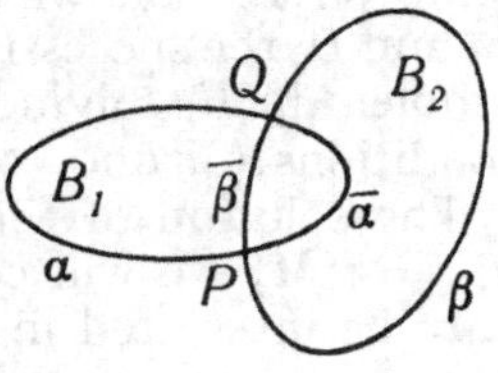

Fig. 65.

We will introduce polar coordinates r and φ with pole at P and direct the polar axis along the tangent to Γ on the side of A. We will take that branch of the angle φ that tends to zero for an approach to P along Γ from the side of A. The direction of reckoning the polar angle will be chosen so that for an approach to the point P along Γ from the side B the polar angle tends to the value $+\pi$.

The function $\frac{\varphi}{\pi}$ will be a regular harmonic function in D and will acquire on Γ continuous values, except at the point P, where there will be a saltus equal to unity. The function $U - \frac{\varphi}{\pi} = v$ will be harmonic in D. Its boundary values are continuous near the point P and tend to zero for an approach to this point. We can accordingly say that for an approach to this point P over any path from D, and along C in particular, v will tend to zero.

But if we approach P along C, then, since the tangent to C at P is different from the tangent to Γ at this point, which latter tangent has been taken as the polar axis, the angle φ will tend to some positive limit less than π. In our figure the limit value of the angle has been called α. Therefore for an approach to P along C, U will tend to the limit $\frac{\alpha}{\pi}$, lying between zero and unity.

The same thing can be said of the point Q, too. And with this the lemma is proved.

Below we will preserve the notation of No. 1. Let there be given two regions B_1 and B_2, bounded by the contours L_1 and L_2 and overlapping each other (Fig. 65). The part of L_1 lying within B_2 we denote by $\bar{\alpha}$, as before, and the remaining part of L_1 by α; the part of L_2 lying within B_1 we will call $\bar{\beta}$, and the remaining part β. To simplify the reasoning we will consider that $\bar{\alpha}$ and $\bar{\beta}$ each consists of a single arc. The ends of $\bar{\alpha}$ we will call $P_{\bar{\alpha}}$ and $Q_{\bar{\alpha}}$, and the ends of $\bar{\beta}$, $P_{\bar{\beta}}$ and $Q_{\bar{\beta}}$. $\bar{\alpha}$ will be some curve lying in the region B_2 and joining the points $P_{\bar{\alpha}}$ and $Q_{\bar{\alpha}}$, and an analogous statement can be made about $\bar{\beta}$.

For the case depicted in the figure, $P_{\bar{\alpha}}$ and $Q_{\bar{\alpha}}$ coincide with $P_{\bar{\beta}}$ and $Q_{\bar{\beta}}$ respectively, and are designated as P and Q.

We will assume that the ends of $\bar{\alpha}$ and $\bar{\beta}$ are simple points of the contours B_1 and B_2 and that at these points there exist tangents to L_1 and L_2 inclined to each other at an angle different from 0 and π.

We will assume that along the boundary $L = \alpha + \beta$ of the region $B = B_1 + B_2$ there are given piece-wise continuous boundary values $f(M)$. Let us supplement them by some piece-wise continuous values $\varphi(M)$ on $\bar{\alpha}$. Following Schwarz, we shall next construct in the regions B_1 and B_2 the successive approximations (5) to the solution of the Dirichlet problem in B, solving the boundary-value problems in accordance with conditions (3) and (4) of No. 1.

The solution of the Dirichlet problem in B for the boundary values $u|_L = f(M)$ we will consider to exist, denoting it by $u(x, y)$ [1]. We shall only be interested in an estimate of the differences $u - u_n$ and $u - v_n$. These differences can be computed one after another by solving the following Dirichlet problems in the regions B_1 and B_2: $u(x, y) - u_1(x, y)$ is regular in B_1 and has on L_1 the boundary values:

$$u - u_1 = \begin{cases} 0 & \text{on } \alpha, \\ u - \varphi(M) & ,, \ \bar{\alpha}. \end{cases}$$

The difference $u - v_1$ is regular in B_2 and on L_2 take the values:

$$u - v_1 = \begin{cases} 0 & \text{on } \beta, \\ u - u_1 & ,, \ \bar{\beta}. \end{cases}$$

The differences $u - u_n$ and $u - v_n$ are regular in B_1 and B_2 respectively and are obtained from the preceding differences by solving the Dirichlet problems under the boundary conditions:

$$u - u_n = \begin{cases} 0 & \text{on } \alpha, \\ u - v_{n-1} & ,, \ \bar{\alpha}, \end{cases} \qquad u - v_n = \begin{cases} 0 & \text{on } \beta, \\ u - u_n & ,, \ \bar{\beta}. \end{cases}$$

Let us apply Schwarz's lemma to the estimation of all the differences. We will choose m so large that we have $|f| \leqslant m$ and $|\varphi| \leqslant m$. By the choice of m, everywhere in B $|u(x, y)| \leqslant m$.

The harmonic function $u - u_1$ acquires on α boundary values equal to

[1] The existence of the solution of the Dirichlet problem in the region B follows from the assumption of the existence of the solution of the Dirichlet problem in B_1 and B_2, as was seen from the reasoning of No. 1 of this section.

zero, and on $\bar{\alpha}$ values estimated as follows: $|u - u_1| \leqslant 2m$. On the basis of the lemma, a number $\vartheta_1 < 1$ can be found, dependent only on the geometric configuration of B_1 and B_2, such that on the arc $\bar{\beta}$ there will hold the estimate

$$|u - u_1| \leqslant 2M\vartheta_1.$$

So also does the harmonic function $u - v_1$ acquire on β values equal to zero; and its values on $\bar{\beta}$ are estimated thus: $|u - v_1| = |u - u_1| \leqslant \leqslant 2M\vartheta_1$. By the same lemma a number $\vartheta_2 < 1$ can be found such that on $\bar{\alpha}$ we shall have:

$$|u - v_1| \leqslant 2M\vartheta_1 \cdot \vartheta_2.$$

Continuing these estimates, we shall have proved the validity of the inequalities

$$|u - u_n| \leqslant 2M\vartheta_1^n\vartheta_2^{n-1} \text{ on } \bar{\beta}, \quad |u - v_n| \leqslant 2M\vartheta_1^n\vartheta_2^n \text{ on } \bar{\alpha}.$$

From these are easily obtained estimates of the differences under study in the regions B_1 and B_2. By the construction, $u - u_n$ assumes on α values equal to zero; its values on $\bar{\alpha}$, by the inequalities just cited, are estimated thus: $|u - u_n| = |u - v_{n-1}| \leqslant 2M\vartheta_1^{n-1}\vartheta_2^{n-1}$. By the principle of the maximum, everywhere within B_1 we must have

$$|u - u_n| \leqslant 2M\vartheta_1^{n-1}\vartheta_2^{n-1}.$$

Analogously, for $u - v_n$ in B_1 we obtain the estimate:

$$|u - v_n| \leqslant 2M\vartheta_1^n\vartheta_2^{n-1}.$$

Both these last inequalities tell us that the differences between the solution u of the problem and the successive approximations u_n and v_n constructed according to Schwarz decrease not more slowly than the terms of a geometric progression with ratio $q = \vartheta_1\vartheta_2$.

3. Reduction of Schwarz's method to the solution of a system of integral equations by successive approximations. Until now we have been engaged in a study of the convergence of the successive approximations constructed in accordance with Schwarz, and have touched not at all on the question of methods for the effective construction of the approximations u_n and v_n themselves. We shall now dwell on one of such methods, which is based on the reduction of our problem to the solution of a system of integral equations of the Fredholm type.

We will preserve here the notation of No. 1. For the construction of the pair of approximations $u_{n+1}(x, y)$ and $v_{n+1}(x, y)$ in the regions B_1 and B_2 by the preceding pair u_n and v_n, as is seen from equations (4) No. 1, it is sufficient to know u_n and v_n on $\bar{\beta}$ and $\bar{\alpha}$ respectively. From this it is evident that for the determination of all the successive approximations u_n and v_n, it is sufficient to know how to find their values on $\bar{\beta}$ and $\bar{\alpha}$ only.

For the regions B_1 and B_2 let Green's functions $G_1(x, y;\ \xi, \eta)$ and $G_2(x, y;\ \xi, \eta)$ be known. In terms of them and the boundary values of the function $u(x, y)$ solving the Dirichlet problem in B, we have the following expression

$$\left.\begin{aligned} u(x, y) &= -\int_{L_1} u(\xi, \eta)\,\frac{\partial G_1(x, y;\ \xi, \eta)}{\partial n}\,ds \text{ in } B_1, \\ u(x, y) &= -\int_{L_2} u(\xi, \eta)\,\frac{\partial G_2(x, y;\ \xi, \eta)}{\partial n}\,ds \text{ in } B_2, \end{aligned}\right\} \tag{9}$$

where n is the external normal to the contours L_1 and L_2 at the point (ξ, η).

Separate the integrals along L_1 and L_2 into the sums of integrals along α, $\bar{\alpha}$ and β, $\bar{\beta}$. We take into consideration the fact that the boundary values of $u(x, y)$ on α and β are known to equal $f(M)$, and consequently the integrals along α and β will give known functions, which we will denote thus:

$$F_1(x, y) = -\int_{\alpha} f(M)\,\frac{\partial G_1}{\partial n}\,ds, \quad F_2(x, y) = -\int_{\beta} f(M)\,\frac{\partial G_2}{\partial n}\,ds. \tag{10}$$

$F_1(x, y)$ is a function harmonic in B_1, the boundary values of which on α are equal to $f(M)$ and on $\bar{\alpha}$ are equal to zero. In exactly the same way, $F_2(x, y)$ is harmonic in B_2 and assumes on β values equal to $f(M)$ and on $\bar{\beta}$ values equal to zero.

We will consider that in the first of equations (9) the point (x, y) lies on $\bar{\beta}$, and in the second, on $\bar{\alpha}$. Then for the unknown values of the function $u(x, y)$ on $\bar{\beta}$ and $\bar{\alpha}$ we obtain the following system of integral equations:

$$\left.\begin{aligned} u(x, y) &= F_2(x, y) - \int_{\bar{\beta}} u(\xi, \eta)\,\frac{\partial G_2}{\partial n}\,ds \text{ on } \bar{\alpha}, \\ u(x, y) &= F_1(x, y) - \int_{\bar{\alpha}} u(\xi, \eta)\,\frac{\partial G_1}{\partial n}\,ds \text{ on } \bar{\beta}. \end{aligned}\right\} \tag{11}$$

By the very meaning of the problem, $u(x, y)$ must take bounded values on $\bar{\alpha}$ and $\bar{\beta}$, and therefore the values we seek must give a bounded solution of system (11). We will now take, conversely, any bounded solution of system (11) and show that it provides the values we require for the harmonic function on $\bar{\alpha}$ and $\bar{\beta}$.

We will denote by $u^*(M)$ and $v^*(M)$ functions defined at the points of $\bar{\alpha}$ and $\bar{\beta}$, satisfying system (11). We will construct by them in the regions B_1 and B_2 respectively the functions

$$u(x, y) = F_1(x, y) - \int_{\bar{\alpha}} u^*(M) \frac{\partial G_1}{\partial n} ds,$$

$$u(x, y) = F_2(x, y) - \int_{\bar{\beta}} v^*(M) \frac{\partial G_2}{\partial n} ds.$$

$u(x, y)$ is harmonic in B and takes on the boundary the values:

$$u(x, y) = \begin{cases} f(M) & \text{on } \alpha, \\ u^*(M) & ,, \quad \bar{\alpha}. \end{cases}$$

Analogously, v is harmonic in B_2 and the values taken by it on L_2 will be:

$$v(x, y) = \begin{cases} f(M) & \text{on } \beta \\ v^*(M) & ,, \quad \bar{\beta}. \end{cases}$$

Moreover, they are bounded, since under the integral sign, as multipliers of the normal derivative of the Green function, stand bounded functions.

It can be seen that in the common part of the regions B_1 and B_2, u and v coincide. To satisfy oneself of this, it is sufficient to establish that u and v acquire identical values on $\bar{\alpha}$ and $\bar{\beta}$. And this will surely be so, by virtue of the fact that u^* and v^* satisfy system (11). Indeed, the values of $v(x, y)$ on $\bar{\alpha}$ are equal to

$$F_2(x, y) - \int_{\bar{\beta}} v^*(M) \frac{\partial G_2}{\partial n} ds, \quad (x, y) \in \bar{\alpha}.$$

But on the strength of the first equation of (11):

$$u^*(x, y) = F_2(x, y) - \int_{\bar{\beta}} v^*(M) \frac{\partial G_2}{\partial n} ds.$$

and accordingly on $\bar{\alpha}$, $v(x, y) = u^* = u(x, y)$. It can also be established that $v(x, y) = u(x, y)$ on $\bar{\beta}$.

Thus in the common part of B_1 and B_2 the functions we have constructed, $u(x, y)$ and $v(x, y)$, coincide, and consequently $v(x, y)$ is the continuation of $u(x, y)$ from B_1 to B_2. In addition, it is seen from the boundary values for u and v cited above that the continued function $u(x, y)$ will acquire the assigned values on the entire boundary L of the region B.

From this it follows, by the way, that system (11) can have only one bounded solution, since if there were several of them, we could have constructed a solution of the Dirichlet problem in B for each of them. And it is clear that solutions of the Dirichlet problem corresponding to different solutions of system (11) would be different. We would have obtained several different bounded solutions of the Dirichlet problem in B, which could not be.

We will remark that in certain cases, besides the indicated bounded solution, system (11) can have an unbounded solution as well. It can be seen — and we shall not tarry on this — that the functions u and v corresponding to such an unbounded solution will have singularities at the ends of the arcs $\bar{\alpha}$ or $\bar{\beta}$.

For practical computations, the fact of the existence of an unbounded solution for system (11) is immaterial. We shall show below that if this system be solved by the method of successive approximations, then whatever bounded initial values we may start with, we shall always obtain in the limit the bounded solution of system (11).

We will now establish that the solution of system (11) by successive approximations is in essence none other than the application of Schwarz's method to the Dirichlet problem in the region B.

To ease the notation we will denote the sought values of $u(x, y)$ on $\bar{\beta}$ by v. The values that it acquires on $\bar{\alpha}$, however, we will call u as before. System (11) will then take the form:

$$\left.\begin{aligned} u &= F_2(x, y) - \int_{\bar{\beta}} v \frac{\partial G_2}{\partial n} ds \text{ on } \bar{\alpha}, \\ v &= F_1(x, y) - \int_{\bar{\alpha}} u \frac{\partial G_1}{\partial n} ds \text{ on } \bar{\beta}. \end{aligned}\right\} \tag{11_1}$$

We will now assign for u on $\bar{\alpha}$ the arbitrary bounded and piece-wise continuous values $u_1 = \varphi_1(M)$. Substitute them for u in the right side of the second equation of (11_1) and adopt the function obtained on $\bar{\beta}$ as the first approximation v_1 for v. From the point of view of the Dirichlet problem, as is seen from (10) and the structure of the second term of the right side, our computations signify that we are finding in the region B_1 the harmonic function

$$u_1(x, y) = F_1(x, y) - \int_{\bar{\alpha}} \varphi(M) \frac{\partial G_1}{\partial n} ds,$$

as the solution of the Dirichlet problem for the boundary condition:

$$u_1(x, y) = \begin{cases} f(M) \text{ on } \alpha, \\ \varphi(M) \;\;,, \;\; \bar{\alpha}, \end{cases}$$

and are then determining its values on $\bar{\beta}$. They will give v_1.

On finding the function v_1, we substitute in the right side of the first equation of (11_1), and the function obtained on $\bar{\alpha}$ after the substitution we adopt as the second approximation u_2 to u.

From the point of view of harmonic functions, our computations signify that in B_2 we are constructing the harmonic function

$$v_1(x, y) = F_2(x, y) - \int_{\bar{\beta}} v_1 \frac{\partial G_2}{\partial n} ds$$

in accordance with the boundary values:

$$v_1(x, y) = \begin{cases} f(M) & \text{on } \beta, \\ v_1 & \text{,, } \bar{\beta}, \end{cases}$$

and are determining its values u_2 on $\bar{\alpha}$, etc. It is perfectly clear that the successive approximations $u_1, v_1, u_2, v_2, \ldots$ to u and v that are obtained in such a process will be exactly those obtained by Schwarz's method described in No. 1.

In Nos. 1 and 2 we have shown that Schwarz's process converges to the solution of the Dirichlet problem, whatever bounded and piece-wise continuous values $\varphi(M)$ for $u(x, y)$ on $\bar{\alpha}$ we may begin the computations with.

Therefore the method of successive approximations for system (11) will converge to the bounded solution of this system, whatever bounded and piece-wise continuous initial approximation $\varphi(M)$ for u on $\bar{\alpha}$ we may begin the computations with.

In practical computations, it goes without saying that there is no need to compute both series of successive approximations $u_1, u_2, \ldots,$ $v_1, v_2, \ldots$ to both the sought functions u and v.

One of these functions can be eliminated, only one of the series of successive approximations thus being computed. Let us eliminate v, for instance. For this purpose we substitute for v its value from the second equation of system (11_1) in the first equation. After the substitution we obtain the following equation, with only the one unknown function:

$$u = F(x, y) + \int_{\bar{\alpha}} uk(x, y; \xi, \eta)\, ds. \qquad (12)$$

Here (x, y) are the coordinates of any fixed point on $\bar{\alpha}$, (ξ, η) the coordinates of a variable point along the curve of integration $\bar{\alpha}$,

$$F(x, y) = F_2(x, y) - \int_{\bar{\beta}} F_1(\xi, \eta) \frac{\partial G_2(x, y; \xi, \eta)}{\partial n} ds,$$

and

$$k(x, y; \xi, \eta) = \int_{\bar{\beta}} \frac{\partial G_2(x, y; \xi_1, \eta_1)}{\partial n} \cdot \frac{\partial G_1(\xi_1, \eta_1; \xi, \eta)}{\partial n} ds_1.$$

Equation (12) is then solved by the usual method of successive approximations. After u is found, the function v can be recovered by using the second of equations (11_1).

§ 2. THE SCHWARZ-NEUMANN METHOD FOR THE SOLUTION OF THE DIRICHLET PROBLEM FOR THE INTERSECTION OF TWO REGIONS

1. Description of the method and investigation of the convergence of the successive approximations. In § 1 we have expounded Schwarz's method for the solution of the Dirichlet problem in a region that is the sum of two regions, if it is known how this problem is solved for each of the component regions separately. As was remarked by C. Neumann [1]), a similar trend of thought can be applied to the solution of the Dirichlet problem in a region that is the common part (the intersection in the set-theoretical sense) of two other regions overlapping one another.

We will now expound the idea of the method. We will not limit ourselves, however, to the consideration of the Laplace equation only, as Neumann did, but will have in view an arbitrary linear homogeneous partial differential equation of the second order that satisfies the requirements enumerated below. We will take an equation of the form:

$$A\frac{\partial^2 u}{\partial x^2} + 2B\frac{\partial^2 u}{\partial x\, \partial y} + C\frac{\partial^2 u}{\partial y^2} + D\frac{\partial u}{\partial x} + E\frac{\partial u}{\partial y} + Fu = 0, \qquad (1)$$

and let D be some region of the xy-plane, bounded by the contour Γ.

We will consider that for equation (1) there are fulfilled the following conditions, analogous to those that we formulated in § 1.

I. *If two bounded functions $u(x, y)$ and $u^*(x, y)$ satisfy equation* (1) *in D and have identical boundary values on the contour Γ of the region D, except, perhaps, at a finite number of points, then they coincide with each other*:

$$u(x, y) = u^*(x, y) \text{ everywhere in } D.$$

II. *A solution of equation* (1) *cannot have within the region of regularity either positive maximum or negative minimum; expressing it differently, if $u(x, y)$ satisfies equation* (1) *in D and if on the contour Γ of the region there is fulfilled the inequality*

$$-h \leqslant u(x, y) \leqslant +g \qquad (h, g \geqslant 0),$$

then everywhere within D, too,

$$-h \leqslant u(x, y) \leqslant +g.$$

We will note one consequence following from this. Let $u(x, y)$ and $u^*(x, y)$ be two solutions of equation (1) and on Γ let $u(x, y) \geqslant u^*(x, y)$. Then everywhere within D, $u(x, y) \geqslant u^*(x, y)$. Indeed, in view of the linearity and homogeneity of (1), the difference $u(x, y) - u^*(x, y)$ will be a solution of the equation, and since on Γ $u(x, y) - u^*(x, y) \geqslant 0$ by assumption, then everywhere in D, too,

$$u(x, y) - u^*(x, y) \geqslant 0.$$

---

[1]) C. Neumann, [1].

III. *The limit of a monotone and bounded sequence of solutions of equation* (1) *is also a solution of equation* (1).

IV. Let $u(x, y)$ be bounded and satisfy equation (1) in region D. Let us consider the arc γ, part of the boundary Γ of region D; assume that on it is given a continuous function $f(M)$, and that P is an interior point of γ. We shall assume that *if* $u(x, y)$ *has at all points* M *of the arc* γ, *excepting, perhaps, the point* P, *boundary values equal to* $f(M)$, *then for an approach of* (x, y) *to* P, $u(x, y)$ *will also tend to a limit value, and this limit value is equal to* $f(P)$.

All the assumptions I—IV that we have formulated up to now coincide with the corresponding assumptions that we made in studying the question of the convergence of the Schwarz process for the sum of two regions. In our present reasoning there is still required one additional assumption, V, which we will formulate later when considering the question of the convergence of the successive approximations.

Let us take two regions B_1 and B_2 in the xy-plane, bounded by the contours L_1 and L_2. Let them overlap one another, and let B' be their common part: $B' = B_1 \cdot B_2$.

With respect to the parts of L_1 and L_2 we will keep the same notation as in § 1. The part of L_1 lying within B_2 we shall call $\bar{\alpha}$, and the remaining part α. The part of L_2 lying within B_1 we shall denote by $\bar{\beta}$, and the remaining part by β. Finally, the common part of β and α, which belongs to the boundary of B', we will call γ. The boundary of B' will obviously be $\bar{\alpha} + \bar{\beta} + \gamma$.

For the sake of simplicity, we will consider that all these parts consist of a finite number of arcs of Jordan curves. We will assume that for each of the regions B_1 and B_2 we can solve the Dirichlet problem for any continuous or piece-wise continuous boundary values of the sought functions, and let it be required to find a function $w(x, y)$ satisfying equation (1) in the region B' and acquiring on its boundary L' the assigned values:

$$w(x, y) \Big|_{L'} = f(M).$$

Here $f(M)$ is a piece-wise continuous function of the position of the point M on L'.

Neumann's idea consists in representing the sought function $w(x, y)$ in the form of the sum of two functions

$$w(x, y) = u(x, y) + v(x, y),$$

the first of which is defined and satisfies equation (1) in B_1, the second being defined and satisfying equation (1) in B_2. Both of the summands $u(x, y)$ and $v(x, y)$ must be so chosen that the sum satisfies boundary condition (2).

The problem as stated, of resolving $w(x, y)$ into summands, contains in itself an element of indeterminacy.

To disclose the degree of it, we will consider a particular case, the region depicted in Fig. 63. In it this indeterminacy is fully revealed; after a consideration of it, the measure in which the problem is indeterminate in the most general case will become obvious.

We recall that, in conformity with the first assumption, to know the function $u(x, y)$ or $v(x, y)$ everywhere in B_1 or B_2 it is necessary to indicate its values on L_1 or L_2 respectively, except, perhaps, at a finite set of points, which for simplicity of exposition we will omit in the reasoning next following.

Let us take any solution $\omega(x, y)$ of equation (1) regular in $B = B_1 + B_2$. It is determined by its boundary values on the contour $L = \alpha + \beta$ of this region. In our problem this contour is the broken line $ABCDEFGH$. We note also that the values of $\omega(x, y)$ on L can be given as any piece-wise continuous ones, since from the assumption of the existence of the solution of the Dirichlet problem in each of the component regions B_1 and B_2 for arbitrary piece-wise continuous boundary values follows, by what was expounded in § 1, the existence of the solution of the same problem in B for any piece-wise continuous boundary values.

Let us represent the sought function $w(x, y)$ in the form:

$$w(x, y) = [u(x, y) + \omega(x, y)] + [v(x, y) - \omega(x, y)].$$

On the part of α not belonging to the boundary L', i.e., on $\alpha - \gamma$, which in our case will be the broken line $BAHG$, where no conditions on the boundary values of u have been imposed, we can so choose the boundary values of ω that the first summand $u + \omega$ has whatever preassigned values we please. Analogously, on the part $\beta - \gamma$ of the contour L_2, which in our problem will be the broken line $DEFG$, one can so manage the choice of the boundary values that the second summand $v - \omega$ has arbitrary boundary values there. On the part γ of the region B, consisting in our example of the broken line BCD and the point G, the choice of ω can be so managed that one of the summands, for instance $u + \omega$, has any values. The second summand, $v - \omega$, however, will then obtain on γ fully determined values $f(M) - (u+\omega)$, since here their sum w is given.

In conformity with this, we can assign arbitrary values on α to the first function u (or to v on β). For the second function v, however, arbitrary boundary values can be assigned only on $\beta - \gamma$, its values on γ being fully determined and equalling $f(M) - u$.

We will now show that by such an assignment of the boundary values, both summands, if they exist, are determined uniquely. Indeed, let us resolve w in two ways:

$$w = u' + v' = u'' + v'',$$

where u' and u'' are regular in B_1 and have identical values on α, and v' and v'' are regular in B_2 and acquire identical values on β. Consider the differences $u' - u''$ and $-(v' - v'')$. They are solutions of the equation in the regions B_1 and B_2 respectively and vanish on the parts α and β of their boundaries. Moreover, as is seen from the last equation, in the common part B' of these two regions they coincide:

$$u' - u'' = -(v' - v'').$$

Therefore the difference $v'' - v'$ is the continuation of the solution $u' - u''$ of equation (1) from B_1 into B_2. Moreover the solution thus continued has null values on the boundary $L = \alpha + \beta$ of the region B. But then, on the basis of Assumption I, it is identically equal to zero everywhere in B and, accordingly, $u' = u''$ and $v' = v''$. Thus the first and second summands into which we resolve w are uniquely determined by the stipulated boundary requirements.

We have reduced the proof of the existence and the construction of the solution we require for equation (1), $w(x, y)$, to the construction of the functions $u(x, y)$ and $v(x, y)$. To determine them we will apply the method of successive approximations in a form somewhat different from that proposed by C. Neumann.

We must now make the choice of the boundary values of $u(x, y)$ on α and $v(x, y)$ on $\beta - \gamma$. On $\alpha - \gamma$ we will assign the function $\varphi(M)$ so that its values will form together with the values $f(M)$ on $\bar{\alpha} + \gamma$ a piece-wise continuous function of the position of the point M on L_1, and shall consider that

$$u(x, y) = \begin{cases} \varphi(M) & \text{on } \alpha - \gamma, \\ f(M) & \text{,, } \gamma. \end{cases}$$

By the choice of $u(x, y)$ on α we have determined the values of $v(x, y)$ on γ; we must have, namely, $v(x, y)|_\gamma = 0$. The values of $v(x, y)$ on $\beta - \gamma$ can be assigned arbitrarily, as we said above. For simplicity we will set them equal to zero, so that for $v(x, y)$ we obtain the following boundary values:

$$v(x, y) = 0 \text{ on } \beta.$$

Such a choice of the boundary values of $u(x, y)$ and $v(x, y)$ on α and β is not, of course, obligatory, and has been made by us for definiteness and to simplify the notation somewhat. Moreover it is evident that for convergence of the successive approximations such a choice of the boundary values of $u(x, y)$ and $v(x, y)$ is inessential: if the successive approximations converge for our choice of boundary values, they will also converge for any other choice of them, provided only that they be piece-wise continuous and such that $u + v = f(M)$ on γ. Indeed, any other manner of choosing the boundary values can be reduced to ours by the substitution

$$u' = u + \omega \text{ and } v' = v - \omega$$

and a suitable choice of the function ω, of which we spoke a few lines above.

The values of $u(x, y)$ on $\bar{\alpha}$ are unknown to us. The problem of constructing $u(x, y)$ reduces, indeed, to finding them. We arbitrarily assign on $\bar{\alpha}$ the values $\bar{\varphi}(M)$, subjecting them to the sole requirement that together with the values of $u(x, y)$ on α, they form piece-wise continuous values on the entire boundary L_1.

In the theoretical investigations we shall not impose on $\varphi(M)$ and $\bar{\varphi}(M)$ any other limitations. In practical computations their arbitrariness must obviously be so utilized as to render as simple as possible the computations required to determine the actual values of $u(x, y)$ on $\bar{\alpha}$ by the scheme to be expounded below.

The first approximation to $u(x, y)$, $u_1(x, y)$, we construct as the solution of the Dirichlet problem for equation (1) under the boundary conditions:

$$u_1(x, y) = \begin{cases} \varphi(M) & \text{on } \alpha - \gamma, \\ f(M) & \text{,, } \gamma, \\ \bar{\varphi}(M) & \text{,, } \bar{\alpha}. \end{cases} \tag{2_1}$$

We find the values that $u_1(x, y)$ takes on β and subtract them from $f(M)$. If $u_1(x, y)$ coincided with $u(x, y)$, we should have found the exact values of $v(x, y)$ on $\bar{\beta}$. But since $u_1(x, y)$ is only an approximation to $u(x, y)$, the values $f(M) - u_1(x, y)$ on $\bar{\beta}$ found by us will not coincide with $v(x, y)$ but will give $v(x, y)$ on $\bar{\beta}$ only approximately. By them and by the known values of $v(x, y)$ on β we construct the first approximation $v_1(x, y)$ to $v(x, y)$, solving the Dirichlet problem for equation (1) in B_2 for the boundary values:

$$v_1(x, y) = \begin{cases} 0 & \text{on } \beta, \\ f(M) - u(x, y) & \text{on } \bar{\beta}. \end{cases} \tag{2_2}$$

Using this first approximation $v_1(x, y)$, we construct the second approximation $u_2(x, y)$ to $u(x, y)$, solving the Dirichlet problem in B_1 under the boundary conditions:

$$u_2(x, y) = \begin{cases} u_1(x, y) & \text{on } \alpha, \\ f(M) - v_1(x, y) & \text{,, } \bar{\alpha}, \end{cases}$$

and so on.

We determine the succeeding approximations $u_{n+1}(x, y)$ and $v_{n+1}(x, y)$ using the preceding ones, as solutions of a boundary-value problem of equation (1) in B_1 and B_2, with the boundary data:

$$\left.\begin{aligned} u_{n+1}(x, y) &= \begin{cases} u_n(x, y) & \text{on } \alpha, \\ f(M) - v_n(x, y) & \text{,, } \bar{\alpha}, \end{cases} \\ v_{n+1}(x, y) &= \begin{cases} 0 & \text{,, } \beta, \\ f(M) - u_{n+1}(x, y) & \text{,, } \bar{\beta}. \end{cases} \end{aligned}\right\} \tag{3}$$

For each of the sought functions $u(x, y)$ and $v(x, y)$ we construct a sequence of approximations

$$\left.\begin{aligned} u_1(x, y),\ u_2(x, y),\ \ldots \\ v_1(x, y),\ v_2(x, y),\ \ldots \end{aligned}\right\} \tag{4}$$

We will now investigate the question of their convergence. The idea of the following reasoning is the same as in No. 1 § 1. For each of the sequences we construct a majorant and a minorant sequence and show that they both converge to one and the same limit. It will then be clear that sequences (4) converge to the same limit. As regards the construction of the majorant and minorant sequences, it is conducted by a method similar to that which we employed in § 1 No. 1, and will differ from it primarily at one point only. Namely, there we constructed simultaneously the majorants u_n^+ and v_n^+ for u_n and v_n. Here, however, the construction of the majorant u_n^+ for u_n will be conducted parallel with the construction of the minorant v_n^- for v_n. The cause of this is obvious: the construction of the majorant for one term of the sum must evidently be connected with that of the minorant for the other term.

We will denote by N a positive number greater than $\max|\bar{\varphi}(M)|$. We will concern ourselves with the choice of it somewhat later. We next construct the sequence of functions $u_n^+(x, y)$ and $v_n^-(x, y)$ solving the Dirichlet problem for equation (1) in the regions B_1 and B_2 respectively for the boundary values:

$$u_1^+(x, y) = \begin{cases} \varphi(M) & \text{on } \alpha - \gamma, \\ f(M) & ,, \ \gamma, \\ +N & ,, \ \bar{\alpha}, \end{cases}$$

$$v_1^-(x, y) = \begin{cases} 0 & \text{on } \beta, \\ f(M) - u_1^+(x, y) & ,, \ \bar{\beta}, \end{cases}$$

. . . .

$$u_{n+1}^+(x, y) = \begin{cases} u_n^+(x, y) & \text{on } \alpha, \\ f(M) - v_n^-(x, y) & ,, \ \bar{\alpha}, \end{cases}$$

$$v_{n+1}^-(x, y) = \begin{cases} 0 & \text{on } \beta, \\ f(M) - u_{n+1}^+(x, y) & ,, \ \bar{\beta}. \end{cases}$$

We will show that the sequence $u_1^+(x, y), u_2^+(x, y), \ldots$ will be majorant for $u_1(x, y), u_2(x, y), \ldots$, and that the sequence $v_1^-(x, y), v_2^-(x, y), \ldots$ will be minorant for $v_1(x, y), v_2(x, y), \ldots$.

Let us consider the differences between the corresponding functions of these sequences. They will be solutions of equation (1) in the region B_1 or B_2, and their boundary values will be interconnected by the following relations:

$$u_1^+ - u_1 = \begin{cases} 0 & \text{on } \alpha, \\ N - \bar{\varphi}(M) & ,, \ \bar{\alpha}, \end{cases}$$

$$v_1^- - v_1 = \begin{cases} 0 & \text{on } \beta, \\ u_1 - u_1^+ & ,, \ \bar{\beta}, \end{cases}$$

. . . .

$$u_{n+1}^{+} - u_{n+1} = \begin{cases} 0 & \text{on } \alpha, \\ v_n - v_n^{-} & ,, \ \bar{\alpha}, \end{cases}$$

$$v_{n+1}^{-} - v_{n+1} = \begin{cases} 0 & \text{on } \beta, \\ u_{n+1} - u_{n+1}^{+} & ,, \ \bar{\beta}. \end{cases}$$

In view of the choice of the number N, it follows from the first of the relations that the boundary values of $u_1^{+} - u_1$ on L_1 are non-negative. By Assumption II, it follows from this that everywhere in B_1 we shall have $u_1^{+} - u_1 \geqslant 0$. But then, as is seen from the second relation, the values of $v_1^{-} - v_1$ on L_2 will be non-positive, which means, on the strength of the same Assumption II, that everywhere in B_2 we shall have $v_1^{-} - v_1 \leqslant 0$. Continuing this reasoning, we convince ourselves that for any n there will be fulfilled the inequalities

$$u_n^{+}(x, y) \geqslant u_n(x, y), \ v_n^{-}(x, y) \leqslant v_n(x, y).$$

The sequences $u_1^{-}, u_2^{-}, \ldots$, minorant for $u_1, u_2, \ldots$, and $v_1^{+}, v_2^{+}, \ldots$, majorant for $v_1, v_2, \ldots$, are constructed analogously. We determine the functions u_n^{-} and v_n^{+} recurrently as solutions of equation (1) in B_1 or B_2 respectively, under the boundary conditions:

$$u_1^{-}(x, y) = \begin{cases} \varphi(M) & \text{on } \alpha - \gamma, \\ f(M) & ,, \ \gamma, \\ -N & ,, \ \bar{\alpha}, \end{cases}$$

$$v_1^{+}(x, y) = \begin{cases} 0 & \text{on } \beta, \\ f(M) - u_1^{-}(x, y) & ,, \ \bar{\beta}, \end{cases}$$

. . . .

$$u_{n+1}^{-}(x, y) = \begin{cases} u_n^{-}(x, y) & \text{on } \alpha, \\ f(M) - v_n^{+}(x, y) & ,, \ \bar{\alpha}, \end{cases}$$

$$v_{n+1}^{+}(x, y) = \begin{cases} 0 & \text{on } \beta, \\ f(M) - u_{n+1}^{-}(x, y) & ,, \ \bar{\beta}. \end{cases}$$

It can be shown that for any n there will be fulfilled the inequalities:

$$u_n^{-}(x, y) \leqslant u_n(x, y), \ v_n^{+}(x, y) \geqslant v_n(x, y).$$

We will now try to so choose the number N that all the auxiliary majorant and minorant sequences of functions we have constructed will be monotonic. We can succeed in doing this with an additional assumption, V, which will relate not only to the differential equation, as did propositions I—IV, but also to the regions B_1 and B_2 under consideration.

V. We will denote by $U_{\bar{\alpha},f}(x, y)$ the function solving the Dirichlet problem for equation (1) in region B_1 under the boundary conditions

$$U_{\bar{\alpha},f} = \begin{cases} 0 & \text{on } \alpha, \\ f(M) & ,, \ \bar{\alpha}. \end{cases}$$

Analogously, we will call $V_{\bar{\beta},f}(x, y)$ the function solving the Dirichlet problem in B_2 under the boundary conditions

$$V_{\bar{\beta},f} = \begin{cases} 0 & \text{on } \beta, \\ f(M) & \text{,, } \bar{\beta}. \end{cases}$$

Let us consider the function solving the Dirichlet problem in B_1 for boundary values equal to zero on α and unity on $\bar{\alpha}$. This is $U_{\bar{\alpha},1}$. We will take the values that it acquires on $\bar{\beta}$ and solve the Dirichlet problem in B_2 for boundary values equal to zero on β and to $U_{\bar{\alpha},1}$ on $\bar{\beta}$. The problem is solved by the function which in the adopted notation is written $V_{\bar{\beta},U_{\bar{\alpha},1}}$.

From the second assumption it is seen that all values acquired by $U_{\bar{\alpha},1}$ in B_1 will lie in the interval [0, 1]. In particular, in this interval will lie all values acquired by $U_{\bar{\alpha},1}$ on $\bar{\beta}$. Therefore all the boundary values for $V_{\bar{\beta},U_{\bar{\alpha},1}}$ will also belong to the interval [0, 1], and accordingly all values that this function acquires in B will lie between zero and unity. Of interest to us now are the values that $V_{\bar{\beta},U_{\bar{\alpha},1}}$ acquires on the part $\bar{\alpha}$ of the contour of the region B_1. They are in any case non-negative and not greater than unity. With respect to them we will require — and in this consists our Assumption V — that none of them shall exceed some proper fraction:

$$V_{\bar{\beta},U_{\bar{\alpha},1}} \leqslant \vartheta < 1 \text{ for } (x, y) \in \bar{\alpha}. \tag{5}$$

Let us now return to our auxiliary sequences, and study the difference $u_1^+(x, y) - u_2^+(x, y)$. Its values on L_1 are the following:

$$u_1^+(x, y) - u_2^+(x, y) = \begin{cases} 0 & \text{on } \alpha, \\ N - f(M) + v_1^-(x, y) & \text{,, } \bar{\alpha}. \end{cases}$$

Clearly $u_1^+(x, y)$ can be represented in the form of the sum of two solutions of equation (1), the first of which, $NU_{\bar{\alpha},1}$, acquires on $\bar{\alpha}$ values equal to N, and on α values equal to zero; the second of them, $U_{\alpha,u}$, acquires on α the same values as does u_1^+ or u, and on $\bar{\alpha}$ values equal to zero:

$$u_1^+(x, y) = NU_{\bar{\alpha},1}(x, y) + U_{\alpha,u}(x, y).$$

In accordance with this, $v_1^-(x, y)$, as is seen from its boundary values, is representable in the form:

$$v_1^-(x, y) = V_{\bar{\beta},f} - V_{\bar{\beta},u_1} = V_{\bar{\beta},f} - NV_{\bar{\beta},U_{\bar{\alpha},1}} - V_{\bar{\beta},U_{\alpha,u}},$$

and therefore the values on $\bar{\alpha}$ of the difference $u_1^+(x, y) - u_2^+(x, y)$ that we are interested in are equal to

$$N - f(M) + v_1^-(x, y) = N(1 - V_{\bar{\beta},U_{\bar{\alpha},1}}) + V_{\bar{\beta},f} - V_{\bar{\beta},U_{\alpha,u}} - f(M).$$

The three last terms are bounded functions of the position of the point

(x, y) on $\bar{\beta}$. The parenthesis standing as the multiplier of N, by our Assumption V, is positive and not less than $1 - \vartheta$. Accordingly the number N can always be chosen so large that

$$N(1 - V_{\beta, U_{\bar{\alpha},1}}) \geqslant |V_{\bar{\beta}, f} - V_{\bar{\beta}, U_{\alpha, u}} - f(M)|.$$

Then the values on α of the difference under study will be non-negative. For such a choice of N, $u_1^+ - u_2^+$ will be not less than zero everywhere on L_1 and by the second assumption we can affirm that everywhere in B_1 we shall have:

$$u_1^+(x, y) - u_2^+(x, y) \geqslant 0.$$

Using this inequality and the following expressions for the boundary values of u_n^+ and v_n^-,

$$v_{n+1}^- - v_n^- = \begin{cases} 0 & \text{on } \beta, \\ u_n^+ - u_{n+1}^+ & \text{,, } \bar{\beta}, \end{cases}$$
$$u_{n+2}^+ - u_{n+1}^+ = \begin{cases} 0 & \text{on } \alpha, \\ v_n^- - v_{n+1}^- & \text{,, } \bar{\alpha}, \end{cases}$$
$$(n = 1, 2, 3, \ldots),$$

inequalities can be proved without difficulty in succession, establishing the monotonic character of the sequences $u_n^+(x, y)$ and $v_n^-(x, y)$ constructed by us:

$$u_1^+(x, y) \geqslant u_2^+(x, y) \geqslant, \ldots$$
$$v_1^-(x, y) \leqslant v_2^-(x, y) \leqslant, \ldots.$$

For the same choice of N is established analogously the monotonicity of the second pair of sequences too:

$$u_1^-(x, y) \leqslant u_2^-(x, y) \leqslant, \ldots$$
$$v_1^+(x, y) \geqslant v_2^+(x, y) \geqslant, \ldots.$$

Now that we have proved the monotonic character of the auxiliary sequences, the further reasoning will in considerable measure repeat the reasoning of the end of No. 1 § 1. First of all, from $u_n^+ \geqslant u_n \geqslant u_n^-$ and the preceding inequalities it follows that

$$u_1^+(x, y) \geqslant u_n^+(x, y) \geqslant u_n^-(x, y) \geqslant u_1^-(x, y).$$

The sequences u_n^+ and u_n^- will be not only monotone but also bounded. By the third assumption the limit functions, which we will call

$$u^+(x, y) = \lim_{n\to\infty} u_n^+(x, y) \text{ and } u^-(x, y) = \lim_{n\to\infty} u_n^-(x, y),$$

will satisfy equation (1) in the region B_1.

From analogous considerations it is seen that the sequences v_n^+ and v_n^-

will also converge, and the limit functions

$$v^+(x, y) = \lim_{n\to\infty} v_n^+(x, y) \text{ and } v^-(x, y) = \lim_{n\to\infty} v_n^-(x, y)$$

will satisfy equation (1) in the region B_2. $u^+(x, y)$ is determined in B_1 and $v^-(x, y)$ in the region B_2. Their sum $w(x, y) = u^+(x, y) + v^-(x, y)$ is determined in $B' = B_1 \cdot B_2$ and satisfies equation (1) there. We shall find the limit values that it acquires on the boundary $L' = \bar{\alpha} + \bar{\beta} + \gamma$ of this region.

Let the point M lie on the part $\bar{\alpha}$ of this boundary, and let the assigned function $f(M)$ be continuous at this point. By the construction of the functions $u_{n+1}^+(x, y)$ and $v_n^-(x, y)$, at it there holds the equation:

$$u_{n+1}^+(x, y) = f(M) - v_n^-(x, y).$$

Starting from this, by quite uncomplicated reasoning it can be shown that for an approach of (x, y) to M, $w(x, y)$ will have a limit value equal to $f(M)$. Indeed, let the point (x, y) approach the point M. From the inequality $u^+(x, y) \leqslant u_{n+1}^+(x, y)$ it is clear that for this we shall have

$$\lim_{(x,y)\to M} u^+(x, y) \leqslant \lim_{(x,y)\to M} u_{n+1}^+(x, y) = f(M) - v_n^-(M).$$

This inequality is true for any n. Let us increase n without limit in it. The left side of the inequality does not depend on n and will not change as n changes. The right side, however, will tend to $f(M) - v^-(M)$, and in the limit we obtain the following inequality:

$$\overline{\lim_{(x,y)\to M}} u^+(x, y) \leqslant f(M) - v^-(M).$$

Let us consider, on the other hand, the function $u_*(x, y)$, solving the Dirichlet problem in B_1 under the boundary condition:

$$u_*(x, y) = \begin{cases} \varphi(M) & \text{on } \alpha - \gamma \\ f(M) & \text{,, } \gamma, \\ f(M) + v^-(M) & \text{,, } \bar{\alpha}. \end{cases}$$

Since $v_n^-(x, y) \leqslant v^-(x, y)$, it is clear that on the boundary L_1, and accordingly everywhere in the region B_1, $u_{n+1}^+(x, y) \geqslant u_*(x, y)$. Therefore

$$\lim_{n\to\infty} u_{n+1}^+(x, y) = u^+(x, y) \geqslant u_*(x, y).$$

If in this inequality the point (x, y) be brought to M, we find:

$$\underline{\lim_{(x,y)\to M}} u^+(x, y) \geqslant \lim_{(x,y)\to M} u_*(x, y) = f(M) - v^-(M).$$

This last inequality is compatible with the estimate from above, obtained earlier, of the limit value of $u^+(x, y)$ at the point M in only one case, when the greatest and least limit values of $u^+(x, y)$ at M coincide and both are equal to $f(M) - v^-(M)$.

From this it is clear that $w(x, y) = u^+(x, y) + v^-(x, y)$ will have for an approach of the point (x, y) to M a limit value equal to $f(M)$.

In the same way it is proved that for an approach of (x, y) to any point M on $\bar{\beta}$ at which $f(M)$ is continuous, $w(x, y)$ will tend to a limit value equal to $f(M)$.

We shall investigate the limit values of $w(x, y)$ on γ. Let M be an interior point of γ, and let the function $f(M)$ be continuous at M. We have to establish that for an approach of the point (x, y) to M, $w(x, y)$ will tend to $f(M)$. We recall that

$$u_1^+(x, y) \geqslant u^+(x, y) \geqslant u_1^-(x, y).$$

Furthermore, by the construction of the functions u_1^+ and u_1^-, when the point (x, y) approaches M, both $u_1^+(x, y)$ and $u_1^-(x, y)$ will tend to one and the same limit $f(M)$. We recall that we have defined both of these functions as a solution of the Dirichlet problem, their boundary values on γ being equal to this $f(M)$. Therefore $u^+(x, y)$ will also tend to the same value $f(M)$.

We will note one more fact, useful to us in what is to follow. Let M be a point on $\alpha - \gamma$ at which $\varphi(M)$ is continuous. When (x, y) approaches M, both $u_1^+(x, y)$ and $u_1^-(x, y)$ will tend to $\varphi(M)$. From the last inequality it follows that $u^+(x, y)$ will thereby tend to the same limit.

From the inequality

$$v_1^-(x, y) \leqslant v^-(x, y) \leqslant v_1^+(x, y),$$

and from the fact that for an approach of (x, y) to any interior point of the part β of the contour L_2, $v_1^-(x, y)$ and $v_1^+(x, y)$ tend to zero, it follows that $v^-(x, y)$ has limit values equal to zero at all interior points of β.

Thus $w(x, y) = u^+(x, y) + v^-(x, y)$ at all points of the boundary L' of the region B', except points of discontinuity and, perhaps, points that are ends of the arcs of the curves figuring in $\bar{\alpha}$ and $\bar{\beta}$ (and of these and the others, there are, on assumption, a finite number), has limit values equal to $f(M)$. To show that $w(x, y)$ solves the posed Dirichlet problem in B', it remains only for us to establish that if $f(M)$ is continuous at a point that is the end of any arc of $\bar{\alpha}$ or $\bar{\beta}$, then $w(x, y)$ will tend to $f(M)$ for an approach to this point.

In view of the fact that there is a finite number of such points, as also of the points of discontinuity of $f(M)$, about each such point there can be indicated an arc on L' containing the point and containing no points of discontinuity of $f(M)$ nor other ends of arcs from $\bar{\alpha}$ and $\bar{\beta}$. For an approach of (x, y) to any point of the arc except that under consideration, $w(x, y)$ tends to $f(M)$, as has been shown above. Moreover $w(x, y)$ is bounded. By the fourth assumption, we can affirm that for an approach to the point under consideration, $w(x, y)$ will also tend to the value $f(M)$ at this point.

By similar reasoning, it can be shown for $u^-(x, y)$ and $v^+(x, y)$ that the sum $u^-(x, y) + v^+(x, y)$ is a solution of the posed Dirichlet problem in the region B'.

Since the Dirichlet problem in B' can have only a single solution, we have

$$u^-(x, y) + v^+(x, y) = u^+(x, y) + v^-(x, y) = w(x, y).$$

On the other hand, the resolution of $w(x, y)$ into two summands $u(x, y)$ and $v(x, y)$ satisfying the requirements we have stipulated can, as is stated above at the beginning of No. 1, only be unique. Therefore we must have

$$u^+(x, y) = u^-(x, y) = u(x, y), \quad v^+(x, y) = v^-(x, y) = v(x, y).$$

Let us now return to the sequences of approximations u_n and v_n constructed by us initially. We have shown that the majorant, u_n^+ and v_n^+, and minorant, u_n^- and v_n^- sequences converge to the functions $u(x, y)$ and $v(x, y)$ solving the problem of the resolution of $w(x, y)$ into two summands regular in B_1 and B_2 respectively. Our sequences u_n and v_n must converge to the same functions:

$$\lim u_n(x, y) = u(x, y), \quad \lim v_n(x, y) = v(x, y), \quad w(x, y) = u(x, y) + v(x, y).$$

With this the converge of the Schwarz-Neumann algorithm is established, under the five assumptions we have made, in the Dirichlet problem for the product of two regions.

A few words yet about practical computations by the Schwarz-Neumann method. For the solution of the entire problem it is sufficient to know how to find either one of the sought functions $u(x, y)$ or $v(x, y)$ on $\bar{\alpha}$ or $\bar{\beta}$ respectively. Indeed, let us have found the values that $u(x, y)$ acquires on $\bar{\alpha}$. The values of $u(x, y)$ on the remaining part α of the contour L_1 have been assigned by us beforehand. We shall know, accordingly, the values of $u(x, y)$ on the entire boundary L_1 of the region B_1. Solving the Dirichlet problem in B_1, we shall, by the known contour values of $u(x, y)$, have recovered it everywhere within B_1.

Moreover, on computing the values that $u(x, y)$ acquires on $\bar{\beta}$ and subtracting them from $f(M)$, we shall have determined the values of $f(M) - u(M)$, which v acquires at the points of $\bar{\beta}$. After this the values of $v(x, y)$ everywhere on L_2 will be known to us, since on the remaining part β of the contour L_2 we have considered the values of $v(x, y)$ to be equal to zero. Having solved the Dirichlet problem in B_2, we shall have found, by the known contour values of $v(x, y)$, its values everywhere within B_2.

Let us now assume that, proceeding from some considerations or other, the character of which is immaterial to us now, we have assigned approximate values $\varphi^*(M)$ to the function $u(x, y)$ on $\bar{\alpha}$. We explained above that these values can be assigned completely arbitrarily, provided the condition of the piece-wise continuity of the boundary values of $u(x, y)$ is fulfilled.

Together with the known exact values of $u(x, y)$ on α, these values will give approximate values for $u(x, y)$ on the entire boundary L_1. Using them we find approximately the values acquired by $u(x, y)$, not everywhere in B_1, but only at the points of the part $\bar{\beta}$ of the contour L_2 lying within B_1. Subtracting from the assigned values $f(M)$ on $\bar{\beta}$ the values of $u(x, y)$ found, we shall have determined approximately the values of $v(x, y)$ on $\bar{\beta}$. Together with the null values on the remaining part β, they will give approximately the contour values of $v(x, y)$ on the entire boundary L_2. Solving by them the Dirichlet problem now in region B_2, we shall find the values that $v(x, y)$ acquires on $\bar{\alpha}$. And finally, subtracting them from the assigned values of the function $f(M)$, we shall have determined corrected values of $u(x, y)$ on $\bar{\alpha}$. Generally speaking they will be closer to the true values of $u(x, y)$ on $\bar{\alpha}$ than those assigned by us.

If the corrected values obtained by us after one cycle of computations coincide, within the specified limits of accuracy, with the initial values assigned by us, we terminate the computation on this, since the initial values have proved to coincide with the true values within the adopted limits of accuracy.

If the corrected values are different from the initial ones, however, then we adopt them as initial values and repeat the entire cycle of computations. These computations are continued until the values of $u(x, y)$ on $\bar{\alpha}$ at the beginning of a cycle coincide with the values at the end of it.

2. An example of the investigation of the convergence of the Schwarz-Neumann method. An estimate of the rate of convergence in the case of the Laplace equation. We will present an example of a differential equation for which are verified the five sufficient conditions of the convergence of the Schwarz-Neumann method enumerated in No. 1. We will take an equation of the form:

$$\frac{\partial^2 u}{\partial x^2} + \frac{\partial^2 u}{\partial y^2} - q(x, y)u = 0. \tag{6}$$

We preserve the previous assumption about the coefficient $q(x, y)$, considering it to be, namely, continuous and non-negative. In the investigations of No. 2 § 1 we showed that for equation (6) are fulfilled the first four requirements that we formulated at the beginning of No. 1. We will now indicate sufficient conditions for the fulfillment of the last assumption, V, stated by us in the preceding No. in the form of inequality (5).

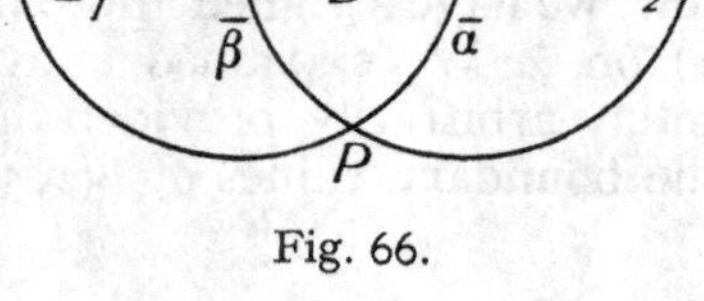

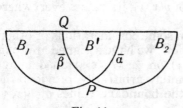

Fig. 66.

We will consider for simplicity that $\bar{\beta}$ consists of one arc of a curve and is some section of the region B_1 between points P and Q.

It may happen that the end of the arc $\bar{\beta}$ is simultaneously an end of one of the arcs figuring in $\bar{\alpha}$. Such is the end P for the case depicted in Fig. 66. We will consider that in this case $\bar{\beta}$ has at its end a tangent not coinciding with the tangent to $\bar{\alpha}$ at this point.

It may also turn out that an end of the arc $\bar{\beta}$ will not be simultaneously an end of $\bar{\alpha}$. Such is the end Q in Fig. 66. We will then impose no limitations on the behavior of $\bar{\beta}$ near the end of the arc.

Let us now consider a solution of equation (6) acquiring the value unity on $\bar{\alpha}$ and the value zero on α (in No. 1 we designated it $U_{\bar{\alpha},1}$). Estimate the values that it acquires on $\bar{\beta}$. We represent $U_{\bar{\alpha},1}$ as the sum of two terms,

$$U_{\bar{\alpha},1} = v_* + w_*, \tag{7}$$

the first of which v_* is a harmonic function having the same values on the contour L_1 of the region B_1 as does $U_{\bar{\alpha},1}$, the second w_* acquiring on L_1 values equal to zero and satisfying within B_1 the Poisson equation:

$$\Delta w_* = q(x, y)\, U_{\bar{\alpha},1}. \tag{8}$$

Since the boundary values of $U_{\bar{\alpha},1}$ on L_1 are non-negative, $U_{\bar{\alpha},1}$ will be non-negative everywhere in B_1. In view of the fact that $q(x, y) \geqslant 0$, the right side of equation (8) will be non-negative everywhere in B_1. It is known, moreover, that a solution of a Poisson equation (8) with a non-negative right side that has on the boundary of the region values equal to zero cannot acquire positive values within the region [1]:

$$w_* \leqslant 0.$$

Therefore

$$U_{\bar{\alpha},1} \leqslant v_*.$$

The harmonic function v_* takes, within the region B_1, values less than unity. This will be so on the arc $\bar{\beta}$, in particular. We shall have shown that its values on $\bar{\beta}$ will be less than a certain proper fraction if we establish that for an approach to the ends P and Q of this arc, v_* tends to values less than unity.

Where we approach an end of $\bar{\beta}$ that is not simultaneously an end of an arc of $\bar{\alpha}$ (for the case depicted in the figure, Q will be such an end), v_* will tend to zero.

If in following $\bar{\beta}$ we approach, however, an end of $\bar{\beta}$ that is simultaneously an end of an arc of $\bar{\alpha}$ (P is such an end), v_* will tend to a

---

[1] For, indeed, if $G_1(x, y; \xi, \eta)$ is Green's function for the region B_1 with pole at the point (ξ, η), then w_* has the following expression:

$$w_*(x, y) = -\iint_{B_1} G_1(x, y; \xi, \eta) q(\xi, \eta) U_{\bar{\alpha},1}(\xi, \eta) d\xi\, d\eta,$$

and since the entire function standing under the double integral sign is non-negative, it is evident that $w_* \leqslant 0$. See, for example, Courant and Hilbert, [1], Chap. V, § 14, No. 5.

limit less than unity, since the tangent to $\bar{\beta}$ at such a point does not coincide with the tangent to $\bar{\alpha}$ at this point. The last easily follows from the reasoning that we carried through in No. 2 § 1 in proving Schwarz's lemma.

Thus there must exist a positive number $\vartheta < 1$ such that at all points

$$U_{\bar{\alpha},1} \leqslant v_* \leqslant \vartheta.$$

The boundary values of the function $V_{\bar{\beta},U_{\bar{\alpha},1}}$ here turn out to be less than ϑ, and accordingly the values that this function takes in B_2 will also be less than ϑ, and inequality (5), given our assumptions about $\bar{\beta}$, is fulfilled for equation (6).

This inequality will obviously also be fulfilled in case $\bar{\beta}$ consists of a finite number of arcs for each of which is fulfilled the assumption stated at the beginning of the No.

Our investigations have established only the fact of the convergence of the Schwarz-Neumann process to the solution of the problem, and have not given an estimate of the difference between u_n and v_n and the limit functions u and v.

We will now make such an estimate for the Laplace equation obtained from (6) for $q(x, y) = 0$. We keep the same trend of thought that we pursued in No. 2 § 1. With respect to the regions B_1 and B_2 we make the previous assumptions. We will consider that $\bar{\alpha}$ and $\bar{\beta}$ each consists of one arc and that both of them satisfy the conditions of Schwarz's lemma (see No. 2 § 1). The existence of the functions $u(x, y)$ and $v(x, y)$ solving the problem of the representation of $w(x, y)$ as the sum of two harmonic functions has been established by us in the preceding investigations. We will now interest ourselves in the estimation of the differences $u(x, y) - u_n(x, y)$ and $v(x, y) - v_n(x, y)$. They can be determined in succession as solutions of Dirichlet problems in B_1 and B_2 for the boundary values:

$$u(x, y) - u_1(x, y) = \begin{cases} 0 & \text{on } \alpha, \\ u - \bar{\varphi}(M) & ,, \ \bar{\alpha}, \end{cases}$$

$$v(x, y) - v_1(x, y) = \begin{cases} 0 & \text{on } \beta, \\ u_1 - u & ,, \ \bar{\beta}, \end{cases}$$

. . . .

$$u(x, y) - u_n(x, y) = \begin{cases} 0 & \text{on } \alpha, \\ v_{n-1} - v & ,, \ \bar{\alpha}, \end{cases}$$

$$v(x, y) - v_n(x, y) = \begin{cases} 0 & \text{on } \beta, \\ u_n - u & ,, \ \bar{\beta}. \end{cases}$$

We will denote by N the upper bound of the absolute value of $u - \bar{\varphi}(M)$ on $\bar{\alpha}$. Since both u and φ are bounded functions, N will be a finite number. N of course depends on the estimation of the values of u on $\bar{\alpha}$, unknown to us, and which we shall not discover. The result that we obtain below will serve, therefore, only for a definition of the order of $u - u_n$ and $v - v_n$, and not for an exact determination of the error of the nth approximations, since it will contain the quantity N that is unknown to us.

The curve $\bar{\beta}$ will divide the boundary L_1 into two parts. On one of them $\bar{\alpha}$ will be located.

The boundary values of $u - u_1$ on α are equal to zero, and on $\bar{\alpha}$ are not greater than N in magnitude. By Schwarz's lemma there must exist a $\vartheta_1 < 1$, depending only on the geometrical properties of B_1 and $\bar{\beta}$, such that on the arc $\bar{\beta}$ we shall have

$$|u - u_1| \leqslant N\vartheta_1.$$

The arc $\bar{\alpha}$ will divide the boundary L_2 of the region B_2 into two parts, on one of which will lie $\bar{\beta}$.

The boundary values of $v - v_1$ on β are equal to zero, and on $\bar{\beta}$ are not greater than $N\vartheta_1$ in magnitude. By the same lemma of Schwarz, there must be a number $\vartheta_2 < 1$ such that at all points of the arc $\bar{\alpha}$ we shall have:

$$|v - v_1| \leqslant N\vartheta_1\vartheta_2.$$

Continuing this reasoning, we obtain the following estimates:

$$|u - u_n| \leqslant N\vartheta_1^n\vartheta_2^{n-1} \text{ on } \bar{\beta}, \quad |v - v_n| \leqslant N\vartheta_1^n\vartheta_2^n \text{ on } \bar{\alpha}.$$

From this are immediately obtained the estimates for the differences under study, $u - u_n$ and $v - v_n$, everywhere in the regions B_1 and B_2 respectively. Let us consider the first of these differences. Its boundary values on α are equal to zero, and on $\bar{\alpha}$, to $v_{n-1} - v$, which in absolute value does not exceed, according to the estimate obtained, $N\vartheta_1^{n-1}\vartheta_1^{n-1}$. Therefore everywhere in B_1 there must obtain:

$$|u - u_n| \leqslant N\vartheta_1^{n-1}\vartheta_2^{n-1}.$$

In the same way, for $v - v_n$ in the region B_2 we obtain:

$$|v - v_n| \leqslant N\vartheta_1^n\vartheta_2^{n-1}.$$

Both the last inequalities show that the departure of the nth approximations u_n and v_n from the limit functions u and v diminishes not more slowly than the terms of a geometrical progression with ratio $q = \vartheta_1\vartheta_2$.

3. Reduction of the Schwarz-Neumann method to the solution of a system of integral equations by successive approximations. In No. 3 § 1 we showed that Schwarz's method for the solution of the Dirichlet problem in the case of the Laplace equation for the sum of two regions can be reduced to the solution by successive approximations of the system of integral equations identified by number (11).

In like manner, the Schwarz-Neumann method for the solution of the Dirichlet problem for the intersection of two regions can be reduced to the solution of some system of integral equations by successive approximations.

During the construction of this system we will keep the notation and assumptions of No. 1 for the regions B_1 and B_2 and their boundaries. We saw there that for the determination of the summand $u(x, y)$ it is necessary to assign its values on the part $\bar{\alpha}$ of the contour L_1. We considered in No. 1 that $u(x, y)$ acquires on γ the values $f(M)$, as does $w(x, y)$, and assigned only its values on $\alpha - \gamma$. We denoted these values by $\varphi(M)$. For the determination of the second summand $v(x, y)$, however, we have to assign its values on the part $\beta - \gamma$ of the boundary L_2. We have taken them equal to zero. The values of $v(x, y)$ on γ are determined by the assigned values of $u(x, y)$, and are equal to $f(M) - u(x, y)$. For our choice of the boundary values of $u(x, y)$, the values of $v(x, y)$ on γ are equal to zero. Finally, the values of $u(x, y)$ and $v(x, y)$ on $\bar{\alpha}$ and $\bar{\beta}$ are subject to determination.

We will denote Green's function for the region B_1 with pole at the point (ξ, η) by $G_1(x, y; \xi, \eta)$. We will call Green's function for B_2 $G_2(x, y; \xi, \eta)$.

Let n be the external normal to the region, and let $u(x, y)$ and $v(x, y)$ have the following expressions in terms of their boundary values:

$$\left.\begin{aligned} u(x, y) &= -\int\limits_{L_1} u(\xi, \eta) \frac{\partial G_1(x, y; \xi, \eta)}{\partial n} ds \text{ in } B_1, \\ v(x, y) &= -\int\limits_{L_2} v(\xi, \eta) \frac{\partial G_2(x, y; \xi, \eta)}{\partial n} ds \text{ in } B_2. \end{aligned}\right\} \tag{9}$$

Here s is the length of the arc of the contour from some fixed point to a variable point of integration $M(\xi, \eta)$.

If we take into consideration the fact that the values of v on β equal zero, and the values of u on α are known, and introduce the designation:

$$F(x, y) = -\int\limits_{\alpha-\gamma} \varphi(M) \frac{\partial G_1}{\partial n} ds - \int\limits_{\gamma} f(M) \frac{\partial G_1}{\partial n} ds, \tag{10}$$

we then obtain from the preceding equations:

$$\left.\begin{aligned} u(x, y) &= F(x, y) - \int\limits_{\bar{\alpha}} u(\xi, \eta) \frac{\partial G_1}{\partial n} ds, \\ v(x, y) &= -\int\limits_{\bar{\beta}} v(\xi, \eta) \frac{\partial G_2}{\partial n} ds. \end{aligned}\right\} \tag{11}$$

$F(x, y)$ is a function harmonic in B_1, acquiring on $\bar{\alpha}$ values equal to zero, on γ the values $f(M)$, and on $\alpha - \gamma$ the values $\varphi(M)$.

We will consider that in the first of equations (11) the point (x, y) lies on $\bar{\beta}$, and in the second, on $\bar{\alpha}$.

Let us take into account the fact that the values of $u(x, y)$ on $\bar{\beta}$ are equal to $f(x, y) - v(x, y)$, and that the values of $v(x, y)$ on $\bar{\alpha}$ are equal to $f(x, y) - u(x, y)$. By $f(x, y)$ here is denoted the value of $f(M)$ at the point (x, y). Equations (11) give the following system of integral equations for the unknown values of $u(x, y)$ and $v(x, y)$ on $\bar{\alpha}$ and $\bar{\beta}$ respectively:

$$\left.\begin{aligned} v(x, y) &= f(x, y) - F(x, y) + \int_{\bar{\alpha}} u(\xi, \eta) \frac{\partial G_1}{\partial n} ds \text{ on } \bar{\beta}, \\ u(x, y) &= f(x, y) + \int_{\bar{\beta}} v(\xi, \eta) \frac{\partial G_2}{\partial n} ds \qquad \text{on } \bar{\alpha}. \end{aligned}\right\} \tag{12}$$

The theory of this system is similar to the theory of the analogous system (11) No. 3 § 1.

By the conditions of the Dirichlet problem posed by us, the sought values of $u(x, y)$ and $v(x, y)$ must yield the bounded solution of system (12). Conversely, let $u^*(x, y)$ and $v^*(x, y)$ be bounded and let them satisfy system (12). We will show that they give precisely these sought values of the harmonic functions u and v on $\bar{\alpha}$ and $\bar{\beta}$.

Using them we will construct, in conformity with equations (11), the harmonic functions $u(x, y)$ and $v(x, y)$:

$$u(x, y) = F(x, y) - \int_{\bar{\alpha}} u^*(\xi, \eta) \frac{\partial G_1}{\partial n} ds,$$

$$v(x, y) = - \int_{\bar{\beta}} v^*(\xi, \eta) \frac{\partial G_2}{\partial n} ds.$$

They are both bounded, since the function $F(x, y)$ is bounded, as are the u^* and v^* which stand under the integral signs as multipliers of the normal derivatives of the Green functions. The first of them is regular in B_1, and the second in B_2. Their boundary values on the contours of these regions are easily discovered if we recall the value of $F(x, y)$ and the properties of the integral terms:

$$u(x, y) = \begin{cases} \varphi(M) & \text{on } \alpha - \gamma, \\ f(M) & \text{,, } \gamma, \\ u^*(x, y) & \text{,, } \bar{\alpha}, \end{cases}$$

$$v(x, y) = \begin{cases} 0 & \text{on } \beta, \\ v^*(x, y) & \text{,, } \bar{\beta}. \end{cases}$$

The sum of these two functions:

$$u(x, y) + v(x, y) = F(x, y) - \int_{\bar{\alpha}} u^*(\xi, \eta) \frac{\partial G_1}{\partial n} ds - \int_{\bar{\beta}} v^*(\xi, \eta) \frac{\partial G_2}{\partial n} ds$$

will be harmonic and regular in the common part B' of the regions B_1 and B_2. It remains only for us to show that $u + v$ acquires on $\bar{\alpha}$ and $\bar{\beta}$ the assigned values $f(M)$. Let the point (x, y) approach a certain point N on $\bar{\alpha}$. The right side of the last equation will thereby tend to the value

$$u^*(N) - \int_{\bar{\beta}} v^*(\xi, \eta) \frac{\partial G_2}{\partial n} ds,$$

but, by the second of equations (12),

$$\int_{\bar{\beta}} v^*(\xi, \eta) \frac{\partial G_2}{\partial n} ds = u^*(N) - f(N),$$

and therefore the preceding expression is equal to

$$u^*(N) - u^*(N) + f(N) = f(N).$$

Thus if $(x, y) \to N$, we have

$$\lim [u(x, y) + v(x, y)] = f(N).$$

Thus also is it proved that for an approach of the point (x, y) to any point N on $\bar{\beta}$, $u + v$ will tend to the assigned value $f(N)$ at this point.

Finding a bounded solution of system (12) is equivalent to solving the Dirichlet problem in the region B'. From this it is evident, by the way, that system (12) can have only one bounded solution, since if it had two different bounded solutions, there would be found by them two different pairs of functions u and v solving the problem posed — of representing w as the sum of two summands regular in B_1 and B_2 respectively and such that the first of them, u, acquires assigned values on α and the second, v, on $\beta - \gamma$. As we revealed at the beginning of No. 1, such a representation is necessarily unique. Both the bounded solutions of system (12) must necessarily coincide, accordingly.

We will now show that the method of successive approximations for system of integral equations (12) is essentially equivalent to the Schwarz-Neumann algorithm described above. Let us arbitrarily assign on $\bar{\alpha}$ a piece-wise continuous function $\varphi^*(M)$. We will adopt it as the initial approximation, $u_1(M)$, to $u(x, y)$, and substitute it in the first equation of system (12). After the substitution we shall obtain some function $v_1(M)$ defined at the points of $\bar{\beta}$. We take it as the first approximation to $v(x, y)$, substitute $v_1(M)$ for $v(\xi, \eta)$ in the right side of the second equation of the system and obtain, after computations, a function $u_2(M)$, defined at the points of $\bar{\alpha}$. This is the second approximation to $u(x, y)$, etc.

All these computations are equivalent to the following: after assigning $\varphi^*(M)$ on $\bar{\alpha}$ we construct in B_1 a harmonic function

$$u_1(x, y) = F(x, y) - \int_{\bar{\alpha}} \varphi^*(M) \frac{\partial G_1}{\partial n} ds,$$

solving the Dirichlet problem in this region by the boundary data:

$$u_1(x, y) = \begin{cases} \varphi(M) & \text{on } \alpha - \gamma, \\ f(M) & \text{,,} \ \gamma, \\ \varphi^*(M) & \text{,,} \ \bar{\alpha}. \end{cases}$$

Next we find its values on $\bar{\beta}$ and subtract them from the assigned values $f(M)$ on $\bar{\beta}$, i.e., we perform all the computations needed for the Schwarz-Neumann algorithm for the determination of the boundary values on $\bar{\beta}$ of the first approximation $v_1(x, y)$ to $v(x, y)$. As the result we obtain:

$$f(x, y) - F(x, y) + \int_{\bar{\alpha}} \varphi^*(M) \frac{\partial G_2}{\partial n} ds = f(N) - u_1(x, y) \text{ on } \bar{\beta}.$$

But this is precisely the result of substituting, in the right side of the first equation of (12), the function $\varphi^*(M)$ for $u(\xi, \eta)$. Thus the first approximation $v_1(M)$ to $v(x, y)$ that we have found by the method of successive approximations will coincide with the boundary value on $\bar{\beta}$ of the first approximation $v_1(x, y)$ in the Schwarz-Neumann algorithm.

Continuing this reasoning, we satisfy ourselves that the successive approximations we construct for the solution of the system of integral equations (12) coincide with the boundary values on $\bar{\alpha}$ and $\bar{\beta}$ of the successive approximations in the Schwarz-Neumann algorithm.

In practical computations, it goes without saying that there is no need to compute the successive approximations for both the unknown functions $u(x, y)$ and $v(x, y)$. System (12) can be transformed so that only one of the unknown functions appears in it. To eliminate one of the unknown functions, for instance $v(x, y)$, one must substitute for $v(\xi, \eta)$ in the second equation of the system its expression in terms of u from the first equation. After the elimination we obtain an equation with one unknown function, next solving it by successive approximations.

§ 3. AN EXAMPLE OF THE APPLICATION OF SCHWARZ'S METHOD

We will consider one example of the application of Schwarz's method to the solution of the Dirichlet problem in the case of the Laplace equation for the sum of two regions. Let the region B be the "corner" $OACDEFO$, the dimensions and position of which relative to the coordinate axes are shown in Fig. 67.

We will seek the function $w(x, y)$, harmonic in B and acquiring on its contour the following values:

$$w(x, y) = \begin{cases} 0 & \text{on } EFOAC, \\ f(x) & \text{,,} \ DC, \\ \varphi(y) & \text{,,} \ DE, \end{cases}$$

where $f(x)$ and $\varphi(y)$ are given functions.

We will adopt as B_1 the rectangle $OACPO$ and as B_2 the rectangle $OQEFO$. The values acquired in B_1 by the function under consideration, $w(x, y)$, we will call $u(x, y)$, and its values in B_2, $v(x, y)$. According to Schwarz, for the determination of $u(x, y)$ and $v(x, y)$ in each of the regions B_1 and B_2 one constructs the sequences of approximations $u_1(x, y)$, $u_2(x, y)$, ... and $v_1(x, y)$, $v_2(x, y)$, ... to them by solving in these regions the Dirichlet problem under boundary conditions that we shall speak about a few lines below.

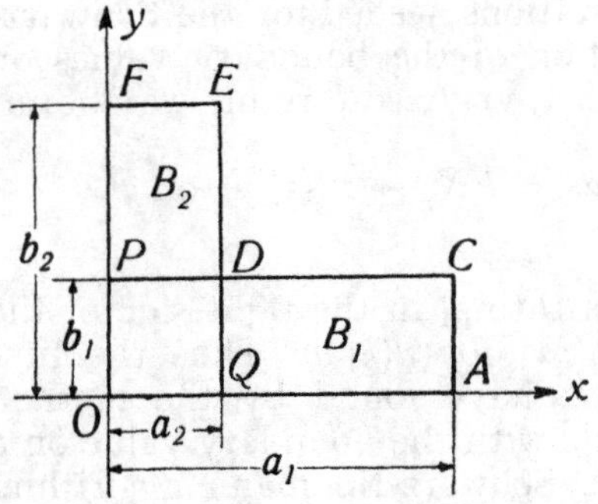

Fig. 67.

Let us begin with B_1. The values of $u(x, y)$ are known only on the part $POACD$ of the contour of this rectangle, equalling zero on $POAC$ and $f(x)$ on CD. On the part DP, however, they are unknown. We will assign them on DP somehow. Expressing it differently, we will continue $f(x)$ from CD to DP, calling the continued function $f_1(x)$. We will consider it to be piece-wise continuous and to satisfy the conditions of the Dirichlet theorem on Fourier expansions. The method of continuation is otherwise arbitrary.

We next construct in B_1 and B_2 the sequences of functions $u_1(x, y)$, $u_2(x, y)$, ... and $v_1(x, y)$, $v_2(x, y)$, ..., solving the Dirichlet problem under the boundary conditions:

$$u_1(x, y) = \begin{cases} 0 & \text{on } POAC, \\ f_1(x) & \text{,, } CP, \end{cases} \qquad v_1(x, y) = \begin{cases} 0 & \text{on } EFOQ, \\ \varphi(y) & \text{,, } DE, \\ u_1(a_2, y) & \text{,, } QD, \end{cases}$$

. . . .

$$u_n(x, y) = \begin{cases} 0 & \text{on } POAC, \\ f(x) & \text{,, } CD, \\ v_{n-1}(x, b_1) & \text{,, } DP, \end{cases} \qquad v_n(x, y) = \begin{cases} 0 & \text{on } EFOQ, \\ \varphi(y) & \text{,, } DE, \\ u_n(a_2, y) & \text{,, } QD. \end{cases}$$

For the construction of each of the functions $u_n(x, y)$, there must be solved the Dirichlet problem in the rectangle B_1 under boundary conditions of the form

$$u_n(x, y) = \begin{cases} 0 & \text{on } POAC, \\ f_n(x) & \text{,, } CP. \end{cases}$$

In our problem $f_n(x)$ $(n \geq 2)$ is equal to $f(x)$ for $a_2 \leqslant x \leqslant a_1$ and to $v_{n-1}(x, b_1)$ for $0 \leqslant x \leqslant a_2$. The solution of such a problem is known, and can be found by means of series by the method indicated in No. 1 § 1 of the first chapter of this book. It has the following form. Expand

$u_n(x, b_1) = f_n(x)$ in a Fourier sine series,

$$u_n(x, b_1) = f_n(x) = \sum_{k=1}^{\infty} \alpha_k^{(n)} \sin k \frac{\pi x}{a_1}. \tag{13}$$

Then $u_n(x, y)$ will be represented by the series.:

$$u_n(x, y) = \sum_{k=1}^{\infty} \frac{\alpha_k^{(n)}}{\operatorname{sh} k \frac{\pi b_1}{a_1}} \operatorname{sh} k \frac{\pi y}{a_1} \sin k \frac{\pi x}{a_1}. \tag{14}$$

To construct $v_n(x, y)$, one must solve the Dirichlet problem in B_2 in accordance with the boundary values

$$v_n(x, y) = \begin{cases} 0 & \text{on } EFOQ, \\ \varphi_n(y) & \text{,, } QE, \end{cases}$$

where $\varphi_n(y)$ is equal to $u_n(a_2, y)$ for $0 \leqslant y \leqslant b_1$, and to $\varphi(y)$ for $b_1 \leqslant y \leqslant b_2$.

If the expansion of $\varphi_n(y)$ in a sine series is

$$v_n(a_2, y) = \varphi_n(y) = \sum_{k=1}^{\infty} \beta_k^{(n)} \sin k \frac{\pi y}{b_2}, \tag{15}$$

then $v_n(x, y)$ will be represented by the series

$$v_n(x, y) = \sum_{k=1}^{\infty} \frac{\beta_k^{(n)}}{\operatorname{sh} k \frac{\pi a_2}{b_2}} \cdot \operatorname{sh} k \frac{\pi x}{b_2} \sin k \frac{\pi y}{b_2}. \tag{16}$$

The construction of $u_n(x, y)$ by use of the known $v_{n-1}(x, y)$ reduces to two operations: 1) the coefficients $\alpha_k^{(n)}$ of the expansion of $f_n(x)$ in series (13) must be found; and 2) series (14) must be formed in accordance with the coefficients $\alpha_k^{(n)}$ found. Analogously, for the determination of $v_n(x, y)$ one must: 1) compute the coefficients $\beta_k^{(n)}$ of the expansion of $\varphi_n(y)$ in series (15); and 2) form series (16) in accordance with the coefficients $\beta_k^{(n)}$ found.

We will now establish recurrence formulas for the computation of the coefficients $\alpha_k^{(n)}$ and $\beta_k^{(n)}$:

$$\alpha_k^{(n)} = \frac{2}{a_1} \int_0^{a_1} f_n(x) \sin k \frac{\pi x}{a_1} dx = \frac{2}{a_1} \int_0^{a_2} v_{n-1}(x, b_1) \sin k \frac{\pi x}{a_1} dx + \\ + \frac{2}{a_1} \int_{a_2}^{a_1} f(x) \sin k \frac{\pi x}{a_1} dx.$$

If in series (16) n be replaced by $n - 1$, b_1 be substituted for y, and the series for $v_{n-1}(x, b_1)$ obtained by this be introduced into the first of the integrals on the right side of the last equation, we shall have reduced it to the form

$$\alpha_k^{(n)} = \sum_{s=1}^{\infty} A_{k,s}\beta_s^{(n-1)} + C_k, \tag{17}$$

where

$$C_k = \frac{2}{a_1}\int_{a_2}^{a_1} f(x) \sin k\frac{\pi x}{a_1}\,dx,$$

$$A_{k,s} = \frac{2}{a_1}\,\frac{\sin s\dfrac{\pi b_1}{b_2}}{\operatorname{sh} s\dfrac{\pi a_2}{b_2}}\int_0^{a_2} \operatorname{sh} s\frac{\pi x}{b_2}\sin k\frac{\pi x}{a_1}\,dx =$$

$$= \frac{2}{\pi}\cdot\frac{s^2a_1^2}{s^2a_1^2 + k^2b_2^2}\cdot\frac{\sin s\dfrac{\pi b_1}{b_2}}{\operatorname{sh} s\dfrac{\pi a_2}{b_2}}\left\{\frac{b_2}{sa_1}\sin k\frac{\pi a_2}{a_1}\operatorname{ch} s\frac{\pi a_2}{b_2} - \right.$$

$$\left. - \frac{kb_2^2}{s^2a_1^2}\cos k\frac{\pi a_2}{a_1}\cdot\operatorname{sh} s\frac{\pi a_2}{b_2}\right\}. \tag{18}$$

Formulas permitting the calculation of the $\beta_k^{(n)}$ from the $\alpha_k^{(n)}$ are obtained in the same fashion:

$$\beta_k^{(n)} = \sum_{s=1}^{\infty} B_{k,s}\alpha_s^{(n)} + D_k,\quad D_k = \frac{2}{b_2}\int_{b_1}^{b_2}\varphi(y)\sin k\frac{\pi y}{b_2}\,dy, \tag{19}$$

$$B_{k,s} = \frac{2}{b_2}\cdot\frac{\sin s\dfrac{\pi a_2}{a_1}}{\operatorname{sh} s\dfrac{\pi b_1}{a_1}}\int_0^{b_1}\operatorname{sh} s\frac{\pi y}{a_1}\sin k\frac{\pi y}{b_2}\,dy = \frac{2}{\pi}\cdot\frac{s^2b_2^2}{s^2b_2^2 + k^2a_1^2}\times$$

$$\times\frac{\sin s\dfrac{\pi a_2}{a_1}}{\operatorname{sh} s\dfrac{\pi b_1}{a_1}}\left\{\frac{a_1}{sb_2}\sin\frac{\pi b_1}{b_2}\operatorname{ch} s\frac{\pi b_1}{a_1} - \frac{ka_1^2}{s^2b_2^2}\cos k\frac{\pi b_1}{b_2}\operatorname{sh} s\frac{\pi b_1}{a_1}\right\}.$$

At the beginning of the computations we assigned the values of $u_1(x, y)$ on PC, considering them to equal $f_1(x)$. Expanding $f_1(x)$ in a sine series,

$$f_1(x) = \sum_{k=1}^{\infty}\alpha_k^{(1)}\sin k\frac{\pi x}{a_1}, \tag{20}$$

we determine the initial coefficients

$$\alpha_k^{(1)} = \frac{2}{a_1} \int_0^{a_1} f_1(x) \sin k \frac{\pi x}{a_1} \, dx. \tag{21}$$

Knowing the $\alpha_k^{(1)}$, and using equations (19), we find $\beta_k^{(1)}$, afterwards determining $\alpha_k^{(2)}$ from equations (17), and so forth.

From the general theory of Schwarz's method, which we expounded in § 1, it is seen that the successive approximations $u_1, u_2, \ldots, v_1, v_2, \ldots$ will converge to the limit functions $u(x, y)$ and $v(x, y)$ solving the posed Dirichlet problem. Moreover, from the estimate of the deviations $u(x, y) - u_n(x, y)$ and $v(x, y) - v_n(x, y)$ of the nth approximations from the limit functions that we have given in No. 2 § 1, it is seen that the convergence will be uniform. Hence it follows, in particular, that the sequences

$$\alpha_k^{(1)}, \alpha_k^{(2)}, \ldots, \alpha_k^{(n)}, \ldots$$
$$\beta_k^{(1)}, \beta_k^{(2)}, \ldots, \beta_k^{(n)}, \ldots$$

obtained in our computations will converge, and that their limit values α_k and β_k will be the Fourier coefficients of the expansions of $u(x, b_1)$ and $v(a_2, y)$ in sine series.

If $f(x)$ and $\varphi(y)$ satisfy the conditions of the Dirichlet theorem on the expansion of a function in a Fourier series, then the series

$$u(x, b_1) = \sum_{k=1}^{\infty} \alpha_k \sin k \frac{\pi x}{a_1}, \quad v(a_2, y) = \sum_{k=1}^{\infty} \beta_k \sin k \frac{\pi y}{b_2}$$

will converge.

Using the α_k and β_k found, we construct the series

$$\left.\begin{aligned} u(x, y) &= \sum_{k=1}^{\infty} \frac{\alpha_k}{\operatorname{sh} k \dfrac{\pi b_1}{a_1}} \cdot \operatorname{sh} k \frac{\pi y}{a_1} \sin k \frac{\pi x}{a_1}, \\ v(x, y) &= \sum_{k=1}^{\infty} \frac{\beta_k}{\operatorname{sh} k \dfrac{\pi a_2}{b_2}} \operatorname{sh} k \frac{\pi x}{b_2} \sin k \frac{\pi y}{b_2}. \end{aligned}\right\} \tag{22}$$

They will also be convergent and will represent the sought functions $u(x, y)$ and $v(x, y)$ in regions B_1 and B_2 respectively.

In practical computations the construction of the sequences $\alpha_k^{(1)}$, $\alpha_k^{(2)}, \ldots, \beta_k^{(1)}, \beta_k^{(2)}, \ldots$ is continued until the values of the coefficients $\alpha_k^{(n)}, \beta_k^{(n)}$ are repeated by the succeeding ones, $\alpha_k^{(n+1)}, \beta_k^{(n+1)}$, within the adopted limits of accuracy.

One more observation on the subject of equations (17) and (19). Consider the following infinite system of linear equations in an infinite number of unknowns α_k and β_k:

$$\alpha_k = \sum_{s=1}^{\infty} A_{k,s}\beta_s + C_k, \quad \beta_k = \sum_{s=1}^{\infty} B_{k,s}\,\alpha_s + D_k. \tag{23}$$

Let us apply the method of successive approximation to it. We will adopt as the initial values for the α_k the coefficients $\alpha_k^{(1)}$ and substitute them in the right side of the second of equations (23). The values obtained after the substitution we adopt as the first approximations $\beta_k^{(1)}$ to the unknowns β_k. We substitute them in the right side of the first of equations (23), take the result of the substitution as the second approximation $\alpha_k^{(2)}$ to α_k, and so on. Equations (17) and (19) will give rules for the computation of the nth approximations $\alpha_k^{(n)}$ and $\beta_k^{(n)}$ to α_k and β_k from the known $(n-1)$st approximation $\beta_k^{(n-1)}$ to β_k. Thus the method of determining the α_k and β_k that we have expounded is essentially none other than the method of successive approximations for the solution of infinite system (23), when (21) are adopted as the initial approximations for the α_k.

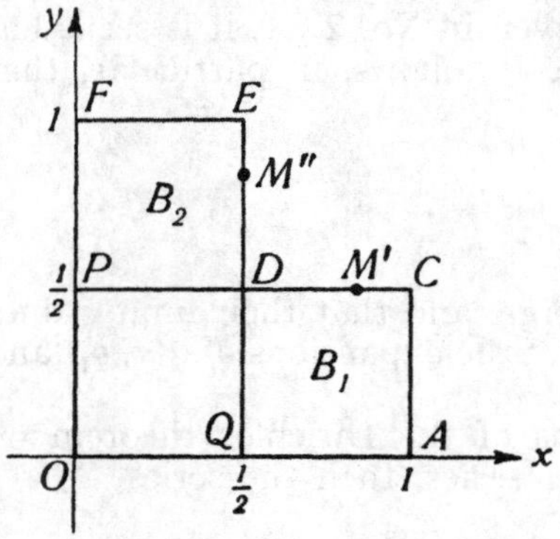

Fig. 68.

We have just expounded the general trend of the idea of solving the Dirichlet problem for a "corner" if series are used. The subsequent computations will depend on the geometrical properties of the "corner" and the boundary values of the sought function $w(x, y)$. We will perform them under the following assumptions. Let the "corner" consist of three equal squares with side $\frac{1}{2}$ (see Fig. 68). The boundary values of $w(x, y)$ we will consider to be the following:

$$w(x, y) = \begin{cases} 0 & \text{on } EFOAC, \\ \sin 2\pi x & \text{,, } DC, \\ -\sin 2\pi y & \text{,, } DE. \end{cases}$$

At the outset we will make several observations which will simplify the computations. Series (22), representing $u(x, y)$ and $v(x, y)$ in the rectangles B_1 and B_2, since $a_1 = b_2 = 1$, $a_2 = b_1 = \frac{1}{2}$, will take the form:

$$\left.\begin{aligned} u(x, y) &= \sum_{k=1}^{\infty} \frac{\alpha_k}{\operatorname{sh} k\dfrac{\pi}{2}} \operatorname{sh} k\pi y \sin k\pi x, \\ v(x, y) &= \sum_{k=1}^{\infty} \frac{\beta_k}{\operatorname{sh} k\dfrac{\pi}{2}} \operatorname{sh} k\pi x \sin k\pi y. \end{aligned}\right\} \tag{24}$$

The region B is symmetric with respect to the bisector $x = y$ of the first quadrant. The boundary values, however, are disposed anti-symmetrically with respect to this bisector in the following sense. If one takes two points M' and M'' on the boundary, symmetric with respect to the bisector $x = y$, then the values of $w(x, y)$ at them are connected by the equation

$$w(M'') = - w(M').$$

From this it is clear that everywhere within B the values of $w(x, y)$ must be arranged anti-symmetrically with respect to the straight line $y = x$:

$$w(y, x) = - w(x, y).$$

This gives the following connection between $u(x, y)$ and $v(x, y)$:

$$v(y, x) = - u(x, y).$$

Therefore the coefficients α_k and β_k in series (24) must satisfy the following relation:

$$\beta_k = - \alpha_k. \tag{25}$$

Let us consider system of equations (17) and (19), serving for the computation of the approximations $\alpha_k^{(n)}$ and $\beta_k^{(n)}$ to α and β. The coefficients $A_{k,s}$ and $B_{k,s}$ in them have been calculated earlier; for our symmetric "corner" they will equal:

$$A_{k,s} = B_{k,s} = \frac{2}{\pi} \cdot \frac{s^2}{s^2 + k^2} \sin s \frac{\pi}{2} \left\{ \frac{\sin k \frac{\pi}{2}}{s} \operatorname{cth} s \frac{\pi}{2} - \frac{k}{s^2} \cos k \frac{\pi}{2} \right\}.$$

When s is even, $\sin s \frac{\pi}{2} = 0$. Therefore in the right sides of these equations there are preserved only terms containing $\alpha_s^{(n)}$ and $\beta_s^{(n-1)}$ with odd indices s. And when s is odd: $s = 2q + 1$, $\sin (2q + 1) \frac{\pi}{2} = (- 1)^q$, and consequently

$$A_{k,2q+1} = B_{k,2q+1} =$$
$$= \frac{2}{\pi} \cdot \frac{(- 1)^q (2q + 1)^2}{(2q + 1)^2 + k^2} \cdot \left\{ \frac{\sin k \frac{\pi}{2}}{2q + 1} \operatorname{cth} (2q + 1) \frac{\pi}{2} - \frac{k \cos k \frac{\pi}{2}}{(2q + 1)^2} \right\}. \tag{26}$$

For our boundary values $f(x) = \sin 2\pi x$, $\varphi(y) = - \sin 2\pi y$, and the free terms C_k, D_k of equations (17) and (19) are

$$C_k = - D_k = 2 \int_{\frac{1}{2}}^{1} \sin 2\pi x \sin k\pi x \, dx = \begin{cases} \frac{1}{2} \text{ for } k = 2, \\ - \frac{4}{\pi} \dfrac{\sin (k + 2) \frac{\pi}{2}}{k^2 - 4} \text{ for } k \neq 2. \end{cases}$$

Equations (17) and (19) themselves take the form:

$$\alpha_k^{(n)} = -\frac{4}{\pi}\frac{\sin (k+2)\frac{\pi}{2}}{k^2-4} +$$

$$+\frac{2}{\pi}\sum_{q=0}^{\infty}\frac{(-1)^q(2q+1)^2}{(2q+1)^2+k^2}\left\{\frac{\sin k\frac{\pi}{2}}{2q+1}\operatorname{cth}(2q+1)\frac{\pi}{2} - \frac{k\cos k\frac{\pi}{2}}{(2q+1)^2}\right\}\beta_{2q+1}^{(n-1)},$$

$$\alpha_2^{(n)} = \tfrac{1}{2} + \frac{4}{\pi}\sum_{q=0}^{\infty}\frac{(-1)^q\beta_{2q+1}^{(n-1)}}{(2q+1)^2+4}, \tag{27}$$

$$\beta_k^{(n)} = \frac{4}{\pi}\cdot\frac{\sin (k+2)\frac{\pi}{2}}{k^2-4} +$$

$$+\frac{2}{\pi}\sum_{q=0}^{\infty}\frac{(-1)^q(2q+1)^2}{(2q+1)^2+k^2}\left\{\frac{\sin k\frac{\pi}{2}}{2q+1}\operatorname{cth}(2q+1)\frac{\pi}{2} - \frac{k\cos k\frac{\pi}{2}}{(2q+1)^2}\right\}\alpha_{2q+1}^{(n)},$$

$$\beta_2^{(n)} = -\tfrac{1}{2} + \frac{4}{\pi}\sum_{q=0}^{\infty}\frac{(-1)^q\alpha_{2q+1}^{(n)}}{(2q+1)^2+4}.$$

In the right side of the equations here α and β appear only with odd indices. Hence it is clear that in computing all approximations $\alpha_k^{(n)}$ and $\beta_k^{(n)}$ except the last, it is sufficient to compute only the coefficients $\alpha_k^{(n)}$ and $\beta_k^{(n)}$ with odd indices k:

$$\alpha_1^{(n)}, \alpha_3^{(n)}, \alpha_5^{(n)}, \ldots$$
$$\beta_1^{(n)}, \beta_3^{(n)}, \beta_5^{(n)}, \ldots$$

The last of the approximations is the exception, of course; in it we take $\alpha_k^{(n)}$ and $\beta_k^{(n)}$ for the actual values of α_k and β_k. By α_k and β_k we must later construct series (24), and for this we shall require not only the odd coefficients α_k, β_k, but the even ones as well.

Setting k odd in the preceding equations: $k = 2p + 1$, we obtain recurrence formulas for the computation of the odd coefficients $\alpha_{2p+1}^{(n)}$ and $\beta_{2p+1}^{(n)}$:

$$\left.\begin{aligned}
\alpha_{2p+1}^{(n)} &= \frac{4}{\pi}\cdot\frac{(-1)^p}{(2p+1)^2-4} + \\
&\quad + \frac{2}{\pi}\sum_{q=0}^{\infty}\frac{(-1)^{p+q}(2q+1)\beta_{2q+1}^{(n-1)}}{(2q+1)^2+(2p+1)^2}\cdot\operatorname{cth}(2q+1)\frac{\pi}{2}, \\
\beta_{2p+1}^{(n)} &= -\frac{4}{\pi}\cdot\frac{(-1)^p}{(2p+1)^2-4} + \\
&\quad + \frac{2}{\pi}\sum_{q=0}^{\infty}\frac{(-1)^{p+q}(2q+1)\alpha_{2q+1}^{(n)}}{(2q+1)^2+(2p+1)^2}\cdot\operatorname{cth}(2q+1)\frac{\pi}{2}.
\end{aligned}\right\} \tag{28}$$

To begin the computations, we must now make a choice of the values of $u_1(x, y)$ on PD.

We will adopt

$$f_1(x) = u_1(x, \tfrac{1}{2}) = \sin 2\pi x.$$

For such a choice of $f_1(x)$

$$\alpha_1^{(1)} = 0, \ \alpha_2^{(1)} = 1, \ \alpha_3^{(1)} = \alpha_4^{(1)} = \ldots = 0.$$

The initial values of all the odd coefficients $\alpha_k^{(1)}$ are equal to zero.

To simplify the computation of the coefficients of system (28), we introduce, instead of the unknowns $\alpha_{2p+1}^{(n)}$ and $\beta_{2p+1}^{(n)}$, new unknowns, setting

$$\alpha_{2p+1}^{(n)} = \frac{4}{\pi(2p+1)} \alpha_{2p+1}^{*(n)}, \quad \beta_{2p+1}^{(n)} = \frac{4}{\pi(2p+1)} \beta_{2p+1}^{*(n)}, \tag{29}$$

whereupon (28) takes the form

$$\alpha_{2p+1}^{*(n)} = \frac{(-1)^p(2p+1)}{(2p+1)^2 - 4} + \frac{2(2p+1)}{\pi} \sum_{q=0}^{\infty} A_{p,q}^* \beta_{2q+1}^{*(n-1)},$$

$$\beta_{2p+1}^{*(n)} = -\frac{(-1)^p(2p+1)}{(2p+1)^2 - 4} + \frac{2(2p+1)}{\pi} \cdot \sum_{q=0}^{\infty} A_{p,q}^* \alpha_{2q+1}^{*(n)}, \tag{30}$$

$$A_{p,q}^* = \frac{(-1)^{p+q} \operatorname{cth} (2q+1)\dfrac{\pi}{2}}{(2q+1)^2 + (2p+1)^2}.$$

In the approximate computation of the $\alpha_{2p+1}^{(n)}$ and $\beta_{2p+1}^{(n)}$, we will take the first 12 of these quantities, $\alpha_1^{(n)}, \alpha_3^{(n)}, \ldots, \alpha_{11}^{(n)}, \beta_1^{(n)}, \beta_3^{(n)}, \ldots, \beta_{11}^{(n)}$ and discard all the rest. In each of relations (30) we will separate out the first six equations, and setting equal to zero all $\alpha_{2q+1}^{(n)}$ and $\beta_{2q+1}^{(n-1)}$ in them with indices greater than 11, we obtain the following recurrence formulas for the calculation of the quantities we need:

$$\left.\begin{aligned} \alpha_{2p+1}^{*(n)} &= \frac{(-1)^p(2p+1)}{(2p+1)^2 - 4} + \frac{2(2p+1)}{\pi} \sum_{q=0}^{5} A_{p,q}^* \beta_{2q+1}^{*(n-1)}, \\ \beta_{2p+1}^{*(n)} &= -\frac{(-1)^p(2p+1)}{(2p+1)^2 - 4} + \frac{2(2p+1)}{\pi} \sum_{q=0}^{5} A_{p,q}^* \alpha_{2q+1}^{*(n)}, \\ p &= (0, 1, \ldots, 5). \end{aligned}\right\} \tag{31}$$

The table of the coefficients $A_{p,q}^*$, computed to four decimal places, follows:

	$q=0$	$q=1$	$q=2$	$q=3$	$q=4$	$q=5$
$p=0$	0.5451	−0.1000	0.0385	−0.0200	0.0122	−0.0082
$p=1$	−0.1090	0.0556	−0.0294	0.0172	−0.0111	0.0077
$p=2$	0.0419	−0.0294	0.0200	−0.0135	0.0094	−0.0068
$p=3$	−0.0218	0.0172	−0.0135	0.0102	−0.0077	0.0059
$p=4$	0.0133	−0.0111	0.0094	−0.0077	0.0062	−0.0050
$p=5$	−0.0089	0.0077	−0.0068	0.0059	−0.0050	0.0041

The successive approximations $\alpha_{2p+1}^{*(n)}$ and $\beta_{2p+1}^{*(n)}$ found from (31) we set forth in the table below. All the computations for the determination of them, with the exception of the operations of addition and subtraction, were performed by us on a 25 cm. slide rule.

n	$\alpha_1^{*(n)}$	$\alpha_3^{*(n)}$	$\alpha_5^{*(n)}$	$\alpha_7^{*(n)}$	$\alpha_9^{*(n)}$	$\alpha_{11}^{*(n)}$
1	0.0000	0.0000	0.0000	0.0000	0.0000	0.0000
2	−0.2651	−0.5834	0.1990	−0.1142	0.0770	−0.0583
3	−0.2781	−0.5805	0.2012	−0.1173	0.0810	−0.0625
4	−0.2800	−0.5794	0.1999	−0.1168	0.0804	−0.0611

n	$\beta_1^{*(n)}$	$\beta_3^{*(n)}$	$\beta_5^{*(n)}$	$\beta_7^{*(n)}$	$\beta_9^{*(n)}$	$\beta_{11}^{*(n)}$
1	$\frac{1}{3}$	$\frac{3}{5}$	$-\frac{5}{21}$	$\frac{7}{45}$	$-\frac{9}{77}$	$\frac{11}{117}$
2	0.2856	0.5759	−0.1980	0.1154	−0.0787	0.0604
3	0.2807	0.5782	−0.1993	0.1164	−0.0804	0.0604
4	0.2802	0.5790	−0.1995	0.1164	−0.0804	0.0604

We have terminated the computations at the fourth approximation, since the next approximation would have repeated it, and have put

$$\begin{aligned} \alpha_1^* &= -\beta_1^* = -0.2802, & \alpha_7^* &= -\beta_7^* = -0.1164, \\ \alpha_3^* &= -\beta_3^* = -0.5790, & \alpha_9^* &= -\beta_9^* = 0.0804, \\ \alpha_5^* &= -\beta_5^* = 0.1995, & \alpha_{11}^* &= -\beta_{11}^* = -0.0604. \end{aligned}$$

Using them we compute α_{2p+1} and β_{2p+1} from the equations

$$\alpha_{2p+1} = \frac{4}{\pi(2p+1)}\alpha_{2p+1}^*, \qquad \beta_{2p+1} = \frac{4}{\pi(2p+1)}\beta_{2p+1}^*,$$

$$\begin{aligned} \alpha_1 &= -\beta_1 = -0.3568, \\ \alpha_3 &= -\beta_3 = -0.2457, \\ \alpha_5 &= -\beta_5 = 0.0508, \\ \alpha_7 &= -\beta_7 = -0.0211, \\ \alpha_9 &= -\beta_9 = 0.0114, \\ \alpha_{11} &= -\beta_{11} = -0.0070. \end{aligned}$$

For the determination of the coefficients α and β with even indices, we utilize equations (27). We drop from them all terms containing the coefficients α and β with odd indices greater than 11, neglecting them in the computations, and apply them to the fourth approximations obtained by us, which are adopted as the actual values of α and β. Considering k even in them: $k = 2p$, we find:

$$\alpha_2 = -\beta_2 = \tfrac{1}{2} + \frac{4}{\pi} \sum_{q=0}^{5} \frac{(-1)^q}{(2q+1)^2 + 4} \beta_{2q+1},$$

$$\alpha_{2p} = -\beta_{2p} = -\frac{p}{\pi} \sum_{q=0}^{5} \frac{(-1)^{p+q}}{(2q+1)^2 + 4p^2} \beta_{2q+1}, \ (p \geqslant 2).$$

We have computed from this $\alpha_2, \ldots, \alpha_{12}, \beta_2, \ldots, \beta_{12}$:

$$\begin{aligned}
\alpha_2 &= -\beta_2 = 0.5638,\\
\alpha_4 &= -\beta_4 = -0.0246,\\
\alpha_6 &= -\beta_6 = 0.0115,\\
\alpha_8 &= -\beta_8 = -0.0061,\\
\alpha_{10} &= -\beta_{10} = 0.0038,\\
\alpha_{12} &= -\beta_{12} = -0.0031.
\end{aligned}$$

Finally, using the α_k and β_k found, we form series (24). The series for $u(x, y)$, for example, has the form:

$$\begin{aligned}
u(x, y) = &-0.3568 \frac{\operatorname{sh} \pi y}{\operatorname{sh} \frac{\pi}{2}} \sin \pi x + 0.5638 \frac{\operatorname{sh} 2\pi y}{\operatorname{sh} \pi} \sin 2\pi x - \\
&- 0.2457 \frac{\operatorname{sh} 3\pi y}{\operatorname{sh} \frac{3\pi}{2}} \sin 3\pi x - 0.0246 \frac{\operatorname{sh} 4\pi y}{\operatorname{sh} 2\pi} \sin 4\pi x + \\
&+ 0.0508 \frac{\operatorname{sh} 5\pi y}{\operatorname{sh} \frac{5\pi}{2}} \sin 5\pi x + 0.0115 \frac{\operatorname{sh} 6\pi y}{\operatorname{sh} 3\pi} \sin 6\pi x - \\
&- 0.0211 \frac{\operatorname{sh} 7\pi y}{\operatorname{sh} \frac{7\pi}{2}} \sin 7\pi x - 0.0061 \frac{\operatorname{sh} 8\pi y}{\operatorname{sh} 4\pi} \sin 8\pi y + \\
&+ 0.0114 \frac{\operatorname{sh} 9\pi y}{\operatorname{sh} \frac{9\pi}{2}} \sin 9\pi x + 0.0038 \frac{\operatorname{sh} 10\pi y}{\operatorname{sh} 5\pi} \sin 10\pi x - \\
&- 0.0070 \frac{\operatorname{sh} 11\pi y}{\operatorname{sh} \frac{11\pi}{2}} \sin 11\pi x - 0.0031 \frac{\operatorname{sh} 12\pi y}{\operatorname{sh} 6\pi} \sin 12\pi x.
\end{aligned}$$

The expression for $v(x, y)$ is obtained from this by changing the signs of all coefficients and interchanging the variables x and y.

To characterize the accuracy of the result we have calculated the values of $u(x, y)$ at four points: $(\frac{1}{4}, \frac{1}{2})$, $(\frac{1}{2}, \frac{1}{2})$, $(\frac{3}{4}, \frac{1}{2})$ and $(\frac{1}{2}, \frac{1}{4})$:

$$u(\tfrac{1}{4}, \tfrac{1}{2}) = \sum \alpha_k \sin k\frac{\pi}{4} =$$

$$= \frac{1}{\sqrt{2}}(\alpha_1 + \alpha_3 - \alpha_5 - \alpha_7 + \alpha_9 + \alpha_{11}) + (\alpha_2 - \alpha_6 + \alpha_{10}) = 0.1122,$$

$$u(\tfrac{1}{2}, \tfrac{1}{2}) = \sum \alpha_k \sin k\frac{\pi}{2} = \alpha_1 - \alpha_3 + \alpha_5 - \alpha_7 + \alpha_9 - \alpha_{11} = -0.0208,$$

$$u(\tfrac{3}{4}, \tfrac{1}{2}) = \sum \alpha_k \sin \frac{3\pi}{4} =$$

$$= \frac{1}{\sqrt{2}}(\alpha_1 + \alpha_3 - \alpha_5 - \alpha_7 + \alpha_9 + \alpha_{11}) + (-\alpha_2 + \alpha_6 - \alpha_{10}) = -0.9999,$$

$$u(\tfrac{1}{2}, \tfrac{1}{4}) = \alpha_1 \frac{\operatorname{sh}\frac{\pi}{4}}{\operatorname{sh}\frac{\pi}{2}} - \alpha_3 \frac{\operatorname{sh}\frac{3\pi}{4}}{\operatorname{sh}\frac{3\pi}{2}} + \alpha_5 \frac{\operatorname{sh}\frac{5\pi}{4}}{\operatorname{sh}\frac{5\pi}{2}} - \ldots = -0.1110.$$

The point $(\frac{1}{2}, \frac{1}{2})$ lies on the contour of the region and, as is seen from the boundary condition of our problem, $u(x, y)$ must vanish at it. The approximate value found by us, $u(\frac{1}{2}, \frac{1}{2}) = -0.0208$, differs from the actual one by 0.0208. The point $(\frac{3}{4}, \frac{1}{2})$ also lies on the boundary, and at it $u(x, y)$ must acquire the value -1. The computed approximate value $u(\frac{3}{4}, \frac{1}{2}) = -0.9999$ is very close to the true one.

On the segment DP the sought functions u and v must be equal to each other. The same thing can also be said of them on the segment DQ.

The approximations to them that we have found will not, generally speaking, coincide on DP and DQ. To characterize how much they differ from each other on these segments, we compute the approximate functions $u(x, y)$ and $v(x, y)$ at the point $(\frac{1}{4}, \frac{1}{2})$ on DP. The value of $u(\frac{1}{4}, \frac{1}{2})$ was found earlier, equalling 0.1122. The value of $v(\frac{1}{4}, \frac{1}{2})$ can be easily determined on the basis of the following considerations. Let us take the point $(\frac{1}{2}, \frac{1}{4})$, symmetric with the point $(\frac{1}{4}, \frac{1}{2})$ relative to the bisector of the first quadrant. From the very construction of $u(x, y)$ and $v(x, y)$ it is clear that their values at points symmetric with respect to the bisector $y = x$ must differ only in sign: $v(x, y) = -u(y, x)$. We must therefore have $v(\frac{1}{4}, \frac{1}{2}) = -u(\frac{1}{2}, \frac{1}{4}) = 0.1110$.

Thus at the point $(\frac{1}{4}, \frac{1}{2})$, $u(\frac{1}{4}, \frac{1}{2}) = 0.1122$ and $v(\frac{1}{4}, \frac{1}{2}) = 0.1110$. The difference between them, 0.0012, constitutes about 1% of their values at this point.

BIBLIOGRAPHY

(The transliteration of the U.S. Library of Congress is followed.)

Abbreviations

M. Moscow
L. Leningrad
G. = GTTI Gosudarstvennoe izdatel'stvo tekhniko-teoreticheskoĭ literatury (State publishing house of technical and theoretical literature). Moscow or Leningrad or both.
DAN SSSR Doklady Akademii Nauk SSSR (Reports of the USSR Academy of Sciences). (Comptes Rendus)
Izd. (or Izd-vo) AN SSSR (Publishing house of the Academy of Sciences, USSR).
Izv. AN SSSR OMEN Izvestiia Academii Nauk SSSR. Otdel' Matematicheskikh i Estestvennykh Nauk (Publication of the Academy of Sciences, USSR. Section of the Mathematical and Natural Sciences.)

A

Akbergenov, I. A. 1. On the estimation of the error of the approximate solution of Fredholm's integral equation of the second kind by E. Nyström's method. (Russian). Trudy vtorogo vsesoiuznogo matematicheskogo c''ezda (1936), 386–7.
— 2. Dissertation. Leningrad Gos. Universitet, Nauchno-issledovatel'skiĭ institut matematiki i mekhaniki im. Bubnova. 1935.
— 3. On the approximate solution of a Fredholm equation and the determination of its proper values. (Russian; German summary). Matem. Sbornik 42 (1935), 6: 679–698.
Akushkiĭ, I. IA. 1. Numerical solution of the Dirichlet equation with the aid of perforated-card machines. DAN SSSR 52 (1946), 5: 375–378.
— 2. The four-counter scheme of solution of Dirichlet's problem by means of punched-card machines. DAN SSSR 54 (1946), 8: 659–662.
— 3. On numerical solution of Dirichlet's problem on punched-card machines. DAN SSSR 54 (1946), 9: 755–758.
Artmeladze, N. K. 1. On the approximate solution of integral equations. (Russian; Georgian summary). Trudy matem. in-ta GrSSR, Tbilisi, 13 (1944), 29–53.
Arutiunian, N. KH. 1. Solution of the problem of the torsion of a bar of polygonal cross section. (Russian.) Priklad. Matem. i Mekh. 13 (1949), 1: 107–112.
— 2. Approximate solution of the problem of the torsion of bars of polygonal cross-section. (Russian.) Priklad. Matem. i Mekh. 6 (1942), 1: 19–30.

B

BAIRSTOW, L. and BERRY, A. 1. Two-dimensional solutions of Poisson's and Laplace's equations. London. Roy. Soc. Proc. (A) 95 (1918), 457–475.

BANIN, A. M. 1. Approximate conformal transformation applied to a plane parallel flow past an arbitrary shape. (Russian; English summary.) Priklad. Matem. i Mekh. 7 (1943), 2: 131–140.

BARI, N. K. 1. Sur les systèmes complets de fonctions orthogonaux. (Russian summary) Matem. Sbornik 14 (56) (1944), 51–108.

BASLAVSKIĬ, I. A. 1. The bending of a rectangular plate of variable thickness. (Russian) Trudy vyssh. inzh.-tekh. uchilishcha Voenno-morskogo Flota 4 (1943), 37–81.

BATEMAN, H. 1. On the numerical solution of linear integral equations. London Proc. Roy. Soc. (A) 100 (1922), 441ff.

— 2. A formula for the solving function of a certain integral equation of the second kind. Messenger Math. 37 (1908), 179–187.

BERGMANN, S. 1. Über die Entwicklung der harmonischen Funktionen der Ebene und des Raumes nach Orthogonalfunktionen. Math. Annalen 86 (1922), 238–271.

BEZIKOVICH, I͡A. S. 1. Ischislenie konechnykh raznosteĭ (The calculus of finite differences). Leningrad. Izd-vo Universiteta. 1939.

— 2. Priblizhënnye vychislenii͡a (Approximate computations). 6 ed, supp. Leningrad. Gos. izd-vo tekhniko-teoret. lit-ry. 1949. 462 pp.

BIEBERBACH, L. 1. Zur Theorie und Praxis der konformen Abbildung. Rendiconti del Circolo Matematico di Palermo, 38 (1914, 2 sem.), 98–112.

BIEZENO, C. B. and KOCH, J. J. 1. On the buckling of a thin-walled circular tube loaded by pure bending. Proc. Akad. Wet. Amsterdam 43 (1940), 783–796, 923–935.

BOCHNER, S. 1. Über orthogonale Systeme analytischer Funktionen. Math. Zeitschrift 14 (1922), 180–207.

BYSTROV, N. 1. Über angenäherte Lösung von partiellen Differentialgleichungen mit drei unabhängigen Variablen. DAN SSSR 3 (1934), 12–16.

C

CARLEMAN, T. 1. Über die Approximation analytischer Funktionen durch lineare Aggregate von vorgegebenen Potenzen. Arkiv för Mat., Astron. och Fys. 17 (1923), No. 9, 30s.

CHEPOV, T. K. 1. The approximate solution of problems of the torsion of certain prismatic bars. Priklad. Matem. i Mekh. 1 (1937), 2: 225–261.

CHRISTOFFEL, E. B. 1. Sul problema delle temperature stazionarie e la rappresentazione di una data superficie. Annali di Matematica (2), 1 (1867–68), 89–104.

CHUFISTOVA, A. M. 1. Approximate conformal mapping by means of the exponential function. (Russian) Uchënye zapiski Leningrad. Gos. Univ-ta, Ser. Matem. Nauk 37 (1939) 6: 119–126.

COLLATZ, L. 1. Schrittweise Näherungen bei Integralgleichungen und Eigenwertschranken. Math. Zeitschrift 46 (1940), 692–708.

— 2. Das Differenzenverfahren mit höherer Approximation für lineare Differentialgleichungen. Schriften des math. Seminars und Inst. angew. Math. der Universität Berlin 3 (1935), 1: 1–34.

— 3. Genäherte Berechnung von Eigenwerten. Z. angew. Math. Mech. 19 (1939), 224–249, 297–318.

— 4. Eigenwertprobleme und ihre numerische Behandlung. Leipzig, 1945.

COURANT, R. 1. Geometricheskai͡a teorii͡a funkt͡sii kompleksnoĭ peremennoĭ (Geometric theory of functions of a complex variable) GTTI, 1934.

—, FRIEDRICHS, K. and LEWY, H. 1. On difference equations ... Uspekhi Matem. Nauk 1941 8:

— 2. Über die partiellen Differenzengleichungen der mathematischen Physik. Math. Annalen 100 (1928), 32–74.

COURANT, R. and HILBERT, D. 1. The methods of mathematical physics. Russian translation: G., 1933.

D

DANILEVSKIĬ, A. M. 1. On the numerical solution of the secular equation. (Russian) Matem. Sbornik 2 (44) (1937), 1: 169—172.

DIXON, A. C. 1. On a class of matrices of infinite order. Cambr. Trans. 19 (1902), 190—233.

E

EGOROV, D. F. 1. Variat͡sionnoe ischislenie (The calculus of variations).

F

FADDEEVA, V. N. 1. Vychislitel'nye metody linearnogo analiza (Computational methods of linear algebra) G., 1950. 240 pp.

— 2. The method of lines applied to certain boundary problems. (Russian). Trudy matem. in-ta im. Steklova, 1949, 28: 73—103.

FIKHTENGOL'T͡S, G. M. 1. Kurs differentsial'nogo i integral'nogo ischisleniia (Course of differential and integral calculus) G., 1947, 1949.

FILONENKO-BORODICH, M. M. 1. Teorii͡a uprugosti (Theory of elasticity). G., 1947.

G

GALËRKIN, B. G. 1. Uprugie tonkie plity (Thin elastic plates). M.-L., 1933.

— 2. Izvestii͡a Politekh. in-ta, 1915, 1916, 1918.

— 3. Vestnik inzhenerov i tekhnikov, 1915.

GEL'FAND, I. M. 1. Lekt͡sii po lineĭnoĭ algebre (Lectures on linear algebra). Ed. 2, GTTI, 1951. (See Appendix 1.)

GERSHGORIN, S. A. 1. On the approximate integration of the differential equations of Laplace and Poisson (Russian). Leningrad. Izv. politekh. in-ta 30 (1927), 75—95.

— 2. Fehlerabschätzung für das Differenzenverfahren zur Lösung partieller Differentialgleichungen. Z. f. angew. Math. 10 (1930), 373—382.

— 3. On the conformal mapping of a simply connected region onto a circle. (Russian; German summary) Matem. Sbornik 40 (1933), 1: 48—58.

GI͡UNTER, N. M. 1. Kurs variat͡sionnogo ischislenii͡a (Course in variational calculus) G. 1941.

GOLDSCHMIDT, E. 1. Numerische Verwendbarkeit der Methoden zur Auflösung unendlich vieler linearer Gleichungen. (Dissertation) Würzburg, 1912.

GOLUSHKEVICH, S. S. 1. O nekotorykh zadachakh teorii izgiba ledovogo pokrova (Certain problems of the theory of the bending of an ice cover). 1947.

GOLUZIN, G. M. 1. Method of variations in the theory of conformal representation. (Russian; English summary) Matem. Sbornik 16 (61) (1946) 2: 203—236.

— 2. Conformal transformation of multiply connected regions into a plane with cuts by the method of functional equations. (Russian) See *Collective Author* 2, pp. 99—110.

GONCHAROV, V. L. 1. Teorii͡a interpolirovanii͡a i priblizhenii͡a funkt͡siĭ (Theory of the interpolation and approximation of functions). G., 1934.

GORBUNOV-POSADOV, M. I. 1. On beams and rectangular plates lying on an elastic half-space. DAN SSSR 24 (1939), 421—425.

GOURSAT, É. 1. Cours d'analyse mathématique. 5 ed., Paris. Russian translation by M. G. Shestopal, ed. V. V. Stepanov. M.-L. GTTI, 1934.

— 2. Sur un cas élémentaire de l'équation de Fredholm. Bull. Soc. Math. de France 35 (1907), 163—173.

— 3. Sur l'équation $\Delta\Delta u = 0$. Bull. de la Soc. Math. de France 26 (1898), 236—237.

GRINBERG, G. A. 1. Izbrannye voprosy matematicheskoĭ teorii elektricheskikh i magnitnykh i͡avleniĭ (Selected problems in the mathematical theory of electrical and magnetic phenomena). Izd. AN SSSR, 1948.

GUREVICH, S. S. 1. Stability of two-dimensional stressed state. (Russian). Uchënye zapiski Leningrad. Gos. Univ-ta, Ser. Matem. Nauk 8 (1939), 137–152.

H

HARDY, G. H., and LITTLEWOOD, J. E. 1. A convergence criterion for Fourier series. Math. Zeitschrift 28 (1928), 612–634.

HECKELER, J. W. 1. Elastostatik. Russian translation: Statika uprugogo tela. 1934.

HELLINGER, E. and TOEPLITZ, O. 1. Integralgleichungen und Gleichungen mit unendlichvielen Unbekannten. Enzyklopädie der Mathematischen Wissenschaften, II.3.2, Heft 13 (1927).

HENCKY, H. 1. Spannungszustand in rechteckigen ebenen Platten. München, 1913.

HILBERT, D. 1. Grundzüge einer allgemeinen Theorie der linearen Integralgleichungen. Leipzig, Berlin, 1912.

HILDEBRAND, FRANCIS B. 1. The approximate solution of singular integral equations arising in engineering practice. Proc. Amer. Acad. Arts Sci. 74 (1941), 287–295.

I

IL'IN, V. P. 1. Estimates of functions having derivatives summable with a given degree on hyperplanes of different dimensions. (Russian) DAN SSSR, 78 (1951), 4: 633–636.

K

KAGAN, V. F. 1. Teoriia͡ opredeliteleĭ (Determinant theory). M. 1922.

KANTOROVICH, L. V. 1. Opredelënnye integraly i riady Fur'e (Definite integrals and Fourier series). Leningrad. Izd-vo Univ-ta., 1940.

— 2. O funktsional'nykh uravneniia͡kh (On functional equations). Uchënye zapiski Leningrad. Gos. Univ-ta. 3 (1937), 17: 51–78.

— 3. Functional analysis and applied mathematics. (Russian) Uspekhi Matem. Nauk, 3 (1948), 6: 89–185.

— 4. On general techniques of improving the convergence in methods for the approximate solution of boundary-value problems of mathematical physics (Russian) Trudy Leningrad in-ta inzh-ov prom. stroit-va, 1 (1934), 2: 65–72.

— 6. On special devices for the numerical integration of even and odd functions. (Russian) Trudy matem. in-ta im. Steklova, 1949, 28: 3–25.

— 7. On the approximate computation of some types of definite integrals and other applications of the method of removing singularities. (Russian; French summary) Matem. Sbornik 41 (1934), 2: 235–245.

— 8. On the method of steepest descent. (Russian) DAN SSSR 56 (1947), 233ff.

— 9. On Newton's method for functional equations. (Russian) DAN SSSR 59 (1948), 7: 1237–1240.

— 10. A utilisation of the idea of the method of Acad. B. G. Galerkin in the method of reduction to ordinary differential equations. (Russian) Priklad. Matem. i Mekh. 6 (1942), 31–40.

— 11. Sur une méthode de résolution approchée d'équations différentielles aux dérivées partielles. DAN SSSR 2 (1934), 532–536.

— 12. Some remarks on Ritz's method. (Russian) Trudy vysshego voenno-morskogo inzhenerno-stroitel'nogo uchilishcha, Issue 3, Leningrad, 1941.

— 13. A direct method of solving the problem of the minimum of a double integral. (Russian) Izv. AN SSSR OMEN 1933, 5: 647–652.

— 14. On the convergence of variational processes. DAN SSSR 30 (1941), 2: 107–111.

— 15. On the convergence of the method of reduction to ordinary differential equations. DAN SSSR 30 (1941), 7: 585–588.

KANTOROVICH, L. V. 16. On conformal mapping. (Russian; French summary) Matem. Sbornik 40 (1933), 3: 294—325.

— 17. On some methods of constructing functions accomplishing conformal transformation. (Russian) Izv. AN SSSR OMEN, 1933, 2: 229—235.

— 18. On the conformal mapping of multiply connected regions. (Russian; French summary) DAN SSSR 2 (1934), 441—445.

— 19. Some corrections to my article "On conformal mapping". (Russian; French summary) Matem. Sbornik 41 (1934), 179—182.

— 20. Effective methods in the theory of conformal mapping. (Russian; French summary) Izv. AN SSSR OMEN, 1937, 1: 79—90.

— *et al.* 21. Konformnoe otobrazhenie odnosvi͡aznykh i mnogosvi͡aznykh oblasteĭ (Conformal mapping of simply and multiply connected regions). M.-L. ONTI, 1937, pp. 5—18.

— and FRUMKIN, P. V. 1. On the application of a method of approximate solution of partial differential equations to the problem of the torsion of a prismatic bar. (Russian) Trudy Leningrad. in-ta inzh. prom. stroitel'stva, 4 (1937), 111—122.

KELDYSH, M. V. 1. On B. G. Galerkin's method for the solution of boundary-value problems. (Russian; English summary). Izv. AN SSSR Ser. Matem. 6 (1942), 309—330.

KELLOGG, O. D. 1. On the existence and closure of sets of characteristic functions. Math. Annalen 86 (1922), 14—17.

K͡HARRIK, I. I͡U. 1. On a problem of the constructive theory of functions, connected with the investigation of the convergence of variational processes. (Russian) DAN SSSR 80 (1951), 1: 25—28.

K͡HAZHALII͡A, G. I͡A. 1. On the theory of the conformal mapping of doubly connected regions. (Russian) Tbilisi. Trudy matem. in-ta Gr. fil. AN, 4 (1938), 123—133.

— 2. On the theory of the conformal mapping of doubly connected regions. (Russian) Matem. Sbornik 8 (50) (1940), 1: 97—106.

KOCH, H. v. 1. Ueber das Nichtverschwinden einer Determinante nebst Bemerkungen über Systeme unendlich vieler linearer Gleichungen. Deutsche Math.-Ver. 22 (1913), 285—291.

KOI͡ALOVICH, B. M. 1. On the theory of infinite systems of linear equations. (Russian) Trudy fiz-matem. instituta im. V. A. Steklova 2; 4 (1932).

— 2. On a partial differential equation of the fourth order. (Russian) St. Petersburg, 1902.

KOVNER, S. S. and ZHAK, D. K. 1. Computation of the powers of the Liebmann and Gershgorin operators and their application to the machine integration of equations. (Russian) DAN SSSR 58 (1947), 5—8.

KRAVCHUK, M. F. 1. Application of the method of moments to the solution of linear differential and integral equations. (Ukrainian) Kiev. Soobs͡hch. AN UkSSR 1 (1932), 168ff.

— 2. On the convergence of the method of moments for partial differential equations. (Ukrainian) Zh. in-ta matem. AN UkSSR 1 (1936), 23—27.

KRYLOV, A. N. 1. O nekotorykh different͡sial'nykh uravnenii͡akh matematicheskogo fiziki (Some differential equations of mathematical physics). Leningrad, 1932.

— 2. Lekt͡sii o priblizhënnykh vychislenii͡akh (Lectures on approximate computations) 3 ed, rev. & supp., L.-M., Izd-vo AN SSSR, 1935. 541 pp.

— 3. On the numerical solution of the equations by which the frequency of small oscillations of material systems is determined in technical problems. (Russian) Izv. AN SSSR OMEN, 1931, 4: 491—539.

KRYLOV, N. M. 1. On the approximate integration of linear integral equations. (Ukrainian: French summary) Kiev. Trudy fiz.-mat. otd. AN UkSSR 3 (1926) 6: 183—208.

— 2. Sur différents procédés d'integration approchée en physique mathématique. Toulouse Ann. 17 (1925), 153—186.

KRYLOV, N. M. 3. Les méthodes de solution approchée des problèmes de la physique mathématique. Mémorial des Sciences Mathématiques 49 (1931).

— 4. The approximate solution of the fundamental problems of mathematical physics. (Ukrainian) Monograph, Kiev, 1931.

KRYLOV, N. M. and BOGOLIUBOV, N. N. 1. La solution approchée du problème de Dirichlet. DAN SSSR 1929, 283—288.

(KRYLOV, N. M. and) BOGOLIUBOV, N. N. 2. Sur l'approximation des functions par les sommes trigonométriques. DAN SSSR 1930, 147—152.

KRYLOV, N. and TAMARKIN, IA. 1. Sur l'application de la théorie des quadratures mécaniques généralisées à l'évaluation par approximations successives de la solution de l'équation intégrale. Kiev. Bull. Ukr. Acad. Science. 1 (1923), 1: 16—21.

KRYLOV, V. I. 1. Application of the Euler-Laplace formula to the approximate solution of integral equations of Volterra's type. (Russian) Trudy matem. in-ta im. Steklova, 1949, 28: 33—72.

KUFAREV, P. P. 1. DAN SSSR 57 (1947), 535—537.

KUROSH, A. G., *et. al.* 1. Matematika v SSSR za tridtsat' let 1917—1947 (Mathematics in the USSR in the thirty years 1917—1947). G., 1948.

KUZ'MIN, R. O. 1. On the theory of infinite systems of linear equations. (Russian) Trudy fiz.-matem. in-ta im. Steklova 2; 2 (1931).

— 2. On a class of infinite systems of linear equations. (Russian; French summary) Izv. AN SSSR OMEN 1934, 4: 515—546.

KUZNETSOV, E. S. 1. Trudy geo.-fiz. in-ta AN, 1949, 4 (131).

L

LAVRENT'EV, M. A. 1. Konformnye otobrazheniia s prilozheniiami k nekotorym voprosam mekhaniki (Conformal transformations, with applications to some problems of mechanics) G., 1946.

LAVRENT'EV, M. A. and LIUSTERNIK, L. A. 1. Kurs variatsionnogo ischisleniia (Course in variational calculus) 2 ed., rev., G., 1950.

LEĬBENZON, L. S. 1. Kurs teorii uprugosti (Course in the theory of elasticity) G., 1947.

— 2. Variatsionnye metody resheniia zadach teorii uprugosti (Variational methods for the solution of problems of the theory of elasticity), G., 1943.

LEKHTIK, S. M. 1. Trudy vysshego voenno-morskogo inzhenerno-stroitel'nogo uchilishcha, Issue 3, Leningrad, 1941.

LEVE, D. 1. Bezbalochnye perekrytia (Beamless floors) Leningrad, 1931.

LEVIN, V. I., and GROSBERG, IU. I. 1. Differentsial'nye uravneniia matematicheskoĭ fiziki (The differential equations of mathematical physics). GTTI, 1951.

LIUSTERNIK, L. A. and LAVRENT'EV, M. A. 1. See vice versa.

LIUSTERNIK, L. A. and PETROVSKIĬ, I. G. 1. Uspekhi Matem. Nauk 8 (1941).

LUCHININ, D. I. 1. The method of infinite systems applied to the solution of the plane problem of the theory of elasticity. (Russian) Matem. Sbornik 3 (45) (1938), 483—508.

LUR'E, A. I. 1. The approximate solution of some problems of the torsion and bending of a bar. Trudy Len. ind. in-ta, 3 (1939), 1: 121—126.

M

MALIEV, A. S. 1. Fourier series of heightened convergence for functions defined in a given interval. Izv. AN SSSR OMEN 1932, 10: 1437—1450.

— 2. On the expansion in Fourier series of heightened convergence of functions defined in a given interval. (Russian) Izv. AN SSSR OMEN, 1933, 8: 1113—20.

MARKOV, A. A. 1. Ischislenie konechnykh raznosteĭ (The calculus of finite differences) St. Petersburg, 1913.

— 2. On a problem of Mendeleev's. (Russian) Memuary Rossiĭskoĭ Akademiia Nauk, 62 (1889), 1—24.

MELENT'EV, P. V. 1. Neskol'ko novykh metodov i priëmov priblizhënnykh vychislenii (Several new methods and devices for approximate computations) L.-M., ONTI, 1937.

MIKELADZE, SH. E. 1. On the numerical solution of integral equations. (Russian; French summary) Izv. AN SSSR OMEN 1935, 2: 255–300.

— 2. On the numerical solution of the differential equations of Laplace and Poisson. (Russian; German summary) Izv. AN SSSR OMEN 1938, 2: 271–292.

— 3. On the numerical integration of partial differential equations. (Russian; French summary) Izv. AN SSSR OMEN 1934, 6: 819–842.

— 4. Chislennye metody integrirovaniia differentsial'nykh uravnenii s chastnymi uravneniiami (Numerical methods for the integration of partial differential equations). Izd-vo AN SSSR, 1936. 108 pp.

— 5. Über die numerische Lösung der Differentialgleichung $\frac{\partial^2 u}{\partial x^2} + \frac{\partial^2 u}{\partial y^2} + \frac{\partial^2 u}{\partial z^2} = \varphi(x, y, z)$. DAN SSSR 14 (1937), 4: 177–179.

— 6. Über numerische integration der Laplaceschen und Poissonschen Gleichungen DAN SSSR 14 (1937), 4: 181–182.

— 7. Über die Lösung von Randwertproblemen mit der Differenzenmethode. DAN SSSR 28 (1940), 5: 400–402.

— 8. On the problem of the numerical integration of partial differential equations by means of nets. (Russian) Soobshch. Gruz. Fil. AN 1 (1940), 4: 249–254.

— 9. On the numerical integration of equations of elliptic and parabolic type. Izv. AN SSSR Ser. Matem. 5 (1941), 57–74.

— 10. Numerical integration. (Russian) Uspekhi Matem. Nauk 3 (1948), 6(28): 1–88.

MIKHLIN, S. G. 1. Integral'nye uravneniia i ikh prilozheniia k nekotorym problemam (Integral equations and their application to certain problems) 2 ed., G., 1949.

— 2. Uchënye zapiski Leningrad. Gos. Univ-ta, Ser. Matem. Nauk, Issue 16 (1949), 167–206.

— 3. On the convergence of Galerkin's method. (Russian) DAN SSSR 61 (1948), 2: 197–199.

— 4. The method of successive approximations applied to the biharmonic problem. (Russian) Trudy seism. in-ta AN 39 (1934), 1–14.

— 5. Priamye metody v matematicheskoi fizike. GTTI, 1950.

MISES, R. v. and POLLACZEK-GEIRINGER, H. 1. Praktische Verfahren der Gleichungsauflösung. Z. f. angew. Math. 9; 58–77, 152–164.

MÜNTZ, G. M. 1. Integral'nye uravneniia (Integral equations). Vol. 1, G., 1934.

MUSKHELISHVILI, N. I. 1. Solution du problème mixte fondamental de l'élasticité pour un demiplan. DAN SSSR 3(8), (1935) 2(62): 51–54.

— 2. Applications des intégrales analogues à celles de Cauchy à quelques problèmes de la physique mathématique. Tiflis, 1922.

— 3. Recherches sur les problèmes aux limites relatifs à l'équation biharmonique et aux équations de l'élasticité à deux dimensions. Math. Annalen 107 (1932), 282–312.

— 4. Nekotorye osnovnye zadachi matematicheskoi teorii uprugosti (Some fundamental problems of the mathematical theory of elasticity). 2 ed., rev. & corr., M.-L. Izd-vo Akad. Nauk SSSR, 1935. 453 pp. [English translation by J. R. M. Radok, Groningen, 1953 (Noordhoff).]

— 5. Singuliarnye integral'nye uravneniia (Singular integral equations). G., 1946. 498 pp. [English translation by J. R. M. Radok, Groningen, 1953 (Noordhoff)]

N

NATANSON, I. P. 1. Konstruktivnaia teoriia funktsii (Constructive theory of functions). G., 1949.

NEUMANN, C. 1. Zur Theorie des logarithmischen und Newton'schen Potentials. Leipziger Berichte 22 (1870), 264–321.

NEWTON, ISAAC S. 1. Philosophiae naturalis principia mathematica. London, 1686.

NIKOLAEVA, M. V. 1. On the approximate computation of oscillating integrals. (Russian) Trudy matem. in-ta im. Steklova 1949, 28: 26–32.

NOETHER, F. 1. Über eine Klasse singulärer Integralgleichungen. Math. Annalen 82 (1921), 42–63.

NYSTRÖM, E. J. 1. Über die praktische Auflösung von linearen Integralgleichungen mit Anwendungen auf Randwertaufgaben der Potentialtheorie. Commentationes Helsingfors: Societas Scientiarum Fennica. Comm. Physico-mathematicae 4, Nr. 15, 52 pp.

— 2. Über die praktische Auflösung von Integralgleichungen mit Anwendungen auf Randwertaufgaben. Acta Math. 54, (1930), 185–204.

— 3. Zur numerischen Lösung von Randwertaufgaben bei gewöhnlichen Differentialgleichungen. Acta Math. 76 (1945), 157–184.

O

OBERG, E. N. 1. The approximate solution of integral equations. Bull. Amer. Math. Soc. 41 (1935), 4: 276–284.

OSTROWSKI, A. 1. Sur l'approximation du déterminant de Fredholm par les déterminants des systèmes d'équations linéaires. Arkiv för matematik, astronomi och fysik 26 (1939), No. 14.

OVCHINSKIĬ, B. V. 1. Trudy Geo.-fiz. In-ta AN 1949 4 (131).

P

PANOV, D. I͡U. 1. On the application of the method of Acad. S. A. Chaplygin to the solution of integral equations. (Russian; German summary) Izv. AN SSSR OMEN 1934, 6: 843–886.

— 2. Spravochnik po chislennomu reshenii͡u different͡sial'nykh uravnenii v chastnykh proizvodnykh (Handbook on the numerical solution of partial differential equations). G., 1949.

— 3. Appendix One to Russian translation of Scarborough's book [1].

— 4. The numerical solution of boundary problems of partial differential equations of elliptic type. (Russian) Uspekhi Matem. Nauk (1938), 4: 34–44.

— 5. On an application of B. G. Galerkin's method for the solution of certain problems of the theory of elasticity. (Russian) Priklad. Matem. i Mekh. 3 (1939), 139–142.

PAPKOVICH, P. F. 1. Teorii͡a uprugosti (Theory of elasticity). L., Oborongiz, 1939.

— 2. Stroitel'nai͡a mekhanika korabli͡a (Ship structural mechanics). Leningrad, 1941.

PELLET, A. 1. Des équations majorantes. S. M. F. Bull. 37 (1909), 93–101.

— 2. Des systèmes infinis d'équations. S. M. F. Bull. 41 (1913), 119–126.

— 3. Sur la méthode des réduites. S. M. F. Bull. 42 (1914), 48–53.

PEREL'MAN, M. I͡A. 1. B. G. Galerkin's method in variational calculus and in the theory of elasticity. (Russian) Priklad. Matem. i Mekh. 5 (1941), 345–348.

PETROV, G. I. 1. Application of Galerkin's method to a problem of the stability of the flow of a viscous liquid. (Russian) Priklad. Matem. i Mekh. 4 (1940), 3: 3–12.

— 2. Trudy T͡SAGI, No. 345 (1938).

PETROVSKIĬ, I. G. 1. Lekt͡sii po teorii integral'nykh uravneniĭ (Lectures on the theory of integral equations). 2 ed., G., 1951.

— 2. Lektsii ob uravneniakh c chastnymi proizvodnymi (Lectures on partial differential equations). GTTI, 1950.

PICONE, M. 1. Nuovo metodo d'approssimazione per la soluzione del problema di Dirichlet. Rom. Acc. L. Rend. (5) 31_1 (1922), 357–359.

PRIVALOV, I. I. 1. Riady Fur'e (Fourier series). 3 ed., G., 1934.

— 2. Integral'nye uravneniiâ (Integral equations). M.-L., ONTI, 1935. 248 pp.
— 3. Vvedenie v teoriiû funktsii kompleksnogo peremennogo (Introduction to the theory of functions of a complex variable). 4 ed., rev. & supp., M.-L. ONTI, 1935. 386 pp.
PROTUSEVICH, ÎA. A. 1. Variatsionnye metody v stroitel'noĭ mekhanike (Variational methods in structural mechanics). G., 1948.

Q

QUADE, E. S. 1. Trigonometric approximation in the mean. Duke Math. Jour. 3 (1937), 529–543.

R

RAFAL'SON, Z. KH. On the question of the solution of the biharmonic equation. (Russian) DAN SSSR 64 (1949), 6: 799–802.
RIESZ, FR. 1. Leçons sur les systèmes d'équations linéaires à une infinité d'inconnues. Paris, 1913.
RITZ, W. 1. Über eine neue Methode zur Lösung gewisser Variationsprobleme der mathematischen Physik. Jour. f. reine und angewandte Mathematik 135 (1908), 1–61.
— 2. Oeuvres complètes. Paris, 1911.
— 3. Theorie der Transversalschwingungen einer quadratischen Platte mit freien Rändern. Ann. der Phys. (4) 28, 737–786.
ROGOV, T. N. 1. The equilateral triangular plate freely supported on the contour. (Russian) Trudy vyssh. inzh.-tekh. uchilishcha Voenno-morskogo Flota, 2 (1940), 34–43.
RUNGE, C. 1. Über eine Methode, die partielle Differentialgleichung $\Delta u =$ Constans numerisch zu integrieren. Zeitschrift für Math. und Phys. 56 (1908), 225–232.
— 2. Graphische Lösung von Randwertaufgaben der Gleichung $\frac{\partial^2 u}{\partial x^2} + \frac{\partial^2 u}{\partial y^2} = 0$. Nachrichten von der Königlichen Gesellschaft der Wissenschaften zu Göttingen, 1911, 431–448.

S

SCARBOROUGH, J. B. 1. Numerical mathematical analysis. Baltimore, 1934. Russian translation: Chislennye metody matematicheskogo analiza, by Gokhman, E. V. and Kontovta, V. I., ed. and appendix by D. ÎU. Panov. M.-L., GTTI, 1934.
SCHMIDT, E. 1. Über die Auflösung linearer Gleichungen mit unendlich vielen Unbekannten. Palermo Rend. 25 (1908), 53–77.
SCHWARZ, H. A. 1. Ueber einige Abbildungsaufgaben. Jour. f. die reine und angew. Math. 70 (1869), 105–120. See also Ges. Math. Abh. 1869, Bd. II, 65–83.
SEMENDIÂEV, K. A. 1. On the determination of the proper values and invariant manifolds of matrices by means of iterations. (Russian; English summary) Priklad. Matem. i Mekh. 7 (1943), 3: 193–222.
SEMËNOV, N. S. 1. Application of the variational method of L. V. Kantorovich to the solution of problems of the bending of thin rectangular plates. (Russian) Priklad. Matem. i Mekh. 4 (1939), 107–116.
SHEVCHENKO, K. N. 1. Application of the variational method to the solution of problems of the theory of elasticity. (Russian) Priklad. Matem. i Mekh. 2 (1938), 219–222.
SHORTLEY, G. H. and WELLER. 1. The numerical solution of the Laplace equation. Journal Appl. Physics. 9 (1938), 334–344.
SLOBODIÂNSKIĬ, M. G. 1. A method for the approximate integration of partial

differential equations and its application to problems of the theory of elasticity. (Russian) Priklad. Matem. i Mekh. 3 (1939), 75–82.
— 2. Spatial problems of the theory of elasticity for prismatic bodies. (Russian) Uchënye zapiski Moskov. Gos. Univ-ta 1940, 39.
SMIRNOV, V. I. 1. Kurs vyssheĭ matematiki (Course of higher mathematics). G., 1941–51.
— 2. Sur la théorie des polynomes orthogonaux. Zhurnal Leningradskogo fiziko-matematicheskogo obshchestva. 2 (1928), 1.
— 3. Kurs vyssheĭ matematiki dli͡a fizikov i inzhenerov (Course of higher mathematics for physicists and engineers).
SMIRNOV, V. I., KANTOROVICH, L. V. and KRYLOV, V. I. 1. Variat͡sionnoe ischislenie (Variational calculus). Kubuch, 1933.
SOBOLEV, S. L. 1. Uravneni͡ia matematicheskoĭ fiziki (Equations of mathematical physics). G., 1947.
— 2. l'Algorithme de Schwarz dans la théorie de l'élasticité. DAN SSSR 4(13) (1936), 6: 243–246.
SOKOLOV, B. A. 1. On the torsion of an axle of variable section. (Russian; German summary) Priklad. Matem. i Mekh. 3 (1939), 3: 153–160.
STEINHAUS, H. 1. Sur les suites complètes. Studia Math., Lwów, 4 (1933), 142–145.
STENIN, N. P. 1. The determination of the parameters in the Christoffel-Schwarz function. (Russian) In *Collective Author* [2], q.v.
STEPANI͡ANT͡S, L. G. 1. Calculation of the laminar boundary layer on solids of revolution. (Russian) Priklad. Matem. i Mekh. 6 (1942), 4: 317–326.
SZEGÖ, G. 1. Über orthogonale Polynome, die zu einer gegebenen Kurve der komplexen Ebene gehören. Math. Zeitschrift 9 (1921), 218–270.

T

TIMOSHENKO, S. P. 1. Teori͡ia uprugosti (Theory of elasticity). G., 1934. (American edition: N.Y., 1934)
— 2. Teori͡ia kolebaniĭ v inzhenernom dele (Theory of vibrations in engineering). 1932.
TREFFTZ, E. 1. Mechanik der elastischen Körper. (In vol. VI of Handbuch der Physik, Berlin, 1928) Russian translation: Matematicheskai͡a teori͡ia uprugosti, L., GTTI, 1934.
TRICOMI, F. 1. Sulla risoluzione numerica delle equazioni integrali di Fredholm. Rend. Accad. dei Lincei 33 (1924, 1 sem.), 12: 483–486.
— 2. Ancora sulla risoluzione numerica delle equazioni integrali di Fredholm. Rend. Accad. dei Lincei 33 (1924), 2 sem., 1: 26–30.

V

VALLÉE-POUSSIN, C. J. DE LA. 1. Cours d'analyse infinitésimale. Paris, 1914. (Russian translation: Kurs analiza, GTTI, 1933.)
VETCHINKIN, V. P. 1. Numerical methods for the solution of non-linear integral equations. (Russian) Trudy T͡SAGI, No. 192 (1935).
VILLAT, H. 1. Sur le problème de Dirichlet dans une aire annulaire. Rendiconti del Circolo Matematico di Palermo, 33 (1912, 1 sem.), 134–174.
— 2. Leçons sur l'hydrodynamique. Paris, 1929.
VLASOV, V. Z. 1. Building mechanics of thin elastic plates. (Russian) Priklad. Matem. i Mekh. 10 (1946), 173–192.
VOLKOV, D. M. and NAZAROV, A. A. 1. On a boundary-value problem and its application to the plane theory of elasticity. (Russian; French summary) Matem. Sbornik 40 (1933), 2: 210–228.

W

WAYLAND, HAROLD. 1. The representation of the secular equation in the form of a polynomial. (Russian) Uspekhi Matem. Nauk 2 (1947), 4: 128 ff. See: Expansion of determinantal equations into polynomial form. Quart. Appl. Math. 2 (1945), 277–306.

WEBSTER, A. and SZEGÖ, G. 1. Partielle Differentialgleichungen der mathematischen Physik. Leipzig-Berlin, 1930. Russian translation by I. S. Gradshtein: Differentsialnye uravneniia v chastnykh proizvodnykh matematicheskoĭ fiziki. Ed. 2., corr., M.-L. Glav. red. obshchtekh. distsiplin, 1934 (on binding: 1935).

WHITTAKER, E. T. and ROBINSON, G. 1. The calculus of observations. London, 1928. Russian translation by V. M. Ozeretskiĭ, N. S. Samoĭlovich and V. P. Tsesevich, ed. N. M. Giunter: Matematicheskaia obrabotka rezul'tatov nabliudeniĭ. M.-L. ONTI, 1935.

WINTNER, A. 1. Ein Satz über unendliche Systeme von linearen Gleichungen. Math. Zeitschrift 24 (1925/26), 266 ff.

Z

ZAGADSKIĬ, D. M. 1. The analog of Newton's method for non-linear integral equations. (Russian) DAN SSSR 59 (1948), 6: 1041–1044.

Collective Author

1. IUbileĭnyi sbornik, posviashchennyi tridtsatiletiiu Velikoĭ Oktiabr'skoĭ sotsialisticheskoĭ revoliutsii (Jubilee symposium, dedicated to the thirtieth anniversary of the Great October Socialist Revolution) Izd. AN (1947).
2. Konformnoe otobrazhenie odnosviaznykh i mnogosviaznykh oblasteĭ (The conformal mapping of simply and multiply connected regions). Sbornik, L.-M, ONTI, 1937.

A CATALOG OF SELECTED

DOVER BOOKS

IN SCIENCE AND MATHEMATICS

Mathematics–Bestsellers

HANDBOOK OF MATHEMATICAL FUNCTIONS: with Formulas, Graphs, and Mathematical Tables, Edited by Milton Abramowitz and Irene A. Stegun. A classic resource for working with special functions, standard trig, and exponential logarithmic definitions and extensions, it features 29 sets of tables, some to as high as 20 places. 1046pp. 8 x 10 1/2. 0-486-61272-4

ABSTRACT AND CONCRETE CATEGORIES: The Joy of Cats, Jiri Adamek, Horst Herrlich, and George E. Strecker. This up-to-date introductory treatment employs category theory to explore the theory of structures. Its unique approach stresses concrete categories and presents a systematic view of factorization structures. Numerous examples. 1990 edition, updated 2004. 528pp. 6 1/8 x 9 1/4. 0-486-46934-4

MATHEMATICS: Its Content, Methods and Meaning, A. D. Aleksandrov, A. N. Kolmogorov, and M. A. Lavrent'ev. Major survey offers comprehensive, coherent discussions of analytic geometry, algebra, differential equations, calculus of variations, functions of a complex variable, prime numbers, linear and non-Euclidean geometry, topology, functional analysis, more. 1963 edition. 1120pp. 5 3/8 x 8 1/2. 0-486-40916-3

INTRODUCTION TO VECTORS AND TENSORS: Second Edition--Two Volumes Bound as One, Ray M. Bowen and C.-C. Wang. Convenient single-volume compilation of two texts offers both introduction and in-depth survey. Geared toward engineering and science students rather than mathematicians, it focuses on physics and engineering applications. 1976 edition. 560pp. 6 1/2 x 9 1/4. 0-486-46914-X

AN INTRODUCTION TO ORTHOGONAL POLYNOMIALS, Theodore S. Chihara. Concise introduction covers general elementary theory, including the representation theorem and distribution functions, continued fractions and chain sequences, the recurrence formula, special functions, and some specific systems. 1978 edition. 272pp. 5 3/8 x 8 1/2. 0-486-47929-3

ADVANCED MATHEMATICS FOR ENGINEERS AND SCIENTISTS, Paul DuChateau. This primary text and supplemental reference focuses on linear algebra, calculus, and ordinary differential equations. Additional topics include partial differential equations and approximation methods. Includes solved problems. 1992 edition. 400pp. 7 1/2 x 9 1/4. 0-486-47930-7

PARTIAL DIFFERENTIAL EQUATIONS FOR SCIENTISTS AND ENGINEERS, Stanley J. Farlow. Practical text shows how to formulate and solve partial differential equations. Coverage of diffusion-type problems, hyperbolic-type problems, elliptic-type problems, numerical and approximate methods. Solution guide available upon request. 1982 edition. 414pp. 6 1/8 x 9 1/4. 0-486-67620-X

VARIATIONAL PRINCIPLES AND FREE-BOUNDARY PROBLEMS, Avner Friedman. Advanced graduate-level text examines variational methods in partial differential equations and illustrates their applications to free-boundary problems. Features detailed statements of standard theory of elliptic and parabolic operators. 1982 edition. 720pp. 6 1/8 x 9 1/4. 0-486-47853-X

LINEAR ANALYSIS AND REPRESENTATION THEORY, Steven A. Gaal. Unified treatment covers topics from the theory of operators and operator algebras on Hilbert spaces; integration and representation theory for topological groups; and the theory of Lie algebras, Lie groups, and transform groups. 1973 edition. 704pp. 6 1/8 x 9 1/4. 0-486-47851-3

Mathematics–Logic and Problem Solving

PERPLEXING PUZZLES AND TANTALIZING TEASERS, Martin Gardner. Ninety-three riddles, mazes, illusions, tricky questions, word and picture puzzles, and other challenges offer hours of entertainment for youngsters. Filled with rib-tickling drawings. Solutions. 224pp. 5 3/8 x 8 1/2. 0-486-25637-5

MY BEST MATHEMATICAL AND LOGIC PUZZLES, Martin Gardner. The noted expert selects 70 of his favorite "short" puzzles. Includes The Returning Explorer, The Mutilated Chessboard, Scrambled Box Tops, and dozens more. Complete solutions included. 96pp. 5 3/8 x 8 1/2. 0-486-28152-3

THE LADY OR THE TIGER?: and Other Logic Puzzles, Raymond M. Smullyan. Created by a renowned puzzle master, these whimsically themed challenges involve paradoxes about probability, time, and change; metapuzzles; and self-referentiality. Nineteen chapters advance in difficulty from relatively simple to highly complex. 1982 edition. 240pp. 5 3/8 x 8 1/2. 0-486-47027-X

SATAN, CANTOR AND INFINITY: Mind-Boggling Puzzles, Raymond M. Smullyan. A renowned mathematician tells stories of knights and knaves in an entertaining look at the logical precepts behind infinity, probability, time, and change. Requires a strong background in mathematics. Complete solutions. 288pp. 5 3/8 x 8 1/2. 0-486-47036-9

THE RED BOOK OF MATHEMATICAL PROBLEMS, Kenneth S. Williams and Kenneth Hardy. Handy compilation of 100 practice problems, hints and solutions indispensable for students preparing for the William Lowell Putnam and other mathematical competitions. Preface to the First Edition. Sources. 1988 edition. 192pp. 5 3/8 x 8 1/2. 0-486-69415-1

KING ARTHUR IN SEARCH OF HIS DOG AND OTHER CURIOUS PUZZLES, Raymond M. Smullyan. This fanciful, original collection for readers of all ages features arithmetic puzzles, logic problems related to crime detection, and logic and arithmetic puzzles involving King Arthur and his Dogs of the Round Table. 160pp. 5 3/8 x 8 1/2. 0-486-47435-6

UNDECIDABLE THEORIES: Studies in Logic and the Foundation of Mathematics, Alfred Tarski in collaboration with Andrzej Mostowski and Raphael M. Robinson. This well-known book by the famed logician consists of three treatises: "A General Method in Proofs of Undecidability," "Undecidability and Essential Undecidability in Mathematics," and "Undecidability of the Elementary Theory of Groups." 1953 edition. 112pp. 5 3/8 x 8 1/2. 0-486-47703-7

LOGIC FOR MATHEMATICIANS, J. Barkley Rosser. Examination of essential topics and theorems assumes no background in logic. "Undoubtedly a major addition to the literature of mathematical logic." – *Bulletin of the American Mathematical Society.* 1978 edition. 592pp. 6 1/8 x 9 1/4. 0-486-46898-4

INTRODUCTION TO PROOF IN ABSTRACT MATHEMATICS, Andrew Wohlgemuth. This undergraduate text teaches students what constitutes an acceptable proof, and it develops their ability to do proofs of routine problems as well as those requiring creative insights. 1990 edition. 384pp. 6 1/2 x 9 1/4. 0-486-47854-8

FIRST COURSE IN MATHEMATICAL LOGIC, Patrick Suppes and Shirley Hill. Rigorous introduction is simple enough in presentation and context for wide range of students. Symbolizing sentences; logical inference; truth and validity; truth tables; terms, predicates, universal quantifiers; universal specification and laws of identity; more. 288pp. 5 3/8 x 8 1/2. 0-486-42259-3

Mathematics–Algebra and Calculus

VECTOR CALCULUS, Peter Baxandall and Hans Liebeck. This introductory text offers a rigorous, comprehensive treatment. Classical theorems of vector calculus are amply illustrated with figures, worked examples, physical applications, and exercises with hints and answers. 1986 edition. 560pp. 5 3/8 x 8 1/2. 0-486-46620-5

ADVANCED CALCULUS: An Introduction to Classical Analysis, Louis Brand. A course in analysis that focuses on the functions of a real variable, this text introduces the basic concepts in their simplest setting and illustrates its teachings with numerous examples, theorems, and proofs. 1955 edition. 592pp. 5 3/8 x 8 1/2. 0-486-44548-8

ADVANCED CALCULUS, Avner Friedman. Intended for students who have already completed a one-year course in elementary calculus, this two-part treatment advances from functions of one variable to those of several variables. Solutions. 1971 edition. 432pp. 5 3/8 x 8 1/2. 0-486-45795-8

METHODS OF MATHEMATICS APPLIED TO CALCULUS, PROBABILITY, AND STATISTICS, Richard W. Hamming. This 4-part treatment begins with algebra and analytic geometry and proceeds to an exploration of the calculus of algebraic functions and transcendental functions and applications. 1985 edition. Includes 310 figures and 18 tables. 880pp. 6 1/2 x 9 1/4. 0-486-43945-3

BASIC ALGEBRA I: Second Edition, Nathan Jacobson. A classic text and standard reference for a generation, this volume covers all undergraduate algebra topics, including groups, rings, modules, Galois theory, polynomials, linear algebra, and associative algebra. 1985 edition. 528pp. 6 1/8 x 9 1/4. 0-486-47189-6

BASIC ALGEBRA II: Second Edition, Nathan Jacobson. This classic text and standard reference comprises all subjects of a first-year graduate-level course, including in-depth coverage of groups and polynomials and extensive use of categories and functors. 1989 edition. 704pp. 6 1/8 x 9 1/4. 0-486-47187-X

CALCULUS: An Intuitive and Physical Approach (Second Edition), Morris Kline. Application-oriented introduction relates the subject as closely as possible to science with explorations of the derivative; differentiation and integration of the powers of x; theorems on differentiation, antidifferentiation; the chain rule; trigonometric functions; more. Examples. 1967 edition. 960pp. 6 1/2 x 9 1/4. 0-486-40453-6

ABSTRACT ALGEBRA AND SOLUTION BY RADICALS, John E. Maxfield and Margaret W. Maxfield. Accessible advanced undergraduate-level text starts with groups, rings, fields, and polynomials and advances to Galois theory, radicals and roots of unity, and solution by radicals. Numerous examples, illustrations, exercises, appendixes. 1971 edition. 224pp. 6 1/8 x 9 1/4. 0-486-47723-1

AN INTRODUCTION TO THE THEORY OF LINEAR SPACES, Georgi E. Shilov. Translated by Richard A. Silverman. Introductory treatment offers a clear exposition of algebra, geometry, and analysis as parts of an integrated whole rather than separate subjects. Numerous examples illustrate many different fields, and problems include hints or answers. 1961 edition. 320pp. 5 3/8 x 8 1/2. 0-486-63070-6

LINEAR ALGEBRA, Georgi E. Shilov. Covers determinants, linear spaces, systems of linear equations, linear functions of a vector argument, coordinate transformations, the canonical form of the matrix of a linear operator, bilinear and quadratic forms, and more. 387pp. 5 3/8 x 8 1/2. 0-486-63518-X